ACPL ITEM

DISCARDED

YO-BWS-117

JUN 13 '69

CEREALS IN THE UNITED KINGDOM

CEREALS IN THE UNITED KINGDOM

Production, Marketing and Utilisation

A report prepared by
DENIS K. BRITTON
DIRECTOR OF A SURVEY COMMISSIONED
BY THE
HOME-GROWN CEREALS AUTHORITY

Associate authors
BASIL E. CRACKNELL
IAN M. T. STEWART

UNIVERSITY OF NOTTINGHAM
DEPARTMENT OF AGRICULTURAL ECONOMICS

THE QUEEN'S AWARD
TO INDUSTRY 1966

PERGAMON PRESS
OXFORD · LONDON · EDINBURGH · NEW YORK
TORONTO · SYDNEY · PARIS · BRAUNSCHWEIG

Pergamon Press Ltd., Headington Hill Hall, Oxford
4 & 5 Fitzroy Square, London W.1
Pergamon Press (Scotland) Ltd., 2 & 3 Teviot Place, Edinburgh 1
Pergamon Press Inc., Maxwell House, Fairview Park, Elmsford,
New York 10523
Pergamon of Canada Ltd., 207 Queen's Quay West, Toronto 1
Pergamon Press (Aust.) Pty. Ltd., 19a Boundary Street,
Rushcutters Bay, Sydney, N.S.W. 2011, Australia
Pergamon Press S.A.R.L., 24 rue des Écoles, Paris 5e
Vieweg & Sohn GmbH, Burgplatz 1, Braunschweig

Copyright © 1969 Home-Grown Cereals Authority
First edition 1969
Library of Congress Catalog Card No. 69—19840

Printed in Great Britain by Watmoughs Limited, Idle, Bradford; and London

08 013896 9

1489002

Contents

Foreword

Letter to the Chairman of the Home-Grown Cereals Authority
(Sir HENRY HARDMAN, K.C.B.)

Department of Agricultural Economics,
University of Nottingham
24th January 1969

Dear Sir Henry,

In 1966 the Authority requested me to undertake a comprehensive survey of the United Kingdom cereals market. I prepared an outline of the points which might be covered in the survey, and after this had been approved a grant towards the cost was obtained from the Agricultural Marketing Development Executive Committee. I now have pleasure in sending you the attached report of the Survey.

In submitting the report I should like to emphasise on behalf of all those who have contributed to it that we regard it as providing a starting-point for further research. We are very conscious of the fact that many topics have been touched upon which are deserving of a much more exhaustive treatment, and we hope that the Authority will initiate further work in a number of the directions indicated.

Yours sincerely,
DENIS K. BRITTON

Acknowledgements

Authorship

The Survey as a whole was directed from the University of Nottingham, where I had the full-time assistance of Dr. Basil E. Cracknell (Senior Research Fellow) and Mr. Ian M. T. Stewart (Research Fellow). Dr. Cracknell has taken a major part in the drafting, co-ordination and editing of the whole work and I gratefully acknowledge the personal debt which I owe to him for his organising capacity, abundant flow of ideas, unfailing patience and complete devotion to the task. He had a hand in the shaping of all sections of the report, but he was particularly associated with the preparation of Part III, Chapter 3 ("The Internal Transport of Grain"), Part IV ("The Importing and Exporting of Grain"), Part VI, Chapter 4 ("Grain Marketing in the U.S.A."), Part VII ("Price Behaviour in the U.K. Cereals Market"), and Part VIII ("U.K. Policy for Cereals"). Mr. Ian Stewart, besides sharing in planning, co-ordination and general editorial work, was largely responsible for preparing the material for Part II, Chapter 1 ("Cereal Growing in the United Kingdom"), Part III, Chapters 1 and 4 ("Introduction to the Merchant Survey" and "Merchants' Dealings in Grain"), Part V(a), Chapter 6 ("Farm-mixing of Feedingstuffs") and Part V(c), Chapter 1 ("Malting by Brewer-maltsters and Sale Maltsters"). It is in recognition of these major contributions that Dr. Cracknell and Mr. Stewart are described on the title page as Associate Authors.

At an early stage in the planning of the Survey I decided that the task was such as to call for a co-operative effort in which a number of universities should take part so that full advantage could be taken of the special knowledge and research experience of various people. The Universities of Reading and Glasgow were invited to combine with Nottingham in the surveys of merchants and of the cereal-using industries. Mr. John Marsh of Reading University, assisted by Mr. David Ansell, was mainly responsible for Part III, Chapters 2, 5 and 6 ("The Structure of the Merchant Sector", "Some Management Problems of Merchant Businesses" and "The Market in Cereals Seed") and for Part V(b) ("Flour Milling"). Mr. Lawrence Smith of Glasgow University, assisted by Mr. Hugh MacLean and Mr.

ix

John Castree, gave particular attention to Part V(a) ("The Animal Feeding-stuffs Industry") and to Part V(c), Chapter 2 ("Malting by Distillers"). At successive stages of the work these colleagues participated fully with the Nottingham team in joint discussions of survey methods and results, and this proved to be a most successful example of close inter-university co-operation.

The University of Cambridge was invited to help on questions of grain storage on farms and the economics of centralised group storage. Mr. G. Davidson undertook the necessary survey work for this and contributed much of Part II, Chapter 5 and Part III, Chapter 7, under the general supervision of Mr. F. G. Sturrock, Director of the Farm Economics Branch at that university. Mr. B. G. Jackson also participated in the planning of the project.

Mr. David Colman, of the University of Manchester, who has done a considerable amount of work on econometric models relating to the supply and demand of grain in the United Kingdom, undertook some special studies of price variations between markets, between seasons and between years which contributed to Part VII ("Price Behaviour in the U.K. Cereals Market"). He also provided useful material about the relationship between weather and grain production (acreage and yield) in this country, which is incorporated in Part II, Chapter 3.

Professor J. T. Coppock, of the University of Edinburgh, provided a substantial amount of material on acreage and yield changes for Part II, Chapter 1 ("Cereal Growing in the United Kingdom") and we much appreciated his help in the drafting of that chapter. He was also responsible for Appendix A ("Factors Affecting the Geographical Distribution of Cereal Acreages and Cereal Yields in the U.K.").

Mr. Michael Butterwick of the University of Oxford (Institute of Agrarian Affairs) contributed the whole of Part VI, Chapter 1 ("Grain Marketing in some Western European Countries"); Mr. Frank Grogan of the Agricultural Adjustment Unit, University of Newcastle-upon-Tyne, was the author of Part VI, Chapter 2 ("Wheat Marketing in Australia"); and Dr. Geoffrey Hiscocks of the Canadian Department of Agriculture (writing purely in a personal capacity) provided the material for Part VI, Chapter 3 ("Grain Marketing in Canada").

Dr. Graham Hallett of the University College of South Wales and Monmouthshire, University of Wales, contributed Part VI, Chapter 5 ("The International Cereal Situation") and Appendix D ("The Future Demand and Supply Position for Cereals").

Market Investigations Ltd. carried out the field work and statistical analysis for the Survey of Farmers and Mr. R. E. Walpole, a Director of that organisation, drafted parts of Chapters 3, 5 and 6 of Part II ("The

Production of Cereals", "Grain Preservation and Storage on the Farm" and "The Marketing of Grain by Farmers") and part of Part V(a), Chapter 6 ("Farm-mixing of Feedingstuffs").

Miss Sonia Lindsay made a thorough review of statistical sources relating to the supply, prices and utilisation of cereals in the United Kingdom and this was a valuable work of reference when the report was in preparation. Copies of this review were made available to the Home-Grown Cereals Authority. Miss Lindsay also assisted in the preparation of Part VIII, Chapter 1 ("Changes in Government Cereals Policy since 1953").

Finally my colleague Mr. Brian E. Hill, in the University of Nottingham, prepared a considerable amount of useful material for Part II, Chapter 4 ("Some Economic Aspects of Cereals Production").

Other Acknowledgements

In the course of preparation of the Report a large number of people were consulted, either in an official or in a personal capacity. The farmers, merchants, manufacturers and N.A.A.S. officers (with their counterparts in Scotland and Northern Ireland) who co-operated in the respective surveys are too numerous to mention by name, but I am happy to be able to express here our thanks to them for all the trouble they took on our behalf and for the information which they provided. A descriptive report of this kind would be inconceivable without the willing co-operation of those who are personally and professionally involved in the various marketing activities under consideration.

The special help which we received from the various national associations connected with cereals marketing is mentioned in the first chapter of the Report. Here again, without such help we could never have effectively undertaken much of the work which the Report entailed.

People in leading positions and other individuals who were of great assistance to us are named below. I am grateful to them all; and to any who have inadvertently been omitted I offer my sincere apologies.

IN CONNECTION WITH PART II (PRODUCERS)

Official Bodies (1) National Farmers' Union: Mr. G. F. Elston, Mr. M. Strauss

(2) Agricultural Co-operative Association Limited: Mr. J. A. E. Morley

(3) Agricultural Co-operative Managers' Association: Mr. T. E. Wilson

Officials of Ministries and Advisory Services:

Mr. D. J. Mitchell, C.B., C.V.O.
Mr. W. E. Jones
Mr. A. C. Sparks
Mr. L. Napolitan ⎬ Ministry of Agriculture, Fisheries and Food
Mr. J. A. Barrah
Mr. E. L. Snowdon
Mr. D. Salton

Mr. O. J. Beilby ⎱ Department of Agriculture and Fisheries for
Mr. P. M. Scola ⎰ Scotland

Mr. H. Shemilt, Ministry of Agriculture, Northern Ireland

Mr. H. E. Evans ⎱ National Agricultural Advisory Service
Mr. A. J. Davies ⎰

Mr. G. A. Catto ⎱ Edinburgh and East of Scotland College
Mr. A. K. M. Meiklejohn ⎰ of Agriculture, Advisory Service

Mr. A. E. Parkinson, West of Scotland Agricultural College, Advisory Service

Mr. A. Howie, North of Scotland College of Agriculture, Advisory Service

Mr. T. Moore, Head of Advisory Service, Northern Ireland Ministry of Agriculture

Individuals:

Mr. P. Savory (Chairman—National Farmers' Union, Cereals Committee)
Mr J. Cossins (Vice-Chairman)
Mr. R. Brown
Mr. D. N. Darbishire
Mr. G. D. French
Mr. J. Macaulay ⎬ Members—National Farmers' Union, Cereals Committee
Mr. L. A. Mason
Mr. R. G. Pike
Mr. A. F. Shaw
Mr. R. Wellesley
Mr. J. B. Eastwood
Mr. J. Parker
Mr. J. Edgar
Mr. R. White   Mr. H. Rice
Mr. J. Fawcett   Mr. S. Miller
Mr. E. Jackson   Mr. Norman Hicks
Mr. J. Jenkins   Mr. J. Sinclair

IN CONNECTION WITH PART III (MERCHANTS)

Official Bodies (1) National Association of Corn and Agricultural Merchants Limited:

Mr. C. G. Metson, O.B.E. (Director-General)
Mr. W. Scott Charles (Secretary, Scottish Council)
Mr. H. S. Leech (Secretary)
Mr. P. J. Ottino (Assistant Secretary)
Mrs. S. Smith
Mr. G. B. Wood, O.B.E. (Past President, N.A.C.A.M.)
Mr. S. T. Skelton (Past President)
Mr. L. F. Ratcliff (Chairman—Cereals Committee, N.A.C.A.M.)

(2) Institute of Corn and Agricultural Merchants: Mr. D. C. Lunn-Rockliffe

(3) Seed Trade Association of the United Kingdom, Inc.: Mr. K. R. Alderton

Individuals:

Mr. E. K. Wherry
Mr. F. H. W. Swallow, O.B.E.
Mr. C. W. Burnett
Mr. D. J. W. Browne
Mr. B. Stevens
Lt. Col. C. A. Brooks, O.B.E.
Col. K. Denham
Mr. A. C. Rose
Mr. A. K. Barby
Mr. G. H. Dann
Mr. T. Loader

Co-operatives:

Mr. T. F. Collis
Mr. D. Small
Mr. L. J. Wright

Transport:

Mr. O. H. Grafton, British Waterways Board
Mr. M. Piggott, British Rail

IN CONNECTION WITH PART IV (SHIPPERS AND BROKERS)

Shippers:

Official Bodies: (1) National Federation of Corn Trade Associations:
  Mr. R. S. Cornelius (President)
  Mr. W. A. Wilson (Past President)
  (2) Chamber of Shipping: Mr. A. S. Wightman

Individuals:

  Mr. L. Betts, Manager P.L.A. Tilbury Grain Terminal
  Mr. R. G. Penderid and Mr. K. J. Slater
  Captain W. A. Adair and Messrs. H. J. Barrett and K. Spence
  Mr. T. G. Johns
  Mr. W. A. Wilson, C.B.E. and Mr. E. M. Hunt
  Mr. T. Naef
  Mr. R. Clark

Brokers:

  Sir Leslie Phillips   Mr. C. Blackmore

Futures Market:

  Mr. A. A. Hooker

IN CONNECTION WITH PART V (MANUFACTURERS)

Compounders and Millers:

Official Bodies: (1) Compound Animal Feedingstuffs Manufacturers
  National Association:
    Mr. C. G. Metson, O.B.E. (Secretary)
    Mr. A. D. Bird (Public Relations and Executive
    Officer)
    Dr. Clare Burgess (Past President,
    C.A.F.M.N.A.)
    The late Rt. Hon. W. Normand (Past President)
  (2) National Association of British and Irish Millers:
    Mr. L. C. Carrington (Secretary)
    Mr. C. L. Copeland (Director of Public Rela-
    tions)
    Mr. B. C. Read (Past President)
    Mr. W. A. George (Past President)
  (3) Flour Milling and Baking Research Association,
  St. Albans:
    Mr. E. N. Greer

Individuals:

Mr. H. R. Philpot, Mr. R. C. Long, Mr. R. S. Jefferies
Mr. A. H. Hunt, Mr. G. A. Watson, Mr. R. Thom, Mr. D. Hutchinson, Mr. P. A. Metaxa
Mr. K. J. Arnott, Mr. J. Fortheringham
Mr. R. O. Prentice, Mr. A. R. Robinson
Mr. J. Weston, Mr. G. Carter, Mr. C. Matthews
Mr. A. J. Fairclough, Mr. E. C. Humphreys
Mr. E. B. Stevenson, Mr. A. W. Brown, Mr. R. E. Atkinson, Mr. C. L. Taylor
Mr. A. G. Gowlett
Mr. W. Grant, Mr. C. G. Goldsmith, Mr. H. N. Weston
Mr. R. H. Towler
Mr. J. M. Heygate

Maltsters and Distillers:

Official Bodies: (1) Maltsters' Association of Great Britain:
Group Captain V. Fairfield, O.B.E. (Secretary)

(2) Brewers' Society:
Mr. D. Horwood (Secretary)

(3) Scotch Whisky Association:
Mr. P. J. Woodhouse (Secretary)

Individuals:

Mr. H. L. Thompson
Mr. T. A. A. Macpherson
Mr. A. G. Hills
Mr. R. Reed

IN CONNECTION WITH PART VI (INTERNATIONAL):

Mr. R. E. Moore (Executive Secretary, International Wheat Council)
Mr. K. Murray (American Embassy, London)
Mr. W. Morgan (Commonwealth Secretariat)

GENERAL:

Home-Grown Cereals Authority:

Sir Charles Norman, C.B.E.
Mr. J. W. Pugsley
Mr. F. T. Rees
Mr. W. J. Hazeldine
Mr. J. Gibbs
Mr. D. Heath
Mr. I. C. Beattie

Others:

Mr. A. D. Clague, Secretary, Economic Development Committee for Food Manufacturing

Finally, a word of thanks is due to those who have helped in the drawing of maps and charts and in the typing of manuscripts. For cartographic work we are grateful to Mrs. M. Sinclair, Mr. K. Wass and Mr. T. Allen. In the typing and checking work we have been ably assisted by Miss I. R. Bruce, Mrs. K. M. Norman, Mrs. B. Jukes, Mrs. K. O. Bradley, Miss A. Bond, Miss P. Harding and Miss H. Gardner.

PART I

Introduction

CHAPTER 1

Background to the Report

1.1. When the Home-Grown Cereals Authority was established in 1965, with a wide ranging responsibility to improve the marketing of home-grown cereals,[1] it found that apart from a few journal articles, and occasional University reports on specific topics, nothing of a comprehensive nature about cereals marketing had been written since the "Report on the Marketing of Wheat, Barley and Oats in England and Wales", prepared in 1928 by the Ministry of Agriculture and Fisheries and published by H.M.S.O. (Economic Series No. 11) and referred to below as the 1928 Report. It was therefore decided that a comprehensive survey should be undertaken covering all aspects of the U.K. cereals market, and work began at the University of Nottingham in the latter part of 1966. An outline of the proposed Research Programme appeared as Appendix III to the first Annual Report of the Home-Grown Cereals Authority 1965–6, published by H.M.S.O. in July 1966.

1.2. The present Report is rather unusual in that it has been prepared by a university at the request of a quasi-government body. It is not therefore comparable with the report of a Committee of Enquiry, or a government-appointed Commission. For example bodies of the latter type are generally set up specifically to examine issues of current political importance; if they collect data it is usually related directly to the particular issues at stake. This report does not deal selectively with issues of this sort but attempts to describe and to document, with statistical material, the whole process of cereals marketing, from the producer through to the user. The bulk of the report is devoted to this kind of documentation, based on various fact-finding surveys, but important policy issues are also taken up.

Method of Working

1.3. It was realised from the outset that the University of Nottingham would not have the staff available to undertake field surveys of cereal

[1] Cereals Marketing Act 1965.

3

producers on the scale required. There are some 175,000 cereal producers in the U.K. and although many of them grow an insignificant acreage it was considered that at least 1000 farmers would be needed to form a reasonable sample. It was therefore decided to enlist the help of a commercial market research agency experienced in agricultural surveys, and Market Investigations Ltd. were appointed to carry out the necessary field work and to analyse the results. The questionnaires were designed in full collaboration with the University and close liaison was maintained at all stages of the work. The surveys carried out by M.I.L. are described fully in Part II Chapter 3 and in Appendix B. This is believed to be the first time that a commercial market research agency has co-operated with a university on an agricultural investigation of this kind. Towards the end of the field surveys meetings with farmers were arranged in different parts of the country to enable the University, together with representatives of M.I.L., to meet groups of cereal farmers face to face. The University also had meetings with members of the National Farmers' Union Cereals Committee.

1.4. The University of Nottingham itself undertook the surveys of merchants and of the cereal-using industries. These involved visits to over 400 firms in all parts of the United Kingdom, and this was a task beyond the capacity of one University Department. It was therefore decided to invite the Universities of Reading and Glasgow to combine with Nottingham in the design and execution of these surveys. The field work for the Survey of Merchants was shared equally by the three universities, and in the analysis and writing up of the results each university concentrated upon a particular feature of the merchant sector. The questionnaires were jointly designed and the writing up was planned so that each part would fit naturally into the whole.

Relations with Trade Associations, Unions and Societies

1.5. Early in the Survey a personal visit was made to each of the main national associations connected with cereals marketing, and an excellent relationship was established. We discussed our proposed questionnaires with them and in each case we received many valuable suggestions and every possible help. In some cases the associations advised their members that the University would be visiting certain of them on a sample basis and asked members to give their full co-operation. As will be shown later, we received a very high degree of co-operation from all sectors of the grain trade and some of the credit for this must go to these central bodies who were as anxious as we were that the issues connected with cereals marketing should be discussed from a firm basis of fact and not, as is so often the case, from supposition.

1.6. Some trade associations took such a keen interest in the Survey that they volunteered to help by sending in statements of their views on topics connected with cereals marketing. We were pleased to have this co-operation and it occured to us that it would add to the interest of the Report to all concerned if we were to include these statements from the associations as an Appendix to the main Report. To help the associations, and to ensure that at least the main issues as we saw them were covered, we sent them lists of questions, inviting them to add others if they wished. All the associations replied, some at considerable length, and their state-ments are included in Appendix G.

Comparison with the 1928 Report

1.7. The 1928 Report covered the whole field of cereals marketing in England and Wales, from the production of the grain and its preparation for market, through the processes of conservation, transport and distribu-tion, to its ultimate use by maltsters, millers or feed manufacturers. As indicated in the following chapter we have adopted the same approach on the grounds that one cannot isolate a particular segment of the cereals market and say where production ends or utilisation begins. The marketing process involves the total relationship between producers, merchants and users.

1.8. At first sight, the position in 1928 seems to bear little resemblance to that of today. Technological progress in the intervening forty years has completely transformed the whole picture of cereals production, marketing and utilisation. In 1928 the grain was cut by binder, put into stacks and threshed later by steam tackle. The combine harvester had not yet come into commercial use, although some trials had been conducted and the report in fact contained a picture of one of the new "combined harvester-threshers". The report commented (p. 56) "The prospects for the machine in this country can thus be regarded as fairly promising", but it adds the prophetic warning: " . . . the general adoption of the machine here would probably accentuate the rush of supplies on to the market at harvest time, unless use were made of storage accommodation and drying facilities for bulk grain" (p. 57). Today combines are of course almost universal in Britain and the marketing problems resulting from their use will figure prominently in this Report.

1.9. Progress has not been confined to new machines but includes also important advances in plant breeding. It comes as a shock to us today, accustomed as we are to increases in average cereal yields, to read in the 1928 Report that yields of all the three cereals had hardly changed between 1885/6 and 1927/8 (Table 11, p. 179). Again, the problems arising out of the rapid increase in production in recent years will occupy much of our attention in this Report.

1.10. In the field of distribution, road transport was still something of a novelty in 1928 and nothing dates the report so much as the picture of a steam-driven bulk grain truck (p. 134). Water and rail transport were used for all but the short farm-to-mills hauls. Today, as indicated in Part III Chapter 3 of this Report, road transport has become universal and has almost entirely superseded water and rail for the transport of grain except for the very long hauls.

1.11. Major changes have also taken place in the utilisation of grain. Whereas in 1928 there were innumerable small flour mills and breweries scattered up and down the country,[1] today, after a long process of concentration, their number is very much reduced. This process was already well-established in 1928, for a footnote on p. 151 of the 1928 Report notes that in 1913 there had been 3643 commercial breweries in England and in 1928 only 1550. In the contrary direction, however, is the growth of animal feed compounding. In 1928 most merchants handled only straight grain, mainly for poultry, or provender such as rolled oats; today nearly 1000 merchants do some manufacturing of compound feeds.

1.12. One striking omission from the 1928 Report is any reference to the role of government policy in relation to cereals. This is because in 1928 there was very little direct government intervention although only four years later, with the passing of the 1932 Wheat Act, the Wheat Commission was established and the system of deficiency payments first introduced. The present Report has to recognise that the Government is deeply involved in all aspects of cereals policy, and that the Home-Grown Cereals Authority is playing an increasingly important role in the marketing of cereals.

1.13. However, in spite of all the changes in technology, business structure and government intervention since 1928, the fundamental problems of cereals marketing have hardly changed at all. It is remarkable how closely many passages of the 1928 Report apply to the present situation, as the following extracts will show.

(a) FROM THE INTRODUCTION

"Among the more important facts to which attention is drawn, is the tendency for the manufacture of cereal products of all kinds, in this country, to become concentrated into larger units. There is no corresponding reduction in the number of farmers selling to merchants or of

---

[1] See the maps on pages 118 and 119 of the 1928 Report. The former has been reproduced as an inset on Map 2.1 in Part Vb of the present report.

merchants selling to manufacturers. In the result, the balance of bargaining power has gone against the seller. This, of itself, suggests the need for some form of organisation among producers to reduce the number of independent sellers on the market. Further, organisation of producers seems to be necessary in order to ensure orderly marketing and avoid the seasonal depression in prices which usually occurs in the early months of each cereal year.

"Then, again, in order to meet the needs of modern industrial conditions, manufacturers require grain—whether wheat, barley or oats— to be delivered in large quantities of uniform quality. To this extent, home-produced grain is at a disadvantage compared with the standardised products arriving from abroad . . .

"Apart from these wider organisational aspects, the Report shows clearly that much can be done in other ways to extend the market for home-produced grain. Indeed, no effort should be spared in this direction in order that cereal growers, in the present state of their industry, may realise the best possible returns for their crops subject to the limitations imposed by world conditions. In this task, there is room for co-operation between farmers, merchants, manufacturers and consumers."

(page 8)

(b) ON FARMERS' SELLING GROUPS

"Although the existing merchant system is economical in its method of working, it does not follow, however, that there is no opportunity for farmers to improve their position as sellers on the market. At present, the seller, whether he be a farmer or a merchant, is placed at a disadvantage in every case—probably not so much with oats as with barley and milling wheat—owing to the fact that sellers are many and buyers are few. The final buyers are very few indeed, and although there is a free market in the sense that each side is free to 'take it or leave it', there is, in fact, no equality of opportunity or bargaining power between buyer and seller. Collective bargaining by farmers on national lines is out of the question as regards cereals in present circumstances, but it is worth considering whether, as a first step in organisation, it is not possible to build up a number of area selling agencies, under farmer-control, which, acting on behalf of members to determine where to sell, would be the sole vendors of grain to merchants and other buyers at country markets in their respective areas."

(pages 162/3)

(c) On orderly marketing and centralised selling

"Orderly marketing is a farmers' problem and no other section of the trade is in a position to change the existing state of affairs. Orderly marketing involves holding back excessive supplies from the market, and this, in turn, involves the questions of credit and storage ... Any organised effort by producers to ensure the more orderly marketing of the crop would need to be of national scope. There would, in practice, be very little hope of making a system of orderly marketing truly effective until control of the whole, or a very large part, of the crop in the country were vested in some central body ... Loyalty would be the very essence of such a movement and, in order to ensure adequate finance for those aspects of the scheme which involve storage off the farm, growers would need to be prepared to sign contracts, not only to sell all their grain through area-selling agencies, such as those referred to above, but to go further and leave the agencies free to control the flow of supplies to market under the direction of a central selling agency."

(pages 164–6)

(d) On use of home-grown wheat by millers

"The difference in price [i.e. between home and imported wheats] is very largely due to the very limited demand for native wheat from the large port millers. This lack of demand has been ascribed to two factors in particular—namely, the lack of 'strength' in native wheat flour and the lack of uniformity in the wheat as delivered ... A marked improvement has certainly been effected, but there is evidently a long way to go before English wheat will make a flour which is anything like as strong and reliable as one containing, say 50 per cent of Manitoba. From the point of view of the ordinary baker, this question of strength becomes almost entirely one of the yield of loaves per sack of flour."

(page 166)

" ... Variations in quality are due to many causes. In the first place, the number of varieties grown is large ... Then again, the condition of the wheat, particularly as indicated by moisture content, varies from season to season, from farm to farm, from one district to another and from time to time according to weather conditions. The amount and character of impurities vary from one consignment to another; within the same consignment, the condition of the grain in the sacks varies according to whether the grain, as threshed, came from the top, middle or bottom of the stack. All these differences give rise to vexatious adjustments in the process of handling and milling the wheat and account for the fact that English wheat is regarded as objectionable by large millers working on factory principles."

(page 168)

(e) On grading

"(5) The disability from which the marketing of home-produced wheat suffers through the absence of defined standards is obvious, but the method of applying a remedy is not. Conditions in this country, as a whole, are not comparable with those existing in the large exporting countries. There, the grain for export must be assembled before it is distributed by buyers; here, the actual assembly of wheat takes place to a limited extent in a few districts only, of which the country in the neighbourhood of the Wash is probably the most outstanding instance . . . In the trade in home-produced grain, the conditions usually associated with systematic grading and standardisation do not exist; the grain does not pass through 'bottlenecks' which might be used as points of inspection; and, except in the case of direct delivery to local mills, the buyer and seller are normally never both represented at any one point along with the grain.

"Any method of determining standards for native wheat would, therefore, need to be on a different basis from the methods commonly employed. When it comes to choosing the type of standard, the possibility of, or necessity for, a wide range of grades can be ruled out at once. From a milling point of view, all *good* English wheat is much on a level as regards quality. The large price range between the highest and lowest grades which exists in Canada, for instance, has no parallel in this country."

(page 170)

B*

CHAPTER 2

Criteria of an Efficient Marketing System

2.1. Although the primary aim of the present study is to provide a comprehensive description of the United Kingdom cereals marketing system as it now exists and to discern the changes which the system is undergoing, rather than to make a critical appraisal of its efficiency, it may be useful to precede the descriptive sections by a consideration of the characteristics of an efficient marketing system.

2.2. What is marketing: where does it begin and end? Conventionally, marketing of an agricultural product is often taken to embrace all the activities relating to the product which take place between the farm gate and the final disappearance (utilisation) of the agricultural product as such. These activities are often labelled "distribution" to distinguish them from the "productive" activities of the farm or the factory, and to emphasise that marketing is associated with the movement of commodities between buyers and sellers rather than with the creation of new products. This gives rise to a certain degree of disparagement of marketing as being a less creative occupation than farming or manufacturing.

2.3. This uncomplimentary attitude to those who fulfil the functions of marketing has been based largely on misconceptions which, though long-established, have been modified to an appreciable extent in recent years, if not dispelled altogether. These misconceptions concern particularly the nature and origin of marketing decisions and the notion of productive and unproductive activity.

2.4. An efficient marketing system will provide a means of determining the needs of consumers (buyers)—the types of commodities they require, the time and place at which they require those commodities and the prices they are prepared to pay; it will transmit this information about requirements promptly and accurately to the producers (sellers) through the mechanism of price quotations (offers and bids); and it will arrange for the transfer of commodities from seller to buyer in the required quantities and at the lowest possible cost. Marketing is thus a "matching" operation, bringing supply into proper relation to demand, and not simply a disposal

11

operation. Whenever a farmer decides to sow a certain acreage of wheat, to select a particular variety, to take measures to exclude weeds from the crop and to store the harvest on his farm for a certain period, these decisions affect his part in the marketing process; and it is to be hoped that in making the decisions he has heeded the signals that came to him, or should have come to him, through the marketing system. Similarly, at the other end of the process, whenever a miller or a brewer decides that he will use only grain of a certain protein content or possessing other specified qualitative characteristics, his decision will eventually have repercussions through the market mechanism which should find a response from farmers.

2.5. The threefold classification of production, marketing and utilisation which has been used in the title of this Report does not signify that these are mutually exclusive activities. It serves only to separate those businesses which are *mainly* concerned with growing grain from those which are mainly concerned with buying and selling it (the merchants) and from those which are mainly concerned with processing it (the end-users). Obviously there are many businesses which operate to an important extent in two of these groups of activities, and some in all three. The whole marketing system can operate efficiently only if the right decisions are taken by people in each of the three groups.

2.6. "Production" can be said to take place whenever resources (including human effort) are used to add value or utility. After grain has been grown in the field there are many ways in which value can be added to it before it comes to its ultimate destination and use. In discussion of marketing it is customary to distinguish at least three kinds of added value, giving the product respectively a higher time-utility, place-utility and form-utility. The creation of these utilities is no less productive than the growing of the crop itself, and it has corresponding claims for reward. Any activity which facilitates the movement of grain of the right kind to the right place at the right time can be called a productive activity; anything which obstructs such movement causes losses which are just as real as the losses represented by crop failure or destruction by pests.

2.7. Because of the value added between the farm gate and the point of delivery to the final user, a gap or "margin" occurs between the price received by the farmer and the price paid by the final user. It is sometimes supposed that the wider is this margin, the more inefficient is the marketing system and the more excessive are the profits of those engaged in marketing. However, profits can only be said to be excessive in relation to the services performed. In modern society the tendency is for more and more treatment to be given to raw materials (manufacturing, purification, standardisation, packaging, storing, sales promotion, retailing, etc.) before final delivery is made. Marketing margins expressed as price differences between the two ends of the chain therefore tend to become wider as time goes on, and this does not necessarily have anything to do with efficiency, which can only be judged after close scrutiny both of the exact nature of the final product and of the resources used to transform it, giving it its own unique specification.

2.8. For a full-scale study of efficiency, detailed analysis of prices in relation to the attributes of the grain at the point when those prices were paid would have been essential. The resources (materials, skill and capital) used in giving the grain those attributes would need to have been listed and costed, and comparisons would then have been necessary with some kind of standard or model of efficient performance. This we have not been able to undertake, nor were we asked to do so; but there is scope for valuable research work of this kind, and examples from the United States have shown what can be achieved in this direction. If the Home-Grown Cereals Authority wishes to promote an effective programme of marketing research this should certainly include price analysis and efficiency studies of the kind indicated. These go far beyond traditional inter-firm comparisons based on accounting data.

2.9. In the very broadest terms, a part of the economic system may be said to be functioning efficiently when it gives the greatest possible satisfaction in return for the resources used. It will be a characteristic of an efficient marketing system that the decisions taken at each point of economic activity between the raw material stage and the final product stage are the "right" decisions, in the sense that in the given circumstances and with the given state of knowledge they incur the least possible waste of resources.

2.10. In cereals marketing, inefficiency could take many forms, such as: oversupply of particular kinds of grain at particular times, leading to its diversion to inferior uses or to its deterioration in store; delivery of grain which did not conform to the specifications sought by the buyer, re-sulting in unjustified transport costs and unforeseen disposal problems; unnecessary haulage over long distances when suitable supplies were available close to the destination; erratic price fluctuations leading to un-expected losses and consequent uncertainty about future production and investment; absence of competition and restrictions on entry into business, resulting in slack performance being unchallenged by efficient operation, and the acceptance of familiar but poor standards of quality and service; failure to offer a price which adequately reflects differences in condition and quality of grain or in loading facilities at the farm; restrictions on the range of choice available to buyers and the provision of goods and services which are not really those which consumers most desire and are willing to pay for (e.g. insufficient product differentiation and inadequate market research); and failures in market intelligence leading to wrong decisions being taken on the basis of outdated or incomplete information.

2.11. In many countries today there is government intervention in the marketing system, designed to impose upon the system certain other functions besides that of matching supply to demand. For example, the Government may decide to use the market mechanism as a means of fulfilling policy objectives such as raising or stabilising farmers' incomes, or reducing imports in order to help the balance of payments. When such objectives are superimposed upon the broader objective of maximum efficiency of operation (greatest total return for the resources used), it becomes considerably more difficult to decide whether the marketing system is "working well"; for it may be achieving one of the objectives at the expense of another.

2.12. There is another important consideration affecting judgements of efficiency. A distinction should be made between the question: is there a reasonable standard of efficient performance within the structure which exists? and the question: is the structure itself an impediment to high performance and therefore to be described as inefficient? Measures to improve efficiency may take the form either of bringing about higher standards of performance within the existing structure or of structural adjustment which itself is conducive to efficiency. Clearly, both kinds of improvement are taking place at the present time in cereals marketing. By "structure" here is meant such features as the number of businesses, their size, their inter-relationships, the legal and commercial framework within which they operate, their geographical disposition and their degree of flexibility or specialisation. Capacity for adjustment is itself an important feature of market structure.

2.13. Short-run changes in supply and demand are normally accommodated by price movements to restore equilibrium to the market. Since demand is always liable to some degree of variation which cannot be foreseen—for example, the weather experienced during the winter and spring in the United Kingdom will, by its effect on the growth of grass, influence demand for manufactured feedingstuffs—and since supply is also liable to be affected unexpectedly by changes in the international market situation reflecting harvest conditions or state trading activities, some movement in prices must be regarded as a *normal* and *purposeful* feature, and need not be attributed either to mismanagement or to irresponsible speculation.

2.14. However, the extent to which it is desirable that all price fluctuations should be transmitted through the various strata within the marketing system or should be absorbed by a particular sector of it or by the Government (e.g. through the operation of deficiency payments) is an open question. When a rough road has to be travelled it is reasonable to ask that some shock-absorbers should be provided; and if the behaviour of prices is very erratic this is a sign that the shock-absorbers are not functioning well, and "over-adjustment" is liable to result. Questions of carryover of stocks and of the operations of futures markets should be considered in this light.

2.15. It can be argued that although producers should be aware of changes in market requirements it may be detrimental for them to experience sharp fluctuations in prices, as these will in no way improve their ability to market grain effectively or lead to more rational long-term decisions. On the other hand, the price system should at all times be able to operate in such a way that it accurately reflects factors to which buyers or sellers attach importance such as quality, timeliness and prompt payment.

2.16. Where economic or social policy requires that there should be a degree of intervention with the free play of the market forces, there is a two-fold obligation: first, to scrutinise the objectives of the policy and their relative position in the order of priorities; and second to consider whether the methods of intervention chosen are appropriate and are efficiently carried out. Frequently it may be found, in any commodity market situation, that the politically imposed goals are acceptable but the instruments chosen are ineffective or harmful.

2.17. The creation of conditions for "orderly marketing" is often held to be a responsibility of governments and of the marketing authorities which they set up, but the concept is seldom clearly defined. No doubt most producers, for their part, would prefer a system which always provided an assured market for whatever is produced, at guaranteed and predetermined prices. A total guarantee of this kind, however, shifts the burden of adjustment and risk-bearing elsewhere; for, as we have seen, there is a perpetual need for adjustment and a perpetual risk of unforeseen price movement. A passion for "tidiness" in prices cannot dispel the realities of changing circumstances. "Stabilisation" is a worthy objective in that it promotes confidence and removes the threat of sudden irrecoverable loss, but when it involves rigidity it exacts a high cost in terms of missed opportunities, which somebody has to pay. While farmers with cereals to sell must naturally be expected to seek the highest possible price, there are other farmers who are concerned with cereals only as a purchased feed. They will expect their merchants to buy in the cheapest market, subject to quality requirements, and they will also want to see a competitive compound feed industry which is constantly on the look out for cheap sources of supply so that the cost of the manufactured product may be kept down to a minimum.

2.18. This discussion has indicated certain principles of efficient marketing and has sought to establish certain theoretical standards or norms against which the realities of the situation as described in subsequent pages may be measured.

PART II

The Structure, Trends and Economics of Cereals Production

CHAPTER 1

Cereal Growing in the United Kingdom

1.1. This chapter describes the acreages, yields and production of cereals in the United Kingdom. It looks at some of the changes which have taken place in the past, and discusses what may happen in the future. Comparisons are made between different regions and counties. Much of the background information used in the chapter has been drawn from comments submitted by the Regional and County Officers of the N.A.A.S. and of the Advisory Services of Scotland and Northern Ireland.

1.2. For the descriptions of past trends, the period covered is from 1926 to 1967 inclusive, so that it is possible to see what has taken place since the last official report on cereals was published in 1928. These 42 years include three quite distinct periods, in each of which British farming was operating under different conditions. First, in the latter part of the inter-war years, farmers were continuing the slow process of recovery from the hard times of the early 1920's. Next, during the Second World War, from 1939 to 1945, came a period when maximum production was the objective and cropping was closely controlled. The distinctive characteristics of the third period have been the importance of government intervention and rapid technical progress. All these changes have been reflected in the acreage and production of cereals in the U.K.

United Kingdom Acreage of Cereals

1.3. In 1967 cereal crops covered some 9·4 million acres in the U.K. This compares with a total of 6·2 million acres in 1926, so that the acreage increased by just over one-half between these dates, despite the loss of more than 2 million acres of land in crops and grass to non-rural uses. The rate of increase, however, has been anything but uniform; Fig. 1.1 shows how U.K. acreages have changed over the period, for total cereals and for wheat, barley and oats separately. The same information is shown in Fig. 1.2 as a percentage of the acreage in crops and grass; subsequent

19

graphs in this chapter have also been drawn in percentage terms, to facilitate comparison of trends of cereal-growing in different parts of the U.K.

1.4. The total acreage under all cereals fell steadily between the two World Wars, reaching a minimum of some 5·2 million acres in 1937.

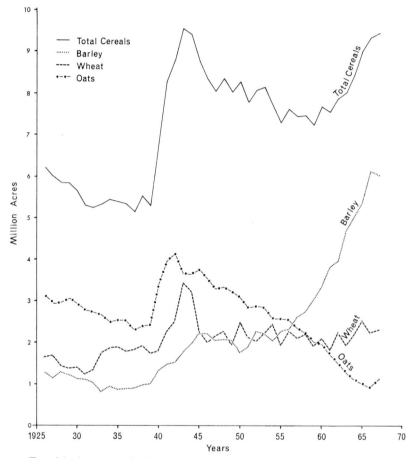

Fig. 1.1. Acreages of wheat, barley, oats and total cereals in the United Kingdom, 1926–67.
Source: Agricultural Statistics.

Following the outbreak of war in 1939 there was a spectacular rise, and the rate of increase continued almost unabated till 1943, when the acreage of cereals in the U.K. reached a peak of 9·6 million acres. There was a sharp decline from 1944 to 1947, followed by a slower and more irregular decrease until 1959—although the acreage in the latter year was still far above its previous low point of 22 years earlier. From 1959 to the present there has been an expansion in acreage nearly as rapid as that of the early war years, and the 1967 acreage is almost, but not quite, as high as the wartime peak.

1.5. In the depressed period between the wars, the low profitability of cereals (as indeed of most farm enterprises at that time) gave little incentive for technical progress. By the late 1930's government intervention was beginning to improve the situation, but there had been only a slight recovery in cereal acreage by the time war broke out. The large increases in acreage which took place during the Second World War were strongly influenced by the compulsory measures made necessary by the needs of the time. When the immediate crisis ended and compulsory cropping ceased,

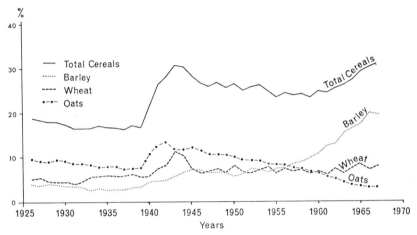

FIG. 1.2. Acreages of wheat, barley, oats and total cereals, as percentage of crops and grass area, United Kingdom, 1926–67.
Source: Agricultural Statistics.

many farmers began to grow fewer acres of cereals, especially in counties with high rainfall and heavy soil. Yet, taking the United Kingdom as a whole, there was no return to the pre-war situation; during the war years, experience had been gained in cereal growing—especially in parts of the country where the emphasis had previously been almost exclusively on livestock farming—the necessary equipment had been acquired, much grassland had been ploughed which in former years had been considered unploughable, and a considerable acreage of land had been upgraded through liming and draining.

1.6. The changes which have taken place since 1945 can be ascribed partly to government policy and partly to technical developments. Probably, especially in more recent years, technical improvements and management factors have been more important than subsidy levels in stimulating the increase in cereal growing. Yet it is doubtful whether technical developments could have gone ahead as quickly as they did if it had not been for the stability offered by government guarantees from the 1947 Agriculture Act onwards, and it is probable that, at least until about 1961, the relative levels of guarantees on cereals as against other enterprises strongly favoured cereal growing in most parts of the country. Cereal growing, especially the production of barley for compounding, was also stimulated by the derationing of animal feedingstuffs in 1954. Better cultivation was encouraged by drainage grants and subsidies on fertilisers and lime, and a direct impetus to crop production (in fact, mainly to cereal production) was given by the grassland ploughing grants. From the late 1950's, too, investment in equipment for cereal production and storage was aided by such measures as the Farm Improvement Scheme and Small Farmers Scheme, as well as by income-tax concessions.

1.7. Another factor, which links government action and technical trends, has been the increasing interest in farm management, and particularly the increased activity of the Advisory Services in assisting farmers with management problems as well as with purely technical affairs. By this means and through the press, radio and television, farmers have become much more aware of the requirements of management and of the advantages of cereal growing.

1.8. Of the technical factors influencing the post-war increases in cereal production, perhaps the most important was the introduction of improved varieties, especially the Scandinavian varieties of barley which began to make their full impact felt in the early 1950's. For oats and wheat as well as for barley, new varieties have made possible greatly increased yields, largely through their resistance to disease and their stiffer straw which discourages lodging and makes harvesting easier. Yields have also been improved by a wider and more scientific use of fertilisers, and by the development and application of new herbicides and agents of disease control. There has been continued progress in mechanisation, both for cultivations (the combine-drill is worthy of special mention) and for harvesting, and in the provision of grain-drying and storage equipment.

1.9. Another important technical development has been the great improvement in grassland management which has taken place in the post-war period. By the use of leys and by better understanding of grass utilisation, farmers have been able to reduce the area of grassland needed to maintain a given number of stock. Furthermore, perhaps as much as 4 million acres of agricultural land (mainly grassland) have been released by the decline in numbers of agricultural horses. Much of the land set free in these ways has been devoted to cereals.

1.10. Perhaps even more important than these technical advances, cereal growing has fitted into the changing patterns of farm management which have taken place since the war. Farms are continually tending to amalgamate and grow larger; and cereals, probably more than any other field enterprise, show economies of large-scale production. Labour difficulties are reduced by growing cereals; their total labour requirement is relatively low, because of the high degree of mechanisation which is possible, and in contrast to many livestock enterprises there is no need for the farmer or his employees to work a 7-day week. For a given financial return, cereals require a relatively low investment in capital as compared to livestock and some other crop enterprises. This has obviously been an attractive feature in the conditions of capital shortage which have faced the industry since the war. At the same time, once a farmer has invested in harvesting or storage equipment, it pays him to spread his overhead costs by expanding his cereal acreage still further.

1.11. In farming, as in other industries, the trend in management is towards specialisation and simplification. Here again, cereal growing scores: the production of cereals is probably the most straightforward type of work which can be undertaken on the farm. Specialisation in cereals has become more and more practicable with the development of fertilisers, selective herbicides and other chemical sprays. Using these, it has been possible to extend traditional rotations to include several successive straw crops, or even on some farms to grow corn continuously, without any serious diminution in yields.

1.12. Given all the advantages which can be gained from cereal growing, it remains to be explained why the cereal acreage did not begin to rise steadily from 1947 onwards. This is a complex question and, without a great deal of detailed research, it is impossible to do more than suggest the most probable reasons. The changes in cereal acreage recorded for the U.K. as a whole were the result of widely varying regional trends (discussed in more detail later in this chapter).

1.13. Immediately after the war there was a rapid fall in the acreage under cereals in those counties where conditions were least favourable, especially where they favoured dairying (which, alone of enterprises based on grazing livestock, has produced comparable returns). Furthermore, the impact of the technical and managerial innovations described above was not instantaneous. They did not occur simultaneously, nor were they adopted immediately or even equally rapidly in all parts of the country; indeed they tended to become accepted first in the main cereal-growing counties and then to spread outwards to other areas. In other words, while forces favouring a decline in cereal growing after 1943 exerted an immediate effect, those favouring its expansion have become progressively more important and began to be fully felt only in the late 1950's.

1.14. In the last decade, farmers have had to use all their management ability to meet steeply rising rents and wages. They needed an enterprise which would provide a good return per acre, using relatively little capital and making modest demands on labour. In many parts of the country, cereal growing fulfilled this need.

Relative Importance of the Three Main Cereals

1.15. As Fig. 1.1 shows, the changes which have taken place in total cereals acreage in the U.K. have been the result of widely differing acreage trends for each of the three main cereals. In examining the reasons for these changes, discussion can be simplified by considering two questions. First, why has the acreage of oats declined? Secondly, why was it barley rather than wheat which increased greatly after the Second World War?

1.16. In the years before 1939, oats could outyield the native barley varieties, such as Common and Spratt Archer, over much of the U.K. In the depressed conditions of the time, farmers had little incentive to apply fertilisers and lime, and soil acidity was often high, a condition which favours oats but not barley or wheat. The barley varieties available were weak-strawed and ran a severe risk of lodging if they were grown on any but the poorest land; furthermore, the main aim of barley growers was to produce a crop of malting quality,[1] so that good weather during harvest was highly desirable. The binder, in general use for harvesting at that time, is usually considered more suitable for oats than for barley (whereas the opposite is true of the combine harvester). Another reason favouring oats was that this was the crop fed to grazing livestock, especially farm horses.

1.17. In the war and post-war periods, subsidies on fertilisers and lime and greater prosperity in the agricultural industry have encouraged farmers to use more lime. As a result soil acidity has been reduced (although there is still very considerable scope for further improvement) and soil conditions have thus become more suitable for barley and less so for oats. Mention has already been made of the Scandinavian varieties of barley which became popular shortly after the war; one of their main properties was that they had stiffer and shorter straw than the old varieties and were much less susceptible to lodging and less vulnerable to bad weather generally, and so they were able to spread to parts of the country which were previously only suitable for oats. The yields from these new varieties of barley often equalled or surpassed those from oats. Combine harvesters were

[1]See the 1928 report, p.30.

being adopted more and more widely and their rise to prominence broadly parallels the emergence of barley as the leading cereal. The virtual disappearance of farm horses removed the need to grow oats for their feed and, with improved understanding of animal nutrition, the traditional reluctance to feed barley to other grazing stock, especially dairy cattle, was broken down. Farmers recognised that barley not only could yield a higher tonnage per acre than oats, but also gave a higher nutritional value in terms of starch equivalent per ton, while better grassland management and conservation techniques meant that the feeding of oat straw became less important than before the war.

1.18. Significant also was the post-war movement towards the growing of cereals as cash crops rather than for feeding on the farm. Whereas barley is in demand as a cash crop, for compounding, malting and for export, the sale market for oats is very small and specialised.

1.19. All these features have combined to bring about the fall in oats acreage and its replacement by barley. Oats persist mainly on higher, steeper land where combining is more difficult, on more acid soils, and in districts of heavier rainfall, where even the new varieties of barley may be at a disadvantage. Plant breeders seem to have had less success in developing improved oat varieties than they have had with barley. Stiffer-strawed types of oats have been introduced, more suitable for combining than the traditional varieties; however, these new varieties have so far been found to be rather late-ripening and this limits their suitability in the main oat-growing areas, where weather conditions late in the season are often bad.

1.20. Undoubtedly the main reason why barley, rather than wheat, has shown such a remarkable increase since the war is the greater suitability of barley for continuous cropping, i.e. for taking successive crops off the same land for several years without a break. If this is attempted with wheat, the crop is likely to suffer from a build-up of diseases such as eyespot, take-all, yellow rust and root-rot, and to become infested with perennial weeds. Barley is much less susceptible to those attacks, and from the experience of many farmers it appears that monoculture of barley can be continued almost indefinitely, provided proper measures are applied to control weeds and diseases. With wheat, on the other hand, few growers would continue taking crops from the same land for more than two successive years and most farmers still consider that wheat can only be grown after a root or grass crop giving a suitable "entry" and that it must be followed by another non-straw crop. In this respect, therefore, barley fits much better than wheat the trends in management towards simplification, specialisation and a quick return on capital.

1.21. In general, wheat is more difficult to cultivate than barley and is less tolerant of unfavourable weather and soil conditions (see Appendix A,

p. 698). This fact alone has restricted the geographical spread of wheat. Time of sowing is much more critical for wheat than for barley. Winter wheat in particular must be sown at precisely the right time; if bad weather prevents sowing at this time, the crop is usually not sown in autumn at all, and the yields from spring-sown wheat are markedly lower than from winter wheat.

1.22. The effect of government subsidy policy has also to be taken into account. Because of the different basis of payment (see pp. 653–5), it is not entirely clear how far the relative levels of subsidy on wheat and on barley have affected the attractiveness of either crop to farmers at different times and in different parts of the country. Probably the truth is that in the areas most suitable for growing wheat, where wheat yields per acre and hence tonnage subsidy payments are relatively high, the balance of advantage from subsidies has been in favour of wheat; in other areas the acreage payments on barley have been more attractive (though this has obviously become less true since the subsidy differential in favour of wheat was widened at the 1967 Price Review).

1.23. It is possible that the difference in methods of subsidy payment on wheat and barley (as distinct from the relative levels of subsidy) has had a specific effect on acreage trends. Because wheat must be sold off the farm to qualify for subsidy, it is in effect restricted to being a cash crop and the return is directly affected by yield. Barley can be a cash or a feed crop as the farmer wishes, and the subsidy payment per acre is the same whatever the yield. It is, of course, a separate question whether significantly more wheat would be retained for feeding on farms if the method of subsidy payment on wheat were changed to an acreage basis (a full discussion of this subject will be found in Appendix E). Quite irrespective of the effects of subsidy, barley is much more diverse in its cash outlets—compounding, straight feed, malting, brewing and distilling and the export market—than wheat, which, apart from the amounts used as a straight feed and in compounding, must go almost entirely for certain restricted uses in the flour-milling industry.

1.24. A feature of Fig. 1.1 which springs to the eye is the break in acreage trends which occurred in 1967 for both barley and oats. The U.K. acreage of barley decreased for the first time since 1954, while oats acreage showed its first increase since 1952. (Provisional acreage figures for the 1968 crop in England and Wales, newly available at the time of writing, indicate that the fall in barley and the rise in oats acreage have continued in that year.) It is too early to say how long these new trends will continue; forecasts made by the advisory officers of the U.K. (see section on Future Developments, pp. 42–46) suggest that during the 10 years from 1967 to 1977 the overall change in barley acreage will continue to be upwards—though at a lower rate than in the 10 preceding years—while oats acreage will level out at around the 1967 figure.

1.25. The decrease in barley acreage must be due in large measure to the changes in the acreage payments which were made in the 1967 and 1968 Price Reviews (see table, p. 816). In addition, the publicity put out by official bodies and others about the development of a surplus in barley must have affected farmers' reactions. Yet this is probably not the whole story. Comments made by the county advisers indicate that other factors are coming into play. Particularly in the more cereal-intensive counties, farmers are beginning to run out of land which can be used for cereal growing—which in effect usually means barley growing; of the acreage which is still not included in the arable rotation, a fairly large proportion is, for one reason or another, unploughable. Apart from this physical restriction, there is an economic restriction operating in counties suitable for arable cropping, in that expansion of cereals would compete with more profitable intensive root or market-garden crops. Most farmers still consider that a certain acreage of break crops must be retained, even in rotations with long runs of cereals, and this also restricts the acreage available for cereals. Many farmers, in any case, are beginning to feel uneasy about the possible long-term results of barley monoculture, and this may be helping to slow down the increase in acreage of the crop.

1.26. The stabilisation in oats acreage can be put down partly to the fact that oats, as has already been mentioned, tolerate soil and weather conditions in which no other cereal will flourish. In the parts of the country least suited to cereals growing, therefore, oats are likely to continue for a long time to come as the only cereal which can be grown for winter feed, and it may be that the substitution of barley for oats has gone as far as it can in these areas. Further, oat varieties are still being developed, and it is possible that plant breeders may be able to make up some of the leeway in development which oats varieties have experienced as compared to barley and wheat. The appearance of a short-strawed, easily combined variety of oat which was early-ripening and disease-resistant might well bring the oats crop back into its own in those districts which were its traditional strongholds before the war.

Cereal Yields in the United Kingdom

1.27. Some of the reasons for increases in cereal yields have already been mentioned. Figure 1.3 shows, for wheat, barley and oats respectively, the changes in yield which have taken place between 1926 and 1967. Also shown is the long-term trend in yield for each cereal, in terms of a 10-year moving average.

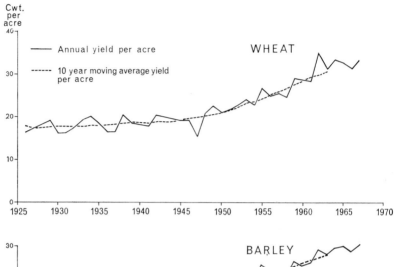

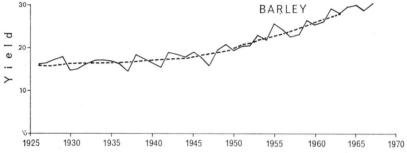

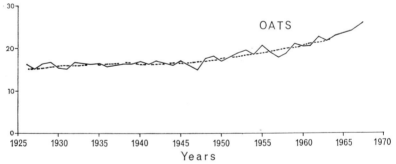

Fig. 1.3. Yields per acre, United Kingdom, 1926–67.
Source: Agricultural Statistics.
Note: 1967 yield figures are provisional.

1.28. Yield trends are similar for all three cereals. Before and during the war, yields rose very slowly. After 1945 the rate of increase became more rapid, and this higher rate has been maintained since about 1950. At first sight, it would appear that the yield of wheat has risen more steeply than that of barley. However, wheat yield started the period at a higher level than barley yield, and in percentage terms the long-term trend for barley shows a slightly greater increase; comparison of the 10-year averages centred on 1963 and 1926 gives an increase of 72% for barley and 70% for wheat. The increase in oats yield, however, has been markedly slower, showing an increase of only 49% between these two 10-year average levels. This reflects the smaller success in developing and adapting new high-yielding varieties of oats and, possibly, the lower application of fertilisers, especially in northern and western districts. Further, the decline in oats acreage has been especially marked in the counties which are more favourable to cereal growing; thus an increasing proportion of total oats acreage has been concentrated into less favourable areas, and this may be a factor contributing to the relative slowness of the increase in average yield of oats.

Cereal Production in the United Kingdom

1.29. Figure 1.4 shows, for wheat, barley and oats, how production has changed between 1926 and 1967. The most obvious feature is the massive

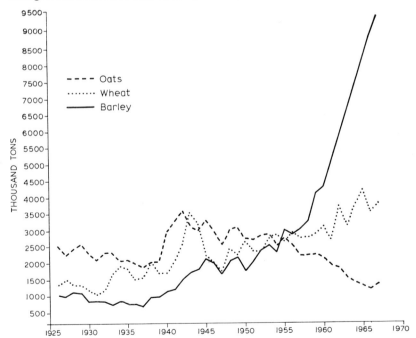

FIG. 1.4. Production of wheat, barley and oats in the United Kingdom, 1926–67.
Note: 1967 figures are based on provisional yield estimates.

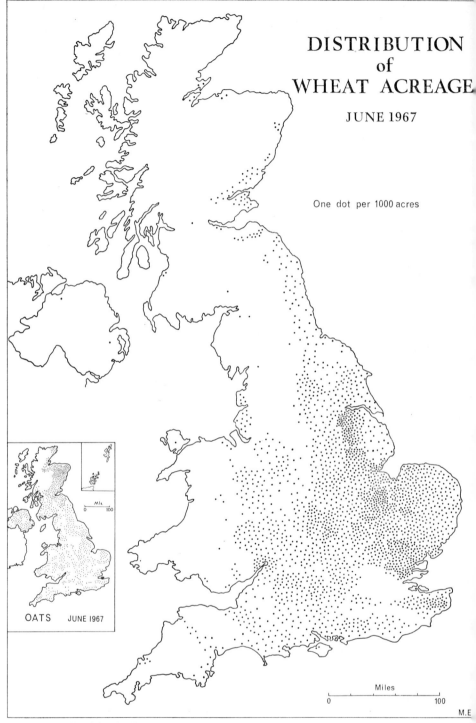

DISTRIBUTION
of
WHEAT ACREAGE

JUNE 1967

One dot per 1000 acres

OATS   JUNE 1967

Miles
0   100

M.E

MAP 1.1. Distribution of wheat and oats acreage in the United Kingdom, 1967.
Source: Agricultural Statistics of acreages by counties (positioning of dots
within counties is approximate only).

DISTRIBUTION
of
BARLEY ACREAGE

JUNE 1967

One dot per 1000 acres

Miles
0   100

M.E.S.

MAP 1.2. Distribution of barley acreage in the United Kingdom, 1967.
Source: Agricultural Statistics of acreages by counties (positioning of dots
within counties is approximate only).

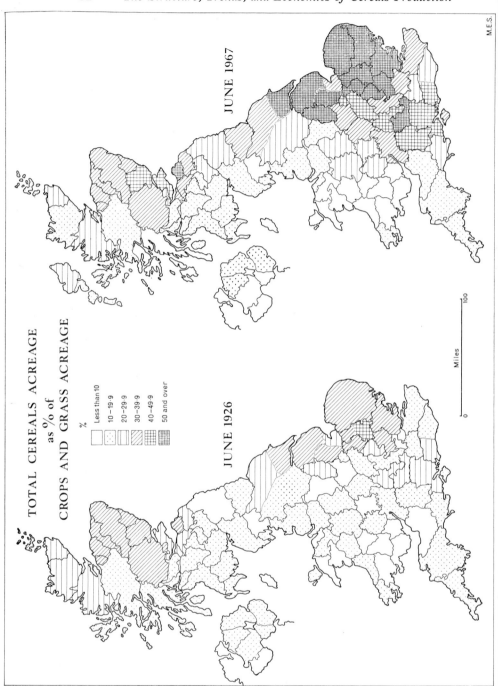

TOTAL CEREALS ACREAGE
as % of
CROPS AND GRASS ACREAGE

%

Less than 10
10 – 19.9
20 – 29.9
30 – 39.9
40 – 49.9
50 and over

JUNE 1926

JUNE 1967

M.E.S.

Miles

0   100

rise in the production of barley. In 1926 less barley was produced than either oats or wheat, and this remained true until 1954. Compared with the 1926 level, the 1967 production of barley showed a ninefold increase, while wheat production was more than two and a half times its 1926 level and production of oats was down by about one-half. These changes can readily be related to the trends in yield already mentioned and to the different acreage trends for the three cereals.

The Geographical Distribution of Cereals Acreage

1.30. Maps 1.1 and 1.2 show the distribution within the U.K. of the acreages under wheat, barley and oats. Cereal growing is concentrated mainly in the eastern counties of Great Britain, especially those extending from the East Riding of Yorkshire to Essex, although the "chalk" counties from Berkshire to Wiltshire are also important. In general, the acreage devoted to cereals declines westward, though both south-west and south-east England are somewhat anomalous. The presence of large upland masses makes the distribution of cereal growing in western and northern parts of Great Britain markedly peripheral, and the small acreages here must be borne in mind in interpreting county maps (Maps 1.3–1.11); this is particularly necessary with the large Highland counties of Inverness and Ross and Cromarty, where most of the cereals are grown in a small area on the east coast.

1.31. While it is possible to suggest fairly obvious reasons for these distributions, the reality is in fact very complex and the factors affecting the distribution of cereal acreages and, *a fortiori,* those of cereal yields, are so numerous and inter-related that no simple explanations can be expected. Appendix A (p. 695) contains a detailed discussion of these factors.

Regional Changes in Cereal Acreage

1.32. The regional concentration of cereal growing, already indicated in the dot maps (Maps 1.1 and 1.2), can be shown in another way by mapping counties according to the percentage of the total crops and grass area occupied by cereals. This has been done for total cereals in Map 1.3 (showing June 1926 and June 1967). Comparison of the maps for these 2 years gives the regional pattern of change in cereal growing over the period; this pattern is shown more clearly in Map 1.4, in which counties have been mapped according to the proportionate changes which have taken place in the importance of cereal growing.

1.33. The picture revealed is an interesting one. The most significant

C

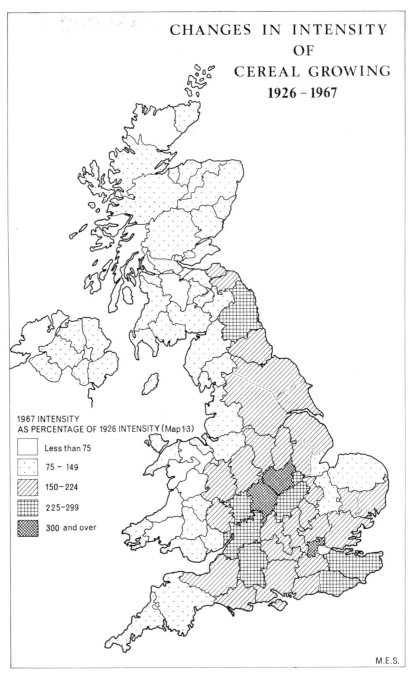

CHANGES IN INTENSITY
OF
CEREAL GROWING
1926 – 1967

1967 INTENSITY
AS PERCENTAGE OF 1926 INTENSITY (Map 1·3)

- Less than 75
- 75 – 149
- 150 – 224
- 225 – 299
- 300 and over

M.E.S.

MAP 1.4. Changes in intensity of cereal growing 1926–67.
Source: Agricultural Statistics.

1489002

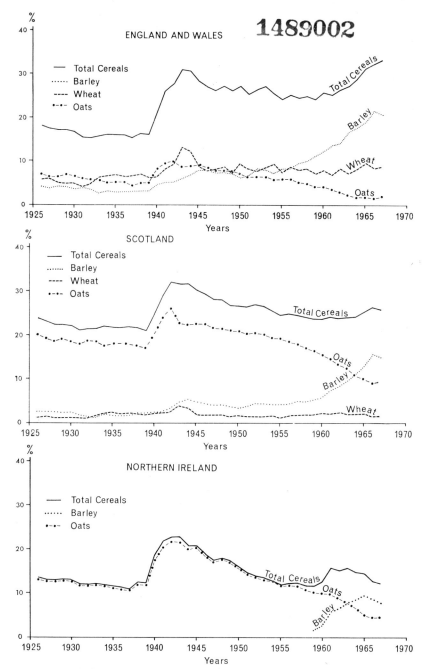

Fig. 1.5. Acreages of wheat, barley, oats and total cereals as percentage of crops and grass area, England and Wales, Scotland and Northern Ireland. *Source:* Agricultural Statistics.

increases in the proportion of cereals to crops and grass have been shown by those counties in the central and south-central parts of England in which cereals were only moderately important in 1926. In Wales and north-west England, on the other hand, there has been little change in the relative importance of cereals compared with 1926. Most noticeable is the fact that over nearly the whole of Scotland and in all six counties of Northern Ireland, there is no significant difference, as between 1926 and 1967, in the proportion of crops and grass accounted for by cereals.

1.34. A clue to the reason for these diverging patterns is given by Fig. 1.5, which shows the changes in the acreage of total cereals, wheat, barley and oats (as percentages of crops and grass) between 1926 and 1967 for England and Wales, Scotland and Northern Ireland.

1.35. In England and Wales the pre-war acreage of cereals was made up of wheat, barley and oats in almost equal proportions (with a small acreage of mixed corn, mainly in south-west England) whereas in Scotland and Northern Ireland oats were far more important than the other cereals. It may be supposed that this was due partly to natural conditions, which at that time made it difficult for any other cereal to compete with oats over most of Scotland and Northern Ireland, and partly to a traditional reliance on farm-grown oats as winter feed for stock in these countries.

1.36. In England and Wales, the increase in the acreage under barley has been much greater than the decrease in oats acreage and, as a result, the total cereals acreage was markedly higher in 1967 than in 1926; in Scotland and Northern Ireland, the increase in barley began later and only just outweighed the decrease in oats. This difference could be explained partly by growing conditions, for England has a relatively large acreage which is suitable for cereal growing and could therefore respond more readily to the various incentives to grow more barley. However, a large part of the explanation must lie simply in the relatively larger pre-war acreages of oats in Scotland and Northern Ireland; when oats declined, these countries were left with a larger gap to fill in total acreage than was the case in England and Wales.

1.37. Regional changes in the relative importance of wheat, barley and oats are shown also on Map 1.5, in which, for 1926 and 1967 respectively, counties have been mapped according to which of the three main cereals took up the largest acreage. In 1926 oats were the principal cereal everywhere in the U.K., except in an area of England extending from Lincolnshire in the north to Somerset in the south-west and Kent in the south-east, where wheat was the leading cereal; barley predominated only in East Anglia. The map for 1967 is in striking contrast. Barley had become the

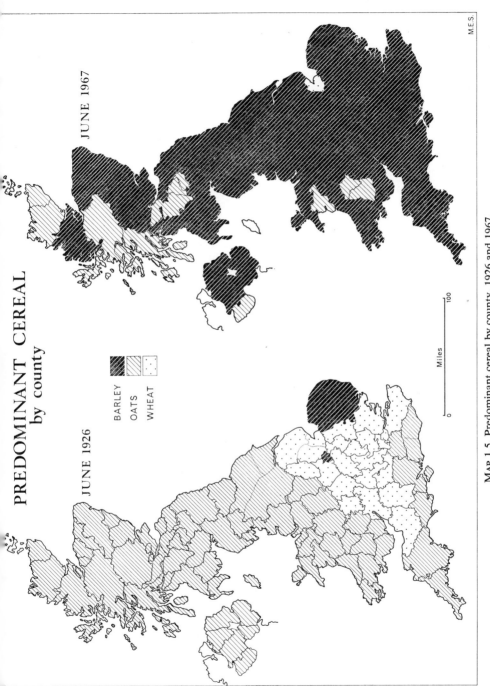

MAP 1.5. Predominant cereal by county, 1926 and 1967.
Source: Agricultural Statistics.

leading cereal in nearly every county in England and in most of the agricultural area in Scotland and Northern Ireland (although mapping on a parish basis would have shown a more restricted area of dominance). Oats had retained their prominence in some Scottish counties, although their share of the cereal acreage had declined considerably. Wheat was the leading cereal only in the Holland Division of Lincolnshire, although it was the second cereal nearly everywhere in the English lowlands.

1.38. These changes, of course, have come about only since the Second World War, and mainly since 1955. Barley extended its dominance from bases in East Anglia and East Lothian, first through other counties of eastern and southern England (excluding the Fenland), then through the Midlands and eastern Scotland to Wales, the Southern Uplands, Northern Ireland and the fringes of the Highlands.

Regional Variations in Cereal Yield

1.39. The relative levels of yield of the three main cereals for the U.K. as a whole have already been outlined. The regional pattern of yield variations for wheat, barley and oats respectively are shown in Maps 1.6, 1.7 and 1.8, in which counties have been shaded according to average yield per acre of wheat, barley and oats respectively over the three crop years 1965/6 to 1967/8.

1.40. Wheat yields are fairly uniform throughout England; they are generally rather lower in Wales, parts of Scotland and Northern Ireland. Yields of barley are, if anything, even more uniform than those of wheat— a result which reflects the relatively high adaptability of barley to different growing conditions. For oats it is difficult to extract a regional pattern of yield, because acreage of this crop is so small in many counties. A more detailed description of regional variations in yields, and a discussion of the causative factors behind these, is given in Appendix A, p. 695.

1.41. The data incorporated in the yield maps show that wheat outyields or equals barley nearly everywhere, and that the advantage of wheat in terms of yield is most marked in the less typically cereal-growing counties of England and also, more surprisingly, in several Welsh and Scottish counties where little wheat is grown. Barley also outyields oats in many counties and especially in the western and northern counties, even in those where oats are most prominent; in part these differences may simply reflect the inadequacy of the county as a basis for estimating yields, but they also suggest that other factors, like eligibility for winter keep subsidy, are also important in the choice of cereal.

1.42. Increases in cereal yields since the 1920's have not been uniform as

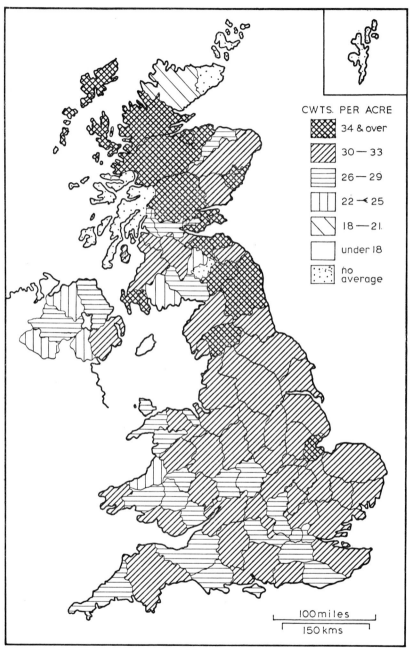

CWTS. PER ACRE

- 34 & over
- 30 — 33
- 26 — 29
- 22 — 25
- 18 — 21
- under 18
- no average

MAP 1.6. Three-year average yields of wheat by county (average 1965–7).
Source: Agricultural Statistics.
Note: The three-year average includes provisional figures for 1967.

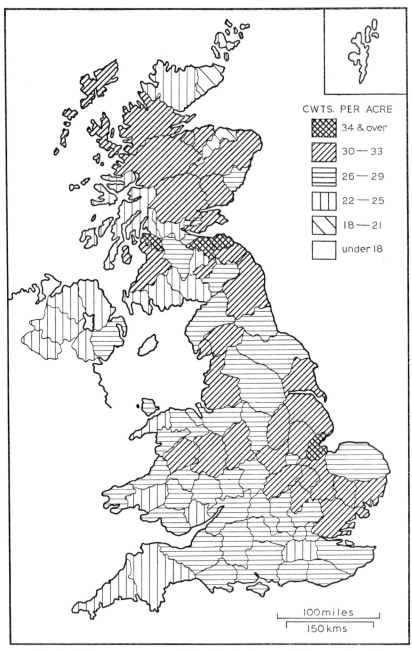

CWTS. PER ACRE

▨	34 & over
▨	30 — 33
▤	26 — 29
▥	22 — 25
▨	18 — 21
☐	under 18

100 miles

150 kms

MAP 1.7. Three-year average yields of barley by county (average 1965–7).
Source: Agricultural Statistics.
Note: The three-year average includes provisional figures for 1967.

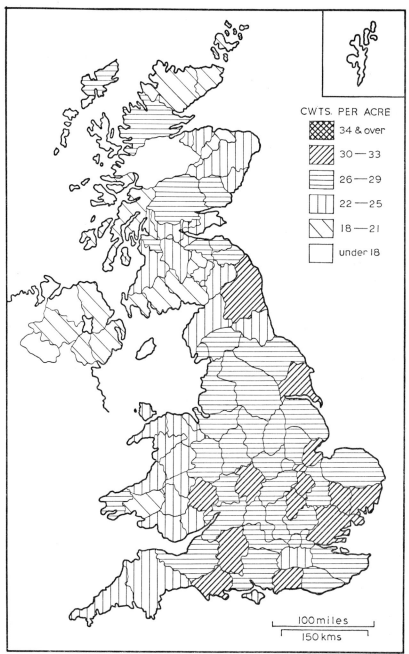

CWTS. PER ACRE

34 & over

30 — 33

26 — 29

22 — 25

18 — 21

under 18

100 miles

150 kms

MAP 1.8. Three-year average yields of oats by county (average 1965–7).
Source: Agricultural Statistics.
Note: The three-year average includes provisional figures for 1967.

C*

between different parts of the U.K. (although the small acreages grown in western and northern counties make comparison hazardous). Increases in yields of wheat have generally been greater in England and Wales than in Scotland and Northern Ireland, possibly because of lower applications of fertilisers and because of the importance of harvesting by binder in these countries. Increases in yields of barley show a similar pattern, and a preference for acid soils for the potato crop may be an additional factor here. Counties with the largest increases in oat yields lie mainly in southern England and in South Wales.

Regional Production of Cereals

1.43. The annual production of each of the three cereals in the counties of the United Kingdom is shown in Maps 1.9, 1.10 and 1.11. In each of these maps, the tonnage production per year (an average for the three crop years 1965/6 to 1967/8) is shown for each county as a circle of area proportional to the tonnage produced. The numerical values of the various circles are the same on all three maps.

Future Developments

1.44. As has already been mentioned, much of the background information in this chapter has been drawn from comments submitted by officers of the Advisory Services. These comments included some forecasts of possible changes in the acreage, yield and production of cereals. It was felt by the Survey team that if the advisory officers' individual forecasts could be collected in a standard form and collated to give predictions for the U.K. as a whole, this would provide a most useful means of forecasting, based on the advisers' intimate knowledge of conditions in their own counties.

1.45. Accordingly, towards the end of 1967 a questionnaire was sent to all county advisers in the U.K., asking specifically for numerical forecasts of acreage and yield for the three main cereals in their counties in the years 1972 and 1977. The advisers were asked to make two separate predictions of acreage; first, assuming that the U.K. would not have joined the Common Market by 1970; and second, assuming that the country would be in the Common Market by this date. For every figure in the questionnaire, the advisers were asked to state not only the forecasts they considered most likely, but also their estimate of the highest possible and lowest possible forecasts

1.46. The collated replies to these questionnaires form the basis of the results given in this section. Forecasts of acreage, yield and production

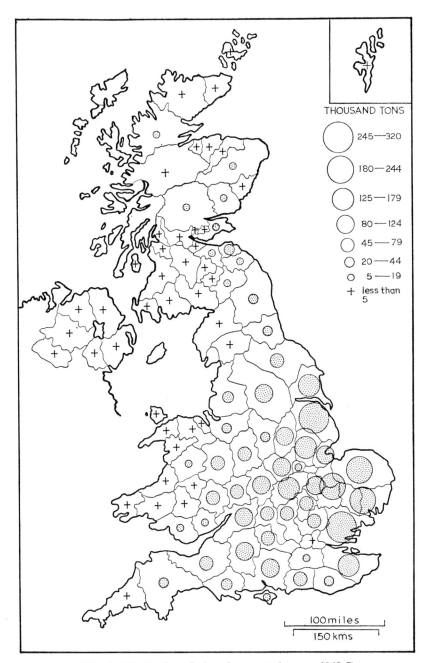

THOUSAND TONS

⬤	245—320
⬤	180—244
○	125—179
○	80—124
○	45—79
○	20—44
○	5—19
+	less than 5

100 miles

150 kms

MAP 1.9. Production of wheat by county (average 1965–7).
Source: Agricultural Statistics.
Note: The three-year average includes 1967 figures based on provisional yield estimates.

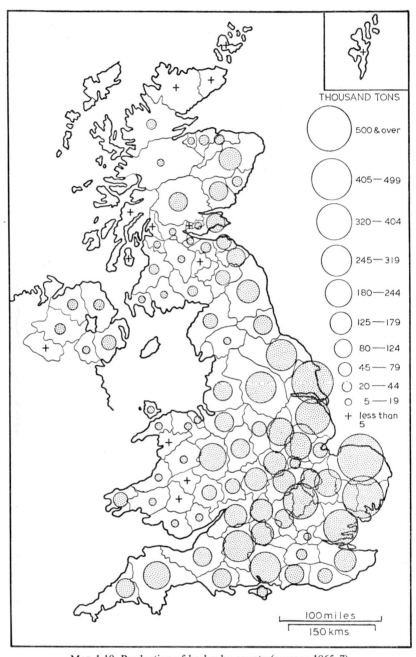

MAP 1.10. Production of barley by county (average 1965–7).
Source: Agricultural Statistics.
Note: The three-year average includes 1967 figures based on provisional yield
estimates.

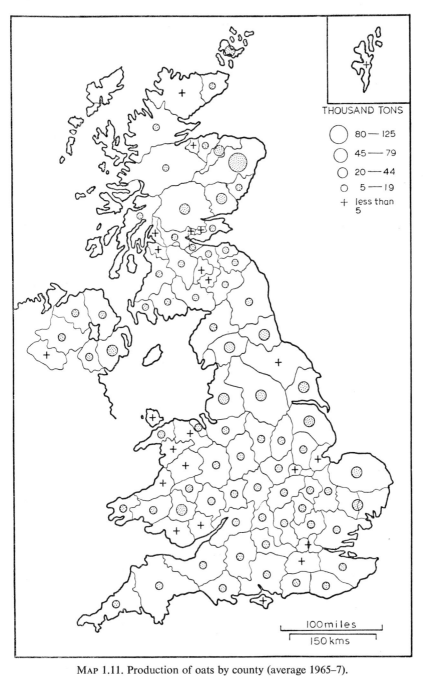

MAP 1.11. Production of oats by county (average 1965–7).
Source: Agricultural Statistics.
Note: The three-year average includes 1967 figures based on provisional yield
estimates.

were worked out for the U.K. as a whole; for England and Wales, Scotland and Northern Ireland; and within each country, for "cereals intensity groups" of counties, i.e. for counties grouped together according to the percentage of crops and grass acreage represented by cereals (see Map 1.3). In all cases, each forecast made for a particular cereal in any given county was weighted into the calculation according to the acreage of that cereal in that county at June 1967.

Acreages

1.47. Table 1.1 shows in percentage terms the forecast changes in cereals acreages calculated in this way for the U.K. Each forecast is shown as an index with 1967 = 100; e.g. the figure 115 means "a rise of 15% on the 1967 acreage"; the figure 91 means "a fall of 9% on the 1967 acreage".

TABLE 1.1

FORECASTS OF CEREALS ACREAGE IN U.K., 1972 AND 1977 FROM COUNTY ADVISERS'
RETURNS (as index with 1967=100)

	1967	1972			1977		
		Maximum	Most likely	Minimum	Maximum	Most likely	Minimum
U.K. outside E.E.C.							
Wheat	100	115	107	101	120	111	104
Barley	100	114	106	100	119	109	100
Oats	100	114	102	91	116	102	87
Total Cereals	100	114	106	99	119	109	100
U.K. In E.E.C.							
Wheat	100	119	112	105	123	114	106
Barley	100	117	110	103	122	113	104
Oats	100	117	105	95	123	105	91
Total Cereals	100	118	110	103	122	113	103

1.48. As Table 1.1 shows, the overall forecast calculated from the advisers' replies was that total cereals acreage in the U.K. in 1972, assuming non-membership of the Common Market, will most probably be 6% greater than in 1967, while by 1972 it will be 9% over the 1967 figure. The most probable percentage changes in wheat and barley acreage are expected to be very much the same as those for total cereals, with wheat showing the slightly higher proportionate rate of increase. Oats acreage, on the other hand, is expected to rise by only about 2% from 1967 to 1972, and, on the basis of the "most likely" forecast, no further change is expected in this crop by 1977.

1.49. For total cereals, and for wheat and barley individually, the "maximum possible" predictions for 1972 and for 1977 are very roughly for increases twice as large as those given in the "most likely" forecasts; the corresponding "minimum possible" figures are for little change on the 1967 level. For oats, on the other hand, there is a much greater range of uncertainty between the maximum and minimum forecasts. No doubt this reflects uncertainty over possible technical developments which might make oats a more attractive crop in relation to barley; it must also stem in part from the fact that the acreage of oats grown in most counties is relatively small, so that a given absolute change in this acreage will represent a relatively high percentage change.

1.50. Many of the advisers said they were very doubtful about the probable effect on cereals acreage of Britain's possible entry into the Common Market, and therefore had reservations about the validity of any forecasts they could make under this assumption. It is, of course, entirely proper to make this point, since at the time of writing it is impossible to say what arrangements would be reached as regards policy on cereal growing should Britain ever join the E.E.C., and the figures given should be interpreted in this light. However, the advisers' collated replies give a consistent picture of the expected effect of Common Market entry on predictions of U.K. acreage: every forecast figure is simply raised by a few percentage points over the corresponding forecast for "U.K. outside E.E.C.". The "most likely" forecasts of total cereals acreage, assuming membership of the E.E.C., are for a gain of 10% by 1972, or 13% by 1977, on the 1967 level. The relationships between forecasts for individual cereals, and between "maximum", "minimum" and "most likely" predictions, remain almost exactly as described in the paragraphs above for "U.K. outside E.E.C.".

1.51. Table 1.2 translates into acreage terms the percentage forecasts shown in Table 1.1. The acreages at June 1967 are given for comparison. According to the advisers' most likely estimates, it is expected that total cereals acreage in 1972 will be about 570,000 acres more than in 1967, and that of this increase roughly 370,000 acres will be in barley and 170,000 acres in wheat. At the highest possible expectation, total cereals acreage might rise by as much as 1·4 million acres between 1967 and 1972, while at the lowest there might be no significant change between the 2 years. Entry into the Common Market, if it takes place, is expected to bring additional rises of about 100,000 to 200,000 acres in each of the three main cereals by 1972, over and above the increases predicted in the absence of Common Market entry. The most likely forecast for 1977 is that total cereals acreage will have gone up by about another 250,000 acres from the level of 1972, with barley contributing about 160,000 acres of this increase.

1.52. The breaks in acreage trends which have taken place since 1966 show how little reliance can be placed on simple projection of past trends as a means of forecasting changes in cereals acreage, and for this reason no attempt is made here to arrive at any projected figures for 1972 and 1977 to compare with the forecasts calculated from the advisers' returns. It is interesting, however, to consider how the trends forecast by the advisers compare with the actual pattern of change in cereals acreage recorded over the period immediately before 1967 (see Fig. 1.1, p. 20). Of

TABLE 1.2

FORECASTS OF CEREALS ACREAGES IN U.K., 1972 AND 1977, FROM COUNTY ADVISERS' RETURNS ('000 acres)

	1967	1972			1977		
		Maximum	Most likely	Minimum	Maximum	Most likely	Minimum
U.K. outside E.E.C.							
Wheat	2305	2647	2475	2326	2765	2566	2385
Barley	6027	6902	6397	6018	7195	6559	6055
Oats	1012	1149	1031	919	1171	1028	881
Total Cereals[a]	9443	10,811	10,009	9362	11,252	10,263	9426
U.K. in E.E.C.							
Wheat	2305	2751	2584	2415	2827	2639	2431
Barley	6027	7077	6637	6215	7360	6828	6268
Oats	1012	1183	1058	959	1245	1063	922
Total Cereals[a]	9443	11,127	10,389	9691	11,552	10,642	9724

[a] Includes rye and mixed corn.

the three main cereals in the U.K., wheat is the only one for which the advisers' forecasts show a trend roughly comparable to that over the 5 years or so up to 1967 (i.e. very slightly upward). Even at the highest estimate from the advisers' returns, the rise in barley acreage is expected to go on much more slowly in the 10 years from 1967 than in the 10 years up to 1966, though it is not thought that the downturn in acreage which began in 1967 will persist to 1972. In the case of oats, the forecast figures represent a levelling-out of the steady downward trend in U.K. acreage which, until 1966, had been going on ever since the last wartime peak in cereal growing. The possible reasons for these breaks in the acreage trends for oats and for wheat have already been discussed (see p. 24).

Yields

1.53. Table 1.3 shows the yield changes forecast by the advisers for wheat, barley and oats in the U.K., again in the form of "maximum", "most likely" and "minimum" predictions for 1972 and 1977. Each forecast is given both as an index with 1967 = 100 and also as a figure of cwt. per acre.

TABLE 1.3

FORECASTS OF YIELD PER ACRE FOR THE THREE MAIN CEREALS IN U.K., 1972 AND 1977, FROM COUNTY ADVISERS' RETURNS

	1967	1972			1977		
		Maximum	Most likely	Minimum	Maximum	Most likely	Minimum
As index (1967=100)							
Wheat	100	111	106	101	117	111	104
Barley	100	111	106	101	118	112	104
Oats	100	113	108	101	120	114	104
Cwt. per acre[a]							
Wheat	33·3	36·9	35·2	33·7	39·0	37·1	34·7
Barley	30·7	34·0	32·6	31·0	36·3	34·5	32·0
Oats	26·5	30·1	28·5	26·8	32·0	30·2	27·8

[a] The 1967 yields shown for the U.K. include provisional figures for certain counties.

1.54. The "most likely" prediction is that both wheat and barley will yield 6% more per acre in 1972 than in 1967, and that a similar rate of increase will be maintained till 1977. For oats, the improvement in yield to 1972 is expected to be slightly greater. At the most, it is forecast, these rates of increase might be almost doubled; at the least, there might be only 1% improvement by 1972 and 4% by 1977. Broadly speaking, the "most likely" forecasts for the three cereals represent a continuation until 1972 of the yield trends recorded in the decade or so up to 1967; in the period between 1972 and 1977, increase in yield is expected to be very slightly slower.

1.55. For all three cereals, it is likely that plant breeders will continue to produce improved varieties; this is the main reason given by the advisers for the expected increases in yield. Development of chemical sprays is expected to continue, bringing better control of weeds, diseases and pests; in particular, a hoped-for possibility is the introduction of a spray which would eradicate wild oats and grass weeds from cereal crops.

Production

1.56. The expected changes in U.K. production of cereals are shown in Table 1.4, first in index terms with 1967/8 = 100, then as tonnages.

TABLE 1.4

FORECASTS OF CEREALS PRODUCTION IN U.K., 1972/3 AND 1977/8, FROM COUNTY
ADVISERS' RETURNS

	1967/8 (a)	1972/3			1977/8		
		Maximum	Most likely	Minimum	Maximum	Most likely	Minimum
(i) *As index (1967/8 = 100)*							
U.K. outside E.E.C.							
Wheat	100	127	114	102	140	124	108
Barley	100	126	113	101	141	122	105
Oats	100	128	110	92	138	115	90
U.K. in E.E.C.							
Wheat	100	132	119	106	144	128	110
Barley	100	129	117	104	143	127	108
Oats	100	132	113	96	147	120	96
(ii) *'000 tons* *U.K. outside E.E.C.*							
Wheat	3836	4867	4359	3919	5374	4760	4142
Barley	9242	11,686	10,413	9301	12,997	11,295	9674
Oats	1342	1716	1469	1233	1846	1547	1214
U.K. in E.E.C.							
Wheat	3836	5057	4556	4068	5502	4907	4221
Barley	9242	11,954	10,791	9578	13,243	11,750	9963
Oats	1342	1766	1520	1286	1970	1613	1286

(a) Based on yields which include some provisional county figures.

1.57. The advisers' forecast of the most likely changes by 1972/3, assuming Britain is not in the Common Market, is that annual U.K. production of wheat will rise by 14%, that of barley by 13% and that of oats by 10% over 1967/8 levels. These figures are equivalent, in approximate terms, to increases on 1967/8 of 520,000 tons for wheat, 1,170,000 tons for barley and 130,000 tons for oats. By 1977/8 annual production with Britain outside the E.E.C. is expected to have risen, as compared to 1967/8, by 24% (920,000 tons) for wheat; by 22% (2,050,000 tons) for barley; and by 15% (200,000 tons) for oats.

1.58. At the highest levels thought to be possible by the advisers, increases for wheat and barley production could be almost twice those given in the foregoing paragraph, while the lowest possible expectation for these crops is of little change by 1972/3 and a rise of 5% or so by 1977/8.

For production as for acreage of oats, the range between the maximum and minimum possibilities is wider than this. Entry into the Common Market, it is forecast, would bring rather larger increases all round; the additional rises would be between about 100,000 and 400,000 tons according to the forecast concerned, as shown in Table 1.4.

1.59. Table 1.5 gives the forecasts of production change by 1972/3 and 1977/8 worked out from the advisers' returns for "cereal intensity groups" of counties (see Map 1.3), and in total for England and Wales, Scotland and Northern Ireland. For simplicity, only the "most likely"

TABLE 1.5

FORECASTS OF "MOST LIKELY" CHANGES IN CEREALS PRODUCTION (ASSUMING U.K. NOT IN E.E.C.) BY CEREALS INTENSITY GROUP AND COUNTRY, 1972/3 AND 1977/8, FROM COUNTY ADVISERS' RETURNS (as index with 1967/8 = 100)

	1972/3			1977/8		
	Wheat	Barley	Oats	Wheat	Barley	Oats
Counties in which Total Cereals as percentage of Crops and Grass (June 1967) was:						
(i) *in England and Wales:*						
50% or over	108	109	113	118	119	119
40%—49·9%	119	117	112	129	123	121
30%—39·9%	118	119	125	130	131	147
20%—29·9%	119	118	130	129	131	143
10%—19·9%[b]	119	104	122	126	106	126
Less than 10%	114	118	111	133	147	126
England and Wales Total	114	113	120	124	122	131
(ii) *in Scotland*[a]:						
30%—39·9%	110	121	96	116	134	95
20%—29·9%	105	113	99	110	124	96
10%—19·9%	106	107	94	107	112	91
Scotland Total	110	115	95	131	125	94
(iii) *in N. Ireland*[a]:						
Northern Ireland Total	108	100	101	101	99	99
U.K. Total	114	113	110	124	122	115

[a] in Scotland, the remaining three groups each contained too few counties for an average to be quoted, and the same is true of the two groups making up Northern Ireland. The total figures for Scotland and Northern Ireland, however, *include* results for all counties.

[b] the unexpectedly low predictions for barley in this group result mainly from the forecasts made by the Devon and Cornwall advisers, who both expected barley acreage in their counties to decline.

predictions have been given, and all the forecasts assume non-membership of E.E.C. The relative rate of production increase in the highest "cereal intensity" group in England—the counties with over 50% of crops and grass acreage in cereals in 1967—is expected to be rather slower than in the rest of England and Wales. Despite this relatively low percentage increase, this group of counties is expected to contribute more to the *total* increase in cereal production by 1972 and 1977 than any other individual group, since its 1967 production figures are comparatively large. While in England and Wales the forecast is for an increase in the production of oats, a decrease in this crop is expected in Scotland. In Northern Ireland, it is forecast, the level of production by either of the two future years covered will not be markedly different from that in 1967/8.

1.60. These production forecasts have, of course, been worked out from the advisers' predictions of acreages and yields. In fact, the variations in production forecasts shown in Table 1.5 result almost entirely from expected acreage changes. The forecast changes in yields show no particular pattern as between "intensity groups" of counties.

1.61. It is necessary to re-emphasise the derivation of the forecasts of acreage, yield and production which have been given in this section. They are not the result of any mathematical technique of projection, but are the weighted averages of informed estimates made by advisers from their own expert knowledge of conditions in their counties. Further, the "maximum", "most likely" and "minimum" forecasts do not express any statistical concept of variation about a mean value; they are worked out directly from the advisers' own estimates of highest, lowest, and most probable changes. There is no reason to believe that this method of forecasting is any more or any less reliable than statistical projection of trends; however, the two approaches are essentially different in nature, and if the results from both are considered jointly they probably give as good an idea of the future as it is possible to arrive at. In the chapter on "The Likely Demand and Supply Situation for Cereals in the Next Five Years" (Part VIII, Chapter 2) the forecasts of production change given above are compared with predictions from other sources.

Technical Forecasts

1.62. In their comments, the advisers mentioned several future possibilities on the technical side of cereal-growing other than those directly connected with acreage, yield and production forecasts, and it is of interest to mention some of these very briefly. It is expected that methods of cultivation will become simpler and will require fewer separate operations. The trend towards larger-capacity machines for cereal harvesting will probably continue, and it is forecast also that still more farmers will install facilities for the bulk handling of grain.

1.63. A question of much current interest concerns the possible developments in break crops. Opinions among farmers differ on this subject, and there are differences also in the comments made by the county advisers. There is general agreement that present break crops are relatively unprofitable, and that more lucrative alternatives are needed. Possible crops mentioned as fulfilling this need in future are beans,[1] oilseed rape, maize, and crops for freezing or canning, such as peas, carrots and brussels sprouts. But each of these crops has its opponents as well as its advocates, and a fairly common opinion is that no strikingly suitable break crop is in view as yet. This particular question, therefore, remains open.

---

[1] In England and Wales the acreage of beans in 1968 (provisional figure) reached some 233,000, compared with about 139,000 in 1967. This rise followed the introduction of an acreage payment of £5 per acre for this crop, announced at the 1967 Price Review as applying over the 3 years 1968–71.

CHAPTER 2

The Structure of Cereals Production

2.1. This chapter will consider how cereals production is distributed between the farms of the United Kingdom. Statistics are given, based on the June returns, of the number of growers of wheat, barley and oats and of all cereals, and of the acreages they grow. The distribution is also studied by size of farm, by region and by type of farm. First the present structure will be described, using the latest available figures, and this is followed by a consideration of the trends which have emerged during the present decade.

Acreage Distribution

2.2. Table 2.1 shows that in 1967 there were some 119,000 growers of cereals in England and Wales[1]. Table 2.2 gives similar figures for Scotland except that no data are available for total cereals, but only for the individual crops. There were nearly 28,000 growers of oats in Scotland in 1967 and, no doubt, there would be some farmers growing barley or wheat who were not growing oats. As a round figure we might, therefore, take 30,000 as the number growing any cereals in Scotland. The number of growers in Northern Ireland in 1967 was 24,725. This makes a total of 174,000 growers in the U.K.

2.3. Table 2.1 shows that over two-thirds of the cereals acreage in England and Wales was grown on only 24,000 farms. At the other end of the scale there were over 50,000 holdings growing less than 20 acres of cereals each and accounting for only 5·2% of the total acreage. There were nearly twice as many growers of barley as of wheat, and one in seven of the barley growers had over 100 acres of this crop. Oats are generally

---

[1] Strictly speaking these figures are not the numbers of individual farmers growing cereals, but the number of holdings for which June returns were made and on which cereals were grown. There are, of course, some farmers or companies who occupy more than one holding and make more than one return. Thus the numbers in the tables somewhat overstate the actual number of persons making decisions to grow cereals.

TABLE 2.1

DISTRIBUTION OF CEREAL GROWING BY ACREAGE GROWN PER FARM
ENGLAND AND WALES, 1967

Acres of the crop	Wheat				Barley				Oats				Total Cereals			
	No. of growers	%	Acres ('000)	%	No. of growers	%	Acres ('000)	%	No. of growers	%	Acres ('000)	%	No. of growers	%	Acres ('000)	%
¼–4¾	6027	11·2	17	0·8	12,836	12·8	38	0·7	10,176	24·1	30	5·7	15,715	13·2	44	0·5
5–9¾	8369	15·6	57	2·5	14,919	14·9	103	2·0	13,404	31·9	92	17·7	16,343	13·7	113	1·4
10–19¾	11,556	21·5	157	7·0	17,807	17·8	247	4·7	11,520	27·3	153	29·3	19,049	16·0	264	3·3
20–29¾	6758	12·6	158	7·1	10,737	10·7	257	4·9	3854	9·1	90	17·2	11,745	9·9	280	3·5
30–49¾	7383	13·8	277	12·5	13,503	13·5	514	9·9	2259	5·3	83	16·0	14,496	12·2	552	6·9
50–69¾	4313	8·0	245	11·0	7942	7·9	462	8·8	548	1·3	31	5·9	9096	7·6	528	6·6
70–99¾	3665	6·9	295	13·3	7590	7·6	626	12·1	248	0·6	20	3·8	8819	7·4	735	9·0
100–199¾	3960	7·4	524	23·6	9981	10·0	1366	26·2	142	0·3	18	3·4	13,622	11·4	1887	23·4
200–299¾	976	1·8	227	10·2	2769	2·8	660	12·6	18	0·1	4	0·8	5331	4·5	1267	15·7
300 and over	567	1·2	266	12·0	2063	2·0	948	18·1	3	—	1	0·2	4867	4·1	2394	29·7
Total	53,574	100	2223	100	100,147	100	5221	100	42,172	100	522	100	119,083	100	8055	100

Source: M.A.F.F.

TABLE 2.2

DISTRIBUTION OF CEREAL GROWING BY ACREAGE GROWN PER FARM
SCOTLAND, 1967

Acres of the crop	Wheat				Barley				Oats			
	No. of growers	%	Acres ('000)	%	No. of growers	%	Acres ('000)	%	No. of growers	%	Acres ('000)	%
¼–4¾	258	9·2	1	1·3	1548	11·3	4	0·6	9683	34·7	17	4·2
5–9¾	446	15·8	3	3·8	1970	14·4	14	2·1	4429	15·9	31	7·6
10–19¾	718	25·5	10	12·5	2583	18·8	36	5·5	6572	23·6	93	22·8
20–29¾	465	16·5	11	13·7	1658	12·1	40	6·1	3350	12·0	80	19·6
30–49¾	499	17·7	19	23·8	1787	12·0	69	10·6	2646	9·5	99	24·2
50–69¾	202	7·2	12	15·0	1179	8·6	69	10·6	739	2·7	43	10·5
70–99¾	131	4·7	11	13·7	1084	7·9	90	13·8	312	1·1	26	6·4
100–199¾	84	3·0	11	13·7	1422	10·4	194	29·7	131	0·5	17	4·2
200–299¾	11	0·4	2	2·5	349	2·5	82	12·4	10	—	2	0·5
300 and over	—	—	—	—	142	1·0	56	8·6	—	—	—	—
Total	2814	100	80	100	13,722	100	654	100	27,872	100	408	100

Source: Dept. of Agriculture and Fisheries for Scotland.

grown in much smaller acreages per farm than barley or wheat, 92% of the growers having less than 30 acres of oats.

2.4. In Scotland (Table 2.2) the growing of oats clearly predominates over the other two cereal crops in terms of numbers of growers, though the total barley acreage exceeds that of oats. More than one-third of the growers of oats in Scotland have less than 5 acres of that crop.

2.5. Table 2.3 deals with the distribution of cereal growing by size of farm in England and Wales. Over one-third of the cereal growers have farms of at least 150 acres in size, and they account for almost 80% of the crop. There are over 30,000 cereal growers on small farms of under 50 acres, but they represent only 3·5% of the total cereals acreage, and even less of the wheat acreage. Wheat tends to be concentrated more on the larger farms than in the case of the other two crops, whereas on the smaller farms the barley acreage is generally three or four times as great as the wheat acreage.

2.6. The distribution of cereals by size of farm may be studied not only by reference to size as measured by farm area, but also to size of farm business. For this purpose it is becoming customary to use "Total standard man-days" as a basis of classification, being an estimate of the labour requirements represented by the crops and livestock on the farm. Table 2.4 shows that in 1966 in England and Wales there were 318,000 holdings in the Agricultural Census, of which 169,000 were of less than 275 standard man-days and generally regarded as part-time holdings. The remaining 149,000 (full-time) holdings are subdivided into three groups according to size of business. In the smallest group (275–599) barley was found on 43·3% of the farms, but wheat on only 19·0%. Among the larger farms (1200 and over) 74·2% were growing barley and 55·7% were growing wheat. Taking the full-time farms as a whole the figures were 55·9% for barley and 31·4% for wheat. The average acreages per grower of these full-time farms were 60 and 45 acres for barley and wheat respectively. Again, as in Table 2.3, it can be seen that wheat growing tends to be associated with the larger farm businesses, one-third of the wheat growers falling within the group "1200 and over" and accounting for 64% of the wheat acreage.

2.7. Table 2.5 shows the distribution of these crops by region and the number of acres per grower in each region. The wheat and barley distributions are quite similar except that more of the wheat is grown in the eastern and south-eastern part of the country, whereas barley is more prevalent in the south-west, the north and Wales. Acreages per grower are highest in the south-eastern region for both crops.

2.8. More detailed information about acreage of cereals per grower is shown in Map 2.1. This emphasises the fact that growers in central

TABLE 2.3

DISTRIBUTION OF CEREAL GROWING BY SIZE OF FARM (ACRES)
ENGLAND AND WALES, 1967

Size of farm (acres of crops and grass)	Total no. of holdings	Number of holdings growing								Total acres grown ('000)							
		Wheat	%	Barley	%	Oats	%	Any cereal	%	Wheat	%	Barley	%	Oats	%	All cereals	%
1–4¾	58,050 (a)	973	1·8	2695	2·7	240	0·6	3854	3·2	2	0·1	6	0·1	1	0·1	8	0·1
5–19¾	68,014	2839	5·3	7525	7·5	1247	2·9	10,597	8·9	11	0·5	45	0·9	4	0·8	61	0·8
20–49¾	51,067	5228	9·8	11,929	11·9	4323	10·2	16,430	13·8	45	2·0	136	2·6	20	3·8	208	2·6
50–99¾	50,441	8708	16·3	21,760	21·7	9815	23·3	27,076	22·8	117	5·3	416	8·0	69	13·2	620	7·7
100–149¾	26,698	7697	14·4	16,500	16·5	7912	18·8	19,021	16·0	147	6·6	514	9·8	79	15·2	755	9·4
150–299¾	30,945	15,438	28·8	24,624	24·6	11,762	27·9	26,608	22·3	541	24·3	1453	27·8	166	31·8	2185	27·1
300–499¾	10,495	7785	14·5	9729	9·7	4382	10·4	10,035	8·4	544	24·5	1186	22·7	94	18·0	1835	22·8
500–999¾	4586	3972	7·4	4417	4·4	2037	4·8	4468	3·8	518	23·3	977	18·7	63	12·1	1565	19·4
1000 and over	1005	934	1·7	969	1·0	454	1·1	994	0·8	298	13·4	488	9·4	26	5·0	817	10·1
Total	301,301 (a)	53,574	100	100,147	100	42,172	100	119,083	100	2223	100	5221	100	522	100	8055	100

(a) Excludes 5322 holdings consisting solely of rough grazings.
Source: M.A.F.F.

TABLE 2.4

DISTRIBUTION OF WHEAT AND BARLEY GROWING BY SIZE OF FARM BUSINESS
ENGLAND AND WALES, 1966

Size of farm business	% of total	Total no. of holdings	Per cent growing		Wheat					Barley				
			Wheat	Barley	Growers	%	Acres ('000)	%	Acres per grower	Growers	%	Acres ('000)	%	Acres per grower
275–599	17	64,674	19·0	43·3	12,294	22·7	216	10·0	17·6	28,030	26·7	727	13·8	25·9
600–1199	26	51,890	31·6	60·2	16,407	30·3	500	23·0	30·5	31,221	29·7	1467	27·7	47·0
1200 and over	50	32,462	55·7	74·2	18,083	33·4	1389	64·0	76·8	24,078	22·9	2824	53·4	117·3
Total (275 or more)	93	149,026	31·4	55·9	46,784	86·4	2105	97·0	45·0	83,329	79·3	5017	94·9	60·2
Less than 275	7	169,027	4·4	12·8	7382	13·6	66	3·0	8·9	21,697	20·7	270	5·1	12·5
Total, all holdings	100	318,053	17·0	33·0	54,166	100	2171	100	40·1	105,026	100	5287	100	50·3

Source: M.A.F.F., *Farm Classification in England and Wales 1966* (Tables 5 and 6), H.M.S.O., 1968.
Note: For parts of this table, comparable data for Scotland (1965) and N. Ireland (1964) are available in *The Structure of Agriculture*, H.M.S.O., 1966, p. 10.

TABLE 2.5

DISTRIBUTION OF WHEAT AND BARLEY GROWING BY N.A.A.S. REGION
ENGLAND AND WALES, 1966

	Wheat					Barley				
	No. of growers ('000)	%	Acres ('000)	%	Acres per grower	No. of growers ('000)	%	Acres ('000)	%	Acres per grower
Eastern	17·1	31·7	763	35·1	44·6	22·6	21·6	1308	24·7	57·9
South Eastern	6·8	12·6	348	16·0	51·2	10·5	10·0	794	15·0	75·6
East Midland	9·3	17·0	433	19·9	47·0	14·4	13·7	828	15·7	57·5
West Midland	6·8	12·6	214	9·9	31·5	12·6	12·0	477	9·0	37·9
South Western	5·6	10·2	197	9·1	35·8	16·8	16·0	750	14·2	44·6
Yorks. and Lancs.	4·8	8·9	141	6·5	29·4	10·1	9·6	537	10·2	53·2
Northern	2·4	4·4	54	2·5	22·5	9·7	9·3	461	8·7	47·5
Wales	1·4	2·6	22	1·0	15·7	8·2	7·8	132	2·5	16·1
Total, England and Wales	54·2	100	2172	100	40·0	104·9	100	5287	100	50·4

Source: M.A.F.F., *Farm Classification in England and Wales 1966*, H.M.S.O., 1968.

TABLE 2.6

DISTRIBUTION OF WHEAT AND BARLEY GROWING BY TYPE OF FARMING
ENGLAND AND WALES, 1966

Type of farming	% of total	Total no. of holdings	Per cent growing Wheat	Per cent growing Barley	Wheat Growers	Wheat %	Wheat Acres ('000)	Wheat %	Wheat Acres per grower	Barley Growers	Barley %	Barley Acres ('000)	Barley %	Barley Acres per grower
Dairy	29	59,348	17·3	46·2	10,267	19·0	290	13·4	28·2	27,393	26·1	880	16·6	32·1
Cattle and/or sheep	11	24,953	18·9	46·0	4716	8·7	138	6·4	29·3	11,471	10·9	415	7·8	36·2
Pigs and/or poultry	6	8744	15·3	36·4	1341	2·5	44	2·0	32·8	3183	3·0	156	3·0	49·0
Cropping, mostly cereals	5	8605	81·8	99·6	7037	13·0	518	23·9	73·6	8567	8·2	1327	25·1	154·9
General cropping	16	17,275	77·5	92·6	13,396	24·7	724	33·3	54·0	15,998	15·2	1290	24·4	80·6
Horticulture	16	14,824	18·3	24·4	2717	5·0	83	3·8	30·5	3620	3·4	116	2·2	32·0
Mixed	11	15,277	47·8	85·7	7310	13·5	308	14·2	42·1	13,097	12·5	833	15·8	63·6
Total (275 or more)	93	149,026	31·4	55·9	46,784	84·6	2105	97·0	45·0	83,329	79·3	5,017	94·9	60·2
Less than 275	7	169,027	4·4	12·8	7382	13·6	66	3·0	8·9	21,697	20·7	270	5·1	12·4
Total, all holdings	100	318,053	17·0	33·0	54,166	100	2171	100	40·0	105,026	100	5287	100	50·3

Source: M.A.F.F., *Farm Classification in England and Wales 1966*, H.M.S.O., 1968.
Note: For parts of this table, comparable data for Scotland (1965) and Northern Ireland (1964) are available in *The Structure of Agriculture*, H.M.S.O., 1966, pp. 31–35.

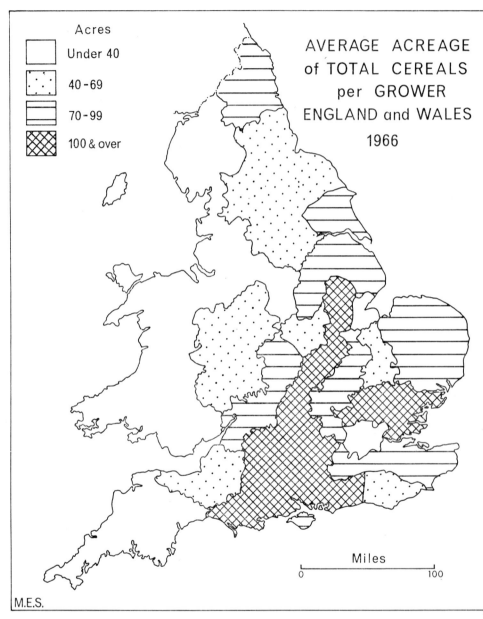

MAP 2.1. Average acreage of total cereals per grower, England and Wales,
1966.
Source: Agricultural Statistics.

southern England, in Essex and in a belt of counties through the Midlands tend to have larger acreages of cereals per farm than in the rest of the country. It was in these counties that the combine harvester first came into general use.

2.9. In recent years the Ministry of Agriculture has analysed the structure of farming in terms of a type-of-farm classification. Table 2.6 shows how many farms were placed in each type-of-farming class in 1966, how many of them were growing wheat and barley, and their acreages of each crop. About one-quarter of the acreage of wheat and barley was on farms classified as "Cropping, mostly cereals". These are farms on which crops are estimated to account for more than 50% of the total labour requirements of the farm and cereals make up 50% or more of the cropping labour requirements. They might be called "cereals specialists". There are only 8605 such farms and numerically they are a minority among cereal-growers, but their relative importance is clear, and comparison with similar data for earlier years shows that they are gradually taking over more and more of the total production. The remaining acreage of cereals is concentrated mainly in general cropping farms, dairy farms and mixed farms. Almost all of the "cereals specialists" grow barley, with an average of 155 acres under that crop, but in 1966 nearly one in five of them were not growing wheat.

2.10. Table 2.7 presents average figures from the financial accounts of thirty-four farms which were placed in the "Cropping, mostly cereals" group and in the largest size-of-business group (1200 or over). It is interesting to see the composition of their total turnover and of their expenditure. Their receipts from the sale of cereals came to over £21,000 per farm out of a total turnover of £35,000. They were spending, on an average, £3,116 a year on fertilisers and £1,517 on seeds. This was on a farm of 859 acres (average size). Although they were mainly engaged in cash cropping, it will be seen that they received over £8,000 from the sale of livestock products.

2.11. Table 2.8 shows corresponding figures for small farms engaged mainly in the production of cereals. Here total turnover was only just over £5000 per year on 135 acres, with sales of cereals accounting for £3248. Expenditure on fertilisers, seeds and other items were correspondingly low compared with the larger farms shown in Table 2.7, but in both cases total inputs absorbed about four-fifths of the total value of output.

TABLE 2.7

AVERAGE OUTPUT, INPUTS AND NET INCOME OF THIRTY-FOUR FARMS IN
"CROPPING, MOSTLY CEREALS" TYPE GROUP AND IN "1800 OR OVER"
SIZE-OF-BUSINESS GROUPS

ENGLAND AND WALES, 1966

Output	£	Inputs	£
Wheat	6879	Labour	6535
Barley	13,904	Rents and rates	4058
Other cereals	549	Machinery and power	6102
(Total cereals)	(21,333)	Fertilisers	3116
Potatoes	1265	Seeds	1517
Sugar beet	1407	Feeds	3079
Horticultural crops	728	Other	2866
Other crops	1842		
Cattle, sheep and wool	3683		
Pigs, poultry and eggs	4186		
Other livestock products	495	Total inputs	27,273
Misc. revenue and production grants	379		
Total gross output	35,317	Net Income[a]	8044

Average size of farm: 859 acres
Estimated value of tenant's
capital: £39,000.

[a] Net income is the amount remaining to remunerate the farmer for his management
and labour and before charging interest on all farming capital other than land and
buildings.
Source: M.A.F.F., *Farm Incomes in England and Wales 1966,* H.M.S.O. 1968, p. 49.

Trends in the Structure of Production

2.12. The foregoing section has described the structure as it existed at
a recent date (1966 or 1967). In the diagrams which follow some significant
trends have been portrayed, based on an analysis of annual figures prepared by the Ministry of Agriculture from the June Returns. The statistics
themselves have not been reproduced in detail, but they are available
from the Ministry and some have been published.

2.13. Figure 2.1 shows that the number of growers of cereals in England
and Wales has been falling steadily since 1960. In the case of oats this
has been largely due to the decline in the acreage of that crop, but with
wheat the reduction in the number of growers is attributable to a concentration of the crop into fewer farms. In the case of barley it will be
noted that the number of growers has shown only a modest increase, in
spite of the remarkable expansion in acreage (see Chapter 1 and Fig. 1.1).
If the trend continues, the number of cereal growers in 1972 in England
and Wales is likely to be less than 100,000, compared with 189,000 in 1954
and 147,000 in 1960.

TABLE 2.8

AVERAGE OUTPUT, INPUTS AND NET INCOME OF TWENTY-ONE FARMS IN "CROPPING, MOSTLY CEREALS" TYPE GROUP AND IN "275–499" SIZE-OF-BUSINESS GROUPS

ENGLAND AND WALES, 1966

Output	£	Inputs	£
Wheat	943	Labour	751
Barley	2280	Rents and rates	593
Other cereals	25	Machinery and power	1256
(Total cereals)	(3248)	Fertilisers	425
Potatoes and sugar beet	157	Seeds	326
Other crops	319	Feeds	504
Cattle, sheep and wool	812	Other	451
Pigs, poultry and eggs	351		
Other livestock products	—		
Misc. revenue and production grants	155	Total inputs	4306
Total gross output	5042	Net income[a]	736

Average size of farm: 135 acres
Estimated value of tenant's
capital: £4000.

[a] Net income is the amount remaining to remunerate the farmer for his management and labour and before charging interest on all farming capital other than land and buildings.
Source: M.A.F.F., *Farm Incomes in England and Wales 1966*, H.M.S.O., 1968, p. 49.

2.14. Figure 2.2 shows the other side of the coin. While the number of growers has been falling, the acreage per grower has risen steadily, with oats again being the exception. This increase in cereals acreage per farm has occurred partly because farms themselves are tending to become fewer and larger, but mainly because of the tendency towards specialisation of cropping—fewer and larger enterprises within the farm. In view of the capacity of the modern combine harvester, it is, however, surprising that the average acreage of cereals per grower is still only about 70 acres. There are, of course, still many growers on the smaller farms who are harvesting with binders.

2.15. In Fig. 2.3 the decline in the numbers growing wheat, already noted in Fig. 2.1, is analysed by size of farm. It is clear that there were farms of all sizes (except the very largest) which tended to give up wheat production during the period shown. Thus, if we take farms of 100 to 150 acres, Fig. 2.3 shows that in 1958, 50% of these were growing wheat, but by 1966 this had fallen to 30%.

2.16. By contrast, Fig. 2.4 shows that for barley the reverse was true. In 1958 only 48% of the farms of 100–150 acres were growing barley, but by 1966 this proportion had risen to 64%. The parallel movement in other size groups is very striking, emphasising that the conditions which made

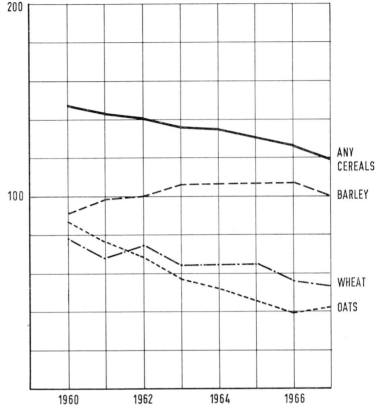

FIG. 2.1. Number of growers of cereals, England and Wales, 1960–7.
Source: Agricultural Statistics.

barley a more attractive proposition for farmers were applicable over a wide range of farm sizes.

2.17. The picture given for oats in Fig. 2.5 is quite different from that of either wheat or barley. The decline in the total number of growers of oats has already been shown in Fig. 2.1, but whereas in the case of wheat the larger farms are still growing that crop despite its declining popularity, in the case of oats many of the large farms have given up this crop. In 1958 oats were to be found on nearly 80% of all farms over 300 acres; by 1966 this proportion had fallen to less than 40%.

2.18. The tendency for cereal production to be concentrated into fewer hands has already been mentioned and the point is further illustrated by Figs. 2.6 and 2.7. Figure 2.6 indicates that those who grow at least 100 acres of a cereal crop are taking on a larger and larger proportion of total

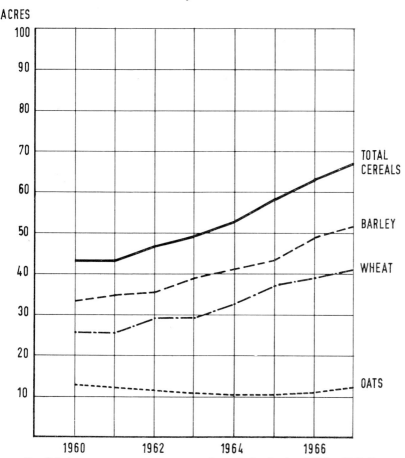

FIG. 2.2. Average crop acres per grower of cereals, England and Wales, 1960–7.
Source: Agricultural Statistics.

production. This is a very steady trend for barley and wheat, and is beginning to show itself also, though to a much more modest extent, in oats. It may well be that by 1972 at least 80% of the total cereals acreage in England and Wales will be on holdings growing at least 100 acres of cereals.

2.19. From Fig. 2.7 it is possible to see what proportion of the total acreage is in the hands of a given number of growers. For example, in 1954 the "top" 30,000 growers accounted for 60% of the acreage; by 1967 they covered more than 75%. This provides a useful indication of the relative importance to the cereals market of the largest producers. However, they are still to be numbered in tens of thousands; the process of concentration has still a long way to go in agriculture by comparison with most other industries.

D

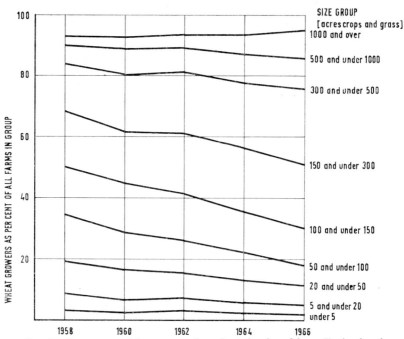

FIG. 2.3. Percentage of farmers growing wheat, by size of farm, England and
Wales, 1958–66.
Source: Agricultural Statistics.

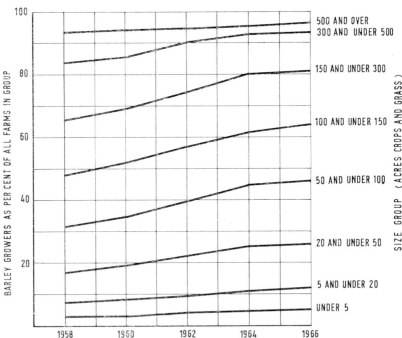

FIG. 2.4. Percentage of farmers growing barley, by size of farm, England and
Wales, 1958–66.
Source: Agricultural Statistics.

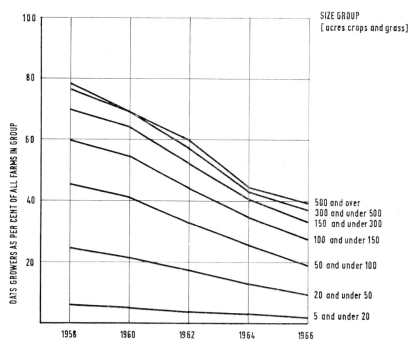

SIZE GROUP
[acres crops and grass]

500 and over
300 and under 500
150 and under 300

100 and under 150

50 and under 100

20 and under 50

5 and under 20

OATS GROWERS AS PER CENT OF ALL FARMS IN GROUP

Fig. 2.5. Percentage of farmers growing oats, by size of farm, England and Wales, 1958–66.
Source: Agricultural Statistics.

PER CENT

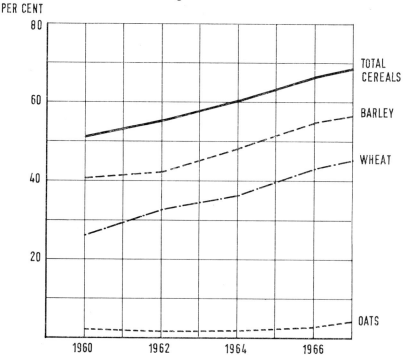

TOTAL CEREALS

BARLEY

WHEAT

OATS

Fig. 2.6. Proportion of total acreage in England and Wales which was grown on farms having at least 100 acres of the respective cereal.
Source: Derived from Agricultural Statistics.

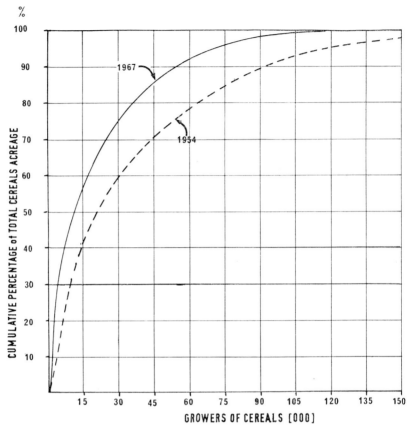

FIG. 2.7. Concentration of cereals production on England and Wales:
percentage of total cereals acreage accounted for cumulatively by various
numbers of growers, arranged in descending order of their cereals acreage, 1954
and 1967.
Source: Derived from Agricultural Statistics.

CHAPTER 3

The Production of Cereals

Introduction

3.1. This chapter, together with Chapter 6, is mainly based on the results of field surveys conducted by Market Investigations Ltd. The way the samples were drawn is described more fully in Appendix B, but briefly the method used was as follows:

NATIONAL SURVEY

3.2. This comprised two randomly-selected samples, each of 600 cereal farmers in the U.K. The first 600 farmers were visited during July/August 1967 (Phase I) and the second 600 during October/November 1967 (Phase II). Most of the following tables have been based on the two samples taken together (called for convenience "the National Survey") but in some cases it has been more appropriate to use one or the other (e.g. when discussing harvest intentions).

AREA SURVEY

3.3. This comprised a total sample of 202 cereal farmers, split more or less evenly between six contrasting areas of Great Britain: Norfolk (around Hickling), Hampshire (around Whitchurch), Yorkshire (East Riding), Devon (South Hams), Northamptonshire (around Thrapston) and East Lothian (around North Berwick). The object of this Survey was to study in some depth the system of cereal production and marketing in contrasting areas, not at one point in time only but over the cereal year as a whole. Thus personal interviews were made in May/June 1967 and again in April 1968, whilst telephone interviews with these same farmers were carried out in July and October 1967 and again in February 1968.

Specimen blank copies of the many questionnaires used in these Surveys will be made available by the University on request.

A. The Characteristics of Cereal Farmers

3.4. A number of questions were put to the farmers interviewed with the object of assessing the extent to which they tried to keep abreast of technological and commercial developments and agricultural matters generally. The replies to these questions, besides being of interest in themselves, served for the construction of indicators of the general "awareness" or "participation" of the cereal farmers interviewed. Further analysis then showed the extent to which this quality was associated with certain production and marketing methods.

PARTICIPATION INDEX

3.5. The following aspects were taken into account for the purpose of classifying farmers according to their general level of "participation":
—membership of farmers' union, club, society or association;
—membership of buying or selling groups;
—participation in other business arrangements with farmers in the neighbourhood;
—attendance at agricultural shows;
—attendance at technical meetings and farming conferences;
—registration of forward contracts with H.-G.C.A.;
—views concerning quality of grain in relation to its price;
—views concerning grading of grain.

3.6. Three categories of "participation" were established (A, B and C) from highest to lowest, and in the sample the dividing lines were drawn so as to give approximately equal numbers in each category. However, after weighting to adjust for the over-representation of the larger farms (see Appendix B) the distribution was as follows:

Participation Index	Weighted proportion
Class A ("above average")	14%
Class B ("average")	34%
Class C ("below average")	52%

In all the following tables the figures have been properly adjusted, or "weighted", so as to be an unbiased representation of all cereal growers in the United Kingdom.

3.7. Table 3.1 shows the proportion of farmers falling in the three Participation Index classes according to their age, size of farm, and size of cereal acreage. It is clear that the larger the size of the farm or the cereal acreage of the farm, the higher was the proportion of Class A farmers; and that farmers up to 50 years of age tended to score higher than those over 50.

TABLE 3.1

DISTRIBUTION OF FARMERS IN EACH PARTICIPATION INDEX
CLASS BY AGE, SIZE OF FARM AND CEREAL ACREAGE

	Participation Index			
	Class A %	Class B %	Class C %	Total %
All cereal farmers	**14**	**34**	**52**	**100**
Age:				
30 and under	11	34	55	100
31–50 years of age	19	38	43	100
Over 50	9	30	61	100
Size of farm (acres):				
Up to 49	—	17	83	100
50–99	12	24	64	100
100–199	16	42	42	100
200–299	18	48	34	100
300–399	20	44	36	100
400 and over	37	36	27	100
Cereal acreage:				
Up to 49	6	30	64	100
50–99	18	36	46	100
100–299	30	45	25	100
300 and over	48	36	16	100

MEMBERSHIP OF FARMERS' ORGANISATIONS

3.8. According to the replies given at interview, 69% of cereal growers were members of one of the Farmers' Unions and there were a few who belonged to other farmers' organisations though not to the Union; so that in all 78% of cereal growers participated in activities of this kind. For cereal growers with less than 50 acres the proportion was only 68%, but for those with over 50 acres it was 87%. As one would expect, the farmers in Participation Class A had the highest percentage membership of farmers' organisations (97%); those in Class B followed with 87%; and in Class C the proportion was 67%.

MEMBERSHIP OF BUYING AND SELLING GROUPS

3.9. Approximately one in five cereal farmers was found to be a member either of a buying group, or of a selling group, or both. In total the membership of buying groups was more widespread (19% of cereal farmers) than membership of selling groups (13%). Very few members of a selling group did not also belong to a buying group (Table 3.2).

TABLE 3.2

MEMBERSHIP OF BUYING AND SELLING GROUPS

	Percentage of farmers who are members of:			
	Buying group only %	Selling group only %	Buying and selling group %	Neither %
All cereal farmers	**7**	**1**	**12**	**80**
Cereal acreage 1967:				
0–49 acres	6	1	10	83
50–99 acres	6	—	12	82
100–299 acres	11	1	18	70
300 acres and over	11	1	22	66
Region:[a]				
Northern	4	—	11	85
East and S.E.	6	2	12	80
Central and Southern	14	1	12	73
Wales and S.W.	5	—	30	65
Scotland	1	—	17	82
Northern Ireland	4	—	2	93
Pattern of sales:				
Forward contract only—				
Any registered	11	4	21	64
Not registered	10	—	14	77
F.C. and spot sales—				
Any F.C. registered	14	—	20	66
Not registered	—	—	3	96
Spot only	6	—	9	85
None sold	2	1	10	87

[a] In this and subsequent tables in this chapter the Regions are the "Farm Management Survey" Regions described in Appendix C.

3.10. Membership of groups is much more widespread among the larger farmers. Among those growing 300 acres or more of cereals in 1967, one in three were members of a buying group and just under one in four were members of a selling group.

3.11. Group membership is above the national average in the Southern and South-Western regions and well below the national average in Northern Ireland. It is also above the national average among those farmers whose pattern of sales included registered forward contracts.

3.12. Farmers who were members of buying groups were asked what products they bought through the group. The wide diversity of practice in purchases made by the farmers through their buying groups is shown by the following table:

	Percentage of farmers in buying groups %
Buying through group:	
Fertilisers	33
Machinery	26
Feedingstuffs	23
Fuel and lubricants	22
Seeds	19
Chemicals	10
Twine and cord	10
Veterinary products	9
Building materials	4

A wide variety of other answers have been omitted from the above table. It is of interest to note that 10% of cereal farmers said they bought all their agricultural needs through the buying group, while 4%, though members, said they bought nothing through the group.

3.13. Members of selling groups were also asked what products they sold through the group. It is perhaps not surprising that in the present Survey the answer most frequently given was grain; 35% of all members of selling groups (who form just over $4\frac{1}{2}$% of the total population of cereal farmers) said they sold some grain through their group. However, the relative lack of interest in selling activities of groups is illustrated by the fact that no less than 41% of those farmers who were members of a selling group said they sold nothing through the group.

OTHER BUSINESS ARRANGEMENTS WITH OTHER LOCAL FARMERS

3.14. One in five cereal farmers said they took part in such arrangements. These took a variety of forms. The sharing of cereal production equipment with other farmers was mentioned by 5% of all cereal growers, this practice being rather more common in Class A (11%) and in Wales and the South-West (13%). The great majority of cereal farmers who shared cereal production equipment were under 50 years of age. In addition to the formal sharing of equipment there was a certain amount of informal borrowing or lending with other farmers in the neighbourhood (2% of all cereal growers), this being rather more common among farms whose total size was less than 50 acres (5%).

3.15. Another 2% of all cereal growers said they shared with other farmers the ownership of equipment *not* related to cereal production, and a further 3% mentioned equipment sharing but provided insufficient information to establish whether or not such equipment was concerned with cereal production. In addition, 1% borrowed or lent equipment not related

D*

to cereal production, and 4% said they borrowed or lent equipment without specifying sufficient detail to establish whether or not it was related to cereal production. Among a variety of other answers 1% of farmers said they did contract work for other farmers.

VISITS TO AGRICULTURAL SHOWS

3.16. In both phases of the National Study farmers were asked how many agricultural shows they had attended in the year 1966. Seventy per cent of cereal farmers had visited at least one agricultural show in the course of that year; 32% had visited more than one show.

TABLE 3.3
VISITS TO AGRICULTURAL SHOWS IN 1966

	Percentage of farmers who visited:				
	no shows %	one show %	two shows %	three shows %	four or more shows %
All cereal farmers	30	38	19	9	4
Class A	6	37	31	17	10
Class B	9	47	27	11	9
Class C	50	32	13	5	—

ATTENDANCE AT TECHNICAL MEETINGS

3.17. Farmers were also asked whether they ever went to technical meetings organised by commercial companies and providing information about the use of fertilisers, crop protection or any other aspects of husbandry. Half said they did attend such meetings, though included among these were 11% who last attended such a meeting over one year ago. Table 3.4 shows that attendance was somewhat below the national average in Scotland and Northern Ireland but well above the national average in Wales and the South-West where nearly three-quarters of the informants claimed to have attended such a meeting, 61% of them within the 12 months prior to interview.

ATTENDANCE AT FARMING CONFERENCES

3.18. Just over a quarter of all cereal farmers (but as many as 69% of Class A) said they attended farming conferences, 9% having done so within the last 6 months, 9% 7–12 months ago and 10% over a year ago.

TABLE 3.4

ATTENDANCE AT TECHNICAL MEETINGS

	Percentage of farmers who:					
	never attend %	have attended				
		in last 3 months %	4–6 months ago %	7–12 months ago %	over 1 year ago %	can't remember %
All cereal farmers	**51**	**11**	**11**	**12**	**11**	**3**
Region:						
Northern	49	9	12	7	17	5
East and S.E.	50	12	11	16	9	2
Central and Southern	45	15	11	8	14	7
Wales and S.W.	26	8	18	35	13	—
Scotland	59	8	14	11	6	2
N. Ireland	65	11	7	4	13	—
Class A	8	24	24	20	17	6
Class B	35	15	16	17	14	4
Class C	74	5	5	8	8	1

Not surprisingly, the Survey also showed that those who attend farming conferences are generally also frequent visitors to agricultural shows and technical meetings.

USE OF N.A.A.S. AND OTHER ADVISORY SERVICES

3.19. Farmers were asked if they had used the N.A.A.S. or other Advisory Services during the previous 6 months, and 29% replied that they had. The proportion for Participation Class A was 50%, for Class B 34% and for Class C 17%. Cereal farmers up to 50 years of age used the Advisory Services more than those over 50 (35% compared with 22%). Cereal farmers in England and Wales used these services (about 35%) rather more than those in Scotland (20%) or Northern Ireland (18%). Table 3.5 shows the topics on which advice was sought.

B. Choice of Cereal Varieties

AVERAGE NUMBER OF VARIETIES

3.20. The analysis of information given with regard to the varieties of cereals grown in 1967 shows that few farmers grew more than two varieties

TABLE 3.5

TYPE OF N.A.A.S. AND OTHER ADVISORY SERVICES USED IN LAST 6 MONTHS

	Total	Participation Class			Region					
		A	B	C	North	East and S.E.	Central and S.	Wales and S.W.	Scotland	Northern Ireland
	%	%	%	%	%	%	%	%	%	%
Percentage of those using N.A.A.S., etc., services in last 6 months who requested advice/service on:[a]										
Soil testing	26	25	29	22	14	27	23	38	43	12
Pest control	14	19	7	19	17	18	9	—	25	—
Animals	12	8	16	10	6	8	20	16	20	—
Cereals	11	13	10	10	17	8	18	9	11	—
General farm management	14	16	14	13	25	16	10	30	4	12
Farm buildings	8	5	11	5	10	6	9	9	3	12
Land drainage and reclamation	8	9	7	7	6	11	7	—	—	—
Accounts/costing	7	5	9	5	1	4	15	7	12	13
Land usage	4	2	7	4	5	5	3	—	1	23
Other (specified)	11	9	10	13	7	16	4	7	1	25
Technical advice (unspecified)	6	8	5	5	6	4	3	7	2	—
Proportion of all cereal farmers who have made any use of N.A.A.S. in last 6 months	29	58	34	17	27	34	37	35	20	18

(a) These percentages do not add to 100 because some farmers sought advice on more than one topic.

of each cereal. It is convenient to express the results in terms of average number of varieties per farm, as in Table 3.6.

3.21. It was naturally to be expected that the larger cereal farmer, or the Class A farmer, would make use of more varieties on his farm, and this is indeed borne out for wheat and barley by Table 3.6. The average barley grower with 300 acres or more of cereals is seen to grow more than two varieties.

TABLE 3.6

AVERAGE NUMBER OF VARIETIES OF CEREALS GROWN

	Wheat	Barley	Oats
All farms growing each cereal	**1·36**	**1·56**	**1·12**
Cereal acreage 1967:			
0–49	1·19	1·28	1·14
50–99	1·24	1·68	1·15
100–299	1·52	1·82	1·07
300 and over	1·73	2·20	1·14
Participation Index:			
Class A	1·56	1·89	1·06
Class B	1·42	1·53	1·09
Class C	1·20	1·43	1·19

VARIETIES OF WHEAT GROWN IN 1967

3.22. It is plain that in 1967, Capelle-Desprez was easily the dominant variety, being found on 69% of all farms growing wheat. A large number of other varieties were mentioned but only the four shown in Table 3.7 achieved any appreciable level of mention.

3.23. Regional differences in the proportion of farms growing particular varieties of wheat are influenced by the extent to which the variety may be suitable for winter as opposed to spring sowing, and by its suitability for various soil types.

VARIETIES OF BARLEY GROWN IN 1967

3.24. While Proctor and Zephyr were the two varieties which were most widely grown, a much wider range of barleys was grown by an appreciable number of farmers (Table 3.8). It is interesting to observe that the Class A farmers appear to have adopted the relatively new variety Zephyr faster than the Class C farmers, a greater proportion of whom still grow Proctor.

VARIETIES OF OATS GROWN IN 1967

3.25. Only four varieties of oats—Condor, Astor, Forward and Blenda —were grown by more than 10% of farms growing oats (Table 3.9).

TABLE 3.7

VARIETIES OF WHEAT GROWN IN 1967

	Percentage of farmers growing wheat who grew some:				
	Capelle %	Kloka %	Opal %	Champlein %	Maris Widgeon %
Total, all farmers growing wheat	**69**	**25**	**14**	**8**	**4**
Cereal acreage 1967:					
0–49	58	27	11	5	3
50–99	69	20	16	7	6
100–299	80	29	16	10	4
300 and over	85	39	7	18	6
Participation Class:					
A	82	31	10	11	7
B	68	26	20	9	3
C	63	22	10	5	5
Region:					
Northern	56	22	25	4	3
East and S.E.	77	25	14	6	3
Central and Southern	63	41	11	12	11
Wales and S.W.	86	14	—	—	—
Scotland	58	1	4	10	—

Other varieties, each grown by less than 4% of farmers growing wheat, are omitted from this table.

CHANGE IN WHEAT VARIETIES

3.26. Of the farmers growing wheat in 1967, 66% were growing only varieties which they had grown in the previous year. Of the remainder, 15% were growing Kloka, presumably for the first time (i.e. they were not growing it the previous year), 5% Capelle for the first time, and 4% Champlein for the first time (Table 3.10). No other variety was mentioned by more than 3% of farmers growing wheat.

3.27. While the issue is not clear-cut, there is certainly some suggestion that at least as far as Kloka and Champlein are concerned, the proportion of Class A and B farmers growing them for the first time in 1967 was higher than the national average. There is also some indication that a greater proportion of large cereal farms—those growing 300 acres and over in 1967—tried Champlein for the first time (8% compared with the national average of 4%).

3.28. The reasons given by farmers for growing particular varieties suggest that a reputation for good yields, and recommendations made to

TABLE 3.8

VARIETIES OF BARLEY GROWN IN 1967

	Percentage of farmers growing barley who grew some:							
	Proctor %	Zephyr %	Impala %	Vada %	Ymer %	Rika %	Dea %	Maris Badger %
Total, all farmers growing barley	**31**	**26**	**17**	**12**	**12**	**9**	**8**	**7**
Cereal acreage 1967:								
0–49	31	17	13	9	12	9	5	4
50–99	31	31	16	14	12	11	8	9
100–299	28	34	26	13	12	3	16	7
300 and over	36	44	23	13	11	4	27	19
Participation Class:								
A	24	36	33	12	12	13	17	11
B	31	26	16	12	12	7	7	6
C	32	22	11	11	12	9	6	6
Region:								
Northern	19	40	17	31	2	10	11	2
East and S.E.	57	26	13	13	—	4	15	9
Central and Southern	29	32	32	12	—	15	4	7
Wales and S.W.	29	29	41	3	—	31	—	3
Scotland	—	11	2	2	67	—	4	11
N. Ireland	—	14	5	—	13	4	—	—

Other varieties, each grown by less than 7% of farmers growing barley, are omitted from this table.

TABLE 3.9

VARIETIES OF OATS GROWN IN 1967

	Percentage of farmers growing oats who grew some:			
	Condor %	Astor %	Forward %	Blenda %
Total, all farmers growing oats	**26**	**20**	**15**	**14**
Region:				
Northern	35	9	16	28
East and S.E.	38	10	17	14
Central and Southern	38	7	11	10
Wales and S.W.	42	1	17	6
Scotland	24	49	20	17
Northern Ireland	5	—	—	9

Other varieties, mentioned only rarely, are omitted from this table.

TABLE 3.10

PERCENTAGE OF FARMERS GROWING VARIETIES OF WHEAT IN 1967
WHICH WERE NOT GROWN IN 1966

	All farmers growing wheat	Participation Class		
		A	B	C
	%	%	%	%
Percentage of farmers growing wheat who grew (in 1967 but not in 1966):				
Kloka	15	18	17	12
Capelle	5	7	2	6
Champlein	4	6	5	1
	(Other new varieties omitted)			
No new varieties grown	66	59	67	69

them personally, are the two most important factors—the yield being particularly significant to Class C farmers. Disease resistance and response to weather were also quoted by an appreciable number of farmers growing new varieties. In addition, 10% said they had grown the new variety purely as an experiment. Certain varieties have particular attractions. It would seem that Kloka is grown for the first time primarily because of its yield and its disease-resistant qualities. The same is equally true of Capelle, but for Champlein, on the other hand, the attraction most often quoted was its suitability for the farmer's land.

TABLE 3.11

PERCENTAGE OF FARMERS GIVING VARIOUS REASONS FOR GROWING VARIETIES
IN 1967 NOT GROWN IN 1966

	All farmers growing new varieties in 1967 %	Participation Class			Growing for first time:		
		A %	B %	C %	Kloka %	Capelle %	Champlein %
Percentage growing variety of wheat in 1967 not grown in 1966 who did so because:[a]							
Good yield	27	22	14	40	27	39	11
Recommended	20	18	24	18	12	—	14
Disease resistant	15	20	20	8	26	29	1
Weather	10	7	12	9	8	13	—
New variety/experiment	10	11	9	9	8	—	4
Suitable for land	5	5	8	3	6	1	27
Hardy/good standing	5	4	5	5	2	2	11

[a] The percentages total less than 100 because reasons only rarely quoted have been omitted.

C. Cereal Yields

3.29. All farmers in the National Study were asked to give details of their cereal yields—for the 1966 harvest in Phase I, and for the 1967 harvest in Phase II. They were then asked, for wheat, barley and oats separately, the methods they had employed in estimating these figures. The replies given were classified as far as possible according to whether the methods used were accurate or inaccurate. Methods regarded as accurate included figures based on actual weighing of the grain, sales figures, or estimates based on the capacity of bins or other containers with specific storage capacities. A small number of farmers gave multiple replies—e.g. yields based partly on sales and partly on estimated bin capacity—but these amounted to only 3% of farmers growing wheat and an insignificant number of farmers growing barley.

3.30. On this issue, as one would expect, there were considerable differences between the two phases of the study. In Phase I, data relating to the 1966 crop were collected at a time when virtually all the crop had been disposed of. In Phase II, information was collected shortly after completion of harvest when an appreciable amount of the 1967 crop had yet to be sold. Thus inevitably the proportion of farmers whose yield was based on sales was much lower in Phase II than in Phase I, while the proportion whose yield was estimated by guess-work was much higher in Phase II.

3.31. Examination of the replies given (Table 3.12) showed that for all three cereals a considerable proportion of farmers were unable to give an accurate estimate of their yield at any stage. Even after disposing of their harvest, only just over half the farmers growing wheat and barley were able to provide accurate figures, and only one-third of those growing oats. The proportion using accurate methods of estimation was, of course, higher among the larger cereal farmers, but even among those growing 300 acres or more of cereals, more than one in four based their yield estimate on "guess-work".

3.32. Just over half of the farmers growing wheat estimated their cereal yields *after drying* (Table 3.13). The corresponding figure for barley was 46% and oats 32%. For all three cereals the proportion of Class A farmers stating yields after drying was very much higher than among Class B, which in turn was appreciably higher than in Class C.

3.33. Rather more than a third of all cereal farmers claimed to know their yield per acre for different fields and 45% claimed to know their yield per acre for different varieties of cereal (Table 3.14). On this aspect, it is interesting to note that in the East and South-East region (where acreages are generally larger) only 35% claimed to know the yield for

TABLE 3.12

PERCENTAGE OF FARMERS USING VARIOUS METHODS OF ESTIMATING CEREAL YIELDS

	Farmers growing:					
	Wheat		Barley		Oats	
	Phase I (July/August) %	Phase II (October/November) %	Phase I (July/August) %	Phase II (October/November) %	Phase I (July/August) %	Phase II (October/November) %
Farmers estimating yields on the basis of:[a]						
Sales	35	10	25	7	13	2
Weighing	15	17	15	13	10	7
Capacity of bins, etc.	3	11	11	14	6	9
Other accurate methods	2	3	2	4	5	3
Total—Accurate methods	55	41	53	38	34	21
Estimate/guess	24	46	30	40	41	52
Capacity of other buildings, etc.	2	2	2	1	3	1
Other inaccurate methods	2	1	3	3	9	4
Total—Inaccurate methods	28	49	35	44	53	57
Unclassifiable as accurate or inaccurate	7	5	4	12	2	9
Yield not given	12	8	9	6	11	13

(a) Percentages total more than 100 because more than one method was used by some farmers.

TABLE 3.13

PERCENTAGE OF FARMERS WHO STATED CEREAL YIELDS
AFTER DRYING

	Wheat %	Barley %	Oats %
All farmers stating yields after drying	**52**	**46**	**32**
Participation Class:			
A	73	67	59
B	57	48	35
C	38	38	24
Region:			
Northern	55	55	51
East and S.E.	45	40	28
Central and Southern	62	54	58
Wales and S.W.	44	39	14
Scotland	50	47	36
Northern Ireland	n.a.	48	11

TABLE 3.14

PERCENTAGE OF CEREAL FARMERS WHO CLAIMED TO
KNOW YIELD PER ACRE FOR SEPARATE FIELDS OR
VARIETIES

	Fields %	Varieties %
Cereal growers knowing yields	**38**	**45**
Region:		
Northern	41	40
East and S.E.	35	54
Central and Southern	49	51
Wales and S.W.	44	70
Scotland	36	41
Northern Ireland	19	20

different fields, but 54% claimed to know the yield for different varieties. In Wales and the South-West as many as 70% claimed to know the yield for different varieties. These findings were confirmed in the Area Studies.

3.34. In the Area Studies farmers were asked at the interview in May/ June 1967 whether they considered their yield for winter wheat or barley differed from their yield for spring wheat or barley. Only 14 farmers of the 234 farmers interviewed were growing both winter and spring wheat, and only ten were growing both winter and spring barley. Of the fourteen wheat growers, eight considered they had achieved a higher yield from winter wheat, two a lower yield and four had found no difference. Of the ten barley growers, seven found no difference, one found a higher yield, one a lower, and one did not know. These results, though for only a few farms, reflect the general experience that yields of winter wheat exceed those of spring wheat, but with barley the yield advantages of winter sowing are much more open to doubt in the farmer's mind.

FACTORS AFFECTING YIELDS

3.35. All informants in the National Study were asked: "Was your yield seriously affected this year by any of the following factors: adverse weather conditions, weed infestation, crop disease or anything else?" The replies in Phase I relate to the 1966 harvest, and in Phase II to the 1967 harvest. Approximately four out of ten informants in both years said that their crop had been affected by adverse weather conditions and one out of ten that the yield had been affected by weed infestation (Table 3.15). There was marginally more mention of crop disease and other factors in relation to the 1966 harvest than to the 1967, which gives a total of 53% of cereal farmers who claimed that some factors seriously affected their yield in 1966 compared with 48% making the same claim in relation to the 1967 harvest.

TABLE 3.15

PERCENTAGE OF CEREAL FARMERS WHO CLAIMED
THAT THEIR YIELDS WERE SERIOUSLY AFFECTED BY
WEATHER, WEEDS, DISEASE AND OTHER FACTORS

	1966 Harvest (Phase I) %[a]	1967 Harvest (Phase II) %[a]
Farmers stating their yields were seriously affected by:		
Adverse weather conditions	41	40
Weed infestation	9	10
Crop disease	10	6
Other factors	4	1
Nothing	47	52

[a] Percentages total more than 100 because more than one factor was mentioned by some farmers.

3.36. Since differences between the two phases are, on the whole, small it is justifiable to combine them in order to study regional variations. Against an overall figure of 49% for the two harvests, the proportion of farmers who stated that their yield had been seriously affected varied by region as follows:

	Proportion of farmers claiming cereal yield seriously affected %
Region:	
Northern	52
East and S.E.	48
Central and Southern	60
Wales and S.W.	41
Scotland	48
Northern Ireland	51

3.37. Adverse weather conditions were mentioned relatively more often by farmers in Scotland (46%) and Northern Ireland (44%), and less often in the Eastern and South-East (36%) and in Wales and the South-West (35%).

3.38. Weed infestation secured hardly any mention in Scotland, but was above the national average in Eastern and South-East (12%), Northern Ireland (12%), and Central and Southern (14%).

3.39. The great majority of mentions of crop disease arose in the Eastern, South-East and Central and Southern regions of the country. In the Eastern and South-East region 11% of cereal farmers named crop disease, while in the Central and Southern region no less than 20% named this factor as having seriously affected their yield.

3.40. Mayweed, chickweed, wild oats and redshank were the types of weed infestation most often mentioned by those who claimed that their yield had been seriously affected by weed infestation:

Proportion of farmers whose yield was seriously affected by weed infestation who named:	%
Mayweed	14
Chickweed	14
Wild oats	13
Redshank	12
Couch	7
Bindweed/black bindweed	5
Other weeds (specified)	38
Weeds (general)	4

3.41. Only one in ten farmers mentioned more than one weed as having seriously affected yields. Mayweed was particularly widely mentioned in the Eastern and South-Eastern region, chickweed in Northern Ireland, Wales and the South-West, and wild oats in the Northern, Eastern and South-Eastern regions.

3.42. The type of crop disease most widely mentioned was rust: no less than 50% of all farmers who said their yield had been affected by crop disease referred to this specifically, the great majority of them being located in the Eastern, South-Eastern and Central and Southern regions. Only take-all (mentioned by 16% of those suffering from crop disease), mildew (11%), smut (9%) and *Rhynchosporium* (7%) secured any appreciable level of mention.

3.43. In Phase II only of the National Study, those informants whose yields were depressed as a result of weed infestation or crop disease were asked whether they planned to take any action to combat this in the future. Of those whose harvest was affected by weed infestation, 58% said that they would spray earlier or more heavily than they had done in 1967. A further 7% commented that they had been unable to spray in 1967 but certainly intended to do so in 1968. The remainder quoted a variety of plans, of which altering the rotation (9%) was the only one to achieve any significant level of mention.

3.44. As far as crop disease was concerned, 29% of those whose 1967 yields were affected by such disease said they did not intend to take any action to combat it in the future. One in four said they would alter their rotation, and a similar proportion hoped that the plant breeders would be able to breed resistant strains. Among the variety of answers given by the remainder, it is worthy of note that 11% confessed they did not know what action to take to combat future attacks of disease.

TRENDS IN YIELDS PER ACRE ON INDIVIDUAL FARMS

3.45. There is some evidence that the difference in yield between the high-yielding and the low-yielding farms may be diminishing. To supplement the results of the Survey of Farmers a study was made by the University of Nottingham (Department of Agricultural Economics) of the records of 127 farms which had been in the East Midlands Farm Management Survey throughout the 10-year period 1955/6 to 1964/5.[1] These records included statements by the farmers of the yields per acre which they obtained each year for their cereal crops. The results for the same farms in successive 5-year periods are summarised in Tables 3.16 and 3.17. In obtaining the group averages, the individual yields were not weighted by acreage.

3.46. The figures for wheat and barley show that although for this group of farms as a whole yields per acre were rising at the rate of about 1 cwt per

---

[1] For an analysis of yields in the Farm Management Survey for England and Wales as a whole see Table 3 of Appendix E.

TABLE 3.16

YIELDS PER ACRE FOR 127 FARMS IN THE EAST MIDLANDS, 1955 TO 1965

Crop	Average yield, 1955–9 cwt per acre	No. of farms	Average yield, cwt per acre		Increase	
			1955–9	1960–4	cwt	%
Wheat	Less than 22	23	19·5	26·2	6·7	34·4
	22·0–24·9	31	23·4	29·2	5·8	24·8
	25·0–27·9	27	26·1	31·9	5·8	22·2
	28·0 and over	33	30·4	34·0	3·6	11·8
	Total	114	25·3	30·6	5·3	20·9
Barley	Less than 21	23	16·8	24·1	7·3	43·5
	21·0–23·9	24	22·3	26·5	4·2	18·8
	24·0–26·9	35	25·5	29·3	3·8	14·9
	27·0 and over	22	28·6	32·7	4·1	14·3
	Total	104	23·5	28·2	4·7	20·0
Oats	Less than 20	33	16·7	22·4	5·7	34·1
	20·0–22·9	26	21·4	25·6	4·2	19·6
	23·0–25·9	20	24·3	27·6	3·3	13·6
	26·0 and over	22	28·5	28·2	−0·3	−1·1
	Total	101	22·0	25·5	3·5	15·9

year, the low-yielding farms were improving at a faster rate than this and the high-yielding farms at a slower rate. In percentage terms (last column) the contrast is more striking. The farms in the lowest-yielding groups improved their yields in five years by 34·4% for wheat and 43·5% for barley. In the case of oats the improvement in yields achieved by the group as a whole was more modest than in the case of wheat or barley, but the gap in yields between the worst and the best groups was halved.

3.47. Using the same data, an attempt was made to discover whether the growers with the larger acreages of cereals obtained higher yields per acre than the smaller growers (Table 3.17).

3.48. For wheat the figures indicate a definite tendency for higher yields to be associated with larger acreages, but for barley this feature is less marked and for oats it does not appear to exist at all. Similar results were obtained when the same analysis was carried out for these farms in the period 1955–9. The rate of improvement in yield between the two periods appeared to be somewhat slower among growers of over 75 acres of cereals than among the smaller growers.

3.49. A further analysis examined the difference between farms which had increased their total farm size (in acres) during the period and those which

TABLE 3.17

AVERAGE YIELD 1960–4 BY ACREAGE OF
CEREALS GROWN

Farm acreage of cereals in 1964 (acres)	Average yield, 1960–4 (cwt per acre)		
	Wheat	Barley	Oats
Less than 15	28·3	27·8	25·5
15·0–29·9	30·0	26·5	25·5
30·0–44·9	30·0	28·5	26·7
45·0–59·9	29·7	29·1	27·3
60·0–74·9	31·7	28·4	24·5
75 and over	31·4	28·9	25·1

had maintained the same total farm acreage throughout. It was found that in both five-year periods the "expanding" farms achieved somewhat higher yields per acre than the "stationary" farms (Table 3.18). The interpretation of this may be that for farms with the higher yields there was a stronger incentive to enlarge the farm boundaries. They may also have accumulated more capital for land purchase out of past profits, or have found it easier to obtain loans. It might also be supposed that the same enterprising and progressive attributes which prompted a farmer to increase his acreage had also influenced his methods of cultivation, fertiliser practices, etc. and that the higher yields and the expansionist tendency had the same basic cause.

D.   Changes in Cereal Acreage between 1966 and 1967

ACTUAL CHANGES

3.50. Detailed examination of data from the Survey of Farmers concerning cereal acreages grown in 1966 and in 1967 showed a wide variety of changes of acreage for each type of cereal on individual farms. An exhaustive study of the nature of these changes would be highly complex, but certain major types of change have been examined in Table 3.19. Particularly interesting is the way in which the balance between winter and spring wheat varied over the two years on farms with different total cereal acreages. It is the larger cereal farms which show the major proportion changing—no doubt in part influenced by the fact that on the larger farms the timing of operations is more crucial and there is a greater risk that the weather will prevent the whole of the desired acreage of winter wheat from being sown. Generally this table (and many others, not all of which are included in this Report) suggest quite clearly that it is the larger cereal farmer who is more flexible in his approach and more likely to take positive decisions to change his cereal-growing policy.

TABLE 3.18

RELATIONSHIP BETWEEN INCREASING FARM ACREAGE AND
AVERAGE YIELDS

	Farms with unchanged total acreage 1955–64	Farms with increased total acreage 1955–64
	(average yield per acre, cwt)	
Wheat:		
1955–9	24·4	25·3
1960–4	30·2	31·2
Barley:		
1955–9	23·3	23·9
1960–4	28·2	28·8
Oats:		
1955–9	22·2	22·1
1960–4	25·4	25·8

TABLE 3.19

CHANGES IN CEREAL ACREAGE BETWEEN 1966 AND 1967

	Cereal acreage in 1967				
	Total %	0–49 %	50–99 %	100–299 %	300 and over %
Percentage of farms growing:					
Less winter wheat 1967	9	6	10	20	27
More winter wheat 1967	8	2	11	18	32
Less spring wheat 1967	4	2	6	9	10
More spring wheat 1967	6	3	9	9	13
Less winter barley 1967	1	—	2	3	3
More winter barley 1967	2	—	3	4	8
Less spring barley 1967	26	23	28	34	39
More spring barley 1967	13	14	30	35	37
Less oats 1967	7	8	6	5	6
More oats 1967	9	5	12	12	17

REASONS FOR CHANGING CEREAL ACREAGE

3.51. On all farms included in the National Survey, a comparison was made between the total cereal acreage in 1966 and that grown in 1967. All told, 22 % were growing an increased total acreage of cereals in 1967 compared with 1966, and 18 % grew less. Reasons for changing the total acreage of cereals from one year to another (defined as a change in the total acreage of 5 acres or more) are summarised in Table 3.20. Overall, rotation in one form or another, or deliberate changes in the proportion of land under the plough, appear to be the two reasons most often quoted for change. But it should be noted that 15% of those growing more cereals in 1967 mentioned profitability and the higher guaranteed price as a reason, whereas in contrast only 5% of those growing less in 1967 mentioned unprofitability as the reason.

TABLE 3.20

PERCENTAGE OF FARMERS QUOTING VARIOUS REASONS FOR CHANGE IN TOTAL
CEREAL ACREAGE BETWEEN 1966 AND 1967

	More cereals grown in 1967 than in 1966 (22% of cereal farms) %	Less cereals grown in 1967 than in 1966 (18% of cereal farms) %
Percentage of farmers changing who did so because of:[a]		
Rotation	21	30
Change in acreage of farm	10	7
Weather	5	1
Other reasons not directly concerned with cereals	1	3
More/less land under plough	27	22
Profitability/higher guaranteed price	15	5
Changes in other crops	13	17
Other deliberate reasons	7	9
Reason not stated	8	8

[a] Percentages add to more than 100 because some farmers gave more than one reason.

3.52. Table 3.21 shows the percentage of farmers giving weather and profitability (or the increase in the guaranteed price) as reasons for changing their total cereal acreage in 1967 compared with 1966, for different regions, cereal acreage groups and classes of farmer. Of particular interest is the extent to which comments on the effect of weather varied by region, and comments on profitability by class and size.

E. Farmers' Attitude to Cereal Growing in the Future

3.53. The Survey showed that a majority of cereal farmers were slightly pessimistic about the future of cereal farming in this country, though this pessimism was concentrated among the smaller farmers. When asked whether they thought that in three or four years' time their cereal acreage would have changed, 10% said they expected to be growing more, 20% to be growing less, and 58% expected to be growing about the same (Table 3.22). However, the greater the present cereals acreage of the farm, the greater was the expectation of an increase in that acreage. Thus those

TABLE 3.21

PERCENTAGE OF FARMERS CHANGING THEIR CEREAL ACREAGE BETWEEN 1966
AND 1967 WHO GAVE WEATHER OR PROFITABILITY AS THE REASON

	Percentage of farmers who gave weather as the reason for change		Percentage of farmers who gave profitability or the increased guaranteed price as the reason for change	
	Cereals grown in 1967 compared with 1966		Cereals grown in 1967 compared with 1966	
	More %	Less %	More %	Less %
All farms	5	1	12	5
Region:				
Northern	2	—	14	—
East and S.E.	6	1	8	3
Central and Southern	3	2	8	4
Wales and S.W.	10	1	19	16
Scotland	5	—	7	7
Northern Ireland	—	—	34	—
Cereal acreage 1967:				
Up to 49	6	—	9	8
50–99	3	—	15	2
100–299	5	4	10	—
300 and over	6	5	5	4
Class of farmer:				
A	4	1	7	7
B	3	1	17	4
C	6	—	10	—

farmers who expected to be growing more represented almost exactly the same proportion of total cereal acreage (about 15%) as those who expected to be growing less.

3.54. This pattern of qualified pessimism at one end of the scale and qualified optimism at the other is repeated in the age breakdown. Younger farmers were more inclined to expect an expansion in cereals acreage than were older farmers. Regional differences also showed that there was slightly more of an expansionist outlook in the Eastern and South-Eastern areas than in the rest of the country—a difference which may be associated with size of farm. The reasons most frequently given by those who expected to be growing more cereals in 3 to 4 years' time were: a general policy of expansion (mentioned by 24% of those who expected to be growing more), the profitability of cereal enterprises compared with others (18%), and

TABLE 3.22

PERCENTAGES OF CEREAL FARMERS WHO EXPECTED IN THREE TO FOUR YEARS'
TIME TO BE GROWING MORE, LESS, OR ABOUT THE SAME ACREAGE OF CEREALS

	Higher cereal acreage %	About the same %	Lower cereal acreage %	Don't know %
All cereal farmers	**10**	**58**	**20**	**13**
Cereal acreage (1967):				
0–49	8	54	24	14
50–99	10	65	16	10
100–299	15	60	13	12
300 and over	22	58	11	9
Age:				
30 and under	24	39	17	20
31–50	11	59	19	11
over 50	6	60	21	13
Region:				
Northern	10	63	13	13
East and S.E.	10	69	13	8
Central and Southern	8	57	25	9
Wales and S.W.	5	55	24	16
Scotland	9	57	21	13
Northern Ireland	13	33	29	24
Class:				
A	18	52	19	11
B	8	64	20	8
C	8	56	20	16

more intensive use of grassland (15%). Among the reasons given for growing less in 3 to 4 years' time, the most common was the expansion of other
enterprises (mentioned by 29% of those who expected to be growing less);
others were that there was more money to be made in other enterprises
(14%) and that the land needed resting (12%). Once again, a wide variety
of other answers were also named, but each represented only a small
proportion of the total sample.

3.55. Two out of three cereal farmers expected that the proportion of the
various cereal crops which they would be growing in 3 to 4 years' time
would be the same as now. This pattern was remarkably uniform by size
of present cereal acreage, by age of farmer and even by region—only Wales
and the South-West (where more oats were expected to be grown) showing
any significant change.

3.56. Those farmers expecting a change were asked in what ways they
thought the proportion of different cereals would change. A variety of

different answers was given, of which Table 3.23 illustrates the most important. In addition to these, a number of other, less specific, answers were also given. The table shows that there is rather greater expectation of an increase in wheat acreage than of barley acreage. On the other hand, the increase in wheat acreage is particularly associated with the larger cereal farmers and the Class A farmers, whereas those whose total cereal acreage in 1967 was less than 50 acres were expecting an increase in the barley acreage.

TABLE 3.23

PERCENTAGE OF CEREAL FARMERS ANTICIPATING CHANGES IN
CEREAL PROPORTIONS IN THREE TO FOUR YEARS' TIME

	Cereal acreage in 1967				
	Up to 49	50–99	100–299	300 and over	Total
	($\%$ of cereal farmers anticipating change)				
Percentage of farmers anticipating change who expect to be growing:[a]					
More wheat/less barley	7	8	12	23	8
More wheat/less oats	2	—	—	2	1
More wheat	12	35	38	32	23
More barley/less wheat	2	—	9	5	2
More barley/less oats	9	3	2	—	6
More barley	23	8	6	11	16
More oats/less wheat	—	3	6	—	2
More oats/less barley	3	3	4	2	3
More oats	3	10	7	4	6

[a] These percentages do not add to 100 because not all possible combinations of answers have been shown.

3.57. As far as the profitability of cereal growing in this country in the future was concerned, there was an atmosphere of uncertainty and wide divisions of opinion. Overall, 33% of cereal farmers took a favourable view of the future against 27% who took an unfavourable view (39% in the N.A.A.S. Eastern Region). The larger cereal farmers took both a more positive and a more favourable attitude (Table 3.24). But the largest proportion (40%) would not express a view.

3.58. Those who took a favourable view often made the spontaneous comment that cereal farmers would be better off if Britain joined the Common Market (17% of all farmers volunteered this comment). On the other hand, 12% expected that the profitability of cereals would decline,

TABLE 3.24

PERCENTAGE OF CEREAL GROWERS TAKING FAVOURABLE OR UNFAVOURABLE
VIEWS ON THE FUTURE PROFITABILITY OF CEREAL GROWING

	Favourable %	Unfavourable %	Don't know %
All cereal farmers	33	27	40
Cereal acreage (1967):			
0–49	32	28	40
50–99	33	27	40
100–299	37	26	37
300 and over	43	25	32
Participation Class:			
A	42	24	34
B	37	28	35
C	28	28	44

and just over 9% specifically mentioned that they expected costs of production to increase. A further 9% (including 13% of farmers growing less than 50 acres of cereals in 1967) feared for the future of the small farmer. Farmers were specifically asked what effect they expected Britain's entry into the Common Market would have on cereal producers. Overall, nearly twice as many farmers expected our entry to be advantageous or profitable to the cereal producers as expected it to be disadvantageous (36% as against 17%). Nearly one-third, however, were unwilling to commit themselves and a further 17% expected it to have no effect (Table 3.25). Once again, it was the large cereal growers who took the most favourable view of the advantages such entry would bring. A secondary analysis relating to cereals density of the farm (i.e. the proportion of the farm acreage down to cereals) showed that whereas only 25% of those who had less than 30% of their land down to cereals thought that Common Market entry would be profitable for them, the proportion among farmers having 71–80% down to cereals was 62%.

F. The Importance of Weather in Cereals Production and Marketing

3.59. The foregoing presentation of some of the results of the Survey of Farmers has indicated the importance of the influence of weather on cereals production. Through its effects on yields, on quality and on the planting of spring or winter varieties, the weather also affects cereals marketing. This section considers these weather effects in more detail.

3.60. Variations in weather conditions affect cereal production in the United Kingdom in three ways. There are the effects on preparing the

TABLE 3.25

PERCENTAGE OF CEREAL GROWERS EXPRESSING VARIOUS VIEWS ON THE EFFECT
ON CEREAL PRODUCERS IF THE U.K. WERE TO JOIN THE COMMON MARKET

	Advanta- geous/ profitable %	Disadvan- tageous/un- profitable %	No effect %	Don't know %	Total %
Total, all farmers	35	17	17	31	100
Cereal acreage (1967):					
0–49	27	22	20	31	100
50–99	41	11	16	32	100
100–299	53	13	9	25	100
300 and over	57	13	10	20	100
Class of farmer:					
A	59	13	19	9	100
B	39	10	33	18	100
C	27	22	33	18	100

seedbed and sowing the crop; on the course of plant growth; and on conditions at harvest.[1]

WEATHER EFFECTS ON SOIL PREPARATION AND SOWING

3.61. Weather conditions for the autumn sowing of wheat have proved highly variable in recent years. When rainfall has been relatively high during the period August to November the acreage of winter wheat has generally been reduced, and vice versa. This can be clearly seen in Fig. 3.1. High rainfall during August and September delays the harvest and thus reduces the opportunity to get on to the land before winter sets in. Moreover, heavy rain during this period makes cultivation and sowing difficult, even when a late harvest does not present a constraint to those operations. Exceptionally, adverse conditions for planting may also be caused by extremely dry weather baking the ground to an unworkable condition. This state of affairs has been rare in the U.K., but it did happen in the autumn of 1955 when dry conditions held up soil cultivations until November.

3.62. Besides this inverse relation between autumn rainfall and autumn sowing, a most interesting feature is the pattern of alternate "dry" and "wet" autumns which prevailed over most of the period 1953 to 1966 with only three minor exceptions in 14 years. As can be seen from Fig. 3.1,

[1] In other countries additional influences may be significant. For example there is the "winter effect" in French cereal production, which is the name given by Oury to the killing of winter cereals by periods of extreme cold. Oury, B., *A Production Model for Wheat and Feedgrains in France (1946–61)*, North Holland Publishing Company, 1966.

December wheat acreages also followed a distinct alternate year pattern, which is the converse of the rainfall pattern. The only year out of phase in this case is 1964, in which acreage rose for the second year in succession, in association with the lowest autumn rainfall recorded during the period. It can also be seen that the June wheat acreages follow the pattern set by

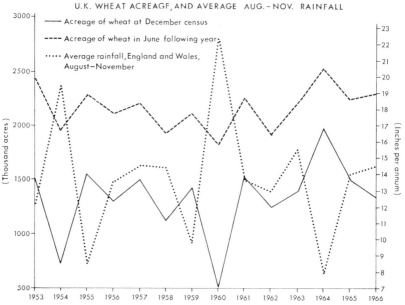

Fig. 3.1. United Kingdom wheat acreage and average August–November rainfall, 1953–66.
Sources: Unpublished Econometric Study of Aspects of the U.K. Cereals Market by D. R. Colman, Department of Agricultural Economics, University of Manchester.

wheat acreage of the previous December, although the variation in the June acreage is less than that at December. This is because the level of spring sowings compensates for those of December to some extent, being larger in years when the December acreage is low; but spring sowings of wheat are never sufficiently large to make up completely for low sowings in the previous autumn.

3.63. Poor weather conditions may also interfere with the establishment of spring crops of cereals, but owing to the longer time available to farmers for spring cultivation they usually manage to establish their planned acreages even in years of adverse spring weather. Certainly, it has not proved possible to identify statistically the effects of adverse conditions upon spring planting of cereals.

WEATHER EFFECTS IN THE GROWING PERIOD

3.64. As a result of agronomic experiments it has been determined that there are optimal conditions for seed germination, tillering, ripening and for all the stages of plant development. These conditions are produced by the existence of a correct balance between such weather factors as moisture availability, temperature, solar radiation, and air movement—factors affecting evapotranspiration and photosynthesis—plus a complex of factors to do with soil structure and the availability of plant nutrients.

3.65. To produce a single index measure of the weather conditions affecting plant growth is difficult, especially when the yield to be explained is of a crop which is dispersed over a wide area. The host of problems involved has been discussed at length elsewhere.[1] The optimal weather conditions may differ at different stages of physiological development, so that aggregation of weather data over time may result in cancelling out the effects of extreme conditions which might occur at some limited and critical period for plant growth. There are important problems in aggregating meteorological and crop data over a wide geographical area, since the weighting given to meteorological observations at different points should depend upon the importance of the crop being studied at the same points. Data availability also creates problems, in that while experiments may determine the importance of a number of interacting weather variables, nationwide data on all but rainfall and temperature are rarely available. Weather indices constructed to take account only of temperature and rainfall have been extensively used in Oury's recent study of French cereals production.[2] Similar indices have been used to test the impact of weather on U.K. cereals yields, but it has been found that changes in the rainfall level dominate the values of the index, and thus rainfall data only are presented in Fig. 3.2, to help explain changes in cereals yields.

3.66. Testing with a number of weather variables has revealed that variations in weather conditions during June, July and August give a better explanation of cereal yields than variables for any other periods. The graph of the average June to August rainfall in England and Wales for the period 1954–66 shows an alternating year-to-year pattern. Over the whole period cereal yields have risen substantially due to the continuous adoption of higher yielding varieties, increased fertiliser application and the generally expanded use of improved technology. Around this rising trend, however, *yields from year to year have usually varied in the opposite direction to rainfall.* The correlation is not a perfect one, but as Fig. 3.2 shows, reported

---

[1] See, for example, Shaw, L. H., The Effect of Weather on Agricultural Output: A Look at Methodology, *Journal of Farm Economics,* Vol. 64, 1964, pp. 218–230.

[2] Oury, B., op. cit.

E

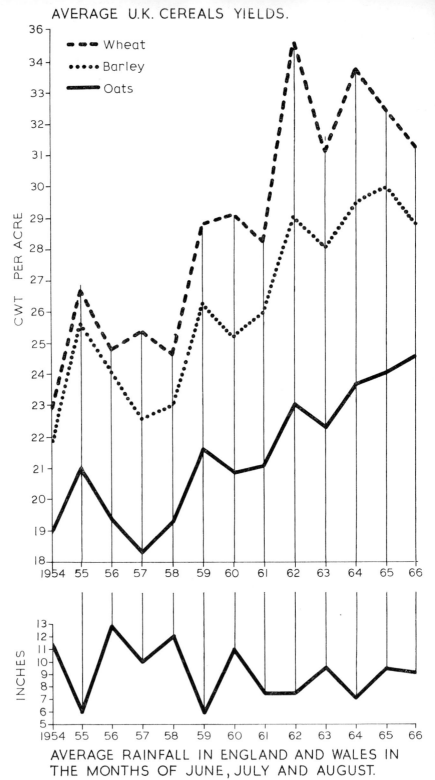

FIG. 3.2. Average United Kingdom cereals yields and rainfall during June–August, 1954–66.
Sources: Unpublished Econometric Study of Aspects of the U.K. Cereals Market by D. R. Colman, Department of Agricultural Economics, University of Manchester.

wheat yields have shown a pattern which is approximately the converse of that of June to August rainfall, after allowance is made for the trend. Barley and oats yields have followed the pattern slightly less closely, but it is clear that an important relationship exists between grain yields, and rainfall during the ripening period.

WEATHER EFFECTS ON HARVEST

3.67. The effects of weather on harvesting are less easy to quantify than the effects previously indicated, for no easily defined indicator is available to measure the outcome of fluctuating rainfall conditions at harvest. That harvesting costs may be greatly increased by heavy and prolonged periods of rain or by high winds in the harvesting period is self-evident, and can be documented by reference to N.A.A.S. Crop Husbandry records. A catalogue of effects attributed to poor harvesting conditions would include lodging of the grain, sprouting of grain in the ear, green ears in the ripe crop, and the delaying of autumn preparation of the land. One of the most important of these effects is that poor weather late in the season may adversely affect the quality of the harvested crop, so that in a year of bad conditions a relatively small proportion of the crop is eligible for human and industrial use and the rest has to be diverted to the animal feed market.

CHAPTER 4

Some Economic Aspects of Cereals Production

Increased Acreages and Receipts per acre with Falling Prices

4.1. During the 1950's the U.K. cereals acreage fell from 8·3 to 7·3 million acres; but after 1959 it increased very rapidly, reaching 9·4 million acres in 1967 (see Appendix H, Table 1, and Fig. 1.1 in Part II). This latter increase occurred in spite of falling producers' prices for cereals and rising wages, rents and prices of means of production. The guaranteed price of barley fell from 29s. 0d. per cwt. in 1959/60 to 24s. 9d. in 1967/8—a decline of nearly 15%—and that of wheat from 27s. 7d. to 25s. 11d. (5%) in the same period, while average weekly earnings of farm workers rose by over 30%.

4.2. The expansion of the cereals acreage during this time of adverse price movements is attributable partly to technological advances which influenced resource use and yield per acre, and partly to a drastic revision of ideas among many farmers about successive cereal cropping as a desirable practice, taking account of new methods of controlling weeds and diseases.

4.3. Rising yields have resulted in a general, though modest, upward trend in gross returns per acre, even in face of the falling prices already mentioned. Table 4.1 shows that in 10 years, average returns per acre of wheat increased from £38 to £41 and for barley from £32 to £36.

4.4. Table 4.1 gives £44 as the average gross return per acre for wheat in 1964/5, and this corresponds quite closely to the figure of £45·4 which resulted from the National Wheat Survey for 1964. Table 4.2 shows that this average return was fairly consistently obtained in all the regions surveyed, except that served by Wye College where much of the wheat crop is grown on poor Weald clay. The table also shows the cost structure for wheat production in 1964. Another source of information on the cost structure is the O.E.C.D. report *Interrelationship between income and supply problems in agriculture* (Paris, 1965), from which the figures in Table 4.3 are taken.

TABLE 4.1

GROSS RETURNS PER ACRE FOR WHEAT AND BARLEY PRODUCTION IN THE UNITED
KINGDOM, 1954/5 TO 1967/8

(£ per acre)

Harvest year	Wheat		Barley	
	Actual	3-year moving average	Actual	3-year moving average
1954/5	36·2	37·8	29·4	32·2
1955/6	40·4	37·9	31·6	30·8
1956/7	37·3	37·9	31·5	32·0
1957/8	36·1	36·0	32·8	32·5
1958/9	34·5	36·7	33·3	34·7
1959/60	39·4	38·0	38·1	35·9
1960/1	38·4	38·5	36·2	36·7
1961/2	37·7	41·2	35·8	37·3
1962/3	47·4	42·3	40·0	37·7
1963/4	41·7	44·4	37·3	38·5
1964/5	44·0	41·9	38·3	37·6
1965/6	39·9	40·9	37·1	36·6
1966/7	38·8	40·6	34·4	36·3
1967/8[a]	43·2	—	37·4	—

[a] Provisional.

Sources: Derived from *Agricultural Statistics*, U.K. various years; *Monthly Digest of Statistics;* and *Annual Review White Papers*. Returns have been calculated from national average yield estimates and average prices received by farmers inclusive of subsidies.

Returns and Costs of Wheat and Barley Production

4.5. Figure 4.1 compares the trend in gross returns per acre with the trend in costs of production per acre, for wheat and barley respectively. The data for returns are taken from Table 4.1. The index of costs has been obtained from published and unpublished reports of surveys carried out by various universities in England and Wales. These surveys have been made on non-random samples of farms and in different years in different regions; year-to-year comparisons of the aggregated results are therefore not very reliable, and for the purpose of this report a three-yearly moving average has been used.

4.6. Figure 4.1 shows that farmers have been able to maintain a fairly constant relationship between costs and returns, though the profitability of both wheat and barley appears to have reached a peak in 1962/3. This was achieved mainly by raising yields (see Fig. 1.3), reducing labour, employing less contract services and cultivating and harvesting more acres per machine. Yield is undoubtedly a major determinant of profitability, as costs per acre are virtually the same for low yields as for high.

TABLE 4.2

NATIONAL AVERAGE COSTS AND RETURNS FROM WHEAT—1964

(£ per acre)

Province	East					West		Aver-age
	Cam-bridge	Leeds and Newcastle	Wye	Nott-ingham	Read-ing	Bristol and Exeter	Man-chester	
Yield (cwt)	36·4	35·8	29·2	36·4	32·2	34·8	35·0	35·2
Seed	3·1	2·8	3·4	3·0	2·8	3·1	3·9	3·1
Fertiliser	3·1	3·4	3·9	3·3	3·7	3·8	3·6	3·5
Sprays	0·7	0·4	0·5	0·7	0·4	0·4	0·5	0·6
Miscellaneous	0·5	0·8	0·4	0·5	0·6	0·8	0·8	0·6
Variable costs	7·4	7·4	8·2	7·5	7·5	8·1	8·8	7·8
Labour	2·4	2·4	2·2	2·3	1·8	2·7	2·8	2·4
Machinery	6·7	6·3	6·6	6·1	6·2	6·9	7·5	6·6
Rent	6·7	4·3	5·6	4·7	4·2	4·7	5·4	5·5
Storage	1·4	1·1	1·8	1·6	1·6	1·6	0·5	1·4
Overheads	3·7	3·2	3·7	3·3	3·2	3·6	3·8	3·6
Fixed costs	20·9	17·3	19·9	18·0	17·0	19·5	20·0	19·5
Total costs	28·3	24·7	28·1	25·5	24·5	27·6	28·8	27·3
Grain sales	47·2	46·3	38·6	46·0	43·1	44·8	44·0	45·4
Gross margin	39·8	38·9	30·4	38·5	35·6	36·7	35·2	37·6
Net margin	18·9	21·6	10·5	20·5	18·6	17·2	15·2	18·1

Source: B. G. Jackson and F. G. Sturrock, "The National Wheat Survey," *Agriculture,* July 1966, p. 310.

Yield Variation between Seasons

4.7. If farmers can both raise yields and reduce the degree of variation in yield from year to year, this will materially increase the attractiveness of cereals as a crop. Improved knowledge of fertilisers and their use in combination with selective herbicides should promote a crop growth which is more independent of climatic conditions. The development by plant breeders of varieties which are disease-resistant might also be expected to stabilise yields over varying seasons. The use of short-strawed varieties has reduced the incidence of lodging, and even when lodging does occur the reels of combine-harvesters pick up a high proportion of the crop, gathering in much that would have been wasted with older methods and equipment.

4.8. However, in spite of all these technological advances there is no clear evidence from national statistics that annual variations in yield have in fact been appreciably reduced in the past twenty years. At the individual

TABLE 4.3

PERCENTAGE DISTRIBUTION OF PRODUCTION
COSTS FOR WHEAT AND BARLEY, ENGLAND
AND WALES[a]

	Wheat	Barley
Labour and management	12	14
Tractor	6	6
Fuel, oil and equipment, depreciation and repairs	15	19
Rent and rental value	13	16
Work by contractors	8	2
Seeds	13	11
Fertilisers	20	17
Other costs	13	15
Total	100	100

[a] The publication does not indicate to what year or years the figures relate, but they may be taken to represent the mid-1960's.

Source: O.E.C.D. *Interrelationship between income and supply problems in agriculture,* Paris, 1965.

farm level, a study of the same 31 farms over the 12 years 1952 to 1963 inclusive indicated a persistently high degree of fluctuation from year to year.[1] Without this element of uncertainty, specialisation in cereal growing would probably have gone even further than it has.

The Importance of Fixed Costs

4.9. Using the customary division into fixed and variable costs, it can be shown that the relative importance of fixed costs (rent, depreciation, repairs and labour) has increased (see Fig. 4.2). A major element in these fixed costs is the machinery which is specific to the cereals enterprise, and notably the combine-harvester. The operation and performance of this machine has therefore had a significant influence on the economics of cereals production over the years.

4.10. Table 4.4 displays the advantages of larger combines when these are used to capacity, while Fig. 4.3 shows the cost reductions to be gained from spreading heavy depreciation costs over the largest possible acreage. In particular, the costs per acre for the farmer with the small machine rise dramatically if he does not use it near to capacity, and if he has to harvest

[1] Allen, P. G., *Yield uncertainty in grain crop production,* unpublished B.Sc. dissertation, University of Nottingham School of Agriculture, 1967.

WHEAT

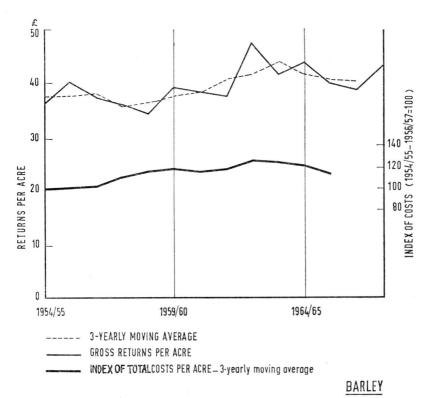

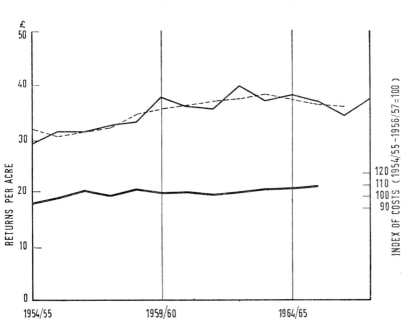

FIG. 4.1. Wheat and barley: gross returns and index of costs per acre.
Note: Gross returns per acre relate to the United Kingdom, index of costs to England and Wales.

E*

TABLE 4.4

FIXED AND VARIABLE COSTS FOR COMBINES OF VARIOUS SIZES AND AT CERTAIN
LEVELS OF USE

	5/6 ft cut bagger	8 ft cut tanker	10 ft cut tanker	12/14 ft cut tanker
Normal use: acres per year	75	200	300	400
	£	£	£	£
Approximate initial cost	850	2100	2850	3500
Fixed costs:				
Depreciation	120	294	399	490
Tax and insurance	5	10	15	18
Total	125	304	414	508
Total per acre	1·67	1·52	1·38	1·27
Variable costs:				
Repairs and fuel per acre	0·45	0·38	0·34	0·31
Total costs per acre for "normal" acreage	2·12	1·90	1·72	1·58

Normal use is the acreage which the combine harvests during the 20 days which on average are available.

Source: Derived from Davidson, J. G., *Farm Planning Data 1967*, University of Cambridge, School of Agriculture, Farm Economics Branch, pp. 68/9.

less than 50 acres it would generally pay him to use a contractor rather than own a machine. In this sense, there is more flexibility with the larger machines, though at full utilisation their fixed costs per acre are not substantially below those of smaller machines.

4.11. On the other hand, the great advantage of large-scale operation is most clearly shown by reference to labour productivity. Suppose that for any of these combine sizes three men were involved in the harvesting operation (with tanker combines two men only are frequently used), that the combine harvests a "normal" acreage and that the value of grain produced is £40 per acre; then the value of grain harvested per man is £1000, £2667, £4000 and £5333 respectively with increasing size of combine. It is obvious that the economic incentives favouring movement towards large scale cereal production are very powerful.

4.12. It is instructive to consider the position of a farmer growing 300 acres of cereals who has the opportunity to expand to 400 acres by purchasing an additional 100 acres of land. His fixed costs should change little. Assuming a gross margin of £30 per acre[1] the farmer could then repay a mortgage loan at 8% in 20 years after paying nearly £300 per acre for

[1] Davidson, J. G., *Farm Planning Data 1967*, University of Cambridge, p. 4.

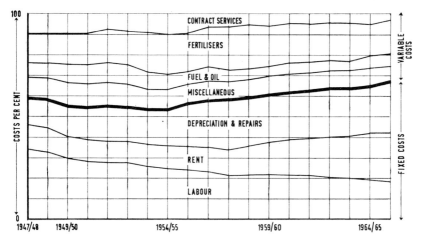

Fig. 4.2. The changing cost structure of cereals production in England and Wales.
Source: Published and unpublished cost investigations carried out by various universities in England and Wales.

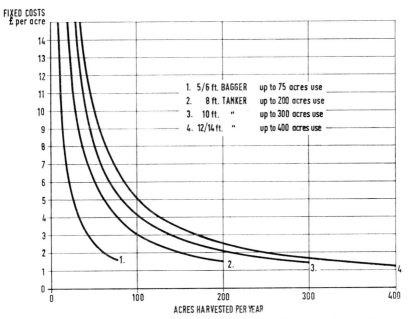

Fig. 4.3. Relationship between combine fixed costs and acreage harvested.
Source: Davidson, J. G., *Farm Planning Data 1967*, Farm Economics Branch, University of Cambridge, pages 68-9.

the land. On fertile land giving a higher gross margin it is not surprising that land prices have sometimes reached this level recently.

4.13. Many farmers, however, have no opportunity to acquire more land, or lack the necessary capital. Consequently much of the expansion of cereals which has occurred has been within the confines of individual farms at the expense of other land-using enterprises. There has been a trend towards specialisation in cereal production. On some farms this has been taken to the extreme of continuously cropping with cereals, especially barley. The farmers consider that any slight reduction in yields consequent upon this practice is more than compensated for by its cost advantages. Reference has already been made to the rising proportion of total farm area which is devoted to cereals (Chapter 1, p. 19 and Fig. 1.2).

Competitive Position of Cereals

4.14. Technological advance and price changes have not, of course, been confined to cereals. Other land-using enterprises have also changed. But comparisons of profitability of alternative enterprises are really meaningful only within a specific set of conditions on an individual farm. There are too many variables such as farm size, soil type, topography and climate for generalisations to be confidently made from diverse cost studies.

4.15. However, a comparison of changes in average producers' prices for various products is of interest in this context. Milk is frequently a major land-using enterprise which is to some extent an alternative to cereals, and Fig. 4.4 shows changes in average prices received for milk, wheat and barley over the years 1955/6 to 1966/7. After 1955/6 the price of barley clearly improved not only in absolute terms but also relatively to the prices of milk and of wheat, and this suggests that price might well have been an important factor contributing to the expansion of the barley acreage by almost 50% between 1955/6 and 1960/1. On the other hand, after 1962/3 the milk price improved while barley and wheat prices declined, yet there was no interruption to the expansion of barley. The conclusion seems to be that the technical reasons which prompted the expansion of barley in the first instance continued to exert a powerful influence long after the price attractiveness of barley had markedly diminished.

Yield Fluctuations and their Implications for Marketing

4.16. The year-to-year fluctuations in total production of cereals are bound to present problems of marketing. Table 4.5 shows the annual changes in production between 1955 and 1967. It will be seen that wheat

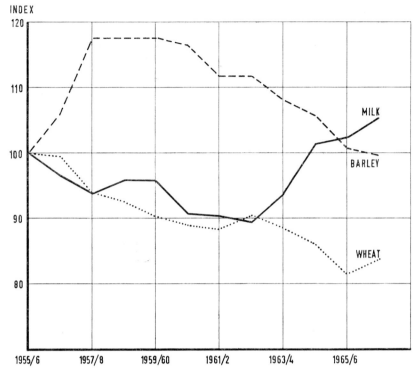

INDEX

FIG. 4.4. Movements in producers' prices for milk, wheat and barley, 1955/6 to 1966/7 (index 1955/6 = 100).
Source: Milk prices from M.M.B. *Dairy Facts and Figures* (annually), net price received by wholesale producers in U.K., ex-farm. Wheat and barley prices, *Annual Review White Papers,* total return including subsidy.

production changed on average by 478,000 tons per annum, the range being from a drop of 685,000 tons between 1965 and 1966 to an increase of 1,338,000 tons between 1961 and 1962. These changes are, of course, much influenced by fluctuations in the acreage of autmun-sown wheat due to weather (see Chapter 3). The production of barley tended upwards for most of the period, the average increase being 548,000 tons per annum. With oats there was a decline in production in most years, the average year-to-year change being 162,000 tons per annum.

4.17. Taking all cereals together, the average change in production per annum was 702,000 tons, but between 1961 and 1962 the increase was over 2 million tons. Two years later there was another very substantial increase of 1·4 million tons. Table 4.6 shows that in four years out of 12 wheat production changed by over half a million tons from the previous year. To the extent that they are unpredictable, year-to-year changes of this

TABLE 4.5

YEAR-TO-YEAR FLUCTUATIONS IN CEREALS PRODUCTION, U.K., 1955 TO 1967

('000 tons)

| Year | Change from preceding year | | | | |
	Wheat	Barley	Oats	Total	%
1956	+246	−136	−223	−113	1·4
1957	−162	+157	−341	−346	4·2
1958	+28	+213	−7	+234	3·0
1959	+74	+846	+49	+969	12·1
1960	+279	+225	−129	+375	4·2
1961	−491	+733	−236	+6	—
1962	+1338	+799	−75	+2062	22·0
1963	−913	+826	−309	−396	3·5
1964	+735	+805	−113	+1427	12·9
1965	+372	+658	−112	+918	7·4
1966	−685	+524	−111	−272	2·0
1967	+416	+656	+238	+1310	10·0
Average variations between years	478	548	162	702	6·9

Source: M.A.F.F. *Agricultural Statistics.*

TABLE 4.6

FREQUENCY DISTRIBUTION OF YEAR-TO-YEAR FLUCTUATIONS IN PRODUCTION, U.K., 1955 TO 1967

| Change from preceding year | Wheat | Barley | Oats | Total |
		(No. of years in period)		
Production UP by:				
0–500,000 tons	6	3	2	3
Over 500,000 and up to 1 million tons	1	8	—	2
Over 1 million tons	1	—	—	3
Production DOWN by:				
0–500,000 tons	2	1	10	4
Over 500,000 and up to 1 million tons	2	—	—	—

order of magnitude must present difficult marketing problems, and usually result in violent swings in market prices and a search for new means of stabilisation.

4.18. The year-to-year change in production is influenced by three factors: acreage changes; long-term trends in yields; and seasonal fluctuations in yields due to weather. An attempt was made to estimate the significance of seasonal fluctuations by eliminating the influence of the long-term upward trend in yields. It was found that the average year-to-year change in

yield attributable to weather was about 2·5 cwt for wheat, 1·5 cwt for barley and 1·0 cwt for oats. Taking the 1967 acreages of each crop, this represents average fluctuations in production due to weather (i.e. in the
absence of trend in acreage or yield) as shown in Table 4.7.

4.19. These are not additive because it does not always follow that when
the yield of one crop is down the yield of another is down also. But these

TABLE 4.7

	Wheat	Barley	Oats
1967 acreage ('000)	2305	6027	1012
Average "weather" fluctuation (cwt/acre)	2·5	1·5	1·0
Average annual fluctuation in production (tons)	268,000	452,000	51,000

figures give some indication of the extent of year-to-year fluctuations in
production to be expected even if the acreage remains unchanged. The
implication of this for grain marketing is that there will continue to be a
need for considerable flexibility in the marketing system. Research into the
causes and means of combating yield variation is desirable. Greater yield
stability would be beneficial nationally; to the individual farmer it would be
a considerable aid when planning the harvesting and disposal of his crop.

CHAPTER 5

Grain Preservation and Storage on the Farm

5.1. The harvesting of grain is concentrated into a period of about six successive weeks once a year. Consumption is spread fairly uniformly throughout the year, and can be envisaged as a flow which gradually uses up the stock of grain produced at harvest. But while consumption is reasonably uniform, the flow of home-grown supplies on to the market is affected by the amount of grain being imported from abroad. To reduce storage requirements, millers and compounders could, in theory, be encouraged to concentrate on using home-grown grain during the British harvest, but even so it would still be necessary to store three-quarters of the home crop for some time. The need for additional storage is, in any event, growing: the acreage of cereals is increasing, yields are rising, and, moreover, obsolete plant must be renewed.

5.2. Traditionally the bulk of the grain was stored near the point of production. Before the combine harvester came into use this was inevitable, because unthreshed grain is bulky and costly to transport. Threshing also provided winter employment for the farm labour force, and the rate at which this was carried out regulated the flow of grain on to the market. With the development of the combine harvester, the grain is still usually stored on the farm but in a grain store rather than a stack. Consequently the situation has altered in three important respects. Firstly, with such large quantities of threshed grain available at harvest time, the farmer is in a weak selling position unless he can store his grain. Secondly, grain straight from the field is usually too moist for safe storage and some method of preservation is necessary. Thirdly, the mechanisation of harvesting with larger machines and fewer men means that proper provision must be made for handling grain in bulk. The use of sacks is becoming uneconomic except in special circumstances.

5.3. The changeover to combine harvesters since 1939 was a natural reaction to rising labour costs. The farmer has, however, been compelled to invest far more capital. Because of this, and as part of a general encouragement of the re-equipment of farming, government grants have

113

been available for farm storage of grain. Assistance was given in two ways, (a) by offering a direct grant of one-third of the approved capital cost of storage installation, and (b) by offering seasonal incentives. These grants and incentives are, however, restricted to farmers—they are not given to merchants—thus helping to perpetuate the practice of storing grain in small lots on the farm. The general lack of success of central storage in the early post-war years reinforced this decision. Whether or not this policy should now be changed is discussed in Part III Chapter 7.

5.4. The Survey of Farmers included questions on the drying and storage of grain on farms. Its results form the main basis of the next two sections of this chapter, which describe the distribution of grain drying and storage equipment on farms in the U.K. The final section goes on to examine the economics of grain preservation and storage on the farm.

Grain Drying on Farms

5.5. At the time of the Survey in 1967/8, 31% of the cereal growers interviewed owned some form of drying facilities. Predictably, this proportion varied very markedly with size of farm: on holdings of under 50 acres, only 6% of farmers owned a drier, while on farms of 400 acres and over the proportion was 83%. There was an even more noticeable relationship between total cereals acreage and the existence of drying facilities: among farmers with 300 or more acres of cereals, 95% had a drier.

5.6. The proportion of cereal farmers owning drying facilities did not vary a great deal from one region of Great Britain to another (considering F.M.S. regions, the maximum of 37% was found in the Central-Southern region and the minimum of 29% in Wales and South-West). In Northern Ireland, however, the Survey indicated much lower levels of ownership, with only 7% of farmers reporting that they had drying facilities.

5.7. In addition, for the sample as a whole, there were 2% of farmers who shared the ownership of drying facilities with another farmer or farmers. There was some indication that a higher proportion of farmers shared driers in the Northern F.M.S. region (9%), though the reason for this is not apparent from the Survey. In view of the relatively high capital cost of a grain drier, there are obvious savings in cost to be had by sharing facilities. This would seem to be specially true of smaller farms, and, indeed, nine out of ten of the farmers reporting that they shared a drier were on farms with between 50 and 200 acres of crops and grass. However, even among these smaller farms the farmers sharing a drier represented only a small percentage of the total, and there may well be scope for an increase in this approach to the ownership of driers.

5.8. The farmers who owned or shared grain driers were asked to describe the drying facilities they had. Just over one in three (36 %) of those with a drier reported having ventilated bins; 26 % said they had a tray drier, 21 % ducted floors, 5 % cascade or vertical drier, and 16 % gave a variety of other answers (the various replies add to over 100 % because some farmers had more than one type of drier).

5.9. Of the ventilated bins and ducted floor installations reported in the Survey, capacities varied as follows:

Capacity (tons)	Percentage of farms with ventilated bins or ducted floors
10 or under	13 %
11–50	23 %
51–200	35 %
201–400	10 %
Over 400	6 %
Capacity not known	13 %
	100 %

The Survey data indicates that the average capacity per farm of ventilated bins and ducted floors together was 115 tons. For ventilated bins only the average was 80 tons, and for ducted floors only the figure was 180 tons.

The throughput of tray driers varied as follows:

Throughput (tons per hour)	Percentage of farms with tray driers
Under 1	5 %
1–2	50 %
Over 2 and under 5	31 %
5 and under 25	8 %
25–50	2 %
Throughput not known	4 %
	100 %

5.10. Farmers owning or sharing drying equipment were asked when it had been installed; their answers can be summarised thus:

How long ago installed	Percentage of farmers owning or sharing equipment
1 year ago or less	17 %
Over 1 and up to 2 years ago	13 %
Over 2 and up to 3 years ago	14 %
Over 3 and up to 4 years ago	14 %
Over 4 and up to 5 years ago	10 %
Over 5 and up to 7 years ago	11 %
Over 7 and up to 10 years ago	9 %
Over 10 years ago	8 %
Not known	4 %

The pattern of answers given above indicates an increasing level of purchases over the 10-year period covered, though it is not possible to be sure what proportion of the purchases in each year represented replacement as opposed to new investment.

5.11. Evidence on the trends in ownership of drying equipment is provided also in the Machinery Censuses carried out by the Agricultural Departments of England and Wales, Scotland and Northern Ireland. Table 5.1 summarises the results of these for the period between 1959 and 1967. As will be seen, the headings under which the various types of machine are recorded are different from those used in our Survey, and so the two sets of results cannot be directly compared. In Scotland and Northern Ireland, the use of drying equipment has risen substantially; in Scotland, the large increase in the number of ventilated silos, bins or ducted floors between 1964 and 1967 is especially noticeable. In England and Wales, the number of continuous grain-flow driers has gone up steadily, while driers described as "tray", "batch" or "platform (in sack)" have stabilised in numbers since about 1964. Because of non-comparability of the 1964 or 1966 figures with earlier years (or with each other), it is not possible to say what changes there have been since 1962 in the numbers of ventilated silos, bins or ducted floors in England and Wales, but a rise is obvious in the years up to 1962.

5.12. Of the farmers taking part in the Survey, 40% dried none of the grain they harvested in 1966, and 55% did no drying after the 1967 harvest. This difference, it can be assumed, is due to the unusually good harvest conditions in most parts of the country in 1967. Of those farmers who did dry grain in either year but did not own or share any drying facilities, about half had their drying done for them by other farmers and half used drying facilities belonging to merchants.

Grain Storage on Farms

5.13. In the analysis of the Survey data, it was found that a high proportion of farmers (87%) had stored some or all of their grain crop on their farms after their latest harvest. Half the farmers in the sample were interviewed before the 1967 harvest and half after, so that their answers referred to the 1966 and 1967 crops respectively; but exactly the same proportion was found to apply in each of the two years.

5.14. The space available for grain storage on farms includes not only purpose-built facilities but also what may be termed general-purpose space—barn floors, lofts and so on, which can be used to store not only grain but also feeds or many other commodities as the need arises. There

TABLE 5.1

GRAIN-DRYING MACHINERY IN THE U.K., 1959–67

	Machinery Censuses								
	1959	1960	1961	1962	1963	1964	1965	1966	1967
England and Wales:									
Types of grain-drier:									
Continuous grain-flow	4670	5210	6100	7260	(a)	9790	(a)	11,110	(a)
Tray or batch	1320	1660	1950	2510	(a)	3710	(a)	6650	(a)
Platform (in sack)	3900	4680	4140	4190	(a)	3680	(a)	5650(b)	(a)
Ventilated silo, bin or floor-forced air flow	4570	4970	5170	5750	(a)	9830(b)	(a)	(a)	(a)
Scotland:									
Types of grain-drier:									
Continuous grain-flow	229	(a)	293	(a)	(a)	678	(a)	(a)	1117
Tray or batch	118	(a)	} 293	(a)	(a)	641	(a)	(a)	} 732
Platform (in sack)	143	(a)	1127	(a)	(a)	1695	(a)	(a)	5551
Ventilated silo, bin or floor-forced air flow	536	(a)		(a)	(a)		(a)	(a)	
N. Ireland:									
Total grain-drying machinery	220	(a)	750	(a)	820	(a)	(a)	955	(a)

Notes: (a) Not recorded, or Census not taken; (b) not comparable with earlier years because of differences in recording.
Source: Machinery Censuses for England and Wales, Scotland and N. Ireland.

can therefore be no precise measure of "maximum available grain storage". Information collected in the Survey in fact suggests that if account is taken of general-purpose space as well as specialised grain storage equipment, the total of storage space for all farms in the U.K. is enough to hold rather more than the total annual production of cereals.

5.15. Among the farms taking part in the Survey, those in the Central-Southern region on the F.M.S. classification showed a rather lower proportion storing grain (80%) than the national average, whereas the percentages recorded were above the national average in Wales and the South-West (92%), Northern Ireland (93%) and the Northern region of England (95%). This is a rather unexpected result; by way of explanation, it is only possible to guess that a relatively high proportion of the crop may be sold off the combine in the Central-Southern region.

5.16. Of the farmers who did not store any of their grain on the farm, the great majority (85%) reported that they also did not store grain anywhere else. Relatively few farmers used storage on neighbouring farms; and even fewer used merchants' storage facilities.

5.17. Many farmers have more than one kind of storage facility for grain, and many combinations of these various types were found in the course of the Survey. The following list shows the proportions of farms with storage facilities on which various combinations were recorded:

Proportion of cereal farmers with storage facilities who have:

	%
Bins/meal store only	10
Bins plus other purpose-built facility	6
Bins plus other non-purpose-built facility	7
Silos/ducted floor only	11
Silos/ducted floor plus other purpose-built facility	3
Silos/ducted floor plus other non-purpose-built facility	5
On floor/sacks	21
On floor only	9
Other non-purpose-built	27
Not stated	1

In other words, 23% of the farms with storage capacity had bins; 19% had silos or ducted floors; and 57% had no specially-installed facilities for the storage of their grain. The Survey showed also that even among farmers who had purpose-built facilities it was very common for part of the crop to be stored in sacks or on the barn floor.

5.18. For storage equipment, unlike drying equipment, the Machinery Censuses do not give enough data to show the trends in ownership which have taken place in past years. However, some evidence is again available from the Survey question in which the date of purchase of equipment was recorded. In this case, the question was asked for storage bins only.

Because the answers were recorded as calendar years (and not as "one year ago", etc.) the figure for 1967 applies to a part-year only; had it covered the whole year, it would probably have been similar to the figures for 1965 and 1966.

Proportion of farmers with storage bins bought in:

	%
1967 (part-year)	5
1966	12
1965	10
1964	17
1963	13
1962	8
1961	5
1960 or earlier	25
At a number of dates	7
Not known	4

Purchases of bins appear to have risen to a peak in 1964 but to have dropped off a little thereafter; the reason for this is not immediately obvious.

5.19. Among the most important incentives to the installation of grain storage facilities on the farm have been the grants available to farmers under the Farm Improvement Schemes. It has been estimated that grant-aided improvements carried out under these schemes since 1957 have included the provision of between 6 million and 7 million tons' purpose-built grain storage capacity on U.K. farms. Since storage equipment is seldom installed without grant-aid, this figure is probably a close approximation to the total increase in purpose-built storage during the period.

5.20. The farmers who stored grain on the farm were asked whether they had experienced any difficulties with storage. Twenty-seven per cent said they had. The most common problem was infestation by rats or by mice (mentioned by 13% and 8% respectively among those storing on the farm). Rather unexpectedly, problems arising from dampness or condensation were reported by only 6% of the farmers who stored grain.

The Economics of Grain Preservation and Storage on the Farm

5.21. This section is concerned with the merits of different systems of preserving, storing and handling grain on the farm. It is the outcome of an investigation into the characteristics of a variety of systems, together with their capital and running costs. Particular attention was paid to the size of installation, and to possible economies of scale. There is in existence a large body of technical information about the various systems, but far less economic data. No doubt this is partly due to the fact that no two farms need exactly the same installation, and costs obtained from one farm cannot be applied indiscriminately to another. While this is true, it is nevertheless quite possible to obtain reasonable average costs from adequate samples and also to analyse these into their components which can be used to estimate the costs of other systems.

5.22. Details of capital costs were obtained principally from agents selling and installing the equipment, with specialist building and machinery costs provided by constructional engineers and manufacturers respectively. All these made a significant contribution to the investigation, both by supplying data and by providing fresh thinking on the problems of grain storage. The acreage of cereals in each N.A.A.S. region was used as a guide to the number of agents to be contacted in different parts of the country. The information collected covered a wide range of systems, and although emphasis was laid on new installations, some data on the cost of expanding existing facilities was also obtained.

5.23. In the course of the Survey detailed costs were collected for 103 complete new installations, with a storage capacity of 46,540 tons and costing a total of £678,500. The average capacity was 452 tons and the average capital cost per ton £14·6 (before grant). A further 176 costings covered (a) extensions of capacity and replacements for existing equipment, (b) conversions of existing buildings, and (c) butyl bags.

METHODS OF PRESERVATION AND STORAGE

5.24. When harvested, each cereal grain is a living embryo and normally it will continue to respire and use up its food store. The rate at which respiration continues depends on the moisture content, the temperature and the supply of oxygen. It follows, therefore, that to reduce storage losses respiration must be reduced by eliminating moisture, by reducing temperature or by restricting the oxygen supply. Grain may deteriorate through the development of moulds or mite infestation, and these are controlled by the same measures. Losses can also be reduced by killing the embryo, a method which is satisfactory as long as germination is of no importance. Normally grain from the field cannot be stored for more than a few days without some form of preservation. At present the four principal methods, which vary widely in importance, are as follows:

(a) drying,
(b) sealing,
(c) chilling,
(d) addition of chemical preservative.

Currently drying is overwhelmingly the most popular system, but the investigation indicated that other methods could be cheaper. Chemical additives are still in the experimental stage, but they may prove a useful alternative on a mainly livestock farm.

5.25. The four possible methods of preservation can be used with a variety of types of storage. The principal types of storage are as follows:

Bins
Ventilated bins inside
Ventilated bins outside
Unventilated bins inside
Unventilated bins outside
Floor
Ventilated on-floor
Unventilated on-floor
Other
Sealed silo—metal towers
Sealed silo—butyl bag
Sacks

5.26. The possible combinations of method of preservation and type of storage are summarised in Fig. 5.1. Dotted lines show the possible storage methods that may be used in conjunction with ventilated bins. Drying is divided into low and high ambient rise methods, although this distinction is not absolutely clear-cut. With low ambient rise drying, large quantities of slightly-warmed air are moved through the grain, whereas with the high ambient rise method the grain is moved through a smaller volume of much hotter air. This distinction is important in a number of ways.

(a) Moving grain is likely to be more costly than moving air, because of the more sophisticated equipment required.

(b) High ambient rise drying is more open to abuse if the temperature is raised above the correct level in an attempt to increase throughput (this is particularly important where the grain is for malting or milling).

(c) The high ambient rise method may be the only practicable method in areas where grain harvested with a high moisture content must be dried rapidly to prevent deterioration.

(d) Low ambient rise drying is a slower process and therefore implies some storage capacity. It is thus less suitable for rapid conditioning of the grain for immediate sale at harvest.

5.27. The chilling of grain as a means of preservation is a recent introduction and although the grain keeps satisfactorily and its germination rate may even be improved, the method is meeting consumer resistance. The grain, however, can be used for milling or malting, whereas grain preserved by sealing or adding chemicals is only suitable for stockfeed. In the case of wheat, subsidy is only payable if the grain is sold and leaves the farm, which effectively precludes the use of sealing or chemicals as methods of

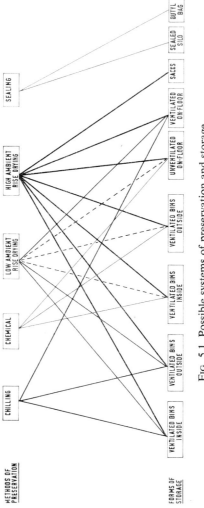

Fig. 5.1. Possible systems of preservation and storage.

Note: Dotted lines show storage methods which may be used in conjunction with ventilated bins.

preservation.[1] This discourages the use of wheat as a component of home-mixed feeds.

5.28. Of the eighteen systems shown in Fig. 5.1, only five were encountered in sufficient numbers to provide a direct comparison between systems. These involved the use of:

Ventilated bins inside
Ventilated bins outside
Ventilated on-floor
Sealed silos
  (a) metal
  (b) butyl

5.29. The first three systems all made use of low ambient rise drying, although in principle they could have been used with chilling units instead of fans and heaters. Many installations made use of high ambient rise drying but there were too few of each of the seven possible systems for effective comparison. Estimates for the remaining systems were synthesised from component costs.

5.30. Ventilated bins indoors may have radial or vertical air flow, with warm air from a static or a mobile fan and heater. Outdoor bins have a vertical air flow, and are usually grouped in a semi-circle with a central intake pit. Most outdoor installations have both ventilated and non-ventilated bins, with the former at least half the capacity. The radial flow ventilation system dries the grain rather more rapidly than vertical flow. It can be used in conjunction with unventilated on-floor storage. Special drying bins can be obtained with a drying rate faster than normal, and these are really midway between high and low ambient rise drying.

5.31. The capital costs of ventilated bin installations are shown in Fig. 5.2. Functions of the rectangular hyperbola type have been fitted to the individual readings, to represent the average relationship. The figures for indoor bins include building costs. Most installations have less than 500 tons capacity (wheat) but as far as can be seen capital costs gradually fall as capacity increases. It is evident that an indoor bin system (including a new building) usually costs about twice as much as the equivalent in outdoor bins, but grain handling is slightly more automatic. Bins placed in an existing building are, of course, cheaper than a complete installation; a reasonable saving is between a quarter and a third of the full cost.

5.32. Ventilated on-floor drying and storage can be likened to a very

[1] A change in the system of payment of wheat subsidy could, of course, materially affect this situation (see Appendix E).

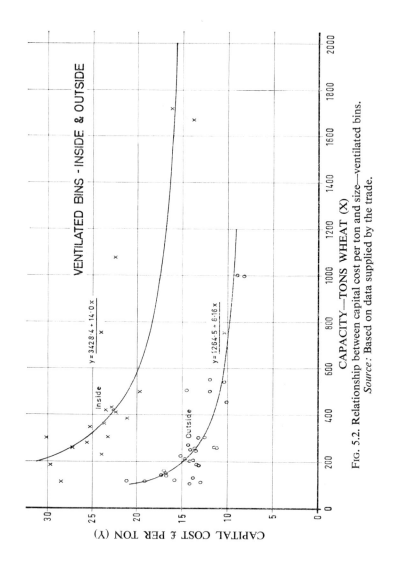

Fig. 5.2. Relationship between capital cost per ton and size—ventilated bins.
Source: Based on data supplied by the trade.

large but shallow vertical flow ventilated bin. The main air flow is through
a large duct (sometimes below floor level) with small lateral ducts at right
angles and 3–4 feet apart. The grain is stored level at 8–10 feet deep,
supported at the side by grain-retaining walls. The capital cost per ton for
this system is summarised in Fig. 5.3. The range of costs for a given
capacity is considerable, but there is evidence of a fall as capacity increases,
although not as pronounced as in the case of bin storage (Fig. 5.2). The
cost curve falls from about £15 at 250 tons to £12·5 at 600 tons and £10

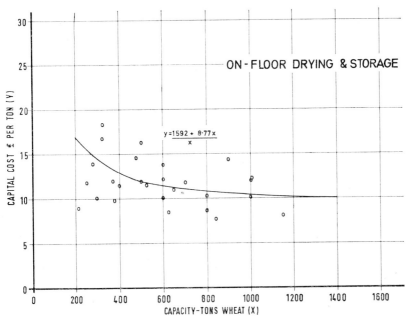

Fig. 5.3. Relationship between capital cost per ton and size—ventilated on-
floor drying and storage.
Source: Based on data supplied by the trade.

at 1300 tons. There are however a number of installations with costs below
£10 a ton for 600 tons or less. The cost pattern is in fact similar to that for
outdoor bins, but labour for handling grain on the floor is likely to be much
greater than for outdoor bins. A conventional building, on the other hand,
has more alternative uses when not required for grain. In wetter areas both
these systems may be unable to reduce high moisture contents sufficiently
rapidly to prevent spoilage (unlike radial-flow indoor bins).

5.33. Sealed silos preserve the grain by excluding oxygen, thereby preventing the growth of aerobic bacteria and fungi. Up to 25% moisture
content is acceptable and grain can be put in direct from the field. The

grain is only suitable for stockfeed and can be harvested earlier than grain for milling or malting. Metal silos are made with a non-corrosive lining, and are usually emptied by a bottom-unloading auger. Capital costs are shown in Fig. 5.4, with capacity measured in tons of wheat. In practice wheat must be sold to earn the subsidy and therefore this system would only be used for barley or oats, which require more storage space.[1] Most of the readings for metal silos are for small capacities, to suit the size of livestock enterprise carried on the farms. Small sizes are not particularly cheap in capital, but running costs are negligible. Such evidence as there is suggests some economies of scale up to about 400 tons, but beyond this

TABLE 5.2

AVERAGE CAPITAL COSTS PER TON OF WHEAT

Type of installation	1 200 tons	2 200–399 tons	3 400–599 tons	4 600+ tons
	£	£	£	£
Ventilated bins inside	29·2	24·9	22·1	16·8
Ventilated bins outside	16·0	13·5	11·9	8·6
On-floor drying and storage	—	13·7	12·6	10·7
Sealed silos metal	13·5	11·9	10·3	—
High ambient rise driers with non-ventilated				
(a) Floor Storage	—	19·8	15·8	10·3
(b) Bins outside	—	16·3	14·3	11·7
(c) Bins inside (steel)	—	20·8	17·9	15·2
(d) Bins inside (temporary)	—	18·0	14·3	11·9

level further savings are likely to be small because a larger plant merely involves multiplying the number of towers.

5.34. Table 5.2 shows the average capital cost per ton for each of the four systems described above, broken down into four size-groups.

5.35. Another form of sealed silo is the butyl rubber bag, supported in a circular wire-mesh container. The usual capacity is about 40 tons. Loading and unloading is by an auger, through the neck and base of the bag, and is slightly more difficult than with the sealed metal silo. The capital cost was found to be about £5 a ton, but it would be prudent to count on a life of no more than five years. As with sealed metal silos, the "shelf life" of the preserved grain is only a few days once it has been removed from the bag.

5.36. As already stated, the costs of some of the less usual systems of

---

[1] Wheat requires 46 ft³ per ton, barley 51 and oats 70, a ratio of 1·0 : 1·1 : 1·5.

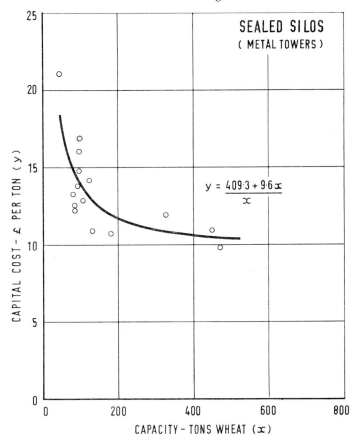

Fig. 5.4. Relationship between capital cost per ton and size sealed silos (metal towers).
Source: Based on data supplied by the trade.

storage were based on quite small samples. Costs could, however, be confirmed by synthesising totals from component costs already known. The same was true for systems not represented at all in the survey.

CHEMICAL PRESERVATION

5.37. Moulds and other harmful organisms found in grain can be killed with a chemical (normally propionic acid) before storing. A special applicator is necessary to ensure that the grain is properly covered with the correct amount of acid. Once treated the grain can be stored in a weatherproof container or building. Steel, zinc and concrete are gradually corroded by the acid, and containers must be proofed by painting or covering with plastic sheet. The treated grain does not, however, deteriorate when taken out of store prior to feeding. This method of preservation is a recent development and was first used commercially during the 1968 harvest. It is therefore too early to reach a firm conclusion about its success on a farm scale. Assuming, however, that the chemical works as well as its manufacturers claim (and experimental results suggest that it does) it can readily be compared with other systems. Apart from the building or bin, the two main items are the chemical and the applicator. The amount of chemical necessary varies with moisture content, from £0·7 per ton at 15% to £1·8 at 30%, at current recommended prices for proprietary propionic acid. A reasonable middle value is £1·25 per ton at 21% moisture content. This is, of course, an annual rather than a capital cost. The applicator is attached to an auger, and the complete unit costs from £150 to £350 according to size.

CHILLING

5.38. Moulds and granivorous insects and mites can be inactivated, and respiration of the grain minimised, by reducing the temperature of grain to between 1° and 5°C (34° and 41°F) depending on moisture content. The system is similar in many respects to low ambient rise drying with floor or bin storage, but the fan and heaters are replaced by a refrigeration unit. As the aim is to cool rather than dry the grain, very low airflows are adequate. Because grain is a good insulator and needs rechilling for only a few hours a week, running costs are well below those for drying. On the other hand the capital cost is greater than for an equivalent drying unit. The chilling process keeps the grain in good condition, and is even claimed to improve germination. The system has not, however, become popular in Britain. It would work well if the farmer intended to keep all his grain for home use. But if he wished to sell part of the crop and the moisture content was above a level acceptable to merchants (16–18%), drying would be necessary. If the farmer buys a drier, however, it would normally be cheaper to use it for all the grain instead of installing chilling plant. Alternatively for small quantities the grain might be dried by contract.

HIGH AMBIENT RISE DRYING

5.39. An important omission from the systems so far considered is the high ambient rise drying method, usually represented by continuous-flow driers. These are the most popular single method of drying in the U.K., as indicated in Table 5.1. They operate by moving a thin layer of grain through a stream of air heated to between 43° and 93°C (110° and 200°F), depending on the desired throughput and the intended use for the grain. Stockfeed grain can be exposed to higher temperatures because germination is not required and the feeding value is not affected. Two advantages of this method compared with low ambient rise drying are firstly, that very damp grain which would spoil easily can be dried rapidly to a safe level, and secondly that it is independent of the method of storage. More labour is, however, needed for supervision, and there is a temptation, when the drying capacity is overloaded, of exceeding the safe temperature limit.

5.40. There are several possible permutations using this method, and the Survey provided too few costs for each system to make direct comparisons worth while. For this reason, capital requirements for driers covering a range of capacities were derived from the installations making use of this method of drying; all driers were of the continuous flow type. The costs include:

> Drier
> Grain-handling equipment
> Foundations (including intake pit)
> Installation and electrical costs (excluding
>   connection charges)

The results are shown in Fig. 5.5. The function can be used to estimate capital costs for any desired drier capacity, and (in conjunction with building and bin costs) to derive costs for systems not well represented in the survey. The economies of scale are quite apparent; two 5-ton-per-hour driers would cost £9600, whereas a single 10-ton-per-hour drier would cost only £7700. Building costs for housing the driers are not included at this stage.

5.41. The combination of this type of drying with (non-ventilated) floor storage over a range of capacities is represented in Fig. 5.6, which is derived from representative building costs and the function in Fig. 5.5. In this system the grain is heaped in the centre of the building to its natural angle of repose (35°) from 8-ft retaining walls, thus making fuller use of the volume enclosed by the span. Unlike the similar but much larger stores discussed in Part III, Chapter 7 it is not considered necessary to "condition" the grain during storage, nor to reduce moisture content as low as 12·5%. Average capital costs for four capacity ranges are included in Table 5.2, for ease of comparison with the systems common in the survey. It can be seen that this is a relatively high-cost system at the small tonnages represented. The other system shown on Fig. 5.6 is the combination of continuous-flow drier and non-ventilated bins outside. The building for the drier is allowed at 20% of the drier cost. Capital costs are rather lower than with floor

F

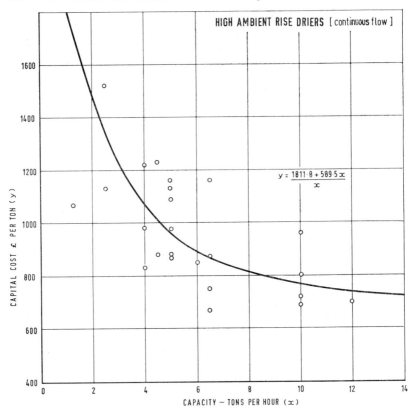

Fig. 5.5. Relationship between capital cost per ton and size—high ambient
rise driers (continuous flow).
Source: Based on data supplied by the trade.

storage at the smaller tonnages, but even so are above those for ventilated
on-floor and ventilated bins outside.

5.42. Figure 5.7 represents a further system, using non-ventilated bins indoors. Steel bins of this kind are clearly an expensive system of storage when
combined with a new building, as to a large extent the weatherproofing is
duplicated. The temporary bins referred to are made of hessian or strong
paper supported by wire mesh, and can be dismantled when not required.
These bins are very cheap (£0·75 per ton capacity) but being circular are
more wasteful of floor space than square bins. Provided they are placed
correctly they can be filled by overhead conveyors and unloaded with an
auger from above, the wall sections being removed as the grain level falls.
Inevitably labour requirements are higher than with more permanent
systems, but not much extra labour is needed during the harvest period
which is usually the busy time. The broken line in Fig. 5.7 shows the cost

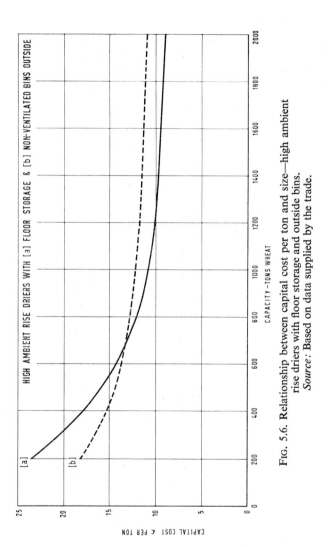

Fig. 5.6. Relationship between capital cost per ton and size—high ambient rise driers with floor storage and outside bins. *Source*: Based on data supplied by the trade.

of the temporary bin system, excluding the building. Some average costs are shown in Table 5.2.

COMPARISON OF SYSTEMS

5.43. Cost must be the main standard for comparison and the method adopted here takes into account both capital and running costs as far as these can be identified. An important difference is that capital costs are generally incurred in the first year, whereas running costs are an annual

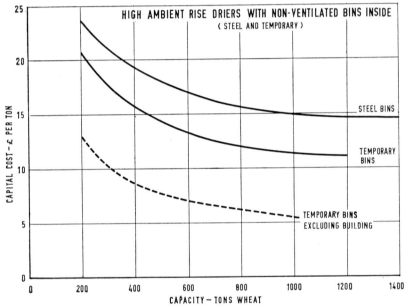

FIG. 5.7. Relationship between capital cost per ton and size—high ambient rise driers with inside bins.
Source: Based on data supplied by the trade.

expenditure year by year. The criterion of minimum present costs is therefore found by discounting future expenditure back to the present at a suitable rate of interest, over a given period of time. Here the discount rate is taken as 7% and the planning period as 10 years. This procedure is described in greater detail in Part III, Chapter 7.

5.44. Running costs are defined here as fuel oil, electricity, repairs and (in one case) chemicals. No charge has been made, however, for the electricity required by handling equipment. Labour charges are excluded as being too arbitrary.

5.45. Tables 5.3A and 5.3B summarise the discounted costs for eight of

TABLES 5.3A AND 5.3B. COMPARISON OF DISCOUNTED COSTS FOR EIGHT SYSTEMS AND THREE SIZES OF GRAIN STORAGE

5.3A

	Continuous flow drying with floor storage			On-floor drying and storage			Chilling with ventilated on-floor storage			Chemical additive with non-ventilated floor storage		
Capacity tons (wheat)	200 tons	400 tons	1000 tons	200 tons	400 tons	1000 tons	200 tons	400 tons	1000 tons	200 tons	400 tons	1000 tons
	£	£	£	£	£	£	£	£	£	£	£	£
Capital cost per ton	23·5	18·4	10·8	17·7	12·8	10·4	19·8	14·0	10·9	8·7	7·2	5·3
Total capital cost	4700	7360	10,800	3540	5120	10,400	3965	5600	10,950	1750	2900	5350
Running costs per ton	0·53	0·48	0·41	0·42	0·38	0·35	0·26	0·20	0·17	1·28	1·27	1·25
Running costs per annum	107	194	408	85	151	354	52	80	169	257	509	1253
10-year total of discounted running costs	751	1362	2865	597	1060	2486	365	562	1187	1805	3575	8800
Plus capital cost	4700	7360	10,800	3540	5120	10,400	3965	5600	10,950	1750	2900	5350
	5451	8722	13,665	4137	6180	12,886	4330	6162	12,137	3555	6475	14,150
Less salvage value at 20% discounted	520	748	1097	360	520	1057	403	569	1113	178	295	544
Total	4931	7974	12,568	3777	5660	11,829	3927	5593	11,024	3377	6180	13,606
Average cost/ton stored per annum	2·46	1·99	1·26	1·89	1·41	1·18	1·96	1·40	1·10	1·69	1·54	1·36

5.3B

	Ventilated bins outside			Temporary bins inside			Sealed silos Metal towers			Sealed silos Butyl bags		
Capacity tons (wheat)	200 tons	400 tons	1000 tons	200 tons	400 tons	1000 tons	200 tons	400 tons	1000 tons	200 tons	400 tons	1000 tons
	£	£	£	£	£	£	£	£	£	£	£	£
Capital cost per ton	14·8	11·5	9·4	20·7	15·7	11·4	11·8	10·6	10·0	5·0	5·0	5·0
Total capital cost	2960	4600	9400	4140	6280	11,400	2360	4240	10,000	935	1870	4675
Running costs per ton	0·40	0·36	0·34	0·64	0·60	0·55	0·00	0·00	0·00	0·00	0·00	0·00
Running costs per annum	80	146	344	129	239	554	0	0	0	0	0	0
10-year total of discounted running costs	562	1025	2416	906	1678	3891	0	0	0	666	1332	3330
Plus capital cost	2960	4600	9400	4140	6280	11,400	2360	4240	10,000	935	1870	4675
	3522	5625	11,816	5046	7958	15,291	2360	4240	10,000	1601	3202	8005
Less salvage value at 20% discounted	150	234	478	210	319	579	120	215	508	—	—	—
Total	3372	5391	11,338	4836	7639	14,712	2240	4025	9492	1601	3202	8005
Average cost/ton stored per annum	1·69	1·35	1·13	2·42	1·91	1·47	1·12	1·01	0·95	0·80	0·80	0·80

the systems discussed above, each with figures for three sizes of store, 200, 400 and 1000 tons (wheat) respectively. The 10-year total of discounted running costs is added to the capital cost, which with one exception is incurred in the first year and is therefore not discounted. The exception is butyl bags, which can be installed immediately before harvest. Another difference is that the bags are assumed to need replacing in 5 years, and the 10-year total of discounted costs here includes the discounted capital cost of replacement in year 6. Buildings are credited with a salvage value of 20% after 10 years on the assumption that they have other uses. Outside bins have been given a salvage value of 10% and butyl bags nil. The costs are summarised as a cost per ton stored per annum.

5.46. Other systems could have been included in the comparison but were rejected because of high capital costs. Sacks have not been included because they are only suitable for very small tonnages or specialised purposes.

5.47. In general the annual cost per ton stored follows the same order as the capital cost per ton, for a given capacity. This reflects the importance of capital relative to running costs (in part because of the discounting procedure). The main exception is chemical preservation, where despite low capital costs the exceptionally high running costs make this system one of the more expensive. In any case the method is only suitable for stockfeed grain. This is also true of the cheapest system, butyl bags. Here the calculations do not allow for the inconvenience of having the storage silos exposed to the weather, and being much less automatic in operation. Metal sealed silos are also relatively cheap, particularly at the smallest tonnage represented here, which is a convenient size for many livestock units.

5.48. Of all the systems suitable for storing milling and malting grain, chilling appears as one with generally low costs, but it does assume that a market is available for grain at a moisture content exceeding about 18%. If this is not so, the advantages of lower running costs are lost. Apart from chilling the cheapest system is the outside ventilated bin. All the costs assume complete installations. Thus if a farm already has a suitable building the capital cost can be reduced accordingly, and this can affect the order of preference between systems. In particular, if a building is already present, a system with outside bins is likely to be relatively less attractive since these are almost bound to be specially constructed.

5.49. As costs for extending only 11 installations were collected during the investigation, referring exclusively to in-bin systems, it is difficult to draw any firm conclusions. The cost of providing this additional capacity varied from £4·8 to £21·5 per ton, depending on the availability of spare drying capacity and the system involved. However, it is interesting to note that in total the 11 extensions gave an extra capacity of 3014 tons at a cost of £42,158, an average of £14·0 per ton—a figure very similar to that recorded for complete new installations.

Conclusions

5.50. (a) The investigation showed a wide variation in capital costs per ton both within and between the various systems in common use. The average cost for complete installations was just over £14 per ton, for 450 tons capacity; if, however, there were a greater concentration on the cheaper systems described here, the average could no doubt be reduced, although it seems unlikely to fall below £10 per ton. The most expensive systems could be made ineligible for grant-aid, or alternatively this could be restricted to a fixed payment per ton of capacity.

(b) Extensions to existing installations are one possible way of increasing capacity at relatively low cost, but the method assumes spare capacity in drying (or other method of preservation) and often means extending the existing system, in some cases tending to perpetuate systems which may be less desirable in the long run.

(c) One of the advantages of systems using a building without major internal structures is that the space is more likely to be usable for other purposes when not needed for grain. Bins may have other uses but generally are less flexible in this respect.

(d) Chilling is a technically satisfactory method of preservation but cannot be widely adopted unless marketing arrangements are changed. The cost figures shown in this chapter provide an indication of the size of benefits obtainable by making such a change.

(e) Similarly, chemical preservation seems likely to be satisfactory technically, as a system for stockfeed grain, but its economic future depends on the price of the acid preservative.

CHAPTER 6

The Marketing of Grain by Farmers

6.1. This chapter is based on that part of the Survey of Farmers which relates to the marketing of grain by farmers. The first part covers such aspects as farmers' attitudes to the agricultural merchants and the H.-G.C.A. and their opinions about the Cereals Deficiency Payments Scheme and the grading of cereals, while the second deals with the number, size, timing and destination of sales of grain. The questions about attitudes and opinions were asked mainly in the Area Studies where there was greater opportunity for interviews in depth, but the other topics were covered in the National Studies.

The Farmer and his Merchant

6.2. While a certain amount of information on farmers' attitudes to merchants emerged from the National Study, where growers were asked their reasons for choosing a particular buyer for their grain, information directly related to their attitude to merchanting was covered solely in Phase 5 of the Area Studies. A certain amount of caution must therefore be used in interpreting these results as being representative of cereal farmers as a whole; nevertheless certain distinct patterns emerge. In the first place, it is plain that the great majority of farmers in all parts of the country considered that the grain merchant, on the whole, served a useful function: 87% of all informants in the Area Studies were of this opinion, the level being consistent over all six areas. The remainder divided almost equally between those who quite positively said they did not feel that the merchant served a useful function and those who felt unable to express any opinion.

6.3. Of all the services provided by the merchant, reciprocal trading was the one most frequently mentioned as the most useful. Among other services commonly mentioned in most areas were the transport of grain, promptness of service and offering the best or fair prices. A number of other services were mentioned but in no case by as many as 10% of all farmers included in the Area Studies. These minority mentions included

F*

advice on marketing, drying facilities, credit facilities, and storage.

6.4. Another aspect of the farmers' attitude to the merchant which was explored was the extent to which farmers believed that merchants quoted the same price to them at a particular point in time as they were quoting to other producers for the same quality of grain. In all areas at least twice as many farmers said they believed merchants quoted the same price to them as believed they quoted different prices—some of those who believed that the merchant was quoting different prices believed that their merchant quoted a higher price to them than to other producers because they were "good customers".

6.5. In view of the increasing tendency in recent years for local merchanting businesses to be taken over by national organisations, it was decided to examine to what extent farmers believed that they got a better or worse deal when such a take-over took place. About one in three felt that such a take-over made no difference, and one in five did not express an opinion; the balance divided evenly between those who believed they got a worse deal and those who believed they got a better deal. Some farmers believed that they received better attention from a small firm compared with the impersonal attitude of the large companies, but others said they received the same treatment irrespective of ownership.

6.6. The majority of farmers considered that an overall reduction in the number of merchanting companies would be to the disadvantage of the farmer. This view was again held uniformly in all six districts in which the Area Studies were carried out. Overall, only 14% of farms taking part in the Area Studies felt that such a reduction would be to the farmer's advantage, as against two-third who felt it would be to his disadvantage.

6.7. In Hampshire and in the East Riding of Yorkshire, two out of three farmers said that they sought advice from one or more merchants about the best time to sell grain and uniformly took the advice they were given. In Norfolk, Northants and East Lothian, farmers divided almost equally between those who did seek advice and those who did not, but in Devon only one in five admitted to seeking advice about the best time to sell grain, a result obviously influenced by the limited selling of grain that takes place in that county.

6.8. A very similar situation applied in respect of advice sought by farmers from merchants on which varieties of grain to sow. Just over three out of four farmers in Hampshire and East Lothian said they did so, two out of three in the East Riding, but rather lower proportions (less than half) in Devon and Northants. On this aspect also advice when sought was almost uniformly accepted.

6.9. Of all farmers who provided information in the Area Studies, approximately one half said that they did at some time engage in reciprocal

trading with one or more merchants. In Hampshire, a much higher proportion did so, (seven out of ten) but in East Lothian a much lower proportion (two out of ten). Over the other districts studied, there was an
almost equal division between those who did or did not engage in this form
of trading. The advantages seen in reciprocal trading by those who did
engage in it were: better prices, the friendly relations which existed with
their merchants, and the fact that reciprocal trading made ready cash unnecessary. Only individual farmers in any district mentioned the disadvantages accruing from being under an obligation to the merchant under
such circumstances or the need to be free to play one merchant off against
another.

6.10. On the other hand, only a relatively small proportion of farmers
said that they regularly used merchant's credit as opposed to aiming to get
full discount on purchases by paying cash. Only in Devon and East Lothian
did the proportion using merchant's credit facilities reach or exceed one in
five, compared with more than two-thirds who said they aimed to get full
discount on purchases for paying in cash. Since the question distinctly
referred to farmers' "usual" procedures, it would not be possible to say
that a higher proportion of farmers did not occasionally make use of such
facilities. But it is clear that overall about four out of five farmers normally
aim to get their full discount on purchases.

6.11. In general farmers were inclined to take a favourable view of
merchants' representatives. At least three out of five in all of the districts
covered in the Area Studies, and three out of four overall, said that
merchants' representatives were useful to them.

Cereal Farmers and the Home-Grown Cereals Authority

6.12. While one might have expected that the great majority of cereal
farmers would have been found to be aware not only of the Authority
itself but also of its functions, the National Survey proved that this was
very far from the case. Overall, only 54% of cereal farmers said that they
had heard of the H.-G.C.A. before the start of the interview (Table 6.1).
Awareness of the Authority was lowest among the smallest farms, only one
in four farms under 50 acres claiming to have heard of it. But, surprisingly
enough, at the other end of the scale, only three out of four farms with
over 300 acres of crops and grass in total had heard of the Authority. Even
among the biggest cereal farmers, those with 300 acres or more of *cereals,*
only 92% had heard of the Authority; even 12% of farmers who had
registered forward contracts said that they had not heard of the Authority.
To some degree, this lack of knowledge at the upper end of the size scale
might be ascribed to the nature of the informants—e.g. farm managers who

TABLE 6.1

PERCENTAGE OF CEREAL FARMERS WHO
WERE AWARE OF THE H.-G.C.A. 1966/7

	%
All cereal growers	**54**
Size of farm (acres):	
0–49	25
50–99	43
100–199	60
200–299	81
300 and over	75
Cereal acreage (1967):	
0–49	36
50–99	67
100–299	83
300 and over	92
Participation class:	
A	95
B	69
C	32
Region:[a]	
Northern	67
Eastern and S.E.	67
Central and Southern	70
Wales and S.W.	53
Scotland	42
Northern Ireland	7

[a] In this and subsequent tables in this chapter the Regions are the "Farm Management Survey" Regions described in Appendix C.

might conceivably have been more interested in managing the farm rather than deciding the disposal of the crop—but this seems improbable. Certainly, awareness was at its highest among Class A farmers (95%) and very low among farmers growing oats only.

6.13. The regional differences show, not surprisingly, that it is in the important cereal growing regions that awareness of the Authority is highest. Contrasted with a level of 67–70% in the North, East, South East, and Central and Southern parts of the country, in Wales and the South-West awareness was only 53%, in Scotland 42% and in Northern Ireland very low indeed at only 7%.

6.14. Not only was the overall awareness of the Authority found to be at a low level, so too was spontaneous awareness of specific functions of the

Authority. When asked, without prompting, to indicate what they understood the functions of the Authority to be, 17% of cereal farmers said its function was to improve trading or marketing conditions, and 12% to standardise prices. No other single function was mentioned by more than 3% (Table 6.2).

TABLE 6.2

PERCENTAGE OF CEREAL GROWERS MAKING
SPONTANEOUS MENTION OF FUNCTIONS OF THE
H.-G.C.A.

	All cereal farmers %
Percentage who named spontaneously:	
To improve trading/marketing	17
To standardise prices	12
To offer bonus on forward contracts	3
Recognition of contracts as collateral	3
To provide market intelligence and guide prices	2
To encourage storage	2
Other answers (relevant)	4
Other answers (irrelevant)	4
Don't know	12
Never heard of H.-G.C.A.	46

6.15. The ratio of favourable to unfavourable comments was greatest in the South-Western (N.A.A.S.) Region—nearly 4 to 1 compared with the national average 2 to 1. Comparison between Phase I and Phase II of the National Study shows a slight hardening of opinion after harvest in 1967, but the overall balance between favourable and unfavourable views was the same on both occasions. Some encouragement for the Authority may be derived from the fact that the level of favourable comment was much higher among those farmers who had sold grain only **with** a forward contract than it was among those who had sold some forward and some spot or spot only (Table 6.3).

6.16. Of those who were aware of the Authority, 9% regarded its formation as highly necessary and desirable and 46% as quite necessary and desirable; 32% said that they felt it was not really needed or desirable, and 13% said that it was totally undesirable. Those who considered that it was necessary and desirable cited most often the need to achieve orderly marketing by the stabilisation of supply and demand. Among the other points mentioned was the need for information on market prices, and for more level prices, and there were also favourable comments on the idea of forward contracts. Those who thought the Authority was unnecessary and

TABLE 6.3

PERCENTAGE OF CEREAL GROWERS COMMENTING FAVOURABLY AND
UNFAVOURABLY ON THE FUNCTIONS OF THE H.-G.C.A.

	Farmers giving comments which were:			
	Favourable only %	Un-favourable only %	Both favourable and un-favourable %	Indeter-minate %
All farmers claiming to know something of the functions of the H.-G.C.A.	**45**	**22**	**3**	**30**
Phase I (before harvest 1967)	42	20	5	33
Phase II (after harvest 1967)	48	24	1	27
Pattern of sales: Sold only with forward contract:				
Registered	52	18	5	26
Not registered	56	15	—	28
Forward contract and spot				
Any registered	46	23	2	29
None registered	49	19	—	32
Spot only	42	23	5	31
None sold	43	24	2	31

undesirable frequently gave as their reason the fact that the Authority had been of little help to them or had added to farmers' costs of marketing. Some mentioned that its existence had depressed prices, and a small minority commented that it lacked powers. These adverse comments, particularly that the Authority was "no use to me," were more widespread among farmers selling spot only.

6.17. When certain specific functions of the Authority were identified to farmers, only the payment of a bonus for registered forward contracts was accepted by the majority as a function of the Authority which they were well aware of (Table 6.4). Among the 54% of cereal farmers who had heard of the Authority, over 90% had heard of the bonus payments. Just about half of those who were aware of the Authority knew of its functions in providing marketing intelligence. Slightly smaller proportions were aware of the scientific functions which the Authority is entitled to carry out, or the influence which it may bring to bear on Price Reviews etc., and the fact that registered forward contracts may be used as collateral. At least half of those who were aware of each function regarded it as necessary and a function which helped the farmers. The balance divided approximately equally between those who considered the function unnecessary, and unhelpful to the farmer, and those who either had not formulated a view on the issue or who felt that the functions were neither necessary nor un-

TABLE 6.4

PERCENTAGE OF CEREALS FARMERS WHO WERE AWARE OF CERTAIN FUNCTIONS
OF THE H.-G.C.A.

	Bonus %	Influence %	Collateral %	Market intelligence %	Scientific functions %
Percentage of all cereal farmers who had heard of the function	**48**	**18**	**17**	**28**	**22**
Participation Class A	94	41	44	66	39
Participation Class B	63	24	23	37	31
Participation Class C	26	8	5	11	12
Percentage who regarded it, from the farmers' point of view, as:					
Necessary	24	11	9	19	17
Unnecessary	11	3	4	4	3
Neither one nor the other	9	3	3	4	1
Could not say	4	1	2	1	1
Percentage who said that it:					
Helped the farmer	26	11	9	19	16
Did not help the farmer	11	3	2	4	2
Did not affect him	7	3	4	3	2
Could not say	4	1	2	2	1

necessary, neither helpful nor unhelpful. Those who sold only on forward contract were particularly inclined to stress the value of the bonus arrangements.

6.18. Only 13% of all farmers—approximately one in four of those who were aware of the H.-G.C.A.—could name any useful function which they thought the Authority should undertake which it does not at present undertake. Approximately one in four of these felt that the Authority should have the power to fix prices, and one in five that it should have powers to control imports. A variety of other suggestions were made by the remainder including the suggestion that the Authority should have the powers of a marketing board, should offer an increased bonus or subsidy, should have export responsibilities and should provide additional information in the form of market research or export and import figures.

H.-G.C.A. BONUS SCHEME FOR FORWARD CONTRACTS

6.19. While 48 % of all cereal farmers (as noted above) had heard of the H.-G.C.A. bonus scheme for forward contracts, when asked more specifically whether they knew anything about it, only 37 % among them claimed to do so. Here the contrast between different participation classes of farmers was most marked. Eighty-six per cent of Class A cereal farmers claimed to know something about the bonus scheme against 47 % among Class B and 17 % among Class C. Of the 37 % of all farmers who claimed to know some details of the bonus scheme however, only 20 % of all farmers gave a correct answer when asked what the bonus level currently was (Class A 53 %, Class B 26 % and Class C 8 %); 6 % gave an incorrect answer but 11 % were unable to give a precise answer at all.

6.20. Obviously, the level of farmers' awareness may be depressed somewhat by the degree to which merchants undertake the function of completing the appropriate returns on behalf of the farmer. Nevertheless, it is plain that the smaller farmer is very much less likely to be aware of the bonus scheme than his larger counterpart. The function of the merchant in completing contract forms, etc. is clearly indicated by the fact that even among those who had registered forward contracts, by no means all claimed to know details of the scheme. Those who did give a correct answer divided almost equally between those who thought the bonus level was about right and those who felt that the level was too low, though the larger the cereals acreage of the farm, the greater was the tendency to regard it as too low. Less than 1 % of cereal farmers in total felt that the current level was too high. On the other hand, 23 % among those who claimed to know something about the scheme considered that the conditions were fair, against 10 % who thought they were not fair, and 4 % who were unable to express an opinion. Among those who felt that the present conditions were unfair, just over one in four put in a plea for a greater tolerance for any discrepancy between tonnage contracted and tonnage delivered, claiming that the current level of 5 % was too low. Approximately one in six commented that storage costs money or that the H.-G.C.A. were getting grain stored at the farmer's expense. Smaller proportions (less than one in ten in each case) felt that there should be greater tolerance on timing, complained about the levy, or felt the bonus conditions were not fair to the farmer using his own grain.

EXPERIENCE OF REGISTERING FORWARD CONTRACTS

6.21. Overall, 27% of cereal farmers claimed to have registered a forward contract with the H.-G.C.A. at some time. It must be borne in mind here that this figure may be somewhat lower than that found in practice because, in some instances, forward contracts may have been entered into without the farmer realising that such contracts had been registered with the Authority in that the merchant may have taken over the responsibility for the documentation and the farmer may have paid little attention to the documents he signed. Accepting these limitations, however, there is no doubt that the vast majority of larger cereal farmers have registered a forward contract with the Authority. Eighty-three per cent of those whose cereal acreage exceeded 300 acres in 1967 claimed to have done so, and 69% of those whose cereal acreage lay between 100–299 acres. On the other hand, small cereal farmers, i.e. those with less than 50 acres, are much less likely to have experience of registering a contract—only 8% claimed to have done so.

6.22. On this issue, as with many others, there is relatively little difference according to the age of farmer, but it is plain that experience with forward contracts, as one would expect, is much greater in the East, South East and Central and Southern areas of the country, rather lower in Scotland and the Northern area, and very low indeed in Wales and Northern Ireland. These regional differences clearly reflect the location of the larger cereal farmers with more marketing experience.

6.23. An appreciable proportion of those farmers who had registered forward contracts claimed at some time to have experienced difficulty in doing so; 6% of cereal farmers in total mentioned such difficulties, though no individual difficulty was mentioned by more than 1%. Among the individual comments an appreciable number were, in fact, irrelevant to the issue, but there was some mention of difficulties with tolerance, and comments on timing and price. About half of those who mentioned the existence of such difficulties claimed to have overcome them, more often by their own efforts than by the merchant's activities on their behalf.

6.24. A wide variety of reasons were given for not registering a forward contract with the H.-G.C.A., though no one reason was mentioned by more than 6% of all cereal farmers. The reasons given included:

1. Farmers did not sell their grain.
2. Farmers sold grain as soon as it was harvested.
3. Farmers always sold to a particular merchant and did not contract forward, or considered that they did not sell sufficient quantities for a forward contract to be worth registering.

A minority (about 2% of cereal farmers in each case) said they disliked being tied in this way, or that they had never thought about taking such action.

THE H.-G.C.A. WEEKLY BULLETIN

6.25. Only 15% of all cereal farmers claimed ever to have seen the H.-G.C.A. Weekly Bulletin. Once again, it must be recognised that this level is probably below the true figure in that many farmers may have seen the Bulletin without recognising it as being a document produced by the Home-Grown Cereals Authority. Once again, there was a steep increase in the proportion who had seen it as size of cereal acreage increased—rising from 8% among those who grew less than 50 acres of cereals in 1967, to 45% of those who grew 300 acres and over. Farmers over 50 years of age were much less likely to have seen it than younger ones. None of the farmers interviewed in Northern Ireland had ever seen a copy of the Bulletin. Approximately one in three of those who had registered a forward contract in relation to their 1966 cereal harvest (Phase I) or 1967 cereal harvest (Phase II) claimed to have seen the Bulletin, a proportion very markedly above that for those who had not registered any forward contract or had sold spot. Only 4% of cereal farmers claimed to receive the Bulletin themselves. Among the remainder, 4% said they had been sent a free copy, 3% had seen insertions extracted from the Bulletin in farming magazines, 2% had been shown a copy by a friend or neighbour and 1% had seen a copy at a meeting. A number of other locations were mentioned by other farmers but in no one case were they mentioned by as many as $\frac{1}{2}$%.

Farmers' Attitudes to Cereal Deficiency Payments

6.26. Just over half the farmers interviewed (52%) expressed satisfaction with the present system of Cereal Deficiency Payments; 38% said that they thought the present system was unsatisfactory, and 10% admitted that they did not know enough about the present system to be able to say whether it was satisfactory or not (Table 6.5). The proportion expressing satisfaction with the present system was higher among those who were not growing barley (66%) than among other cereal farmers. Among Class A farmers slightly more found the present system unsatisfactory (49%) than found it satisfactory (45%); in contrast, among Class C farmers 55% found it satisfactory and 33% unsatisfactory.

6.27. When asked to state in what ways they found it unsatisfactory, or to suggest how it could be improved, the most frequent comment was a plea for higher subsidy levels, especially among the smaller farmers and among the Class B and C farmers. A somewhat smaller proportion suggested abolishing subsidies altogether, this view being more widely held among the larger cereal farmers and among Classes A and B. Overall 7% of cereal growers (rising to 12% of those not growing barley) stated that they felt all subsidies should be geared to acreage. A much smaller proportion—

TABLE 6.5

PERCENTAGE OF CEREAL GROWERS EXPRESSING VARIOUS COMMENTS ON THE
CEREALS DEFICIENCY PAYMENTS SCHEME

	Total all farms	Cereal acreage 1967				Participation class		
		0–49	50–99	100–299	300 and over	A	B	C
	%	%	%	%	%	%	%	%
Percentage finding the C.D.P. Scheme unsatisfactory because of delay in payments	6	6	5	6	6	4	7	5
Percentage suggesting that the C.D.P. Scheme would be improved by:								
Higher subsidies	22	26	20	15	11	10	21	27
Abolishing subsidies	9	8	8	14	12	12	11	6
Securing stable prices	9	8	10	2	7	11	8	10
Gearing subsidy to acreage	7	7	8	4	5	12	9	4
Gearing subsidy to tonnage	2	1	3	5	4	8	3	1
Rewarding quality	2	2	3	2	2	5	1	2

(Other answers omitted from this table.)

only 2%—felt that subsidies should be geared to tonnage, and all of these were barley growers.

6.28. Lack of knowledge of the present system was highlighted by the fact that no less than 38% of all farmers were unable to detail any way in which they found the present system unsatisfactory or any way in which it could be improved. The Area Studies threw further light on this aspect in that certain questions asked in the last phase of the Area Studies showed very clearly the wide lack of knowledge among farmers of the details, method and administration of the present system. Here six out of ten said they would like the present system to continue; half of them because without it, their profit would be small, and the other half because the system helps the farmer. Among the 25% who were in favour of discontinuing the present C.D.P. system the majority were in favour of the removal of the deficiency payment and fixing instead a fair price to the farmer. A few suggested that the present system should be replaced by an import levy on grain.

6.29. When these farmers were asked whether they saw any advantage in the Ministry of Agriculture continuing to administer the wheat and barley incentive payments, the replies very clearly indicated a wide lack of knowledge or appreciation of the present system. Approximately four out of ten claimed that there were advantages but a similar number admitted that they did not know enough about the system to comment; when those who stated there were advantages in the Ministry administering the incentive payments were asked what the advantages were, the great majority gave answers which were irrelevant—only a very small minority remarking that since the Ministry already had the records available they were the obvious people to do it.

Effect of the Annual Price Review on Cereal Farmers' Plans

6.30. Only 4% of farmers said that the 1967 Price Review had had any effect on their cereal growing in 1967. Though the differences between large and small farms were minor, a rather greater proportion of cereal farmers growing 300 acres and over in 1967 (7%) had made some changes as a result of the Price Review. The 40% of farmers who said that the Review had affected their cereal growing in 1967 generally specified an increase in their wheat or bean acreage and a reduction in barley.

6.31. Among the much larger group whose growing plans had not been affected by the Price Review, the reason for this most commonly mentioned, especially by Class A farmers, was that the results of the Review had been published too late to affect their plans (28% gave this reason). A further 20%, most of whom had only a small cereals acreage, said that they had made no changes because they did not sell their grain but used it only for feed; 16% said their cereal acreage was not sufficient to warrant a change; and 12% had made no change because of the extent to which their cereal acreage was controlled by crop rotation plans.

6.32. When farmers were asked whether the 1967 Price Review had had any effect on their intentions for 1968, 10% said that it had. This was particularly evident with the large cereal farmers, 19% of those growing 300 acres and over saying that the 1967 Review had had an effect on their 1968 plans, against only 7% of those growing less than 50 acres of cereals. As with 1967, the change named most often was an increase in wheat acreage (31% of farmers whose 1968 plans had been affected): another 15% said they would increase their bean acreage, 8% that they would increase their barley acreage (though this was mainly confined to small farmers) and 4% that they would increase their oats acreage; 8% anticipated an increase in cereal acreage without specifying the type of cereal that would be affected. On the other hand, 18% planned a reduction in cereal acreage overall, and 9% a reduction specifically in their barley acreage.

6.33. Those who were not changing their 1968 plans as a result of the Price Review gave as the most common reason, that cereals were not the main business of the farm (mentioned by 20% of those whose plans were unchanged); 19% said that there was nothing in the Review to suggest a change of plan and that the Review substantially was the same as in 1966, 18% said that they planned several years ahead and had already formulated their policy, 13% felt no need, or were unwilling, to "chop and change", whilst 11% quoted their rotation system as a reason for not incorporating any changes. A further 7% said that they were already growing what was best for the land they farmed.

Attitudes of Cereal Growers to the Relationship between Market Prices and the Quality of Grain

6.34. Considerable dissatisfaction was expressed by certain farmers interviewed during the Survey with regard to the extent to which the prices they received for their grain reflected its quality. This feeling was largely concentrated among the large cereal farmers and among Class A farmers. Overall 39% of cereal growers expressed themselves as not satisfied on this issue, but this figure rose as the cereal acreage grown increased: 33% of those growing less than 50 acres of cereals were dissatisfied; 44% among those growing 50–99 acres; 49% among those growing 100–299 acres; and 60% among those growing 300 acres or more. Only 27% of Class C were dissatisfied as against 72% in Class A. However, it is noteworthy that less than half of those who were dissatisfied on this issue felt that a grading scheme would help (Table 6.6); among this minority there was considerable diversity of opinion as to the method under which they would like to see a grading scheme operating, grading by some government body or by the merchant being the two most frequently named. Once again it was the Class A farmers who were both more widely in favour of a grading scheme and more explicit about the way in which they would like to see it operate.

6.35. In considering the extent to which farmers consider that the price they get for their grain reflects the quality of the grain, one should also consider the extent to which cereal farmers currently aim to produce the highest possible yield per acre as opposed to the best possible quality, possibly at the expense of yield. This particular issue was explored only in the Area Study (Phase 5) and unfortunately it is therefore not possible to produce a definitive national figure. But of all the districts in which the Area Study was carried out, only in East Lothian did the number of farmers who said they aimed for quality exceed the number who said they aimed for yield. In all five of the other areas at least six out of ten said they aimed for yield—in Devon, Hampshire, and Northants eight out of ten said so. Of all informants who provided information on the Area Studies two out of three said they aimed for yield, one out of five that they aimed

TABLE 6.6

PERCENTAGE OF CEREAL GROWERS EXPRESSING VARIOUS VIEWS ON THE
POSSIBILITY OF A GRADING SCHEME

	All farmers growing cereals	Cereal acreage 1967				Participation class		
		0–49	50–99	100–299	300 and over	A	B	C
	%	%	%	%	%	%	%	%
Percentage of cereal growers who were dissatisfied with the extent to which the market reflected the quality of grain who:								
Thought a grading scheme would help	16	13	20	20	19	35	17	11
Did not think so, did not know	23	20	24	29	41	37	27	16
Percentage who were satisfied	61	67	56	51	40	72	44	27

for quality and one in nine that they aimed for both yield and quality. To the extent which these figures may be taken as typical of the country as a whole it is obvious that those who aimed for quality in their production were considerably outnumbered by those who were dissatisfied with the extent to which the price reflected quality.

6.36. The Area Study showed considerable variation from one district to another in respect of screening out impurities and tail corn before selling grain. In East Lothian nearly three out of four followed this course, in Hampshire and the East Riding of Yorkshire just over one out of two; in Northants on the other hand only one in three did so and less than one in five in Norfolk and Devon. Those farmers who adopted this practice were rather more likely than not to say that the price they got for the grain had an effect on the care they took in screening. Obviously employment of this procedure must be dependent to an appreciable extent on the use to which the grain is to be put, but it is plain that there was a minority of farmers—perhaps one in five overall—who considered it worthwhile in view of the financial reward.

The Pattern of Sales

6.37. The average number of sales from the 1966 harvest was 2·6 per farm selling grain (21% of cereal farmers sold no grain). Table 6.7 shows the number of sales by cereal acreage grown.

TABLE 6.7

NUMBER OF SALES OF CEREALS FROM THE 1966 CROP BY CEREAL ACREAGE GROWN

	All farms	Cereal acreage grown, 1967			
		0–49	50–99	100–299	300 and over
	%	%	%	%	%
Percentage of cereal farmers making:					
No sales	21	33	14	5	—
One	22	28	21	12	3
Two	25	24	28	21	19
Three	15	8	8	25	16
Four	7	4	9	12	14
Five	5	1	5	15	19
Six–eight	4	1	5	8	17
Nine or more	1	—	—	3	11
Average number of sales per farm	2·6	1·8	2·7	3·6	4·7

6.38. Table 6.7 shows that the larger the cereal acreage grown the larger was the number of sales. However, the tonnage per sale also increased as size of cereal acreage grown increased (Table 6.8).

6.39. Table 6.9 shows that the average tonnages per sale for grain sold on forward contract were about half those for all sales (Table 6.8). It also shows that the larger the cereal acreage the higher was the proportion of the crop sold on forward contract. The great majority of sales on forward contract were registered, especially in the largest cereal acreage group. Oats are not included in Table 6.9 because only 9% of oats growers sold any on forward contract.

6.40. As to the timing of cereal sales, Phase II of the National Study, conducted in October/November 1967, showed that 45% of cereal farmers had not sold any of their grain at that time. Of the sales completed at the time of the interview Table 6.10 shows the percentage that had been sold spot and on forward contract.

TABLE 6.8
AVERAGE TONNAGE PER SALE (SPOT AND FORWARD
SALES), 1966 CROP

	Wheat	Barley	Oats
All farms	**60**	**74**	**23**
Cereal acreage:			
0–49	24	33	14
50–99	39	58	18
100–299	103	105	51
300 and over	220	390	71

TABLE 6.9

AVERAGE TONNAGE PER SALE, GRAIN SOLD ON FORWARD CONTRACT, 1966 CROP
(Percentages of total sales in brackets)

	Wheat				Barley			
	All sales on forward contract		Registered sales on forward contract		All sales on forward contract		Registered sales on forward contract	
	tons	%	tons	%	tons	%	tons	%
All farms	**35**	**(57)**	**30**	**(49)**	**37**	**(51)**	**34**	**(46)**
Cereal acreage 1967:								
0–49	8	(34)	6	(25)	7	(20)	4	(15)
50–99	19	(47)	13	(33)	25	(44)	23	(40)
100–299	64	(62)	61	(59)	63	(60)	58	(55)
300 and over	161	(73)	155	(70)	282	(73)	276	(68)

6.41. Among those who had intentions to make further sales, many intended to sell only one lot, though many were undecided (Table 6.11). For barley, in particular, the proportion planning to sell in more than one lot was greater among the larger farms.

6.42. Plans to make further sales in particular months appeared to be well formulated, as Table 6.12 indicates.

6.43. For wheat, 61 % of lots to be sold by farms growing 300 acres or more of cereals were intended to be sold over the period March–June 1968. The pattern for barley also shows larger farms intending to sell later, but with a much greater proportion undecided.

6.44. Even among those whose forward planning left them uncertain in how many lots they would sell their remaining cereals, many had formulated an expected time of sale (Table 6.13).

TABLE 6.10

PERCENTAGE OF PRODUCTION SOLD SPOT OR FORWARD AT
OCTOBER/NOVEMBER 1967

	Wheat	Barley	Oats
	%	%	%
Percentage of total production sold spot:			
All farms	**26**	**23**	**19**
(percentage at end of 1966 season)	(39)	(34)	(32)
Cereal acreage 1967:			
0–49	38	18	12
50–99	50	32	25
100–299	16	17	24
300 and over	14	17	12
Percentage of total production contracted forward:			
All farms	**15**	**16**	(a)
(percentage at end of 1966 season)	(54)	(36)	(4)
Cereal acreage 1967:			
0–49	—	6	—
50–99	6	11	—
100–299	23	21	(a)
300 and over	21	26	(a)
Percentage of total production registered with H.-G.C.A.:			
All farms	**13**	**13**	—
(percentage at end of 1966 season)	(47)	(32)	(—)
Cereal acreage 1967:			
0–49	—	2	—
50–99	3	8	—
100–299	21	19	—
300 and over	19	22	—

(a) Less than 0·5%.

6.45. Table 6.14 shows that a month or two after the harvest farmers have made firm plans for the disposal of only a relatively small proportion of their wheat on forward contract (about a third) and even less of their barley (about a quarter). At this point in the cereal year it would seem that most farmers are keeping their freedom of manoeuvre so that they can take

TABLE 6.11

PERCENTAGE OF FARMERS INTENDING TO SELL IN ONE LOT OR MORE

	Wheat	Barley	Oats
	%	%	%
Percentage of farmers intending to make further sales (as at Oct./Nov. 1967) who planned to do so in:			
One lot	57	33	33
Two or more	8	21	14
Don't know	35	46	53
Total	100	100	100

TABLE 6.12

PERCENTAGE OF LOTS OF GRAIN INTENDED TO BE SOLD IN PARTICULAR MONTHS

	Wheat	Barley	Oats
	%	%	%
Percentage of lots, intended for future sale (as at Oct./Nov. 1967) which will be sold in:			
October 1967	9	9	7
November 1967	12	25	7
December 1967	9	9	10
January 1968	9	12	14
February 1968	9	7	19
March 1968	6	8	22
April 1968	9	9	1
May 1968	17	4	1
June 1968	3	2	—
Undecided	17	15	19
Total	100	100	100

advantage of what they think—or hope—the price situation will be. Farmers who had decided on the number of lots in which they were going to sell their remaining grain were asked to whom they intended to sell. The results are also shown in Table 6.14.

6.46. The reasons for choosing a particular buyer emphasise the personal relationship which so often exists between the farmer and his merchant and the limited amount of "shopping around" which farmers engage in (Table 6.15). The dominant reason for choosing a particular buyer is clearly that the farmer knows and trusts him from past experience.

Reasons for Selling

6.47. Farmers were asked why they sold when they did in relation to each sale made. Considering first the overall pattern emerging from the disposal

· TABLE 6.13

PERCENTAGE[a] OF FARMERS EXPECTING TO SELL THEIR GRAIN IN PARTICULAR MONTHS

	Wheat	Barley	Oats
	%	%	%
Percentage of farmers intending sales of each cereal (as at Oct./Nov. 1967) who were uncertain how many lots they would sell, but expected to sell in:			
October 1967	4	2	—
November 1967	18	15	7
December 1967	10	5	3
January 1968	16	9	3
February 1968	27	24	11
March 1968	13	12	18
April 1968	10	5	1
May 1968	3	4	5
June 1968	1	2	—
Don't know	40	45	67

[a] These percentages add to more than 100 because some farmers mentioned more than one month.

TABLE 6.14

PERCENTAGE OF LOTS OF CEREAL INTENDED FOR SALE TO BE SOLD TO DIFFERENT BUYERS AND IN DIFFERENT WAYS

	Wheat	Barley	Oats
	%	%	%
Percentage of lots of cereal intended for sale (as at Oct./Nov. 1967) to be sold to:			
Merchant	78	70	60
Miller	3	5	34
Compounder	5	11	1
Seedsman	1	2	—
Other farmer	—	2	—
Other and undecided	13	10	5
Total	100	100	100
Through selling group	6	1	25
Under forward contract	33	26	4

of the 1966 harvest, three reasons emerged clearly as being of greater importance than any others (Table 6.16). Lack of storage was the most widespread of all, reflecting the lack of adequate on-farm storage on certain types of farm. The next most widespread was "to get a good (or better) price" reflecting the degree of concern among farmers who can store to

TABLE 6.15

PERCENTAGE OF LOTS OF GRAIN INTENDED FOR SALE WHICH WOULD BE SOLD
TO PARTICULAR BUYERS FOR PARTICULAR REASONS

	Wheat	Barley	Oats
	%	%	%
Percentage of lots of cereal intended for sale (as at Oct./Nov. 1967) which would be sold to specified buyers because:			
"We always deal with them"	22	26	47
"We know them well"	11	8	4
"They are good to deal with"	8	12	—
"They made best offer"	15	21	12
"We have a mutual arrangement"	4	6	3
"We have a contra-account"	2	—	—
Other, including "No particular reason", etc.	40	27	41

(N.B. Other answers for oats include 14% who said "Not much choice here".)

achieve the best return they can for their efforts. Third—almost equal with price—was "we needed the money". There is also some evidence of pressures from the buyers from such comments as "the buyer needed it then".

6.48. The proportion who had no specific reason to sell at a particular time was not insignificant, and this probably reflects the relative unimportance of cereals on some farms. Particularly interesting is the relative lack of regional variation in the proportion who answered "to get a good or better price". Class A farmers gave this reply more often than others. Only in the Northern Region did the proportion who quoted the need for cash fall notably below the national average; in that region lack of storage capacity was more widely mentioned than anywhere else. Table 6.17 presents similar data relating to sales already made from the 1967 harvest by October/November 1967.

6.49. Certain major differences stand out. Half the spot sales effected by October/November had been made because of storage problems. This applied almost equally to all classes of farmer and showed little variation by size of cereal acreage. The next most important factor was the need for money. This was quoted in relation to 31% of spot sales by Class A farmers, against 19% for all farmers. The position here may have been influenced by informants' reluctance to admit this need; the fact that it was quoted more often by Class A than Class B or C farmers suggests that this may indeed be so. But it may also reflect the higher bank borrowing often associated with the larger cereal farmers and the pressure on them to reduce overdrafts. Third in importance for spot sales was the chance of a good price—a reason quoted more often by Class A farmers than by

TABLE 6.16

PERCENTAGE OF FARMERS SELLING CEREALS AFTER THE 1966 HARVEST GIVING
DIFFERENT REASONS FOR SELLING WHEN THEY DID

	No room to store	Good/ better price	Needed funds	Buyer needed it then	Other reasons	No particular reason
	%	%	%	%	%	%
Percentage of those selling cereals giving each answer:						
All farms	**40**	**30**	**25**	**9**	**17**	**6**
Participation class:						
A	38	40	32	10	17	—
B	43	28	20	15	16	7
C	38	27	25	6	18	8
Region:						
Northern	60	36	14	10	13	—
Eastern and S.E.	35	29	31	11	15	3
Central and Southern	43	31	24	9	16	6
Wales and S.W.	35	28	26	15	33	—
Scotland	37	29	23	7	20	14

(N. Ireland data were insufficient to justify analysis.)

other farmers. There were also minority mentions, in relation to spot sales, of the need to dry the grain (4%) and of its being surplus to the farm's requirements (4%).

6.50. As far as forward contracts are concerned, price seems to be the dominant factor—virtually as dominant as the storage problem is in regard to spot sales. Storage problems, however, were mentioned by some farmers (presumably reflecting the knowledge that some at least of their grain could not be satisfactorily stored over the winter) as was also the need for money. Among other factors some farmers sold because they thought this was the best time to make the contract, and some because the market seemed to be in a secure state at the time the contract was made.

6.51. One of the reasons why the Area Studies were undertaken was to check that replies given in the National Studies to questions relating to the reasons for the timing of sales were not clouded by a failure of farmers to recall the circumstances and reasons affecting their decisions. In the event, there is no reason to believe that this occurred; the Area Studies in general confirm the findings quoted above from the National Studies—in particular the dominant importance of price in relation to forward contracts—not only those made immediately after harvest but also those relating to later contracts from the 1967 harvest. In the Area Studies price was quoted in

TABLE 6.17

PERCENTAGE[a] OF FARMERS SELLING CEREALS AFTER THE 1967 HARVEST GIVING DIFFERENT REASONS FOR SELLING WHEN THEY DID

	No room to store	Good/ better price	Needed funds	Buyer needed it then	Secure market then	Thought it was best time	Other reasons
	%	%	%	%	%	%	%
Proportion of sales for which each answer was given:							
All farms:							
Spot sales	50	17	19	7	2	3	13
Forward contract	12	47	8	3	10	10	16
Class A:							
Spot sales	46	24	31	6	—	2	9
Forward contract	8	43	3	6	2	13	25
Class B:							
Spot sales	53	18	11	14	1	3	11
Forward contract	11	44	7	3	17	9	15
Class C:							
Spot sales	48	15	21	3	3	3	13
Forward contract	19	50	20	—	7	4	1

[a] These percentages add across to more than 100 because some farmers mentioned more than one reason.

relation to about half the sales on forward contract as the main reason for selling at a particular time.

Forward Contracts

6.52. Over two-thirds (69%) of all farmers making forward contracts for the sale of their 1966 cereal crop made fixed-price contracts only; 25% made open-price contracts only, and the remainder (6%) made both open and fixed-price contracts. Table 6.18 shows the regional pattern. There is some indication that a higher proportion of Scottish farmers made contracts of both types, but those who made all contracts at a fixed price predominated in all areas.

6.53. It is interesting to note that in the majority of cases where farmers contracted some of their crop forward, contracts had been made in respect

TABLE 6.18

PERCENTAGE OF FARMERS SELLING GRAIN FROM THE 1966 HARVEST UNDER FORWARD CONTRACTS WHICH WERE FIXED OR OPEN

	All fixed-price	All open-price	Some fixed/ some open
	%	%	%
Percentage of farmers making contracts	69	25	6
Region:			
Northern	65	29	6
Eastern and S.E.	70	24	6
Central and Southern	64	33	3
Scotland	72	16	11

(The number making forward contracts in Wales and S.W. and in N. Ireland was not large enough to justify analysis).

TABLE 6.19

DISTRIBUTION OF FORWARD CONTRACTS 1966 BY TOTAL NUMBER OF SALES
MADE PER FARM

| | Number of forward contracts | | | | | |
	None	One	Two	Three	Four	Five or more
	%	%	%	%	%	%
Total number of sales per farm:						
One	32	50	—	—	—	—
Two	41	12	52	—	—	—
Three	12	24	28	53	—	—
Four	7	6	10	22	80	—
Five or more	8	8	10	25	20	100
Total	100	100	100	100	100	100

of the majority of their sales (Table 6.19). Overall 54% of farmers making one or more sales under forward contract, made all their sales under forward contract. A further 17% made only one sale not under forward contract.

6.54. In Phase I of the National Study conducted in July/August 1967 all farmers were asked whether they had entered into any contracts for the forthcoming harvest; 7% said they had done so. It is not surprising to find that those who had sold their 1966 harvest only under forward contract and registered at least some of their forward contracts, showed a proportion (13%) greater than the national average, but it is perhaps somewhat surprising to find that 7% of those who sold spot only in 1966 had already entered into a contract for their 1967 crops. Rather more than three out of four farmers who had already made a contract for their 1967 crop were located in the Eastern, South-Eastern, Central and Southern areas of the country. Just over 40% of these farmers had entered into a normal forward contract for the supply of grain, these dividing almost equally between those who had made a forward contract at a fixed price, those who had done so at an open price and those who did not indicate how the price was to be decided. Another 27% had entered into a forward contract for seed grain whilst the balance gave a variety of other answers which could not be classified under any of these headings.

6.55. Virtually all the forward contracts for seed were in the Eastern, South-Eastern, and Central and Southern regions; virtually all the fixed-price forward contracts were in the South-Eastern region, while the majority of those at an open price were in the Central and Southern regions. Most contracts were with merchants but 6% of farmers in Phase I mentioned contracts with compounders and 5% contracts with other farmers.

6.56. When asked why they had entered into these contracts, 44% said they had done so because they had been offered a good price and this was the only way of getting the bonus, financed by a levy on producers. The only other reason at all widely quoted was "habit" (27%). The influence of habit in entering into such contracts was equally widespread among those who had contracted with a seedsman as among those who had contracted with a merchant. Only 23% of farmers said that the contract into which they had entered had had any effect on their choice of variety of cereal. Virtually no-one said that it affected his choice either of crop protection chemical or of fertiliser.

6.57. In addition to the 7% who had already made a forward contract by July/August 1967 were those 20% of cereal farmers who said they were considering doing so. There were considerable regional differences in that only 9% of farmers in Wales and the South-West said they were thinking of doing so, and none in Northern Ireland. Among those who had registered forward contracts in relation to the sale of their 1966 harvest, 50% said they were considering entering into a contract for their 1967 crop. In contrast to this, only 4% of those who sold spot only in 1966 were considering entering into a contract for their 1967 crop. About a third of those who had already entered into some contract for their 1967 crop were considering a further contract.

6.58. About a third of the cereal growers who were not considering entering into a contract in July/August 1967 expected to have little or no cereal left to sell after meeting their own feed requirements. Among the remainder, the reasons most often mentioned for not considering a contract were that they found the price unattractive (14%), that they sold regularly to one particular merchant (7%), that it was too early to decide (6%) or that they were unable to store (4%). In addition, 9% expressed generally unfavourable views regarding forward contracting. The great majority of this latter group were farmers who had in 1966 only sold spot, but nearly one in four of the farmers who expressed this view had in fact made sales under forward contract registered with the H.-G.C.A. in 1966. An appreciable proportion (slightly more than one in three) of those who had found forward contract prices unattractive, were also farmers who had sold in 1966 under registered forward contracts. Obviously, therefore, there is some evidence here of unfavourable views to forward contracting among those who had in 1966 sold in this way.

6.59. There seems very little evidence from the Survey that the identity of the buyers of grain (see Table 6.14) differed to any great extent according to whether the sale was under forward contract or spot. Sales to other farmers were almost exclusively made spot—presumably because of the nature of the sale. Sales to millers were proportionately more numerous among forward contract sales than among spot sales. It is perhaps somewhat surprising that the proportion selling to seedsmen did not differ appreciably between those who sold spot and those who sold under forward contract; it might have been expected that a higher proportion would have sold under forward contract. Both among those selling forward and those selling spot, higher than average proportions of Class A farmers sold to compounders, and of Class C to millers (Table 6.20). But the merchants predominated as purchasers among all classes of farmer.

TABLE 6.20

PERCENTAGE[a] OF CEREAL GROWERS SELLING SPOT AND FORWARD TO CERTAIN BUYERS

	Spot sales Participation class			Forward contract Participation class		
	A	B	C	A	B	C
	%	%	%	%	%	%
Percentage of cereal farmers who sold to:						
Merchant	77	71	80	85	75	64
Miller	5	7	10	5	12	25
Seedsman	9	10	5	10	9	6
Compounder	13	4	3	14	5	5
Maltster	1	2	1	4	—	—
Other farmer	6	7	6	4	—	—
Other	1	1	2	—	2	—

[a] Percentages total more than 100 because some farmers mentioned more than one type of buyer.

6.60. In Phase I of the National Study each informant who had made a forward contract was asked if he had previously approached anyone else with a view to entering into a similar sale. Seventy-eight per cent had contacted no other potential purchaser, 5% had contacted more than one potential buyer in respect of some but not all their contracts, while the balance (17%) had contacted more than one potential buyer in respect of all contracts. This emphasises the strength of the personal aspect of cereals marketing. Nearly four out of five selling under forward contract made no effort to "shop around" before making a sale—even among the Class A farmers 74% made no alternative contacts.

The Bulk Handling of Grain

6.61. Questions relating specifically to sales of grain made in bulk rather than in sacks were confined solely to the Area Studies. Approximately four

G

out of ten cereal farmers sold none of their 1967 crop in bulk, whilst a similar proportion sold all of it in bulk (Table 6.21). In Devon, Norfolk, Northants. and East Lothian, there seemed to be a clear dichotomy on this issue—a farmer either sold all his grain in bulk or none of it in bulk. In

TABLE 6.21

PERCENTAGE OF CEREAL FARMERS SELLING GRAIN IN BULK

	Areas in the Area Study						
	All areas	Devon	Hants.	Norfolk	North-ants.	Yorks. (E.R.)	East Lothian
Number of informants	202	37	37	34	29	40	25
	%	%	%	%	%	%	%
Percentage of 1967 harvest sold in bulk:							
None sold	5	11	3	—	10	5	—
None sold in bulk	37	70	27	59	24	10	32
Up to ¾ sold in bulk	7	—	11	6	3	20	4
¾ and over sold in bulk	10	3	22	3	—	24	—
All sold in bulk	40	16	38	32	62	41	64
Percentage of farmers having problems in the loading of grain into vehicles	5	—	5	9	3	5	8
Percentage of farmers who had taken steps to improve their bulk handling of grain	15	5	22	6	17	20	20
Percentage of farmers intending to take steps to improve their bulk handling of grain	13	8	14	12	7	22	12

Hampshire and the East Riding of Yorkshire the issue was not quite so clear-cut; in both these areas six out of ten of informants made sales in bulk covering three-quarters or more of their total harvest. Only one in eleven of all farmers who had made any sales in bulk said they had experienced problems in loading grain into the vehicles which collected it. This proportion was too small for any useful information to be derived concerning the nature of the problems which they encountered. About one in five of farmers in Hampshire, Northants, East Riding and East Lothian, said they had already taken steps to improve their bulk handling of grain; hardly any informants in Devon or Norfolk had done so. Over all areas a further small proportion said they did intend to take further steps to improve their bulk handling of grain.

PART III

The Role of the Merchant

CHAPTER 1

Introduction to the Merchant Survey

1.1. The merchant sector is the most vital part of today's cereals market in the United Kingdom, and is at the same time the area of the market about which least is known. This became clear at a very early stage of the Cereals Survey. To supply some of the missing information, a major study of the merchanting industry was carried out by the Survey team.

1.2. Before survey work could begin, a problem of definition had to be solved. Everyone knows, in rough terms, what a corn merchant does—he buys corn (from farmers or others) and sells it to cereal users of one sort or another, perhaps making it into feed first. But this rough definition is of no use for the purposes of a systematic survey; for the results to have any value, one must know exactly what section of the cereals industry is being dealt with, and it must be possible to choose a sample of firms covering the whole of that section and no other.

1.3. The first step, therefore, was to find a list of names which could fairly be taken as including all the firms in the merchanting industry—or, in statistical terms, as defining the *universe* from which the sample was to be drawn. One possibility would have been to use the membership lists drawn up by the merchant trade associations. This was decided against, however, because it was felt that there might possibly be firms who could reasonably be called "merchants" but who, for one reason or another, did not appear on the trade lists. The sampling list eventually adopted was the list of Registered Wheat and Barley Buyers compiled by the M.A.F.F., which gives the names and addresses of all those currently authorised to buy wheat and barley under the terms of the Subsidy Schemes. In brief, the advantages of using this list were these:

(a) By definition, the lists must give complete coverage of all non-farm purchases of grain (apart, of course, from transactions not qualifying for subsidy, which are likely to be an insignificant part of the total).

(b) In particular, all sales of wheat should be covered (since the subsidy on wheat is paid only if the grain is sold through an authorised

165

merchant), making possible a statistical check on the results of the survey.

(c) Each address on the list represents a precisely defined unit—namely, a single buying point for grain. This avoids one great difficulty of using trade lists, on which an address may represent anything from a single buying point to a whole chain of branches, each of them operating independently but shown in the trade list under the address of their head office.

1.4. The original sample of firms to be surveyed, therefore, was drawn from the M.A.F.F. list of about 2600 names. The procedure followed was that of "systematic random sampling", taking the first address at random from the first ten firms on the M.A.F.F. list and taking every tenth address thereafter, i.e. about 260 addresses.

1.5. Some of the addresses on the M.A.F.F. list were found on investigation to refer to firms which were clearly not "corn merchants" in any acceptable sense of the word, even though they bought grain from farmers —such as maltsters, distillers, flour millers and national compounders. Then again, some firms which were authorised to buy grain did not in fact do so—either because they had closed down or had ceased their merchanting activities, or because they kept their buying licence in reserve but did not use it. In order to give a large enough final sample of merchant firms for results to be reliable, these non-merchant businesses had to be replaced. The replacement sample was drawn from the M.A.F.F. list by taking the fifth address after each one drawn in the original sample. Whenever an address in the original sample was found to refer to a non-merchant business, it was replaced by the address immediately following in the replacement sample; if the latter was also a non-merchant business, the next address in the replacement sample was taken, and so on until a merchant business was found. Replacement was necessary also in the very small number of cases where firms refused to give information.

1.6. The merchant survey was carried out in the early Summer of 1967. Members of the Survey team visited each address on the sample list (replacing as necessary in the manner described) and asked a set questionnaire, the summarised answers to which form the main basis of the chapters which follow. In addition, a set of tables was left with each interviewee for return through the post; as was to be expected, the rate of response for these tables was noticeably lower than for the interview. Where a respondent failed to return the tables, no replacement was taken.

1.7. In practice, there was very little difficulty in deciding which businesses visited were merchants and which were not. For precision, certain standards were adopted to define the "merchant business", though there turned out to be very few borderline cases in which these standards had to be invoked. The rules adopted were that a firm on the M.A.F.F. list was to be regarded as a "merchant business" unless:

(a) It bought no grain from farmers.[1]

(b) It was one of the few large firms producing compound feeds on a national scale.

(c) More than 75% of its turnover came from malting, brewing, distilling or flour milling.

1.8. Another potential difficulty—which likewise proved to be unimportant in practice—was that of defining the commercial units represented by the buying points in the M.A.F.F. list. It was known before the Survey began that some of the addresses listed were separate branches of a larger concern, or subsidiaries of other companies; and, conversely, that some other firms listed had branches which were not shown separately, the address referring to the head office only. This problem was met by deciding that the merchant sample should be a sample of *buying points* and not a sample of firms; in other words, the address on the M.A.F.F. list was to be regarded as the basic unit of sampling, irrespective of whether it represented a whole firm or several firms, or a branch or branches of a firm. To avoid double-counting in the final results, interviewers sought to ensure that the information collected from each buying point referred to the commercial unit whose grain purchases from farmers were made at that point, and to no other unit. In practice, this standard was found to reflect very accurately the commercial structure of the merchanting industry. Most buying points do represent individual firms; and where a firm with several branches has more than one address on the M.A.F.F. list, each buying point usually corresponds to a distinct grain-buying unit within the firm.

1.9. Where the term "merchant" is used without qualification in the following chapters, therefore, it refers to a buying unit corresponding to an address in the M.A.F.F. list and meeting the definition of a "merchant business" given in paragraph 1.7 above.

1.10. Even within this definition, one finds a vast diversity of firms. There can be few industries so heterogeneous as agricultural merchanting. Enterprises range in size from a small operator working part-time from his garden shed, or an old water-mill turning out a few dozen tons of feed per year, to large concerns with an annual turnover of many million pounds. Some merchants deal only in grain, or grain and feeds; others trade in the whole range of agricultural goods, and some may have a

---

[1] The only exceptions to this rule were a few small businesses which dealt only in purchased feeds, but which were agreed to have all the other characteristics of the typical merchant. There were four such businesses in the original sample (i.e. before replacement) and five in the final sample—all of the latter being in Northern Ireland.

TABLE 1.1

Region	(1) Businesses in original sample found to be merchants	(2) Universe estimates (col. 1 × 10)	(3) Merchant businesses interviewed in final sample[a]
England and Wales:			
South-west	25	250	32
South-east	27	270	29
East	32	320	45
East Midland	17	170	22
West Midland	23	230	28
Wales	11	110	11[b]
Yorks. and Lancs.	16	160	24
North	11	110	14[b]
Scotland:			
East	5	50	8[b]
West	7	70	8[b]
North	7	70	8[b]
Northern Ireland	24	240	31
United Kingdom	205	2050	260

[a] The numbers returning information in the postal tables were rather lower than the numbers of completed interviews. [b] Because of the small samples in the regions so marked, they were merged with other regions for the purposes of raising: Wales was merged with W. Midland; North England with Yorks. and Lancs.; and the three Scottish regions were combined.

Note on "raising"

Some of the statistical material in the chapters which follow (notably for Figs. 2.1 to 2.3) was prepared by "raising" results from the merchant survey—that is, by taking the result concerned from the merchant sample and multiplying it by an appropriate factor so as to give an estimate of that result for the whole merchanting industry. The raising procedure adopted for each item was to find the average of that item over the number of firms returning information, separately for several regions of the United Kingdom; then, within each region, to multiply the per-firm figure by the total number of merchant businesses in that region. The regions used were, for England and Wales, those adopted by the N.A.A.S., and for Scotland the Scottish College areas; Northern Ireland was treated as a single region. The United Kingdom total for each item was arrived at by adding the regionally raised figures. An obvious complication was that the total "number of merchant businesses" in a region could not be found directly, since the M.A.F.F. list contains many entries for other types of business; instead, the number of merchant businesses in each region was estimated by taking the number of such businesses in that region in the *original* sample and multiplying it by ten, since one firm in ten was the number originally sampled.

large trading interest outside agriculture. Some firms use most of their grain to make feeds, others make none. Despite all these variations, each firm is unmistakably an agricultural merchant business. In the analysis of the merchant survey, firms were categorised in several ways according to size and activities; the results of this part of the study are given in Chapter 2 below.

1.11. Table 1.1 shows in column 1 the numbers of merchant businesses found for each region in the *original* sample, and in column 2 the resultant estimates of the *total* numbers of merchant businesses by region. Column 3 shows the number of interviews with merchants which the Survey team completed in each region. Because of the system of replacement which has been described, the number of completed interviews in most regions was higher than the number of merchants in the original sample. In fact, for the United Kingdom as a whole, roughly one in eight of all merchant businesses were interviewed.

Classification of Merchant Businesses

1.12. In the description of the Merchant Survey results in the chapters which follow, merchant firms have been classified into groups according to various criteria. No single classification system was suitable for all the analyses which were carried out, and firms have been grouped in the ways which seemed most relevant to the topics under discussion. Since an understanding of the groupings used is essential in the interpretation of tabular results, a detailed description of the standards of classification is given in this section.

1.13. Two of the most important groupings used are based on the total turnover of firms (namely "turnover group" itself and "type of business"). It should be borne in mind that in most cases the turnover reported by a firm must only be regarded as correct within rather wide limits. It became clear to interviewers during the Survey that merchants were very often quoting an approximate rather than an exact figure of total turnover— occasionally because they preferred not to disclose the precise amount, but most often because records of the exact figure were not available. In the case of the "type of business" classification, which depends on a breakdown of turnover between various activities, additional difficulties arose where either the turnover figure, or its subdivisions between activities, were not reported by a firm. In these cases, an estimated classification was made by the interviewer who had visited the firm. For these reasons, the "turnover" and "type of business" groupings should be regarded as aids to description rather than as precisely defined sub-divisions. They can, however, be taken as very close approximations to the groupings which actually exist.

1.14. The various classifications used are as follows:

(a) *Country:* England, Wales, Scotland, Northern Ireland.

(b) *County.*

(c) *Cereals Intensity:* the percentage of total crops and grass made up by cereals in the county in which the business is situated (see Map 1.3, page 32).

(d) *Advisory Regions:* these are, for England and Wales, the regions distinguished by the N.A.A.S., and for Scotland the regions used for the advisory work of the Scottish Agricultural Colleges (see Appendix C). Northern Ireland is treated as a single region.

(e) *Surplus/Deficit Area:* this describes the balance in a county between production of cereals for livestock feed and their estimated consumption. A detailed description is given in Chapter 3.

(f) *Company Status:* in this classification, firms are grouped as follows, according to answers given at the Survey interview:

    (i) Subsidiary (or branch) of a national company;

    (ii) Co-operative, or branch of co-operative;

    (iii) Firms, other than in (i) or (ii) above, having links with any other business (by ownership, management or financial arrangements);

    (iv) Firms having no links with any other businesses.

(g) *Type of Business:* here, firms are divided into several broad groups according to the importance of certain main activities, as measured by percentage of total turnover.

The groups are as follows:

Specialist grain merchant:	75% or more of total turnover in sales of straight grain.
Mainly grain merchant:	more than 50% and less than 75% of total turnover in sales of straight grain.
Specialist feed merchant:	75% or more of total turnover in sales of purchased feeds.
Mainly feed merchant:	more than 50% and less than 75% of total turnover in sales of purchased feeds.
Specialist compounder:	75% or more of total turnover in sales of own-manufactured feeds.
Mainly compounder:	more than 50% and less than 75% of total turnover in sales of own-manufactured feeds.

General Agricultural
  Merchant: no single main activity exceeds 50%
                            of total turnover.
Other Merchant: more than 50% of total turnover is
                            in some activity other than straight
                            grain or feeds (e.g. fertilisers, hay,
                            seeds).

In carrying out this classification, the value of sales for straight grain, provender and compounds was estimated by applying standard value factors to the tonnage sales figures reported by the firms in the Survey.

(h) *Labour Force:* for the purposes of this classification, any part-time workers employed are taken as equivalent to one full-time worker, and firms are then grouped on the resulting total labour force figures as follows: 1 only; 2–5; 6–10; 11–20; 21–50; 51–100; and over 100. These totals include managers and working principals.

(i) *Turnover:* the following classification is used, on the basis of the annual value of turnover reported (after rounding to the nearest thousand): under £75,000; £75,000–£149,000; £150,000–£499,000; £500,000–£1,249,000; £1,250,000 and over.

1.15. In the description of Survey results, these classification systems have often been simplified by merging some of the groups within a particular classification. For example, under "Type of Business" the "Specialist Grain" and "Mainly Grain" classes may be described together; similarly with "Specialist Feed" and "Mainly Feed", and so on. But although the groups described above may be merged, they are never sub-divided in the analysis of Merchant Survey results.

CHAPTER 2

The Structure of the Merchant Sector

2.1. This chapter describes the structure of the merchant sector. It examines both the current pattern of economic activity and changes which are taking place within the sector. Finally, it relates the structural characteristics of the merchant trade to its function in marketing home-grown grain. It is against this background that judgements must be made concerning the overall economic efficiency of the merchant community, and in particular concerning the oft-expressed contention that there are too many merchant businesses.

The Number and Types of Merchants

2.2. The evidence of the Survey suggests that there are some 2050 agricultural merchanting businesses in the United Kingdom. These firms include a diversity of legal and organisational types. The most common legal form was the private limited liability company, 55% of all firms taking part in the Survey. Sole traders and partnerships came next. Public limited liability companies (9%) and co-operatives (8%) were less common, but enjoyed in many cases a relatively large share of the trade.

2.3. Differentiation by legal status owes more to taxation rules and regulations than to important economic characteristics of the firms concerned. A more useful classification is by "Company Status" as defined in Chapter 1, p. 170. This divided firms into subsidiaries of national companies, co-operatives and independent firms. The latter were further subdivided into those having links with any other business (by ownership, management or financial arrangements) and those having no such links. Table 2.1 shows the numbers of these firms and their distribution by type of business, as defined in Chapter 1.

2.4. The most numerous group is "independents, no links". Together with the "independents with links" these account for 84% of all the firms visited. This corresponds well with the notion of the agricultural merchant as an essentially independent trader, relying on his knowledge of the market

173

TABLE 2.1

NUMBERS OF MERCHANT BUSINESSES IN SURVEY SAMPLE, BY COMPANY STATUS AND
TYPE OF BUSINESS

	Company status				
	Subsidiaries of nationals	Co-op-eratives	Independents with links	Independents, no links	Total sample
Type of business:					
Specialist/mainly grain	3	—	10	28	41
Specialist/mainly feed	4	3	1	28	36
Specialist/mainly compounding	3	—	6	18	27
General merchant	14	15	19	77	124
Other merchant	1	2	6	22	31
Total sample	25	20	42	173	260

and the farming community more than on massive commercial strength. As will be seen, this numerical superiority overstates the importance of independents in relation to the proportion of the market they control. Nevertheless, they form a very vital part of the total merchanting sector.

The Size of Firms

2.5. Here, three measures are used: turnover (or total sales); labour force; and aggregate purchases of grain. A fourth, capital, would have been very valuable, but differences of accounting procedure and an understandable reluctance to disclose such confidential information made it impossible to gather satisfactory data for the sample as a whole.

2.6. The measure of turnover ignores important differences in the value added by the businesses concerned. For example, a turnover of £100,000 may be achieved by one man and a telephone where the business is limited to grain buying and selling: a manufacturer who sells £100,000 of his own compounds will need more men, more capital, and will earn a higher margin. Nevertheless, the considerable range in turnover of the merchant businesses incorporated in this sample indicates real differences in size. Table 2.2 shows the number of merchant businesses in the Survey by turnover and type of business. This table indicates the considerable numerical importance of relatively small businesses. One-third of the firms had a turnover of less than £150,000 and two-thirds of less than £500,000. This suggests that, at least up to now, economies of scale have been of insufficient importance in any branch of the merchant sector to eliminate or reduce to insignificance the number of small and medium-sized firms.

2.7. Nevertheless, Table 2.3 indicates that in terms of total turnover the small firms are collectively of relatively little importance. The table is based on 196 firms who provided actual turnover figures for their last year of operation, as distinct from those who merely placed their business in one of the turnover categories used in Table 2.2. At the extremes, 21% of the firms produced only 2% of the turnover, while 3% of the firms controlled 28% of turnover. Most significantly, 59% of the trade in the Survey firms was in the hands of the top 14% of firms with turnover in excess of

TABLE 2.2

NUMBERS OF MERCHANT BUSINESSES IN SURVEY SAMPLE, BY TYPE OF BUSINESS AND TURNOVER GROUP

	Type of business					
	Specialist/ mainly grain	Specialist/ mainly feed	Specialist/ mainly com- pounding	General merchant	Other merchant	Total sample
Turnover group (£'000):						
Under 75	8	12	3	7	12	42
75–149	8	9	2	20	5	44
150–499	11	12	12	51	8	94
500–1249	7	3	9	23	4	46
1250 and over	6	—	—	19	2	27
Turnover not known	1	—	1	5	—	7
Total sample	41	36	27	125	31	260

TABLE 2.3

DISTRIBUTION OF 196 MERCHANT BUSINESSES BY TURNOVER, AND THEIR SHARE OF THE TOTAL TURNOVER OF THE SAMPLE

Range of turnover (£'000)	Number of firms	% of firms	% of total turn- over of sample
0–99	41	21	2
100–199	35	18	4
200–299	22	11	4
300–399	25	13	7
400–499	9	4	3
500–999	38	19	21
1000–1999	13	7	14
2000–3999	7	4	17
4000 and over	6	3	28
Total sample	196	100	100

£1,000,000. In national terms this suggests that under 300 firms account for around 60% of the trade of the merchant sector.

2.8. The other measures of size confirm this general picture. Tables 2.4 and 2.5 show the relationship between turnover and labour force, and between type of business and labour force, for the 260 merchant businesses in the sample. Part-time workers have been adjusted on to a full-time basis, using the rough and ready assumption that each works half his time for the firm which employs him.

TABLE 2.4

NUMBERS OF MERCHANT BUSINESSES IN SURVEY SAMPLE, BY TURNOVER GROUP AND NUMBERS ENGAGED

	Turnover group (£'000)						
	Under 75	75–149	150–499	500–1249	1250 and over	Turnover not known	Total sample
Numbers engaged:							
1	11	—	1	—	—	—	12
2–5	22	12	4	—	—	1	39
6–10	1	17	23	1	—	—	42
11–20	1	4	27	14	—	1	47
21–50	—	—	14	16	8	—	38
51–100	—	—	1	6	8	1	16
Over 100	—	—	—	1	4	—	5
Not available	7	11	24	8	7	4	61
Total sample	42	44	94	46	27	7	260

TABLE 2.5

NUMBERS OF MERCHANT BUSINESSES IN SURVEY SAMPLE, BY TYPE OF BUSINESSES AND NUMBERS ENGAGED

	Type of business					
	Specialist/ mainly grain	Specialist/ mainly feed	Specialist/ mainly compounding	General merchant	Other merchant	Total sample
Numbers engaged:						
1	3	6	—	1	2	12
2–5	11	7	4	10	7	39
6–10	6	9	—	23	4	42
11–20	6	8	4	26	3	47
21–50	5	5	10	15	3	38
51–100	5	—	3	6	2	16
Over 100	—	—	—	4	1	5
Not available	5	1	6	40	9	61
Total sample	41	36	27	125	31	260

2.9. Of the 199 firms giving exact information on the size of their labour force, twelve (6%) relied solely on the services of the proprietor, and close on half employed ten men or fewer. Even the largest firms are relatively small in terms of numbers employed compared with many other industries.

2.10. A closer examination of the data revealed that among the small number of firms who derived a relatively large proportion of turnover from seed, labour forces tended to be well above average for the turnover category concerned.

2.11. From the point of view of the marketing of home-grown cereals, the distribution of firms according to their total purchases of grain (Table 2.6) is most significant.

TABLE 2.6

GRAIN PURCHASES BY 193[a] MERCHANT BUSINESSES BY TOTAL ANNUAL VOLUME OF GRAIN PURCHASED

Total purchases of grain (tons)	Number of firms	% of firms	% of purchases by all firms
Less than 1000	55	29	1·4
1000–4999	69	36	9·4
5000–9999	27	14	11·0
10,000–19,999	22	11	19·0
20,000 and over	20	10	59·2
Total	193	100	100

[a] Data were not available for sixty-seven firms in the sample.

2.12. Thus concentration has proceeded to the point where one-third of the firms account for nine-tenths of total purchases of grain by merchants in the sample. This would represent about 700 firms in the U.K. as a whole.

2.13. Many, but not all, of the small businesses are independents. Table 2.7 shows that such firms form the largest single category of merchant businesses in all but the largest turnover groups. Despite this, the fact is that the relative importance of co-operatives and of subsidiaries of national companies far exceeds their numerical ratio to the total merchant population.

The Activities of Merchant Businesses

2.14. Merchant businesses are in many cases the "general shop" of the farming community. Grain purchases and sales form a part, and in some cases only a small part, of their total activities. The role which they play in marketing home-grown grain is, however, conditioned by the full range of their activities. It is at this level that competition for capital within the merchant sector takes place. It is because of the facilities existing to deal

TABLE 2.7

NUMBERS OF MERCHANT BUSINESSES IN SURVEY SAMPLE, BY COMPANY STATUS
AND TURNOVER GROUP

	Company status				
	Subsidiaries of nationals	Co-op-eratives	Independ-ents with links	Independ-ents, no links	Total sample
Turnover group (£'000):					
Under 75	1	3	1	37	42
75–149	3	1	2	38	44
150–499	4	8	18	64	94
500–1249	6	1	13	26	46
1250 and over	9	6	6	6	27
Turnover not known	2	1	2	2	7
Total sample	25	20	42	173	260

with these non-grain activities that some firms are prepared to handle grain for very modest margins. It is because of outstanding debts associated with some of these activities that merchants may take a farmer's grain rather than risk non-payment. We shall describe first the grain-handling activities, then the other agricultural activities, and finally the non-agricultural activities of merchants.

Grain-Handling Activities

2.15. The merchant discovers suitable supplies of grain on farms and arranges for its physical collection and transport at appropriate times to those who will pay most. He keeps in touch with the needs of a wide range of grain users so that grain flows to those destinations where its value is greatest. To some extent these activities may be conducted on the basis of the small group of farmers and end-users known to the merchant. Over and above these, he may sell grain to other merchants or buy from them according to prevailing price differentials. The more sensitive his appreciation of small variations in supply and demand conditions, the more likely he is to make a profit for himself and to offer good prices both to farmers and to the users of grain.

2.16. The 199 merchants who provided information about grain sales and purchases in the course of the Survey bought some 1·6 million tons of grain. They sold 1·3 million tons. Table 2.8 shows the proportionate breakdown of these transactions between major sources of supply and destination of sales.

2.17. The pattern of sales demonstrates one important feature of the

TABLE 2.8

PERCENTAGE DISTRIBUTION OF MAIN SOURCES OF SUPPLY AND DESTINATIONS OF
SALES OF WHEAT, BARLEY AND OATS (RAISED RESULTS FROM 199 MERCHANT
BUSINESSES)

	Wheat	Barley	Oats		Wheat	Barley	Oats
Bought from:				*Sold to:*			
Farmers	84	74	77	Farmers	17	15	39
Other merchants	11	25	19	Other merchants			
Imports	5	1	4	(incl. exports)	19	36	37
Other sources	(a)	(a)	(a)	Industrial			
				users (b)	64	49	21
				Other sales	(a)	(a)	3
All sources	100	100	100	All destinations	100	100	100

(a) = less than 0·5%. (b) Industrial users = maltsters, distillers, compounders, millers.

structure of the merchant sector. By far the most important outlet is that for industrial uses. Such sales to compounders, millers, maltsters and distillers dominate the market for home-grown grain. Even though the proportion of firms in the merchant sector owned by end-users is relatively small, less than 10%, the influence on trade of the large firms which dominate the milling and compounding sector is much more far-reaching. Many "independent" firms, especially those who do not make their own compounds, depend on the buying policy of these large public companies, which can change the whole pattern of trade. One clear example of this is the variation in standard moisture content for barley sales (16%), and for wheat (18%), which seems to owe more to the accepted standards of compounders and millers respectively than to any technical merit of the standard applied.

2.18. As might be expected, the specialist grain firms, some 10% of the sample, handled a rather greater proportion of grain purchases (22%) than their numbers might suggest. For many of the firms classified as "general" and "other merchants" grain is just one more feature of their comprehensive and reciprocal trade with farmers. Its contribution to profits is probably very small, but given the need to sell compounds, fertiliser, etc., it is important to offer an outlet to the farmer for his grain. Manufacturing merchants, on the other hand, must buy grain. Quite apart from margins they may make on grain transactions, grain itself forms the most important raw material they use. Although their aggregate consumption of the home-grown cereal crop, some 15%, is relatively small, the fact that they provide an alternative outlet to the major national users introduces an important competitive element into the market.

2.19. An examination of the volume of grain bought by firms in different "company status" groups confirms the view that the influence of the national firms is more through the pressures they create for independent firms, than through their own ownership of a share of the merchant sector.

TABLE 2.9

PERCENTAGES OF TOTAL GRAIN PURCHASES
(VOLUME) BY FIRMS IN DIFFERENT COMPANY
STATUS GROUPS

	% of total grain purchases
Company status:	
Subsidiaries of nationals	22
Co-operatives	6
Independent with links	22
Independent, no links	50
Total purchases	100

2.20. The oft-expressed contention of merchants' trade associations that the independent business still performs a major role in marketing grain is well demonstrated. Together the "independents with links" and the "other independents" buy about 70% of all grain purchased by the sector. Equally, the rather small share of the trade handled by co-operatives is striking. It seems that the attractions of co-operative trading have been insufficient to wean many farmers away from their traditional dependence on the merchant sector. Such a conclusion must not, however, ignore the competitive effect of co-operatives, which may well have helped to keep private traders on their toes.

2.21. Figures 2.1, 2.2 and 2.3 illustrate the different flow patterns for wheat, barley and maize. Being drawn to scale they show the relative importance of the various types of merchant businesses in the handling of each type of grain. Figure 2.1 also illustrates the importance of inter-merchant trade in wheat. These charts have been based on statistics derived from the Survey of Merchants and included in Appendix H (Tables 7–11).

2.22. Table 2.6 on p. 177 has made it clear that many of the firms in the merchant sector buy only small quantities of grain. Table 2.10 reinforces this picture and shows that the basic pattern of purchase is similar for individual grains, though few firms bought oats at a rate of more than 1000 tons per year. In contrast, more than half of the firms bought more than 1000 tons of barley in the last accounting year.

2.23. If we assume a price of £20 per ton and a gross margin of 3% (rather more generous than many quoted by merchants in the course of the Survey) a firm buying 1000 tons of grain will make a gross margin of only

THE FLOW OF WHEAT (U.K.) 1966/7

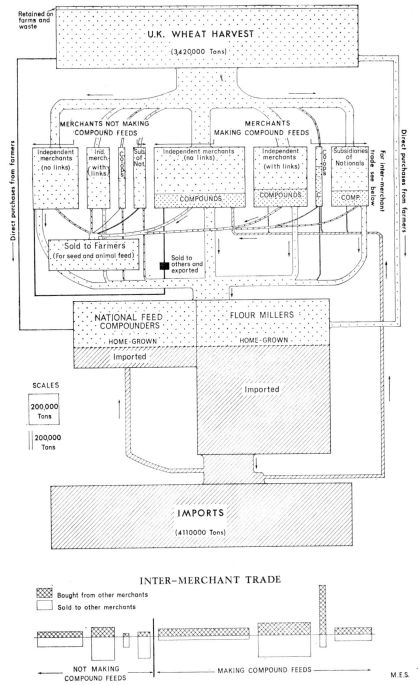

FIG. 2.1. The flow of wheat, U.K., 1966/7.
Source: Based on agricultural statistics, and results of the merchant survey—
see Statistical Appendix.

THE FLOW OF BARLEY (U.K) 1966/7

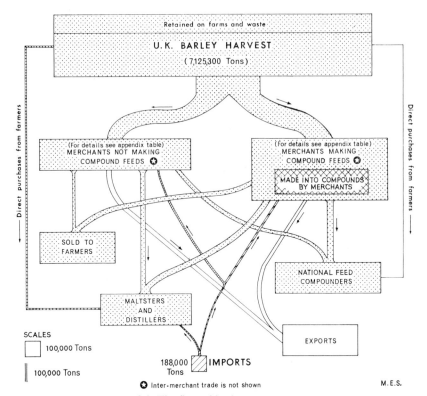

FIG. 2.2. The flow of barley, U.K., 1966/7.
Source: Based on agricultural statistics, and results of the merchant survey—
see Statistical Appendix.

£600. From this, labour, handling, and administrative costs have to be deducted. It seems very unlikely then that the basic commercial *raison*

TABLE 2.10

DISTRIBUTION OF 200 MERCHANT BUSINESSES[a] BY ANNUAL TONNAGES OF WHEAT,
BARLEY, OATS AND TOTAL CEREALS PURCHASED

	Number of firms buying:					
	Less than 100 tons	100–999 tons	1000– 4999 tons	5000– 9999 tons	10,000 tons and over	Total
Wheat	58	64	56	14	8	200
Barley	33	58	71	16	22	200
Oats	108	83	8	1	—	200
Total cereals	13	48	68	29	42	200

[a] Data were not available for sixty firms in the sample.

THE FLOW OF MAIZE (U.K.) 1966/7

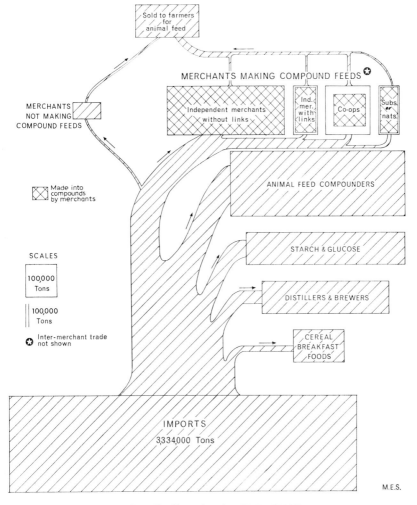

FIG. 2.3. The flow of maize, U.K., 1966/7.
Source: Based on agricultural statistics, and results of the merchant survey—
see Statistical Appendix.

d'être of many of these small buyers can be grain marketing. Even where
very much more grain is bought it still seems probable that other activities
must play an important part in the overall economy of the firm.

Other Agricultural Activities

2.24. Table 2.11 indicates the range of agricultural activities other than grain handling undertaken by the firms visited in the course of the Survey. Merchants suggested four main reasons for this wide range of trade: shared overheads; diversified risks; the need to restrain farmers from dealing with competitors; and the desire to provide a complete service for the farmer.

TABLE 2.11

AGRICULTURAL ACTIVITIES (OTHER THAN GRAIN DEALING) UNDERTAKEN BY MERCHANT BUSINESSES IN SURVEY SAMPLE, BY TYPE OF BUSINESS (PER CENT)

	Percentages of firms trading in:					
	Own compounds	Purchased compounds	Fertiliser	Spray	Seed	Other ag. activities
Type of business:						
Specialist/mainly grain	19	59	73	49	76	32
Specialist/mainly feed	42	100	69	42	56	56
Specialist/mainly compounding	100	66	56	26	44	19
General merchant	67	97	88	74	80	54
Other merchant	16	58	58	35	58	55
Total sample	53	83	76	56	70	47

2.25. Many of the overheads involved in handling grain can simultaneously serve a number of other items of agricultural trade. By increasing the volume of trade in relation to these overhead costs, unit cost can be lowered and profitability increased. Economies arise in the use of labour, of transport and of premises. Although contact with farmers for the purpose of buying grain can be achieved simply by telephone conversation and an occasional visit from the proprietor of the merchant business, few firms work in this way. Especially where animal feed is sold, a more elaborate and continuing pattern of visits by the firms' representatives is needed. Such salesmen will be less costly the greater the volume of trade they transact, at given margins, regardless of the nature of the trade concerned. At an administrative level the merchant has to sustain an office staff whose size will not bear any fixed relation to the volume of business handled. Within fairly wide limits trade may be increased, particularly trade with the same group of farmers, without adding to the number of people employed in the office.

2.26. Transport is a major element in the economy of most merchant businesses. Its efficient use involves keeping lorries loaded and running for the maximum possible time. By dealing in the whole range of bulky goods which have to be moved from or to farms (feedingstuffs, fertilisers, seed and grain), the possibilities of efficient usage are enhanced. The evidence of the Survey suggests that where such diverse trading possibilities are not open, the merchant may prefer to rely on hired transport rather than incur this heavy overhead cost which he is unable to spread over an adequate volume of business. Of the specialist grain merchants, 75% of whose business was in grain, three-quarters relied entirely on hired transport.

2.27. Premises form a rather less important element in many merchant businesses, but again it is clear that the more intensively stockholding capacity and storage facilities can be used, the lower their impact on the costs per unit of the business.

2.28. Diversification is a common way of handling the risks associated with commercial activity. Within the context of state support for farmers, the threat of a collapse in demand for farmers' requisites is greatly reduced. Nevertheless, changing policy and technical development on and off the farm may in time tend to erode some parts of the merchant's business and to increase the scope for others. By involvement across the board the merchant reasonably hopes to secure his long-term position.

2.29. Many merchants were conscious of the keenly competitive nature of their trade. They argued that if a rival firm gained entry to a farm through supplying some item which they did not stock themselves, then the farmer might choose to deal with the rival concern not only for that item but also for goods which the firm did stock. Though there may be "rules of the game" against poaching of customers, merchants often feel safer if they can provide the whole range of farmers' requirements.

2.30. To wish to provide a "complete service" is in many respects a reflection of the desire to exclude competitors, but it goes further. The merchant's two-way traffic with the farmer builds up a situation of interdependence. The merchant's long-term prosperity depends upon that of the farmers with whom he deals. He is often keenly interested in animal and crop husbandry, and he or his representative is a friend and consultant of the farmer. By providing a high proportion of the farmer's requirements, he can develop and sustain this relationship; even more, he can judge better the extent to which credit facilities can safely be granted.

2.31. The most important activity of many merchants was the sale of animal feedingstuffs. In many cases, feed made by the merchant was sold at prices lower than those of national feedingstuffs manufacturers. The price differential, often said to be about £1 a ton, was usually based on lower raw material cost rather than on economies in production. Some merchants were very substantial country compounders, and it seems possible that their products were, from the farmers' point of view, of equal quality to those of national firms but cheaper. That these businesses also sold the nationally known products confirms that this was not the view of all farmers, and reflects the desire of merchants to retain business and sustain turnover even where margins were lower than on their own products.

2.32. Three out of four merchants sold fertiliser, but most did so without enthusiasm. The discount system of the fertiliser manufacturers leaves very small margins for traders who sell only small quantities. Such business was only worthwhile where its overhead costs could be shared with those of other activities.

2.33. This examination of other agricultural activities makes it clear that grain marketing is only one element in an interlocking network of trading activities carried out by merchants. In many cases it is a subsidiary element, in others the merchant is more concerned with meeting his own manufacturing needs than in making a margin on grain. The merchant is concerned with the overall profitability of his business and within this context may well continue to handle grain even when, by itself, the income it yields is very small.

Non-Agricultural Activities

2.34. Diversification extends in an appreciable number of cases beyond agriculture itself. The types of non-agricultural activity most commonly discovered are summarised below:

	No. of firms		No. of firms
Garden sundries	41	Fuel and oil	24
Coal	30	Pet food	17
Builders' materials	24		

Table 2.12 shows the importance of such non-agricultural activities in relation to turnover. Non-agricultural activities are more common among the medium-sized firms, but their importance to turnover is relatively small. Although fewer small firms had non-agricultural activities, where they did exist they often formed a fairly substantial part of the whole business.

2.35. Garden sundries and pet food represent an extension into a different and often very rewarding market. To some extent they may stem from trade in seed, sprays, fertilisers, etc., with farmers, but the scale of operation and the need to carry a range of goods not needed in agriculture means that extra and different forms of capital are needed. The profitability

TABLE 2.12

	Percentage of firms having:				
	No non-ag. activities	Non-ag. activities less than 10% of turnover	Non-ag. activities more than 10% of turnover	Non-ag. activities —% of turnover unspecified	Total sample
Turnover group (£'000):					
Under 75	67	2	24	7	100
75–149	66	7	23	4	100
150–499	53	23	16	8	100
500–1249	76	11	11	2	100
1250 and over	56	15	22	7	100
Turnover not known	71	29	—	—	100
Total sample	62	14	18	6	100

of this investment is likely to depend on the site of the business concerned.

2.36. Coal and builders' materials, on the other hand, can utilise some of the facilities which are needed for trade with farmers. Transport, access to railway sidings and storage space may serve the needs of each type of trade. In so doing, overheads may be effectively reduced, and the competitiveness of the small business enhanced in each direction.

The Geographical Structure of the Merchant Sector

2.37. Very few farmers are far from a merchant. Most have the choice of several. This dispersed distribution of merchant businesses is well demonstrated in Table 2.13 and also in Map 2.1. The fact that the same area may be covered by two or more merchants is not necessarily wasteful. Although in purely physical terms one firm might deal with all the trade, the existence of competition helps to ensure that costs are kept down and farmers offered a fair price for the grain they sell.

2.38. The regions used in Table 2.13 are not of equal area, nor do they contain the same number of farmers. However, it is clear that farmers in Wales and Scotland are likely to be more distant from a merchant than those in the south-west of England or in Northern Ireland. Specialist grain and mainly grain merchants were heavily concentrated in eastern England and in Yorkshire.

2.39. The distribution of merchant businesses in terms of the surplus/

TABLE 2.13

NUMBERS OF MERCHANT BUSINESSES IN SURVEY SAMPLE, BY TYPE OF BUSINESS AND
ADVISORY REGIONS

	Type of business					
	Specialist/ mainly grain	Specialist/ mainly feed	Specialist/ mainly comp'ing	General merchant	Other merchant	Total sample
Advisory Regions:						
South-west	—	5	5	20	2	32
South-east	—	2	—	25	2	29
East	19	1	6	18	1	45
East Midland	4	4	1	9	4	22
West Midland	1	4	1	19	3	28
Yorks. and Lancs.	8	4	2	8	2	24
Northern England	4	3	—	5	2	14
Wales	—	7	1	3	—	11
Scotland	4	3	5	6	6	24
Northern Ireland	1	3	6	12	9	31
Total sample	41	36	27	125	31	260

deficit areas described in Chapter 3 (see Table 2.14) shows that in the surplus areas, specialist grain and mainly grain merchants predominate, whereas in the deficit areas most of the firms are general merchants. It is also noticeable that of the merchants dealing mainly in purchased feed or in their own compounds most are to be found in the areas of greatest deficit, probably because such firms rely heavily on imported grain and hence have to be near the ports.

TABLE 2.14

NUMBERS OF MERCHANT BUSINESSES IN SURVEY SAMPLE, BY TYPE OF BUSINESS AND
SURPLUS/DEFICIT CLASSIFICATION

	Type of business					
	Specialist/ mainly grain	Specialist/ mainly feed	Specialist/ mainly comp'ing	General merchant	Other merchant	Total sample
*Surplus/deficit classification**						
S3	11	2	—	4	—	17
S2	7	2	1	6	2	18
S1	2	—	2	6	3	13
B	10	—	4	15	1	30
D1	4	7	1	27	6	45
D2	3	4	—	9	2	18
D3	4	21	19	58	17	119
Total sample	41	36	27	125	31	260

* See Chapter 3, page 241.

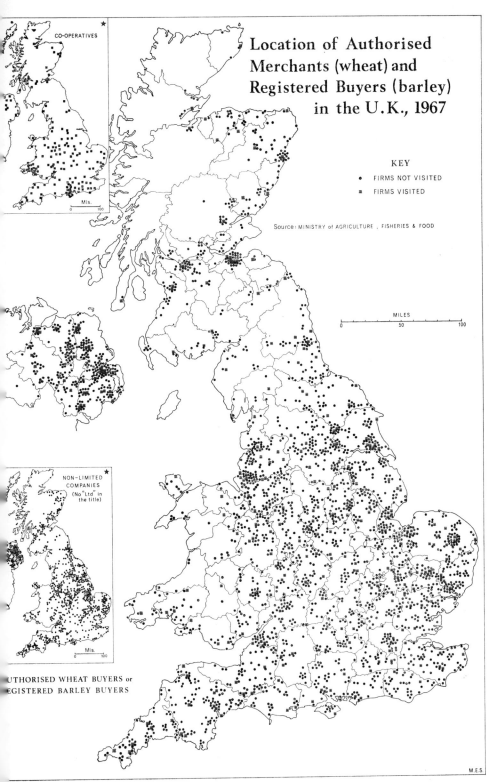

Location of Authorised Merchants (wheat) and Registered Buyers (barley) in the U.K., 1967

CO-OPERATIVES

Mls.
0   100

KEY

● FIRMS NOT VISITED
■ FIRMS VISITED

Source: MINISTRY of AGRICULTURE , FISHERIES & FOOD

MILES
0   50   100

NON–LIMITED
COMPANIES
(No "Ltd" in
the title)

Mls.
0   100

UTHORISED WHEAT BUYERS or
EGISTERED BARLEY BUYERS

M.E.S.

MAP 2.1. Location of wheat and barley buyers in the U.K., 1967.
Note: The buyers shown on the two inset maps are also included in the main maps.

2.40. It seems clear from this table that the number and type of merchant businesses in an area is closely related to the local farm economy. This impression has been further analysed in terms of four factors which seem likely to influence the prospects for profitable merchant trade. These are:

1. The number of farms in the area.
2. The size of farms in the area.
3. The acreage of crops and grass in the area.
4. The total trade potential in terms both of aggregate cereal production and aggregate demand for animal feedingstuffs.

These factors have been worked out for sixteen counties in England and Wales. The choice of the county as a unit was governed by the availability of official statistics. Considerable differences of farming exist within counties. The analysis is also inexact because merchants situated in a particular county do not deal exclusively with farms in that county. However, the effect of this overlap is to a considerable extent neutralised by the fact that merchants in adjacent counties will trade into the county in question.

2.41. The results of this analysis are set out in Table 2.15. It is clear from the table that the number of merchant businesses in an area bears no fixed relation to the number of farms. There is a range in the estimated number of full-time holdings per merchant from forty-one in Hertfordshire, to 450 in Carmarthenshire. Difference in size of farm could not wholly account for this.

2.42. A better explanation may be derived in terms of the "intensity" of business in the county, i.e. the potential trade per acre. This divides the total production of cereals together with the total feed requirements in each county by its acreage. The last two columns of Table 2.15 show a very close relationship between potential trade per acre and the number of merchants per 100,000 acres. The lower the level of potential trade per acre, the lower tends to be the number of merchants, the wider the area over which they have to operate, and the greater the aggregate volume of trade which is needed to sustain their business. Merchants in an area where there is a high level of trade per acre derive less benefit from extending their trading areas. To do so will add to transport costs and lead to a lower level of trade per unit of overhead charges than would arise from expanding trade closer to home.

2.43. Two exceptions to the general conclusions to be drawn from Table 2.15 must be noted. Cornwall and Sussex both have relatively low potential trade per acre yet a fairly high number of merchants. The evidence of the Survey suggests at least one possible explanation: in both

TABLE 2.15.

NUMBER OF FULL-TIME HOLDINGS, ACREAGE OF CROPS AND GRASS AND POTENTIAL TRADE PER MERCHANT FOR FIFTEEN SELECTED COUNTIES IN ENGLAND AND WALES.

County	Surplus/ deficit area	Number of merchants	Full-time holdings 1966	Full-time holdings per merchant	Acreage of crops and grass per merchant ('000 acres)	Estimated potential[a] trade per merchant ('000 tons)	Potential trade, tons per 100 acres	Merchants per 100,000 acres
Bedfordshire	S3	22	1268	58	11	13	118	9·6
Lincolnshire	S3	92	7052	77	17	18	106	6·0
Northumberland	S1	18	2400	133	33	21	64	3·0
Essex	S1	44	3308	75	15	21	140	6·6
Yorkshire	S2 D2 D1	143	15,023	105	17	17	150	6·0
Suffolk	B	75	3394	45	10	15	150	10·2
Hertfordshire	B	30	1216	41	9	12	133	11·0
Berkshire	B	24	1141	48	13	13	100	7·6
Herefordshire	D2	12	2533	211	37	29	78	2·7
Dorset	D3	21	2325	111	22	21	95	4·7
Carmarthenshire	D3	7	3150	450	54	25	46	1·9
Shropshire	D3	32	4157	130	23	22	96	4·6
Cornwall	D3	45	5348	119	15	13	87	7·1
Somerset	D3	39	5569	143	20	17	85	4·9
Sussex	D3 D2	42	3190	76	14	12	86	7·3

[a] Estimated feed requirements and grain production in tons.

counties there exist an unusually high proportion of small merchant business.

Radius of Trading Area of Merchants

2.44. Merchants who took part in the Survey were asked to indicate the approximate radius of their trading area. Clearly, there could be no precision about such a concept. Merchants usually have a few customers who are many miles outside their normal trading area. These may be long-standing customers who have moved away but continue to trade with the merchant, or a small group of customers who farm close to the home of a representative who lives outside the main trading area. The exact boundaries within which trade is carried out are seldom defined very clearly. They tend to be irregular, extending further where transport facilities are good and being curtailed where the coastline or a major town imposes extra transport costs.

2.45. Table 2.16 shows that more than half of the merchants traded within an area of 20 miles from base. Larger firms, in turnover terms, tended to have a more extensive trading area. Even so, it is only among the merchants whose turnover was over £1,250,000 that more than half had trading areas reaching beyond 30 miles. This evidence is in broad agreement with the analysis of Table 2.15, which led to the conclusion that merchants would generally be able to operate most successfully in an intensive agricultural area within which transport costs could be kept low in relation to the volume of trade.

2.46. Independent firms worked in smaller trading areas than subsidiaries of nationals. More than four out of every ten firms in the latter group had a radius in excess of 30 miles. Only 18% of the independent firms fell into this category. Again, part of the explanation must be the larger turnover characteristic of the national subsidiaries, but part of it may be due to the greater dependence of the independent firm on personal contact with the farmer. Some of the restrictions on trading area may be overcome where independent firms form links with other independents. Thus, Table 2.16 shows that 47% of the "independents with links" traded over areas in excess of 30 miles.

2.47. The siting of merchants' premises often owed more to historical accident than to the economic logic of the current pattern of trade. Many of the firms visited occupied the site of former wind- or water-mills. Sometimes these were in congested town areas, inconvenient both for vehicular access and for expansion.

TABLE 2.16. PERCENTAGE OF AGRICULTURAL MERCHANT BUSINESSES WITH VARIOUS TRADING AREAS
(based on a sample of 260 merchant businesses)

Annual turnover	Radius of area within which most purchases from, and sales to, farmers took place (miles):						
	0–10	11–20	21–30	31–40	41–50	51 and over	Total
1. By Annual Turnover							
	%	%	%	%	%	%	%
Under £150,000	41	37	13	5	3	1	100
£150,000 to £499,999	12	53	19	9	5	2	100
£500,000 to £1,249,999	7	16	41	16	5	15	100
£1,250,000 and over	—	7	13	20	24	36	100
Total	19	36	20	10	7	8	100
2. By Company Status							
Subsidiaries of nationals	—	19	38	10	19	14	100
Agricultural co-operatives	10	60	—	10	—	20	100
Independents with links	4	30	19	16	12	19	100
Independents, no links	26	35	21	9	5	4	100
Total	19	36	20	10	7	8	100
3. By Region							
East, S.E. and E. Midlands	20	35	20	11	7	7	100
South West and W. Midlands	19	34	12	14	7	14	100
Wales	—	73	27	—	—	—	100
Yorkshire, Lancashire and North	16	39	19	8	10	8	100
Scotland	—	33	42	8	4	12	100
N. Ireland	43	23	20	14	—	—	100
United Kingdom	19	36	20	10	7	8	100

H

Date of Establishment of Firms

2.48. Table 2.17 shows the date of establishment of the 260 firms who took part in the Survey. The date of establishment does not indicate the age of the premises currently occupied, but it does indicate the period during which the firm has existed in its current form.

TABLE 2.17.

THE LENGTH OF ESTABLISHMENT OF 260 MERCHANT BUSINESSES IN
SURVEY SAMPLE

Period of establishment	Number of firms	Percentage of firms
Before 1850	32	(13)
1851–1914	105	(40)
1915–1939	68	(26)
1940 or later	55	(21)
	260	(100)

2.49. Merchants were asked if they had changed their business location at some time in the past. Only 20% said they had done so; 15% failed to answer the question, but the implication of their replies together with the information in Table 2.17 suggests that locational disadvantages had not prevented the survival of many existing firms which were established fifty or more years ago. Of course, we have no comparable data showing the characteristics of those firms which did fail, but it seems unlikely that their situation differed markedly from many who have survived. This is remarkable in view of the drastic changes in power supplies, transport, and farming which have taken place over the past century. Today the advantage of water power and access to rail transport have largely disappeared, and the old site might be expected to be less well placed and less competitive than newer sites chosen for their ability to handle modern trade. However, only one-third of the firms admitted that they had a locational problem. The problems named were traffic congestion—twenty-five firms; labour scarcity—twenty-four firms; inadequate premises—sixteen firms; and railway closure—six firms. A diversity of other problems were mentioned by one or two firms, among them flooding and urban expansion.

2.50. In general, however, the merits of specific locational sites were of relatively little importance. More significant were the general characteristics and especially the intensity of farming in the area, and the flexibility and ingenuity of the merchant in adapting his business to changing circumstances.

The Labour Force of the Merchant Sector

2.51. Table 2.4 on p. 176 has already demonstrated that the majority of merchant businesses are small employers of labour. In aggregate terms the results of the Survey suggest that employment in the merchant sector is about 46,000. This figure includes workers who are engaged in non-agricultural and non-grain activities.

2.52. The difference between the function of manufacturing, specialist grain, and other merchants is well demonstrated by the differing average levels of turnover per man (Table 2.18).

2.53. Turnover figures do not measure value added by businesses. Thus, although specialist grain and mainly grain firms appear to achieve high

TABLE 2.18

TURNOVER PER MAN FOR 197 MERCHANT BUSINESSES[a] BY TYPE OF BUSINESS

Type of business	Number of firms	Average turnover per man (£)
Specialist/mainly grain	36	39,298
Specialist/mainly feed	34	20,103
Specialist/mainly compounding	21	16,633
General merchant	84	29,662
Other merchant	22	11,790
Total sample	197	26,757

[a] Data were not available for sixty-three firms in the sample.

levels of turnover per man, the value added by each man may well be no greater than among compounding merchants. The greater value of work done in making and distributing animal feeds means that the higher labour force of manufacturing merchants can be borne profitably. The wide differences of turnover per man between businesses within each category will be discussed more fully in Chapter 5 on Management.

2.54. Table 2.19 illustrates well one aspect of the structure of the industry: the strong element of family control and the relatively low numbers of hired managerial and executive staff. Many firms are too small to hire expensive managerial resources. Even in firms which might do so, control of the business, in so far as it is distributed among several individuals, often remains within the family. The strong family tradition in agricultural merchant businesses has probably enabled some of the smaller firms to weather periods of economic stress.

TABLE 2.19

THE STRUCTURE OF EMPLOYMENT IN 193 MERCHANT BUSINESSES[a]

Category of employment	Number	% of total employment
Working principals	324	7·8
Departmental managers	180	4·3
Representatives	565	13·6
Office clerks	673	16·2
Foremen	203	4·9
Mill operatives	1007	24·2
Lorry drivers	766	18·4
Other	442	10·6
	4160	100·0

[a] Data were not available for sixty-seven firms.

2.55. The number of people employed as office clerks is undoubtedly increased by the amount of paperwork involved in the administration of the Cereals Deficiency Payments Scheme, and more recently the Home-Grown Cereals Authority Forward Contract Bonus Scheme (see Chapter 4, p. 260). This is one of the hidden costs of official intervention in agriculture. Several firms interviewed claimed that they had had to augment their office staff as a direct result of the Bonus Scheme.

2.56. Mill operatives are the largest single category of worker—almost a quarter of all workers, despite the fact that 50% of the merchants did not manufacture. This illustrates in yet another way the importance of the "fixed cost element" in manufacture, and the need to seek higher levels of turnover.

2.57. In many respects the most important members of a merchant's staff are his representatives and his lorry drivers. Both are in direct contact with the farmer. In each case the reputation of the firm can be enhanced or damaged by the care and consideration they display. The representative often acts as counsellor and guide well beyond the role of a mere "order-taker". The lorry-driver who is skilful in negotiating awkward farmyards, who is prepared to do a fair share of the loading and unloading of goods on the farm, and who observes such elementary principles as closing gates behind him, helps to create goodwill. It is because of the need to sustain this goodwill that many merchants own transport rather than rely exclusively on contract haulage.

2.58. The wage rates paid by merchant firms are generally determined by prevailing wages in the locality. Thus, wages paid by a firm in an agricultural area are likely to be significantly lower than those paid where competition from other industry is more pronounced. Representatives were commonly paid on a commission basis.

Links, Groups and Integration Between Businesses in the Merchant Sector

2.59. Something like a quarter of all transactions in grain handled by the sector are between merchant businesses. To facilitate this flow of trade there exists a very close network of information between merchants. By informal telephone contact, meetings at corn exchanges and in trade gatherings, merchants gain a much wider picture of the national market in grain than could be derived from their local transactions with farmers, but in many cases more formal arrangements have developed. These links are discussed under three heads—links which imply no change in the ownership status of the business; links through which independent businesses are wholly or partly owned by other independents; and links through which merchant businesses are owned by national companies.

2.60. Merchants who export grain are seldom in a position to provide a whole shipload. Problems of moving grain for shipment in the time necessary to avoid high port charges mean that more lorries are needed than most merchants possess. The volume of grain to be moved involves collection from several farms in a short period. By acting together merchants can make up a shipload without overstraining their facilities or driving up the local price of grain. This form of joint action has undoubtedly made possible the very considerable export of barley in recent years. It represents a very valuable feature of the merchant's activities.

2.61. Fertilisers and feedingstuffs are sold to merchants on a discount basis. The greater the volume of purchases by the merchant, the more the discount. The range of these discounts is considerable. Their effect is to enable the large buyer to undercut the small in selling to farmers. A partial solution is for a number of small merchants to join together in placing a large order through one firm. By so doing they can enjoy much of the discount associated with the larger transaction. Of the merchants interviewed, twenty-four reported making joint purchases of this type, and others may have done so although they did not specifically refer to it. This type of link stems from the policies of national feed and fertiliser manufacturers. It does not affect other aspects of the merchants' transactions, and has very little influence on the marketing of grain.

2.62. An extension of these arrangements of a more elaborate and permanent nature is the "merchant group". Two such groups were reported in the Survey. One, in the south of England, consists of twenty-four merchants within a radius of some miles who buy a wide range of requisites jointly; the group's membership has been fairly constant since its formation. Raw materials for compound feed manufacture, including imported grain, were bought jointly. Each firm bought its own grain from farmers, and sold in competition with all the others. A smaller group of seven or eight merchants stretched from the Midlands to the South Coast. The group meets to exchange information and to agree on a buying policy for the requisites purchased jointly. These include raw materials for compounding, bags, and motor-vehicle accessories. The attractions of better bargaining powers and pooled information had to be set against the possibility that decisions taken by the group would not fit the needs of every business. As one merchant declared: "There are different opinions among the members, and the advantages of buying big are too small to offset the hazards of buying at the wrong time." The existence of groups of this type can have only marginal effects on the marketing of home-grown cereals. They may achieve some slight improvement in the already excellent network of inter-merchant communication, but this is unlikely to be of great importance. The fact that in grain marketing so much depends on judgement, as distinct from information about the market, suggests that most businesses will continue to exercise an independent judgement rather than pool this aspect of their business.

2.63. A potentially greater influence on grain marketing stems from the fact that of the 260 firms in the sample, thirty-eight were owned by or owned other businesses. This figure excludes those firms which were the wholly-owned subsidiaries of national feed manufacturing companies. Two types of link may be distinguished. In seven cases the firm visited was owned by a holding company which possessed assets far removed from agricultural merchanting. The other interests of these companies ranged from import/export brokerage to the manufacture and sale of swimming pools. Two of the holding companies were property and finance organisations. In the bulk of the cases, however, association was with a firm which was also engaged in some or all aspects of the agricultural merchanting trade.

2.64. A wide range of specific reasons had led individual businesses to purchase other firms. However, the major general reason for buying another merchant firm is to extend the market and increase turnover. This reason is of great importance where fixed capital in transport or compound manufacture can be used more fully over the larger amount of business. To extend the market by price competition or by establishing representatives in new areas is a slow and expensive process. To buy another firm ensures immediately a certain volume of trade which, with good management, may be retained and even expanded after the takeover. In addition, some of the businesses bought may bring with them skilled staff or well-sited premises which will strengthen the company as a whole. Some merchants have sought to be taken over. For them it may represent the most satisfactory way of realising their assets in the business before retirement. In other cases, shortage of capital has prevented expansion which, as members of a larger group, they have been able to undertake. It is noticeable that many former owners were retained as managers of subsidiary companies.

2.65. A group of firms owned in this way, although small by national standards, often has an important share of the local market. This must make it more difficult for farmers to play off one merchant against another and must tend to limit price competition. The Survey provided no evidence of effective monopoly positions of a regional character. Even where an independent group had secured a substantial fraction of local trade an alternative buyer could always be found. Grain marketing can be only marginally affected by the existing degree of horizontal integration in the merchant trade.

2.66. Vertical integration—the ownership of merchant businesses by end-users—may be more important. Some merchants who manufacture compounds are vertically integrated within their own businesses. For them, grain-buying policy must be influenced by their own manufacturing needs as well as by the margins to be derived from buying and selling cereals straight. Of greater interest is the ownership of merchant businesses by large national compounders or millers. In the Survey, twenty-five of the firms visited were national subsidiaries of this type.

2.67. The influence of national companies on merchant businesses is already considerable. Apart from their importance as a major outlet for grain bought by the merchant, they also determine the profitability of the compound sales he makes by regulating the discount they offer on their own products.

2.68. National companies differ in the extent to which they trade through merchants, but for those who do rely on the merchant sector the continued existence and prosperity of merchant businesses is of central importance. Efficiency may be fostered through advice, assistance in technical matters, and the provision of credit for modernisation. In some cases, merchant businesses develop close relationships with one particular firm which gives them preferential treatment. Thus considerable influence may be exercised short of actually owning the merchant firm. However, where influence alone is insufficient to sustain a satisfactory pattern of outlets for manufactured compounds, outright ownership may take its place. The advantage of this is a greater degree of control over management, and a greater certainty that market shares will not be eroded by competition from other national companies who offer better terms to the merchant, or from compounds manufactured by the merchant.

2.69. The Survey showed that 22% of the grain purchased by merchants was bought by the subsidiaries of national companies (Table 2.9). We were assured that in buying grain, national companies worked on the basis of offering the same price to their own merchants as to others. Two exceptions were mentioned. First, when the market was flooded with grain, preference would be shown to wholly-owned firms in that their grain was bought at the prevailing price, in preference to that of firms which were not owned. Second, a merchant who needed to offer particularly good terms to a farmer in order to retain his custom (for compounds) might be offered a slightly better price by his national parent company. Over time, it seems probable that these policies will favour the wholly-owned subsidiaries and increase the proportion of the grain which they handle.

The Relationship of the Structure of the Merchant Sector to Grain Transport, Storage and Credit

2.70. The role of the merchant in transporting grain is conditioned by the structure of the merchant sector. First, the diffuse location, often in relatively remote rural communities, of many merchant businesses limits the type of transport which may be employed. Railway closures have reduced the number of merchants who could make use of rail transport, but the short-haul nature of journeys in a small trading area, the fact that goods have to be loaded or unloaded on the farm, and the flexibility of lorries, means that road is likely to be preferred in most cases. The effect of pressure to use rail services could well be to diminish the profitability of some merchant businesses very greatly, and even in some instances to force them out of the trade. Second, the reciprocal nature of the merchants' trade makes the use of lorries much less costly than if grain alone were to be handled.

TABLE 2.20

NUMBERS OF LORRIES OWNED BY MERCHANT BUSINESSES IN SURVEY SAMPLE, BY TYPE OF BUSINESS.

| | Number of lorries owned: | | | | | |
	None	1–5	6–15	More than 15	Not known	Total sample
Type of business:						
Specialist/mainly grain	20	12	6	2	1	41
Specialist/mainly feed	7	21	6	1	1	36
Specialist/mainly compounding	2	15	10	—	—	27
General merchant	5	72	32	14	2	125
Other merchant	5	21	4	—	1	31
Total sample	39	141	58	17	5	260

2.71. The importance of reciprocal trade is clearly demonstrated in Table 2.20. Half of the "specialist/mainly grain" merchants own no transport. For them grain movement is simply a question of choosing the lowest-cost method: rail, road, or water. At the other extreme, only two "specialist/mainly compounding" merchants manage without a lorry fleet. Here, the need to take compounds to the farm at the time and in the volume which is needed dominates policy. Grain may often be transported when the lorry is not needed for other jobs or as a return load from a delivery journey. The real cost to the community and to the merchant of such grain movement is very low. Alternative transport would only be worthwhile if the price at which it was offered fell below the running cost of the lorry already owned by the merchant. So long as the structure of the merchant sector remains as complex as it is now, the advantages of specialist grain transport by rail, or by bulk lorries, are unlikely to offset the benefit the merchant derives from using his own fleet more intensively.

2.72. Storage problems, like transport, are discussed elsewhere in this report. Here, it is only those aspects which flow from the structure of the merchant sector which are considered. The smallness of many businesses precludes storage in quantities likely to achieve economies of scale. The need to increase turnover by moving quickly a commodity on which margins are low conflicts with a commitment to tie up capital in grain stores or in stored grain. Seventy-seven of the firms in the Survey had no storage. They moved grain from its point of purchase to its point of sale with no intermediate period in store. Only sixty merchants had storage capacity in excess of 500 tons, and more than half of these were manufacturers. For manufacturers, storage represents security against shortfalls in grain supplies for the mills rather than a function of marketing grain. If merchants were to undertake extensive storage the structure and capital base of the sector would have to be changed.

2.73. Credit is one of the important services which merchants provide for farmers. The structure of the merchant sector emphasises the two-way relationship between farmers and merchants. If this were destroyed credit provision would become much more difficult and hazardous. The merchant who extends credit for feedingstuffs or fertilisers can have much greater security if he knows that the farmer will, in due course, produce and sell through him a satisfactory cereal crop. Even where no formal agreement exists to sell grain to the merchant, farmers will often choose to do so. The existence of credit can thus influence the channels through which grain is moved to its final use. No evidence of abuse of power derived from a strong creditor position was discovered during the Survey. Indeed, more than one merchant confessed that he accepted grain of lower quality from some farmers in order to liquidate a protracted credit position.

H*

Changes in the Structure of the Merchant Sector

2.74. Three types of change call for special comment: the continuing decline in the number of merchants; the tendency for farmers to establish trading groups; and the growing concentration of the trade, associated with the takeover of merchant businesses by national manufacturers of animal feedingstuffs.

The Decline in the Number of Merchants

2.75. In order to attempt to measure the decline in the number of merchants, comparison was made between the membership of the National Association of Corn and Agricultural Merchants (N.A.C.A.M.) in 1947 and 1967. Not all merchant businesses belong to N.A.C.A.M., but in England and Wales most do. It seems improbable that the proportion belonging to the Association declined between the years considered. Table 2.21 sets out the results of this comparison (see also Map 2.2).

2.76. The table shows that half of the firms who were members of N.A.C.A.M. in 1947 had ceased to be so in 1967, but new entrants reduced the net disappearance to one-third.

2.77. The reasons for the disappearance of particular businesses must be related to the individual characteristics of the firm concerned. In an enterprise which depends heavily on the energy and skill of the proprietor there must be a business life cycle closely related to that of the owner and his

TABLE 2.21

THE NET DISAPPEARANCE OF FIRMS FROM THE LIST OF N.A.C.A.M. MEMBERS
1947–67.

	No. of firms 1947	Gross disappearance 1947–67		New entrants 1947–67	Net disappearance 1947–67	No. of firms 1967
		No.	%			
South-west	369	182	49	55	127	242
South-east	392	218	56	83	135	257
East	355	154	43	90	64	291
West Midlands	285	155	54	49	106	179
East Midlands	252	115	46	61	54	198
Wales	127	83	65	19	64	63
Yorks. and Lancs.	374	214	57	69	145	229
North	87	46	53	18	28	59
North Scotland	38	16	42	5	11	27
East Scotland	67	33	49	20	13	54
West Scotland	103	57	55	16	41	62
All firms	2449	1273	52	485	788	1661

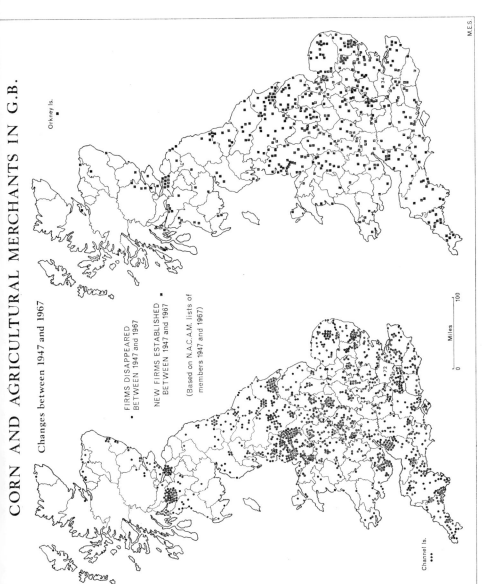

MAP 2.2. Changes in numbers of merchants in Great Britain between 1947 and 1967.

family. Changes of the magnitude revealed in Table 2.21 suggest that some more general adverse pressures must have affected the merchant sector since the end of the Second World War. Three of these may be distinguished: competition for resources; the differential effect upon merchants of greater home-grown cereal production; and changes in technology.

2.78. From 1947 to 1967 aggregate value of sales from farms rose by 258%. In the same period the value of cereal production increased 390%. By themselves these figures might suggest a prosperous time for agriculture and for all who serve the industry. The facts are somewhat different. Increased production in other sectors has meant keener competition for the resources which may be used in agriculture or elsewhere. Labour and capital costs have risen so that although the volume of trade may have grown and selling prices be sustained by government action, there has been a persistent tendency for margins to be squeezed by rising costs. Those businesses who are unable to participate in the areas in which the volume of trade is growing most rapidly are likely to suffer most.

2.79. Merchants who took part in the Survey were asked whether further investment in their businesses would be justified. Their answers reflect the stringencies outlined in the previous paragraph. Only 115 believed that more investment would be worthwhile.[1] Of these, thirty-four had links with other companies through which capital could be made available. Where investment was contemplated it was often associated with compound manufacture. One merchant in this category described his predicament as investment for "survival". He was planning handling and production facilities which would use less labour and increase his turnover. Most merchants, however, believed their profit margins were so low that more investment would be foolish.

2.80. Merchants who find their position deteriorating do not necessarily wind up their business or sell its assets to another firm. In many cases a solution is sought through diversification. Tables 2.11 and 2.12 have given some indication of the existing diversity of merchant businesses. Activities such as garden sundries and pet foods are likely to share more fully in the expenditure pattern of consumers as their incomes rise. A small agricultural merchant suitably situated may thus transform his business until his dependence on farming becomes very small.

2.81. The increased production of home-grown cereals has been of great benefit to the merchant situated in a grain-producing area, and to the merchant who is also a manufacturer of compounds. The effect of narrowing margins can be offset if the number of transactions is sufficiently

---

[1] Independent firms sixty-five, independent with links eighteen, national subsidiaries sixteen, and co-ops sixteen. The proportion of the last three categories who thought investment worthwhile is much higher than the first.

increased, or if by adding the extra function of manufacture a double margin can be gained. Closeness to the source of supply has improved the position of the country compounder *vis-à-vis* the port compounder, and the higher the proportion of home-produced grain used in mixes, the greater this advantage becomes.

2.82. The effect on non-manufacturing merchants outside the main grain-producing areas is not simply a relative deterioration in their position. In absolute terms, they may suffer as production of compound-consuming livestock moves towards those areas where grain and compounds are more cheaply available. Further, in areas where no surplus of grain has to be moved from farms, the manufacturing merchant may find the profits he makes on compounds enable him to offer rather better prices to farmers, or to accept lower margins. Again, the effect may be an absolute decline in the proportion of the trade left in the hands of non-manufacturing merchants.

2.83. Changes in technology have tended to increase economies of scale in the merchant sector. Because larger firms can often attain lower costs, there is an inevitable pressure towards fewer firms for any given level of trade. The technological changes which seem to have given most impetus in this direction are improvements in transport and sophistication in the manufacture and use of compound feeds.

2.84. Improvements in transport to which the Survey drew attention covered both the bulk movement of grain in specialist lorries, and the improvement in handling grain on the merchants' premises. Bulk lorries are not adaptable to other uses, and unless the amount of grain to be moved is fairly large they are likely to have relatively long idle periods. Improved handling of grain, automated control systems for milling, and low-cost bulk storage all affect the manufacturing merchant. They require considerable investment, and can only be justified where there is a fair volume of business. The new techniques set the pace of competition. Firms who have invested will be prepared to sell at below prevailing prices in order to boost the volume of their sales. Firms who are already small find it difficult to resist this pressure.

2.85. The growing emphasis on more complex farming methods affects the merchant in two ways. Within the business the merchant who manufactures needs a growing array of machinery for cubing, pelleting and mixing in exact proportions. In his contact with farmers he must be able to give advice on a wide range of pesticides, herbicides, feeds, fertilisers, etc. Although small manufacturing firms usually buy the most complex parts of their rations ready prepared by specialist manufacturers, the ability to vary formulation in response to raw material prices, while retaining a product of equal acceptability to the farmer (and his stock), requires flexibility. The larger manufacturer scores here: he can afford to pay for specialist services necessary to achieve flexibility. Equally, merchants can draw on specialist advisers to help farmers sort out problems which arise in using sophisticated products on the farm. Farmers, however, often require technical advice at the time the goods are bought. The representative is becoming less of an itinerant salesman and more of an advisory agent. The small firm can neither afford highly skilled representatives nor readily keep abreast of current developments in a multitude of applied sciences.

2.86. Yet there are some advantages in small-scale operation. It has long been accepted that the relationship between farmer and merchant is closer than pure business logic would suggest. Merchants often know their farmers personally, and the smaller the merchant the better he knows his customers—their managerial ability, the type of land they are farming and the kind of grain they deliver.

2.87. Despite these advantages, the economies associated with largeness of scale seem to have grown and will probably grow further. The long-run effect must be a further decline in the number of merchants.

Farmers' Trading Groups

2.88. As a middleman, the merchant exists under the constant threat that either of his customers will decide to by-pass him. From the farm side, this threat has to some extent materialised in the form of farmers' trading groups. Among end-users there has been a tendency to buy or set up merchant businesses belonging to the national manufacturers of compound feedingstuffs. In this section the threat of group trading by farmers is discussed.

2.89. Groups established for the purpose of selling grain are relatively rare. Much more common are farmers' buying groups, some working independently and some through Agricultural Central Trading, an N.F.U.-promoted organisation for combining the bargaining power of all farmers' groups. A.C.T. is seeking to expand its role in cereal marketing, but at the time of the Survey only a very small fraction of the total trade in home-grown cereals went through this channel.

2.90. The direct impact of group trading on cereal marketing has so far been small, but the further sharpening of competition in sales of requisites to farmers associated with farmers' buying groups has threatened the merchant's position and it affects grain marketing inasmuch as it impairs the reciprocal nature of his trade, makes it more difficult to utilise fully the equipment he possesses and reduces his turnover. As a result, further investment is likely to be curtailed.

2.91. In all, the operation of groups must diminish the demand for merchants' services. The individual merchant faces a complex choice. If he trades with a group he may increase his turnover. Group trade may be carried out at narrower margins, but it is carried out in greater volume than trade with individual farmers. On the other hand, the merchant may conclude that any advantage to be derived from dealing with farmers' groups is more than offset by the irksome nature of negotiation and the uncertainty that an organisation with no final directing authority will in fact deliver the goods. Merchants who participated in the Survey exhibited the full range of possible attitudes. A few had promoted groups. Most said that they were prepared to deal with groups on the same terms as any other buyer, i.e. to offer discounts for quantity. Some were resolutely opposed to group trading, regarding it as an unwarranted invasion of their own territory.

2.92. The extent to which group trading will grow in the future depends on a number of commercial and technical factors. On the commercial side, farmers who deal with merchants are often paying for a range of services, credit, advice, the delivery of urgently needed small quantities of farm requisites, etc. As farming itself becomes more specialised the advantages of this bulk package are likely to decline. Groups of farmers who are not using these services will find it cheaper to make special terms with the merchant or to by-pass him altogether. The merchant may respond by making the terms on which he deals more selective. This will simultaneously reduce the costs of his services to well-placed farmers and increase them to farmers who have to use the credit which he can provide. The most important technical factor is the growing need for storage associated with increased production of home-grown cereals. The merits of large-scale storage are discussed in Chapter 7. Here we should note that if, to attain these advantages, favourable terms are offered to farmer group storage arrangements through government subsidies, then the probability of group selling of grain is likely to increase.

2.93. These considerations might lead to an expectation of drastic contraction in the grain marketing aspects of merchant businesses. However, two fundamental restraints on this development must be noted. First successful group trading depends upon a homogeneity of interests among farmers. This is likely to be rather unstable. The varying circumstances of individual farms make different courses of action appropriate to each. At one time a farmer may find it no problem to accept the discipline of a group, at another it may be a costly and frustrating experience. Second, as trading organisations grow in size they find it increasingly unsatisfactory to rely on the voluntary part-time efforts of enthusiastic members. There is a natural tendency for fixed costs in premises and staff to be incurred.

Trading groups handling only one commodity are less well equipped to spread these fixed costs over a range of goods, and so face the danger of higher costs. It is notable that co-operatives, which are farmer-controlled bodies, find it difficult to offer better terms than many private merchants, and that their share of the grain trade has remained very modest. Finally, it should be noted that the merchant is a skilled operator. The range and degree of skill vary among merchants, and there is no reason why the necessary knowledge should not be acquired by outsiders. However, trade which ignores these skills is likely to be unprofitable. If knowledge of the market is inadequate, or the most suitable method of transport not chosen, farmers who have by-passed merchants may find themselves worse off.

2.94. On balance it seems likely that further developments in group trading will take place, but improbable that they will command a very large proportion of the national trade in cereals.

The Acquisition of Merchant Businesses by National Compounders

2.95. The survey of merchants showed that about 10% of the firms were owned by national companies, most of whom were manufacturers of animal feedingstuffs. They bought 22% of the cereals purchased by the sample and were larger both in turnover and in trading area than the majority of merchant businesses.

2.96. The association with national firms was often of recent origin. Merchants who were not taken over frequently commented on the penetration of their area by subsidiaries of national companies, and indicated how greatly this had developed in recent years.

2.97. Although no calculations of return to capital in merchanting were attempted as part of the Survey, the attitude of merchants to new investment suggests that it cannot be high. Three main reasons were offered to explain why national firms were investing in so apparently unrewarding a direction: the growth of country compounding; the possible effects of membership of the European Economic Community; and the need to protect outlets for animal feedingstuffs.

2.98. Increased domestic production of grain has to some extent diminished the advantages of port locations. Country compounders can offset the higher costs of smaller scale and the cost of moving imported raw materials to their premises by lower costs of home-grown grain and shorter delivery hauls for finished compounds. In siting manufacturing capacity, national firms now have to pay attention not only to the ease with which imports may be brought in but also to accessibility to home-produced grain. This shift in emphasis is far from rendering existing port locations uncompetitive, but it does provide a reason why national companies should buy country compounding capacity. In this way, the risks of a further swing in favour of home-grown grain can be hedged. Rather than establish new mills, the tendency has been to purchase existing compounding capacity. By further investment the equipment may be modernised, where necessary, but the business brings with it an existing body of customers, and a staff locally known and equipped to deal with the exigencies of the country compounding trade.

2.99. Successive applications to join the European Economic Community have so far failed. The urgency has to some extent departed from arguments about the consequences of membership, but the eventual accession of the United Kingdom remains a possibility. If we were to join domestic grain prices would rise, and so would the cost of some imported raw materials for animal feeds. The effect of these changes would be to further reinforce the benefits of country compounding. Hence, it seems plausible that the possibility of Common Market membership was one factor encouraging the acquisition of merchant businesses by national manufacturers of animal feeds.

2.100. Port manufacturers have relied, in the main, on the independent merchant trade to market their products. So long as sufficient independent businesses remain to carry out this job, it avoids the necessity to establish an elaborate and expensive organisation for selling goods to farmers. The fact that some merchant businesses have been bought by national companies disturbs the situation. Although wholly-owned subsidiaries have continued to sell the products of companies outside the organisation to which they belong, uncertainty must exist concerning how long this policy will be continued. A very strong probability exists that over time great emphasis will be placed on selling the products of the group. This poses a threat to national concerns who do not own merchant businesses. It is intensified by the low profitability of many merchant businesses, and the consequent long-run decline in their numbers.

2.101. Two courses of action may be pursued. Selected independent merchants may be assisted through specially favourable terms, or through injections of capital, to resist the encroachment of wholly-owned national subsidiaries. In return for such support, the merchant will ensure an outlet for the firm's compound feed. Alternatively, national firms whose judgement on other grounds did not encourage the acquisition of merchant businesses, may now do so as a means of defending their outlets. The more the second policy is pursued, the more urgent it becomes for those who have not bought companies to do so. The danger is that the better businesses will already have gone to rival firms, leaving an expensive uphill struggle to upgrade less thriving firms which may now be purchased.

2.102. The implication of this need to protect outlets for animal feed products is that there is likely to be a spate of takeovers once the equilibrium of trade through merchants has been upset. Each of the national firms is likely to proceed either by direct purchase, by establishing new subsidiaries or by providing support for independent merchants, until a new balance is achieved. Not all independent merchants will be bought up or controlled; only enough to protect outlets regarded as adequate to maintain the compounding firms' share of the market. It is difficult to be certain when this point will have been reached, but the recent slowdown in the rate at which merchant businesses have been acquired by national firms suggests that it may be close at hand.

2.103. The effects of this process of concentration on the marketing of cereals must not be exaggerated. The large end-users of cereals already possess very considerable power. Merchants depend on the larger compounders and flour millers as an outlet for their grain. The prices offered and the quality standards enforced set the tone for the whole market. The changes which flow from a greater degree of direct control are more diffuse but still of great importance.

2.104. Association with a large national company is likely to improve access to capital, to specialist technical skills and to expert management. In the longer run this will improve the competitive position of national subsidiaries and make the lot of the remaining independents more difficult. In so far as preference is given, at times when the market is overloaded, to grain supplied by firms controlled by national companies, this will diminish the outlets available to the independent firm and increase the risks it faces. The reduction in the number of merchants who are able to come to an independent view of the market will reinforce the already strong tendency for smaller independent merchants to follow prices set by the national companies.

2.105. Even if the recent spate of takeovers is complete, it is clear that the consequential effects have not yet worked themselves out. The full effects will depend on a wide range of issues which cannot be foreseen, but it seems reasonable to expect that the already concentrated nature of the final market for grain will be brought closer to the farmer. The merits of this trend will be discussed in the next section.

The Effect of the Structure of the Merchant Sector on the Efficiency with which Grain is Marketed

2.106. Three potential advantages possessed by merchants for grain marketing have been discussed: specialist knowledge, shared overheads and flexibility. The structure of the sector is relevant to their attainment by the industry as a whole.

2.107. At first sight, the size structure appears to be incompatible with the development of new techniques, services and skills which will be necessary. There is, however, an important degree of specialisation of function even among the small firms within the merchant sector. For technical advice the small firm can, and often does, call on the resources of large compounding companies, or upon its 'trade association'. New markets are served by working with specialised firms who can keep abreast of changing opportunities. By making use of hired transport, where special vehicles offer significant economies over flat trucks, many of the problems of evolving transport systems are removed from the merchants' shoulders.

2.108. The small merchant survives in part because despite his relative disadvantage in technical knowledge, and the second-hand nature of much of his market knowledge, he has a close personal knowledge of many of his farmers. So long as small-scale farming remains important, a substantial task remains to service these farms in terms of the small quantities they require and produce. It is here that the small merchant scores.

2.109. Among the 300 or so large merchants who purchase most of the home-grown cereal crop, specialist knowledge of an expert type is more clearly evident. The interconnections, commercial and organisational, between merchant businesses are effective. They mean that there is a continual pressure to move grain into uses where its value is highest.

2.110. The second potential advantage of the merchant in marketing home-grown grain is the possibility of sharing overhead costs over a large number of transactions. The structure of the trade is well adapted to the realisation of this goal. It is clear from Tables 2.11 and 2.12 that most merchants engage in a diversity of activities. The possibility of shared overheads is greatly enhanced by the reciprocal nature of much of the merchant trade. The optimum sizes of unit for handling grain, feed, and fertiliser as separate activities are likely to be different. Selective analysis based on considering each one alone might well lead to the conclusion that a radical reorganisation of the merchant sector was needed. However, taken together, where overhead costs can be shared, and allowing for inter-merchant co-operation through groups, it seems probable that a far less clear-cut range of optima would exist. This may go a long way to explain why the merchant sector contains a very wide range of firms both in terms of their size and their combination of activities. Many firms will continue to function although they fall below the optimum indicated for any one activity considered alone. The diversity of farming, the variety of talent and wealth within the merchant sector and variations in local transport costs associated with topography all contribute to the diversity of firms. The conclusion to be drawn when considering the most appropriate system for marketing home-grown cereals is that cereals must not be considered alone, but account must be taken of the impact of any change on the overall efficiency of the firms concerned.

2.111. The third potential merit of the merchant system is its flexibility in handling grain. It is clear that the structure of the trade assists this in so far as it fosters inter-merchant movements of grain. Something like a quarter of all transactions fell into this category, and there is no evidence that any structural change would result in a better regional distribution of grain. The structure of the trade means, however, that the merchant has relatively little influence over the timing of the disposal of grain. Few merchants have much storage capacity, and most lack the resources necessary to create new stores. The most important contribution merchants can make in current circumstances to a good distribution of the crop through time is by offering forward contracts. Of much greater influence is the policy of the Government in relation to cereal and farm improvement subsidies and of the Home-Grown Cereals Authority's Forward Contract Bonus Scheme. If merchants are to play a greater part in achieving an orderly disposal of a larger home-grown cereals crop, it is necessary that they should have access to capital for storage on the same terms as farmers, and that market prices should reflect more faithfully to farmers the changing value of their crop over time. In current circumstances, the fact that merchants have a good knowledge of farmers assists them to locate grain where it is needed and to buy it from the farmers' store. This is one of the features of a varied and competitive merchant sector. To some extent at least it compensates for the uncertainty over the timing of grain marketing by farmers, which is likely to persist in some degree so long as most of the crop is stored on the farm.

CHAPTER 3

The Internal Transport of Grain

Introduction

3.1. Transport plays a vitally important part in the operations of all agricultural merchants, and it is frequently said that the skill with which the merchant operates his transport activities largely determines the profitability of his business. The whole transport situation is undergoing rapid change, and will continue to do so in the years ahead; these changes present merchants with a number of very difficult decisions to take.

3.2. There are two main kinds of movement of grain. First, there are the short-distance movements from farms to end-users located in the same area, e.g. country compounders, local flour millers, seed merchants and other farmers. These movements are seldom more than 20–30 miles. Second are the medium to long-distance movements of grain. Sometimes these involve malting and distilling barley moving from the cereal-growing areas to the localities where the breweries or distilleries have to be located, e.g. because of their special water requirements. But mostly these movements are of feed grains and milling wheat from the cereal-growing areas into the great ports where most of the large animal feed and flour mills are located. A substantial part of this movement is from the grain-surplus areas (mostly in the eastern parts of Britain) to the grain-deficit areas (mostly in the west). As a background to the study of these medium to long-distance movements, and as an aid to the study of how merchant businesses have developed in different ways in different regions, we have attempted a classification of all the counties of the U.K. into grain-surplus or grain-deficit areas.

The Concept of Grain-surplus and Grain-deficit Areas

3.3. The basic concept, which is defined in terms of animal feed requirements only, is that if the tonnage of cereals suitable for livestock feed in a county is equal to the estimated tonnage of concentrated feed the livestock in that county are estimated to require, then the county is in balance, if it is

213

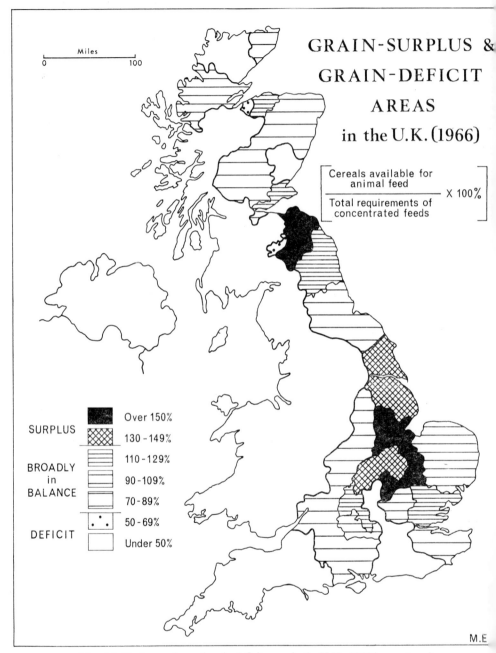

MAP 3.1. Grain-surplus and grain-deficit areas in the U.K., 1966 (percentages).
Source: Derived from agricultural statistics.

greater the county is in surplus, and if it is less the county is in deficit. The tonnage of cereals suitable for livestock feed is defined as total cereals produced less the estimated tonnage of wheat for milling, barley for malting and distilling, and grain required for seed or industrial use. A note on the methods of calculation used will be found at the end of this chapter.

3.4. It was to be expected that the U.K. as a whole would emerge as a deficit area, since imported maize, sorghum and wheat comprise a substantial proportion of total concentrates for livestock feed, and since the "requirements" figures include protein and cereal by-products, whereas these are omitted from the "cereals available" figures. Table 3.1 shows that, in fact, for the U.K. as a whole the index is 59, which reflects a substantial deficit.

TABLE 3.1

INDICES OF SURPLUS OR DEFICIT

Country	Cereals available as % of concentrates required[a]
England and Wales	64
Scotland	64
Northern Ireland	14
U.K.	59

[a] After deducting grain for human, seed and industrial use.

3.5. Map 3.1 shows the results of this analysis. It shows that in general the eastern parts of the country are in surplus and the western and northern are in deficit. However, it shows that quite important cereal-growing areas like Sussex, Dorset, Shropshire and Staffordshire are actually in deficit, whilst other more important cereal areas, like the Eastern Counties, Hampshire and Wiltshire, are only broadly in balance. The main cereal surplus areas are the counties stretching northwards from Cambridgeshire through Rutland and Lincolnshire to the Yorkshire Wolds, and the Tweed Valley further north.

3.6. Map 3.2 shows that the surplus tonnages are relatively small compared with the huge deficits in Lancashire and the North-west, the South-west Peninsula and Northern Ireland. Domestic grain cannot supply all the needs of the great factories in these deficit areas; they must continue to rely heavily on imports. But there is a substantial movement of grain across the country, e.g. from Lincolnshire and the surrounding grain surplus areas to the mills of Manchester and Liverpool (see Map 3.3A), and there is also a growing grain export trade developing in these surplus areas (see Map 3.3c and Part IV, Map 3.3).

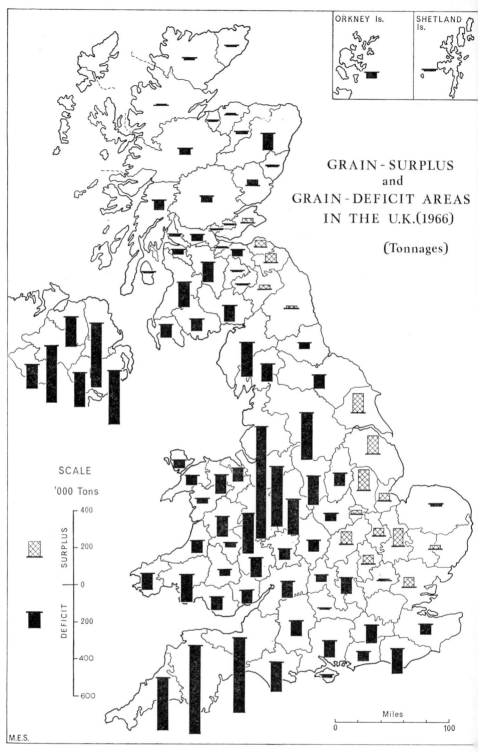

GRAIN-SURPLUS
and
GRAIN-DEFICIT AREAS
IN THE U.K.(1966)

(Tonnages)

MAP 3.2. Grain-surplus and grain-deficit areas in the U.K., 1966 (tonnages).
Source: Derived from agricultural statistics.

3.7. Map 3.3B shows that there are also many movements of grain from the main ports to destinations in their hinterlands; these will mainly comprise imported maize and wheat offals going to country compounders and manufactured feeds going to merchants generally. Map 3.3D shows that there are also many cross-hauls of grain other than the movements to and from the ports. These arise partly because of local surpluses or deficits of grain, and also because there are a number of maltings, flour mills and other grain-using factories located away from the main ports. The nature of this inter-merchant trade is discussed more fully in Chapter 4.

Road Transport

To own or hire?

3.8. A major decision a merchant has to take is whether to own his own lorries or to hire some or all of the transport he needs. The Survey showed that merchants have reacted to this choice in different ways. The majority of them (89%) hired all or part of their road transport requirements, and the remainder (11%) did not hire road transport at all. Table 3.2 shows that the smaller the firm, the more likely it is that either all the transport will be hired, or none, because a lorry represents a large investment for a small business and such a merchant may hire for all his needs or, if he has bought a lorry, he will try to keep it occupied all the time and not hire at all.

3.9. There is a marked difference between the various regions of the U.K. in this matter of owning or hiring vehicles. As Table 3.3 shows, a higher proportion of merchants in the Eastern Region, Yorks./Lancs. and Northern England and Scotland used all hired transport than is the

TABLE 3.2

PERCENTAGE OF FIRMS HIRING ROAD TRANSPORT BY SIZE OF BUSINESS

	Annual turnover					
	Under £75,000	£75,000– £149,000	£150,000– £499,000	£500,000– £1,249,000	£1,250,000 and over	All firms
	(%)	(%)	(%)	(%)	(%)	(%)
All hired	29	20	13	11	11	16
Some hired	52	66	76	83	89	73
Probably most	(21)	(23)	(31)	(29)	(36)	(27)
Probably little	(19)	(18)	(17)	(16)	(10)	(17)
No indication given	(12)	(25)	(28)	(38)	(43)	(29)
None hired	19	14	11	6	—	11
	100	100	100	100	100	100

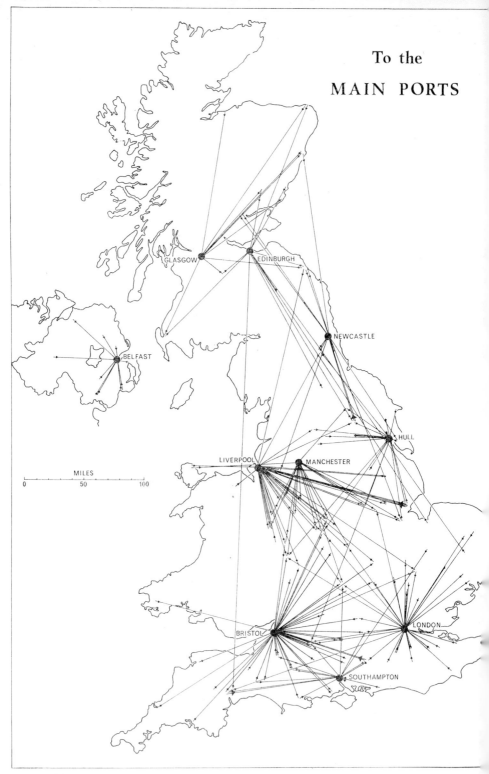

To the

MAIN PORTS

MAP 3.3A. Main movements of grain reported by 260 merchants in the U.K., 1967.
Note: In Maps 3.3A–D no attempt was made to assess the tonnage involved and
some movements, e.g. seed oats from Scotland to England, represent small
tonnages. Short-distance movements were excluded.

From the

MAIN PORTS

MAP 3.3B. Main movements of grain reported by 260 merchants in the U.K., 1967.

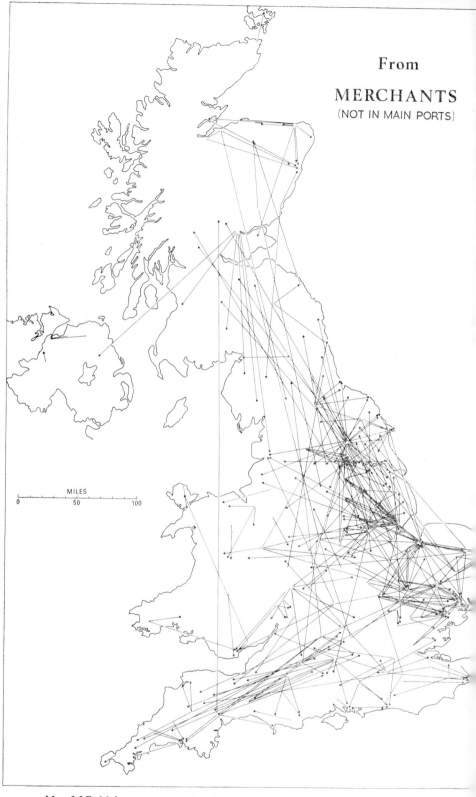

From

MERCHANTS
(NOT IN MAIN PORTS)

MILES
0   50   100

Map 3.3C. Main movements of grain reported by 260 merchants in the U.K., 1967.

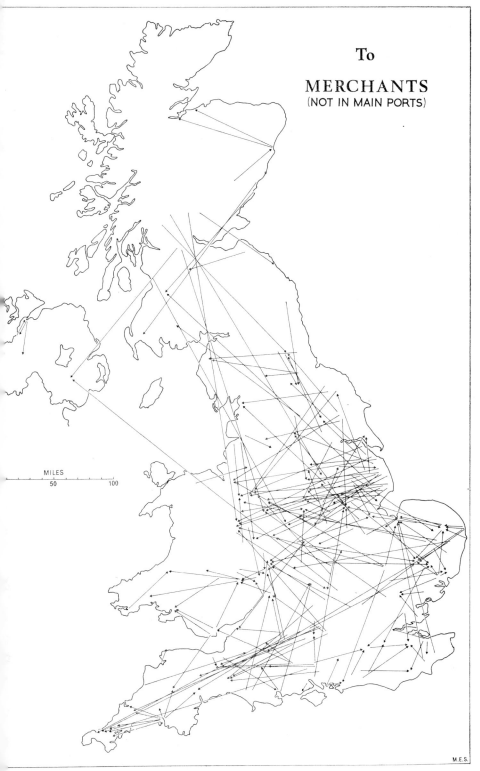

MILES
50   100

M.E.S.

MAP 3.3D. Main movements of grain reported by 260 merchants in the U.K., 1967.

TABLE 3.3

HIRING OF ROAD TRANSPORT—REGIONAL ANALYSIS[a]

(% HIRED WITHIN EACH REGION)

	S.E.	S.W.	E.	E.M.	W.M.	Y. & L.	N.	Scotland	Wales	N. Ireland	U.K.
	%	%	%	%	%	%	%	%	%	%	%
All road transport hired	14	9	24	9	11	28	22	25	9	3	16
Some road transport hired:	86	82	69	77	75	65	56	75	91	58	73
Probably most	(70)	(50)	(22)	(23)	(21)	(14)	(22)	(8)	(36)	(6)	(27)
Probably little	(16)	(6)	(20)	(18)	(—)	(7)	(12)	(12)	(27)	(45)	(17)
No indication given	(—)	(26)	(27)	(36)	(54)	(44)	(22)	(55)	(28)	(7)	(29)
No road transport hired	—	9	7	14	14	7	22	—	—	39	11
	100	100	100	100	100	100	100	100	100	100	100

[a] These are Advisory Regions—see Appendix C.

case in the other regions. The main reasons for this are that grain from these areas often travels long distances (malting barley northwards, Scottish oats southwards), and, in the Eastern region, a high proportion is handled in bulk and many merchants do not have bulk grain vehicles. On the other hand, only Northern Ireland has a significant proportion of firms not relying on hired transport at all. Even those firms that use hired transport in Northern Ireland generally do so only very rarely, e.g. for a long haul to Belfast. One explanation for this is that there is very little grain grown in Northern Ireland and hence little harvest-time pressure and little need for bulk lorries. The main traffic is in animal feeds and this can be adequately handled by a merchant's own lorries. The probable reason why firms in Scotland and Wales, in other respects similar to Northern Ireland, do not dispense altogether with hired transport is that long hauls are necessary, if only occasionally, and for these hired transport is used; in Northern Ireland, on the other hand, the longest journey is still within a day's travelling distance.

3.10. A study of the comments on this problem by the merchants interviewed shows that a high proportion (90%) tended to be in favour of hiring. This represented about three-quarters of merchants interviewed since some merchants made more than one comment. The reasons given for or against hiring, with an indication of their relative importance, are listed in Table 3.4.

Advantages of hiring

3.11. Undoubtedly the main reason why most merchants use hired transport is that there is a marked seasonal element in most of the goods they transport. The most usual practice is to buy enough lorries to meet the demand during the slackest period and to use hired transport for the rest. Another reason for using hired transport is that merchants are often less able to arrange for a return pay-load on long hauls than is a haulage contractor, and it is usually more economical to use hired transport for these journeys. A third reason for using hired transport is that a merchant will often not have sufficient grain or bulk feed trade to justify purchasing a bulk grain lorry or a blower-fitted bulk-feed lorry of his own and he has no choice but to use a haulier for this traffic.

3.12. Less importantly, hired transport is sometimes used in emergencies when a merchant's own transport is not available, and for special classes of traffic, such as hay and straw. Finally, an advantage of using hired transport is that the transport cost of each journey is known exactly and the profit on a deal can be assessed more readily—with his own transport a merchant seldom knows what the true transport cost will be when he

TABLE 3.4

COMMENTS ON HIRING

Comments favourable to hired transport	No. of times mentioned	% of all favourable mentions
		%
Hired transport is used at peak times when a merchant's own lorries cannot cope, or because trade is so seasonal that it would not pay the merchant to have his own lorries	65	30
Hired transport is used for longer journeys where no return payload is possible, e.g. to port mills, to export wharves, etc.	64	29
Hired transport is used because of the increasing difficulties of running a transport fleet of one's own, e.g.		
Problem of finding drivers 13		
High cost of repairs 12		
Problem of keeping a lorry busy in a small merchant business—or of buying the more economic large lorries 6		
The recent M. of T. Plating Regulations 5		
Traffic delays—especially at the ports 2		
Conflict with the sugar-beet season in arable areas 2		
Lack of space at premises 1		
—	41	19
Hired transport is used because the merchant has no:		
Bulk grain vehicle 22		
Blower lorry (bulk feed) 14		
—	36	16
Hired transport is used for emergencies, e.g. vehicle breakdowns, holidays or illness, or for rush jobs occasionally	7	3
Hired transport can be costed exactly and this helps the merchant in negotiating a deal	5	2
Hired transport is used for special classes of traffic, e.g. straw or hay	2	1
Total favourable mentions	220	100

Comments unfavourable to hired transport		% of all unfavourable mentions
		%
Hired transport often represents a poorer service to the farmer or end-user, e.g.		
Little control over when the haulier will deliver 6		
Haulier's labour less sympathetic to the farmer and less tolerant of poor loading or unloading facilities 3		
Hauliers are not always reliable 2		
Hauliers give poorer service generally 6		
—	17	65
Hired transport is often unobtainable when wanted, sometimes a haulier will refuse to handle sacks	8	31
Hired transport can be more expensive	1	4
Total unfavourable mentions	26	100

negotiates for the grain, or even what the true profit was when the whole transaction has been completed. Often merchants have only a hazy idea what the true transport costs are.

3.13. Apart from these positive advantages of using hired transport, the sheer difficulties and anxieties of running a transport fleet at a profit are also causing many merchants to turn more to hired transport. The problem of finding suitable drivers was frequently mentioned, especially in areas where there is alternative employment in other local industries. Contractors are better able to meet this difficulty because they can offer more overtime working and can pay higher rates because they operate the larger vehicles, which are more economic. Maintenance and repairs are a constant headache, and again the contractor is better able to cope with these because he often has his own workshop and does his own maintenance. Often a small merchant finds that he cannot run even one lorry economically, his business is not large enough to justify it, and he tends to switch to hired transport. The recent Ministry of Transport Plating Regulations, which are designed to ensure that all lorry operators adhere strictly to the licensed tonnage for each vehicle, are adding to the merchants' problems and in some cases have been the last straw, causing merchants to cease running their own vehicles.

3.14. As one would expect, the importance of hiring is related to the type of business being conducted. Table 3.5 shows that over two-thirds of the merchants with over 75% of their turnover in straight grain hired for all their transport needs and probably the rest of the firms of this type hired for most of their needs. There would obviously be little for lorries owned by such firms to do during the summer months. On the other hand, none of the specialist country compounders or general merchants, and very few of the "mainly" country compounders, hired for all their needs, since their feed business (and fertilisers, sprays and other goods in the case of general merchants) would provide employment for their lorries the whole year round.

DISADVANTAGES OF HIRING

3.15. Of the 246 comments analysed, twenty-six (11%) were unfavourable to hiring. Most of these unfavourable comments stemmed from the merchant's fear that the use of a haulier would upset the harmonious relationship between himself and the farmer or end-user. Many of those who used hired transport made a point of emphasising that they often use it only for long hauls from their premises to a mill, or for deliveries to farms, but not for collecting grain on the farm; this is because the farmer may have very inadequate methods of loading and they do not want to

TABLE 3.5

HIRING OF ROAD TRANSPORT—PERCENTAGE DISTRIBUTION BY TYPE OF BUSINESS

	Type of business							
	Grain merchants		Feed merchants		Compounders		General agricultural merchants	Other merchants
	Specialist	Main	Specialist	Main	Specialist	Main		
	%	%	%	%	%	%	%	%
All hired	68	25	30	20	—	6	5	17
Some hired	32	69	60	60	79	88	85	67
Probably most	(12)	(31)	(40)	(16)	(37)	(25)	(32)	(20)
Probably little	(4)	(6)	(20)	(20)	(21)	(25)	(19)	(13)
No indication given	(16)	(32)	—	(24)	(21)	(38)	(34)	(34)
None hired	—	6	10	20	21	6	10	16
	100	100	100	100	100	100	100	100

offend him (or be in trouble with the haulier) by sending a hired lorry along. Moreover, if the haulage contractor is operating a large number of lorries it is likely that some of the drivers will be antagonistic to having to handle agricultural goods and unfamiliar with the farms and their layout; this often causes trouble for the merchant. One merchant who had switched entirely to hired transport was reverting to his own again because of the loss of trade as a result. Often a merchant will hire only from a small haulier, who is either an owner-driver or operates only a few vehicles and has reliable drivers. Another difficulty with hired transport is that it is usually outside the merchant's immediate control, and thus he is often unable to assure the buyer exactly when delivery will be made, or the seller when goods will be collected. Sometimes, when there is seasonal pressure for transport services, he may even have difficulty in getting hired transport at all.

3.16. Attempts have been made to overcome these difficulties of using hired transport in several ways. Some firms have set up a separate transport company under the umbrella of the parent merchant business so that it enjoys most of the advantages of being a separate haulier whilst it is still within the full control of the merchant and must put the needs of that business first. In others there is an arrangement with a haulier that he will give priority to the merchant firm above other clients, and often some of his lorries are attached to the merchant business either more or less permanently or for long periods of the year. Often a merchant will have a formal contract with a local haulier in which he undertakes to provide a minimum value of haulage business during the year in return for the exclusive use of the vehicles involved. These are all devices to get the best of both worlds, the increased flexibility and cheaper transport resulting from the hiving off of road transport from the merchant's primary responsibility, and the maintenance of an effective service to farmers and end-users.

FLATS OR BULKERS

3.17. If a merchant decides to run his own transport fleet he must decide what type—and size—of vehicles he needs. The larger he is the easier this decision becomes, because he has considerable flexibility; but for the smaller and medium-sized merchants it is often a very difficult decision to take. Nor is it solved once and for all. Lorries have only a limited life and replacements are necessary at regular intervals. Whilst this gives some flexibility, it also means that a transport-owning merchant is never free of the problem of having to decide what type and size of transport fleet is most economic. Flats, or platform lorries, are the usual choice for most of the smaller merchants and their sizes range from 3 tons or so to 20 tons carrying capacity. The large merchants tend to have more bulk grain and bulk feed lorries. Of the estimated total of 11,600 lorries owned by all corn and agricultural merchants in the United Kingdom, 8700 (75%) are flats, 2520 are bulk lorries (22%) and 385 (3%) are vans. The percentage distribution by size of lorry and turnover of the business for the 219 merchants in the survey who owned lorries is given in Table 3.6. This shows how the proportion of small flats (under 5 tons) declines as size of business increases, whilst the proportion of the large flats (11 tons and over) increases. Similarly with bulkers; the few large businesses with over £1,250,000 p.a. turnover owned nineteen out of the twenty-two bulkers of 16 tons and over.

3.18. As one would expect, the bulkers are concentrated in the grain-growing areas, but the increasing use of bulk lorries for animal feeds is reflected in the considerable number of bulk lorries in the South-west, Northern Ireland and Scotland (Table 3.7).

3.19. As with the decision whether to own or hire, there are a number of compromises available to the merchant which make the choice between a flat and a bulker, and between one large lorry or two or three smaller ones, less final than would otherwise be the case. For example, seventy-four of the flats owned by the merchants in the sample (7%) were convertible to bulk (e.g. by bolting on to the platform a large bulk grain tank), whilst 14 (5%) of the bulk lorries were convertible for sacks (e.g. a door is let into the side to facilitate the handling of sacks). When a bulk lorry has been fitted with a blower, however (as 24 (8%), had been among the survey firms), it became an expensive piece of equipment, costing in the region of £8,000, and it would not be economic to use it for other purposes. The problem of whether to buy one large lorry or two or three smaller ones has been met in some cases by using articulated vehicles. Thus several trailers will be bought for each "tractor" unit. There was a total of forty-six such vehicles (4%) among the sample firms. Articulated lorries were particularly common in Northern Ireland. They have considerable advantages in flexibility, since a trailer can be unhitched for loading and an already loaded trailer taken away, so keeping the tractor and its driver busy. Several merchants thought that articulated lorries would become more widespread in the future, but others believed they had a tendency to jack-knife. No firm in the sample with a turnover less than £150,000 had a blower lorry, and only two had articulated lorries; however, a higher proportion of the flats owned by these firms (12%) were convertible to bulk, than was the case for the firms with a turnover exceeding £150,000 (6%). A convertible lorry is often a necessary compromise for a small merchant.

TABLE 3.6

DISTRIBUTION OF FLATS AND BULKERS BY SIZE OF VEHICLES AND TURNOVER OF THE BUSINESS

Range of turnover	No. of firms owning transport in sample	Flats					Bulkers			
		Tons carrying capacity					Tons carrying capacity			
		0–5	6–10	11–15	16 and over	Total	6–10	11–15	16 and over	Total
£'000		%	%	%	%	%	%	%	%	%
Under 75	30	23 (50)	22 (48)	1 (2)	— (—)	46 (100)	2 (66)	1 (34)	— (—)	3 (100)
75–149	35	17 (20)	64 (75)	4 (5)	— (—)	85 (100)	3 (75)	1 (25)	— (—)	4 (100)
150–499	82	37 (13)	231 (79)	17 (6)	7 (2)	292 (100)	33 (77)	9 (21)	1 (2)	43 (100)
500–1249	41	21 (9)	173 (73)	29 (12)	13 (6)	236 (100)	44 (69)	18 (28)	2 (3)	64 (100)
1250 and over	25	6 (2)	295 (78)	62 (16)	15 (4)	378 (100)	84 (53)	56 (35)	19 (12)	159 (100)
Not known	6	1 (2)	49 (96)	1 (2)	— (—)	51 (100)	42 (100)	— (—)	— (—)	42 (100)
Total	219	105 (10)	834 (77)	114 (10)	35 (3)	1088 (100)	208 (66)	85 (27)	22 (7)	315 (100)

Source: Sample of 219 merchant businesses owning lorries.

TABLE 3.7

REGIONAL DISTRIBUTION OF VEHICLES OWNED BY CORN AND AGRICULTURAL MERCHANTS IN THE U.K.

Region	Flats	%	Bulkers	%	Vans	%	All vehicles	%
South-east	1164	(64)	614	(34)	28	(2)	1806	(100)
South-west	1710	(72)	562	(24)	102	(4)	2374	(100)
East	1195	(70)	469	(27)	57	(3)	1721	(100)
East Midlands	688	(72)	247	(26)	23	(2)	958	(100)
West Midlands	1491	(92)	131	(8)	—	(—)	1622	(100)
Yorks. and Lancs.	754	(73)	263	(25)	21	(2)	1038	(100)
Northern	128	(62)	64	(31)	14	(7)	206	(100)
Scotland	467	(79)	56	(10)	64	(11)	587	(100)
Wales	358	(95)	17	(5)	—	(—)	375	(100)
Northern Ireland	743	(81)	93	(10)	77	(9)	913	(100)
U.K.	8698	(75)	2516	(22)	386	(3)	11,600	(100)

Source: Raised results from the Survey of Merchant Businesses.

RETURN LOADS AND THE ECONOMICS OF TRANSPORT

3.20. The possibility of getting a return load is of vital importance to the merchant in operating his transport at a profit. All merchants who run their own lorries try to get a return load for journeys exceeding a few miles if they possibly can, and the longer the journey the more important this factor becomes. Up to 30 miles it is sometimes questionable whether the difficulty of trying to find a return load, and the complications involved, are worthwhile. But for the journeys of 30–100 miles all merchants would look for a return load, especially if the journey is a repetitive one over a period of time. Beyond 100 miles the return load usually becomes an absolute necessity if road transport is to be profitable. Merchants do not usually keep separate accounts of the journeys made with a double pay-load, but a common estimate was about 50% of all lorry journeys, with a range down to only 5% at one extreme to 75% at the other.

3.21. For short journeys of only a few miles the cost per ton is relatively small (assuming that unproductive time is kept to a minimum) but for journeys of over, say, 30 miles the transport cost becomes important. Take, for instance, a journey of 60 miles from Newbury to Avonmouth at 2s. 6d. per mile[1], the round trip would cost about £15, or about £1 per ton for grain, assuming there was no return load. However, if the same lorry can bring back a load of feedingstuffs the journey would take a little extra time, but the net cost for the round trip expressed per ton of grain would have been reduced to, say, 12s. There is nothing a merchant or haulier likes better than this kind of transport arrangement, and very few of the lorries bringing grain into the feed factories at the main ports leave without taking back a load of feeds. As a next best arrangement a merchant will collect a load of fertilisers from a factory as near as possible to the place where the grain has been delivered. Often the possibility of return loads is in any case small, e.g. in the livestock areas where there is no grain leaving the farm, and a considerable quantity of feed to be brought in. Frequently a contractor is better placed to obtain return loads and there is a substantial movement, for example, of coal outwards from South Wales, and grain and feed inwards. Merchants who are primarily seedsmen or who handle only straight grain are denied the opportunity of return loads and seldom run their own transport fleets. Movements of grain to export ports seldom have return loads unless the small ports now in use happen to be near compound and fertiliser factories. The more remote areas, such as the north of Scotland and the south-west Peninsula, are particularly badly off for return loads, and one merchant in Caithness commented that there were

[1] Based on a paper by W. Lindsay on "Transport Costs" delivered to an A.C.C.A. Conference in October 1958, updated to October 1967; and on the *Commercial Motor* Supplement, "Farm Transport", 1967. See Table 3.8.

three loads going out for every one coming in. One of the problems of buying a special-purpose vehicle like a bulk feed blower is that it may not be suitable for the carriage of other commodities on return loads.

3.22. Apart from the possibility of return loads, a merchant is anxious to keep his vehicles busy for as long as possible. A lorry standing idle, especially if the driver is idle too, is costing the merchant a lot of money. Putting a sack hoist on a lorry not only saves the driver a lot of back-ache—it is good economics too. Better still, of course, is a blower lorry that can discharge 10 tons of grain, the equivalent of 400 56-lb sacks, in 45 minutes.

TABLE 3.8

COSTS OF LORRY OPERATION

	Small lorry (6 tons)			Large lorry (22 tons)		
	200 miles	400 miles	600 miles	500 miles	600 miles	700 miles
	£ s. d.	£ s. d.	£ s. d.	£ s. d.	£ s. d.	£ s. d.
Standing costs per week	21 16 6	21 16 6	21 16 6	51 17 6	51 17 6	51 17 6
Running costs per week	10 9 6	17 2 6	25 5 6	27 7 11	32 17 6	38 7 1
Total costs per week	31 6 0	38 19 0	47 2 0	79 5 5	84 15 0	90 4 7
Costs per ton per mile[a] (shillings)	0·52	0·32	0·26	0·14	0·13	0·12

[a] Assuming loading to capacity.

Source: see footnote, p. 231.

This desire to keep lorries busy is one reason why corn merchants have often had a subsidiary coal business—the peak activity for coal was in the winter months, whilst the peak activity for grain was in the late summer months; the two were therefore complementary. Merchants are particularly glad to have customers who order and take delivery of their fertilisers during the summer months, because they usually have idle lorries available at that time. More usually, however, the rush for fertilisers coincides with a very busy marketing period for grain, i.e. at the back-end of the year. If more efforts could be made to encourage farmers to take delivery in the summer it would ease the grain transport problem considerably.

3.23. Another important economic consideration which influences the merchant in the organisation of his transport fleet is the rising cost of labour and the need to spread this cost over a larger pay-load. The significance of this may be illustrated by Table 3.8.

3.24. By having a larger lorry the cost of the driver and the other running costs are spread over more tons and this more than outweighs the heavier standing charges. Many merchants are too small to afford these large lorries (and in any case many of the smaller farms could not accommodate them) but the larger merchants and the hauliers are increasingly switching to larger lorries and the smaller flats are fast becoming uneconomic to run. The Ministry of Transport's Plating Regulations are lending impetus to this trend. Under these regulations a vehicle is only allowed to carry the tonnage for which it is licensed regardless of the fact that there may be space for a larger tonnage (vehicles are designed to carry a certain tonnage but some classes of commodity are more bulky in relation to their weight than others and this is allowed for in the design. Thus with a relatively dense commodity like bulk grain the plating capacity may represent only 70% of the actual physical capacity, or even less). The plating capacity is related to the number of axles and wheels, so that the larger vehicles—and especially the special-purpose ones—are less likely to be loaded above their plated limit than are the smaller ones.

3.25. The bigger and more expensive the lorry the more vital it is to keep it running, and the more expensive are delays at ports, in crowded cities, at animal feed factories, wharves or farms. The Road Haulage Association recently launched a "Turn that Lorry Round" campaign in recognition of this fact. Contractors are particularly sensitive to these delays because it is usual for hired transport to be paid on a "per ton" basis and the haulier has to meet the costs of any delays. Often they will charge demurrage (say, 15s. per hour) where their vehicles are unreasonably delayed (e.g. because farms are inadequately equipped to load bulk grain). Increasingly merchants and hauliers are trying to avoid making journeys into the great cities because of the delays involved. Often drivers attempt to get round this problem by driving into the cities at night and a few of the big feed factories now remain open all night, but the new Transport Act may make this more difficult if it becomes law. Another way of avoiding the cities is to collect feed and fertilisers from railside depots established at strategic points. This system can be expected to increase in the years ahead, but only slowly unless the manufacturers can find a way of collecting grain from railway depots also, thus providing opportunities for double pay-loads.

3.26. The increasing tendency for animal feeds to be moved in bulk presents the merchant with one of his most difficult economic problems. A blower lorry must work full time for most of the year to cover the high capital cost, but during the interim period when farmers are slowly making the change, a merchant cannot expect to be able to run a blower lorry full-time until sufficient farmer customers have equipped themselves for bulk. Yet if he delays too long those of his customers who want feed in bulk may take their business elsewhere. He may hire for a while, but eventually he will often decide to buy his own vehicle and hope that the trade will expand to justify it. The big national feed firms have stationed blower lorries at suitable localities throughout the country, so that the smaller merchants (and their own subsidiaries) can hire them as necessary. The introduction of the blower lorry is yet another factor adding to the transport problems of the smaller merchant and the small country compounder.

3.27. A merchant's lorries are often used very inefficiently, not through any fault of his own but because his farmer customers are frequently not prepared to order their feeds, fertilisers, etc., in advance or in reasonable quantities. Thus a merchant's lorry will typically have as many as seven different "drops" of small lots of feed or fertilizers on a round trip of, say, 80 miles. True, the merchants charge special rates for small lots, and some of them find these very profitable, but the persistence of this kind of trading represents a cost to the nation which in many instances could well be avoided. Some merchants refuse to deliver in lots of less than 10 cwt. Often the smaller farmers suddenly find they are running out of feed and the merchant has to send a lorry out especially—often with an uneconomic small load. It may be that the rising costs of keeping a small lorry for this kind of trading may force the farmers concerned to change their ways.

3.28. One side effect of the trend towards large bulk lorries is a change in the pattern of transport that is slowly developing. Whereas hitherto the pattern has been for a merchant to collect feed from the factory and deliver it direct to the customer, there is a tendency now for the merchant to bring the grain from the factory to his premises in large bulk carriers and to tranship it in smaller vehicles to the farms. This system is not as yet widespread, because of the high cost of transhipment, but it is likely to gain momentum as the economic scales turn increasingly in favour of the large bulk lorry.

3.29. Although a few merchants were able to quote detailed costings for their own transport (and often admitted that the smaller lorries did not pay their way) there were as many who confessed that they had no idea what the true costs of their transport were. Often transport was regarded as a necessary service that the merchant had no choice but to give, and there was no point in working out the cost. Many merchants themselves commented that if some of their competitors had any idea of the true cost of their transport they would realise that they could not make a profit out of some of the prices they were quoting. We are satisfied that this is often the case, and that more attention to the costing of transport, as indeed of the other departments of a corn merchant's business, would be desirable.

COLLECTION FACILITIES ON FARMS

3.30. The merchant and haulier cannot always be blamed for the inefficiencies of transport that exist—these are often caused by inadequate collection (and delivery) facilities on farms, particularly the smaller farms.

3.31. The general question was put to all merchants interviewed: "Are collection facilities on the farm adequate?" Of the 212 who replied (merchants who used hired transport did not usually comment), 45% thought they were adequate and 55% that they were inadequate. These percentages however, do not signify a great deal because the usual response was "Some farmers are good, some are bad—on the whole they are all getting better". What is of interest are the comments on the particular inadequacies of the present collection facilities, and these have been summarised in Table 3.9. Several merchants made more than one comment and the 172 comments represent the views of 123 merchants. The three major criticisms are small augers and slowness of loading generally; lack of assistance on the part of the farmer in the loading or unloading of vehicles; and inaccessible and ill-equipped buildings. These comments were made by all sizes of merchants, and in all regions. In the western parts of the country, however, there were more complaints about the lack of elevators than about the inadequacy of augers, since most of the grain in these parts is still handled in sacks; some of it indeed goes off straight from the combine or is delivered by the farmer in his own vehicles, and the problem of collection facilities does not arise in these cases.

3.32. Several merchants commented on the fact that the farmer's only concern is to get his grain safely into store; he spares little thought to the problem of getting it out again! Perhaps the merchant is himself partly to blame for this since it is very rare that a merchant will explicitly penalise a farmer for having inadequate facilities, although many of them undoubtedly quote a lower price in such circumstances. There are signs that some merchants are seriously considering imposing more direct penalties for unreasonable delays at farms; there were eight merchants who were in this position among those interviewed.[1] The merchant is understandably reluctant to take this step because he is afraid that another merchant will take the trade if he does. However, with the increasing size of lorries coming into use, rising costs of labour and falling profit margins, there is little doubt that more merchants will be forced to charge for delays in the future. In our view it is desirable that this should be done openly rather than disguised in the price, so that the farmer may be encouraged to improve his facilities.

[1] One large merchant is reported to have made an agreement with the local N.F.U. that a penalty will be imposed if the rate of loading is less than 12 tons an hour (*Farmer and Stockbreeder*, 7th April, 1967).

TABLE 3.9

COMMENTS ON THE INADEQUACY OF COLLECTION FACILITIES ON FARMS

A. *Inadequacies of Equipment*

		No. of comments recorded	
			%
Elevator—too few		6	(3)
Augers —too few	5		
—too small	31		
—need standardising	1		
—lack of electricity to drive them	2		
	—	39	(22)
Bulk hoppers—too few		13	(8)
General comments:			
General lack of mechanical equipment		12	(7)
Lack of facilities for handling bulk grain		19	(11)
Loading too slow		25	(14)
Total		114	(65)

B. *Inadequacies of Buildings*

Poor buildings			
Badly adapted	4		
Low outlets for grain	4		
Inaccessible	16		
No cover from the rain	2		
	—	26	(15)
Poor layouts and approaches			
No space for lorries to turn	2		
No hard standings or adequate access roads	6		
	—	8	(5)
Total		32	(20)

C. *Poor Liaison between Farmers and Merchant or Contractor*

Lack of farm labour to help load or unload the lorry	19		
The farmer is not there when vehicle arrives	7		
	—	26	(15)
Total		26	(15)
Total (all comments)		172	(100)

Rail Transport

3.33. The decline in the use of railway transport by corn merchants is a feature of the last 15 to 20 years. Only just over a third of merchants visited used rail transport, and a third of these said they used it very rarely, perhaps for only one or two consignments a year. In Northern Ireland not one of the thirty-one merchants visited used rail transport.

3.34. Nearly a third of the merchants said that they had stopped using the railway fairly recently. The reasons they gave, in order of frequency of mention, were:

	% of total reasons mentioned
Rates too high	32
Stations closed or lines removed	31
Double handling costs	19
Unreliable service	11
Too slow	5
Inadequate bulk handling facilities	2
	100

3.35. The reasons all stem from the same fundamental cause—the lack of flexibility of rail transport. A lorry can pick up a load where it happens to be and deliver it direct to where it is wanted. But with the railway the load has first to be collected at the farm or wharf and taken to the railside; then, at the other end of the rail journey, it has to be hauled to the final destination. This double-handling generally more than outweighs the lower running costs of rail transport. The closing of railway lines and stations was not, of course, the cause of the decline in rail traffic but the effect of it; many of those who referred to the closing of lines or stations said that, in fact, it made little difference since they had already stopped using the railway. It is significant that those merchants who most regretted the closing of the railways were those with stores or silos adjacent to the tracks, i.e. they had minimised the inconvenience of double handling.

3.36. The relative advantage of road transport diminishes as the distance travelled increases, particularly if the commodity being carried is not particularly valuable and can be carried in bulk. It is still cheaper in many of the more remote places to buy a trainload of fertiliser delivered to a convenient rail-head than to haul it by lorry. Where long hauls on a regular basis from the same origin to the same destination are involved, then rail transport can still be cheaper than road. The prime example of this is the movement of distilling barley and malt from the marshalling yards of eastern England (e.g. Whitemere, near March) to Dufftown in Banffshire to serve the many distilleries of nearby Speyside. A weekly trainload of up to thirty wagons of 20 tons each travels from Doncaster to Dufftown for this purpose. The regular movement of animal feeds from Avonmouth down to Cornwall is another example. However, in spite of the advantages of rail for long hauls, there are some merchants who put such store by the reliability and convenience of door-to-door transport that they will send barley from Norfolk to Scotland by road. They may often be able to offset the cost to some extent by arranging to pick up a load of feedingstuffs on the return journey, e.g. from the feed mills at Selby or Gainsborough.

3.37. What is needed now is a simple and cheap means of getting the grain from the farm to the railway. British Rail have recently introduced such a system. It consists of an independent grain container, capable of holding 8 tons and with its own in-built discharging system, which can be carried by any vehicle of the required size.

3.38. There can be no simple answer to the question of where the break-even point lies between road and rail transport; it depends upon so many variables, particularly, so far as road transport is concerned, on the availability of a return pay-load. But one large grain merchant who makes extensive use of rail transport and has his own railway siding said that in his case it would usually be cheaper to use rail beyond 100 miles. However, our impression is that most merchants would probably choose road transport for journeys of up to 150 miles provided that a pay-load can be obtained for at least part of the return journey.

3.39. A further consideration in this connection is that British Rail usually require a minimum lot of 100 tons (equivalent to five railway wagons), although in one merchant's experience it was 150 to 200 tons. Not many merchants dispose of such quantities over a long distance. It is therefore not surprising that the lowest proportion of firms using rail traffic was among those in the £75,000 per annum turnover group (20%) and the highest in the £1,250,000 and over per annum group (53%), the intermediate group being 37%. Grain was never mentioned in connection with rail traffic by the small merchants, except occasionally for Scottish oats, and it was only the firms with turnovers of £500,000 per annum or more that made any significant use of rail for moving grain.

3.40. The economic advantages of road transport for all but the longest hauls are so overwhelming that we cannot see any likelihood of a major return to rail transport. However, apart from the effects of the government's decision to use legislative powers to force long distance traffic back to the railway (discussed later), we can see the following factors that might favour a limited return to rail transport for middle- to long-distance bulk traffic:

   (a) The concentration of agricultural merchants into fewer and larger businesses, many of them controlled by national companies. This brings more opportunities for the bulking of grain and other commodities, and the negotiation of attractive rates.
   (b) The trend towards bulk grain, bulk feeds, etc.; especially if British Rail improve their facilities for handling bulk grain and other commodities at the rail siding (e.g. tipping facilities), and introduce new systems to offset the disadvantages of double-handling.
   (c) The H.-G.C.A. Forward Contract Scheme, which has spread the period of marketing and is conducive to rail transport (i.e. less

harvest-time rush to dispose of grain quickly—more time to consider and arrange for the use of rail).

(d) The rapid rise in costs of road transport. Already a substantial number of merchants are disposing of their lorries and switching to haulage contractors; this, of course, leaves them more free to consider alternative systems of transport.

(e) The increasing delays due to traffic congestion on many roads, especially in and around the big ports.

(f) The new grain terminals at Tilbury (which will handle bulk grain only) and Liverpool may act as a focus for an extensive use of rail transport to take imported grain to inland mills and to bring grain for export to the large silos there.

(g) British Rail are trying to offer a more reliable service and guaranteed delivery times, although they do not appear to be anxious to encourage more bulk grain back on to the railways.

Water Transport

3.41. Although canals are still used to carry large tonnages of imported grain on the last stage of its journey (see Part IV, Table 3.3.) water transport is now seldom used by corn and agricultural merchants for the transport of home-grown grain. Only twenty-one of the 260 firms visited (8%) were using water transport, and these only rarely as a rule. Three had to use water transport in any case to communicate with nearby islands (e.g. Isle of Wight), and ten were using water transport purely for the export of grain, or occasionally for the import of basic slag and fertilisers from the Continent, i.e. not for coastwise trade. Of the remaining eight firms, one was a large merchant-compounder in the Midlands who was still making intensive use of the canal, bringing several thousand tons of grain up from London each year in 40-ton barges; one was using the Mersey and another the Trent. The other five were in Scotland (4) and Lincolnshire (1), and the cargoes traded were bulk commodities such as sugar-beet pulp, fertilisers, distilling grain and oats.

3.42. The decline in both canal traffic and coastwise trade in the last 50 years has been due to the greater speed and flexibility of the lorry. Once a canal or a harbour ceases to carry much traffic it begins to silt up, warehouses and wharves fall into disrepair, railway and road bridges are built impeding navigation, small sailing barges capable of navigating creeks and inlets rot away and are not replaced, and firms vacate their wharf-side sites and develop in more spacious surroundings elsewhere. Thus of the 239 firms not using water transport twenty-four (9%) said that they had used water transport in the past, and several still had wharves on rivers or creeks that were not now navigable.

3.43. Water transport is still generally uneconomic in competition with motor transport, but there are signs of a renewal of interest in this form of transport in view of the delays being experienced on crowded roads and the rapidly rising costs of motor transport. The tendency for new compound feed mills to be established in inland areas may lead to more bulk commodities being moved by canal, provided facilities are available. The new Transport Act is likely to add to the difficulties of road hauliers and merchants. Moreover, the rapid growth of the grain export trade in recent years, mostly conducted in small coasters, had led to the improvement of many coastal waterways and the refurbishing of wharves. Although as yet this does not seem to have led to any significant increase in coastal trade between British ports there is a possibility that it will do so in due course.

The Transport Act, 1968

3.44. The basic purpose of the new Transport Act is to encourage integrated use of railway and road transport and to ensure that goods are carried by rail "whenever this is efficient and economic". Certain features of the Act will undoubtedly ease the difficulties of agricultural merchants and hauliers. For example, small vans and lorries of under 30 cwt unladen weight will be freed altogether from the necessity to have a carrier's licence of any kind, whilst operators of all other vehicles will obtain a carrier's licence provided they can assure a minimum quality of service, and undertake to run their vehicles safely. The present distinction between haulage on own account ("C" licences) and haulage for hire or reward ("A" or "B" licences) will disappear and few will mourn its passing. These restrictions frequently prevented vehicles from being used to their optimum potential, and the new situation should lead to substantial improvements in efficiency all round. Even the new regulations restricting the length of the drivers' working day, although they are certain to add to operating costs, may stimulate more efficient use of the driver's time.

3.45. The provisions in the Act aimed at encouraging more bulk traffic on long hauls to use the railways are unlikely to have much impact on corn and agricultural merchants. British Rail do not have an adequate fleet of bulk grain wagons and are therefore unlikely to object to merchants transporting grain over long distances. In practice most merchants will probably receive a general authorisation from the licensing authority, valid for 5 years, and covering not individual journeys but a "transport service" as a whole. British Rail is mainly concerned with freightliner traffic, i.e. large-scale movement between the big cities, and agricultural merchants are unlikely to handle much traffic of this kind.

3.46. The original proposal in the Act to increase road haulage taxes on vehicles with an unladen weight exceeding 3 tons by from £50 to £200 per annum depending on the type of vehicle, was superseded by the Chancellor of the Exchequer's announcement of increased vehicle licence rates in the 1968 Budget. The effect, however, is the same, namely a substantial increase in the cost of running heavy road vehicles.

3.47. Whether all these measures will succeed in transferring any substantial volume of agricultural traffic from road to rail transport is doubtful. Certainly the general trend in recent years has been in the opposite direction. Table 3.10 shows that the trend over the past few years for inland freight of all kinds has been steadily against rail transport and coastal shipping and in favour of road.

TABLE 3.10

INLAND FREIGHT TRANSPORT TRENDS 1962–6

	Ton-miles ('000 mill.)					% Share				
	1962	1963	1964	1965	1966	1962	1963	1964	1965	1966
Road	33·6	35·0	39·0	41·0	41·5	55	57	59	60	61
Rail	16·1	15·4	16·1	15·4	14·8	26	25	24	23	22
Coastal shipping	10·8	10·6	10·7	10·9	10·6	18	17	16	16	16
Inland waterways	0·2	0·1	0·1	0·1	0·1	—	—	—	—	—
Total (incl. pipeline)	61·1	61·6	66·6	68·2	67·9	100	100	100	100	100

Source: The Transport of Freight, Cmnd. 3470.

3.48. Farms are geographically scattered, and tonnages of grain and other agricultural goods to be moved from individual farms are usually relatively small, compared, for instance, with a train load representing 500/600 tons. It is unlikely that there would be a substantial change to rail unless grain could be bulked at some intermediate stage, as is done in fact with barley and malt destined for Scotland. We have suggested elsewhere in this report that a move in this direction is worth trying but it would be over-optimistic to expect it to have much impact on the transport problem immediately. The main effect of the Act will therefore be to force vehicle operators to look for further economies in their transport operations to offset the increased operating costs.

A Note on the Methods used in Calculating Surplus and Deficit Areas

A. CONCENTRATED FEEDS AVAILABLE

3.49. The acreage of each class of cereal in each county in June 1966 was multiplied by the average yield in that county (using Ministry of Agriculture statistics of yield) to obtain total production for each class of cereal.

From each county total the estimated percentages of wheat used for milling, and barley for malting or distilling, and of grain used for seed or industrial use, were deducted. U.K. percentages had to be used for this purpose, and it was assumed that the remainder was available for animal feed. The results for all cereals are given in Appendix H, Table 13, cols. 6 and 7. The U.K. percentages were derived as in Table 3.11.

TABLE 3.11

HUMAN, INDUSTRIAL AND SEED USES OF GRAIN FROM THE U.K. CROP 1966/7[a]

(mill. tons)

	Human	Seed	Industrial	Total non-feed uses	Production	% Non-feed uses	Balance is for animal feed, i.e. %
Wheat	1·624	0·181	0·005	1·810	3·383	53·5	46·5
Barley	1·328	0·393	—	1·721	8·578	20·1	79·9
Oats	0·099	0·089	—	0·188	1·110	16·9	83·1
Rye	0·007	0·001	—	0·008	0·011	72·7	27·3

[a] *Sources:* Wheat: Commonwealth Secretariat, *Grain Bulletin,* Vol. 13, No. 7, Dec. 1967. Barley, Oats, Rye: *ibid.,* Vol. 14, No. 1, Jan. 1968.

B. CONCENTRATE REQUIREMENTS BY LIVESTOCK

3.50. To estimate total requirements of concentrates the estimated average requirements per head of each class of livestock were multiplied by the number of livestock in each category in each county at June 1966. The source of the per-head feeding rates was *Concentrated Feedingstuffs for Livestock in the United Kingdom, 1960/1 to 1965/6,* by Paul W. H. Weightman, Department of Agricultural Economics, Cornell University, Ithaca, New York. Mr. Weightman collected information from a wide variety of sources, including sample surveys, and the farm accounts collected by various university departments and farm institutes in the U.K. His figures include, in addition to cereals, an unknown proportion of cereal by-products like wheatfeed and brewers' grains, and high-protein feeds such as oil cake, oil seed residues and fish meals. Table 3.12 shows the main figures used in this analysis; they are taken from Table 3 of the above report. The detailed results are given in Appendix H, Table 13. The total estimated requirements of just over 16 million tons compares very closely with the total consumption of concentrated feedingstuffs on farms in the U.K. in 1965/6 (16·5 million tons) and 1966/7 (15·6 million tons), as given in the Annual Review White Paper, Appendix I, Table 9 (Cmnd. 3558).

TABLE 3.12

FEEDING RATES PER CLASS OF LIVESTOCK IN THE U.K.

Class of livestock	1965/6
	cwt per head
Cows and heifers in milk and in calf (milk types)	24·7
Cows and heifers in milk and in calf (beef types)	8·7
Heifers in calf (first calf, milk and beef types)	5·3
Other cattle 1 year and over	7·8
Cattle under 1 year	6·3
Sows and gilts for breeding (including litters)	31·0
Baconers	5·1
Pigs used in part for bacon (mainly heavy pigs)	8·2
Other pigs (mainly porkers)	3·7
Upland ewes, shearlings and rams	0·5
Lowland ewes, shearlings and rams	1·1
Laying hens	0·91
Growing pullets	0·23
Broilers and table birds	0·089

Source: Concentrated Feedingstuffs for Livestock in the U.K. 1960/1 to 1965/6, Paul W. H. Weightman, Cornell, N.Y.

Note. Adjustments were made to the "per head" figures for pigs and poultry to put them on to a "per annum" basis.

C. INDEX OF SURPLUS OR DEFICIT

3.51. This was obtained by expressing the cereals available for livestock feed as a percentage of concentrated feeds required. The results are shown by individual counties in Appendix H, Table 13. It is recognised that these indices to some extent overstate the deficit position, as wheat used for milling and barley used for malting and distilling yield by-products which become available for livestock feed in the form of wheatfeed, brewers' grains and malt culms. But the quantities involved are relatively small, and moreover it would be wrong to assume that they return to the county of origin. For these reasons it is suggested that the proportion after deducting grain for human use is the more appropriate for the purpose in hand.

CHAPTER 4

Merchants' Dealings in Grain

4.1. This chapter describes the ways in which merchants buy and sell grain (forward as well as spot dealings); how they arrive at prices, and in particular what sources of price information they use; and what their attitudes are on certain topics connected with the purchase and sale of grain.

4.2. Some closely related topics are covered elsewhere in the Report. In particular, the quantitative importance of merchants' grain dealings in the U.K. cereals market is discussed in Chapter 2, the place of grain dealing in individual merchant businesses is described in the same chapter, and Chapter 3 covers geographical aspects and the transport of grain by merchants.

The Mechanisms of Buying Grain from Farmers

4.3. It is not a serious exaggeration to say that in grain merchanting there are as many systems of buying as there are firms. Each merchant's buying practices are developed to suit the requirements of his business and its customers; these requirements in turn will be affected by the size of the business, its pattern of activities and its situation. Buying practices in any one business may vary from one time of year to another. The approach taken in the survey was to ask specific questions about a few of the most important aspects of grain-buying methods. They were:

(a) Who in the firm is directly concerned in grain buying?
(b) Where is the initial contact made?
(c) Is the purchase made immediately, after sampling, or subject to instructions from head office?
(d) What degree of initiative is allowed in fixing price?

This scheme is also followed in the description below.

4.4. Each time a parcel of grain is bought, the transaction can be said to follow a certain "buying path". Most firms have more than one possible buying path. For example (to quote a typical situation among medium-sized businesses), the main buying path might be that a representative, while calling round farms on his normal selling trips, bought feed grain on the spot at a price dictated to him by the firm's grain buyer. For the occasional lot of malting barley, instead of buying on the spot, he might take a sample back to head office, where the buyer would evaluate it and quote a price to the farmer. A subsidiary buying path might be for the buyer to travel round farms, solely with the purpose of buying grain, and buy on the spot with full discretion on price.

4.5. Clearly, it is not possible to summarise all the different buying paths used; the possible variations run into hundreds. Instead, in the tables which accompany this section, the four aspects of buying mentioned above have been treated separately, in terms of the numbers of firms which dealt with that aspect in a particular way. If, at a particular stage of grain buying, a firm followed more than one alternative practice, all the practices reported were counted. For instance—using the example given in the previous paragraph—in the section on "initiative in fixing price" there would be a count of "one" for "representative buys at fixed price"; another count of "one" for "principal, etc., buys with full discretion"; and another count of "one" for "representative refers back sample, principal quotes to farmer"—this last referring to the malting barley bought by the representative. Thus in nearly all the tables, the numbers of businesses shown as reporting each feature will total more than the numbers of businesses sampled. The degree of this double-counting is, in fact, a rough measure of the variability of practice within individual firms at that stage of the buying process.

4.6. *Who are concerned in buying?* This first question can be summarised more straightforwardly than most of the others in the section. A firm may have no representatives buying grain, all buying being done by the principal or by a director, buyer, manager or other executive (for brevity's sake, all these possibilities will from now on in this section be expressed in the text by the single word "principal", except where otherwise stated—this applies even where more than one executive buys grain in the same firm). There may be one representative buying grain, or several.[1] Where one or more

---

[1] A slight difficulty of definition arises in cases where representatives do not buy immediately but refer back to head office, sometimes taking back a sample, so that the principal can approve the purchase and quote a price. In this case, who is "concerned in buying"? The solution adopted here has been to include anyone who has direct contact with the farmer at any stage of that particular buying path. Thus if the principal telephones the quotation back to the farmer and closes the deal, both "representative" and "principal" are counted. But if the principal sends the representative back to the farm to quote the price and buy the grain, only "representative" is counted. This may seem a rather artificial distinction; but if it were not made, other difficulties would arise— for example, in many firms the representatives do all the day-to-day buying, working to

representatives buy, the principal may or may not also buy. Table 4.1 shows how often these possibilities were recorded for the various sizes and types of merchant firms.

4.7. The pattern of results by size-group is very much as might have been expected. The larger the firm, the less common it was for the principal alone to do all the grain buying, and the more frequently were several representaives employed. The use of one representative only was uncommon among large firms, because of their large sales forces, and again among the smallest firms because in most cases all buying was done by the principal. The low figures for "principal buys as well as representatives" in the two smallest size-groups was, of course, due to the high proportion of firms in these groups which had no representatives buying; the count was higher for medium-sized firms, and dropped again among the larger businesses, presumably because of greater delegation of duties by principals. No very striking differences in practice were noticeable as between different types of business, except that among the firms depending mainly on dealings in straight grain or purchased feed the majority (about two-thirds in each case) reported that all buying was done by the principal, whereas among general merchants and those dealing mainly in compounds there was a similar bias towards the use of several representatives in buying. This is a predictable result in the case of businesses dealing mainly in grain, but it is rather unexpected at first sight to find so many purchased-feed businesses who do not use representatives to buy. The clue probably lies in the size structure of the "specialist/mainly feed" group: it is biased towards small businesses to a markedly greater extent than are the "specialist/mainly compounding" or the "general" type-groups.

4.8. *Where is first contact made?* Here the question referred to the contact which initiated each individual deal, not to the first contact with a new customer (indeed, many merchants remarked that all or most of their grain was bought from customers of long standing who had sometimes had connections with the merchant firm for generations). A special case, included as a separate heading in the question, was the growing of grain on contract for the merchant. The breakdown of answers for the whole merchant sample is shown in Table 4.2. The proportion of businesses which *never* bought by visiting farms was greater among smaller firms than among large; of the firms with a turnover of less than £75,000 a year, one-third did not mention that they ever bought on farm visits, while in the largest group—firms with more than £1,250,000 of annual turnover—all reported

---

prices quoted to them by the principal. In such a situation, the principal is clearly not "concerned in buying" in the present sense (though he is, of course, very much concerned in *pricing*), and the only usable demarcation line is that of coming in direct contact with the farmer.

TABLE 4.1. RESPONSIBILITY FOR BUYING

	Turnover group (£'000)							Type group					
	Under 75	75–149	150–499	500–1249	1250 and over	Turnover not known	Total sample	Specialist/ mainly grain	Specialist/ mainly feed	Specialist/ mainly com-pounding	General	Other	Total sample
(i) Numbers of firms:													
1. No representatives buy—all buying done by principal (or buyer, director, manager)	30	28	33	6	3	—	100	24	19	6	38	13	100
2. One representative buys (see also 4 below)	2	5	13	1	1	—	22	5	3	2	8	4	22
3. Several representatives buy (see also 4 below)	—	4	42	33	22	6	107	10	7	16	67	7	107
4. Principal (or buyer, director, manager) buys as well as representative(s)(a)	2	2	42	16	10	3	75	6	7	10	49	3	75
5. No information on topic(a)	10	7	6	6	1	1	31	2	7	3	12	7	31
(ii) As percentages of firms answering in each group(b):													
1. No representatives buy—all buying done by principal (or buyer, director, manager)	94	76	38	15	12	—	44	62	66	25	34	54	44
2. One representative buys (see also 4 below)	6	14	15	2	4	—	10	13	10	8	7	17	10
3. Several representatives buy (see also 4 below)	—	11	48	82	85	100	47	26	24	67	59	29	47
4. Principal (or buyer, director, manager) buys as well as representative(s)	6	5	48	40	38	50	33	15	24	42	43	12	33

(a) This line includes firms which bought no grain from farmers, or only insignificant amounts (seven firms for the sample as a whole). (b) Percentages calculated separately for each group, after *excluding* returns with no information on topic.

that farm visits were used for at least some purchases of grain. This presumably reflects the greater use of representatives in buying among larger firms.

4.9. The use of markets or exchanges in buying from farmers was reported by relatively few firms (less than one-sixth of those replying to the question). The proportion was rather higher for the "specialist/mainly grain" group of firms—one in three of whom bought some grain from

TABLE 4.2

WHERE IS INITIAL CONTACT MADE IN GRAIN PURCHASE
TRANSACTIONS FROM FARMERS?

	Number of firms	Percentage of firms[a]
Principal, etc., or representative visits farm	200	(84)
Farmer comes to merchant's office	40	(17)
At market or exchange	32	(14)
Grain grown on contract	12	(5)
No information on topic	23	(—)[a]

[a] Percentages calculated after excluding "no-information" returns, i.e. on a total of 237 firms.

farmers at markets—than for any other type-group. The use of markets as a centre for transactions in grain between merchants and farmers has been declining for many years, and seems likely to go on doing so. No specific question on the reasons for this decline were included in the Merchant Survey, but comments made by individual merchants give some clue on the subject. As compared to the situation even immediately before the war, merchants now can visit farms much more easily and, of course, many more farms are on the telephone. With the rise of combine harvesting and on-the-farm drying and storage, many merchants now like to see the bulk of the grain on the farm and buy in large lots, rather than buying small parcels of grain on sample, as was the usual practice at local markets. A few merchants alleged that, in the farming and the merchanting industries alike, the tempo of business was quicker now than formerly, so that neither merchants nor farmers any longer had time for regular visits to markets.

4.10. Only twelve firms (5%) reported having grain grown for them on contract. In all except one of these cases, the only grain grown on contract was grain destined for sale as seed. For feed grain or malting barley, as far as direct purchase by the merchanting industry is concerned, contracting for the growing of grain is virtually nonexistent. (The position in the malting industry is different: see Part Vc, Chapter 1, p. 475.)

4.11. *Is the deal negotiated immediately, after sampling, or subject to instructions from head office?* The proportion of firms reporting that at least some purchases were negotiated immediately was very similar as between different size-groups and type-groups. About 15% of all firms surveyed *never* bought grain from farmers straightaway, but delayed the purchase pending examination of a sample or for some other reason; and the proportion of firms following this policy did not vary much with type or size. (See Table 4.3.)

4.12. Often, a firm would report that some of its grain was bought straightaway, but that on other occasions the purchase was delayed until a sample had been taken back to the merchant's office (usually by a representative, but sometimes by the buyer) and evaluated. Many firms commented that their usual policy was to buy immediately when the purchase was of feed grain, but to bring back a sample before purchase

TABLE 4.3

NEGOTIATION OF GRAIN PURCHASE DEALS WITH FARMERS

	Number of firms	Percentage of firms[a]
Purchase negotiated immediately	198	(84)
Sample taken back to merchant's office	74	(31)
Representative negotiates after instructions	18	(8)
Representative passes information back to principal	7	(3)
No information on topic	23	(—)[a]

[a] Percentages calculated after excluding "no-information" returns, i.e. on a total of 237 firms. The answer "purchase negotiated immediately" refers both to buying by principals and buying by representatives; where immediate negotiation both by principal and representative was reported by a given firm, a count of "one" only was entered for that firm.

when buying grain for malting or flour milling. As Table 4.3 shows, about one-third of the firms interviewed said that, at least on some occasions, a sample was brought back for examination before grain was bought. The percentage of firms reporting this practice increased with increasing average size, presumably because among the bigger firms more buying was done by representatives. Variations on this theme arise where the representative, instead of taking a sample back, refers back to the principal with a description of the grain and buys on the principal's instructions, or where he simply passes back the information that there is grain available and leaves the principal to complete the deal. As the table shows, these practices were reported by relatively few firms in the sample as a whole, and they were not significantly more important for any one type or size of firm.

4.13. *What degree of initiative is allowed in fixing price?* The answers to this final question in the section on buying practices are summarised in Table 4.4. The answer "buys with full discretion on price" applies to purchases by either principal or representative, and, just as in the case of the first answer in the previous table, if both principal and representative buy with full discretion in a firm, a count of "one" only has been entered for that firm. This answer was recorded for around three-quarters of the firms interviewed in the sample as a whole. There was a clear variation by size of firm: of the smallest firms, nearly all reported that principal or representative bought with full discretion on price, and hardly any other arrangements were noted in this group. It is clear from the data on "who buys?" (Table 4.1) that this reflects the small proportion of these firms who employed representatives to buy. Going up the size scale, the percentage of firms reporting this answer declines steadily; again, this reflects the relative importance of representatives in the buying system. The remaining three answers to the question refer to buying by representatives only, and correspond to successively lessening degrees of freedom allowed to representatives to fix the price of grain bought. Representatives may be allowed to negotiate around a stated price, within a set range, or, less often, to buy at whatever price they can get up to a stated maximum; the price limits in question are notified to the representatives by principals, either daily or at longer intervals. Comparing the various size-groups this system was most popular among firms of above-average turnover (though rather less frequently reported by the very biggest firms). The largest firms tended more often to restrict their representatives' pricing powers even further, and to dictate a rigid price at which the representative could either take the grain or leave it. The lowest degree of pricing initiative occurs when the representative cannot quote a price immediately to the farmer, but must either take a sample back to the principal, or send back a description of the grain, and leave the principal to quote a price. As has already been mentioned, this arrangement is frequently (though not exclusively) used for buying malting barley and wheat for flour milling, rather than when buying feed grain; it is a relatively important method among larger-than-average firms.

Variations in Grain Purchase Price between Customers

4.14. The sources of information merchants use in determining the "going price" for grain purchases are described in the last section of this Chapter (pp. 272-7). The present section discusses the extent to which

TABLE 4.4. DEGREE OF INITIATIVE IN FIXING PRICE

	Turnover group (£'000)							Type group					
	Under 75	75–149	150–499	500–1249	1250 and over	Turnover not known	Total sample	Specialist/ mainly grain	Specialist/ mainly feed	Specialist/ mainly com- pounding	General	Other	Total sample
(i) Numbers of businesses:													
1. Principal or representative buys with full discretion on price	35	34	66	20	11	4	170	32	25	15	78	20	170
2. Representative buys in a price range (or to a maximum price)	1	1	11	20	6	1	40	8	5	4	20	3	40
3. Representative buys at a fixed price	1	3	13	1	11	2	31	4	1	1	20	5	31
4. Representative refers sample or information back to principal, who quotes to farmer[a]	—	—	26	23	11	4	64	8	5	11	39	1	64
5. No information on topic[a]	6	6	4	5	4	1	26	1	4	4	11	6	26
(ii) Percentages of businesses[b]:													
1. Principal or representative buys with full discretion on price	97	89	73	49	48	67	73	80	78	65	68	80	73
2. Representative buys in a price range (or to a maximum price)	3	3	12	49	26	17	17	20	16	17	18	12	17
3. Representative buys at a fixed price	3	8	14	2	48	33	13	10	3	4	18	20	13
4. Representative refers sample or information back to principal, who quotes to farmer	—	—	29	56	48	67	27	20	16	48	34	4	27

[a] and [b] as for Table 4.1.

merchants, when buying grain from farmers, deliberately vary the purchase price as between one customer and other. It had been suggested to us that at least some merchants varied their buying prices between customers not in response to the dictates of demand and supply nor in recognition of variations in grain quality, but as a result of pressures exerted on them by those customers—for instance, that the price might often be set higher than demand warranted because the farmer owed the merchant money and the merchant wanted to "contra" his account. In so far as these practices exist, they are, of course, quite rational from the point of view of the individual merchant, who must try to minimise the amount of capital he has outstanding with farmers and at the same time stay in business. But they reduce efficiency in the cereal market, to the extent that they prevent the play of demand being signalled accurately back to the cereal producer.

4.15. The results of this question were among the least conclusive to emerge from the Survey of Merchants. It very soon became obvious that day-to-day pricing is a complex process; for many merchants, it is so much a part of everyday life that they do not normally stop to deliberate on the motives for their decisions. The technique of depth interviewing could have thrown much more light on the question, but this takes time which was not available at Survey interviews. Therefore, the results given below should be regarded as broad indications and guides to possible further research, rather than as definite answers.

4.16. Table 4.5 shows, for the 260 merchant businesses covered in the Survey, the numbers of firms which varied purchase prices for grain between one customer and another for the reasons stated. The numbers shown add to more than 260, because some firms reported more than one reason for variation. About one-third of all firms interviewed did not vary prices between one customer and another (though price will, of course, usually be varied according to the quality of grain a customer offers). About one firm in five gave favourable prices to big buyers of compound feeds, and the same proportion favoured good or old-established customers. Rather fewer firms said that they varied prices where farmers owed them money; merchants' comments confirmed that when price is varied for this reason, the usual practice is to give a rather better price than the average in order to get the farmer's grain, then to credit his account with the purchase value. This picture of the relative importance of reasons for price variation did not differ noticeably as between different sizes or types of merchant firm.

4.17. Interpretation of these figures would be much easier if the size of price variations were known exactly. Here again, it was not practical to ask a specific question in the Survey interview. However, from

TABLE 4.5

VARIATIONS IN GRAIN PURCHASE PRICE BETWEEN CUSTOMERS

	Number of firms	Percentage of sample
No variation between customers	82	(32)
Variations made:		
To large buyers of feeds	46	(18)
For customer loyalty	46	(18)
Where the customer owes money	39	(15)
To attract new customers	27	(10)
To large buyers of fertiliser	23	(9)
According to quantity of grain supplied	20	(7)
All other reasons	22	(8)
No definite answer	31	(12)

comments made by merchants, our impression is that, in general, price variations are relatively small; where figures were quoted, the deviation from the average price was hardly ever greater than 10s. per ton, and usually less than this. It was pointed out to us that word of excessive price variation gets round a merchant's customers and can very quickly lose him goodwill; and that margins in merchanting are usually too small to allow any large variations in price.

4.18. A suggestion made to us before the Survey began was that small merchant firms varied grain price more than the larger concerns, since the smaller firms were supposedly in a weaker buying position. But the Survey results gave no indication whatsoever that small firms did, in fact,

TABLE 4.6

COMPARISON OF NUMBERS OF PRICE VARIATIONS AND
"NO VARIATION" REPORTS BY TURNOVER GROUP

Turnover group (£'000)	(1) Total instances of variation	(2) "No variation"	(3) Ratio (1) to (2)
Under 75	16	16	1·0 : 1
75–149	34	13	2·6 : 1
150–499	98	24	4·1 : 1
500–1249	40	18	2·2 : 1
1250 and over	33	8	4·1 : 1
Turnover not known	2	3	0·7 : 1
Total sample	223	82	2·7 : 1

vary price more than large. A very rough idea of the prevalence (though not the size) of price variation in each size-group can be gained by adding up all the instances of price variation reported by firms and comparing this figure with the number of firms reporting "No variation" (Table 4.6). The prevalence of price variation seemed if anything to be relatively low among smaller firms, and to be higher among firms in the medium and large size-groups. A similar analysis was done for the various types of merchant firm; the indication was that general merchants (those with no single activity accounting for more than 50% of turnover) tended on average to vary prices more often than any other business type, having a ratio of 4·5 : 1 between instances of variation and "No variation" reports.

4.19. The Survey has thus confirmed that price variations not occasioned by the interplay of supply and demand are common in cereal merchanting. But, in our view, any recommendation of ours that this practice be discouraged would not lead to much gain in efficiency. The reasons for this view are as follows:

(a) A large number of merchant firms already follow a policy of not varying prices, and many more vary prices only occasionally; when variations are made, they are generally small.

(b) Favourable treatment of large customers and good customers is part and parcel of trading practice in any competitive industry. We think it would be naïve and ineffectual to recommend that merchants should cease trading in this way if they prefer to do so.

(c) There are obvious objections to the situation in which a farmer gets a higher price for grain simply because he owes the merchant money—though it is worth repeating that only 15% per cent of the firms who answered the question said that this happened in their business. Here again, however, we feel that any direct recommendation could have little effect; the question is bound up with the much wider issues of merchant credit and the buying position of the merchant industry, both of which topics are discussed elsewhere in the Report.

4.20. Though for these reasons we do not recommend any direct attempt to discourage variation in pricing as between customers, we would draw attention to the fact that one merchant firm in three manages to carry on trade without making such variations. We found no evidence to show that these firms were losing customers, nor that they were less successful than firms which did vary prices. We suspect that some merchants have exaggerated fears of what would happen if they ceased to favour individual customers, and suggest that more firms might profitably follow the example of the one-third who make no such variations in price.

Buying Grain from Farmers' Groups

4.21. For the merchant sample as a whole, about one firm in every six bought some grain from farmers' selling groups. There was a clear variation in this proportion as between different sizes of firm; the bigger the firm, the more common was it to buy from groups. In particular, among the largest concerns (over £1¼ million turnover annually) more than half the firms interviewed bought some grain in this way. The detailed figures are given in Table 4.7. This pattern of results by size-group is, of course, bound up with the fact that the main purpose of group trading is to gain the advantages of bulk dealing; the bigger the merchant, the more easily can this be done. The results for the various types of merchant firm support

TABLE 4.7

REPLIES TO QUESTION, "DO YOU BUY GRAIN FROM FARMERS' GROUPS?"

Turnover group (£'000)	"Yes"	"No"	No answer	% of turnover-group buying from farmers' groups[a]
Under 75	1	38	3	(3)
75–149	3	40	1	(7)
150–499	13	79	2	(14)
500–1249	9	36	1	(20)
1250 and over	16	10	1	(62)
Turnover not known	1	5	1	(17)
Total sample	43	208	9	(17)

[a] Percentages calculated after *excluding* "No answer" returns.

this interpretation: the highest proportion of firms buying grain from farmers' groups was found among merchants dealing mainly in grain (34%), while the lowest was reported by firms who dealt mainly in purchased feeds (3%).

4.22. Because relatively few merchants bought from farmers' groups, not many comments were volunteered on these groups. The general feeling, more especially among smaller merchant firms, was one of disapproval. In many cases, the merchant did not particularise further. Of the specific allegations which were made, the following were the most common (numbers in brackets show how often the comments were recorded): groups do not provide the personal service that the merchant can provide (11); groups are incompetent to run a grain business (10); groups force the merchant to operate on low margins (10). Some merchants said that groups of their acquaintance were running into difficulties, or soon would do so, because farmers would not keep to the rules or because, as the groups increased in size, they ran into administrative troubles and found that overheads had to rise considerably (17 comments in all).

4.23. Clearly, if group selling were to expand markedly among farmers, many smaller merchants would find themselves in difficulty. As far as the present position is concerned, our impression is that group selling by farmers has had little effect on the merchanting industry in general. Large merchant firms, in areas where groups operate, find them a convenient source of bulk purchases; most smaller merchants are unaffected by the operations of groups—often simply because there is no selling group nearby—while those who are affected consider groups a nuisance, but not up to now a serious threat to their own existence.

Buying on Behalf of Other Principals

4.24. It had been suggested to us before the Survey that many merchants nowadays were acting virtually as brokers—that is, that a large part of their business consisted in buying grain on commission for another principal. A question on this topic was included in the merchant survey. The results show that far from being a widespread practice, buying for another principal is virtually unknown in the merchanting industry. Of the total of 260 merchants interviewed, only eight (3 %) reported this activity. Three of these merchants said that commission buying only took place very occasionally; another two only bought in this way for some small parcels of specialised quality; and another was semi-retired. It is clear, therefore, that only negligible amounts of grain are bought by merchants on other principals' behalf.

Statutory Grading Scheme for Cereals

4.25. Merchants were asked whether they thought a statutory grading scheme for cereals would be desirable or possible. The answers, for the 260 firms in the sample, are shown in Table 4.8. While one merchant in every three interviewed thought a scheme would be desirable, only about one in six considered it to be possible. About half the firms surveyed thought a scheme would be undesirable. The figure for merchants considering a scheme "impossible" may well be an understatement, since the question on desirability of a scheme was asked first in the interview, and when a merchant said a scheme would be undesirable, interviewers did not press him on the question of possibility. This accounts for the large proportion of "No answer" returns in the second half of the question. It may be assumed that most of the merchants who had said a scheme was undesirable would also have said it was impossible if pressed to answer this second part of the question.

4.26. A more instructive picture of merchants' attitudes to a statutory

K

TABLE 4.8

MERCHANTS' OPINIONS ON A STATUTORY GRADING SCHEME

	Number of firms	Percentage of sample
(a) *Would scheme be desirable?*		
Desirable	80	(31)
Undesirable	138	(53)
Don't know	25	(10)
No answer	17	(6)
	260	(100)
(b) *Would scheme be possible?*		
Possible	41	(16)
Impossible	121	(46)
Don't know	43	(17)
No answer	55	(21)
	260	(100)

grading scheme is given by the comments they volunteered to interviewers while this question was being dealt with. These comments are summarised in Table 4.9; as will be seen, there was a very strong leaning against the introduction of statutory grading. In the table, the numbers of comments shown relate to the sample total of 260 firms; but they add to more than that total, because some merchants made more than one of the points listed. It is worth noting that, in total, 300 comments unfavourable to a statutory grading scheme were recorded, as against twenty-five comments in favour.

4.27. There is thus no doubt of the prevailing opinion among merchants as regards a possible statutory grading scheme. Our own opinion is equally unfavourable. Primarily, this view is based on sheer physical impracticability. We suspect that many advocates of a grading scheme have had in mind the successful operation of such schemes in the U.S.A., Canada and other overseas countries. But growing conditions in these countries are vastly different from those in the U.K. There, very large acreages are grown under virtually uniform conditions of climate, soil and harvesting technique; here, as merchants pointed out, holdings are relatively small, soils and harvest conditions vary from farm to farm and even from field to field, and weather may change greatly between seasons. For all these reasons, we accept that the chief advantage of grading—namely the ability to buy on description—could not be achieved in the U.K. We believe that if a statutory mechanism for grading were set up, it could at best not achieve more than is done at present by merchants and end-users, and would probably do it at greater expense and with much less flexibility. A large part of the merchant's function in the market is to judge the quality

TABLE 4.9

MERCHANTS' ATTITUDES TO A STATUTORY GRADING SCHEME

	Number of comments	Percentage of sample
(a) *Scheme unnecessary or of no benefit:*		
Reason not specified	27	(10)
Merchants already grade grain	26	(10)
End-users already grade grain	21	(8)
Not justified by increase in price	14	(5)
Other reasons	21	(8)
(b) *Scheme unworkable:*		
Reason not specified	22	(8)
Impossible to decide who would do the grading	13	(5)
Would cause disputes and loss of goodwill	13	(5)
Too much variation between samples	12	(5)
Too much regional variation	11	(4)
Too much seasonal variation	10	(4)
Other reasons	26	(10)
(c) *Scheme too costly:*		
Reason not specified	34	(13)
Other reasons	10	(4)
(d) *Scheme otherwise undesirable:*		
Would complicate the merchant's job	10	(4)
Would mean more regulation of the industry	10	(4)
Other reasons	20	(8)
(e) *Scheme desirable:*		
Would match price more closely with quality	6	(2)
Would help to improve quality	6	(2)
Other reasons	13	(5)

of grain and direct it to the outlet for which it is best suited. We have found no evidence to suggest that this function would be better performed under a statutory grading scheme. It is important to bear in mind that these answers relate to a *statutory* (i.e. a compulsory) grading scheme. The possibilities of a *voluntary* scheme are discussed in Chapter 7.

Forward Buying

4.28. Merchants were asked whether they bought any grain from farmers on forward contract. Their answers are summarised in Table 4.10, for the whole sample of 260 merchants and for each size-group separately. For the sample as a whole, just over three-quarters of all merchants bought

TABLE 4.10

ANSWERS TO QUESTION "DO YOU BUY GRAIN FORWARD?"

Turnover group (£'000)	"Yes"	"No"	No answer	Percentage of firms buying forward[a]
Under 75	18	24	—	(43)
75–149	29	12	3	(71)
150–499	79	14	1	(85)
500–1249	36	8	2	(82)
1250 and over	27	—	—	(100)
Turnover not known	7	—	—	(100)
Total sample	196	58	6	(77)

[a] Percentages calculated after *excluding* "No answer" returns.

some grain forward. The prevalence of forward buying increased, on average, with size of firm; among the smallest concerns, less than half bought any grain forward, while all the very large firms interviewed did some forward buying (this was, of course, not to say that all their purchases were brought forward). From the comments made by merchants on forward buying—summarised in more detail below—it appears that small firms often do not think it worth their while to contract forward; this may be because they do not have staff time to spare for the paperwork required, or because, having a relatively small amount of grain to buy, they feel they can get it satisfactorily on the spot market only. Among the various types of firm, those dealing mainly in straight grain used forward contracting more frequently than any other business type (95% of firms in this group reported buying some grain forward).

4.29. The working of the H.-G.C.A. Forward Contract Scheme is discussed in another chapter of this Report. The present section sets out only to describe merchants' *attitudes* to the Scheme, as revealed in the Merchant Survey interview. Merchants were asked "What is your opinion of the H.-G.C.A. Forward Contract Scheme?". Their answers are summarised in Table 4.11. It is important to bear in mind that interviewers did not prompt the merchant in any way, nor suggest a list of topics which the merchant was to cover in this connection. This obviously affects the interpretation of the figures; if a certain comment was made by, say, 10% of merchants, it means that this number of merchants thought the point important enough to raise without prompting. Further, it does *not* mean that the remaining 90% of merchants would necessarily have disagreed with that comment, had they been specifically asked.

4.30. The percentages shown in Table 4.11 all refer to the total sample of 260 merchants. They add to more than 100, because many merchants made several points. For convenience, the comments made have been grouped into four categories: (a) adverse comments on the Scheme's effect on the market; (b) favourable comments on the same topic; (c) specific suggestions as to changes needed in the working of the Scheme; and (d) other comments. The four classes are mutually exclusive. It will be seen

TABLE 4.11

MERCHANTS' ATTITUDES TO THE H.-G.C.A. FORWARD CONTRACT SCHEME

	Number of comments	Percentage of sample
(a) *Adverse comments:*		
Too much administrative work for merchants	33	(13)
Shortages now arise in the spot market	23	(9)
Farmers lose (or do not gain) from contracting	21	(8)
Scheme has not improved seasonal flow of grain	14	(5)
Farmers may fail to honour contracts	11	(4)
Other adverse comments	67	(26)
(b) *Favourable comments:*		
Scheme has evened out seasonal fluctuations	33	(13)
Scheme makes it easier to plan buying	14	(5)
Other favourable comments	21	(8)
Reason for approval not specified	23	(9)
(c) *Comments on the working of the scheme:*		
The 5% delivery tolerance causes difficulties	14	(5)
The contract period is too long	13	(5)
Open-priced contracts are preferable to fixed-priced	12	(5)
Fixed-price contracts are preferable to open-priced	11	(4)
Other comments on working of scheme	15	(6)
(d) *Other comments:*		
Scheme is not relevant to the small buyer of grain	15	(6)
Customers are not interested in contracting forward	11	(4)
Other comments	16	(6)

that adverse comments on the effects of the Scheme outnumbered favourable comments by almost two to one. The most common complaint was that the Scheme created too much paperwork and other administrative tasks for merchants; many of those interviewed pointed out that the merchant gets no pay for this work. It was suggested to us several times in the course of the Survey that such payment should be made. However, we do not feel that it would be realistic to recommend this. Almost certainly, any payment to merchants would have to be financed out of the levy on farmers (which might have to be increased) and it is very unlikely that this arrangement would be acceptable to the majority of farmers. We would, however, draw the Authority's attention to the need to keep paperwork on the Scheme as light as possible; otherwise its benefits may be lost to many merchants—particularly the smaller firms—and their customers.

Forward Selling

4.31. Table 4.12 shows, for each size-group and for the whole merchant sample together, the numbers of firms which sold some grain forward in the Survey year. The table also gives the same information for the various types of business. Rather more than half the merchants interviewed said that they sold some grain forward (as in the case of forward purchases, it was not practicable to record the *amounts* of grain entering into forward dealings). Generally speaking, the smaller a firm was, the less likely was it to sell any grain forward: only a quarter of the firms in the smallest size-group sold grain in this way, while among the largest firms the great majority did so. As might be expected, the breakdown by type of firm showed that forward selling of grain was much more common among

TABLE 4.12

ANSWERS TO QUESTION "DO YOU SELL GRAIN FORWARD?"

	"Yes"	"No"	No answer	Percentage of firms selling forward[a]
(i) *By turnover group* (£'000)				
Under 75	10	30	2	(25)
75–149	15	26	3	(37)
150–499	56	35	3	(62)
500–1249	25	19	2	(57)
1250 and over	23	1	3	(96)
Turnover not known	4	2	1	(67)
Total sample	133	113	14	(54)
(ii) *By type group* Specialist/mainly grain	37	3	1	(92)
Specialist/mainly feed	6	21	—	(22)
Specialist/mainly compounding	12	23	1	(34)
General	71	46	8	(61)
Other	7	20	4	(26)
Total sample	133	113	14	(54)

[a] Percentages calculated after *excluding* "No answer" returns.

merchants who dealt mainly in grain than for any other type. Where the business depends mainly on compounding or purchased-feed sales, forward selling of grain is less common—very largely because in these types of business little or no straight grain may be sold in any case.

4.32. Merchants were next asked what proportion of grain bought forward was covered by selling forward. It very soon became obvious to the Survey interviewers that relatively few merchants work to a specifically thought-out policy of covering forward buying; most often, forward sales and forward purchases are made independently of each other, to suit the requirements of a particular customer or in response to the merchants' own assessment of the market. Therefore, while Table 4.13 shows the extent to which forward purchases were reported as being covered by forward sales, it must not be assumed that deliberate covering operations were taking place in all the cases reported; indeed, it is safe to say that firms following such a deliberate policy—even for part of the grain they buy forward—are in a small minority. Because merchants so seldom covered forward buying specifically, they often found it difficult to give more than a rough estimate in answer to the question, and this also should be borne in mind when looking at the figures. As Table 4.13 shows, about half the merchants in the sample reported that no forward purchases were covered by forward selling.[1] The differing patterns of results between different sizes and types of merchant firm are as might be expected from what has already been said about the prevalence of forward selling: on average, larger firms covered more of their forward buying than did smaller firms, and "specialist grain" or "mainly grain" businesses covered a much larger proportion of forward purchases than firms mainly dealing in feed or manufacturing compounds. Again it should be pointed out that businesses of the latter types will usually buy forward with the main object of ensuring supplies for compounding, so that the question of forward selling hardly arises.

Speculation

4.33. Even in the planning stages of the merchant survey, it was realised that "speculation" would be a difficult concept to deal with. This indeed turned out to be the case. Originally, interviewers were instructed to ask the merchant "To what extent do you engage in speculation? Up to what proportion of your total purchases of grain would you be prepared to

---

[1] This number, 113 merchants in all, corresponds to those reporting "no forward sales" in Table 4.12; the "percentage of the sample" represented by this numbe differs slightly as between Tables 4.12 and 4.13, because of differing numbers of "No answer" returns to the two questions.

TABLE 4.13. PROPORTION (BY WEIGHT) OF FORWARD PURCHASES COVERED BY FORWARD SALES

Turnover group (£'000)

Proportion of forward purchases covered	Under 75	75–149	150–499	500–1249	1250 and over	Turnover not known	Total sample
Number of businesses							
Nil	30	26	35	19	1	2	113
Up to half	3	3	12	1	4	1	24
Half up to 9/10ths	3	4	11	14	5	1	38
9/10ths and over	3	8	24	8	8	1	52
No answer	3	3	12	4	9	2	33
Percentage of businesses[a]							
Nil	77	63	43	45	6	40	50
Up to half	8	7	15	3	22	20	10
Half up to 9/10ths	8	10	13	33	28	20	17
9/10ths and over	7	20	29	19	44	20	23
Total	100	100	100	100	100	100	100

Type group

Proportion of forward purchases covered	Specialist/ mainly grain	Specialist/ mainly feed	Specialist/ mainly compounding	General	Other	Total sample
Number of businesses						
Nil	3	21	23	46	20	113
Up to half	3	2	3	15	1	24
Half up to 9/10ths	10	3	2	21	2	38
9/10ths and over	13	1	6	28	4	52
No answer	12	—	2	15	4	33
Percentage of businesses[a]						
Nil	10	78	68	42	74	50
Up to half	10	7	9	14	4	10
Half up to 9/10ths	35	11	6	19	7	17
9/10ths and over	45	4	17	25	15	23
Total	100	100	100	100	100	100

[a] Percentages calculated separately for each group, after *excluding* "No answer" returns. Percentages rounded where necessary to add to 100.

speculate?" It quickly became obvious that this wording could be interpreted by merchants in many different ways. A major difficulty was that the term "speculation" has a derogatory sense in many people's minds. No such suggestion was intended by the question. The figure wanted was the proportion of total grain purchases which the merchant held for resale in the hope of benefiting from an upswing in price. Interviewers added this explanation whenever merchants expressed doubt about the meaning of the question. Even so, it was obvious that different merchants were interpreting it in different ways. Some quoted a percentage without further ado; others said they could not quote any definite figure. Many qualified their answers with doubts about the appropriateness of the concept of "speculation" as applied to a grain merchanting business; a frequent comment was that the merchant had to assess the future state of demand whenever he bought grain—to use the merchant's phrase, he had always to "take a view of the market"—but that this was not the same thing as speculation. As one merchant put it: "The answer to this question could be 'all' or 'none'."

4.34. These difficulties should be kept very much in mind when considering Table 4.14, which gives the numbers of businesses quoting various percentage ranges in answer to this question. The numbers, in fact, show the extent to which merchants consider themselves to be engaging in speculation, rather than any more objective measurement. As some merchants pointed out, the scope for speculation is restricted where there is not much spare storage for grain, or where capital is limited and the possible effects of losing some of it on a speculative deal are, therefore, relatively more serious; it is fair to suppose that both these considerations will tend to apply to the small man more often than to the large. The variations between the other size-groups cannot be considered significant, in view of the relatively high rate of non-response and the difficulties in interpreting the question, though the results for the largest firms do not, at least, contradict the supposition that large firms consider themselves to be speculating more often than do smaller firms.

4.35. The pattern of results as between the various types of firm is determined mainly—as it was in the case of forward selling and covering—by the fact that merchant-compounders or firms dealing mainly in purchased feeds will often buy grain for their own use rather than for sale, so that to this extent the possibility of speculating on straight grain does not arise for these firms.

K*

TABLE 4.14. ANSWERS TO QUESTION "UP TO WHAT PROPORTION OF YOUR GRAIN PURCHASES WOULD YOU BE PREPARED TO SPECULATE?"

Turnover group (£'000)

Per cent of purchases	Under 75	75–149	150–499	500–1249	1250 and over	Turnover not known	Total sample
	Number of businesses						
Nil	30	15	38	25	3	2	113
Up to 10%	6	7	29	10	12	3	67
10% up to 30%	2	7	5	3	2	1	20
30% and over	—	—	4	—	2	—	6
No answer	4	15	18	8	8	1	54
	Percentage of businesses[a]						
Nil	80	52	50	66	16	33	55
Up to 10%	15	24	38	26	63	50	32
10% up to 30%	5	24	7	8	10	17	10
30% and over	—	—	5	—	11	—	3
Total	100	100	100	100	100	100	100

Type group

Per cent of purchases	Specialist/ mainly grain	Specialist/ mainly feed	Specialist/ mainly com-pounding	General	Other	Total sample
	Number of businesses					
Nil	16	18	16	44	19	113
Up to 10%	10	5	2	47	3	67
10% up to 30%	5	3	2	9	1	20
30% and over	1	1	—	3	—	5
No answer	9	9	7	21	8	54
	Percentage of businesses[a]					
Nil	50	67	80	42	83	55
Up to 10%	31	18	10	45	13	32
10% up to 30%	16	11	10	9	4	10
30% and over	3	4	—	4	—	3
Total	100	100	100	100	100	100

[a] Percentages calculated separately for each group, after *excluding* "No answer" returns. Percentages rounded where necessary to add to 100.

Straight Grain Dealings with Other Merchants

4.36. As Table 4.15 shows, about five in every six of the merchants interviewed in the sample as a whole bought some grain from other merchants, and this proportion did not vary much as between different sizes or types of firm. If anything, the very largest firms showed a tendency to

TABLE 4.15

ANSWERS TO QUESTION "DO YOU BUY STRAIGHT GRAIN FROM OTHER MERCHANTS?"

	"Yes"	"No"	No answer	Percentage of firms buying from other merchants[a]
(i) *By turnover group* (£'000)				
Under 75	32	9	1	(78)
75–149	31	13	—	(70)
150–499	77	16	1	(83)
500–1249	40	6	—	(87)
1250 and over	26	1	—	(96)
Turnover not known	7	—	—	(100)
Total sample	213	45	2	(83)
(ii) *By type group*				
Specialist/mainly grain	35	6	—	(85)
Specialist/mainly feed	29	7	—	(81)
Specialist/mainly compounding	21	6	—	(78)
General	107	18	—	(86)
Other	21	8	2	(72)
Total sample	213	45	2	(83)

[a] Percentages calculated after *excluding* "No answer" returns.

buy more often from merchants than did the average size of business.

4.37. However, these percentage results, based on numbers of firms, give a much-overstated idea of the importance of grain purchases from other merchants as compared to those from farmers. For most merchant firms, the amount of grain bought from merchants makes up rather a small proportion of total grain purchases. This is demonstrated in Table 4.16, which has been worked out in terms of tonnages of the three main cereals purchased, and shows for each cereal what percentage of total purchases (grain and seed) came from merchants. On average over all the firms surveyed, only about one-quarter of the total barley purchased was bought from other merchants, with smaller proportions of wheat and oats. The

TABLE 4.16

PERCENTAGES OF TOTAL STRAIGHT GRAIN PURCHASES
REPRESENTED BY PURCHASES FROM OTHER MERCHANTS
(total purchases of each cereal = 100)

	Wheat	Barley	Oats
(i) *By turnover group* (£'000)	%	%	%
Under 75	23	32	28
75–149	16	16	26
150–499	11	16	21
500–1249	15	17	14
1250 and over	7	33	20
Turnover not known	25	8	4
Total sample	11	25	19
(ii) *By type group*			
Specialist/mainly grain	9	31	15
Specialist/mainly feed	16	19	40
Specialist/mainly compounding	45	55	38
General	9	14	17
Other	20	32	25
Total sample	11	25	19

only noticeably higher figures, not surprisingly, were for merchants whose main interest was in making their own compounds.

4.38. It proved unexpectedly difficult, in the time available at the Survey interview, to get any well-defined picture of what prompts firms to buy from other merchants. A firm may be temporarily short of grain; or other merchants may have a temporary surplus. It may be that the merchant takes advantage of an unexpected price difference. Some firms may follow a policy of buying as much as possible of their grain from farmers; others may prefer to buy a specific proportion from merchants, for reasons of quality, price or convenience. All these considerations of demand, supply and price are closely bound up one with another, and to disentangle them would mean much closer examination than was possible in the Survey. However, the general impression left with interviewers was that the majority of firms tried to buy as much of their grain from farmers as was practical, and only resorted to buying from merchants when supplies from their farmer customers were not enough to meet their needs for grain, either for compounding or to meet sale commitments. This reason was mentioned specifically by about fifty of the merchants interviewed, or almost one in four of the total number who bought grain from other merchants. Since this was a freely volunteered comment, and not an answer to a set question, it can be assumed that many more merchants would have made the same statement had they been asked. A frequent additional comment was that purchases from merchants were concentrated towards the end of the season, because supplies of grain from farms tended to fall off at that time. Relatively few firms in the Survey reported a continuous policy of dealing with other merchants. In fact, this latter type of trade is mainly the province of a few large and well-known firms—often called "wholesalers"—who specialise in moving grain between merchants, often over long distances, in response to changes in price.

4.39. Merchants were also asked whether they sold any straight grain to other merchant businesses. For the sample as a whole, about one firm in three did so. As might be expected, the proportion of firms selling grain to other merchants was highest in the "specialist/mainly grain" group (88%) and lowest among the feed merchants and country compounders (46% and 38% respectively). As in the case of grain purchases, the proportion of total grain sales going to other merchants will be much lower than these figures would suggest; in fact, forty firms—about one in four of all those selling to other merchants—volunteered the comment that these sales were only occasional. Of the specific reasons given, the most frequent was that grain was only sold to other merchants if it was surplus to the firm's own requirements for compounding or sale.

4.40. First and foremost, therefore, the average merchant bought his grain direct from farmers, and either sold it to end-users or utilised it in feed manufacture. The reason for this is not difficult to suggest: margins in grain-handling are so low that it cannot normally pass through more than one pair of hands. Grain as a rule moved between merchants only when temporary stresses of demand or supply caused it to do so. As one merchant said: "Merchant dealings are the safety-valve of the grain trade." This in our view is the correct function of inter-merchant deals, and one which they are performing with great efficiency: movements between merchants are kept at a low level, yet they occur when necessary to even out variations in demand and supply. Not only for the wholesalers but also on occasion for most merchant firms, this function can be performed over long distances; the geographical pattern of inter-merchant trade is discussed in Chapter 3 (pp. 213–17).

Price Decisions in Selling Grain

4.41. The Merchant Survey questionnaire included a question in which merchants were asked to state which of various possible alternative methods of pricing they used for sales of straight grain to end-users, to other merchants and to farmers respectively. However, it soon became apparent that many firms' pricing policy was too variable, too complex or too vague to be recorded adequately on a set coding system. Another difficulty was that many firms either sold no grain, or only insignificant amounts, to any or all of the three main classes of buyers. For these reasons, it would be misleading to quote specific numbers of firms against the results of this question. The main conclusions are clear, however, and have been arrived at on the basis of merchants' comments as well as on the coded answers.

4.42. As a rule when selling grain to end-users, and particularly when selling to the big animal-feed manufacturers, most merchants had simply to take or leave the price quoted to them by the buying firm. Even in this situation, of course, the merchant has the theoretical alternatives of storing his grain or selling it to someone else if he does not think the price offered is high enough; how far he is actually able to follow these alternatives will depend on his storage facilities, his position in relation to alternative outlets, and the state of demand and supply at the time. Sometimes a certain amount of negotiation is possible around the quoted price. Merchants stated that maltsters and millers were usually more open to negotiation than the large compounding firms. Even the big compounders, however, might engage in a small degree of haggling over price if they happened to be short of grain. Some firms said that compounders were more ready to negotiate price on large lots than on small lots of grain; but it must be pointed out that the Survey's evidence is not consistent enough to allow firm conclusions on the way in which lot size affects pricing, as regards sales to compounders or indeed any other transactions. The large merchant firms, however, reported much more frequently than the smaller firms that they had scope for some price negotiation in grain sales to end-users. In some cases, this might be because the larger merchants could offer larger parcels of grain, but large merchants also gain bargaining strength by virtue of their purchases of feedingstuffs from the compounding firms concerned.

4.43. Merchants realised that the big firms of end-users would inevitably quote identical or very similar prices on a given day, because their intelligence services would come to the same conclusions about the state of the market. However, none of the 260 merchants visited voiced the opinion that collusion took place between these large buyers, and the few merchants who volunteered specific comments on the subject felt that nothing in the nature of a price ring existed. Especially near the main port centres, merchants said that to all intents and purposes the "going price" for grain was the price offered by the national compounding firms. But the nationals' quotations themselves depended largely on import prices, and the level of the import price, rather than any deliberate action on the part of the large buyers, was considered to be the main determinant of the feed grain price in the port areas.

4.44. Not surprisingly, the Survey showed that when one merchant sold grain to another, the price was almost always arrived at by negotiation between the two parties. This accords with the fact that, for most firms, merchant-to-merchant grain deals normally take place only to clear temporary surpluses or to meet a temporary shortage. The scope for negotiation will depend on how serious is the surplus or shortage concerned, on the general state of the market, and on the bargaining skill of the merchants; but both sides are well aware of the going price for grain, and variations from it are usually small. At the same time, of course, even such small price differences can decide whether a merchant profits or loses on a deal.

4.45. Pricing policy on sales of straight grain to farmers was less well defined than in the case of sales to end-users or to other merchants. A fair number of firms said that when selling grain to farmers, as to merchants, the price was arrived at simply by negotiation. (It might have been expected that this report would occur relatively more often among smaller merchant firms, but the survey showed no evidence of this.) The majority of merchants said that, in theory at least, their policy was to sell grain to farmers at a fixed margin over the buying price. This margin was quoted inclusive of transport charges by some firms, and after transport by others. However, it became obvious that in practice the grain merchant can very seldom work to a definite margin on sales in the same way as, say, the retail shopkeeper. For one thing, merchants often sell grain forward (or accept forward orders) before that grain has been bought. Even where the purchase has been made and the buying price is known, the selling price the merchant can ask is set between very narrow limits by the going price for grain at that particular time and place, and farmers are usually as well acquainted with this price as are merchants themselves. Perhaps the only circumstance in which a merchant may be able to apply a fixed margin in the usual sense is when he is selling small lots of grain to buyers who use only very small amounts, for instance farmers who run a small poultry flock. Otherwise, when merchants refer to a "margin on sales", what virtually always happens in practice is that the merchant has in mind a certain margin which he would like to achieve, and tries by skilful buying and selling to ensure that on average over a period of trading this margin is reached. On some deals, he will get more; often, he must be content with less.

4.46. Merchants were not asked specifically what margins they made on grain sales, since it was realised that many firms might regard this as confidential information. However, some fifty merchants volunteered figures. They did not all quote their margins in the same form: some gave the figure they aimed to get, while others quoted their estimate of the margin which was actually made on average. Some quotations were before allowing for transport charges, others after. Therefore, no precise average can be given. But broadly speaking, where merchants quoted the margin they *aimed* to get on grain sales, the figures mentioned were almost all between 10s. and £1 per ton clear of transport charges. Figures quoted for the margin actually *averaged* over a period ranged more widely, from about 5s. to 25s. per ton clear of transport. It is not correct to assume, of course, that these quotations necessarily reflect the size of margins for the merchant sample as a whole, although the fifty merchants who did volunteer figures were not concentrated in any one type or size of business.

4.47. In selling grain, then, is the merchant a "price-maker" or a "price-taker"? Broadly speaking, as this section has shown, the answer depends on whom the grain is sold to. When selling to end-users, the merchant must usually take or leave the price offered to him. In merchant-to-merchant deals, in so far as a price-maker and price-taker relationship exists at all, the role taken by the selling merchant will depend on his bargaining position at each individual deal. When a merchant sells to farmers, he is usually in the position of price-maker to a greater or less extent. These results are what might have been expected from the relative sizes and degrees of bargaining power of the buyers concerned. But the most important point to note is that the whole concept of "price-maker and price-taker" is one which only applies in a very restricted sense to the selling of grain. In the cereals market, as in the markets for most other agricultural commodities, the actions of any individual firm can influence the "going price" to only a very small degree, and indeed will not influence it at all unless the firm concerned is one of the very largest operators in the market. First and foremost, the price of grain at any one time is determined by the relationship of supply and demand; knowledge of the "going price" is spread through the market by a sensitive information system, and grain quickly moves to even out any imbalance. The pricing policy of everyone in the market—even the large compounding firms—is governed by this situation.

The Information Used in Grain Pricing

4.48. In the grain trade, where prices continually fluctuate and margins are small, market information is of obvious importance. Merchants were asked what sources of information they used in arriving at the prices they quoted when buying grain. It was not practicable at the Survey interview to enquire deeply into the relative importance of the various sources mentioned by each merchant. However, to gain a very rough indication of this, the answers to the question were recorded in two parts: first, the interviewer noted the sources which the merchant quoted spontaneously, without prompting. Next, a set list of sources was read out to the merchant, and the interviewer noted any of these which the merchant agreed he used, other than those already mentioned spontaneously. Table 4.17 shows the results of these two parts of the question. The various sources are set out in the same order as in the list read out at interview, with the addition of a few other headings volunteered by merchants. Of the total sample of 260 merchants, eleven gave no answer to the question; the numbers of firms listed under "spontaneous answers" have been shown in the table as percentages of the 249 firms who did reply. For "answers after reading out list" the number not replying was much higher, for the obvious

TABLE 4.17

SOURCES OF PRICE INFORMATION MENTIONED BY MERCHANTS

	Spontaneous answers		Answers after reading out list	
	Number of firms	Percentage of sample[a]	Number of firms	Percentage of sample[a]
Sources quoted on interview list:				
M.A.F.F. notices	32	(13)	23	(9)
H.-G.C.A. notices and bulletins	87	(35)	55	(22)
Farming press	112	(45)	60	(24)
Trade press	98	(39)	48	(19)
Prices at corn exchanges or markets				
—found by visit	92	(37)	18	(7)
—found otherwise	30	(12)	3	(1)
Futures prices	48	(19)	40	(16)
Direct contact with other merchants	159	(64)	30	(12)
Price information from end users				
—price lists	31	(12)	5	(2)
—direct contact	162	(65)	15	(6)
Prices mentioned by farmers	65	(26)	35	(14)
Other sources mentioned:				
Newspapers	28	(11)	—	(—)
Other written sources	15	(6)	—	(—)
Brokers—direct contact	16	(6)	—	(—)
Other direct contacts	12	(5)	—	(—)
Radio	13	(5)	—	(—)
No answer	11	(—)[a]	140	(—)[a]

[a] Percentages calculated after excluding the eleven firms who gave no answer to either part of the question, i.e. 249 on firms.

reason that many merchants mentioned all their sources of information before the list was read to them. In this case also, the numbers of replies are shown in the table as percentages of the 249 firms who answered the first part of the question. It will be obvious that not all the headings in the table are mutually exclusive; for example, futures prices might be found out from the farming press or trade press, or by direct contact with brokers or merchants. Often, market intelligence issued by the M.A.F.F. or H.-G.C.A. is seen by a merchant in farming or trade journals, even though he himself does not take the notices concerned. There is thus a certain area of approximation in some of the figures in Table 4.17, but this does not affect the main conclusions.

4.49. The most outstanding feature of these results is the importance placed on direct contact with others in the trade. Without prompting, two in three of the merchants interviewed quoted contact with end-users as a source of price information, and a similar number mentioned contact with other merchants. More than one in three said they visited corn exchanges or markets to gather price information, and contact with farmers, brokers and others (e.g. grain importers and exporters) was also frequently mentioned. Apart from visits to markets, day-to-day contact is primarily kept up by telephone, though visits paid between firms by travellers and buyers are also important. Contacts will, of course, not always be concerned merely with passing on information; often, an actual deal in grain will be done, and the discussion of price which goes on at this time is an important source of information for most merchants.

4.50. In contrast, published market intelligence was used less frequently than might have been expected. Rather fewer than half the merchants interviewed mentioned without prompting that they consulted the farming press for price information; just over one in three used trade publications, and a like proportion mentioned the market intelligence issued by the H.-G.C.A.[1] It is interesting to note that when interviewers went on to read out the list of possible sources of price information, the published sources were recalled relatively more often by merchants than were

---

[1] At 30th June 1967, the end of its first year of publication, the *H.-G.C.A. Weekly Bulletin* had 1338 subscribers; of these, 799 were farmers, 532 were merchants or end-users, and seven were of other descriptions. (In addition, around 1000 free copies of each week's bulletin were being sent to official bodies, etc.). By 30th June 1968 the number of subscribers had risen to 1444; this increase was the net result of a gain of 465 subscribers, and a loss of 359 by cancellations, during the year. Exact figures by type of subscriber at June 1968 are not available at the time of writing, but it is estimated that the total of subscribers at this date included about 830 farmers and 600 merchants or end-users. Some of the cancellations by merchants in 1967/8 arose from the fact that a number of merchants, previously subscribers, were appointed by the H.-G.C.A. to report market information, and thenceforward received free copies in place of subscription copies.

sources depending on direct contact. In other words, for most merchants day-to-day contact is the source which most readily springs to mind in a discussion of price information; in comparison, published market intelligence plays rather a background role. Indeed, a fair number of merchants volunteered comments to this effect.

4.51. From these and other remarks made by merchants in the course of answering this question, it became clear that published sources of information were of even less relative importance than would appear from the figures in Table 4.17. Interviewers recorded almost seventy comments on the lack of usefulness of published market intelligence—a very substantial figure when one considers that these comments were unprompted. The most frequent criticism was that published prices were almost invariably received too late to be of any use in pricing decisions; one or two merchants said that information could not be any more than a day old if it was to be of any use. Another complaint was that the price information in published sources did not reflect price levels in a merchant's own locality. This could apply even to sources which quoted the prices ruling at various centres throughout the country; unless a merchant was virtually on top of one of the reporting centres (and not always then) the effects of haulage costs and local variations in supply and demand could make the published price quite irrelevant to his own business. Some merchants alleged that farmers sometimes took advantage of this, using the published price as a bargaining weapon when it was higher than the going price for grain in the area. These reasons help to explain why most merchants only use published sources as a general background to their grain pricing; their day-to-day decisions on price are based above all on continual contact with other traders.

4.52. The results shown for the whole sample in Table 4.17 did not differ much between sizes or types of firm. There was a slight tendency for all sources of information to be more frequently reported with increasing size of firm. In particular, H.-G.C.A. notices, published information in general, and futures prices were mentioned relatively often among larger firms, but in no case did any of these sources exceed direct contact in importance.

4.53. The conclusions drawn from this question have obvious relevance to the market intelligence work of the H.-G.C.A. and other bodies. It is usually taken for granted that more market intelligence would be a good thing for the cereals market. Leaving aside for the moment the question of whether more intelligence is in fact needed, the point arises of how official bodies can dispense information which will be of any practical use to the majority of merchants in their pricing decisions. As has been mentioned, some merchants remarked that information more than a day old would be out of date; even assuming that this lag could be extended to 2 or 3 days (though, in fact, it is just as likely that information less than a day old could be effectively out of date) it is clear that a list of prices received a week after they applied can have very little effect on most merchants' pricing decisions. This objection could in theory be met by putting reports each day in the daily press or broadcasting them on the radio, though relatively few merchants reported using the existing intelligence from these sources (some who did consult daily newspapers commented that the prices in them were up to date). More serious is the difficulty that published prices often do not reflect the situation in a merchant's own area. It is not easy to see how any reporting service could deal with the countless variations which arise over distances of a few miles, but which make all the difference to the merchant's decisions on price.

4.54. In fact, it is possible to call in question the assumption that more market intelligence is needed in the merchant sector. To prove that the need existed, it would be necessary to show that the provision of more intelligence would enable grain to flow through merchants' hands in more accurate response to price changes than it does at present. A conclusive answer on this point would only come from a much more detailed investigation than has yet been made. But the present survey certainly gives no reason to suppose that merchants' pricing decisions are significantly hampered by lack of information. The key to the whole matter is the importance of direct contact. For any given merchant business, the extent of this contact will naturally correspond with the extent of the trade carried out. For example, a merchant who sells almost all his grain to a few firms of compounders will ring them up daily, and this may be his only source of price information; but it is perfectly adequate for rational decision-making in his day-to-day trade. Merchants who sell for export will be in continual contact with brokers and exporters. Firms who deal only within a small area will get their price information mainly from inside that area, whereas if a merchant deals with others a long way off, his information sources will be correspondingly far-flung. In other words, the information flow within the merchant sector corresponds with the flow of trade. Any information coming from outside a merchant's field of action will be irrelevant as far as his day-to-day pricing decisions are concerned: for example, it will do a merchant no good to know grain prices in Scotland if he has neither the immediate intention of trading in Scotland nor the organisation to do so; nor to know malting barley prices if his local farmers and merchants have no malting barley to sell. Of course, such information may be relevant to longer-term decisions (e.g. to begin selling to Scotland) but such decisions are bound up with many other questions of management and capital availability, and there is again no evidence in any case that the necessary information is lacking. The most important conclusion to be drawn on this topic is that, as far as grain merchanting is concerned, it is very doubtful whether any increase in the amount of market intelligence issued by official bodies would bring any improvement in the efficiency of marketing. Further, merchants' comments indicate that thought needs to be given to the form and usefulness of existing market intelligence.

CHAPTER 5

Some Management Problems of Merchant Businesses

5.1. In the present chapter the following problems of the management of merchant businesses are discussed:

1. The problem of delegation and co-ordination.
2. The problem of recruiting and training workers.
3. The problem of choosing the right activities.
4. The problem of credit.
5. The problem of stocks.
6. The problems of transport management.

Finally, the success of management is discussed in terms of profitability and in particular of the low level of monetary reward which many merchants derive from their activities.

Types of Merchant Businesses

5.2. All the tables in this chapter relate to a total of 250 merchants—being those who gave full information on the topics concerned—out of the 260 covered in the merchant sample as a whole. They have been divided into three functional groups, as follows:

(i) *Manufacturing Merchants,* i.e. all those who manufactured any feed (irrespective of the proportion of turnover represented by feed sales).

(ii) *Specialist Grain Merchants,* i.e. those with 75% or more of annual turnover in sales of straight grain.

(iii) *Others,* i.e. the remainder of the 250 merchants covered in the analysis of management problems. This group should not be confused with the "Other Merchant" type-group used in other chapters (see definitions, p. 171).

The Problem of Delegation and Co-ordination

5.3. Merchants seldom operate as one-man businesses; generally, they employ a staff to assist them. This means that there must be some measure of delegation and "remote control". The lorry driver who calls on a farmer is acting for the merchant. His activities cannot be completely controlled. Secondly, merchants seldom confine themselves to one type of business activity. This is bound to lead to divided attention and a need for co-ordination. The sale of pesticides and the purchase of grain involve different areas of knowledge. Some division of time and effort is involved, and because of the variety of firm sizes and types, it is likely that no single system of organisation will be right for every business.

5.4. In an attempt to find out how merchants tackled this problem, they

TABLE 5.1

EXISTENCE OF DEPARTMENTAL ORGANISATION AMONG THE FIRMS IN THE SAMPLE

Turnover (£'000)	No. of firms	Firms with departments			
		Manufacturing merchants	Specialist grain merchants	Others	Total
Less than 75	40	1	—	1	2
75–149	44	—	—	5	5
150–499	94	12	1	8	21
500–1249	45	9	1	7	17
1250 and over	27	15	1	2	18
Total	250	37	3	23	63

were asked whether their business was divided into departments, and, if so, how these departments were chosen. Inevitably, there was a variety of interpretation concerning what constituted a department. In some firms, each department has its own executive head, in others it amounts to little more than an accounting convention and involves no separate managerial control. With this qualification in mind, Table 5.1 sets out the number of firms having departments in terms of their function and their turnover.

5.5. Three impressions emerge from this table. First, that departments are used by a minority of merchants. Second, that departmental structures are more used by large than by small firms. Third, that manufacturing merchants make greater use of departmental organisation than merchants of other types.

5.6. In the smaller firms no real division of responsibility is possible. The merchant himself deals directly with his customers. Decisions affecting every aspect of the business are in his hands, and he can act quickly. No difficulty exists in ensuring that every aspect of the business is dominated by the needs of the firm as a whole. On the other hand, small scale imposes some limits on managerial performance. Many smaller merchants are deeply immersed in the day-to-day clerical and manual work of their firm. Little possibility exists for a detached view of the firm as a whole. Because turnover is small, specialist employees can seldom be justified, so that the efficiency with which individual tasks are carried out may be impaired. Because the merchant is so busy, he may find great difficulty in keeping abreast of technical information.

5.7. Larger scale provides more opportunity for departmental organisation. Increased size often involves a range of dissimilar activities. The problems of each of these can be regarded separately and provide a basis for delegated authority. Where separate activities exist, some degree of delegation may be appropriate at relatively small levels of turnover. For example, the smallest firm with a recognisable departmental organisation had only £50,000 turnover: its two departments dealt with agricultural and horticultural trade respectively. More typically, larger firms divided their activities into categories covering grain, fertilisers, seed, milling, and feed. Firms sometimes used departments based on the function of the staff employed. Thus, one firm, with a turnover of £650,000, reported departments covering management, mill, sales, and office administration. In some cases departments were designated partly on the basis of business activity and partly on the function of the staff. For example, a firm with £800,000 turnover had departments covering grain, horticultural seeds. and sales and purchases.

5.8. The greater use of departmental organisation among manufacturing merchants reflects the distinctive character of their business. The organisation of a compound mill requires skills different from those involved in trading. Work routines and production patterns approximate more closely to factory conditions than in other aspects of the merchant trade. In addition, the opportunity to create meaningful divisions is greater among manufacturing merchants because of their larger average size. In this group, too, a somewhat higher proportion of firms owned their own transport fleet and thus became involved in an additional and different range of managerial opportunities and problems. Finally, as will be seen later in this chapter (p. 288) manufacturing merchants had credit outstanding to farmers for rather longer periods than had other types of merchant business. The conjunction of size, diversity of activities, transport, and credit facilities all afford a reason and basis for departmental organisation.

5.9. Some idea of the diversity of departmental organisation among manufacturing merchants firms is given in Table 5.2.

TABLE 5.2

DEPARTMENTAL NAMES USED BY 37 MANUFAC-
TURING MERCHANT BUSINESSES (I.E. THOSE
HAVING ANY MANUFACTURING ACTIVITY)

Title	No. of times reported
Mill	27
Grain	20
Administration	17
Transport	12
Sales	8
Other departmental names	50

5.10. Despite this profusion of titles, the fact remains that only 63 of the firms used any recognisable form of departmental organisation. Many merchants felt that their business was too small to divide into departments; and even the larger merchants had often given the matter no active consideration. Even where departmental labels do exist, the boundaries they delineate represent spheres of interest rather than autonomous areas of responsibility. Four main reasons were given for this reluctance to fragment the business. First, the merchant has transactions with an individual farmer over a range of activities. If the farmer is to be passed from one specialist to another his time and patience may be strained. Second, separation of buying and selling can have far-reaching effects on the whole business of the firm. To lose a good customer for feed by applying arbitrary standards in buying grain may damage the competitive position of the firm as a whole. Third, the trade of the agricultural merchant is seasonal. If employees are unable to switch from grain buying to fertiliser selling, or from feed sales to sales of sprays, it is more difficult to keep them fully occupied over the year. Finally, most firms aim at increasing turnover. To do so each of their representatives and executives must be fully aware of the total range of services the firm can provide. Ignorance or lack of interest about the activities of a separated department can lead to loss of trade.

5.11. Nevertheless, good management requires the identification and control of key elements in the merchant business. Decentralisation might be expected to play a greater role in this in the future. The growing complexity of agricultural and administrative techniques suggests that even small businesses may need the services of some experts whose field of knowledge is specialised. The growing volume of the grain trade and the greater possibility of matching home-grown grain to consumer needs may make specialist grain departments more valuable. The overhead costs of such departments become easier to justify because grain is marketed throughout the year. Finally, decentralisation may assist in identifying more closely the costs and returns from particular lines of business activity. In a changing and competitive world this may well help the firm to adapt more successfully to new opportunities. On the other hand, although many merchant businesses might improve management techniques through a more systematic approach to departmental organisation, the need to ensure a co-ordinated policy related to the whole range of activities in which the firm is involved demands that the central command structure of the firm retains a close degree of overall control.

The Problem of Recruiting and Training Labour

5.12. The structure of the merchant's labour force has already been described (Chapter 2, p. 195). Here, we are concerned with those issues which relate to finding the right man for the job and ensuring that his talents are well used. The difficulty of separating different tasks within the merchant business, already noted above in the context of departmental organisation, makes the task of choosing the right workers more difficult and more important. Two types of basic quality are demanded: technical competence, and an acceptable personality. In the past, emphasis has been more heavily placed on the second: in the future, technical proficiency may become more important.

5.13. The range of technical skills involved includes:

1. Skill in salesmanship.
2. Knowledge of the grain trade.
3. Knowledge of each of the other activities in which the firm is involved—usually at least feed, fertiliser, and sprays.
4. Skill in organising transport, both with local farmers and with large-scale suppliers and customers who may be 100 or more miles distant.
5. Skill in administration—covering, for example, work routine in an office, control of purchases and supplies, and supervision of credit.
6. Ability to work in a mill in which a variety of feeding stuffs are made from a range of raw materials.

5.14. The range of these skills makes it most unlikely that any one individual would excel in them all, or could readily be transferred from one to another. It is a particular handicap of the small business that it has to seek to combine a large number of skills in one individual. Larger firms can justify a greater degree of specialisation, and so might be expected to achieve a greater technical competence in each direction.

5.15. The Merchant Survey did not provide evidence upon which technical performance could be objectively assessed. The specific problems of each business are sufficiently different to make simple comparisons dangerous.

However, some impression of the range of technical skills was forthcoming. At one extreme were found relatively large-scale businesses which used such sophisticated techniques as computerised accounting and computer-controlled mills. A few of these firms employed specialists in formulation and in transport. At the other extreme existed a number of small firms whose business practices were largely traditional. Decisions were made on a rule-of-thumb basis, supplemented by advice from trade journals and from large-scale suppliers of feedingstuffs.

5.16. Recruitment to many small merchant businesses is influenced to a considerable extent by their family basis and relatively rural location. Because the business is owned by the family, sons tend to join the firm, and the prospects of executive responsibility are often remote for non-relatives. Because the business is sited away from major centres of population, there may be relatively few applicants from whom to choose when appointments are made. Changes in the structure of the merchanting sector, the growing size of firms, and the tendency for firms to link with other businesses, may reduce the impact of these factors, but they remain of importance.

5.17. Merchants who were asked what qualities they looked for in staff stressed ability to sell, experience, and integrity. The accent on experience reveals some of the problems of training. In a small business inexperienced employees may be a distinct liability, yet the cost of training them, in terms of time spent by senior staff, may be considerable; and in the end the trained man may leave for another firm where his experience commands a higher wage. The larger firms are in a better position to undertake training. Some the subsidiaries of national companies provide training schemes for all grades of staff. The better career prospects which large businesses can offer trained men afford a powerful reason for remaining with the firm after training. Even in companies whose resources are more modest, training in the elements of sales technique, pooling of experience within the firm and familiarisation with the firm's products may well be worthwhile.[1]

5.18. The Industrial Training Act 1964 attempted to remedy some of the deficiencies in the training of manpower. In particular, it sought to improve the supply of trained men and women, to improve the quality of training, and to share the costs of training more evenly among the firms who participate in the industry. In the agricultural merchant sector a good foundation for this activity already existed in the activities of the Institute of Corn and Agricultural Merchants. Before 1964 this body, to which

[1] See R. Tuck, Training is Your Business, *Agricultural Merchant*, Vol. 46, No. 7, July 1966.

firms contributed on a voluntary basis, already provided training for all levels of staff. Under the new legislative framework this activity has continued, but the programme has been co-ordinated under the Food Processing Industrial Training Board, and the cost of its services will henceforth be borne very largely by levies paid to the Board. The merits of this Scheme are evident. The large number of small firms who have little interest in or capacity to provide training can now participate in the benefits of a higher quality labour force without bearing an undue proportion of the cost.

5.19. The quality of the labour force is more than a question of capacity. It demands, too, a willingness on the part of those who work for the firm to use their talents to the full on its behalf. Management may influence this by providing jobs in which talent can be recognised. Some degree of independence of action, freedom to try new ideas and to seek new custom may help to engage the enthusiasm of staff. A further factor which is likely to call forth a good response from employees is the opportunity for promotion, but the most obvious and commonly accepted way of inducing a satisfactory pattern of motivation was through pay. In a large number of cases, salesmen were paid a basic wage plus a commission related to sales. Some merchants dislike this system because undue emphasis on volume of sales could lead to too little regard for the terms of sales, the credit liabilities involved or the need to promote more difficult products which earned more per unit sold for the firm. Even here, however, some system of relating remuneration to the profitability of the business was commonly discovered.

5.20. Representatives were the "front-line" workers of most merchant businesses. In the Survey of Merchants firms were asked if they used representatives for selling compound feeds, and the replies were supplemented by subsequent discussion of the mechanism of grain buying which revealed that representatives often played a key role here (see Chapter 4).

5.21. It is clear that the principal function of representatives is selling.

TABLE 5.3

FIRMS USING REPRESENTATIVES TO SELL FEEDINGSTUFFS

	No. of firms	No. who use representatives	as % of firms in group
Manufacturing merchants	117	90	78
Specialist grain merchants	24	6	25
Others	109	58	49
Total	250	154	62

The high proportion of manufacturing merchants who use representatives (see Table 5.3) reflects not only their greater average size, but also the need to find an outlet for the produce of the mill. (The importance attached to this is evidenced by the comment of several firms that they believed an increased team of representatives was a pre-requisite to increased turnover.) Specialist grain merchants, in contrast, have less need to employ representatives, as they have little or no feed to sell. Their major customers are the large millers and compounders.

The Problem of Choosing the Right Activities

5.22. To ensure maximum profitability it is axiomatic that the firm's resources should contribute as much to profits in their present use as they would in any other. If, for example, by moving resources from fertiliser sales to compound feed production they can earn a higher reward, the profitability of the firm cannot be maximised until this change has taken place. Good management involves the discovery of such inequalities in return and decisions which will reduce them. For this to be possible, the manager must first have a sufficient awareness of what is the return on resources used in various ways, and second, must assess the probabilities of this pattern persisting into the future. It did not seem that many of the firms visited in the Merchant Survey had adopted a systematic approach to this problem.

5.23. This impression was to some extent confirmed when merchants were asked, for instance, why they undertook non-cereal activities. Table 5.4 sets out the results of this enquiry, which include some multiple answers.

5.24. A variety of techniques exist for guiding decisions of this type. At their simplest, they involve setting out the prospective costs and returns from various activities. More elaborately, the problem may be explored in detail using the techniques of linear programming and relying on a

TABLE 5.4

REASONS FOR NON-CEREAL AGRICULTURAL ACTIVITIES

	No reason given	To even out the work-load	To provide a total service	To keep out com- petition	To spread risks	Other
Manufacturing merchants	39	26	48	2	1	15
Specialist grain merchants	13	6	5	2	—	—
Others	51	19	49	14	7	15
Total	103	51	101	18	8	30

computer to do the formidable amount of arithmetic involved.

5.25. Within the merchant sector it seemed reasonable to anticipate a wide range of methodological approaches to this problem associated with the diversity of firm size and type represented. In fact, reasons given for choosing activities suggest that in this respect decisions were more often based on tradition and intuition than upon any formal analysis of the situation.

5.26. Such an informal decision-making process may well produce good results. Evening out the work-load through the year adds to the return on capital at times when it might otherwise be idle and unprofitable. "Providing a total service" or pursuing an activity in order to "keep out the competition" may well be consistent with maximum return on the assets of the firm. Strictly, the lower return on the activities pursued to retain business should be regarded as a cost incurred in prosecuting the more profitable activities of the firm. Risk-spreading, too, is consistent with a good use of resources. By diversification many of the firm's fixed capital assets may continue to earn a profit despite changes in the relative profitability of the constituent activities.

5.27. The difficulties of a more systematic approach to choosing enterprises should not be underestimated. Determining the past contribution to profits of various activities is difficult where several activities share the same resources. Conventionally this may be done by examining comparative gross margins and assuming their shared resources are in fact fixed costs. As a practical expedient this may often be justified, but in some cases it could be misleading. The representative who is selling goods which have a low margin, as well as those which are more profitable, is, on this basis, regarded as a fixed cost to each activity. In fact, part of the problem of good enterprise selection is precisely to identify products which will make good use of representatives' time. To do so, it is necessary to know not only the gross margin on each, but also the rate at which representatives can sell them. A low gross margin per unit may still yield a satisfactory profit overall if sufficent units are sold in a short time. Even where the contribution of each enterprise is known, considerable difficulties exist in relating this to the closely-knit structure of the merchant business. Although compound feed sales may offer a better return than grain purchases, it may still be necessary to buy the farmer's grain in order to sell feed to him. Finally, the advocate of a systematic approach, as well as assuming that the past achievements of the business are known and that the interrelationships of the activities have been fully understood, has still to establish what future costs and returns will be, and to discount these for inescapable elements of risk.

5.28. Enough has no doubt been said to explain the preference of many merchants for rule-of-thumb methods. Complexity makes the task difficult: uncertainty makes its value less clear-cut. Despite this, some merchant businesses which were visited did approach the problem in a systematic manner. Two types of approach were adopted. Some of the larger firms, particularly those who were subsidiaries of national companies, called in expert management advice. Others took part in a co-operative programme of management analysis in conjunction with the Centre for Interfirm Comparisons Ltd. Participating firms provide data in a standard form and receive a series of ratios based on this data, and on comparative data for other anonymous merchants. These help to indicate not only how well the firm is doing in comparison with others, but also to suggest where the strong points and weaknesses exist in the merchant business. The interpretation of this data requires some care, but there is no doubt that it does provide a valuable aid to good management. It seems probable that the fees charged are fully recouped through more effective business planning.

5.29. The fact that only a minority of the merchant businesses visited had taken part in this project is disappointing. In a rapidly changing technical and economic environment, traditional rules-of-thumb are of diminishing relevance. Greater recognition of the need for sound business analysis and planning techniques could improve the efficiency with which existing tasks are performed and help businesses to adapt to new circumstances. It is paradoxical that while great effort is sometimes exerted to plan feed formulation correctly or to use transport wisely, less thought seems to be given to the overall direction of the buiness.

The Problem of Credit

5.30. It was made clear during the course of the Survey that one of the biggest managerial problems confronting agricultural merchants was the successful control of the credit which traditionally is supplied to farmer customers.

5.31. Any estimate of the total amount of such credit supplied to farmers by merchants must be more or less an informed guess, and indeed some merchants only have a hazy idea of the extent to which their own firm is committed in this respect. In 1958 the Ministry of Agriculture supplied the Radcliffe Commission of Inquiry with the figure of £120 million in Great Britain. This figure covered credit extended both by agricultural merchants and by some other businesses such as livestock dealers and auctioneers. At the time, this represented some 40% of total short- or medium-term loans to farms. (For Northern Ireland a sum of about £5 million should be added.) In a survey of 2460 farmers in Bucks. (Table

5.5) it was estimated that credit outstanding to merchants represented 4 %
of total liabilities of farmers (this includes all long- and short-term borrow-
ings)—as much as was borrowed from all private sources. The figure most
often quoted in trade circles for merchant credit outstanding is now about
£130 million.

5.32. A considerable amount of controversy exists among interested
parties as to the place of merchant credit in agricultural finance, and it
might be worthwhile restating the view recorded by the Radcliffe Commission in its Final Report. Paragraph 914, page 317, in the Report,
contains the following view:

"We believe that merchants' credit has a natural, important, and proper

TABLE 5.5

2460 FARMERS IN BUCKINGHAMSHIRE, 1962 — DISTRIBUTION OF TOTAL
LIABILITIES BY SOURCE

Size group (acres)	Banks £'000	A.M.C. £'000	Private £'000	Family £'000	Merchants £'000	Other £'000	Total £'000
5–99	801	—	141	72	48	114	1175
100–299	1287	364	176	252	90	—	2168
300–699	873	242	63	65	70	47	1359
Over 700	557	20	10	—	—	—	587
Total	3518	626	390	387	207	160	5289
	66%	12%	7%	7%	4%	3%	100%

Source: A. Harrison, "Some Features of Farm Business Structures", *Jour. Agric.
Econ. Soc.*, Vol. XVI, No. 3, 1965 (Table 9, p. 34).

part to play in the provision of finance for agriculture, and we think it
likely that the choice between merchants' credit and bank credit is determined much less by considerations of comparative cost than by considerations of convenience, and in some cases by the feeling that indebtedness to a bank is in some way reprehensible and disadvantageous in a way
which indebtedness to a merchant is not."

5.33. The above paragraph goes some way towards explaining why merchant credit is so important. It is important, firstly, because it is available,
although at a fairly high price as will be discussed later. It is available
because there are severe pressures on merchants to maximise their sales
even if this is at the cost of a higher credit commitment; and also because
merchants, together with the livestock dealers and a few other groups, are
in a unique position to judge the creditworthiness of clients who ask for
financial assistance. The merchants, particularly the smaller ones, move
in the same social circle as many farmers, and the relationship is often
sufficiently intimate to enable informality to be achieved. In circumstances
such as these, continuity is common in trade between a particular farmer

L

and a merchant. These links, valuable in many respects, make it difficult for a merchant to refuse credit, and equally difficult for him to call money in if the farmer is reluctant or unable to pay. Apart from these considerations, it is natural to expect the suppliers of agricultural requisites to be asked for credit, and it is natural that in a highly competitive situation most of them will be persuaded to supply it.

5.34. The farmers' attitude to finance from this quarter is mixed. Whilst they make use of it, they often suggest that it is extremely expensive. In a statement to the Radcliffe Commission, the N.F.U. cited the extent of merchants' credit as evidence of a fundamental gap in the short-term credit structure. The interest rate, it is alleged, is between 15% and 20% per annum, but the Report points out (para. 903, p. 314) that "it seems

TABLE 5.6

PERCENTAGE DISTRIBUTION OF FIRMS ACCORDING TO AVERAGE
NUMBER OF WEEKS CREDIT OUTSTANDING TO FARMERS

	Weeks of credit				
	0–4	5–8	9 or over	Not available	Total
Manufacturing merchants	7	44	48	1	100
Specialist grain merchants	29	21	38	12	100
Others	17	35	44	4	100
All firms	13	38	45	4	100
(No. of firms)	(33)	(95)	(113)	(9)	(250)

unlikely that the cost of merchants' credit presents itself to a farmer's mind in terms of a rate of interest".

5.35. The Survey investigated the question of merchants' credit not from the point of view of total amount of capital outstanding, but rather in weeks of credit, i.e. what was the average time which elapsed between the date of the invoice and the settlement of the account. Table 5.6 indicates the proportion of firms in the three principal functional categories falling into three credit periods.

5.36. The amount of credit outstanding to farmers was much larger than was expected when the questionnaire was originally planned, so that a large number of firms fall into the open-ended category of nine weeks or over. Many firms in this category estimated that the average period was three months or more.

5.37. It can be seen from Table 5.6 that manufacturing firms tended to have credit extended longer than either other merchants or specialist grain merchants, and that specialist grain merchants were less extended in this respect than other merchants. The specialist grain merchants, of course, purchase more from farmers than they sell to them, and thus are not confronted with the problem to the same extent as firms which are principally suppliers of farm requisites. Other merchants are probably forced to be more severe in their credit policy than manufacturing merchants, in that a greater proportion of their sales is likely to be made up of purchased compounds, for which they have to pay promptly themselves.

5.38. Table 5.7 shows the distribution of credit period amongst firms in the various turnover categories.

TABLE 5.7

PERCENTAGE DISTRIBUTION OF FIRMS ACCORDING TO AVERAGE CREDIT
PERIODS AND TURNOVER CATEGORIES

Turnover group (£'000)	Weeks of credit			Not available	Total
	0–4	5–8	9 or over		
Under 75	22	33	30	15	100
75–149	12	34	52	2	100
150–499	9	48	43	—	100
500–1249	13	25	62	—	100
1250 and over	15	41	37	7	100

There is no particular association between size of business and extension of credit. The category "under £75,000" has the highest proportion in the shortest credit period, and the smallest proportion in the longest of the three credit periods, but also a relatively high number where the firm was unable to give an answer. This seems likely to mean that the credit period is relatively long, which would put the smallest firms in a closer position to the others.

5.39. There is little evidence to suggest that the extent of credit commitments by merchants is decreasing. In answer to the question "Is the period for which credit is extended tending to lengthen?", 20% of all firms admitted that it was, 41% said that there was no significant change, 24% said it was decreasing, and 15% did not know or were unable to answer. Assuming that it is the aim of most firms to reduce the amount of working capital loaned to farmers it would thus appear that only a quarter of all firms are being successful in this.

5.40. Farming is a seasonal industry, and consequently the amount of credit outstanding to farmers from merchants also has a seasonal aspect. Twenty-two per cent of merchants said that March/April were their worst months from this point of view, 21% named May/June, and 22% July/August. This pattern, which was found in most regions, is what one would expect to emerge in most farming systems. Feedingstuff requirements reach a peak during the winter; fertiliser and seed are purchased during the early spring, and for many farms little money comes in until the cereal harvest at the end of the summer. The months when the amount of credit outstanding is lowest follow harvest, although in the Eastern region September/October was reported as being a relatively extended time. The reason for this is not clear, but it may be that in this area storage facilities are better and cereals contracts used more heavily than elsewhere, so that a lower proportion of the total crop is sold at harvest.

5.41. In conclusion, it must be emphasised that the figures quoted as weeks of credit were average figures for each merchant, and that each merchant has a range of customers, from those who invariably pay within the month to those who do not pay at all. One medium-sized firm, for example, stated that approximately one-third of customers settled within one month, a third within 6 months, and a third between 6 months and 3 years. Another merchant, however, claimed that 50% paid within 28 days, 25% in 2 months, and the rest within 12 months; and a compounding firm estimated that 75% of farmers paid within one month.

Credit Control

5.42. The amount of credit outstanding to an agricultural merchant is largely a function of his own credit control policy. The farmer is encouraged to pay early by the granting of "discounts" for prompt payment, or occasionally by charging for credit at a certain rate—usually 1% or 2% above bank rate. There is, however, a wide divergence of terms and conditions among different firms, and the degrees of success of their credit control policies. This can range from having 80% of customers paying cash, and no credit at all allowed over 6 months, to the situation of the merchant who stated that most of his farmers only paid once or twice a year, and that he only bothered to send out statements every 3 or 4 months.

5.43. The "discount" system of credit control is widely used by merchants, though the word "discount" is not a correct indication of the actual procedure which takes place. If the farmer pays the merchant on the same day that he receives the produce he does not get a discount off the net price; rather it is the practice for merchants to add on to the net price of compounds and fertilisers a surcharge of £1 or £2 per ton. Then if the farmer pays within the stipulated period this surcharge is removed. In fact, farmers are not rewarded for paying promptly, but are charged for not paying promptly. It is, of course, quite proper that this should be the case, but farmers may get the impression that they are buying feed at less than the net price when this is not true. Similarly some farmers who are tardy in settling accounts might be less so if the full extent of the cost of merchant credit was made more easily apparent to them.

5.44. There is a good deal of variation in the magnitude of the prompt payment incentive, and there is clearly a substantial difference in the cost of credit from different merchants. For example, some merchants add £2 per ton on to the invoice for a feed delivery, and refund the whole of this if the account is paid within 28 days, 30s. of it if the bill is paid within 2 months, £1 if paid within 3 months, and 10s. if within 4 months. This represents a rate of interest of approaching 20% per annum, if we assume the price per ton of the compound to be about £30. In other cases, the initial credit surcharge may be 30s. or only £1, and a similar sliding scale applies. However, in some cases the initial surcharge is reclaimable wholly if paid within the month, or within 28 days, but no part of it is paid back if this payment is not made. Thus, the pressure on the farmer to pay promptly is increased, but if he goes beyond the loss of the £2 surcharge in the first month then there is relatively little incentive for him to pay after that. In other cases, the period for maximum "discount" is by the end of the month following the month of delivery. This gives the farmer a longer period in which to pay and yet avoid a credit charge, but some merchants reported that it resulted in a bunching of orders for feed during the first week in each month because farmers were anxious to have as long to pay as possible. In a few cases, however, the period for which full discount could be claimed was substantially less than one month—in one case 14 days.

5.45. The weakness of the "discount" method of charging for credit is that once the discount has been lost, there is no incentive for the farmer to pay at an early date. If the merchant adopts a sliding scale of diminishing "discounts" over time, then the impact is not so great as if a sum of £2 per ton is suddenly lost overnight. The choice may well be between having a substantial proportion of customers paying within the month, and the remainder paying after a relatively long time lapse, or a pattern whereby the flow of payments gets off to a slower start but continues more steadily.

5.46. The alternative method of charging for merchants' credit is to apply a straightforward rate of interest. At its simplest, it can be illustrated by a firm in Scotland which allows three months' credit without charge, and thereafter charges 1% above bank rate. In some cases a variable rate of interest is employed. It may be $2\frac{1}{2}$% per month for 1 to 4 months, and 3% for 4 to 6 months. After this, some other action will be taken. The number of times when an explicit credit charge was used as opposed to a "discount" offered were relatively few.

5.47. There are aspects of credit control other than monetary incentives to prompt payment. When the farmer has lost his "discount" and is committed to paying an additional credit charge, the merchant or his representative has to coax or threaten the farmer until he pays. Obviously, it is against the merchant's interests to make an enemy of a farmer. The threat of legal action is used only as the last resort. When farmers fail to pay because they are immersed in the day-to-day running of the farm and regard any kind of paper-work as anathema, enterprising but tactful representatives can do much to hasten the payment of outstanding accounts. A personal visit is often more effective than a postal communication in achieving this end.

5.48. One large firm in Scotland provided representatives with an incentive to persuade farmers to pay by deducting $\frac{1}{2}d$. per month from the representative's commission for each £1 overdue from that particular representative's customers.

5.49. There is then a wide divergence both in the amount of credit which merchants are prepared to offer, and in the terms on which they are prepared to advance it, and the farmer can go "shopping around" to find the terms most beneficial to him. Not only are the formal terms offered by individual merchants often substantially different, but they are often not adhered to on a rigorous basis, and a prospective customer, particularly if he has a reasonably big order, will be in a position to negotiate terms with some merchants.

5.50. It is alleged that merchant credit is expensive to the farmer, and it is difficult to refute this on a simple calculation of the rate of interest. However, the provision of this type of credit is also costly to the merchant. He must face the fact that capital could be employed elsewhere at a profit, and he must incur the expense of maintaining ledger accounts for each customer and the cost of envelopes and postage when statements are sent out. To send salesmen to collect debts involves both direct expense and an indirect loss of other business. In some cases, where other methods have failed, the costs of a solicitor must be paid. All these expenses are additional to the immediate cost to the merchant of borrowing the money himself.

5.51. Doubtless, most merchants would prefer to trade solely for cash if it were possible, but the tradition and the competitiveness of the sector force the great majority of merchants to offer credit terms, and so involve themselves in one of the most delicate and troublesome aspects of the merchanting trade.

Storage—the Problem of Carrying the Right Stock

5.52. Apart from capital involved in buildings, transport, and other relatively long-lived assets, merchants require to keep a stock of the goods in which they deal. They must be able to deliver to farmers within a short time goods which are ordered. Failure to do so may lead to lost custom. The more varied the character of the business, the more complex becomes the problem of stock-carrying. For manufacturing merchants, stocks must be carried sufficient to ensure that they are able to keep their milling equipment running on an economic basis.

5.53. The Survey of Merchants showed that very few firms kept substantial stocks of grain. In part this was due to physical limitation of storage capacity, in part to the existence of farm storage facilities and storage subsidies for farmers. Above all, however, it reflects the essentially trading nature of the merchant's job in marketing grain. To store grain on a speculative basis not only exposes the merchant to risk; it also calls upon his capital reserves. A safer and more productive use of capital requires investment in goods with a more rapid rate of turnover. Only where grain was needed for processing did storage form a common feature of the merchant's business. In a changed market situation with parity of storage costs and rewards between farmers and merchants, it might become worthwhile to invest in silos. In the current situation it is not.

5.54. One of the potential economies of scale which larger merchant businesses may enjoy is in terms of stocks. Particularly for items which move relatively slowly or irregularly a minimal stock may be sufficient to service the needs of a relatively large business. Where seasonal shifts in demand take place, the opportunity to use storage space for diversity of purposes may reduce the incidence of its cost on individual items. The preoccupation with increased size which is so much the concern of some merchants is well founded in this respect.[1] On the other hand, mutual assistance by small merchants could eliminate at least part of this advantage.

5.55. The best answer to the problem of stock-carrying must vary from firm to firm, reflecting the differing activities and access to capital which each enjoys. What is clear is that the modest buildings and limited stocks which many firms possess are not in themselves signs of bad management. Provided that an adequate job can be done without more costly investment, there is no reason to suppose that a superficially more impressive array of resources would contribute to economic efficiency.

The Problems of Transport Management

5.56. Managing the transport requirements of a merchant business is in itself a complicated and specialised problem. The existence of separate transport departments, and even in some cases of wholly-owned but separate companies whose sole function is transport, provides *prima facie* evidence of the truth of this assertion.

---

[1] See, for example, Granum Mercandum, *Agricultural Merchant,* Vol. 46, No. 11, p. 65.

5.57. It is in no sense the object of the following observations, derived from the Survey of Merchants, to establish criteria for a good transport policy. Rather, the evidence is used to illustrate some of the problems of using transport, and especially to draw attention to those which affect the movement of grain. The internal transport of grain has already been covered in Chapter 3.

5.58. The first problem concerns the choice of means of transport. For movements off the farm road transport is universal. Here, problems of accessibility for lorries and the use which can be made of the same lorries for deliveries to farms must be weighed against the advantages of using very big bulk vehicles. The merchant who sacrifices some technical economies in grain movement in order to use a smaller, less specific, vehicle, which can also be used to deliver requisites, may well be making a good decision. The extra costs of moving grain in such general-purpose vehicles may be less, both for the merchant and the community, than the losses associated with under-utilised vehicles which are better designed simply from the point of view of transporting grain.

5.59. The choice of transport for movements from ports or from other merchants is more complex. The range of possibilities may include rail and water, as well as road. The benefits of purpose-designed vehicles are more evident as distances increase. Where the delivery is simply to the merchant's premises, this may determine the issue in favour of one particular type of specialist vehicle. On the other hand, if grain has to be moved direct to farms large vehicles may not be suitable, and rail or water may involve transhipment costs. Again, a total view of the merchant's business may lead to the use of vehicles which differ from those which would be chosen were grain haulage costs considered alone.

5.60. Size of business itself must influence the best choice of transport. For a small firm to buy a bulk lorry may be as foolish as for a 25-acre farmer to equip himself with a 14-foot combine harvester. In this context large scale enhances the probability that technical and economic optima will coincide leading to low unit costs of transport.

5.61. The second problem which must be faced is whether to buy or hire transport. Hired transport may enable a wider range of effective choice to be made—although in this case the choice will be in the hands of the contractor. The merchant escapes the capital cost and the worry of organising his own fleet, but he does so at a price. First, he must pay a transport charge which will leave a profit for the haulage contractor. Second, his control over the behaviour of lorry drivers on farms is much diminished. Third, he loses the publicity value of having his own lorries seen on the road and at the farm gate. For merchants whose trade is irregular, or whose operations are on a very small scale, these disadvantages are outweighed by the savings a haulage contractor makes possible. In most cases, as can be seen from Table 3.2, Chapter 3, a mixture of owned and hired vehicles is employed. Most merchants regarded their own fleet as providing the base load of transport requirements. Hiring was resorted to when lorries owned by the firm were fully committed. In a small number of cases, however, the reverse philosophy was adopted. A fleet of small lorries was owned to handle urgent deliveries, but the bulk of the business was based on hired vehicles.

5.62. A third management problem which occurs when lorries are owned is the routing, maintenance, and replacement of the fleet. Good routing demands planning journeys which may involve a whole series of pick-up and delivery points. This must be done within the restrictions imposed by permitted driving hours and licensing arrangements. Maintenance is of increasing importance, not only to keep the lorry operational, but also to observe the stringent safety requirements imposed by Parliament. Again, the advantages are with the larger firms, who can afford a specialist maintenance unit capable of keeping vehicles in good condition. Replacement problems face the firm afresh with decisions about the size of fleet, type of lorry, and role of hired tranport. Changing circumstances demand that these issues be continually reviewed. What is clear is that once investment in a lorry has been undertaken, it pays to use it to the full. Again, this may mean that grain is not always moved in the cheapest possible way if one journey alone is considered, but that the overall economies associated with the remainder of the merchant's activities more than offset this disadvantage.

The Profitability of Merchanting

5.63. In the course of the Merchant Survey no data about profits was collected, but many merchants commented on the low level of profits in the sector as a whole. Their attitude to investment indicated that for a majority profits were insufficiently attractive to justify extensive commitment of new resources. Further, the reduced number of merchants between 1947 and 1967 is consistent with a situation of low profitability (see also Part III, Chapter 2, Table 2.21 and Map 2.2).

5.64. Three points must be made before any conclusions can be drawn from this evidence. First, a high proportion of merchants regarded low margins as synonymous with low profits. This is not so. Even where very small margins per transaction are involved, sufficient transactions can lead to a good return on capital. Discussion of profit as a percentage unit of turnover can lead to the wrong conclusion.

5.65. Second, comparisons of the rate of return on capital in the

L*

merchant sector compared with other industries must take into account the relative levels of risk involved. Merchants are certainly very conscious of the risks attaching to their business. Their reaction, in general, to speculation was adverse. They seemed reluctant to abandon unprofitable activities because of the risk that some other business might also be lost. However, the degree of risk involved should not be exaggerated. The demand for the services of merchants is related to the level of farm activity. Within the framework of agricultural support, this has remained relatively steady. Trading risks can be avoided by hedging operations—though this is seldom done—and while compound manufacturers may face price uncertainties, especially for imported raw materials, these firms account for only half the merchant sector. It remains true that catastrophes such as the recent foot-and-mouth outbreak may severely damage a merchant's business, but the frequency of such occurrences is, fortunately, very rare. Overall the level of risk involved in merchant trade does not seem to be inordinately high, and a fairly modest return on capital would not be inconsistent with the overall health of the sector.

5.66. Finally, before interpreting reports of low profits as indicative of too large a number of firms, we must notice the range of profitability among the firms concerned. The existence of competition and the pressure of newer more successful methods of trading must, of course, bring this about. The contribution of merchants to economic progress may demand that some firms cease to trade, but their business may well be taken over by the expansion of existing businesses or by new entrants.

5.67. In an attempt to throw more light on the variability of the returns experienced by merchants, the data relating to 25 co-operative societies was analysed. These were the societies which dealt in grain and provided information. Co-operative societies have different goals and different business procedures from private firms. It is, therefore, somewhat dangerous to compare their net profit with that of the rest of the merchant sector.

TABLE 5.8

THE FINANCIAL RESULTS OF 25 AGRICULTURAL CO-OPERATIVE SOCIETIES IN 1966

Turnover group (£'000)	No. of societies	Average turnover (£'000)	Average margin per £100 turnover	Average net profit per £100 assets	Average net profit per £100 turnover
			£	£	£
Less than 75	2	61	6·0	3·2	0·2
75–149	1	119	10·9	0·4	0·1
150–499	4	337	8·7	4·8	1·1
500–1249	5	857	8·2	4·5	1·3
1250 and over	13	7524	11·4	5·9	1·6

However, in our experience the characteristics of private merchants and of agricultural co-operative societies were sufficiently similar to make data relating to the co-operative section of the trade of interest and relevance. Table 5.8 sets out the results of this enquiry.

5.68. This evidence supports the notion that the level of profitability is low. Three aspects of these low returns must be mentioned. First, neither the profit per unit of capital employed, shown by the average net profit per £100 assets, nor the profit in relation to turnover can be closely related to the average margin on sales. This emphasises the danger of too great a concern with margins. Second, the bigger societies tend to be more profitable. The very small number of societies in the smaller size groups makes this an impression rather than a conclusion. The impression, however, is consistent with the view commonly expressed among merchants that the smaller businesses were most vulnerable. Third, the average

TABLE 5.9

TRENDS IN TURNOVER, ASSETS AND NET PROFIT OF SIX AGRICULTURAL CO-OPERATIVE SOCIETIES (1966 AS PERCENTAGE OF 1956)

	Size range (1966 turnover £'000)					
	150–499			500–1249	1250 and over	
	Society A	Society B	Society C	Society D	Society E	Society F
Turnover	136	107	140	203	188	340
Assets	116	100	140	251	212	268
Net profit	46	9	62	85	31	215

rate of return on capital earned in these businesses was lower than that which prevailed for gilt edged securities during 1966. On the basis that the risks involved in merchanting exceed those of holding government stock, the firm conclusion may be drawn that the assets of some of these societies could have earned higher profits elsewhere—always remembering, however, that profit was not their sole objective.

5.69. The figures of one particular year may prove a misleading basis for judgement of the profitability of any group of firms. For this reason, data relating to six societies were analysed for a 10-year period, 1956–66. These societies were chosen to give a fair range of size of turnover but should not be regarded as being necessarily representative of the normal range of profitability among agricultural co-operative societies generally. Table 5.9 gives the result of this analysis in respect of turnover, assets employed, and net profit. It is immediately clear that while each society has substantially increased its turnover, only one has managed to increase its profitability. Both as a percentage of turnover and of capital, the profitability of these societies has declined.

5.70. The economic environment within which these societies operate has become more rigorous during this 10-year period. This climate is common to all merchant businesses, and there is no reason to ascribe the difficulties demonstrated in Table 5.9 to the co-operative characteristics of the societies involved. Equally, within this adverse environment the performance of individual societies varies greatly. The most successful society has more than tripled its turnover and has doubled the value of its assets. The least successful had about the same turnover and level of assets in 1966 as in 1956. At the heart of these differences lies variation in the quality and capacity of management.

5.71. Private businesses as a group may be more successful, although there is little evidence to support such a claim. More certainly, the pressure on poor-quality management is very real.

CHAPTER 6

The Market in Cereals Seed

6.1. About 5% of the grain produced in the United Kingdom is used for seed. Some of this is retained on the farm of origin but a large proportion passes through various trade channels before it is used to produce a new crop. This chapter describes the market in seed and discusses some of the special characteristics of seed transactions. The production and development of new seeds is a highly complex and specialist operation, and no attempt is made here to evaluate either the technical problems or the technical opportunities associated with the introduction of new varieties. Discussion is limited to those aspects of the seed trade which impinge on the marketing of the home-grown cereal crop.

6.2. Table 6.1 sets out an estimate of the total volume of seed used in the United Kingdom. Our Survey suggested that a high proportion of this passes through the trade, but the sample results did not provide sufficient basis for a tonnage estimate. A survey by the National Association of Corn and Agricultural Merchants indicated that over 400,000 tons of wheat, barley and oats seed were handled by its members in 1967/8. (See also *The Agricultural Merchant,* March 1968, p. 42.) A very large proportion of seed purchases are made from general agricultural merchants. There are also seed wholesalers and specialist seed breeders.

Purchase of Seed from Farmers by Merchants

6.3. Some 63% of the firms visited in the Merchant Survey buy seeds direct from farmers. Those who do not buy from farmers meet their requirements for seed by purchases from other merchants or from specialist seed houses. Seed bought from farmers is frequently inspected during growth and requires extensive cleaning, drying and dressing procedures when it comes into the merchant's hands. Facilities to carry out such operations would place a considerable strain on the resources of merchant businesses whose dealings in seed were on a small scale. Hence many of

TABLE 6.1

SEED USED ON AGRICULTURAL HOLDINGS IN
THE U.K., 1962/3–1966/7 ('000 TONS)

	1962/3	1963/4	1964/5	1965/6	1966/7
Wheat	150	170	194	173	181
Barley	307	330	351	400	393
Oats	114	99	90	81	89

Source: M.A.F.F.

the smaller businesses rely on trade suppliers to meet their needs. In fact, as Table 6.2 makes clear, few merchants buy large quantities of seed from farmers.

6.4. Two points about Table 6.2 should be noted. First, a fairly high proportion of those merchants who do buy seed purchase only one or two of the three principal cereals produced in the United Kingdom. Second, although 63% of the merchants were involved in buying seed (Table 6.3),

TABLE 6.2

SEED PURCHASED FROM FARMERS BY 199 AGRICULTURAL MERCHANTS

(% of merchants buying the following tonnages)

	Nil	1–99	100–199	200–499	500–999	1000 and over	Total	Total bought (tons)
Wheat	65	14	8	6	3	4	100	23,720
Barley	60	14	5	10	6	5	100	37,103
Oats	72	18	5	2	2	1	100	9556

Note. This table refers to 199 merchants who provided tonnage information on seed purchases from farmers.
Source: Survey of Merchants.

a very much smaller number of firms accounted for the bulk of purchases from farmers. Further analysis of the results of the Survey clarifies this point. The 10% largest buyers of seed accounted for 66% of the wheat, 59% of the barley and 77% of the oats bought for seed from farmers.

6.5. Table 6.3 shows the regional distribution of merchants who bought seed from farmers. The dominant position of the Eastern region in seed production reflects its position as the major cereal-producing area of the U.K. In Scotland the high proportion of merchants buying seed is more to be explained in terms of the local importance of oats and of the tradition of seed production.

6.6. Seed may be bought from farmers on a spot basis or subject to contract. Spot transactions for seed differ little from those for other

TABLE 6.3
REGIONAL DISTRIBUTION OF 164 MERCHANTS
WHO BOUGHT CEREAL SEED FROM FARMERS

Region [a]	No. of firms in sample	No. buying seed	% Buying seed	% of those buying seed who bought on contract
South-west	32	23	72	26
South-east	29	21	72	29
East	45	39	86	33
East Midlands	22	12	55	50
West Midlands	28	21	75	24
Yorks. and Lancs.	24	10	42	30
Northern	14	7	50	42
Wales	11	4	36	25
Scotland	24	17	71	35
N. Ireland	31	10	32	30
	260	164	63	32

[a] These are Advisory Regions (see Appendix C).
Source: Survey of Merchants.

cereal purchases. Provided a satisfactory crop is available, a price is negotiated between merchant and farmer. Contract arrangements are more complex. Many contracts will specify the field in which seed is to be grown, the husbandry processes to be undertaken and the method of harvesting, drying and cleaning the crop. Clauses empowering the merchant to inspect the growing crop are generally a feature of such contracts. Prices are fixed under the terms of the agreement, commonly on the basis of an addition between 2s. 6d. and 5s. per cwt to the ruling market price on the day the seed is collected by the merchant. Table 6.3 shows that about a third of the merchants who bought seed did so on contract. Only half of these merchants relied on contracts alone. It is not possible to estimate from these figures the relative amount of seed corn bought on contract. The impression gained in conversation with merchants was that contracts tended to be used by the larger firms but that, despite this, only a minor proportion of seed was in fact grown on contract.

6.7. The merits of a contract system of trade are more evident when grain is grown for seed, than when the market is less clearly specified. Seed production demands especial care on the part of the farmer. For this extra trouble and cost his reward depends on actually selling the crop as seed. Its value for other purposes is no greater than grain produced in a less costly manner. A contract ensures that a seed market will be available. For this security he has to sacrifice the possibility of an advantageous spot deal when the crop is ultimately harvested. The value of security depends on the circumstances and attitudes of the farmers concerned but it seems reasonable to assume that it will be fairly high for those farmers whose income is largely derived from seed crops or who are regular producers of seed.

6.8. Merchants, too, may derive benefit from contract arrangements. In the short term they ensure the availability of seed for the next season. In the longer term they may help to build a continuing relationship with farmers who produce good-quality seed. One of the larger seed-buying merchants stressed the importance of this. For him it was essential that farmers should have sound technical knowledge. In addition they should grow a large enough area of seed, 500 to 600 acres, to justify special equipment to deal with seed, regular visits of inspection from the firm and should have the ability to build up an accurate history of the fields upon which seed is grown. This firm buys all its cereal seed on contract, and contracts with the larger producers are negotiated by management executives rather than by representatives.

Sale of Seed to Farmers by Merchants

6.9. The Survey of Merchants showed that 78% of the firms sold cereal seed (Table 6.4). The smallest businesses were less concerned in the trade, but even among the group with less than £75,000 turnover, 40% had some seed business.

TABLE 6.4

CEREAL SEED SALES AMONG 260 AGRICULTURAL MERCHANT BUSINESSES

Turnover (£'000)	No. of firms	No. selling seed	Percentage of firms selling seed, %
Less than 75	42	17	40
75–149	44	32	73
150–499	94	88	94
500–1249	46	38	83
1250 and over	27	27	100
Not known	7	4	57
Total	260	202	78

Source: Survey of Merchants.

6.10. Table 6.5 shows that there is less specialisation in the sale of seed than in its purchase from farmers (Table 6.2). Merchants who sell seed at all usually sell all sorts of cereal seed. A considerably higher proportion of merchants sold cereal seed to farmers than purchased it from them. These factors demonstrate the essentially retail character of seed sales by many merchants.

6.11. In Chapter 2 of this Part the extent to which merchants attempt to offer farmers a comprehensive service was discussed. Inter-merchant

TABLE 6.5

SEED SOLD TO FARMERS BY 199 AGRICULTURAL MERCHANTS

(% of merchants selling the following tonnages)

	Nil	1–99	100–199	200–499	500–999	1000 and over	Total	Total sold (tons)
Wheat	49	27	10	8	3	3	100	21,185
Barley	41	26	11	13	4	5	100	35,379
Oats	56	32	7	4	1	—	100	6041

Note. This table refers to 199 merchants who provided tonnage information on volume of sales.

Source: Survey of Merchants.

transactions make up any deficiencies in a merchant's own purchases from farmers. In contrast some of the larger seed buyers are very much in the position of wholesalers whose sales are aimed as much at other merchants as directly to farmers. Some indication of the extent of inter-merchant transactions in seed can be derived from Table 6.6.

TABLE 6.6

PURCHASES AND SALES OF SEED FROM OR TO OTHER MERCHANTS BY 199 AGRICULTURAL MERCHANTS (tons)

	Bought from other merchants			Sold to other merchants		
	Quantity	No. of merchants buying	Av. annual purchase per merchant buying	Quantity	No. of merchants selling	Av. annual sale per merchant selling
Wheat	6797	79	86	4408	28	157
Barley	10,102	90	112	8737	43	203
Oats	3827	70	55	6003	21	286

Source: Survey of Merchants.

6.12. The network of inter-merchant communication assists seed trans-actions, but, in addition to this, there exist a small number of seed brokers. These firms do not physically handle seed. Their function is to exploit a very full knowledge of the market in order to arrange transactions between merchants who have a relative surplus of some variety and merchants who are facing shortage. Essentially this is an inter-regional trade and a supplement to the communication system which exists in the merchant sector.

6.13. Such firms cannot properly be described as wholesalers since they do not own the goods at any stage. Very few firms in fact sell exclusively to merchants and can thus properly be regarded as wholesalers. More typical are the quasi-wholesalers who have extensive direct dealings with farmers but also derive a considerable proportion of their trade from sales to merchants. Such firms include specialist seedsmen as well as agricultural merchants who have a large business in seed. Compared with the general merchant, specialist seed firms have a very acute seasonal problem so far as their labour force is concerned. During the growing season large numbers of men are needed to inspect and certify seed crops. The general merchant can redeploy his staff into other actitivites during the remainder of the year but for the specialist seedsman it may be difficult to find worthwhile employment. On the other hand, the seed trade is becoming more complex and demanding an increased degree of technical skill. Such sophistication is probably less important for cereal seed than for herbage or horticultural seeds. In current circumstances the balance of advantage appears to lie with the general merchant who can spread his overhead costs over a larger volume of trade. Discussion with several specialist seeds firms emphasised the economic pressures under which they were working. As evidence attention was drawn to the tendency towards concentration in the seed trade in recent years.

6.14. The production and dissemination of new seed varieties is of immense importance to cereal production. One authority[1] has estimated that something like half the total improvement in cereal crops may be directly attributable to the use of new and better varieties. Pedigree seed houses play a major role in the discovery and importation of new varieties. New varieties have also been developed by the Plant Breeding Institute at Cambridge, and the work done by the Welsh and Scottish Plant Breeding Stations should also be mentioned. Royalties are now payable on new cereal varieties. The *Financial Times* stated that "by the time a new variety is offered to farmers it will have cost its breeder anything from £20,000 to £50,000 and involved a fantastic gamble in the process".[2] For a really successful variety, returns might reach well over £100,000 a year.

6.15. The value of improved seed varieties for the production of home-grown cereals depends not only on their relative merits compared with established seeds but also on the speed with which they are adopted by farmers. The process of sifting new varieties so that those which have most to offer are accepted by farmers is greatly facilitated by the work of the National Institute of Agricultural Botany. The N.I.A.B. tests all new varieties before they appear on the Index of Varieties,[3] and issues a recommended list. Its recommendations cover yield, resistance to disease,

---

[1] F. R. Horne, Prospects for crop improvement through new varieties and better seeds. *Journal of Farmers' Club,* Pt. I, 1961.

[2] Reported in *Agricultural Merchant,* Oct. 1966.

[3] Only varieties which appear in the Index may be sold as seed.

grain quality and suitability for different regions. The list is published each year and widely reported in the farming press. Similar recommendations are made in Scotland by the Agricultural Colleges. This impartial and informed volume of advice helps farmers to distinguish which of the new varieties are most likely to prove successful in their own environment.

6.16. The information provided by the N.I.A.B. is not the only influence on farmers to adopt new seeds. Seed breeders advertise their products in the trade and farming press. Merchants advise their customers on new varieties. Inevitably in a market in which so much choice is possible and which may be greatly affected by weather conditions at seed-time there is some uncertainty concerning the uptake of new varieties. Horne[1] estimated that an intensive advertising campaign might raise the up-take of a new and little-known variety to as much as 10% of total national usage before trials can be completed. The records show that a variety is not likely to be maintained at the 10% unless it does well in trials and is placed on the recommended list. Very successful varieties such as Cappelle-Desprez wheat or Proctor barley have risen to as much as 40% of the total acreage within 5 years of being placed on the recommended list.

6.17. Uncertainty about the reaction of farmers to new varieties and the possibility that weather conditions may curtail winter wheat and barley acreages adds to the complexities and costs of the seedsman's trade. If a variety is popular he may be short of supplies. If farmers fail to buy, stocks of expensive seed may remain unsold. Such seed may, if it has not been dressed for planting, fetch feed grain prices as opposed to prices ranging from 45s. to 90s. per cwt which may have been expected for seed. If the seed has been dressed the price at which it can be sold may fall to 8s. per cwt.

6.18. Trade in seed is regulated under two Acts of Parliament, the Seeds Act, 1920, and the Plant Varieties and Seeds Act, 1964. In addition extensive modifications in existing practice were made under the Seeds Regulations, 1961, and corresponding regulations for Scotland. All sellers of seed must give a declaration that seed has been tested for purity and germination. The name and address of the seller of seed and the variety of seed and the country of origin must be included. The name and number of injurious weed seeds[2] must be declared. Merchants may sell seed as reaching a "minimum declarable percentage of germination". For wheat and barley this is 90%, for oats 85%. Seeds may be sold at lower rates of germination but, in this case, the actual percentage must be declared.

---

[1] F. R. Horne, op. cit.

[2] Wild oats, dodder, docks and sorrels, black grass and couch-grass are classed as injurious weed seeds.

Because seed dressing to prevent damage by pests or soil-borne diseases may adversely affect germination the seedsman is required to state whether the seed was tested before or after treatment and the nature or name of the treatment. Tests of purity and germination must be carried out at official seed-testing stations or at a private licensed station.

6.19. The emphasis of these statutory arrangements is to safeguard the buyer of seed. Seed producers adopt a much more positive attitude to quality. Because cleaning operations are costly and because as much as 30% of a crop bought from farmers may have to be discarded in the cleaning process,[1] buyers attempt to ensure that good quality is attained when the crop is grown rather than by extensive cleaning operations after harvest.

6.20. The National Institute of Agricultural Botany distinguishes four grades of quality:

(a) *Basic seed* which is produced under the responsibility of the breeder and is intended for the production of Certified seed.

(b) *Certified seed* which is intended for producers of Certified or Multiplication seeds.
Both Basic seed and Certified seed meet international standards established under the O.E.C.D. Cereal Seed Scheme.

(c) *Multiplication seed* which may be used for the production of further Multiplication seed and for the production of Field Approved seed.

(d) *Field Approved seed* which is intended for use by farmers for commercial crop production. It may not be used for further multiplication as seed.

6.21. The first three grades are intended for use by farmers who produce seed. Field Approved seed in contrast provides the ordinary cereal-growing farmer with a reliable officially sponsored seed. The Cereal Seed Field Approval Scheme, administered by the N.I.A.B. in England and Wales, is being extended in 1968 as the British Cereal Seed Scheme, to include Scotland and Northern Ireland. To qualify for Field Approved status the crop must be inspected, in England and Wales, by qualified seeds inspectors trained by the N.I.A.B. and employed by participating firms or regional seed growers' organisations. Similar provisions apply in Scotland and Northern Ireland. In 1967 some 130,000 acres reached Field Approved standards, of which 91,000 qualified for the more exacting standards of Multiplication seed.[2]

---

[1] S. T. Skelton, The Trade in Agricultural Seeds, *Agriculture*, Vol. 68, No. 7, Oct. 1961.

[2] An outline of the standards for Certified seed, Multiplication seed and Field Approved seed is set out in a note at the end of this chapter.

6.22. The Scheme aims at providing a quality seed which is true to type and free from disease or weeds. It offers the farmers seed of higher standards than those required under the Seeds Acts, for example varietal purity must attain a standard of 1 in 1000 compared with the 98 % needed by law (for wheat and barley). It provides the merchant with seed which needs less cleaning and from which therefore a smaller proportion of the grain brought from farmers will have to be discarded.

6.23. Three main price categories of seed may be distinguished: pedigree seed, which is usually supplied by the national seed houses and corresponds to the upper categories of the N.I.A.B., retailing at 80s. to 90s. per cwt; Field Approved seed, sold by merchants at around 50s. per cwt; and commercial seed which meets the minimum requirements of the Seeds Acts and sells at about 35s. per cwt. Approximately 30 % of the cereal seed purchased by farmers is of Field Approved origin. A much smaller proportion is pedigree seed and the remainder commercial.

6.24. The greater reliability of Field Approved seed helps to explain the price differential between it and commercial grades of seed. The price gap between Field Approved seed and pedigree seed seems less easy to understand. Pedigree seed may be needed where farmers wish to produce seed crops or to experiment with a new variety, but for other purposes its advantages are less evident. Undoubtedly the operation of the Cereals Field Approval Scheme has tended to improve both the quality of seed available to farmers and to provide a mechanism whereby farmers can with greater confidence aim at the seeds market. There may in fact be some further hidden benefits. Not all the seed which qualifies as Field Approved is sold as such to farmers. The remainder is likely to find its way into the commercial market where it will help to improve the quality of cheaper seeds. To some extent this may have made the position of specialist seed houses more difficult.

6.25. There is very little that the trade can do to increase the domestic seed market as a whole. The total volume of seed used is closely related to the aggregate acreage sown. In recent years farming practice has tended towards lower seed rates. Exports of cereal seed are very small and although they may grow as quality improves and the new royalty scheme fosters the development of new varieties, they are unlikely to constitute a large market in the near future.

6.26. The inflexible nature of the total demand for seed means that, given the cereals acreage, firms with a major interest in seed can only expand their volume of sales by increasing their share of the market. Two factors which derive from the nature of the merchant trade are of great importance here. First, the margin per unit on grain sold for seed is far higher than that for other grain transactions. Seed may be bought from farmers at 27s. per cwt and sold at 50s. per cwt. Even when due allowance has been made for the cost of transport, cleaning, wastage, drying and dressing, the profit margin still appears relatively attractive compared with many other parts of the merchant business. Second, the major national companies, who now own an important share of the merchant trade, are capable of providing capital in order to exploit the more promising prospects for seed among their merchant businesses. The effect of these factors has been to increase the interest of merchants in seed and to add to their share of the market at the expense of more specialised seed firms. One of the large national companies, which owns merchant businesses, now has a national seeds organisation operated through eight of its merchants who serve the remainder so far as seed is concerned. The ability to provide capital, to finance advanced technical programmes which can assist the group as a whole and to develop a nationally known reputation in seed, represents an important commercial advantage.

6.27. Farmers who buy seed through the trade are accepting a higher level of cost than those who rely on their own previous crops for seed. The extra cost may be justified because of the greater production potential of the bought seed. As a result of cleaning, careful drying and systematic testing, farmers may expect a more reliable version of existing varieties than they would derive from their own stock. Further, new varieties represent improvements over existing types, and they must be bought via the trade if the farmer is to gain maximum benefit from the other factors of production employed. Seed costs usually form a relatively small part of the total cost of seed production; the difference between commercial seed and Field Approved seed represents only about 21s. per acre. A larger gap exists between farmers' own seed and Field Approved seed, perhaps of the order of 37s. per acre. These extra costs can be justified if yields are 2 cwt per acre higher or if the bought seed results in a weed-free crop which will minimise the need for cleaning cultivations at a later stage.

6.28. Specialised seed-cleaning equipment and the availability of expert technical knowledge on seed varieties and husbandry seem likely to give greatest advantage in future to farmers who produce relatively large volumes of seed and to merchants whose trade is on a sufficient scale to justify the capital cost of the equipment involved. These factors should encourage farmers to buy at least the current proportion of their seed through the trade. If, through quality improvements and new varieties, the gap in yield between bought and home-produced seed can be extended, the market for seed produced on a relatively few specialised farms and sold through the trade may tend to grow.

Note on British Cereal Seed Scheme

The detailed arrangements for The British Cereal Seed Scheme are set out in *Seed Production,* N.I.A.B. Publication No. 7. The following table, which derives from this publication, indicates the different standards required for Certified seed, Multiplication seed and Field Approved seed.

A. Tested by field inspection

	Certified seed	Multiplication seed	Field Approved seed
Varietal purity %	99·97	99·95	99·8
Species purity %	99·996	99·99	99·99
Loose smut %	99·98	99·97	99·96
Other smuts %	100	99·999	99·999
Wild oats in oats (plants per acre)	Nil	Nil	Nil
Wild oats in barley or wheat (plants per acre)	3	3	3
Cleavers (plants per acre)	50	50	50
Runch (plants per acre)	3	3	3
Wild onion	Nil	Nil	Nil

B. Tested by sample (Certified and Multiplication seed only)

For each 20 ton lot a 4-lb sample is examined and must not contain more than:

	Certified seed	Multiplication seed
Wild oats	Nil	Nil
Other weeds	1	2
Other cereals and obvious other varieties[a]	1	2
Other varieties[b]	2 in 200 grains	2 in 200 grains

[a] e.g. black oats in white and vice versa.

[b] This standard relates to those characters which can only be detected by a hand lens, microscope or other laboratory test.

CHAPTER 7

The Economics of Centralised Group Storage compared with On-the-farm Storage

7.1. Cereal yields and total acreage in Britain are both rising, and with only a relatively small amount exported at harvest, it follows that a larger grain-storage capacity is needed year by year. Taking into account also the annual replacement of a proportion of existing storage as it becomes obsolete, there is a continuous requirement for new capacity. Traditionally, storage capacity for home-grown grain has been located on the producing farms, and therefore has consisted of relatively small units, e.g. a few hundred tons per installation. This pattern has been reinforced by the government policy of restricting grant-aid (for capital expenditure), and the seasonal guaranteed price differential, to farmers alone. Despite this a number of large commercial grain stores have been constructed recently without government financial assistance. Other central silos built during the war years were, after two relatively unsuccessful years as a government agency, leased to commercial firms and the fact that they are still operating suggests that they have not been without success.

7.2. The purpose of this chapter is to examine the relative merits of different scales and systems of operation in providing the extra storage requirement. The problems are to decide what form it should take, and who should be responsible for operating it. Thus there are two aspects, the technical-economic, and the institutional. For example, a large-scale store might prove satisfactory technically and economically, but be unsatisfactory in practice because of institutional difficulties, e.g. farmers requiring that their grain should be kept in separate lots. In the first section, attention is concentrated on the technical and economic aspects; any conclusion on these grounds can then be considered in relation to other aspects discussed in the second section.

Technical and Economic Aspects of Centralised Storage

7.3. The criterion for choosing between different forms of drying and storage is the cost of each, expressed on a per-ton basis. Therefore the objective is to determine how the additional storage capacity can be provided, with the minimum use of scarce resources, i.e. at minimum annual cost per ton. Here it is possible to exclude from consideration methods of preservation and storage using sealing or chemicals. These are only suitable where the grain is for livestock feeding, preferably near the point of storage. Although capital costs may be low, these systems are impracticable for storing the bulk of the country's grain production. Drying or possibly chilling are the only methods suitable for a saleable product.

7.4. The basic idea behind centralised grain storage is that additional capacity could be provided by a large central store instead of a much larger number of small installations on individual farms. Such a scheme would inevitably increase transport costs, and also require additional labour at the central store. The question is whether these are likely to be outweighed by savings in capital costs. To answer the question it is necessary to know how each of these items is affected by changes in the size of store. As in Part II, Chapter 5, store size is measured here in terms of wheat; barley requires about 10% more storage space for the same weight.

7.5. Considering first the variation in capital cost per ton with total capacity, the costs for relatively small tonnages have already been summarised in Part II, Chapter 5. It was concluded that the capital cost of farm storage was unlikely to fall below £10 per ton of wheat; the actual average was more than £14 per ton, for an average capacity of 450 tons of wheat. For larger tonnages, particularly those above 2000 tons, there is little empirical information but what there is indicates a progressively lower capital cost per ton. This is supported by "synthesised" cost estimates for a range of tonnages not covered by actual data. The figures refer to the system using a continuous flow or batch drier in conjunction with floor storage, with grain heaped in the centre of the building to its natural angle of repose (35°) from 10-ft retaining walls. This system makes fuller use of the volume enclosed by the span than does on-floor drying, where the grain is normally level at 8 ft deep. However, during storage it is necessary to "condition" the grain by low-volume air flow, so that ventilating equipment on the floor is required.

7.6. Three features of this system need special mention. Firstly, the capital cost per ton depends to some extent on the shape of the building, because the grain must find its natural angle of repose at each end as well as at the sides of the building. From this point of view a square building would be least efficient as the grain would form a pyramid. On the other hand, this may be offset by lower construction costs for a fairly square building. The second point is that the grain must be dried down to 12–13% moisture content instead of the 15% usually considered adequate for long-term storage. The lower moisture content is necessitated by the depth of grain in the centre, which may be 30 to 40 ft. It is also a wise precaution against trouble developing in these very large amounts of grain. Extra fuel and drying capacity is therefore required for this system, but the cost is usually outweighed by savings in other capital expenditure, although in individual cases this may not be so. The third point is that the system is relatively little used at present, probably because it is more suitable for large installations. A recent report by Davey and Ross[1] also considered this to be the cheapest system but their conclusion on economies of scale is different. When considering the merits of central storage much depends on the extent of the economies of scale in capital and operating costs. In calculating the extent of scale economies Davey and Ross make two important assumptions which although they may be justified in the particular context of their study do not appear to be realistic as a general rule. These assumptions are (a) that building costs are a constant £3·6 per ton stored, and (b) that the particular range of drying equipment they use as an example is representative of the equipment available. It has been found, for the type of building considered here, that the cost per ton stored becomes progressively less as the capacity of the store increases. Also when different manufacturers' drier costs are used in the calculations, economies of scale can be achieved to almost 10,000 tons and do not end, as they suggest, at approximately 6000 tons.

7.7. Estimates of capital costs for various sizes of large store, using the system just described, indicate that the cost falls from about £8 per ton at 2000 tons to £3·5 at 25,000 tons. There are probably slight economies of scale even beyond this size. Although these estimates are synthesised from manufacturers' and agents' quotations for buildings, driers, etc., they are confirmed by the case of a recent large installation, holding approximately 25,000 tons. The costs have been built up from the following components:

    building
    foundations
    grain walling
    drier(s) and handling equipment
    pre-drying storage
    floor ventilation
    cleaner
    weighbridge.

[1] B. H. Davey and G. Ross, *Economic Aspects of Co-operative Grain Drying, Storage and Marketing,* Report 168G, University of Newcastle upon Tyne, Department of Agricultural Economics, April 1968.

Drying capacity is sufficient for 10% moisture extraction. The cost estimates for various capacities are summarised in Table 7.1. The figure for a 2000-ton building is included because it is a common size for farmer-syndicate stores. These capacities will be used as examples for comparisons with farm storage.

7.8. A central store would need transport and labour additional to that required by farm storage. If haulage were direct from field to central store,

TABLE 7.1

CAPITAL COSTS FOR LARGE STORES

	Store capacity (tons wheat)					
	2000	5000	10,000	15,000	20,000	25,000
	£/ton	£/ton	£/ton	£/ton	£/ton	£/ton
Building, foundations and grain walling	3·27	2·37	1·98	1·83	1·71	1·69
Other equipment	4·82	2·60	1·88	1·88	1·70	1·55
Total	8·09	4·97	3·86	3·71	3·41	3·24

the extra transport would be the difference between the average distances to the central and the farm store. Direct haulage may be possible where the grain is nearby but to err on the cautious side it is assumed that the grain is hauled first to a (farm) holding store, before being taken to the central store. If the latter is to cope with only the additional grain resulting from higher yields and cereal acreages, the effective amount of grain available nearby is small, e.g. if 10% of the wheat in a given area goes to the central store, with a yield of 1·5 tons per acre and 25% of the land in wheat, there would be only 300 tons within a radius of 2 miles.

7.9. A further reason for assuming double handling of grain stems from weather variation at harvest. Growers naturally wish to harvest as much as possible while the weather is suitable, and with combine capacities steadily increasing, direct haulage to a central store is likely to overwhelm the intake capacity, since every lorry or trailer load has to be weighed and unloaded. In practice it can be assumed that existing farm capacity will act as holding storage and there is no need to allow for any further capital expenditure for this purpose.

7.10. The cost of haulage to the central store is here based on contractors' charges, since this is the most likely method of transport. Also farm transport could be used during bad harvesting weather, and in the early morning and late evening during good weather. Contractors' charges are often a flat rate per ton within certain mileage limits, but this is only for administrative convenience. To reflect resource-use more accurately, the cost per ton has been varied for each of the six sizes of store, as shown in Table 7.2. The average distance of haul is based on the assumption of a circular catchment area around each central store. This area is found from:

$$\frac{\text{store capacity}}{\text{yield per acre} \times \text{proportion of land in cereals}} \qquad ^1$$

The average haulage distance is two-thirds the radius of the catchment area. To reduce haulage distances the stores could be constructed to take two types of cereal, by siting the pre-drying storage bins across the centre of the building to form the main part of a dividing wall. The costs per ton shown in Table 7.2 are based on actual transport charges for sugar beet, and are considered adequate to allow for (a) haulage distances being somewhat greater than the theoretical average, because of winding roads, and (b) haulage requirements being slightly larger than the dried weight, because of the higher moisture content at that stage.[2]

TABLE 7.2

HAULAGE CHARGES AND DISTANCES[a]

	Store capacity (tons wheat)					
	2000	5000	10,000	15,000	20,000	25,000
Contract haulage, cost per ton (£)	0·336	0·412	0·437	0·462	0·483	0·500
Average haulage distance, miles	2·6	4·1	5·8	7·0	8·1	9·1

[a] Based on (dried) yield per acre of 1·5 tons, and *effective* proportion of cereals of 5%.

7.11. A central store will require labour for its operation, but not in direct proportion to the capacity. A manager and at least one other full-time man will be needed to cope with the store during the year, with additional labour at harvest, depending on the capacity. The suggested staffing and costs are summarised in Table 7.3. The 2000-ton store is excluded because it is assumed that it is operated by farm labour with no

[1] The proportion of land in cereals is that contributing grain to the central store, not the overall proportion; e.g. if the latter is 50%, and 10% of the grain goes to the central store, the *effective* proportion in cereals is 5%.

[2] The haulage costs in Table 7.2 are considerably higher than those quoted by Pratten and Dean in their study of scale economies in manufacturing industry, e.g. for road transport of 50 tons per hour of granular material, the cost per ton is quoted as £0·225 for a six-mile haul. On the other hand these figures refer to a constant route, throughout the year. See C. Pratten and R. M. Dean, The Economies of Large-scale Production in British Industry, *Occ. Paper 3*, University of Cambridge, Department of Applied Economics, 1965.

opportunity cost. In fact, syndicate stores of this size often do have some additional paid labour.

7.12. Fuel costs for central drying and storage are based on a 7·5% moisture extraction (i.e. from 20% to 12·5% moisture content) and 3·5 gallons of oil per ton of grain stored, at 1s. 3d. per gallon, or £0·22 per ton. Electricity costs are derived from the horse-power requirements assuming 85% efficiency and are estimated at a constant £0·10 per ton for all instal-

TABLE 7. 3

LABOUR REQUIREMENTS AND COSTS

	Store capacity (tons wheat)				
	5000	10,000	15,000	20,000	25,000
Intake per day (tons)	100	200	300	400	500
Hours per day[a]	4	8	12	16	20
		(1 short shift)	(1 long shift)	(2 short shifts)	(2 medium shifts)
Salaries and wages per annum:	£	£	£	£	£
Manager	1500	1750	2000	2000	2000
Assistant manager	—	—	—	1 at 900	1 at 900
Weigher(s)	$\frac{1}{2}$ at 450	1 at 900	1 at 1000	1 at 1000	1 at 1000
				1 at 200	1 at 250
Unloader(s)	$\frac{1}{2}$ at 450	1 at 200	1 at 350	2 at 250	2 at 300
Floormen	—	2 at 240	1 at 360	1 at 160	1 at 220
		1 at 160	1 at 280	1 at 240	1 at 300
Secretarial	—	1 at 50	1 at 100	1 at 100	1 at 100
Total	2400	3300	4090	5100	5370
Labour cost per ton (£)	0·48	0·33	0·27	0·25	0·21

[a] Assumes 25 tons per hour intake.

lations. The total fuel and electricity costs for central drying and storage are therefore £0·32 per ton. For farm drying, only a 5% moisture extraction is assumed, or £0·15 per ton for fuel, plus £0·1 for electricity, a total of £0·25 per ton.

7.13. The total capital and operating costs included here are summarised in Table 7.4 for the usual six sizes of store. The comparison of these large stores with farm storage is based on a capital cost per ton for farm storage of £10. It should be noted that this is well below the average which has been found in practice, although rather higher than some individual cases which could be quoted.

7.14. It is evident from Table 7.1 that large-scale storage could give capital costs below £10 per ton; the question is whether these savings outweigh the additional transport, labour and drying costs involved in a

TABLE 7.4

TOTAL CAPITAL AND OPERATING COSTS[a]

	Store capacity (tons wheat)					
	2000	5000	10,000	15,000	20,000	25,000
	£	£	£	£	£	£
Capital cost	16,180	24,845	38,600	55,648	68,262	80,962
Haulage	672	2060	4370	6930	9740	12,500
Labour	—	2400	3300	4090	5100	5370
Fuel oil	440	1100	2200	3300	4400	5500
Electricity	200	500	1000	1500	2000	2500
Repairs[b]	162	248	386	556	683	810
Operating costs per annum	1474	6308	11,256	16,376	21,923	26,680

[a] See Appendix F for full details.   [b] 1% per annum of total capital cost.

central store. In answering this question it is relevant that capital costs are incurred in the first year, whereas operating costs are spread more or less equally over the life of the plant. In choosing between alternatives the minimum-cost solution is found by discounting future costs back to the present, at an appropriate rate of interest (r). This approach emphasises the greater "cost" of expenditure now rather than in the future. As an example of the method the procedure for the comparison of farm storage with the 10,000-ton store is set out in detail (Table 7.5).

7.15. The two capital costs are found as already described (see Table 7.1). Annual operating costs are set out for each year of the discounting period, and discounted by dividing by $(1+r)^n$, where r is the discount rate and n is the year in question; e.g. in year 10 the annual costs are divided by $(1+r)^{10}$. (Operating costs for farm storage consist only of fuel and electricity.) To the sum of the discounted costs is added the capital cost, which is not discounted because it is all incurred in year 0. From this is subtracted a salvage value for the installation, allowed at 20% of the discounted value of the plant in the tenth year. Since the equipment, as distinct from the building, can be given a physical life of about ten years, and the building and equipment are roughly equal in initial cost, the procedure amounts to crediting the building with a salvage value approximately 40% of its discounted value in the tenth year. The choice of discount rate is open to argument but here a rate of 7% is used. This is probably less than the current rate at which money could be borrowed for this type of investment, but the market rate of interest contains an element to allow for inflation. If this were explicitly included it would be logical to increase future operating costs also.

TABLE 7.5
DISCOUNTING PROCEDURE USED FOR COMPARISON[a]

	Central store (10,000 tons)		Farm stores (10,000 tons)	
Capital cost	£38,600		£100,000	
	Operating cost per annum	Discounted at 7%	Operating cost per annum (see Table 7.7)	Discounted at 7%
	£	£	£	£
Year 1	11,256	10,524	3500	3272
Year 2	11,256	9826	3500	3056
Year 3	11,256	9185	3500	2856
Year 4	11,256	8588	3500	2670
Year 5	11,256	8026	3500	2496
Year 6	11,256	7496	3500	2331
Year 7	11,256	7012	3500	2180
Year 8	11,256	6551	3500	2037
Year 9	11,256	6123	3500	1904
Year 10	11,256	5718	3500	1778

10-year total of discounted operating costs		79,049	10-year total of discounted operating costs	24,580	
Plus capital cost		38,600	*Plus* capital cost	100,000	
		117,649		124,580	
Less salvage value at 20% (£7720 discounted)		3922	*Less* salvage value at 20% (£20,000 discounted)	10,160	
10-year total discounted costs		113,727	10-year total discounted costs	114,420	

[a] In practice the 10-year total can be found directly, by the use of annuity tables.

7.16. The two alternatives in Table 7.5 are compared in terms of the total of their discounted operating costs, initial capital and salvage value. It can be seen that in this case central and farm storage have very similar total costs (about £114,000). Similar calculations for the other five sizes of central store gave the results shown in Table 7.6. The 2000-ton central store appears a better proposition than the 5000-ton store, but it should be remembered that no labour has been charged. If it were more realistic to charge labour for a particular situation, a medium-sized store of this type could offer a useful alternative. On the other hand the existence of a full-time store manager (at larger capacities) should enable more efficient marketing arrangements to be made. The competitive position of the 15,000-ton central store would be rather better if a more convenient size of drier were manufactured. On the whole, however, discontinuities of this kind were not a major problem.

TABLE 7.6

COMPARISON OF FARM AND CENTRAL DRYING AND STORAGE

	Store capacity (tons wheat)											
	2000		5000		10,000		15,000		20,000		25,000	
	central	on-farm	central	on-farm	central	on-farm	central	on-farm	central	on-farm	central	on-farm
	£	£	£	£	£	£	£	£	£	£	£	£
(a) Capital cost	16,180	20,000	24,845	50,000	38,600	100,000	55,648	150,000	68,262	200,000	80,962	250,000
(b) Operating costs per annum	1474	700	6308	1750	11,256	3500	16,376	5250	21,923	7000	26,680	8750
(c) Total discounted operating costs	10,352	4916	44,301	12,290	79,049	24,580	115,009	36,871	153,965	49,161	187,374	61,451
(d) (a)+(c)	26,532	24,916	69,146	62,290	117,649	124,580	170,657	186,871	222,227	249,161	268,336	311,451
(e) Salvage value	1644	2032	2524	5080	3922	10,160	5654	15,240	6935	20,320	8226	25,400
(f) Total discounted costs (d)—(e)	24,888	22,884	66,622	57,210	113,727	114,420	165,003	171,631	215,292	228,841	260,110	286,051

M

7.17. To give an indication of the level of annual costs per ton, the different store sizes are also compared with farm storage in the more conventional way using amortisation charges. With this method there is a constant annual charge covering capital repayment and interest. The results are shown in Table 7.7, where the total capital is written off in 10 years, and other assumptions are as before. The conclusion is similar to that already reached, because although operating costs are not discounted, this is balanced by the *total* capital being written off, with no salvage value.

TABLE 7.7

ANNUAL COST PER TON STORED

| | Store capacity (tons wheat) | | | | | | On-farm |
	2000	5000	10,000	15,000	20,000	25,000	
Capital cost per ton	£ 8·09	£ 4·97	£ 3·86	£ 3·71	£ 3·41	£ 3·24	£ 10·00
Operating costs:							
Haulage	0·34	0·41	0·44	0·46	0·49	0·50	—
Labour	—	0·48	0·33	0·27	0·25	0·21	—
Fuel	0·22	0·22	0·22	0·22	0·22	0·22	0·15
Electricity	0·10	0·10	0·10	0·10	0·10	0·10	0·10
Repairs	0·08	0·05	0·04	0·04	0·03	0·03	0·10
Sub-total operating costs	0·74	1·26	1·13	1·09	1·09	1·07	0·35
Capital cost (amortised at 7% over 10 years)	1·15	0·71	0·55	0·53	0·48	0·46	1·42
Total annual cost per ton	1·89	1·97	1·68	1·62	1·57	1·53	1·77

7.18. Sensitivity analysis was carried out to assess the effect of changing labour and haulage costs and the capital cost of farm storage. Haulage charges are affected by changes in the average proportion of cereals and in the cost per ton at any given distance. Here it is assumed that it is the proportion of cereals which is changing, thereby altering the average haulage distance. Labour costs are varied on the assumption that only part of the total cost is chargeable to the central store, e.g. a proportion might be charged to the marketing function. The results in terms of 10-year totals of discounted costs are shown in Table 7.8. It is evident that relatively small changes in the capital cost of farm storage have a big impact on total costs.

TABLE 7.8

COMPARISON OF TOTAL DISCOUNTED COSTS
(£'000)

Store capacity (tons wheat)	Central store costs with effective proportion of land in cereals of:					Central store costs with proportion of labour charged to storage of:				Total cost of farm storage with capital cost per ton of:			
	0·5%	1%	2·5%	5%	10%	100%	75%	50%	25%	£8	£10	£12	£14
5000	72	70	68	67	66	67	62	58	54	48	57	66	75
10,000	128	120	117	114	112	114	108	102	96	96	114	132	150
15,000	191	181	172	165	161	165	158	151	143	145	172	201	226
20,000	255	238	223	215	208	215	206	197	188	193	229	265	301
25,000	318	292	271	260	251	260	251	241	232	241	286	331	376

7.19. A number of conclusions can be drawn from this stage of the analysis.

(a) On-farm and central storage are fairly evenly balanced in economic terms, but central storage appears to have a slight advantage at higher capacities, particularly since so many assumptions rather unfavourable to it have been incorporated. From this it follows that

(b) if the Government wishes to subsidise grain storage, subsidising on-farm storage is not necessarily unreasonable but consideration of costs would not appear to justify restricting this encouragement to farm storage only, as at present.

(c) If the subsidisation of farm storage *is* to be reasonable, it should be fixed in amount, i.e. a flat rate payment per ton of capacity and not a percentage of the total cost.

(d) Any proposals for syndicate or other central storage arrangements should be examined on their merits, and in the light of local conditions, e.g. the amount of grain for which new storage will have to be provided in some form or other.

(e) Any farmer-controlled scheme of this kind would have to include provision for equalising transport costs between farms using the store, otherwise its location could give rise to dispute and dissatisfaction.

Other Aspects of Centralised Storage

7.20. In addition to the comparative economic costs of building and operating on-the-farm or centralised stores, there are a number of other important factors to be considered. These may be summarised as follows:

(a) Centralised drying and storage could lead naturally to co-operation in marketing. Deals can be negotiated with buyers of grain for tonnages much larger than any individual farmer can usually offer, and because the grain is bulked in one place it becomes a more feasible proposition to grade the grain and to offer the quality and type of grain required by the customer. The individual farmer rarely sells a sufficient quantity of grain to justify such treatment. If the centralised drying and storage organisation is large enough it may adopt a different trading relationship with the agricultural merchants than is normal, e.g. instead of the merchant buying the grain and thus acting as a principal, he may sell it for the group on a commission basis. Although the centralised drying and storage organisations already in existence do not appear to have obtained any substantial price advantage through group selling it is possible that as the buyers become aware of the potential benefits of buying a graded commodity in greater bulk they may be prepared to pay a slight premium.

(b) The experience of co-operating together to build and operate a centralised grain store might well lead on to co-operation in the production
of grain as well as its disposal. For example there might be more
sharing of equipment, more uniformity in choice of fertiliser, or
sprays and in their application so as to produce the more uniform
product usually required by processors.

(c) Some of the financial problems experienced by growers can be eased,
e.g. some groups operate in such a way that they pay the farmer on
delivery to store rather than when the grain is eventually sold.

(d) Grants might be available through the Central Council for Agricultural and Horticultural Co-operation for centralised grain drying and
storage installations as well as for other forms of co-operation in
production. Such grants might widen still further the economic
advantage in favour of centralised stores.

(e) Centralised stores would make possible a more economic system of
transporting grain. At present grain is moved from a multitude of
individual farms, often in small 5–10 ton lorries, and, as Part III,
Chapter 3 shows, there are many problems involved in arranging for
the grain to be collected, providing labour to load, and unload, and
dealing with inefficient loading equipment. Whilst improvements are
being made all the time, it would be unrealistic to expect this system
ever to be very efficient. Many farms will always have inefficient
grain-delivery equipment, or inadequate turning space for long lorries
or too small access roads, whilst such practical difficulties as fixing
suitable arrival times for each individual collection, and arranging
for labour to be available, are inherent in the system, and there is no
way round them provided the present system remains.

The centralised store however opens out the possibility of a new
and streamlined system of transport from store to end-user. The grain
would be taken straight from the combine, or from the barn-floor, to
the store. On arrival at the store it would be weighed and tipped into
a below-ground reception chamber—the whole operation taking only
a few minutes. Bulking at the store would enable economical transport schedules to be worked out with a shuttle service of huge 25–30
tons bulk grain carriers capable of being filled in a matter of minutes
from bulk overhead hoppers, going straight from the store to the end-
user with the minimum of administrative effort and at a cost per unit
considerably less than at present. If such a store were located by a
railway siding or by a wharf, some of the grain intended for longer
distances could move by rail or water.

It has been suggested to us that 7 miles might be the furthest distance grain could be carried direct from the combine to the store, otherwise the harvesting schedule would be thrown out of gear. Experience at R. and W. Paul's new malting and grain store at Mendlesham in Suffolk shows that up to 7000 tons can be successfully moved direct from the combine to the malting over the space of about 6 weeks. This installation is located, of course, in an intensive cereal growing region. An alternative to moving grain direct from the combine to the grain store would be to deposit it temporarily for a few hours, or even a day or two, on a barn floor somewhere on the farm, whence it could be taken by lorry to the store. This alternative was the one assumed in the costings given in the earlier section. This method would involve double-handling, which would add to the cost, but it would enable the radius of the "catchment area" of the store to be increased considerably. The costings show that the costs of double-handling are more than outweighed by economies in the building and operating of the drier and store.

There seems to be no reason why the centralised grain store should not serve a wider purpose, i.e. as a distribution depot for bulk feeds, fertilisers or imported maize. Thus the bulk grain carriers could return to the store with these commodities ensuring a double payload all the time. With bulk feed, where transhipment into the store and out again might cause damage by separation of the particles, the return journey could be made direct to the farm provided it were capable of receiving a vehicle of this size.

(f) If the number of centralised stores operated by farmers became substantial, and the grain involved amounted to a reasonable proportion of the total coming onto the market, the farmers might for the first time have some collective strength as sellers of grain. In the sense that increasingly the merchant and manufacturing sectors of the grain trade are coming to be dominated by a few large national companies some degree of countervailing power of this kind might not be undesirable.

(g) The existence of a number of centralised grain stores might facilitate the operation of a system of season-to-season carryover of grain should this become necessary (see Part VIII, Chapter 3).

(h) One of the problems of the grain export business is to concentrate speedily a large tonnage of grain within a few hours to enable a ship to be filled between one tide and the next. The existence of centralised grain stores, provided they were accessible to export ports, would facilitate this process.

(i) A group operating a centralised store on a large enough scale will find it worth while, and indeed essential, to employ specially skilled staff to deal with the processes of drying and storage and with the commercial aspects of marketing the grain.

7.21. The question arises—who would be likely to establish and operate these grain (or multi-commodity) stores? There seems to be no logical reason why farmers alone should operate such centralised stores although they obviously have a vital interest. Firms of maltsters, shippers, merchants and co-operatives are already successfully operating schemes roughly similar to what has been proposed above. What is important is the function—not the particular institution that is performing it. We see no reason why centralised stores, if they develop, need be channelled into any particular mould. In our view any organisation proposing to establish a scheme approved by some central body such as the Central Council or the H.-G.C.A. should qualify for the same benefits, in particular, equality with the farmers as regards:

(a) the de-rating of agricultural premises;

(b) qualifying for Improvement Grants for the erection of the buildings.

7.22. Some would go further and have suggested to us that all grants to farmers to erect individual farm grain stores should immediately be stopped since there is now an adequate provision of this kind of grain storage. The money saved in this way could then be diverted to assist the building of centralised stores of the kind described earlier. We would however, prefer for the time being a trial period when experiments could be made with the new approach, but we would certainly not rule out the possibility in a few years' time of the flow of government money for individual farm storage being discontinued.

PART IV

The Importing and Exporting of Grain

This Part deals with the role of the shippers in the U.K. grain trade (Chapter 1), the role of the brokers, port merchants and futures markets (Chapter 2), and finally with the sea transport of grain and the significance of the new grain terminals, at Tilbury and elsewhere, for the U.K. grain trade (Chapter 3).

CHAPTER 1

The Role of the Shippers

1.1. The imported grain trade in the U.K. is handled by a small number of shippers, five of whom are very large businesses with world-wide interests not only in grain and other commodities, but also in such activities as ship-owning, manufacturing, banking and insurance. Visits were made to the London houses of the five big international shippers, but not to their headquarters offices which in each case lie outside the U.K. What follows therefore relates to the activities of these firms in the U.K. rather than internationally. Some of the smaller shippers in the U.K. were also visited.

The "Big Five" Shippers

1.2. Although some shippers run their own ships, the freightage of goods is not their primary function; they are primarily traders, buying usually in one country and selling in another. The shippers are thus comparable in function, though often vastly bigger in size, to corn and agricultural merchants in that they act as principals in the grain trade: and just as the corn merchant sometimes uses his own transport and sometimes hires it, so the shipper sometimes uses his own ships ("house cargoes") and sometimes charters freight through a freight broker (not to be confused with a grain broker) on the Baltic or New York Exchanges. At one time it was thought advantageous for shippers to carry their goods in their own ships, but today the more general view is that it is more economic to charter freight. A shipper will usually operate his ships as completely independent units and he will often transport grain in ships belonging to someone else if he can obtain a cheaper freight that way. In short, despite his title, the shipper is primarily a trader in grain and only secondarily a mover of it.

1.3. The skilful chartering of freight is nevertheless of vital importance to the shipper. A potential profit on the grain buying side can easily be turned into a loss if no freight is readily available to shift the grain except at a high rate. Freight rates vary quite remarkably according to the availability of ships and cargoes to fill them,[1] and they constitute an element of risk in the shipper's assessment of the likely profit on a transaction, although a certain amount of freight hedging is done to offset these risks. When freight rates are tending to rise the shipper may charter vessels for an extended period, or for a number of consecutive voyages, whilst if they are falling he will charter only on very short term. In a sense the shipper treats freight as a commodity and adopts a "long" or a "short" position.

1.4. Because of the necessarily large scale of operations involved when individual shipments of thousands of tons of grain are being negotiated, the five international shippers are powerful concerns with huge capital resources. One shipper's representative visited in the Survey ventured the opinion that no firm with assets of less than £50 million could be successful in the international grain trade. Since the taking of some risks is unavoidable, a firm with inadequate resources could easily succumb if it made a serious mistake. This is one reason why the shippers tend to be closely associated with banking firms. None of the "Big Five" firms of international shippers is British. Bunge is an Argentine firm, Louis Dreyfus is French, Cargill (or Tradax as it is known in Europe) is American, Wm. H. Pim Junior, once a British firm, is now a subsidiary of the Continental Grain Company (U.S.A.), and Garnac (known in the U.K. as the European Grain Company) is a Swiss firm. In addition there are a number of British firms that between them handle a substantial tonnage of grain imports although they do not rank with the "Big Five" in international trade generally.

Role of the Shipper as a Grain Importer

1.5. Traditionally the shippers in the U.K. grain trade have been mainly concerned with the import trade, i.e. selling imported grain to the large port-based millers, compounders and port merchants. More recently the shippers have taken an interest in the U.K. export trade, and this has brought them into closer touch with the corn merchants, which in turn has led them to take a greater interest in the internal U.K. grain market; but still their primary role is that of grain importers.

---

[1] For example, between November 1967 and January 1968 the rate for wheat from United States Gulf to U.K. fell from £3 18s. per long ton to £2 16s., a decline of 28% in two months. See Commonwealth Secretariat, *Grain Bulletin*, Vol. XIV, No. 1 and No. 3.

1.6. Outside Australia and Canada,[1] where the grain trade is centrally controlled, the big international shippers have many offices and often own interior elevators in the countries where the grain is grown—particularly the U.S.A. and Argentina. They are thus able—in the shippers' terminology—to "originate" grain, move it to the seaboard, clean and grade it for export, and ship it to the buying country. Often the shippers also have an extensive selling organisation within the country of origin. Similarly, each of the shippers has its own widespread selling organisation covering the main buying countries. These various parts of a shipper's chain of operations are fairly loosely tied together, and the selling companies, for instance, may not trade exclusively with the buying organisations in their own group, although it is usually advantageous for them to do so.

1.7. Each of the big shippers maintains an intricate network of communications, with each office in each country acting as a listening post and passing on vital information to the others. The shipper is thus enabled to trade more effectively in what is probably one of the most competitive markets in the world. A shipper's "market room" is a busy place—it is the nerve centre of an intricate communications system stretching all over the globe. These tele-communication systems are expensive to instal and maintain and this is another reason why the largest companies can operate most effectively in this field. Up-to-the-minute information is one of the shipper's most vital requirements, and one of the reasons why shippers continue to operate their own ships is that this gives them vital information on freights that might make the difference between a profitable and an unprofitable transaction.

1.8. Most grain imported directly into the U.K. (i.e. other than that transhipped via Europe) arrives in ships of 10,000–25,000 tons. Whilst there is no rigid rule, a typical shipment would comprise one or two parcels of about 5000 tons going to one or other of the big national millers, compounders (e.g. Ranks, Spillers or B.O.C.M.), or port merchants, possibly a few smaller shipments of, say, 2500 tons for the smaller national firms, and a few "small lots" of a few hundred tons. Usually the smaller merchants in the U.K. buy their imported grain through the large national firms or port merchants, but sometimes they buy small quantities direct from a shipper. However, a shipper will seldom accept orders for parcels of grain of less

---

[1] Australian wheat is sold only by the Australian Wheat Board through their counterparts in London, the Australian Wheat Committee. Similar arrangements exist for barley. The big five shippers therefore play little part in this trade although they sometimes trade in Australian sorghum or oats. Canadian wheat is sold by the Canadian Wheat Board through their agents domiciled in Canada. The big shippers handle some Canadian grain but some is bought direct from the Board's agents by the big national buyers in the U.K.

than 250 tons, and then only if he is already working a ship to the port in question and there happens to be space available. A shipper will not "start" a cargo with a small lot like 250 tons, although he has to consider his competitors and may be willing to start a cargo for a relatively smaller load than he might wish in the hope that he will find other buyers in due course. Rather curiously, in view of the risks involved, shippers do not seem to charge significantly more per ton for small lots than for large. A few of the biggest U.K. customers will order a full cargo at a time; thus a big starch manufacturer in Manchester takes a 10,000–12,000 ton shipment of maize at regular intervals throughout the year. Also, when world prices appear to be very low and manufacturers in the U.K. are stocking up, the big national buyers may order whole cargoes at a time.

1.9. Nearly all the grain imported into the U.K. (one shipper put the proportion as high as 98%) is already sold when it arrives in this country and goes immediately to the buyer. On the other hand, a shipper tries to avoid "dead freight", i.e. unused capacity in the ship's hold, and sometimes he will complete a cargo with a small proportion of unsold grain which has to be disposed of whilst the ship is afloat or on arrival. This tends to be a risky business because most of the main buyers will have already purchased their requirements and shippers naturally try to keep the tonnage of this kind of business down to a minimum. The shippers very seldom store grain and they are weak sellers if the demand for grain is slack on the ship's arrival. Often, as will be indicated later, firms of specialist port merchants help dispose of these "unsold balances".

1.10. One reason why few of the big shippers have invested capital in grain handling facilities in U.K. ports is that the big buyers have generally installed their own grain handling equipment at their premises; another is that with seven major ports in the U.K. importing grain, there is an insufficient tonnage handled by any one shipper at any one port (except possibly London) to justify elaborate new equipment, and U.K. port authorities have not in general encouraged private investment of this kind. Another reason is that grain entering the U.K. is usually at the end of its journey, whereas much of the grain entering Rotterdam, Amsterdam or Antwerp is intended for transhipment elsewhere—hence there has been no call for large transit silos such as have been erected in these three ports. The Port of London might have justified such a project; and, in fact, a consortium of shippers was interested in the possibility of Tilbury as one of a number of alternative sites, some years ago, but the P.L.A. stepped in and the shippers abandoned their plans. These issues are discussed at greater length in Chapter 3.

1.11. The shipper usually sells grain to the U.K. buyer on a c.i.f. (cost, insurance, freight) basis, i.e. the shipper pays all the costs of transporting the grain to the U.K. port and ownership passes to the U.K. buyer when he has paid for the shipping documents, and his expenses begin "at ship's rail".[1] The smaller British shippers, and the larger shippers when engaged in the U.K. grain export trade, usually buy grain on an f.o.b. (free on board) basis, i.e. the shipper assumes responsibility for all costs incurred once the grain has been delivered into the hold of the ship. Imports tend to follow a steady pattern over the year, although the summer months tend to be a little quieter than the others. The St. Lawrence is frozen over for the winter months but the grain continues to move through other ice-free ports. Changes in relative price levels over the year have most effect on the seasonality of grain flows. The shipper seldom sells grain direct to country merchants or smaller inland compounders or millers, although in recent years there has been a tendency for the larger country compounder to buy direct from the shipper (usually through a broker or port merchant) rather than through a national compounder.

1.12. Imported grain is nearly always bought forward two or three months, and longer still if prices are specially attractive. The shipper has to take a view of the way the international grain market is moving and to back his judgement; if price levels are tending to rise he will take a "long" position, i.e. he will buy in anticipation of being able to sell at a profit, but if prices are falling he will adopt a "short" position, i.e. he will restrict his buying as much as possible to what he knows he can sell. In the last analysis every shipper is a risk-taker and he must decide for himself the extent of the risk he is prepared to carry at any given moment. By acting in this way a shipper is performing a valuable role in the market, since by the exercise of his skill and foresight, supply is matched more equally to demand.

1.13. The bigger the shipper, and the more widespread the markets in which he is trading, the better is he able to spread his risk; there is a reasonable hope that if the price is low in one market it may be higher in another. This is yet another factor making for large size in the international grain trade. The big five shippers take full advantage of the hedging facilities offered by the Kansas City and Minneapolis and other futures markets, but the London Futures Market is too small to play an important part in these operations. No shipper can avoid taking a long or a short position based on the view he takes of the market, but equally no shipper would survive for long if he did not hedge some of these risks. The shipper has to strike

---

[1] Practice varies between the big five shippers as to the point at which the responsibility for grain exported from the country of origin passes to the branch of the company in the country to which the grain is exported. The latter will often assume responsibility at the exporting country's seaboard, but sometimes the responsibility remains with the originating side of the business throughout and it will pay a handling charge to the branch in the country to which the grain is sent.

a balance between risk and profit. Thus the profit on "back-to-back" business (when the grain is sold as soon as it is bought—see 2.4 below) is relatively small because the risk is small. But where the shipper originates the grain (i.e. buys, grades and moves it to the seaboard before selling it) the risks are higher and the rewards more attractive.

1.14. No information about profits was collected in the Survey, but it seems to us that in such a highly competitive market as this, it is unlikely that high profits would be enjoyed by any firm for any length of time. What has to be remembered is that a firm engaged in an operation usually spanning a period of months, and between widely separated countries, must build up reserves sufficient to meet emergencies. The big question is whether there is any "understanding" between the big shippers which could indicate that the market is less than fully competitive. Although the shippers have formed consortia for the erection and operation of grain silos in European countries, we regard this as comparable to the consortium of U.K. animal feed manufacturers to build a factory in Belfast, i.e. a means of securing the full economies of scale, and it does not of itself signify any loss of competition. We have not become aware of any collusion among the shippers in grain trading, either in talks with the shippers themselves or with those who trade with them, and we have no reason to believe that this is other than a highly competitive market.

The Baltic Exchange, the Futures Markets and the Role of the Brokers, Port Merchants and Shipper-owned Merchants

The Baltic Exchange

2.1. The Baltic Exchange, or more accurately the Baltic Mercantile and Shipping Exchange, is not so much one market as several. For instance, it contains not only the freight market, where ship brokers and chartering agents find cargoes for ships and ships for cargoes (mostly for the deep sea trade), but also the commodity markets, which cover oil and oilseeds in addition to grain, and the various futures markets, of which the London barley futures market is the most important.

2.2. The freight market is very active, and the floor of the Exchange is usually thronged with freight brokers and shipowners' representatives discussing ships, cargoes and journeys, and, of course, rates. In contrast to the freight market, however, there has been a steady decline in the use made of the grain market. One reason for this is that there is more direct buyer-to-seller trading today than was once the case. Another is the vast improvement in communications; it is more important these days for a shipper to be within a few minutes' access of his Telex or telephone than to be on the floor of the Exchange since most trade is now done by telephone rather than by personal contact at the Exchange. A further reason is the decline in the number of buyers and sellers. As one shipper expressed it: "Ten years ago I would have had to 'phone 20 people to sell a parcel of wheat and it was easier to meet them on the Exchange, but today I need only contact 6 people and it is easier by 'phone." Nevertheless, the Baltic Exchange, like the various country corn exchanges, is still a place where information is exchanged and where at times one can gauge the "feel" of the market better than one can on the telephone.

2.3. The London Futures Market is located at the Baltic, although it now deals largely with grain of British origin. There are several reasons for this location: the Baltic is open all the week, whereas Mark Lane is only

339

operative for two days a week; international communications are very much better at the Baltic; and the shippers, who account for a sizeable proportion of the trade done on the London Futures Market, are more accustomed to the Baltic's pattern of trading. It is also noteworthy that samples are never seen on the Baltic, whereas at Mark Lane most of the trading is done by sample. The operation of the futures markets in the U.K. is discussed at the end of this chapter.

The Role of the Broker

2.4. Grain imports are often negotiated by the selling agency or shipper direct with the end-user, but sometimes the transaction will be negotiated through a firm of grain brokers. Although the broker usually acts only as the intermediary, and passes the name of his principal to the other party after completing the deal, he occasionally has to act as a principal. For example, when a buyer in the U.K. will buy only from someone domiciled in Britain, the broker may import in his own name and then sell to his client. It is a distinguishing feature of a broker's business that if he trades as a principal it is always on a "back-to-back" basis, i.e. as one of the leading London grain brokers expressed it, "A broker never buys grain that he does not immediately sell, or sells grain that he does not immediately buy". In short, a broker never speculates in grain or "takes a position". The grain broker performs the very important function of putting sellers into touch with buyers; he uses his intimate knowledge of how the grain market is moving from day to day, even from hour to hour, to further the interest of his client. In this way he is helping to keep the grain market as competitive as possible. Grain brokers are to be distinguished, of course, from freight brokers whose task is to arrange the actual freightage.

2.5. The tendency for quasi-Government bodies like the Canadian Wheat Board to deal, through their agents, direct with the main buyers in the U.K. has diminished the role of the broker in recent years. This tendency has also been accentuated by the process of take-overs and amalgamations among the big buyers—the smaller the number of buyers and shippers the less obvious the role of the broker becomes. Moreover, we were told that the big buyers tend to be more concerned these days about their forward stock positions for grain than they once were. This may be a reflection of the increasing influence of the scientists, and those responsible for formulating animal feed rations or milling grists, on the buying policies of these firms. Another indication of this may be the increasing quantity now being bought forward and the decline in spot purchases. Despite these developments, however, a small number of long-established firms of grain brokers continue to flourish, and those engaged in the grain trade who use their services clearly find that the benefits more than cover the $\frac{1}{4}\%$ to $\frac{1}{2}\%$ commission usually charged by brokers. Contrary to the impressions conveyed to us by some of the shippers and national buyers that the role of the broker was diminishing, one of the leading London grain brokers said that a higher proportion of grain business is done through brokers today than was the case 20 years ago. Nevertheless, the number of small (virtually one-man) brokers has declined—the broking business, like most other branches of the grain trade, has had to achieve maximum efficiency to survive. The shippers themselves sometimes use brokers when it suits them, e.g. if they want to "cover in" a position without wishing it to be known that they are doing so. As to the export of U.K. grain and the internal U.K. grain market, where a sizeable proportion of the grain is handled by a relatively large number of sellers and buyers, the grain broker, as will be shown later, can meet an important and growing need. There are some, indeed, who believe that the pattern for the future will be for the U.K. grain producers, acting in groups, to sell their grain within the U.K. through a brokerage system (in which case the "broker" would usually be a merchant working on a commission basis) rather than through a merchant trading as a principal. Grain brokers frequently act for agricultural merchants and manufacturers in the U.K. (for a commission of about 1*s*. 6*d*. to 2*s*. 6*d*. per ton) but they seldom act for farmers direct.

2.6. An interesting development in the last year or two has been the setting up of Eurograin, an organisation of European co-operatives aimed at improving the facilities for inter-co-operative trading among European countries. This organisation has acquired one of the leading firms of grain brokers in Europe, which now acts primarily as a broker for the large co-operatives. The U.K. was at a disadvantage compared with the Continental co-operatives because there was no single organisation capable of acting on behalf of all co-operatives in the U.K. handling grain, the A.C.A. being barred from undertaking trading functions. It was because of this that a new company was set up, Farmers Overseas Trading, to co-ordinate the activities of co-operatives in the international grain trade. Since few U.K. co-operatives are engaged in compounding, F.O.T. has only a minor role to play in grain importing. But it has an important function in the grain export trade and was the main avenue by which the member co-operatives (and 30 or so private merchants who are also members) exported over 50,000 tons of grain in 1966/7. But F.O.T. may have an even greater significance than this, for it represents probably the first serious attempt in the U.K. to organise a "co-operative of co-operatives". As will be shown in Part VI, the lack of anything of this sort in the U.K. before F.O.T. was set up was one of the main differences between co-operatives in the U.K. and those in Europe. With the success of F.O.T. as an example, it may be that the U.K. co-operative movement will advance further along the road of inter-co-operative co-operation.

2.7. Some buyers and sellers seem to regard grain brokers as an unnecessary intermediary to be short-circuited whenever possible, but we do not share this view. The broker sells his skill in understanding the grain market, and since no-one is compelled to use his services, he survives because both parties, buyers and sellers, find it worth their while to buy his skill. The brokers play a useful part in grain marketing and the market as a whole would be less efficient without them.

The Port Merchants

2.8. Acting as intermediaries between shippers, brokers, other merchants, farmers and end-users are a small group of merchants who have specialised over the years in the grain importing business. These are sometimes known as port merchants.

2.9. Before the war the number of these businesses was greater than it is today. There were located in the ports, particularly Liverpool, Hull, Avonmouth and London, and they acted as importers on behalf of manufacturers and the smaller inland merchants. During the war these firms were forced to deal mainly in domestic grain, and after the war they never quite regained their previous importance in the international grain trade. At one time the big national users of grain would import some of their requirements through these merchants, but now they usually buy direct (e.g. Canadian wheat) or through shippers. This method of trading has reduced the importance of the corn exchanges in the big ports so far as grain imports are concerned (particularly the Liverpool Corn Exchange), and the port merchants have suffered accordingly. However, a number of these firms still exist in the main ports and they perform a distinctive role as grain importers, one of their functions being, as mentioned earlier, to help the shipper by disposing of the "unsold balance" of a shipment, i.e. that part of the cargo that has not been sold before the ship docks.

Shipper-owned Merchants

2.10. An important development within the last year or two has been the buying up of agricultural merchant businesses by the big shippers. The best known example of this is the acquisition of an important merchant business in the North of England by one of the big shippers. The business has branches in several parts of Britain and is building up an important position in the export trade, and as a "merchant's merchant", i.e. it will buy grain from merchants and sell to end-users in large quantities. This system is useful to the end-users, since it saves them having to contact a number of merchants to obtain a sufficient quantity of grain of the required type. This firm is well placed to perform this role because its link with the shipper also enables it to import the grain which its customers require and, equally, to export if this is the best outlet. In this sense the firm is operating as a port merchant, although most of its sites are inland.

2.11. Other shippers have acquired corn merchant businesses in Britain and it is likely that further acquisitions are pending. Since it is very unlikely that a shipper would acquire a business capable of handling more than a small fraction of its throughput of grain, imported or exported, one might wonder why they have branched out in this direction. One answer is the over-riding need all the time for reliable information. By owning a merchant business of his own a shipper has a valuable listening post in the internal grain trade, particularly so far as grain exports are concerned. Another possible explanation is that if a shipper is contemplating investing capital in grain-handling equipment for exports, as some of them certainly are, he wants to be assured of a minimum throughput of grain to justify the investment—he can move in this direction by buying a merchant business operating in the hinterland of the ports he is using. Perhaps a third reason is that as the U.K. grain import business declines, consequent upon the rising output of U.K. cereals, so the shippers begin to think in terms of retaining their interest in the grain trade generally by participating in the internal U.K. grain market.

2.12. The trend towards buying up merchant businesses is being resisted by some of the shippers. They can see that an agricultural merchant business typically deals with a wide range of goods, many of which have little to do with the grain trade, and they are reluctant to get involved in what is virtually a new activity so far as they are concerned. Then again, they realise that the competition among the big nationals, and now the shippers, to buy up the most efficient merchant businesses tends to put up the price of these businesses, and this affects the economics of the whole operation. They see this as the thin edge of the wedge—if you become involved with merchanting you may also find yourself in animal feed compounding or the seed trade, or owning a fleet of lorries, and your specialised expertise and skill in the grain trade is of little value in these untried spheres.

2.13. However, for good or ill, the die has been cast by some shippers and as a result there is already a rapid infusion of capital in the "shipper-cum-merchant" side of the industry. These shippers, with their great capital reserves, can command resources quite outside the scope of most independent merchants—nor are they accustomed to operating on a small scale. The firms acquired by shippers are likely to take the lead in investing capital in grain-handling facilities for export; provided they do not acquire too dominating a position in the grain market, their advent is to be welcomed in an industry where innovation has tended to be held back for lack of capital resources.

Importing and Exporting of Grain by Agricultural Merchants

2.14. The foregoing has been based mainly on the visits to shippers and brokers, but the results of the Survey of Merchants confirmed the impressions received from these sources. Of the 260 firms in the merchant survey, 71 imported grain (mostly maize) but did not export any, 40 exported grain but did not import any, and 41 both imported and exported grain. Thus 152 of the Survey firms (60%) participated in the importing and exporting of grain. The exporting firms were located mainly in the eastern parts of the country, but the importing firms were widely spread over the U.K. The firms that did no importing or exporting were mostly in the lower turnover ranges.

2.15. The methods used for importing and exporting were very varied, and many firms used more than one method. Importing grain direct through shippers was relatively rare; it was more usual for brokers to be used, or for merchants to import through other (usually larger) merchants or through national compounders. Similarly, exports were usually arranged through other merchants, generally those situated at or near a port who would make the necessary arrangements for the ship to be loaded; brokers were also sometimes used and some merchants relied on their direct contacts with buyers abroad. The bigger merchants tended to use several methods, both for importing and exporting grain. The detailed results of the Survey are given in Table 2.1.

2.16. Although relatively few merchants had direct contacts with shippers the latter in fact handle the bulk of the U.K. import/export trade in grain. The trade has a pyramid structure—at least so far as the agricultural merchants, small country millers and compounders, farmers' groups and co-operatives are concerned. At the base of the pyramid are the thousand or so small country merchants and other relatively small country businesses who do some importing or exporting of grain. Their contacts, as shown above, are primarily with the few larger merchants specialising in grain-importing and exporting, including the port merchants, or with the national compounders or brokers. These constitute the middle band of the pyramid and they in turn work through a handful of shippers who constitute its apex.

2.17. This pyramid structure has evolved because it admirably suits the circumstances of this sector of the grain trade. Few of the small country merchants or millers would need to import or export a sufficient quantity of grain to justify arranging shipments themselves. For example, the average tonnage of wheat and maize imported, spread over the 900 or so

TABLE 2.1

METHODS OF IMPORTING AND EXPORTING GRAIN
USED BY MERCHANTS IN THE U.K.

(Results of a random sample of 260 merchants)

Of the 260 merchants, 108 did no importing or exporting of grain. The methods used by the remaining 152 firms were:

Merchants engaged in:	No. of firms
A. *Exporting grain only* (i.e. no importing done)	
Through other merchants only	24
Through brokers only	5
Through other agents (not specified) only	3
Through both brokers and other merchants	2
Through brokers and direct	5
Through shippers and direct	1
Total	40
B. *Importing grain only* (i.e. no exporting)	
Through other merchants only	17
Through brokers only	17
Through national compounders only	14
Through shippers only	4
Through others (not specified) only	2
Through brokers and shippers	4
Through brokers and national compounders	4
Through brokers and merchants	3
Through merchants and national compounders	4
Through brokers, merchants and national compounders	2
Total	72

C. *Both Importing and Exporting*

The 41 firms in this group used a variety of means of importing and exporting, but the most frequent were brokers (27 firms), other merchants (23 firms), shippers (11 firms) and national compounders (8 firms).

merchants importing grain, was about 1000 tons per firm per annum. This would represent only a small part of a shipment. By importing through intermediaries, the country merchant for example is able both to offer a service to farmers and to obtain the grain he needs for his own compounding activities, without incurring the overheads that would be necessary if these intermediaries were not available. With exports the same arguments apply, although with rather less force, since the size of vessel used in the export trade is so much smaller. This explains why there is a tendency among grain exporters to look for ways of exporting direct rather than through intermediaries. The pyramid structure still remains, however, the distinguishing characteristic of the grain import/export trade so far as the agricultural merchants, country millers, etc., are concerned.

The Cereal Futures Markets in the U.K.

2.18. A futures contract, like a forward contract, is a legally binding agreement by which one party undertakes to deliver grain at a specified date in the future, and the other party agrees to take delivery at that date. However, the futures contract, unlike the forward contract, is negotiable— it can itself be bought or sold at the currently ruling price for such transactions. This means that futures trading need not necessarily involve transfer of the physical commodity; instead, a trader who has made a contract to sell can offset it by purchasing a contract to buy the same amount of grain (or vice versa). If this is done, the trader is said to have "closed out"; he has cancelled his liability to handle the grain itself. If a trader does not thus offset his commitment to sell or to buy (or, to use the technical terms, if he remains "short" or "long" respectively), he must deliver or take delivery of the grain at the date specified in the contract.

2.19. Three distinct types of operation can be carried out in a futures market: hedging, speculation and procurement. There are many possible variations in the methods and objectives of each of those main classes of operation, and for a description of these and of the complex theory behind them, the reader should refer to specialised publications on the subject.[1] The description which follows summarises the basic nature of each operation.

Hedging involves balancing of price change on spot or forward dealings in a commodity by a simultaneous deal in futures. To quote a simplified example: if a trader bought grain spot at say £20 per ton, and sold it later at £19 per ton, he would have suffered a loss of £1 per ton. But if at the time of his spot purchase he had also sold a futures contract at say £22 per ton, and if when he sold the physical grain he

---

[1] As, for example, the following:

Hooker, A. A., *The London "Home-Grown" Grain Futures Markets* (London, A. A. Hooker & Co. Ltd., 1968).

Kohls, R. L., *Marketing of Agricultural Products* (New York, Macmillan, 3rd edition, 1967).

Working, H., New concepts concerning futures markets and prices, *Amer. Econ. Rev.* **52**, No. 3, June 1962.

Phillips, J., The theory and practice of futures trading, *Rev. Mktg. & Ag. Econ.* **34** (2), June 1966.

Hieronymus, T. A., Uses of grain futures markets in the farm business, *Illinois A.E.S. Bull.* 696, Sept. 1963.

closed out this futures commitment by buying another futures contract at £21 per ton, the profit on the futures deal would balance the loss on the spot deal. In this example, spot prices and futures prices have been pictured as moving exactly together (i.e. both dropping by £1 over the period concerned). In practice, of course, futures prices do not move precisely in line with spot or forward prices, but nevertheless the general movement of prices in all these markets will usually be roughly parallel; therefore a hedging operation, while not eliminating all the risk arising from price fluctuation, will reduce this risk to a greater or less extent. It will be obvious that hedging must not only reduce the risk of loss arising from price fluctuation, but also correspondingly reduce the chance of making a profit from this source.

Speculation in a futures market is the converse of hedging: the speculator, instead of avoiding risk, deliberately leaves himself open to price changes and attempts to take advantage of them by skilful buying and selling of contracts. Some speculators may deal in physical grain as well as in futures; others do not deal in the physical commodity, and may have no facilities for doing so. The word "speculation" has a derogatory sense in some people's minds; but as far as the futures market is concerned, speculation is a legitimate and necessary part of trading. Without the activities of speculators—who virtually underwrite the risks avoided by hedgers and keep up the flow of business on the market—futures trading could not be maintained; and if the futures market were to go out of existence, some potential benefits (discussed later in this section) would be lost to the industry.

Procurement implies use of a futures contract in virtually the same way as a forward contract, delivery of the grain being the intention from the outset.

2.20. In the U.K. there are two separate and independent cereal futures markets, at London and at Liverpool. The London market allows trading in contracts for home-grown barley and wheat, while Liverpool contracts relate to home-grown barley. For the purposes of this Survey, the respective Corn Trade Associations provided figures of total turnover in futures contracts at the two futures markets. These are shown in Table 2.2, which has been compiled for the five crop years 1963/4 to 1967/8; in fact, the Liverpool market for home-grown barley began operations in 1959/60, and the London contracts for barley and for wheat were opened in 1964/5 and 1965/6 respectively. Though no formal forecast estimates are available, the Associations suggest at the time of writing that in the crop year 1968/9 the turnover of contracts on the London market will be up considerably for both cereals (to about 500,000 tons for barley and about 300,000 tons for wheat), but that turnover at Liverpool will show a

TABLE 2.2

TOTAL TURNOVER OF HOME-GROWN
CEREAL FUTURES CONTRACTS
(TONNAGE), 1963/4 TO 1967/8

	London barley (tons)	London wheat (tons)	Liverpool barley (tons)
1963/4	(a)	(a)	203,375
1964/5	236,500	(a)	121,500
1965/6	381,375	38,125	94,875
1966/7	321,625	66,500	55,525
1967/8	294,625	129,625	94,125

(a) Market not operating.

substantial fall. There is no immediate explanation for this difference in the trends in futures turnover at the two trading centres.

2.21. These figures, of course, do not represent tonnages of grain actually delivered; a proportion of the amounts shown would be represented by contracts which changed hands several times and were, therefore, counted more than once. On U.K. production estimates for 1967/8 of some 9·2 million tons for barley and 3·8 million tons for wheat, it will be seen that the total turnover of futures contracts in that crop year was equivalent to about 4% of U.K. production for barley and about 3% for wheat. These percentages are very much lower than in cereal futures markets in other countries; for example, it is not unusual for total production of grain in the U.S.A. to be covered five or six times by futures market turnover.

2.22. When the delivery date of a futures contract arrives, provided that the contract has not been "closed out", the grain must be delivered—in the technical phrase, "tendered"—at a specified place known as a "tendering point". Grain can be tendered at stores belonging to a number of "Approved Service Operators" (A.S.O.'s), who are chosen by the Futures Associations from the list of dryer-operators authorised by the Ministry of Agriculture under the Cereals Deficiency Payments Scheme. At the time of writing, the London futures contract allowed tendering at about ninety stores belonging to A.S.O.'s, and the Liverpool contract allowed use of about thirty such stores (many of these were tendering points for both London and Liverpool contracts). A.S.O.'s premises were situated throughout England, as far north as Yorkshire. The choice of store used for tendering is at seller's option. The grain, of course, does not retain its identity in store, and delivery is effected by the A.S.O. signing over a "warrant of entitlement" to the holder of the contract concerned.

2.23. Trading at both London and Liverpool is in terms of a standard contract unit of 125 tons. All direct trading on each market is done among a small group of operators—elected by Grain Futures Associations at the respective Exchanges—known as the "ring". Trading is subject to strict rules and regulations and there are provisions against default on contracts; at both exchanges, the Associations maintain insurance funds to back futures trading. Members of the public wishing to trade in futures may do so through members of the Associations, who act as brokers.

2.24. Detailed statistics about the operations of the U.K futures market—figures, for example, of the types of firm who made contracts (shippers, merchants, farmers, etc), or of how many contracts were speculative in nature and how many were made by hedgers—are virtually nonexistent, and any description of how the British cereal futures markets perform their economic functions must be based mainly on consultation with knowledgeable operators in the market.

2.25. The bulk of the hedging on the U.K. market (in terms of tonnage represented by the contracts) is done by shippers—though the number of firms represented is of course very small (see Part IV, Chapter 1). In addition, a number of merchants deal in futures; at the London market at the time of writing, this number was about fifty, and a smaller number traded at Liverpool. Some farmers also traded, but their futures dealings accounted for only a small proportion of the total. There is no information on what proportions of merchants' and farmers' trading consisted of speculation and of hedging, but it is fairly safe to guess that hedging was the more important activity.

2.26. As has been seen, the outstanding feature of the U.K. cereal futures market at present is its relatively small size, in terms both of the proportion of total production represented by futures contracts and of the numbers of firms dealing in futures. In any futures market, there is a certain minimum level of turnover below which trading can no longer continue. A thin market obviously gives little scope for the operations of speculators. Further, it has been suggested that when activity in a market falls below a certain critical level, a vicious circle sets in whereby speculators lose interest and withdraw from trading: this makes hedging more difficult and activity falls still further. This is not to say, of course, that the initial thinness of the market can always be ascribed to low levels of speculation; it may equally well arise from lack of hedging interest. Another obvious result of low activity in a market is that brokers find it difficult to make an adequate total net return. It is fairly certain that the U.K. futures markets for home-grown cereals, up to now at least, have been no more than large enough to maintain themselves in existence, though the current expansion of trade at London is an encouraging sign for that market.

2.27. In fact, it is not easy to find an obvious cause for the relatively low level of activity on the U.K. grain futures markets. Farmers, of course, are partly protected against price variation by the guarantee arrangements, and so have relatively little incentive to hedge. But this is not true of merchants, relatively few of whom use the futures market. Certainly there are many merchants who pride themselves on their ability to "take a view" and extract the maximum advantage from all price changes. It is understandable that these merchants might scorn to engage in hedging. All the more reason, however, why they might be expected to enter the market on the other side, as speculators. Here, perhaps, lack of capital is the clue; not many merchants have enough spare capital to sustain the risk of futures speculation. More than for any other reason, however, the low level of interest in futures trading in the U.K. would seem to be caused simply by lack of knowledge—if not of the existence of the market, then of the ways it can be used and the method of dealing on it. This is supported by our own impressions at interviews with merchants, farmers and other cereal traders, and by the opinions of people we have spoken to within the futures market itself. Farmers and merchants may often observe futures prices as a guide to forward contracting; but in most cases they simply have not thought of trading in futures as a possibility for their own business. It is considered too abstruse, too risky or too time-consuming. Indeed, speculative trading may be all these things to anyone except a professional speculator in futures. Hedging, however, need not be; and if hedging interest were to increase, one may take it that the volume of speculative trade would expand in sympathy.

2.28. We believe that there is considerable potential for use of the futures market—particularly by merchants—as insurance against price risk, and that this at the moment is not being exploited to anything like its full advantage. The main responsibility for remedying this situation must lie with the Futures Associations themselves, though official assistance, perhaps through H.-G.C.A. or A.M.D.E.C., would seem well justified. The Associations have already made efforts to spread information about the value of futures trading. But in our view, more can still be done in this direction, and we consider that priority should be given to a programme of information and education directed at all potential users of the cereal futures markets, showing how to go about trading in the market, its possible modes of use and its benefits. We would suggest also that the two separate Futures Associations should begin producing much more comprehensive and detailed statistics on the operation of their markets; we would hope that in this project they would co-operate not only with each other, but also with other interested bodies outside the trade. This would make possible further study in a field where it is much needed.

CHAPTER 3

The Sea Transport of Grain

3.1. There are basically three types of sea transport of grain relevant to the U.K. situation. The first is the importing of grain into U.K. ports, either direct from the country of origin or transhipped via Europe; second is the export trade from the U.K., and third is the coastwise movement of grain between U.K. ports.

Movement of Grain into British Ports

3.2. Grain is imported into the U.K. from all over the world, but as Table 3.1 and Map 3.1 indicate, the most important sources of wheat are Canada, the U.S.A., Australia, Argentina and Eastern Europe (grain from the Netherlands is mostly American or Australian grain transhipped there); of maize and sorghum, the U.S.A., South Africa, France and Argentina; and of barley, Canada. As indicated already in Chapter 1, the grain imported into the U.K. direct from the country of origin comes in bulk grain ships of medium size, i.e. 10,000–25,000 tons. Vessels using Leith, Manchester and Belfast are usually about 12,000 tons, those using Liverpool, Avonmouth (Bristol), Hull, Glasgow and London up to about 25,000 tons. Table 3.2 and Map 3.1 show that these eight ports dominate the situation so far as grain imports are concerned, together accounting for 85% of total imports.

3.3. The restrictions on size of grain ships using U.K. ports are due partly to physical factors (e.g. a harbour bar, depth of water, size of berth, or the problem of passing under the bridges across the Manchester Ship Canal) and partly to economic factors, e.g. vessels of larger size would not necessarily be used, even if they could be accommodated, because the users do not require more grain at a time than they can cope with. Most of the ships are bulk carriers of various kinds, not only for grain but also for other bulk commodities like phosphate or coal. Converted oil tankers are less used today than they were a few years ago and with the tonnage of purpose-built bulk carriers steadily rising, the use of oil tankers is likely to continue

351

TABLE 3.1

UNITED KINGDOM: IMPORTS OF GRAIN IN 1966 AND 1967—BY COUNTRY OF ORIGIN
(thousand tons)

	Wheat		Wheat flour[b]		Maize		Barley		Oats		Sorghum	
	1966	1967	1966	1967	1966	1967	1966	1967	1966	1967	1966	1967
Australia	521	384	27	14	—	—	19	28	17	16	8	7
Canada	1694	1620	159	64	112[a]	8	99	186	11	4	—	—
Malawi	—	—	—	—	25	28	—	—	—	—	—	—
Argentina	179	202	—	—	93	155	—	—	—	—	94	85
Belgium	173	54	—	—	74[a]	76	—	—	—	—	37[a]	18
France	198	170	—	—	68	260	—	—	—	—	—	—
Mexico	—	—	—	—	169	95	—	—	—	—	—	—
Netherlands	499[a]	543	—	—	422[a]	541	2	—	1	1	132[a]	74
South Africa	—	—	—	—	17	537	—	—	—	—	58	113
Eastern Europe	49	286	—	—	43	133	—	—	—	—	—	—
United States	594	437	8	4	2206	1773	1	—	—	—	190	30
Other countries	98	78	2	11[c]	7	57	4	4	—	—	3	—
Total	4005[d]	3774[e]	196	93	3236	3663	125	218	29	21	522	327

[a] Believed to be mainly transhipments.   [b] Actual weight.   [c] Fed. Rep. of Germany.   [d] Including denatured feed wheat 287, of which Belgium 105, France 83, Sweden 61.   [e] Including denatured feed wheat 117, of which Belgium 1, France 38, Netherlands 78.

Source: Commonwealth Secretariat, *Grain Bulletin*, Vol. XIV, No. 2.

to fall off—U.K. importers will not accept grain imported directly in oil tankers.

3.4. Usually a grain ship will carry only one or two types of grain in any one cargo, although there is no difficulty about carrying more types. This homogeneity of many cargoes derives largely from the fact that wheat is not grown in the same geographical regions as maize, though, in contrast, because the Great Lakes provide access to different geographical regions the ships from this area often carry more than one type of grain. Most journeys by grain ships are direct from port of origin to destination, with the ship either going back in ballast or taking back another bulk cargo if one is available; there is very little "triangular trade" these days, although occasionally a ship will return to the Gulf (U.S.A.) via South America. One reason for this is that the ships are specially equipped to handle grain, and certain other bulk cargoes, and are best employed in these particular trades; another is that a bulk carrier is very expensive to operate, in relation to the value of the cargo carried, and it must be kept at sea for as many days in the year as possible, i.e. once it has discharged its cargo it cannot afford to wait for an outward cargo to be assembled.

Transhipment via Europe

3.5. During the last few years the pattern of grain imports has changed significantly. With the introduction, a few years ago, of large bulk carriers capable of carrying up to 60,000 tons of grain, and also oil tankers temporarily converted into grain-ships, the practice developed of taking grain

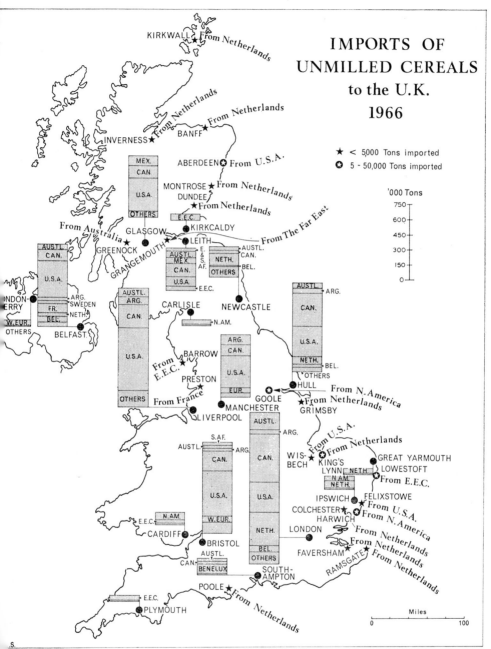

MAP 3.1. Imports of unmilled cereals into the U.K., 1966.
Source: Statistics supplied by H.M. Customs and Excise.
Note: Statistics sometimes relate to a group of ports as indicated on Map 3.3.

N

TABLE 3.2

IMPORTS OF GRAIN INTO THE U.K. (1967)
(thousand tons)

Port of entry	Wheat	Wheat flour[a]	Maize	Barley	Oats	Sorghum and millet	Total[e]
London	828·6	20·7	533·6	6·8	2·9	49·1	1450·9
Liverpool	575·3	4·7	481·1	20·1	1·4	38·1	1122·6
Southampton	131·1	0·3	56·7	—	—	0·6	118·8
Bristol	331·0	5·9	430·6	35·4	2·0	56·2	866·1
Cardiff	120·8	0·1	17·9	—	—	0·4	139·2
Manchester	244·7	14·5	455·6	—	—	11·0	737·7
Hull	502·0	1·0	362·0	7·6	0·1	44·5	919·8
Ipswich	73·8	—	56·6	—	—	15·9	146·3
Newcastle[b]	182·8	1·1	43·8	—	—	0·6	228·8
Glasgow	142·5	21·4	423·9	15·1	11·6	14·2	637·1
Leith	166·4	11·3	124·5	48·4	0·5	6·6	362·2
Belfast	260·1	6·4	403·6	75·9	1·2	73·7	823·5
Other ports	213·4	5·4	273·2	8·8	1·1	24·3	527·9
Total	3773·5[c]	92·8	3663·1	218·1	20·8	335·2[d]	8150·9

[a] Actual weight (i.e. not wheat equivalent).   [b] Including N. and S. Shields.
[c] Including denatured feed wheat 117·4, of which Belfast 40·5.   [d] Of which millet
0·8.   [e] Includes rye.

Source: Commonwealth Secretariat, *Grain Bulletin,* Vol. XIV, No. 4.

in these vessels to Rotterdam, Amsterdam or Antwerp and there transhipping it into very much smaller vessels for the last stage of its journey. The saving in bringing Canadian grain to Europe in a large 50,000 ton bulk carrier, compared with a 20,000 ton ship, is estimated to be about £1 per ton, and this more than compensates for the cost of transhipment. During the last 5 years the tonnage of grain entering the U.K. after transhipment has risen from about 100,000 tons to over one million tons in 1967/8. Without doubt, the 50,000 ton bulk grain ships are certain to oust many of the smaller 10,000–20,000 ton ships, which are now relatively uneconomic and are mainly used because the buyers often do not have sufficient depth of water to receive the larger ships. It is doubtful, however, whether the trend towards even larger vessels will continue much further. Ships of over 50,000 tons could not be accommodated in most loading or discharging ports, nor could they use the Panama Canal. It seems likely, therefore, that for the time being we have reached the limit of size at 50–60,000 tons. The U.K. trade has tended to use Rotterdam, Antwerp and Amsterdam (the "ARA" ports as they are sometimes called) in preference to U.K. ports for several reasons; no U.K. port could accommodate grain vessels of 50,000 tons and over; port handling charges at the main U.K. ports are higher than in Europe, whilst the grain-handling equipment generally available is less efficient and there is a dearth of storage facilities; dock labour problems tend to lead to more delays and frustrations in U.K. ports than in Europe. Another factor is that certain U.K. importers are located on tidal creeks, or by docks and wharves capable of taking only small ships, and if grain is to be delivered there by sea it has to be transhipped in any event. Thus in existing circumstances it is cheaper to bring grain in a bulk carrier to Rotterdam and then ship it by coaster back across the North Sea, then to land it at London in a smaller ship and tranship it to the East Coast.

3.6. The ARA ports—Rotterdam in particular, and the other two ports to a lesser extent—have the great advantage over any U.K. port that they lie on the mainland of Europe, at the mouths of rivers giving excellent access by land or water to the heart of the Continent; they are also as well placed for coastwise movement as any U.K. port. Moreover, because they are well-equipped with silos capable of simultaneously unloading a bulk carrier and loading the grain into coasters they are better placed for the transhipment trade than any U.K. port at present. Because of their geographical location grain can more readily be "buried", to use the shippers' jargon (i.e. dispersed to a variety of destinations), from the ARA ports, and by a greater variety of routes, than is possible at any U.K. port. The large grain silos at the ARA ports, some of which are operated by the shippers and municipalities jointly, are mainly transhipment silos, the grain staying in them for only a week or two at the most. Silos of this kind have not been built at any U.K. port for the simple reason that the bulk of the grain is not usually imported for transhipment but for use at the point of unloading. For reasons of geography, therefore, it seems unlikely that any U.K. port will acquire a position in the international grain trade comparable to that of Rotterdam. This is not to say, however, that the U.K. cannot compete at least for its own share of Rotterdam's trade. Of the U.K. ports London is the best placed, by virtue of its short distance from the mouths of the Rhine and Scheldt, to challenge the ARA ports, and the new grain terminal at Tilbury gives it an opportunity to win back some of its lost trade. Between 1955 and 1967 the volume of all cargo handled by London (other than fuels) dropped from 24·7 million tons to 20·5 million tons at a time when world trade was increasing. This is a vicious circle; as trade declines so it becomes more difficult to justify installing the up-to-date equipment which alone can attract trade back again. The Tilbury terminal is an attempt to break this vicious circle by putting London back on the map as one of the best equipped grain ports in Europe.

The Tilbury Grain Terminal

3.7. The Port of London Authority's new grain terminal at Tilbury will be in operation early in 1969. The scheme includes a deep water jetty capable, after the river has been dredged, of accommodating vessels of up to 60,000 tons. There will be grain discharging equipment capable of handling 2000 tons per hour (the fastest rate of discharge in the world), and grain silos totalling 105,000 tons capacity. These facilities will permit ships to be turned round much more quickly, greatly reducing the cost of the whole operation.

3.8. Some members of the grain trade are frankly sceptical of Tilbury's chances of success. They consider that the largest bulk carriers will continue to use Rotterdam because the London market could not absorb 50,000 tons of grain arriving at one time, and although the surplus could be stored in the P.L.A. silo, this would be very expensive. Moreover, they are afraid that the gains from the acknowledged efficiency of grain handling at the new terminal will be partly offset by the continuation of high port charges and the labour troubles that have bedevilled the Port of London in the past. They forecast that the London millers and compounders will continue to import direct to their mills further up-river from Tilbury in medium-sized ships.

3.9. There are some powerful arguments on the other side, however. First and foremost is that fact that the P.L.A. will take all possible steps, having invested £5 million in this project, to ensure that it is fully used, They could do this by such means as charging very competitive rates, allowing their grain-handling equipment at the docks up-river to run down. and encouraging grain millers and compounders to locate their new factories adjacent to the terminal. One large flour milling concern has already announced plans for a new flour mill there. Occupying a 5-acre site adjacent to the terminal this will be the most modern flour mill in Europe—the grain will be fed direct from the terminal to the mill by a conveyor system capable of delivering 1000 tons per hour. Negotiations are currently in progress with four other millers and grain processors for neighbouring sites. As in other Continental ports, an industrial complex will undoubtedly grow up at Tilbury based on the direct ship-to-factory discharge of a bulk commodity. For a while the big millers and compounders with heavy capital investments at docks up-river may resist this trend, but in the long run they will have to decide whether to follow suit, write off their older mills, and redevelop further downstream. In the short term the P.L.A. may introduce a system of large powered-barges, similar to those seen on the Rhine, carrying grain from Tilbury to up-river sites on a daily schedule. As to the problem of disposing of 50,000 tons of grain at one time, this need present no insuperable difficulty provided all the importers use the Tilbury terminal.

London as a whole imports about 1½ million tons of grain per annum, or about 30,000 tons per week—thus a large bulk carrier arriving, say, once every three weeks could supply a good proportion of London's needs, leaving some to be imported in smaller ships. Of course it is not really as simple as this because grain comes to London from different countries of origin, and different types of grain often come from different ports. Moreover, of the 1½ million tons nearly 400,000 tons are transhipped via the ARA ports and it cannot be assumed that this will come to an end—the ARA ports may well compete successfully with Tilbury to keep some at least of the present transhipment business. On the other hand, the above figures make no allowance for the extra demand likely to arise from the new mills being built at Tilbury itself, or for grain moving through London to destinations inland. It is relevant in this connection that the P.L.A. expect the terminal to handle up to 3 million tons of grain a year. Thus the problem is not insoluble, but it must certainly involve far-reaching changes in current patterns of importing. Our view is that in the short term probably not many ships in the 50,000–60,000 ton class will use Tilbury but it will be used mainly by vessels in the intermediate size range—25,000–35,000 tons. In the long run, however, the economic advantages of moving a bulk commodity like grain over long distances in big ships are so overwhelming that the larger bulk carriers will become increasingly important and the cereal using industries will have to fall in with whatever location pattern is most appropriate. In short, the mills will have to move to the ships and not vice-versa.

3.10. There can be no doubt that the Tilbury terminal will indeed be used, and that it will have a very significant impact on the future development of grain imports. For the first time the trade will have port storage facilities in the U.K. comparable with those provided in European ports, and undoubtedly they will be used for short-term storage. There will probably be less transhipment via Europe, and a new complex of grain-using industries has already begun to develop round the terminal; the terminal will also play an important part in our grain export trade, as indicated below. Because Tilbury is developing rapidly as Britain's largest container port, road and rail communications inland are certain to be vastly improved and this also will be to the advantage of the grain terminal compared with congested up-river sites.

Developments at Other Ports

3.11. London is not alone in attempting to improve its grain-handling facilities; Liverpool and Glasgow for instance are making similar improvements and the largest bulk carriers will soon be able to use these ports.

Because of the desire of each port to retain its competitive position *vis-à-vis* other grain importing ports there is indeed some risk of over-provision of grain-handling facilities. As in London, these developments in West Coast ports are likely to lead to a rationalisation of milling and compounding activity within the port areas, with the older mills, or those located on small docks, closing down and new sites being developed adjacent to the new terminals. Thus at the same time as some animal feed compounding is tending to move out of the great ports into the country areas, there is also beginning a movement towards the concentration of grain-using industries within those ports. This two-fold development will tend to crystallise more precisely the present rough distinction between the "country compounders" and the "port compounders" (now useful only as a geographical not an ownership distinction) as each seeks to maximise its particular advantage in relation to access to its raw material. So far as the U.K. as a whole is concerned, if these various developments succeed in their objective the future is likely to see a reduction in the proportion of grain imports into the U.K. transhipped via Europe, and possibly even some reversal of this role.

Inland Waterways

3.12. The inland waterways perform a useful function in enabling grain transhipped from ocean-going vessels at the main ports to continue their journey by water to the compound feed and flour mills, or merchant premises located inland. The British Waterways and the Manchester Ship Canal Company have supplied statistics on which Table 3.3. and Map 3.2. have been based.

The U.K. Grain Export Trade

3.13. One of the most remarkable features of the U.K. grain trade during the last few years has been the sudden expansion of exports. British malting barley has long been in demand in Europe and until 1964/5 there was an annual export of about 100,000 tons of barley. In 1965/6, however, the export of barley, including now a high proportion of feed barley, rose to 667,000 tons, and in 1966/7 to 1,091,000 tons. In 1967/8 total exports of barley fell to 782,000 tons. West Germany and Denmark were among the principal customers in all three years with Belgium, Italy, the Netherlands and Spain (to 1966/7) also taking substantial quantities (Table 3.4). Most of the grain exported from the U.K. went out in small coasters, usually between 200 and 1000 tons, operating from many small ports along the east and south coasts, some of them, like Gunness in Lincolnshire, hardly even names on the map (Map 3.3). Usually grain-handling equipment was

TABLE 3.3

TONNAGE OF GRAIN AND GRAIN PRODUCTS MOVING TO INLAND DESTINATIONS
IN THE U.K. BY WATER 1967

(A) British Waterways Waterway	From	To	Mileage by water	Tonnage ('000 tons)
1. Aire and Calder Navigation	Hull	Wakefield	59	37
	Hull	Knottingley/ Castleford	42/49	18
	Hull	Selby	39	10
				65
2. Trent	Hull	Gainsborough	43	52
	Hull	Carlton	56	28
	Hull	Newark	75	9
	Hull	Nottingham	98	8
				97
3. Sheffield and South Yorkshire	Hull	Sheffield	65	18
	Hull	Doncaster	47	11
	Hull	Rotherham	59	11
	Hull	Mexborough	55	12
				52
4. River Severn	Avonmouth	Gloucester	40	48
	Avonmouth	Tewkesbury	52	23
	Sharpness	Bristol	29	23
	Various	Various	—	8
				102
5. Fossdyke and Witham	Hull	Lincoln	65	10
6. London	Thames	Wellingborough	90	8
	Maypole Dock	Within London area	—	27
				35
7. Scotland (Caledonian Canal and Coastwise)				0·5
Total				351·5
(B) Manchester Ship Canal Tonnage of grain imported through the Port of Manchester				752
Grand Total				1103·5

Sources: British Waterways Board. Manchester Ship Canal Co.

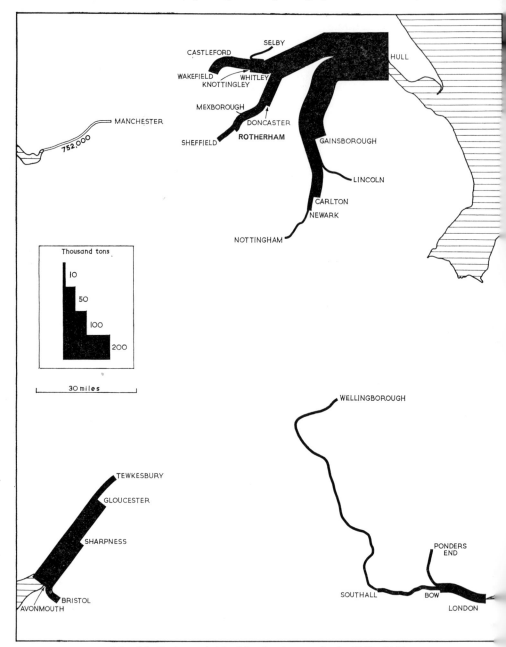

MAP 3.2. Grain carried by inland waterways in the U.K., 1967.
Source: British Waterways Board and Manchester Ship Canal Company.
Note: The tonnage carried on the Manchester Ship Canal (752,000 tons) is too great to be shown on the map.

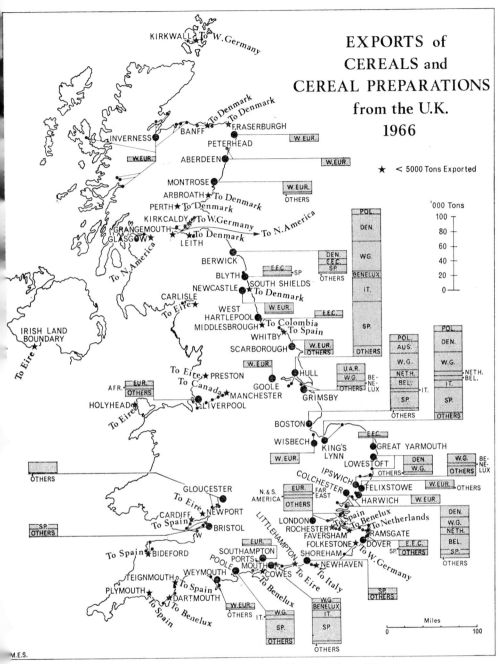

MAP 3.3. Exports of cereals and cereal preparations, 1966.
Source: Statistics supplied by H.M. Customs and Excise.
Note: Statistics sometimes relate to a group of ports as indicated on the map.

N*

primitive in the extreme and there were generally no grain storage facilities at the wharf-side. Most of the ships are "tramp" ships i.e. they move about as and when their services are required, as opposed to "liner" ships which follow scheduled runs. Many are owner-operated and may be British, Dutch or French. Coasters are always chartered through freight brokers.

TABLE 3.4

EXPORTS OF BARLEY FROM THE U.K.

('000 tons)

Destination	July/June 1965/6	July/June 1966/7	July/June 1967/8
Belgium	63	100	36
West Germany	201	183	349
Italy	21	125	139
Netherlands	47	58	55
Denmark	130	166	129
Irish Republic	10	8	5
Norway	1	24	5
Poland	22	33	27
Eastern Germany	30	27	5
Egypt	18	—	—
Spain	114	327	23
Other countries	10	40[a]	9
Total	667	1091	782

[a] Of which Algeria 10 and Israel 11.

Source: Commonwealth Secretariat: *Grain Bulletin,* Vol. XIV, No. 7.

3.14. Exporters have tended to concentrate on a particular region, and have exported mostly through the ports serving its region—some of them have already made improvements in these port facilities and others have plans for limited new investment. The exporting of such a large tonnage was a triumph of ingenuity and expediency, and it is a tribute to the flexibility of the grain merchants, brokers and shippers that they were able to dispose of so much grain in altogether new markets. Had it not been for their success the domestic grain market would have had the greatest difficulty in absorbing such a large additional tonnage without a collapse in prices.

3.15. In the longer term, however, the problem has to be faced of whether or not to "tool up" for a permanent export trade. This involves heavy capital investment in wharf-side silos to enable grain to be concentrated prior to the ship's arrival to ensure rapid loading and turnround, overhead protection to enable loading to continue in wet weather, and bulk handling equipment. Demurrage charges for keeping a vessel in port beyond an allotted period are very heavy, for instance the daily charge for a vessel of 1000 tons is £144. The process of "tooling up" may also mean finding customers abroad and devoting time and expense to foreign travel. So far as the ship-owners are concerned, it is probable that more of the trade will be handled in the slightly larger coaster, e.g. 500/600 tons minimum, since the crew is virtually no larger than is required on a 200 ton ship and the cost can be spread over a larger tonnage. However, an important factor inhibiting the use of larger vessels is the capacity of the loading berth in the U.K. and the discharging berths abroad, (many buyers in Europe are situated on creeks or inlets only capable of accommodating 200 ton ships). Is the future for the export trade sufficiently promising to justify additional investment of this kind? The experience of 1967/8, and the existence of substantial carry-over stocks in European countries, leads one to doubt whether the circumstances are such as to justify much investment on the export side. If the U.K. eventually joins the Common Market there is likely to be a rapid increase in grain trading between the U.K. and the other Common Market countries—this would be in both directions and would partake of the character of normal "internal" trade, being conducted by water rather than by land. But this is a possibility for the future; the present likelihood is that the big shippers will make some limited investments in those ports that are well situated to handle both exports (middle-distance as well as short-sea) and domestic grain, but that caution and flexibility will be the order of the day. With the uncertainties surrounding the U.K. grain export market at present this seems the only wise policy.

3.16. Because of the small scale of most of the grain export trade the big international shippers do not have as dominant a role here as they have acquired in the importing business. It is not usual, for instance, for the shippers to own any coastal vessels, nor did they have, until recently, any direct contacts with agricultural merchants in the grain-growing areas of the U.K. The international shippers have used their great experience in the grain trade and their connections in the European markets to secure a substantial share of total exports, but still quite a lot of British grain is exported by the smaller shippers and by merchants, either selling direct to buyers abroad or working through the grain brokers. Merchants in the exporting areas of the U.K. are tending to look for direct outlets abroad, whilst the co-operatives have already acquired direct links with the co-operatives in Europe. Farmers' producer and marketing groups are also looking for direct outlets abroad. The brokers, in particular, have played a very important role in the export trade, and their importance is likely to increase rather than diminish since the export market is likely to remain a rather volatile one.

3.17. There are some people who criticise the shippers for competing with each other to sell U.K. grain abroad. They argue that this is bound to reduce the price and that it would be better if the exporters stopped competing in this way but agreed to offer British grain at a higher price to the benefit of this country's export earnings. This attitude ignores the fact that the price of grain on world markets is a function of the supply and demand situation at any time and it would make no difference to the total supply of grain if U.K. grain were offered for export by a central export agency or by many sellers acting independently. With the present system each exporter tries to get the best price he can. If free competition were excluded then either a central selling organisation would have to be envisaged, or some system of "allocating" exports among the shippers and others trading in overseas markets would have to be introduced. Neither of these is likely to be as efficient as the present system; moreover the buying countries might react adversely to the imposition of a united selling "front" and take their custom elsewhere, or play one exporting country off against another. We would be reluctant to see any such intervention in this highly competitive and efficient market.

3.18. There is one important new development which may change the character of the U.K. export trade in the years to come. Hitherto relatively little grain has been shipped to the East Mediterranean countries, or countries further afield, because no U.K. port has adequate facilities for the regular loading of ships of more than about two thousand tons, (although in special circumstances ships of up to 5000 tons have been loaded). This is because equipment designed to discharge grain cannot, without prohibitive expense, be set in reverse to be used for loading. A few coasters have shipped grain to Italy and North Africa, but the longer the journey time the less economic it becomes to ship grain long distances in small ships. Once the new Tilbury terminal is operational, the situation could be transformed, since the terminal incorporates facilities for the rapid loading of vessels of up to 7000 tons—either from the silo or direct from a larger bulk carrier alongside. The cost of moving grain from the growing areas to Tilbury has to be considered, but it is quite possible that Tilbury could become an important grain export port for the middle-distance trade, and could thus fill a major gap in our existing facilities.

The Coastwise Movement of Grain

3.19. Bearing in mind the remarkably low cost of carrying grain long distances by sea (one firm of shippers recently exported grain to Poland for less than £1 per ton) it is at first sight curious that there appears to be very little coast-wise movement of grain between British ports, although there is a substantial movement from Britain to Northern Ireland. When this point was discussed with the shippers they all blamed this situation on the activities of what they called the "Coastal Conference" (more correctly the Coastal and Short Sea section of the Chamber of Shipping). During the 1939–45 War all coastal freight rates were tightly controlled by the Government and when the War was over and the Government withdrew, the ship owners apparently decided to keep in being this system of standardised freight rates on a voluntary basis. Coastal freight rates have been determined in this way ever since. To quote an example, there is a freight rate fixed for carrying 600 tons of barley from the Thames to Ipswich (19*s*. 3*d*. per ton)[1] or King's Lynn (23*s*. 6*d*. per ton), or Leith (23*s*. 9*d*. per ton); rates are quoted similarly from the other main ports to all likely destinations. The effect of this system has been that coastal freight rates for grain are now higher than those for road or rail for all except the longest journeys, and there is little coastwise movement of grain between U.K. ports less than 100 miles apart. British coastal freight rates also appear to be less competitive than those charged by Continental countries. In our view the present system of standardised freight rates is unlikely to lead to maximum efficiency in short sea transport. It seems to us that with the rapidly rising costs of road transport and congestion on the roads, there are opportunities in the near future for a resurgence of coastwise movements of grain transhipped via Tilbury, Liverpool and other major U.K. ports, to lesser ports around our coasts, as well as grain shipments from growing areas in the U.K. to the consuming areas. This is unlikely to happen unless rates are allowed to be as competitive as possible and we would welcome any movement in this direction.

---

[1] This is about the same as the cost of bringing grain across the North Sea from Rotterdam.

PART V

Utilisation of Grain in the U.K.

(a) The Animal Feedingstuffs Industry

Introduction

THE approach adopted in this Part is to take a broad look first at the industry as a whole, beginning with the demand for feedingstuffs, e.g. how the pattern of demand has changed seasonally, regionally and by class of livestock over the last ten years, and then at the supply side where the emphasis is mainly on the sources of raw materials for animal feed manufacture. After this broad look comes a more detailed study of the structure of the industry (Chapter 3) and its division into the seven national compounders and hundreds of smaller firms. The focus of interest is progressively narrowed and Chapter 4 deals with the grain buying arrangements of animal feed manufacturers, Chapter 5 with the process of integration within the industry, Chapter 6 with home mixing on farms, and Chapter 7 with the special problems of Northern Ireland's animal feed industry.

CHAPTER 1

The Demand for Animal Feeds

Trends in Livestock Output and Feed Requirements[1]

1.1. The livestock population of the United Kingdom has created over the decades a demand for feedingstuffs of many different types including grass in its various forms, cereals and straw, root crops and other farm products. The demand for the various types of feed obviously depends on the type of animal, its purpose, and the system of production used. With beef and dairy cattle, a producer has a wide range of choice between cereal-based feed, grazing, hay or silage, root crops and by-products such as bran, beet pulp, and brewers' and distillers' grains, and specific management decisions relating to the profitability of the operation will decide in what ratios some or all of these types of feed are used. Similarly with pigs, poultry and sheep there is a degree of choice as to the feed constituents to be used. Geographical location, however, with its associated ecological, climatic and market factors will greatly influence the type of animal production to be pursued and the relative suitability of the various feed alternatives available. Technological advances, also, and improving feed conversion ratios have contributed greatly to changes in animal feeding practices and requirements over the years.

1.2. In Table 1.1 indices of total sales of fatstock and livestock products at constant prices are compared with indices of total concentrated feed[2] consumption. This shows a fairly close parallel between total livestock production over the years and the use of concentrated feedingstuffs.

---

[1] This Part relies for a great deal of statistical material on *Developments in the feeding-stuffs manufacturing industry and the production and utilisation of concentrated feeding-stuffs since 1953*, Economic Trends No. 130, August 1964, and subsequent statistics published by the Ministry of Agriculture, Fisheries and Food in August 1967 (Press release titled *Developments in the Animal Feedingstuffs Industry*) and more recent statistics provided by the Ministry for inclusion in this Chapter. These three sources are collectively referred to as M.A.F.F. *Developments*.

[2] See 1.3 for definition of concentrated feed.

371

TABLE 1.1

A COMPARISON OF LIVESTOCK AND LIVESTOCK PRODUCT SALES
(AT CONSTANT PRICES) AND CONSUMPTION OF CONCENTRATED
FEEDINGSTUFFS, U.K.

Year	Livestock and livestock product sales at constant prices (£ million) (1954/5–1956/7 average prices)[a][c]		Total consumption of concentrated feeds[b][c]	
	Value of output (£ million)	Index 1954/5–1956/7 average = 100	Million tons	Index 1954/5–1956/7 average = 100
1954/5	1008·4	99·9	12·0	99·7
1955/6	974·2	96·6	12·0	99·7
1956/7	1043·3	103·4	12·1	100·6
1957/8	1093·1	108·4	12·3	102·2
1958/9	1101·4	109·2	13·7	113·9
1959/60	1116·5	110·7	13·4	111·4
1960/1	1154·5	114·5	13·6	113·0
1961/2	1246·2	123·6	14·5	120·5
1962/3	1267·9	125·7	15·1	125·5
1963/4	1290·7	128·0	14·8	123·0
1964/5	1308·5	129·7	15·9	132·1
1965/6	1329·4	131·8	16·5	137·1
1966/7	1330·7	131·9	15·6	129·6

[a] *Source:* M.A.F.F. *Output statements.* [b] *Source: Annual Review and Determination of Guarantees,* various years. [c] On holdings over 1 acre.

1.3. In this chapter of the Report we are mainly concerned with the utilisation of cereals as animal feed, but for most of the time we will have to discuss *concentrated feedingstuffs* in general. Cereals are used in a variety of forms in livestock production and it may be as well to establish some terms before we continue. In Ministry of Agriculture definitions[1] "concentrated" (i.e. not bulky) feedingstuffs consist of compounded rations, processed straights (provenders), cereal livestock mixtures and straights. The latter include feed grains, oil cakes, dried sugar beet pulp, etc. It is recognised that within the compound feed trade "concentrates" are defined more narrowly to mean a protein-mineral-vitamin mixture intended for blending with cereals to form a complete food. In this report we will call these mixtures a *concentrate supplement.* Compounded rations consist of a mixture of cereals and other ingredients which have undergone some form of processing. Processed straights or provenders are normally considered to be cereals which have not been mixed but which have been processed; they may include some other processed materials. Cereal livestock mixtures are mixtures of feed grains which have not been processed.

[1] M.A.F.F. *Developments.*

TABLE 1.2

UTILISATION OF CEREALS AND
CEREAL PRODUCTS IN CON-
CENTRATED FEEDS, U.K.
1955/6–1966/7

| Year | Utilisation of cereals and cereal products | |
	Million tons	As % of concentrated feed consumption
1955/6	9·5	80·5
1957/8	10·3	83·7
1959/60	10·9	81·3
1961/2	12·3	84·8
1962/3	13·3	88·1
1963/4	12·9	87·2
1964/5	13·5	84·9
1965/6	14·4	87·3
1966/7	13·2	84·6

Source: Home-grown cereals used for stock feed: M.A.F.F. Utilisation statements. Imported cereals, imported cereal offals, home-produced offals from imported materials and offals produced from home-grown materials: M.A.F.F. *Developments* (Table 6).

Manufactured feedingstuffs are those undergoing some form of processing in a place classified by the Ministry as an animal feedingstuffs factory. These definitions may appear to be complicated but it is necessary to understand them in order to interpret statistics which are provided by the Ministry of Agriculture.

1.4. As Table 1.2 shows, cereals have for a long time been a major item in concentrated feed consumption and have increased in importance in the last ten years.

1.5. From these tables it might be surmised that there has been a slight increase in overall concentrate and cereal feed requirements *per unit of livestock output,* whereas improving conversion ratios and improved grass management might lead one to expect that concentrate feed requirements per unit of output would be on the decline. The obvious explanation of this paradox lies in the different rates of growth of various sectors of the livestock industry in recent years. Intensive poultry production, which has increased very rapidly, relies largely on cereal-based feed rations, and the same may be said of pig production. On the other hand, although concentrated feeds for cattle have a much higher cereal (and hence energy) content than 10 years ago, cattle and sheep production utilises considerably more grass in its various forms, as well as other crops and by-products such as root crops, sugar beet pulp and pulses.

1.6. From Table 1.3 and Fig. 1.1 it can be seen that the expansion of poultry production relying heavily on concentrated feed was much more rapid than that of other products. These differential growth rates are also reflected in the demand for total concentrated feeds (expressed in terms of starch equivalent) (Table 1.4) and in indices of sales of the main types of manufactured compounds (Table 1.5).

TABLE 1.3

UNITED KINGDOM OUTPUT OF LIVESTOCK AND LIVESTOCK PRODUCTS AT CONSTANT
PRICES (prices = 1954/5–1956/7 average)

	Average output 1954/5–1956/7		Average output 1964/5–1966/7		Average of output 1964/5–1966/7 as a % of 1954/5–1956/7
	£ million	% of total output	£ million	% of total output	
Cattle and calves	185·4	18·4	217·4	16·4	117·3
Sheep and lambs	59·3	5·9	82·7	6·3	139·5
Milk and milk products	337·8	33·5	376·4	28·4	111·4
Wool (clip)	15·7	1·6	18·3	1·4	116·6
Pigs	185·3	18·4	235·4	17·8	127·0
Poultry	41·8	4·1	127·2	9·6	304·5
Eggs	179·0	17·7	254·0	19·2	141·9
Other	4·4	0·4	12·0	0·9	272·7
Total	1008·7	100·0	1323·4	100·0	131·2

Source: M.A.F.F. *Output and Utilisation of Farm Produce in the United Kingdom.*

TABLE 1.4

ESTIMATED UTILISATION OF CONCENTRATED FEEDINGSTUFFS IN U.K. BY PRINCIPAL
LIVESTOCK CLASSES[a] (all holdings)

(million tons starch equivalent)

	1955/6	1959/60	1963/4	1965/6	1966/7	1966/7 as % of 1955/6
Cattle	2·9	3·4	3·6	4·5	4·0	138
Pigs	2·7	2·7	3·1	3·2	3·0	111
Poultry	2·5	3·3	3·8	3·8	3·9	156
Other	0·4	0·4	0·3	0·4	0·4	100
Total livestock	8·5	9·8	10·8	11·9	11·3	133

[a] Excluding dried grass products.

Source: M.A.F.F. *Developments.*

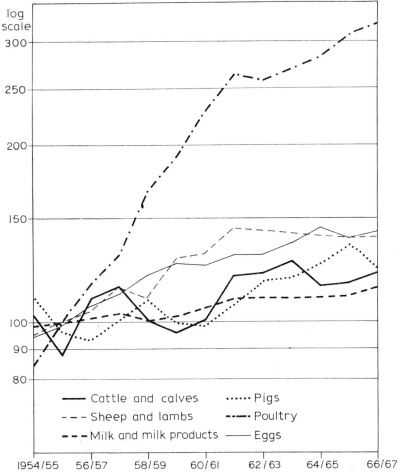

FIG. 1.1. Indices of livestock and livestock product sales at constant prices
(1954/5–1956/7 average prices) 1954/5–1956/7 av. = 100.
Source: M.A.F.F. *Output and Utilisation of Farm Produce in the U.K.*

1.7. In the last column of Table 1.4 it can be seen that the increase in the utilisation of concentrated feedingstuffs by both cattle and poultry has been rapid, but Table 1.5 suggests that the demand for poultry *compounds* has outstripped the demand for cattle and calf compounds. In both tables, the figures for pigs need careful interpretation because of the fluctuating pig population between one year and another—utilisation and sales have been higher in some years than in those selected for the tables.

1.8. The increase in the utilisation of concentrated feedingstuffs for cattle—38 % in 11 years (Table 1.4)—appears to have exceeded the increase in sales of manufactured compounds for this class of animals (Table 1.5) and also the increase in output of cattle and cattle products (Table 1.3). This would suggest a swing away from grazing towards the use of concentrated feedingstuffs, but accompanied by a rather greater reliance on the use of home-mixed rations.

1.9. Feedingstuffs are the major item in the cost of production of intensive livestock products and because of this one would expect changes in the price of feedingstuffs to have an effect on their demand. That is, if the price

TABLE 1.5

INDICES OF SALES OF THE MAIN TYPES OF MANUFACTURED COMPOUNDS[a] IN U.K.
(1955/6 = 100)

Type of compound	1955/6	1959/60	1963/4	1966/7	Percentage of total tonnage 1966/7
Cattle and calf	100	120·6	119·0	130·9	35·8
Pig	100	91·7	104·6	96·4	20·0
Poultry	100	145·5	162·5	163·2	42·4
Other	100	126·6	147·9	180·6	1·8
Total	100	121·2	130·3	133·2	100·0

[a] Note that this table refers to sales of *compounds,* not to utilisation of *concentrated feeds,* as a whole.

Source: M.A.F.F. *Developments,* and Quarterly statements of compound manufacture.

of feedingstuffs declined relatively to the price of the product one would expect that it would become more profitable to keep livestock and hence the demand for feeds would expand.

1.10. This price effect is nullified in the United Kingdom for pigmeat and eggs by the operation of the feedingstuffs formula in the guaranteed price scheme. Essentially, under this arrangement, any change in the price of certain standard feedingstuffs rations is matched by changes in the guaranteed price of the products so that the profitability of these enterprises is largely unaffected by feed price changes. Thus in 1968/9, if the price of the ration for pigs goes up by 1% the guaranteed price for pigmeat will rise by 0·53%, and similarly for eggs by 0·56%. So long as feedingstuffs represent at least 53 and 56% of the total costs of pigmeat and egg production respectively, the producer has no incentive to expand or contract production with changing feed prices. Of course, individual producers using different rations, or buying materials at different prices and with cost structures varying from those assumed in the standard ration, will find their profitability marginally altered by changes in feed prices;

but it remains true to say that the feed formula arrangement operates in the direction of stabilising pigmeat and egg production. At the same time it must be supposed to have some destabilising effect on grain prices, as it counteracts the tendency for demand for feed grains to alter in response to changing grain prices resulting from a changing supply situation. It is therefore apparent that government policy relating to livestock production and feed price formulae is extremely relevant to the future demand for cereals in the U.K.

Regional Aspects of Livestock Production and Compound Feed Consumption

1.11. The different rates of increase of output of the various livestock and livestock products have had corresponding effects on the demand for different types of concentrated feedingstuffs and also on their regional demand. Some statistics of regional production of feedingstuffs for the various categories of livestock and livestock products are available, but only for those concentrated feedingstuffs which are in the form of manufactured compounds.[1] The Ministry of Agriculture, Fisheries and Food uses for this purpose a regional classification on the basis of six port areas in Great Britain which are respectively the hinterlands of London, Hull, Liverpool, Bristol, Glasgow and Leith as shown in Map 1.1. Northern Ireland is regarded as one port area.

1.12. Table 1.6 provides a breakdown of the country's production of manufactured compounds by port areas and by type within each port area, expressed as percentages of area totals. If it is assumed that sales and consumption occur mainly within the area of production, it is seen that in the Liverpool, Bristol and Glasgow areas, sales of cattle and calf compounds are of considerably greater importance than in any of the other areas. The London, Leith and N. Ireland areas, and to a lesser extent the Hull area, on the other hand, are characterised by a high demand for compounds for intensive livestock (pig and poultry) production. This confirms the well-known fact that the western areas of Great Britain have a much stronger emphasis on dairying and cattle production than the Eastern counties. (See also J. T. Coppock, *An Agricultural Atlas of England and Wales,* 1964.)

1.13. Because of the slower rate of growth in extensive livestock output than in intensive livestock, the regional pattern of demand for feedingstuffs has changed over time. A breakdown of regional statistics of concentrated feedingstuffs sales some 10 years ago is not readily available. Instead, we have made use of feedingstuffs requirement factors as calculated in Part III, Chapter 3 (Table 3.12), these being applied to the respective regional

[1] See 1.3 for definition.

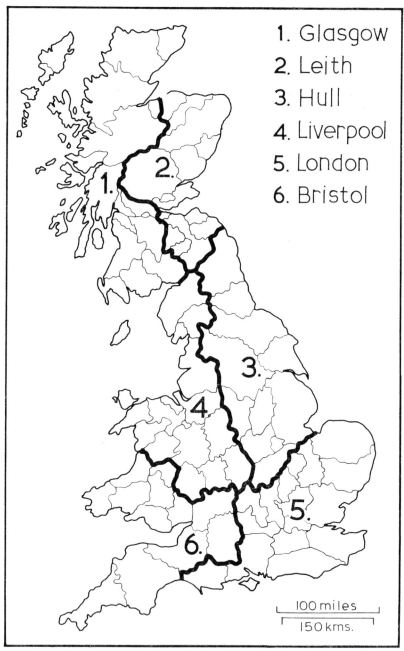

1. Glasgow
2. Leith
3. Hull
4. Liverpool
5. London
6. Bristol

100 miles

150 kms.

MAP 1.1. Port areas in Great Britain (as used for M.A.F.F. statistics production of concentrated feedingstuffs).

TABLE 1.6

THE RELATIVE IMPORTANCE OF TYPES OF MANUFACTURED COMPOUNDS BY
PORT AREAS, 1966/7

Port area	Cattle and calf %	Pig %	Poultry %	Other %	Total %	Total tonnage produced ('000 tons)	As per-centage of total %
London	21·6	24·2	52·5	1·6	100·0	1966·9	21·1
Hull	32·3	20·1	45·3	2·3	100·0	1569·0	16·8
Liverpool	46·3	13·3	38·6	1·8	100·0	2110·5	22·7
Bristol	44·0	17·5	36·7	1·7	100·0	1978·7	21·2
Glasgow	53·0	8·9	35·0	3·1	100·0	504·7	5·4
Leith	24·1	19·2	53·1	3·6	100·0	194·3	2·1
N. Ireland	23·9	36·2	39·1	0·9	100·0	911·7	10·7
Total	35·8	20·0	42·4	1·8	100·0	9235·8	100·0

Source: M.A.F.F. Quarterly statements of production of compounds.

livestock populations in 1956 and 1966. The 1966 estimates of regional concentrated feed requirements have been expressed as indices of the 1956 figures. Table 1.7 therefore indicates the estimated changes in concentrated feedingstuff requirements, based solely on livestock population changes, which occurred between 1956 and 1966, by type of livestock and by port areas.

1.14. One can immediately see that the eastern port areas, London and Hull, have shown larger increases in demand than the other areas of Great Britain, even though their cattle and calf populations have fallen (in the London area) or barely changed (Hull). The large increases in these areas are therefore due entirely to the rapid rise in intensive livestock production and especially to the rise in poultry feed requirements.

1.15. As far as future domestic production of feed grains is concerned, the main question to resolve is the extent to which the production of different types of livestock will change. When we consider the marketing of cereals, however, it is obvious that not only the size and composition of the livestock population but also its regional dispersion is of great importance, as it could have a tremendous impact on the required flow of grain and the future location of processing facilities. It could be that the rapid growth in intensive livestock production which has been observed in the Eastern areas constitutes a movement towards the source of raw materials, that is, towards the major centres of cereal production. More recently it has been suggested that broiler production is moving to the peripheries of the major conurbations.[1] The same analysis is not so strictly applicable to

[1] M.A.F.F. *A Century of Agricultural Statistics*, p. 54. (1968).

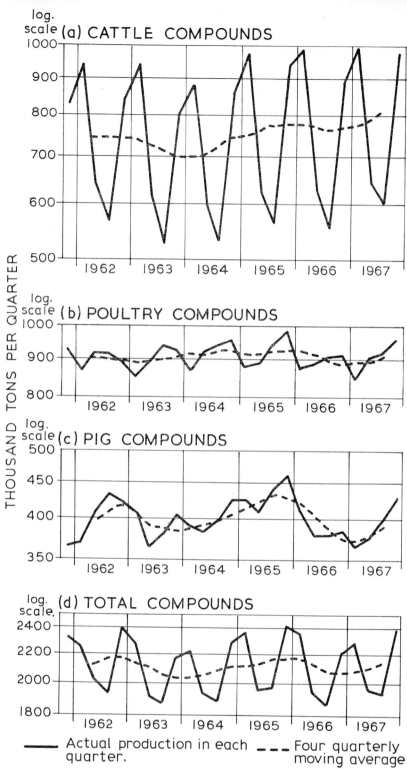

FIG. 1.2. Seasonal fluctuations in Great Britain compound manufacture (by type of feed).

Source: M.A.F.F. quarterly statements of production of compounds.

TABLE 1.7

INDICES OF ESTIMATED CONCENTRATED FEED REQUIREMENT CHANGES BETWEEN 1956 AND 1966. (1956 requirements = 100)

Port area	Type of feed				All feeds
	Cattle and calf	Pig	Poultry	Other	
London	96	141	168	135	131
Hull	104	132	145	125	122
Liverpool	114	125	115	133	117
Bristol	116	125	118	150	119
Scotland	120	116	96	111	114
N. Ireland	130	162	94	121	126
U.K.	112	135	126	127	122

pigs (Map 1.2) where Cornwall and Lincolnshire remain important areas, although the major area is still East Anglia and the Home Counties.

1.16. To summarise, providing we accept the hypothesis that in the future there is likely to be a greater relative growth in poultry and pig production than in dairying and cattle production, we would expect there to be a further shift of emphasis towards heavier feed requirements in the East of the country with consequent effects on cereal marketing.

The Seasonal Demand for Feedingstuffs

1.17. Another aspect of the demand for feedingstuffs is its seasonal variation. Again, we have detailed figures for the production of manufactured compounds on a quarterly basis only from 1962, but it may be assumed that these are good indicators of the overall seasonal demand for feedingstuffs, as it is very unusual for compounds to be stored for any appreciable length of time.

1.18. As Fig. 1.2 shows the production of cattle compounds is some 50 to 60% higher in the 6 months October–March than it is in the 6 summer months, with a marked peak in the January–March period and a trough in July–September. (There is a similar pattern for sheep compounds, but these are relatively unimportant and are not shown here.) There is also a seasonal pattern in the production of poultry compounds but two features distinguish it from the pattern for cattle compounds. First, the trough is in the January–March period, that is, exactly the reverse of the cattle pattern, and the peak in recent years has been in October–December. The most likely reason for this seasonal pattern is the killing of poultry for the Christmas trade coinciding with the culling of poultry flocks. The second difference is that the seasonal fluctuation is much smaller than that for cattle, so that these contrasting patterns do not by any means cancel each other out. There is no evidence of a seasonal pattern in the manufacture of pig compounds. Production does fluctuate widely from quarter to quarter, but this appears to be mainly associated with cyclical fluctuations in the pig population extending over several years and not to any regular pattern of seasonal changes within the year in pig numbers or feeding habits.

1.19. The overall effect (Fig. 1.2(d)) is that total compound production shows a marked seasonal pattern, being high in the winter months of October to March and low in the summer months. Relating this pattern to the date of harvesting the home cereal crop (August and September) it is clear that the amount of storage required is less than would have to be provided if there was a constant demand throughout the year combined with a constant degree of reliance on the utilisation of home-grown crops. (See also Part II, Chapter 5.) However, this seasonal pattern does lead one to question the desirability of providing large storage incentives for coarse grains beyond the end of March and it also creates serious problems of seasonal under-utilisation of capacity for the compound manufacturing industry.

1.20. If we combine the seasonal patterns of compound demand by different types of livestock with the respective regional distributions of live-stock as shown in Table 1.6, we find that the relatively greater importance of cattle in the western areas shows clearly in the seasonal pattern of com-pound production in the Liverpool, Bristol and Glasgow areas (Fig. 1.3A), with marked peaks in the winter months and corresponding troughs in the summer. This pattern is particularly marked in the Glasgow region where cattle and calf compounds accounted for 53% of total compound produc-tion in 1966/7. These violent seasonal fluctuations are even more marked when we compare them with the pattern in the adjacent eastern area of Leith (Fig. 1.3B), where cattle and calf compounds accounted for only 24% of total production in 1966/7 whilst poultry compounds accounted for 53%.

1.21. For all eastern areas it is true to say that normally the seasonal variations in production are less marked than in the western areas,[1] but the seasonal pattern in the western areas is more regular than that in the eastern areas which are more dependent on the relatively volatile annual production of pigs and poultry. Production in Northern Ireland does show a seasonal peak in January–March in most years with a trough in July–

---

[1] The graphs have been drawn on a log scale so that one can compare the proportional seasonal fluctuations by measuring the distance between the dotted line and the con-tinuous line.

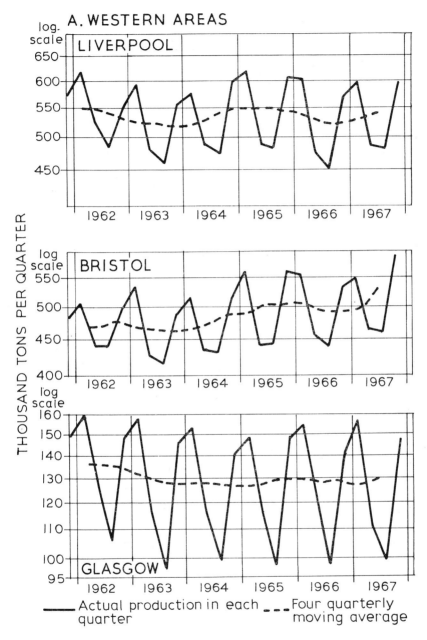

FIG. 1.3A. Seasonal fluctuations in Great Britain compound manufacture (by region).
Source: M.A.F.F. quarterly statements of production of compounds.

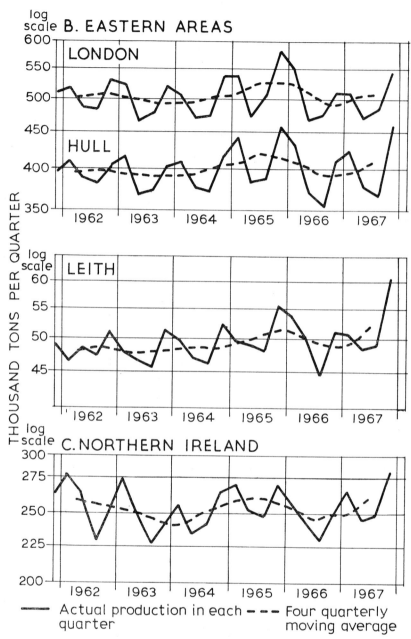

FIG. 1.3B. Seasonal fluctuations in Great Britain compound manufacture (by region).

Source: M.A.F.F. quarterly statements of production of compounds.

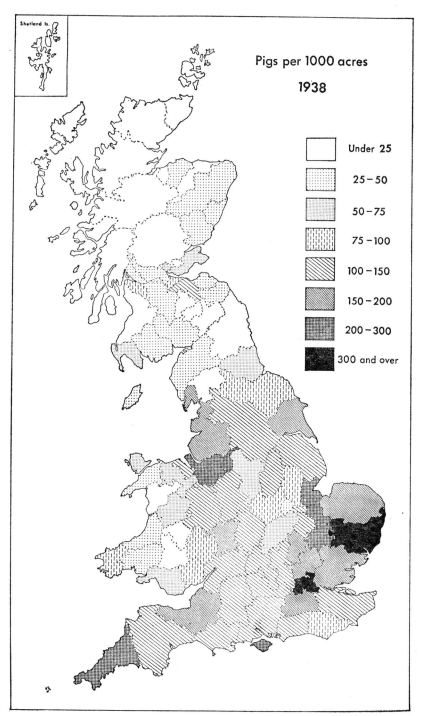

MAP 1.2A. Number of pigs per thousand acres of agricultural land (including common rough grazing) in the counties of Great Britain, 1938. *Source:* M.A.F.F. *A Century of Agricultural Statistics* (map xv).

o

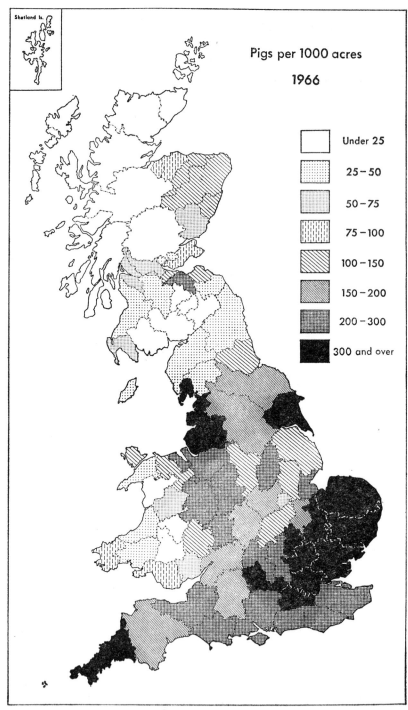

MAP 1.2B. Number of pigs per thousand acres of agricultural land (including common rough grazing) in the counties of Great Britain, 1966.
Source: M.A.F.F. *A Century of Agricultural Statistics* (map xv).

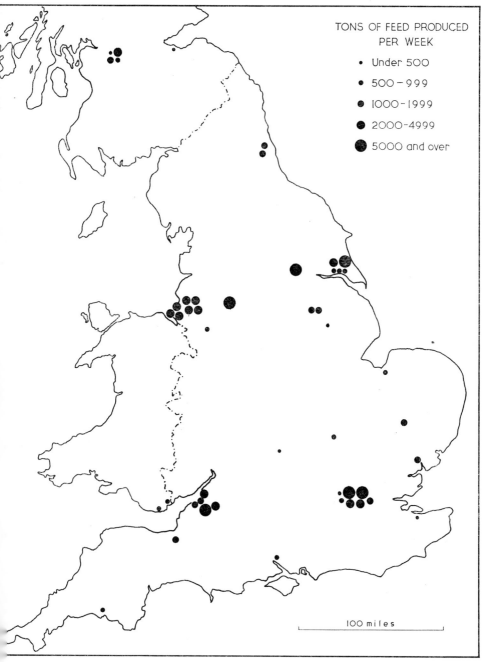

MAP 1.3. Location and size of animal feedingstuffs factories owned by national compounders in the U.K., 1968.
Source: Information supplied by the trade.

September (Fig. 1.3B). Despite the relatively low proportion of cattle, the overall pattern of production shows slightly more similarity with the western areas than the eastern areas.

1.22. In summary, we have shown in this chapter that the demand for feed grains is closely linked to the production of livestock, and that the regional distribution of types of livestock has a marked effect on the seasonal and annual production of compounds and hence on the entire cereal marketing sector.

CHAPTER 2

The Supply of Concentrated Feedingstuffs

2.1. As described in the preceding chapter (Tables 1.1 and 1.2), the livestock industry has created a growing demand for cereal-based feedingstuffs. Indeed by far the largest outlet for home-grown cereals is as feedingstuffs in livestock production. Over 45% of the 1966/7 wheat crop was used directly for animal feed, and another 13% was used for similar purposes in the form of wheat offals. About 66% of the barley crop and over 81% of the oat crop is used directly or indirectly for animal feed. Even though in 1966/7, the latest year for which full utilisation figures are available, approximately 8·7 million tons of home grown cereals or cereal by-products were used for animal feed, another 3·4 million tons of imported cereals were also required, together with over 1 million tons of cereal offals produced in this country from imported cereals.

2.2. Table 2.1 suggests that total consumption of feedingstuffs on farms has increased steadily since decontrol in the middle 1950's, and in 1965/6 accounted for 16½ million tons, falling back to 15·6 million tons in 1966/7. In that year, 13·5 million tons were delivered to the farm which suggests that 2·1 million tons were consumed on the farms on which they were grown. These would be mainly cereals. Of the 13·5 million tons of delivered feedingstuffs, 9·3 million were in the form of compounded rations. The M.A.F.F. estimate[1] that a further 3·7 million tons were delivered straight, just under half a million tons were in the form of processed cereals and a much smaller quantity, only 55,000 tons, were in the form of cereal livestock mixtures. Information on the extent of concentrated feed manufacture, relating mainly to the same year, which we collected from the Survey of Merchants and from interviewing other animal feedingstuffs manufacturers, suggests that manufactured concentrated feedingstuffs sales to farmers may be some half a million tons greater than the Ministry estimates. Moreover, our results would suggest that of the manufactured concentrated feed sales nearly one-quarter is provender compared with the

[1] M.A.F.F. *Developments* (Table 5).

TABLE 2.1

CONSUMPTION AND PURCHASES OF CONCENTRATED FEEDINGSTUFFS IN THE
UNITED KINGDOM, 1955/6–1966/7 (June/May years) (million tons)

	1955/6	1957/8	1959/60	1961/2	1963/4	1965/6	1966/7
1. Total consumption of concentrated feeds[a]	12·0	12·3	13·4	14·5	14·8	16·5	15·6
2. (a) Total deliveries of concentrated feed[b]	10·1	11·0	12·1	13·3	13·1	14·6	13·5
(b) As percentage of 1.	84·2	89·4	90·3	91·7	88·5	88·5	86·5
3. (a) Total deliveries of compounds	7·0	7·3	8·5	9·2	9·1	9·5	9·3
(b) As percentage of 1.	58·3	59·3	63·4	63·4	61·5	57·6	59·6
4. (a) Estimate of total conc. feeds retained on farm of origin[c]	1·9	1·3	1·3	1·2	1·7	1·9	2·1
(b) As percentage of 1.	15·8	10·6	9·7	8·3	11·5	11·5	13·5

[a] Statistics of total consumption are for farms of one acre or more. *Source:* Item 3, Table G, Appendix I, *Annual Review and Determination of Guarantees 1968.* [b] Statistics of total deliveries are for all farms, so item 4 "feeds retained on farm of origin" is somewhat understated. [c] This should not be confused with Item 2, Table G, Appendix I of *Annual Review and Determination of Guarantees* which is misleadingly titled "Home-grown concentrated feeds retained on farm of origin", but which in fact refers to cereals not qualifying for delivery bonuses and includes farm-to-farm sales of barley, farm-to-farm sales of oats and also farm-to-merchant-to-farm sales of oats.

Source: M.A.F.F. *Developments* (Table 5).

Ministry's figure of one-twentieth, although some of this difference could have been caused by misunderstandings about the term "provender" by interviewed merchants. Our Survey suggests that only 1·6 million tons of grain were sold direct to farmers by non-national compound feed manufacturers and merchants. We believe that some grain is sold direct by national compounders to farmers but we are uncertain of the quantities involved. The difference between these cereal sales and the Ministry's estimate of delivered straights of 3·7 million tons is presumably non-cereal materials such as sugar beet pulp and proteins for home mixing. However, we feel that the Ministry figures may overstate the extent of straight deliveries, and that some is processed by merchants not classified by the Ministry as animal feedingstuffs manufacturers. This could explain some of the discrepancy between our own estimates of provender sales and those made by the Ministry.

2.3. It is seen that the greater part of consumption of concentrated feeds is in the form of purchased compounds. Although the quantity of compounds purchased has been rising in absolute terms, they have remained a constant proportion of total purchases of concentrated feed (70%) (see Table 2.1, 3(a) as a percentage of 2(a)) and since 1959/60, a somewhat reduced percentage of total *consumption* of concentrated feed. However, the above table does not include those compounds produced by farmers or by large organisations which do not resell the compounds that they themselves manufacture. It is known that many large broiler and egg producers manufacture compounds from cereals grown by themselves or, more generally, from purchased provenders and straights. The extent of their compound manufacture is not known, but it has been estimated that annual production by one of the largest of these organisations exceeds that of some of the large national compounders. Production by these larger organisations could account for much of the apparent decline in compound feed deliveries as a percentage of total consumption of concentrated feed on farms since the early 1960's.

2.4. Despite the large increase in cereal production, the amount of home-grown concentrated feeds retained on farm of origin has shown only a small increase in the last few years. There are several reasons which can be given for this situation. The first is that in order to obtain the *full* benefits from the cereal deficiency payments system, cereals must be sold off the farm of origin. Another explanation may be that cereal growing farms are not usually cereal-using farms. But perhaps the major reason is the realisation by farmers that home mixing is not always technically justified for sophisticated rations and in many cases is not economically justified either. It has been suggested that the growth in the popularity of home mixing coincided with the boom in barley beef production. Most merchants interviewed in the survey suggested that although a few years ago home mixing was beginning to reach sizeable proportions, in recent years this has again declined.[1]

2.5. Table 2.2 shows the main constituents of delivered concentrated feedingstuffs and whether they are home-grown or imported. It indicates the growing importance of cereals as a major component of delivered concentrated feedingstuffs, a situation almost certainly due to the increasing proportion of feedingstuffs being sold to producers of poultry and eggs, with their requirements for rations of a high energy content. The table also reveals the increasing contribution of the United Kingdom cereal grower to the total cereal content of delivered feedingstuffs, and this trend is reinforced if we consider the total *consumption* of feedingstuffs, whether delivered or retained on farms, as virtually all concentrated feedingstuffs retained on farms will be home-grown cereals.

2.6. Imported wheat and barley have been increasingly replaced by the home-grown products, and it is also now apparent that the imported maize content is being partly replaced by home-grown cereals. After decontrol of

---

[1] This topic is discussed more fully in Chapter 6.

TABLE 2.2

ESTIMATED RAW MATERIAL COMPOSITION AND ORIGIN OF ALL CONCENTRATED FEEDINGSTUFFS PURCHASED IN THE UNITED KINGDOM
('000 tons)

Raw material	1955/6		1963/4		1965/6		1966/7	
	Home	Imported	Home	Imported	Home	Imported	Home	Imported
Wheat	930	569	1060	481	2013	672	1489	516
Barley	885	668	2726	240	2781	143	3106	68
Oats	318	27	201	7	149	8	136	9
Maize	—	1275	—	2454	—	2363	—	2150
Other cereals	11[a]	444[b]	5[a]	281[b]	2[a]	539[b]	5[a]	500[b]
Total cereals	2144	2983	3992	3463	4945	3725	4736	3243
Cereal offals[c]	421	1526[g]	420	1573[g]	423	1370[g]	465	1305[g]
Proteins[d]	173	1528	226	1813	240	1879	244	1655
Beet pulp	310	—	387	—	593	—	538	—
Others	172[e]	489[f]	143[e]	571[f]	283[e]	572[f]	212[e]	560[f]
Total[i]	3220	6526	5168	7420	6484	7546	6195	6763
Cereals as % of total[h]	52·6		59·2		61·7		61·6	
Cereal offals as % of total[h]	20·0		15·8		12·8		13·7	

[a] Mixed corn and rye.   [b] Sorghums.   [c] Mostly wheat offals.   [d] Oil cakes, fish, blood, bone meals, etc.   [e] Mostly grass meal, brewers' grains.   [f] Mostly molasses.   [g] Mainly home-produced from imported wheat.   [h] This relates to delivered feedingstuffs: Table 1.2 relates to *all* concentrated feedingstuffs whether delivered or retained on farms.   [i] If due allowance is made for mineral additives and packaging material, the totals reconcile with those in Table 2.1.

Source: M.A.F.F. *Developments* (Table 6).

the feedingstuffs industry in 1954 there was a phase during the rapid expansion of the industry when wheat and barley, the main cereal constituents, were overtaken by maize, and in 1962/3, maize alone contributed a peak 39% of the cereal content of delivered concentrated feedingstuffs. However, after that year the proportion of maize in the total cereal content of delivered concentrated feedingstuffs declined steadily to 27% by 1966/7, and home-grown barley had become the main cereal constituent at almost 39% of total cereal content in 1966/7. Home-grown wheat has contributed in recent years up to 23% of the total cereal content.

2.7. This is the broad picture of requirements and sources of supply of cereal-based feedingstuffs. We now wish to consider, briefly, to what extent the feedingstuffs industry succeeds in matching demand and supply, and to examine some of the problems which it may encounter. Some of our earlier conclusions may be summarised as follows:

(a) The demand for feedingstuffs in the U.K. is derived from the demand for livestock and livestock products and from British farmers' share in supplying that demand.

(b) Where the cost of feedingstuffs is a major item in the total cost of livestock products, the price of feedingstuffs may affect the quantity demanded by livestock producers, unless there is government intervention to relate receipts to feed costs.

(c) At the moment, more raw materials are required for feedingstuffs than are produced in this country.

Given these conditions, the question we may ask is whether there is some optimal method of using these home-grown raw materials throughout the year? Three possible attitudes to this question may be considered.

2.8. First, the animal feedingstuffs industry might rely almost entirely on home-grown supplies for the months immediately after harvest, in this way reducing the amount of storage on U.K. farms, releasing farmers' capital for short-term use elsewhere and ensuring that the home supplies were all utilised before moving on to imported supplies. Second, if the animal feedingstuffs manufacturers were attempting to make the maximum possible contribution to import-saving they might use home-grown supplies at those times when imported supplies would be most expensive. Under certain conditions, manufacturers might minimise the import bill by being continuously in the market for a relatively small quantity of supplies, rather than entering the market from time to time for large quantities of grain which might cause the world market price to rise. Third, to maximise their own profits, animal feedingstuffs manufacturers would presumably prefer to purchase their supplies in the lowest-priced market regardless of whether these were home-produced or imported supplies. At the same time this course of action would be most beneficial to the livestock sector of

O*

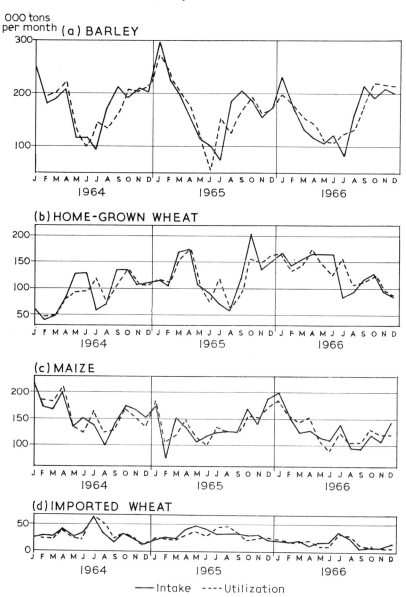

FIG. 2.1. Monthly cereal intake and utilisation by animal feedingstuffs processors in Great Britain, 1964–6.
Source: M.A.F.F. (unpublished).

the agricultural community as it would lead to lower costs of production.

2.9. It should be apparent that the likelihood of these three attitudes coinciding is not very high. To some extent the animal feedingstuffs industry is prevented from adopting the first course, because the storage incentive element in the C.D.P. scheme has given farmers ample compensation for grain storage. As Fig. 2.1 shows, we find as a result that the peak month for home-grown barley intake (and utilisation) by feedingstuffs manufacturers is January and for home-grown wheat the peak in recent years has been as late as March, April and May. However, unless 1968 is to prove an exception, the stocks of home-grown grain have always been cleared by the end of the cereal year so that we have never experienced an occasion when the use of imported supplies has left home supplies unmarketed. The change in the seasonal incentive scales announced for the 1968/9 cereal year for wheat should lead to a more even monthly flow of home-grown wheat on to the market. Nevertheless, the present storage incentive scheme does relieve both the cereal producing and using sectors of much of the cost of financing storage between harvest and the time the cereals are used.

2.10. It is impossible to judge whether the animal feedingstuffs industry has minimised the import bill by judicious purchasing of imported materials, because the prices on world markets are affected by so many different factors. Nevertheless it is true to say that as the price of imported materials has risen relative to the price of home-grown supplies, imported supplies have become a relatively less important source of raw materials. This has occurred continuously through the 1960's, and this is even more true for the cereal content of feedingstuffs than total raw material content of feedingstuffs (Figs. 2.2 and 2.3).

2.11. At the same time the feedingstuffs industry shows a marked ability to alter the relative importance of its raw materials in the face of altering circumstances such as variations in the size of the home-grown cereal crop or the export of barley from the U.K. To some extent this may be due to the changing importance of different types of compounds during the year. Table 2.3 shows the combinations of total cereals used by animal feedingstuffs processors in Great Britain for the months of February and June of the three years 1964–6.

2.12. The patterns of utilisation change so rapidly from month to month and year to year that it is impossible to generalise on the topic. For instance, for several months in 1966, home-grown wheat was a more important cereal than maize, whereas only a few years ago it was a relatively minor raw material in manufactured concentrated feedingstuffs.

2.13. Despite technical constraints as to the degree of raw material substitution that the feedingstuffs industry can accommodate, its extreme

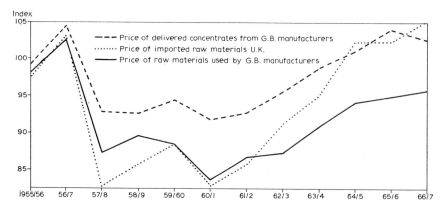

FIG. 2.2. Indices of prices of delivered concentrated feeds; of raw materials used by manufacturers in Great Britain; and of imported raw materials in the United Kingdom (1954/5–1956/7 = 100).
Source: M.A.F.F. *Developments in the Animal Feedingstuffs Industry.*

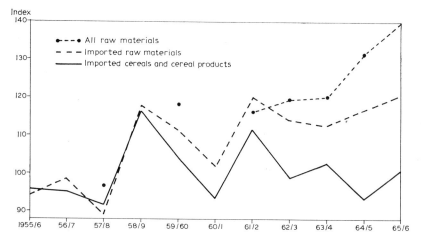

FIG. 2.3. Indices of values of all raw materials; of all imported raw materials*, and of imported cereals and cereal products,† used in delivered concentrated feedingstuffs (1954/5–1956/7 = 100).‡
Source: M.A.F.F. *Developments in the Animal Feedingstuffs Industry.*
Notes: * Quantity of raw materials multiplied by average price per ton.
† Excluding home-produced offals from imported raw materials.   ‡ For all raw materials 1955/6 = 100.

TABLE 2.3

ESTIMATED RELATIVE USE OF DIFFERENT CEREALS BY ALL ANIMAL FEEDINGSTUFFS
MANUFACTURERS IN GREAT BRITAIN (per cent)

(a) February

Year	Barley	Home-grown wheat	Imported wheat	Maize	Others	Total
1964	41·9	10·0	5·5	39·4	3·2	100·0
1965	47·6	22·4	5·4	21·2	3·4	100·0
1966	34·8	27·4	4·6	31·0	2·2	100·0

(b) June

Year	Barley	Home-grown wheat	Imported wheat	Maize	Others	Total
1964	28·6	27·1	6·0	36·0	2·3	100·0
1965	20·4	26·8	11·1	36·1	5·6	100·0
1966	30·8	37·2	4·7	26·4	0·9	100·0

Source: Derived from M.A.F.F. statistics (unpublished).

flexibility in the use of supplies suggests that the industry is quite adept in purchasing its raw materials in the lowest-priced markets, and it has this as one of its main objectives. By varying its use of raw materials in this way, the feedingstuffs industry is able to reduce its stockholding to a relatively low level. More importantly, this policy of flexibility is one reason why the industry has been so successful in containing the rise in raw material prices of concentrated feedingstuffs (Fig. 2.2).

2.14. To summarise, unless 1968 proves an exception, the animal feedingstuffs industry has adjusted its purchasing pattern over the years in order to absorb most of the increased cereal production in the United Kingdom. At the same time it has maintained sufficient flexibility to purchase its requirements in the lowest-priced markets.

CHAPTER 3

The Structure of the
Feed Manufacturing Industry

The Present Situation and its Background

3.1. The United Kingdom's feed manufacturing industry, at present producing concentrated feeds to the approximate value of £350 million annually ex-mill and ex-merchant's store, comprises around 850 to 900 factories ranging in size from a handful of large factories which each produce over 200,000 tons annually to a much larger but a declining number of factories[1] whose annual production does not exceed 1000 tons.

3.2. Since decontrol of the animal feeds industry in 1954, a marked redistribution of factory sizes as measured in terms of annual production has accompanied an overall expansion. In Great Britain alone, during the period 1952/3 to 1966/7, total manufacture rose from 5.4 to 9.0 million tons, but the total number of factories is estimated to have fallen by 303, from 1085 to 782. Table 3.1 shows the change in distribution of factories by annual deliveries during that period and the changing contribution to total manufacture of the different factory size categories.

3.3. It is apparent that this decline in factory numbers has been concentrated in those with the smallest throughput, with increases in the number of factories with a throughput of over 5000 tons per annum.

3.4. Historically, the industry has consisted of two distinct groupings of factories, one in the country areas and the other in and around the major ports.[2] The "country" group has developed mainly from the old country flour milling industry which used to consist of numerous small rural mills utilising locally produced cereals and marketing flour and processed corn locally. Since the end of the last century, when large flour mills developed

---

[1] The M.A.F.F. uses the term "factory" in this context, but many of the firms concerned are primarily merchants rather than manufacturers.

[2] See M.A.F.F. *Developments*. In the M.A.F.F. statistics, port factories are those located in London, Southampton, Bristol, Cardiff, Swansea, Barry, Merseyside (including Manchester), Humber Ports, Tyneside, Clyde Ports and Leith.

at ports and in large cities, many of the small country mills have turned to manufacturing—sometimes putting compensation payments received from the Millers Mutual Association to this purpose—and have become the nucleus of the country feedingstuffs industry. Today, this industry comprises a large number of small firms relying substantially on home-grown grain and selling most of their manufactured feeds and much provender and straights direct to farmers.

3.5. The other group of factories, located in the ports, has its origins in the flour-milling and oilseed-crushing industries which developed during

TABLE 3.1

DISTRIBUTION OF G.B. FACTORIES BY ANNUAL DELIVERIES OF MANUFACTURED CONCENTRATES

Volume of deliveries (tons)	Number of factories			Percentage of deliveries		
	1952/3	1961/2	1966/7	1952/3	1961/2	1966/7
0–999	634	395	296	4·4	1·8	1·4
1000–4999	319	292	239	13·4	7·2	6·3
5000–19,999	87	159	154	15·5	17·3	17·2
20,000–49,999	21	45	59	12·3	15·1	20·8
50,000–99,999	15	19	18	20·0	14·8	13·0
100,000–199,999	7	9	10	19·2	15·2	16·8
200,000 and over	2	7	6	15·2	28·5	24·5
	1085	926	782	100·0	100·0	100·0

Source: M.A.F.F. *Developments.*

the latter half of the last century following increased availability of cheap imported grain from the Americas and oilseeds from the tropics. Wheat offals and protein by-products produced by these mills were further processed into animal feedingstuffs by new factories built close to the parent mills. By the outbreak of the Second World War their distribution had expanded to include all the major ports.

3.6. The distinction between "port" and "country" manufacturers is becoming less clearly defined, due partly to the larger port firms expanding into smaller port areas and even into country areas by opening up new factories or taking over or merging with smaller port and country manufacturers. Also there has emerged, through a process of take-overs, expansion and mergers, a category comprising the seven largest feedingstuffs companies who have become known as the seven "national" compounders, operating as they do on a nationwide basis. Although these companies are still vastly stronger, from the point of view of factory capacities, in the port areas than in the country areas, they have in recent years expanded and spread their manufacturing and distributive activities into the country areas.

3.7. Present-day classifications tend therefore to be on the basis of "national" and "non-national" manufacturers and both categories can occur in port and country areas. Among the latter are some very large flour-milling, food-processing and other companies as large as if not larger in terms of total turnover than some of the seven national compounding firms; but because their actual animal feed compounding activities do not match any of the seven national compounders, they have not come to be regarded as nationals in the animal feedingstuffs trade.

3.8. The firms who are commonly recognised as national compounders are the Unilever companies (B.O.C.M., Silcocks and Levers); Bibby's; Spillers; Ranks; Pauls and Whites; Crosfields and Calthrop; and the Co-operative Wholesale Society. B.O.C.M. is by far the largest single animal feedingstuffs manufacturer, and together with Silcocks and Levers, accounts for a substantial share of the total concentrated feedingstuffs manufactured in the United Kingdom. B.O.C.M., Bibby's, Spillers and Ranks between them account for about one-half of the total, while Crosfields and Calthrop, Pauls and Whites, and the C.W.S. together account for under one-tenth. Between them the seven largest firms produce over 60 % of the concentrated feedingstuffs manufactured in the United Kingdom.[1]

3.9. Our Survey would suggest that none of the non-national animal feedingstuffs manufacturers compares in size with even the smallest of the seven major firms. The largest non-national firm in the Survey produced only one half of one per cent of the total United Kingdom manufacture of concentrated feedingstuffs. It should be remembered, however, that, as stated in paragraph 2.3, compound manufacture by some large-scale egg and broiler producing organisations exceeds that of some of the smaller nationals. Statistics of this form of compound production are not available.

3.10. Table 3.2 shows that in 1966/7 the national companies owned 59 factories in Great Britain and delivered 60 % of all the compounds manu-factured. Our Survey results showed a similar percentage. These factories were specifically excluded from our Survey of Merchants and were all covered by a separate Survey. In our Merchant Survey sample, however, we identified six smaller subsidiary firms owned directly or indirectly by the national compounders in Great Britain and producing more than 1000 tons of manufactured concentrates each (Table 3.3). This would suggest that there are probably around 45 to 50 such firms in total. Thus the seven national companies would own up to 110 animal feedingstuffs factories in Great Britain out of the total of 486 producing more than 1000 tons of concentrates per annum (see Table 3.1).

---

[1] All these figures relating to the seven "nationals" exclude concentrated feeds manu-factured by their subsidiary merchant businesses.

TABLE 3.2

DELIVERIES OF COMPOUNDS BY THE SEVEN LARGEST
COMPANIES IN GREAT BRITAIN

Year July/June	Number of factories	Compound deliveries (million tons)	% of G.B. deliveries
1957/8	56	4·2	64
1961/2	56	5·0	61
1965/6	59	5·1	61
1966/7	59	4·9	60

Source: M.A.F.F. *Developments.*

3.11. In Britain the major companies own all the factories producing more than 100,000 tons of concentrated feedingstuffs per annum and these 16 factories alone account for over 40% of the Great Britain deliveries of manufactured concentrates (Table 3.1.) The Survey suggests that factories owned by national compounders but excluded from Table 3.2 probably produce only about 3 to 4% of the total deliveries in Great Britain. The future significance of these smaller factories is, however, likely to be much greater than at present.

3.12. There does appear to be some difference between the estimates of the number of animal feedingstuffs manufacturers derived from our Survey results and the M.A.F.F. statistics of feedingstuffs factories. The Survey results suggest that there are about 970 feedingstuffs factories in Great Britain excluding the major factories owned by the nationals (Table 3.3), compared with a Ministry total in 1965/6 of 803 including major factories.[1] However, some of the factories in our Survey produced provender only (e.g. barley meal) and when we allow for these we estimate that there are about 785 factories producing compounds, excluding the major factories. Given the nature of our sampling frame these figures are in sufficient agreement with the Ministry figures. The presence of around 185 firms producing provender and not compounds, and probably missed from the Ministry statistics, reinforces our view that the M.A.F.F. statistics understate the amount of provender produced (see paragraph 2.2).

3.13. Table 3.1 shows that the number of factories with a delivery between 5000 and 20,000 tons per annum of manufactured concentrates has remained almost constant since 1961/2, as has their share of the trade. It is important to note, however, that factories delivering 20,000 to 100,000 tons annually have increased both in numbers and in relative and absolute share of the trade, at the expense of both the smallest and the very largest

[1] *Source:* M.A.F.F. *Developments.*

factories. This trend may be a reflection of the policy of some of the national companies to increase production in country factories, but it also indicates growth in the strength of the larger non-national compounders who are benefiting from the increasing availability of home-grown cereals.

Regional Aspects of the Structure of the Industry

3.14. One question worth considering is whether or not the national manufacturers are more important in some areas of the country than in others. Our Survey figures suggest that the national companies' share of the total production in each of the various port areas is as follows:

London:	57%	Liverpool:	64%
Hull:	60%	Bristol:	69%
Leith:	26%	Glasgow:	67%

These figures suggest that the national companies are still strongest in the western port areas. It should be noted, however, that in Scotland the port areas of Glasgow and Leith (see Map 1.1) do not correspond at all closely to the trading pattern based on these ports. It is known that manufacturers based in the Glasgow area, particularly, trade in many parts of the Leith port area, and the reverse situation is also true to a lesser extent. The national companies, traditionally stronger in the ports of the west, i.e. Liverpool, Manchester, Bristol and Glasgow, catered in the earlier days for the more important dairy cattle feed trade, utilising by-products of the port-based flour and oilseed mills, which relied on imported raw materials, and utilising also large tonnages of wheat, barley and maize imported expressly for the animal-feed industry.

3.15. More recently, the increasing availability of home-grown cereals combined with the rising production of poultry products in particular in the eastern areas have caused the feedingstuffs industry to expand its production in that part of the country. Traditionally, these areas were mainly served by non-national compounders in country areas relying to a large extent on home-grown cereals. In effect, both the market and the sources of raw materials have tended to become less concentrated in the western side of the country. If home cereal production continues to expand, and if transhipment of imported grain from the Continental ports becomes more commonplace, and also if the rising trend of intensive poultry production continues as contrasted with the more static dairy industry associated with the west, there will be a continuing effort by the national firms to expand their share of the market in the eastern areas.

3.16. There are indications[1] that expansion by the large national companies and by the large non-national firms, traditionally situated in ports,

[1] See Part III, Chapter 2.

TABLE 3.3

REGIONAL LOCATION OF SUBSIDIARIES OF NATIONALS, AND NON-NATIONAL
FIRMS PRODUCING CONCENTRATED FEEDINGSTUFFS[a]

Port area (see Map 1.1)	Subsidiaries of nationals			Non-nationals			Total	Estimated raised sample total
	Production less than 1000 tons p.a.	Production 1000 tons p.a. or over	Production unknown	Production less than 1000 tons p.a.	Production 1000 tons p.a. or over	Production unknown		
	(Number of firms in Merchant Survey sample)							
London	—	3	—	10	28	1	42	331
Hull	—	—	—	8	7	5	20	158
Liverpool	1	—	—	10	7	5	23	181
Bristol	1	2	—	5	16	—	24	189
Glasgow	—	—	—	2	6	—	8	63
Leith	—	1	—	1	3	1	6	47
Great Britain	2	6	—	36	67	12	123	969
Northern Ireland	—	—	—	3	7	3	13	102
United Kingdom	2	6	—	39	74	15	136	1071
Estimated raised sample total	16	47	—	307	583	118	1071	

[a] Some of these firms produce provender only, not compounds.
Source: Sample Survey of Merchants.

has tended in recent years to be into "country" areas or at any rate away from the actual port precinct due to the increasing use by the industry of home-grown cereals. Large companies, traditionally associated with certain of the large ports, have now expanded their manufacturing outlets to many parts of the country, selecting "country" locations to take advantage of the changing situation in which a port location is no longer so important a prerequisite for successful operation. It is likely that by that means, and by the buying out of existing country compounding businesses and diversifying into retailing of feedingstuffs and other agricultural requisites, expansion by the nationals will continue and their areas of sales will become less and less reliant on their original port factories.[1]

3.17. Despite the fact that the non-national companies are proportionately more strongly represented in the eastern areas of Great Britain, the

[1] This is discussed more fully in Chapter 5.

Survey results show that factory sizes of the non-national firms are, on average, smaller in the eastern areas than in the west. Of the non-national firms, the Survey estimated that the average annual production of concentrated feed in a sample of 98 factories in Great Britain was some 4500 tons, but in the London and Hull port areas the corresponding figures were 3250 and 1950 tons respectively. In the Liverpool port area, on the other hand, annual average production per factory was estimated to be 9300 tons. The marked difference between the east and the west can be largely explained by the fact that the sample included not only country compounders but also non-national port compounders, which are considerably larger than most of the country factories. These non-national port factories, more prevalent in the west, owe their larger size to their port locations and primary reliance on imported raw materials. These factories are better able to expand to larger sizes than country compounders as they do not encounter diseconomies of rising haulage costs as their intake of raw materials grows. The country compounder, on the other hand, has to go increasingly further afield for home-grown raw material as his requirements increase and rising costs of transport may limit the optimum size of plant. The sample reflects the greater occurrence of the larger port facilities in the west and the smaller number of country factories, the latter type of factory not having evolved to the same extent in the grain-deficient west.

3.18. On the other hand, the Survey has shown that the total turnover of non-national firms manufacturing compounds is on average considerably higher in the western port areas (Table 3.4).

3.19. This clearly shows the minor part that compound manufacture plays in the activities of the average firm producing less than 1000 tons annually. With firms producing larger tonnages of compounds, compound manufacture tends to assume a more dominant role in the overall business, particularly in the areas to the west. The firms to the east are appreciably

TABLE 3.4

AVERAGE ANNUAL TURNOVER PER FIRM OF NON-NATIONAL COMPOUNDERS AND THE VALUE OF COMPOUND MANUFACTURE[a] PER FIRM AS A PERCENTAGE OF TURNOVER

Compounds manufactured (tons)	* Average turnover (£'000) † Compounds as % of turnover	Port area						
		London	Hull	Liverpool	Bristol	Glasgow	Leith	N. Ireland
50–999	*	591	448	282	211	219	..	206[b]
	†	2·3	2·6	4·2	6·6	4·3	..	6·0[b]
1000–10,000	*	846	792	500[b]	600	261	1667	290[b]
	†	15·7	12·5	48·7[b]	21·7	37·3	10·2	54·1[b]
over 10,000	*	3375[b]	..	2500[b]	..	1230[b]	..	720
	†	21·5[b]	..	50·5	..	41·0[b]	..	95·3

.. Sample too small to yield realistic results.   [a] Compounds valued at £33 per ton.   [b] Subject to error due to smallness of sample.

Source: Sample Survey of Merchants.

larger, on average, than those in the west despite their compounding activities being smaller as stated earlier. This is mainly because of their greater involvement in grain merchanting activities as shown in Table 3.5.

3.20. It is worth noting that the firms in the Bristol port area show some similarities to those of the eastern areas and this can be largely explained by their proximity to the grain-growing areas of Wiltshire and eastward to Oxfordshire. Table 3.5 also indicates that those firms making less than 1000 tons of compounds annually tend to sell considerably larger tonnages of straight grain than of compounds regardless of location. A possible explanation for such uniformly widespread grain sales at the lower level of

TABLE 3.5

AVERAGE WHOLE GRAIN SALES (TONS) PER FIRM MANUFACTURING COMPOUNDS
AS A PERCENTAGE OF COMPOUNDS MANUFACTURED (TONS)

(excluding major factories of national compounders)

Compounds manufactured (tons)	Port area						
	London	Hull	Liverpool	Bristol	Glasgow	Leith	N. Ireland
	(%)	(%)	(%)	(%)	(%)	(%)	(%)
50–999	2800	610	300	550	120	..	140[a]
1000–10,000	240	440	2[a]	240	10	120	1[a]
over 10,000	240[a]	..	20[a]	20[a]	10[a]	..	0·1

.. Sample too small to yield realistic results.   [a] Subject to error due to smallness of sample.

Source: Sample Survey of Merchants.

the business turnover range is that these firms are more concerned with selling grain as feed to farmers than in buying grain from farmers to sell to other merchants and end-users. In the western port areas and in Northern Ireland much of the grain is either imported or at any rate bought from larger merchants or brokers.

3.21. It is seen, therefore, that those firms manufacturing smaller tonnages of compounds tend to diversify their activities to a much greater extent than the larger compounders. In the eastern areas, grain merchanting is a major activity of many such firms. As discussed in Section 5.3, retailing of feeds for national and other large manufacturers is a major activity of country compounders. The Survey found that specialisation in the manufacture of compounds was a minor occurrence among non-national firms and was more prevalent in the non-cereal areas of Great Britain and in Northern Ireland. In the cereal-growing areas it was apparent that animal feed compounding was more frequently a subsidiary enterprise associated with grain merchanting. In all areas, feed manufacturers were under certain pressures from their farmer-customers to diversify their stocks and services, and it was indeed the policy of many firms to diversify their trading activities. This latter aspect of merchant trade in general is discussed more fully in Part III, Chapter 2.

Port versus "Country" Location

3.22. When the railways were more prepared to handle small packages of goods, and a higher proportion of raw materials for compound manufacture was imported, it was sensible to locate large factories in the port areas. Since the mid-1950's, however, economic considerations have tended to favour the location of new plant away from the major port areas. To a large extent this has been due to the increased emphasis on home-grown raw materials, but it is also a reflection of the larger companies' policies to fill in the interstices in their distributive systems by building new plants some distance away from their existing factories. Despite this tendency, it certainly cannot be said that port location is no longer justified, and companies might still find it economically desirable to locate mills in port areas if they had to replace their existing capacity.

3.23. The discussion on the decline of the port compounder is perhaps seen in better perspective after studying Table 3.6. This shows that manufacture in the port areas has declined from a peak production in 1962/3, but since 1963/4 it has maintained its output in absolute terms whilst taking a declining share of the trade. The country factories have shown a steady rise in absolute tonnage and share of the trade since the early 1960's. Nevertheless, it should be noted that over 60% of total deliveries are still in the hands of port compounders, regions showing variations between just under 50% in London to nearly 80% in Leith in 1966/7 (Table 3.7).

TABLE 3.6

DELIVERIES OF MANUFACTURED CONCENTRATES FROM PORT AND
COUNTRY FACTORS IN GREAT BRITAIN

Year (July–June)	Port factories		Country factories		Total '000 tons
	'000 tons	% total	'000 tons	% total	
1954/5	4416	71·0	1807	29·0	6223
1957/8	4880	66·5	2462	33·5	7342
1961/2	5975	66·8	2976	33·2	8951
1962/3	6040	66·3	3069	33·7	9109
1963/4	5701	64·2	3182	35·8	8883
1964/5	5815	63·4	3364	36·6	9179
1965/6	5709	62·0	3505	38·0	9214
1966/7	5455	60·9	3506	39·1	8961

Source: M.A.F.F. *Developments.*

Utilisation of Grain in the U.K.

3.24. An examination of the regional trends shown in Table 3.7 throws more light on the port/country controversy. In terms of absolute tonnage produced, most of the decline in port compounding over the past 5 years has occurred in the London region. Most other port areas have shown smaller declines, but Bristol and Leith actually showed an increase. However, as there was a more rapid increase in country area production in the

TABLE 3.7

PRODUCTION OF COMPOUND FEEDINGSTUFFS IN GREAT BRITAIN BY PORT AREA
AND BY PORT/COUNTRY LOCATION WITHIN THE AREA 1961/2 TO 1966/7

Port area	Year	Production			Decline in production located at ports in each area as a percentage of the area's production 1961/2–1966/7	Change in area's percentage share of total trade 1961/2–1966/7
		Location				
		Port	Country	Total		
		('000 tons)			%	
London	1961/2	1109·5	871·3	1980·8		
	1966/7	978·9	988·0	1966·9	−6·25	−0·30
Hull	1961/2	1024·6	525·2	1549·8		
	1966/7	983·0	586·0	1569·0	−3·46	+0·12
Liverpool	1961/2	1535·6	664·2	2199·8		
	1966/7	1336·7	773·8	2110·5	−6·47	−1·22
Bristol	1961/2	1402·5	432·8	1835·3		
	1966/7	1436·3	832·4	1978·7	−3·83	+1·60
Glasgow	1961/2	421·3	116·5	537·8		
	1966/7	400·3	104·4	504·7	+0·98	−0·43
Leith	1961/2	131·3	43·5	174·8		
	1966/7	154·5	39·8	194·3	+4·4	+0·22
Total	1961/2	5624·8	2653·5	8278·3		
	1966/7	5289·7	3034·4	8324·1	−3·40	

Source: M.A.F.F. Quarterly statements of compound production.

Bristol region, Bristol port area trade accounts for a declining percentage of the region's trade.

3.25. Ministry statistics show a steady overall increase in average factory size and a decline in total numbers (see Table 3.1). It has already been noted that firms with small manufacturing capacities are more prevalent in the eastern areas and the large port factories in the west (Section 3.17). It might be assumed, therefore, that the incidence of closure would be greatest in the eastern areas. On the other hand, it has been shown that in the eastern areas the market is in fact less static than in the west, due to the increasing role of intensive poultry and pig production in these areas, and the increasing availability of home-grown cereals. Furthermore, because the smaller manufacturers in the east do not rely on compounding as their major activity, they are more able to accommodate lower margins on their compound production, at least temporarily.

3.26. It is obvious that various pressures in opposing directions are making themselves felt on the industry's structure and the final balance of these forces is not at all clear. It is known that, except in Scotland, country compounding is increasing in all regions, but as far as available statistics show the increase has been greatest in Bristol, London and Liverpool port areas since the early 1960's. In other words, there is no clear regional pattern to be seen in the expansion of country compounding and it would be difficult to forecast future trends. Port compounding is, in relative terms, on the decline, but in absolute terms any noticeable decline is confined to the Liverpool and London port areas. Again, any clear regional trends are therefore difficult to identify.

CHAPTER 4

Grain Purchasing Arrangements
by Animal Feedingstuffs Manufacturers

4.1. If we assume that cereals constitute 60% of delivered concentrate feed, then we can estimate from Ministry statistics[1] that approximately 6 million tons of grain were used for compound and provender production in the United Kingdom in 1966/7. From our surveys of merchants and end-users, we have tried to establish the relative importance of different purchasing paths for nationals and non-nationals. Unfortunately this has not been very successful as merchant compounders resell a large proportion of the grain they purchase. For national compounders in Great Britain the exercise was more successful and the results are recorded in Table 4.1. This shows, as is to be expected, that the national compounders rely mainly on merchants for their purchases of home-grown supplies, though it is known that one national compounder relies mainly on brokers. With regard to imports, the national compounders generally do not operate through merchants but directly.

4.2. The Survey of Merchants suggests that non-national compounders used almost 2·7 million tons of grain for concentrated feed production in 1966/7. When the patterns of grain usage by nationals and non-nationals are compared, a marked similarity emerges (Table 4.2), although national compounders do appear to use relatively more maize in *concentrated* feed production than non-nationals. This difference would be greater if the Northern Ireland utilisation of maize were included in the "nationals" column. The figures may, however, overstate the maize component of *compound* feeds as some nationals have a large trade in processed maize, i.e. flaked maize, part of which is sold to other compounders or to merchants. Another interesting feature is the almost complete lack of oats utilisation by national compounders.

---

[1] The M.A.F.F. estimate that 9·3 million tons of compounded feed and 0·5 million tons of provender were produced in 1966/7 (see paragraph 2.2).

TABLE 4.1

GRAIN PURCHASES BY THE SEVEN NATIONAL COMPOUNDERS IN GREAT BRITAIN
FOR FEEDINGSTUFFS MANUFACTURE 1966/7

('000 tons)

Purchased from	Wheat	Barley	Oats	Maize	Total
Farmers	33	56	1	—	90
Merchants	503	847	23	[a]	1373
Imported direct	239	83	[a]	1093	1415
Other sources	39	66	1	1	107
	814	1052	25	1094	2985

[a] Less than 1000 tons.

TABLE 4.2

GRAIN UTILISATION FOR CONCENTRATED FEEDINGSTUFFS BY
NATIONAL COMPOUNDERS (IN GREAT BRITAIN) AND NON-
NATIONAL COMPOUNDERS (IN THE UNITED KINGDOM) 1966/7

	National compounders		Non-national compounders	
	'000 tons	%	'000 tons	%
Wheat	814	27·3	674	25·4
Barley	1052	35·2	1003	37·7
Oats	25	0·8	138	5·2
Maize	1094	36·6	843	31·7
Total	2985	100·0	2658	100·0

4.3. Although it was not possible to identify all the sources of grain actually used for compounding by the non-national manufacturers, it is possible to study the relative importance of direct purchases from farmers in different regions. The results of this analysis from the Survey of Merchants are shown in Table 4.3 where grain purchases from farmers are expressed as a percentage of total grain purchases by manufacturers of animal feedingstuffs.

4.4. From this it is apparent that in general the smaller compounders rely more on direct purchases of grain from farmers than do larger non-national compounders, and this reliance is much stronger in the east of the country than in the west.

4.5. The Survey material was also analysed to see whether maize purchases were more important in the cereal-deficient areas of the west than in the east. The results of this analysis are shown in Table 4.4 and suggest

TABLE 4.3

GRAIN PURCHASES FROM FARMERS AS A PERCENTAGE OF TOTAL GRAIN PURCHASES
BY MANUFACTURERS OF COMPOUND FEEDINGSTUFFS

Feeds manufactured p.a. (tons)	Port areas						
	London	Hull	Liverpool	Bristol	Glasgow	Leith	N. Ireland
	(%)	(%)	(%)	(%)	(%)	(%)	(%)
50–999	92	97	73	63	69	..	99
1000–10,000	69	86	31	79	22	7	26
over 10,000	..	..	9	11	12	..	5

.. Sample too small to yield realistic results.

that maize purchases are higher in Liverpool and Northern Ireland than in the other regions. The higher percentages shown for the small firms, especially in the western regions, probably signify the absence of a whole-grain trade by merchants in these areas rather than a higher percentage of maize in compound rations.

4.6. The actual mechanism of grain purchasing by the non-national compounders is described in Chapter 4 of Part III. The national compounders invariably have some form of centralised grain buying policy. In some cases this takes the form of advice to their local buyers on what prices to pay and on the future prospects for the market, in other cases it is more rigorous and quantities to be purchased are actually specified. In all cases, however, the existence of a centralised buying policy is evidence of the large economies of scale which can be obtained in collecting, interpreting and using market intelligence. Most national compounders buy some part of their grain supplies from their own subsidiaries but none appear (at the moment) to give their subsidiaries preferential treatment over other merchants.

TABLE 4.4

MAIZE PURCHASES (GROSS OF RESALE) AS A PERCENTAGE OF COMPOUNDS MADE
BY NON-NATIONAL COMPOUNDERS 1966/7

Feeds manufactured p.a. (tons)	Port areas						
	London	Hull	Liverpool	Bristol	Glasgow	Leith	N. Ireland
	(%)	(%)	(%)	(%)	(%)	(%)	(%)
50–999	24	17	51	19	30	..	0
1000–10,000	17	18	24	15	13	24	26
over 10,000	19	..	33	..	11	..	52

.. Sample too small to yield a realistic result.

4.7. No compounder was at present interested in quality in feed grain beyond certain minimal standards such as moisture content, but two firms mentioned the possibility that more attention might be paid to protein content in the future. Most firms admitted that it was technically possible to substitute more home-grown grain for imported grain, but that for high energy rations in particular it was rarely economical to do so.

4.8. It has been suggested at various times that there is collusion between the large users of grain to fix prices, but we found no evidence to support this. It is also sometimes asserted that they depress home-grown grain prices by importing quantities of grain when the home market is already weak. Whether or not this is so we cannot say; but even without any deliberate policy of this kind, the largest compounders know full well that their activities can affect the price of grain, especially in the short run when they wish to purchase "spot" supplies, and they frequently mask their purchasing to prevent any undue reaction on the market, e.g. by using brokers. Similarly, it is probable that they buy large quantities of imported grain mainly because they expect that it will be cheaper than home-grown grain, and any subsequent decline in home grain prices, though it may have been caused by their action, was not necessarily the object of it.

CHAPTER 5

Competition and Integration in the Industry

Advantages and Disadvantages

5.1. One of the most interesting factors in studying the relationship of the animal feedingstuffs industry to cereal marketing is the question of the future share of the market to be obtained by the national compounders. Potentially they have a large bargaining power *vis-à-vis* the producer attempting to sell his grain and the livestock producer buying animal feedingstuffs. The national compounder's main advantages lie in the economies of scale he may obtain. Though we have no proof of this it is probable that these are mainly in the fields of buying raw materials and in the overheads of marketing operation. Most of the national compounders now engage to some extent in their own shipping and broking activities for imported feed grains and other raw materials. The national compounders do not appear to have any advantages in the purchasing of home-grown cereals other than their access to much better market intelligence than many of the small compounders. National compounders are also able to spread the overheads of advertising, promotion, research and development activities over a much larger turnover. They therefore claim to be providing the livestock producer with a much more sophisticated product than many smaller compounders can produce.

5.2. Despite their apparent disadvantages in these respects, the smaller compounders do have some means of redressing the balance. It has been suggested that they are not able to afford the capital investment in machinery necessary to make pelleted and cubed feeds. The results of our Survey showed that nearly all the firms producing more than 2000 tons of compounds per annum are producing both pellets and cubes, and even in the 1000 to 2000 ton range 50% of the firms had either pelleting or cubing facilities or both. Most firms with even a small throughput have the ability to produce a wide range of compounds. We do not know how important economies of scale in production are but it is conceivable that full utilisation of plant throughout the year is more important in minimising costs than scale of plant and thus those manufacturers with small

415

seasonal fluctuations in output are probably in a better position than those firms which have marked seasonal fluctuations. Smaller firms also sometimes benefit from the lack of overheads. For instance they do not engage in any research or development activities or advertise their products. A common feature throughout this study has been the use by the smaller firms of expertise provided by the concentrate supplement firms, who supply both technical and economic information on ration formulation. It is ironical to find that some of this advice is given by the national compounders who also produce concentrate supplement mixtures. It emphasises the need to ensure that a competitive environment is maintained in this sector of the industry. The smaller compounders also frequently benefit from a low level of transport cost if they are able to buy their raw materials and to sell their finished products within a very small radius. They are, however, losing this relative advantage as the national compounders spread their production facilities more widely. Many smaller compounders also claim that they have the advantage of being able to provide individual mixes for some of their customers and some also lay great stress on other factors, such as goodwill, the freshness of their compounds and personal contact with farmer customers.

5.3. The survey also revealed clearly the almost universal practice among the smaller manufacturers of selling large tonnages of compounds and provender manufactured by larger manufacturers, usually nationals. Of the 113 firms sampled in the United Kingdom who manufactured compounds and/or provender and who provided information, 100 also sold compounds and/or provender manufactured by other firms, and the tonnage of the latter was not, in fact, far short of that tonnage produced by the sample firms themselves. Each of the 13 firms not selling other manufacturers' feeds produced at least 1000 tons annually.

5.4. The smaller compounder, therefore, has a useful means of increasing his total turnover without necessarily materially increasing his plant size. As pointed out in Chapter 2 of Part III, the smaller compounder is increasingly under pressure to supply the much more widely advertised national compounds and he is compelled to diversify his stocks to hold on to his traditional clientele. He is, however, able to provide the more diversified service merely by acting as a distributor for one or more of the national firms. In this way three objectives are fulfilled, namely: a fuller more sophisticated service is provided to his customers; his turnover is expanded without the necessity to expand his manufacturing plant; and a distributive outlet is provided for the national firms in that area. In addition it may be that the non-national compounder is thereby protecting himself from direct competition from the national compounders, provided that he is still able to dispose of his own plant's production on which he is able to obtain a similar or better trading margin. (The Survey revealed that most non-national manufacturers adopted a policy of pricing their manufactured feeds at 10s. to £1 a ton below the equivalent feedstuff produced by the national compounders.)

5.5. Nevertheless there is little doubt that in the U.K. in general the larger compounder has an overall advantage over the smallest of the country compounders, and notwithstanding the various advantages of the smaller compounder, particularly when servicing the smaller farmer,[1] the present trend for the average size of the manufacturing unit to grow will continue and production will be concentrated in the hands of fewer firms. In a situation where the pressure to expand arises from increasing costs per unit of production and a relatively static demand rather than from a situation of an expanding market, it is inevitable that expansion on the part of the large firms will be at the direct expense of the smaller manufacturer and the latter's outlets will be under direct threat as a result. An accompanying expansion in the size of farming units and more specialised labour-saving use of bulk feed will further decrease the demand for the more personal, diversified, services of the small compounder.

5.6. Because small-scale farming is likely to be with us for a long time to come there will continue to be a livelihood for many small merchant-compounders, but it is difficult to visualise other than a diminishing role for their services. Government policy is very strongly orientated towards increasing the size of the farming unit, fostering co-operative and group trading activities and giving strong financial assistance to small-farm amalgamation projects. All of these policy measures are counter to the type of farming that relies primarily on the services of the smaller country compounder or merchant.

Process of Integration

5.7. Earlier sections of the Report have indicated the steady process of rationalisation that has been taking place in the industry, a trend which is best illustrated by the continuing decline in factory numbers coupled with rising production and greater throughput per factory. Much of the change has been brought about by the integration of manufacturing firms, some merging with one another or being absorbed into the structure of larger companies. Smaller factories have closed down completely in the face of growing competition from expanding larger competitors, or have been closed down following a take-over by a larger production unit that may

[1] See Part III, Chapter 2.

P

have been more interested in the former's retail outlets than in their manufacturing capacity.

5.8. What are the reasons for the various processes of integration, what patterns if any do they follow, and what are the likely implications for the many firms within the industry and for the farming sector that utilises the manufactured feeds? These issues are now examined more closely. Let us first examine the basic forms which integration has taken.

5.9. One form of integration was the purely horizontal type where similar large manufacturing firms combined their productive capacities under one company. Such integration serves primarily to reduce unit costs by enabling the sharing of fixed manufacturing costs such as research and administration, but it also greatly facilitates increased economies of scale in matters of handling larger shiploads of imported raw materials, more fully utilising port handling and storage facilities and many other such opportunities that might not be directly concerned with feed manufacture. It is interesting that, in general, such rationalisation has not gone the full length of specialisation by the various factories in certain limited lines of product. In particular, the Unilever feeds firms have continued for many years to produce full ranges of animal feeds under their original trade names; and for the purposes of distribution they have continued to operate almost as separate firms each with its own sales force and distribution network.

5.10. The above type of integration has been the basic background of the seven national firms as they are known today, and nearly all have expanded by buying out existing manufacturing capacity as well as by constructing their own new plants. More recently, integration has tended to veer more to the vertical type, and the larger manufacturing firms have become increasingly involved in retailing of compounds to farmers. Traditionally, the port and large country manufacturers sold their product wholesale to a large number of retailers or distributors who themselves might or might not be country compounders, and even to this day, the commonest distributive network for national compounds is through the county merchant. Indeed our Survey estimates that merchants sold 4·1 million tons of purchased compounds in 1966/7, which would represent around 80% of the nationals' production. The Merchant Survey estimated that of over 2000 merchants in the United Kingdom, some 1700 sold compounds or provender manufactured by other firms, and that of these 1700 over half sold more than 1000 tons annually. Most of these firms also sold their own compounds but over 700 merchants, it was estimated, sold only those feeds manufactured by other, larger firms. In the early 1960's there was an upsurge, which has for the present died down, of the national or larger companies buying out country merchant businesses and the Survey suggests that some 86 country compounders or distributors have been bought out by one or other of the nationals since 1960.

5.11. A third stage in integration is also taking place in the country, whereby feedingstuffs firms are becoming involved through subsidiary companies in the production of livestock, in particular pig, egg and broiler production. This latter development is still relatively insignificant, but as discussed later on in this section, it may become the most common form of integration and rationalisation within the industry in the future.

5.12. There are several reasons why integration has occurred, some being more relevant at particular times than others. Even before the Second World War, the large port compounding companies were expanding their manufacturing facilities to meet the requirements of a growing market. Expansion was by means of new factory construction, absorption of or merging with other manufacturers and their capacity. Following de-control of the animal feeds trade in 1953, there was revival of competition within the industry, the more aggressive firms with large productive capacity endeavouring to expand their territorial coverage, forcing the less aggressive companies to take steps to protect their existing outlets. Consolidation and expansion of retail outlets took more than one form, and the different national companies tended to adopt their own particular approaches to the task, depending largely on their prevailing retailing system. More generally, the national compounders had relied on whole-saling to many private agricultural merchants who might or might not be "key" merchants with special arrangements with the national company to sell only that company's compounds and thereby secure more favourable trading terms. Less common then was the practice of retailing direct to farmers through distribution points, but it is known that at least one national compounder has always favoured this form of trading.

5.13. In the late 1950's some firms systematically aimed for complete national coverage of retail outlets and constructed distribution depots in strategic locations. Such expansion had the effect of causing other firms, who might themselves be similarly expanding, to strengthen their hold in existing sales areas. Where the main outlets were merchants, or key merchants, who were still sufficiently autonomous to be tempted by attractive trading terms or take-over bids from other nationals or who might not appear as enterprising as they might be from the point of view of sales promotion, the national company frequently made a successful take-over bid, thereby securing most of that firm's existing goodwill and ulti-mate control of its management. As well as taking over some key merchants to protect their existing outlets, national companies also bought out firms with a view to expanding their coverage into new areas. Some companies favoured taking over moribund private businesses that were strategically situated. Others bought out firms that were thriving but whose main activity might not be the sale of animal feedingstuffs. Some seedsmen, grain merchants and other specialists were thus taken over by the animal feeds nationals and although it is likely that the take-over companies visualised, primarily, an expansion of feed sales from these outlets, it is also likely that they intended to exploit the specialised activities of these firms to provide a more complete service to farmers, thereby attracting more feeds trade. It is well-known that farmers welcome a diversified service from a merchant, and seeds and fertiliser sales and grain purchases, particularly, are important activities in a feed merchant's range of trade. These items, as well as being "services" which attract feeds trade, can themselves be remunerative, and can share many of the fixed costs of the feeds trade.

5.14. A very consistent feature of the management of firms taken over by the nationals was the lack of any sudden change in the taken-over firms' activities, and in this respect, policies of all seven national companies tend to be similar. The taken-over firms were allowed to continue largely as before, and in several cases, continued to produce their own brands of compounds. Some even continued to sell the compounds of national competitors, but it has to be said that such practices are being increasingly discouraged. It was apparent that the parent companies were anxious to protect the existing goodwill of the taken-over firms and furthermore, care was taken not to give unfair trading advantages to the taken-over firms that were competing with neighbouring key merchants or agents of the parent firm. For that reason, new subsidiaries were invariably offered identical trading terms to the key merchants. There were indications that, due to increasing costs and competition and reduced trading margins, the national companies were becoming more concerned to rationalise their production and distributive methods, and in this process, it is likely that more influence will be brought to bear on the activities of subsidiary companies, even to the extent of some being closed down. On the other hand, the manufacturing facilities of some of the larger subsidiaries are seen by some national companies as potential assets for further development if the case for utilisation of home-grown cereals strengthens further. Transport savings in the utilisation of local, as opposed to imported grain, could make ownership of such establishments an important advantage in the face of increasing competition and narrower margins in the feeds trade.

5.15. There are clear indications that economies of scale in compound manufacturing are more quickly overtaken by transport diseconomies as increasing reliance is placed on home-grown raw materials. Unlike imported raw materials, which are normally delivered to one particular port, regardless of quantity, home-grown raw materials, i.e. cereals, produced over a wide area, require to be transported over ever-increasing distances as requirements increase. Although no precise information has been collected as to the effect that distance from production area to point of end-use has on home-grown cereal cost, it is obvious that there must be reached, at some stage, a point where it is worthwhile to establish a factory nearer to the growing area. One manufacturer suggested that a 50 mile radius is the maximum which can be economically serviced by a factory relying mainly on home-grown supplies. This factor has been, and will continue to be, important in influencing the large national manufacturers to diversify their manufacturing points away from the ports and into the heart of the cereal-growing areas. However, whilst overall manufacturing capacity has been expanding in this way, the demand for compound feedingstuffs has been relatively static in recent years. There is therefore some surplus capacity at present in the industry. In these circumstances it is inevitable that to secure outlets for their new factories, the nationals must encroach on the markets of competitors, and one means of doing so, and at the same time securing established factories in these areas, is to buy out smaller manufacturers.

5.16. In the last three or four years, the momentum of take-overs has slowed down, due to several factors, the most important being the lack of available uncommitted capital on the part of the national compounder and the characteristically low rate of return on capital in the feeds trade. The almost static demand for animal feeds, increasing production costs and lower trading margins do not justify a deliberate expansion programme at present. As already mentioned, there are signs that future expansion will be directed at vertical, rather than horizontal, integration of the feeds industry and that the large national companies will become increasingly involved in animal production, e.g. broiler, pig weaner and pig fattening activities to gain greater control of and to increase the outlets for their manufactured feeds and to spread the overhead costs of their activities. It is this form of expansion, therefore, that is more likely to give rise to take-overs in the future. If this leads to the spreading of overheads through a larger tonnage of compounds being produced and therefore to lower prices to other producers, then there is little to criticise in this development. Indeed it is preferable for these firms to undertake the risks than to engage in contract farming with producers tied to certain feed outlets. It would be detrimental if feed companies used this diversification to charge higher prices to other broiler producers. However, given the existing competitive structure of the market, this is very unlikely to occur.

5.17. It is obvious that firms have experienced great variations in the success of their diversification activities. The full significance of these developments in recent years are only just beginning to be seen as the national compounders integrate the newly purchased firms into their overall pattern, introduce new managers and management techniques and remove some of the autonomy which these firms have enjoyed. It is possible that in the early 1970's the national compounders will have a much stronger selling network than they have at the moment with a much larger proportion of sales going through their own companies.

CHAPTER 6

Farm-mixing of Feedingstuffs

6.1. Of the total tonnage of feedingstuffs used on farms in the U.K., a certain proportion is prepared by farmers themselves on their own milling and mixing machinery. The grain used for farm-mixing may be grown on the farm where it is used, or bought in; and other ingredients—for example proteins, minerals and vitamins—are often included in the mixture.

6.2. The Survey of Farmers indicated that in 1966/7 over 2 million tons of cereals were processed on milling or mixing machinery on farms. An unknown but probably small quantity would have to be added to this estimate to allow for the fact that the Survey did not cover milling or mixing on farms not growing cereals.

6.3. Of the farms covered by the Survey, as shown below, two out of every three owned some form of milling or mixing machinery. The proportion of farms owning such machinery increased with average farm size up to about 200 acres and then tended to level off. The proportion of sample farms owning milling and mixing equipment was relatively high in areas where stock farming predominates, and lower in areas which specialise more in cereals (e.g. over 80% in the Northern and West Midland regions of the N.A.A.S., but 60% in the Eastern region).

Farm size	Proportion of farms owning milling/mixing machinery
Less than 50 acres	20%
50–99 acres	63%
100–199 acres	77%
200–299 acres	84%
300–399 acres	84%
400 acres and over	83%
All farms in sample	66%

6.4. Overall, 13% of the farms in the sample owned more than one type of milling or mixing machinery. The roller-grinder was the type most often

423

encountered; next in importance came the hammer mill. The results for the full list of machines covered in the Survey were as follows:

Type of machine	Proportion of farms owning
Roller-grinder	29%
Roller	11%
Grinder	5%
Hammer mill	22%
Roller-grinder mixer	4%
Roller mixer	1%
Hammer mill mixer	5%
Pelleter and/or cuber	1%

The more sophisticated machines—those designed to mix the feed as well as grinding the cereals—were relatively seldom reported, and only one in a hundred of the farmers taking part in the Survey as a whole had equipment for pelleting or cubing the mixed feed. The indication is that most of the feed produced on farms is provender, or fairly simple mixes fed in the form of meal.

6.5. All types of machine tended to be more frequently reported on large farms than on small, and, as might be expected, the more complicated machines were found relatively more often on larger holdings. This was especially marked on farms of 400 acres and over, where the number of reports of machinery with mixing facilities equalled 30% of the sample number of farms. In this group, too, 4% of farms had pelleting or cubing equipment—higher than for the sample as a whole, but showing that even among these larger holdings pelleting and cubing of feed are done by only a small minority.

6.6. In an attempt to assess current changes in farm-mixing, farmers were asked whether they made more, less or about the same amount of feed on the farm in 1966/7 as compared to 1965/6. As stated above, about one farmer in three had no facilities for making feed in either year. Of those who did make feed, about 75% said that there had been no significant change in the amount made between these two years; 18% reported an increase, and 7% a reduction. No data are available on the size of these changes, but it is safe to guess that farm mixing in the U.K. as a whole increased slightly in 1966/7 as compared to the previous year (this conclusion is supported, with certain qualifications, by the results of the Survey of Merchants given in paragraphs 6.14-6.16 below).

6.7. Of the farmers who increased their farm-mixing in 1966/7, 55% said they had done so because they had more stock. Among those who mixed less, the most frequently quoted reason was similarly that stock numbers had gone down. An increase in the amount of corn available was quoted by 15% of the farmers who had increased mixing; a variety of other reasons were given both for increase and decrease, but none of these was mentioned often enough to be significant. Many farmers could give no definite reason for the changes (12% of those increasing farm mixing and 22% of those cutting it down). It seems, therefore, that considerations of relative price and quality of farm-mixed as opposed to bought-in feeds do not play much part in determining farmers' year-to-year decisions on how much feed to make on the farm. This is not to say, of course, that farmers disregard these features when they are considering whether or not to install milling and mixing equipment in the first place.

6.8. Farmers were asked whether they would use more wheat in farm-mixing if the basis of subsidy were changed from tonnage to acreage (if at present they used no wheat, they were asked if they would begin doing so). Not surprisingly in view of the complex issues involved, almost one in three of the farmers interviewed gave no definite answer. Only 15% said they would use more wheat, while 57% said they would not. Among the farmers who were already using some wheat for farm-mixing in 1966/7, 28% said they would use more if the basis of subsidy were changed; but here also a larger proportion of those interviewed (53%) said that their usage would not increase. These results obviously have relevance to the question of changing the basis of the wheat subsidy, discussed in detail in Appendix E. At the same time, they do not make it possible to work out with any degree of precision what would be the actual change in the total retention of wheat on farms; the proportion of indefinite answers is too high, and, more especially, the question does not cover retentions of wheat for feeding as straight grain.

Feed Production on Large Poultry Units

6.9. An interesting development in recent years has been the emergence of a few very large enterprises which produce eggs, broilers or both, and themselves make the feed they require. At its most advanced, such a unit will integrate all the stages of the production process—the building of the poultry houses and provender mills; the growing of a large part of the cereals needed and the purchase of the remainder; the maintenance of a breeding flock to provide rearing fowls and replacement laying birds; and the packaging and sale of eggs and broilers. Units of this sort have, of course, arisen principally because of the economies of scale to be gained in poultry and egg production, but to these they have added important economies in the production of feed.

P*

6.10. There are no national statistics on the cereal usage of these large units. Neither are there any data to speak of on their economic performance. This section has been written on the basis of information provided in the course of Survey interviewing.

6.11. It is clear that a unit of this kind can produce its feeds much more cheaply than they could be produced by a compounding firm. The saving, it was suggested, can amount to as much as £4–£5 per ton, and arises in several ways:

(a) provender mills can be built relatively cheaply, to the exact specification and capacity required for the poultry unit concerned;

(b) only a few different kinds of ration are produced, and the mills can turn them out at a steady rate (whereas a compounder must make a much wider range of feeds and vary his output of each according to demand);

(c) all handling is in bulk.

6.12. Up to now there is no evidence that the operations of these very large units are causing any significant change in the marketing of grain in the U.K. One firm providing information was a very substantial buyer of grain; and, like very many other large buyers throughout the cereal market, it bought its grain from merchants. Buying from farmers was not favoured because it was considered to cause too much administrative difficulty. It is impossible to say whether the other large units also take this attitude to buying; but it is instructive that the merchants interviewed in another part of the Survey (see Part III) made no mention of being by-passed in the purchase of cereals by these large poultry enterprises.

6.13. Expansion of these large units is almost certain to continue, and to all intents and purposes may be limited only by saturation of the market for eggs and broilers; whether this stage will ever be reached is not a matter for discussion here. In itself, this expansion is not likely to have any very significant effect on the cereals market, so long as the producers concerned continue to buy a large part of their grain through merchants rather than buying direct from farmers or greatly stepping up their own cereal production. As far as can be predicted, no substantial changes of this sort are likely in the near future.

Farm-Mixing and the Merchant

6.14. As part of the Survey of Merchants (the main results of which are given in Part III), agricultural merchants were asked whether there had been any noticeable increase in the farm-mixing of compound feeds in their area in the last few years. Of the firms answering the question, 60% reported an increase in farm-mixing. However, comments volunteered by merchants made it clear that in many cases a simple choice of answer between "Increase" and "No increase" was an over-simplification, and that what had happened in fact was that farm-mixing had been on the increase until a fairly recent point of time—most merchants making this comment thought that this was around 1964 or 1965—but that since then the trend had levelled off. The explanation of this, according to the few merchants who commented further, was that the idea of farm-mixing had gained favour among farmers from about 1960 onwards; but that experience had shown many farmers that they could not make feed as cheaply or as satisfactorily as the compounder. Merchants thought that this applied particularly to smaller farms; it was only on the larger units that farm-mixing had made headway in the last two or three years.

6.15. Merchants were also asked whether their sales of compound feeds had been affected by farm-mixing. Of the merchants who gave a definite answer, only about one in three said that farm-mixing had reduced their compound sales. Even this is almost certainly an over-estimate, since merchants who had previously reported "no increase in farm-mixing" were often not pressed to answer this latter question; if they had been, their answers would obviously have been in the negative. Very many merchants, whether or not their compound sales had been affected, had turned the situation to their own advantage by increasing sales to farmers of the materials needed for farm-mixing—of straight grain, of provender, but, above all, of concentrates and protein and other supplements.

6.16. By and large, therefore, the mixing of feed on farms has not had an adverse effect on agricultural merchants' trade. Neither is it an anxiety for the future as far as most merchants are concerned. In fact, it is difficult to predict the changes which will take place in the prevalence of farm-mixing in the next few years (leaving aside the effects of a possible change in the basis of subsidy, already discussed). If, as is probable, farm-mixing is more suited to large than to small farms, it may tend to increase slowly as average farm size goes up. But this gradual upward trend could equally well be counteracted by any technical or economic changes, whether in animal husbandry or in compound production, which made purchased compounds more attractive to the farmer.

CHAPTER 7

The Animal Feedingstuffs
Industry in Northern Ireland

7.1. Although all the previous chapters in this Part have been written on a U.K. basis there are certain distinctive features about the situation in Northern Ireland which justify a separate look at that country.

7.2. Northern Ireland, producing some 8 to 9% of the U.K.'s gross output of livestock and livestock products, has an important place in U.K. agriculture, and its contribution is guided by the same overall governmental policies that prevail over the country as a whole. But because it is cut off by water from the mainland of Britain, with the transport problems this poses, the grain merchanting and animal feedingstuffs industries in particular have been prevented from becoming fully integrated into the U.K. market and have developed along fairly independent lines.

The Demand for Animal Feedingstuffs

7.3. With a climate not well suited to crop production, Northern Ireland has developed an intensive pig, poultry and dairying industry, and cereal farming is of relatively minor importance. In Northern Ireland a higher proportion of total compounds manufactured goes into pig feed than anywhere else in the U.K. (Table 1.6) and the increase in pig feed requirements since 1956 has also been higher than elsewhere in the U.K. (Table 1.7). Because of the importance of intensive livestock feed production the seasonal fluctuations in output are less in Northern Ireland than they are in the western parts of Britain (Fig. 1.3c). As indicated in Part III, Chapter 3, and illustrated in Map 3.2, Northern Ireland is one of the major grain-deficit areas in the U.K. The region consumes annually almost 1 million tons of compound feedstuffs compared with a total production of grain for all purposes of not more than 300,000 tons (less than a third of which is sold off farms) and all other raw materials have to be shipped to Northern Ireland, either from Great Britain or from abroad.

The Supply of Animal Feedingstuffs

7.4. Since the early 1950's barley has replaced oats as the main cereal crop in Northern Ireland and much of the barley grown is utilised by the producers themselves for incorporation in home-mixed pig-feed. The general practice of using cereals as feed on the farm of origin is a characteristic feature of Northern Ireland, and it is estimated that only 25 to 28% of all cereals grown are sold to registered buyers.

7.5. The Survey of Merchants revealed the virtual absence in Northern Ireland both of large cereal growers offering substantial tonnages for sale,[1] and, as one would expect, of specialist grain merchants. As noted by the Ulster Farmers Union in their statement at Appendix G, the acreage of wheat grown is insignificant because the subsidy is only paid if the crop is sold, and Northern Ireland farmers prefer to feed their grain to their own livestock. The small parcels of grain sold on the feed grain market are generally bought by the many small country merchants who manufacture their own feedingstuffs for sale, by the few larger specialist country compounders, or by the distributors of compounds manufactured by the large Belfast or Londonderry port compounders. It is probably true to say that only the non-port compounding firms make deliberate efforts to obtain as much home-grown grain as possible to gain the advantage of lower transport costs. The port compounders, or the distributors of their compounds, engage in local grain buying primarily as a service to farmer customers, and to use it as a "contra" item for settling outstanding accounts. So small is the tonnage of home-grown grain coming onto the market, and so variable is it in quality, dryness and cleanness, that it is not surprising that large grain users such as the port compounders find it more satisfactory to rely on imported grain.

7.6. Recent developments in farm drying, farm storage, forward contracting and the bulk transport of home-grown grain have had considerably less impact in Northern Ireland than in Great Britain, but it would be wrong to suggest that there are no exceptions to this. In the eastern counties of Antrim and Down particularly, parts of which are relatively well suited to grain farming and are situated near to the Belfast markets, a few farmers have adopted modern handling and marketing techniques with some success, and country compounders have been quick to engage in forward contracting arrangements with such producers. In the Downpatrick area of County Down a small grain storage and marketing group has been formed, probably the only one in Northern Ireland.

---

[1] The average acreage of cereals grown per cereal grower, in Northern Ireland, is about 12 acres.

7.7. Because of the over-riding reliance on imported cereals, the producer price for home-grown grain is related to the port price for imported grain of the same type, which in effect causes the average producer price in Northern Ireland for barley or wheat to be some £2 higher than in Great Britain. Also, there is considerably less within-season fluctuation in producer prices than in Great Britain for the reason that home-grown supplies have such little impact on the overall supply situation and landed import prices dictate prices paid to local producers.

The Structure of the Feed Manufacturing Industry

7.8. Northern Ireland's animal feedingstuffs requirements are manufactured within the country, and the structure of the industry differs from that of Great Britain in several respects. The influence of the seven national companies is less apparent. Northern Ireland's requirement of manufactured concentrates of over 1 million tons per year is produced by 78 firms, but 77% of that tonnage is produced by the eight largest factories (as shown in Table 7.1) and 48% by the three largest.

TABLE 7.1

DISTRIBUTION OF ANIMAL FEEDINGSTUFFS FACTORIES IN
NORTHERN IRELAND BY ANNUAL DELIVERIES OF
MANUFACTURED CONCENTRATES 1966/7

Volume of deliveries (tons)	Number of factories	Deliveries (tons)	Percentage of total deliveries
0–499	15	2248	0·2
500–999	7	5081	0·5
1000–1999	9	12,768	1·2
2000–4999	22	76,754	7·1
5000–9999	12	84,957	7·9
10,000–19,999	5	67,522	6·3
20,000–200,000	8	821,716	76·8
	78	1,071,046	100·0

Source: Ministry of Agriculture, Northern Ireland.

7.9. Because almost all raw materials for feed compounding have to be shipped to Northern Ireland from Great Britain or from overseas, port compounding is of even greater significance than in Great Britain, and the country compounding firms, who rely to a declining extent on locally grown cereals for a part of their raw material requirements, are in general considerably smaller than the large manufacturers of Belfast, and, to a lesser extent, Londonderry. Although the Survey was not able to give conclusive indications, it is unlikely that any country compounding firm produces more than 30,000 tons of compounds in a year.

7.10. Of the United Kingdom's seven large national feed companies described earlier, only five manufacture feeds in Northern Ireland and two of these have co-operated to construct and operate one factory jointly, to produce their respective brands of feeds. In general, the five nationals represented in Northern Ireland commenced production there relatively recently, constructing their own factories or taking over existing private compounding businesses. Their impact on the compounding industry is not yet as great as in the rest of the United Kingdom, but we estimate that they now produce over 40% of the compounds manufactured in Northern Ireland. One interesting feature of the Northern Ireland industry is that several of the largest firms are still privately owned.

Grain Purchasing Arrangements

7.11. Unfortunately we have no statistics of the purchases of grain by the national compounders in Northern Ireland, but as we estimate that they control 40% of concentrated feed sales in Northern Ireland, we would expect them to use approximately 240,000 tons of grain, most of which would be imported maize and barley (see Table 4.4). The animal feed compounders in Northern Ireland give no preferential consideration to British sources of feed barley and it is for this reason that maize forms such a high proportion of the cereal content of animal feeds. The landed price of maize in terms of units of starch is invariably lower than the landed price of British barley, which contrasts strikingly with the normal relationship between maize and home-grown barley at ports elsewhere in the United Kingdom. It was frequently stated in the course of Survey enquiries that the high cost of shipping grain from Great Britain to Northern Ireland prevented Northern Ireland's compounders from buying more of their grain in the British feed barley market and that shipping costs from North America were little higher than those from Great Britain. Devaluation, while raising the total cost of Northern Ireland's feed grain imports, will tend to alter the balance in favour of the British market. Moreover, it does not appear unreasonable to assume that reductions could be made in the cost of shipping grain from Great Britain to Northern Ireland.[1] On an annual tonnage of 600,000 tons the savings could be considerable. Importing grain from Britain would help Northern Ireland to offset the effects of devaluation on its imported grains bill, and equally important, it would provide an additional outlet for Britain's surpluses of feed grain. There

---

[1] This issue is also discussed in Part VIII, Chapter 3.

might also be a case, on balance-of-payments grounds, for encouraging more cereal production in Northern Ireland itself (see the Ulster Farmers' Union statement, Appendix G). It is interesting in this context to note that the Report on "Agriculture's Import-Saving Role" by the E.D.C. for Agriculture envisages "cereal growing reaching out towards the far west and south" (i.e. of the United Kingdom) (page II,10).

Competition and Integration

7.12. In Northern Ireland there are as great pressures for rationalisation and for improving efficiency by increasing scale of operation as in the rest of the country. The feed industry which developed to meet an expanding market for concentrated feedingstuffs is now facing the same levelling off in demand as is occurring in the United Kingdom as a whole, and adequate returns on investment are becoming increasingly difficult to realise. The larger Northern Irish feedingstuffs firms are still apparently competing successfully with the large British national companies operating in Northern Ireland, but should competition become keener, or even continue at its present level, there will be pressure on the local firms to combine resources with each other or to ally themselves with the nationals with their vastly stronger financial backing and greater range of activities.

7.13. In the case of Northern Ireland's larger country compounding firms, the Survey revealed a general buoyancy of activity and a widespread intention to increase their hold on the market. With Northern Ireland's peculiarly over-riding reliance on imported cereals and other raw materials, the position of the country compounder would not appear to be as favourable as in Great Britain, but it has to be said that the large country compounding firms appeared to be well able to continue to hold at least their present share of the market.

PART V

Utilisation of Grain in the U.K.

(b) Flour Milling

Introduction

FLOUR is an intermediate product. Before it is consumed it must be processed—by industry or by housewives—into bread, biscuits, confectionery, or one of a diversity of minor uses. Changing requirements at this level, together with changes in the techniques of flour milling, determine how much and what types of grain will be bought by the milling industry.

This Part is concerned with the use of home-grown grain for flour milling. It examines first the aggregate demand and supply situation for flour. Second, it explores ways in which structural and technical characteristics of the milling industry affect its utilisation of home-grown cereals. Finally, it discusses the significance of a number of changes which are taking place in the industry. In addition to using published sources, the chapter is based on Survey data and upon conversations with leading firms and institutions involved in the manufacture of flour. Five firms who were flour millers were visited in the course of the Merchant Survey. An additional Survey was made of eighteen independent flour millers, twelve of whom were able to help. Postal enquiries were sent to forty firms of whom twenty-four replied.

CHAPTER 1

The Supply and Demand Situation for Flour

1.1. Table 1.1. demonstrates that flour milling is a relatively static industry whose overall production is tending to decline. Between 1956 and 1967 flour production fell some 7% despite an increasing U.K. population and a rise in *per capita* income levels. At the same time, the industry increased its use of home-grown wheat by about a third. The market is almost entirely domestic, exports representing a very small fraction, less than 1%, of total production. Although more use is made of home-grown wheat than formerly, two-thirds of the grain used in flour milling is imported. In 1966 imports of wheat for all uses cost £101 million; most of this was used for flour milling.

1.2. Most of the wheat bought by flour millers is used to make bread flours. An estimate published by Mr. Greer of the Flour Milling and Baking Research Station shows the use of wheat for different types of flour.[1]

Flour type	Annual wheat usage (million tons)	% home grown
Bread	3·6	20
Cake and biscuit	0·65	90
Household	0·7	50
	4·95	33

Source: E. M. Greer, The wheat variety Maris Widgeon, *Ceres,* Vol. 1, H.-G.C.A., 1968.

1.3. This estimate gives a broad impression of the use of wheat by the industry. Individual firms have different patterns of use and in particular years the proportion of home-grown grain will vary as a result of changes in quality and in its price compared with that of similar imported wheats.

---

[1] The figures are described as a "rough estimate". No year is specified but the aggregate corresponds to wheat milled in 1967.

1.4. Table 1.2 indicates how the downward trend in flour consumption reflects changing patterns of consumer demand. Although the total consumption of the major cereal products has fallen by 17% since 1956, consumption of some of the more expensive items has in fact increased. Cakes, buns, biscuits and brown bread were all more heavily consumed in 1967 than in 1956. The effects of these increases were outweighed by reduced purchases of white bread, which, as Table 1.2 demonstrates, remains the most important form in which flour is consumed.

1.5. This picture is consistent with a rising level of affluence in the community as a whole. As people become better off they substitute other more expensive and more attractive foods for bread. The estimates of the National Food Survey Committee suggest that a 1% rise in income levels is associated with a 0·5% decline in expenditure on large white, wrapped loaves, and a 0·5% fall in the quantity purchased. In contrast, the demand for chocolate biscuits rises by 0·5% for each 1% rise in income.

1.6. In a stationary market the use of home-grown wheat can only be increased at the expense of imported wheat. It may do so because the combination of price and quality which it offers to the flour miller and the flour user makes it more attractive. For bread flours, which still constitute more than two-thirds of all flour produced, home-grown wheat is at a technical disadvantage. It lacks the characteristics of hard imported wheat, which are required to produce the conventional type of loaf. A fuller description of these technical limitations is given later in this part: here it is sufficient to note that they form an effective barrier to the substitution

TABLE 1.1

WHEAT MILLED AND FLOUR PRODUCED: U.K. 1956–67
('000 tons)

	Wheat milled	Of which home-grown	% home-grown	Flour produced
1956	5391	1202	22	3910
1957	5232	1245	24	3720
1958	5145	1075	21	3672
1959	5041	1100	22	3632
1960	5112	1242	24	3693
1961	5005	1409	28	3592
1962	4870	1192	24	3530
1963	5038	1561	31	3645
1964	4967	1630	33	3589
1965	5067	1614	32	3675
1966	5063	1602	32	3645
1967	4968	1583	32	3644

Sources: Annual Abstract of Statistics. Monthly Digest of Statistics.

TABLE 1.2

HOUSEHOLD FOOD CONSUMPTION OF FLOUR AND FLOUR PRODUCTS MEASURED IN OUNCES PER WEEK

	1956	1958	1960	1962	1964	1966	1967	% change 1956-67	1967 % of total consumption of flour and flour products
White bread	44·36	38·43	36·64	36·07	36·02	32·57	33·84	−24	59
Brown bread	2·38	1·89	2·42	2·44	2·64	3·41	3·38	+42	6
Other bread	4·34	6·89	6·41	5·08	3·30	2·66	2·80	−35	5
All bread	51·08	47·21	45·17	43·59	41·96	38·64	40·02	−22	70
Household flour	7·89	7·75	6·76	6·22	6·07	5·95	5·79	−27	10
Buns, scones, teacakes, cakes and pastry	5·67	5·82	6·31	6·61	6·47	6·46	6·04	+7	10
Biscuits	5·30	5·58	5·67	5·75	5·73	5·60	5·87	+11	10
All products	69·94	66·36	64·21	62·17	60·23	56·65	57·72	−17	100

Sources: Annual Reports of National Food Survey Committee, *Household Food Consumption and Expenditure*, M.A.F.F. *Board of Trade Journal*, 10th May 1968.

of home-produced grain for imported, even when price differentials favour the domestic product. For biscuits, on the other hand, the softer flour produced from home-grown grain is suitable. In so far as the biscuit trade has grown, it represents an attractive market for home-grown wheat. However, as Table 1.2 shows, biscuits form only 10% of the total consumption of cereal products in the U.K.

1.7. Confronted by falling demand, the flour-milling industry faces a potentially low-profit situation. Flour-milling machinery is relatively long-lived, and a surplus of capacity is likely to arise. To the extent that technical developments in milling permit existing equipment to produce more flour, the problem of excess capacity is intensified. These pressures have already led to a reduction in the number of mills, to concentration in the industry and, in the past, to the operation of a scheme within the industry[1] to minimise the effect of falling demand or prices. In their submission to the authors the National Association of British and Irish Millers[2] say that it would be a simple matter for the industry to produce approximately 25% more flour than is at present consumed in the U.K.

1.8. Table 1.3 shows that the U.K. is a small net importer of wheat-based cereal products. About £8 million worth of flour was imported in 1966. The existence of potential imports, however, sets a ceiling on prices which may be charged for domestically produced flour.

TABLE 1.3

U.K. TRADE IN WHEAT-BASED CEREAL PRODUCTS

	IMPORTS			EXPORTS	
	Quantity, tons	Value, £'000		Quantity, tons	Value, £'000
Wheat flour	195,588	7719	Wheat flour	7123	384
Groats and meal			Other flour	2956	184
(incl. semolina)	613	29			
Other flour	1708	62			
	197,909	7810			
Bread, rolls, etc.	4440	848	Bread, rolls, etc.	3038	416
Biscuits	1864	505	Biscuits	26,788	7587
Cakes	1690	418	Cakes	1243	342
Other	1436	381	Other	1891	588
Total value		9962	Total value		9502

Source: Annual Statement of Trade, 1966.

[1] See Chapter 2, page 446.

[2] See Appendix G.

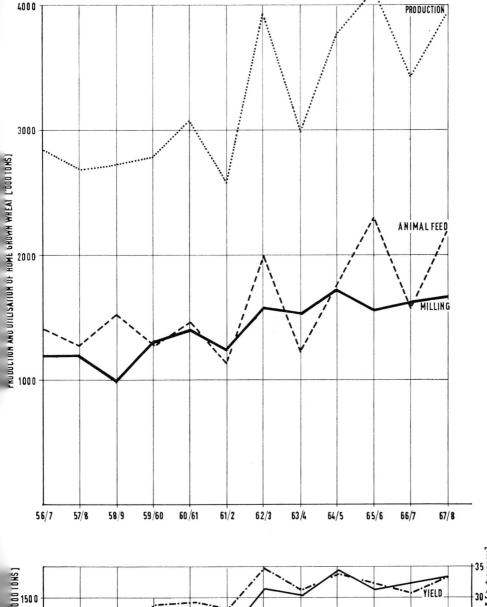

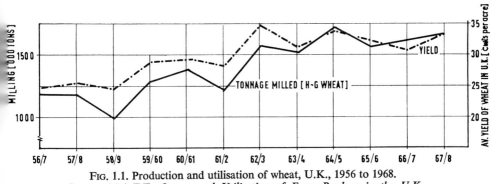

FIG. 1.1. Production and utilisation of wheat, U.K., 1956 to 1968.
Source: M.A.F.F. *Output and Utilisation of Farm Produce in the U.K.*

1.9. Finally, these trends in the flour-milling industry must be set in the context of a basically prosperous and growing national economy. The effect of expansion elsewhere is to intensify competition for the labour and capital used by millers. Rising wage levels may make new labour-saving techniques imperative, despite the unpropitious economic environment for milling, whilst the need to maximise return on capital may demand new and larger mills where lower costs associated with large-scale production can be achieved. Combined with the pressure on existing millers, deriving from the fall in flour consumption, this tends towards a greater degree of concentration, the elimination of small or badly-placed mills and the reduction of the level of employment in milling.

1.10. Figure 1.1 indicates that with the long-run increase in U.K. wheat production there has been an increase in millers' uptake. It might be supposed that a heavy crop would tend to increase pressures on the milling industry to use more domestic wheat, but the figure shows that there has been no consistent tendency for uptake to be greater in years of higher production. Much close correlation is to be seen between production of wheat and its use for animal feed; in other words, fluctuations in the total crop are absorbed by feed use rather than by use for milling.

1.11. An interesting positive relationship, however, appears to exist between the quantity of home-grown wheat used for milling and the yield per acre (Fig. 1.1). The explanation of this would need further investigation, but it seems likely that in years of high yield per acre quality has also been good from the millers' point of view. If this is so, the inference would be that in any particular year it is yield rather than acreage which has the greater influence on the amount which will be taken up by millers. It is noticeable, for instance, that although there was a record acreage in 1965 and a very high production figure, the uptake by millers in 1965/6 was actually lower than in the previous year. The yield per acre in 1965 was somewhat lower than in 1964 and it also appears to have been less acceptable in terms of quality.

CHAPTER 2

The Structure of the Flour Milling Industry

The Process of Concentration

2.1. Flour milling is characterised by a great degree of industrial concentration. Although there are some sixty-nine firms which belong to the Incorporated National Association of British and Irish Millers, four of these (Associated British Foods; The Co-operative Wholesale Society; Ranks Hovis McDougall; and Spillers) produce over two-thirds of all flour used in the United Kingdom.[1] Concentration is not simply a question of ownership; it is evident too in the size distribution of productive capacity. Replies from the survey of flour millers and by the large national companies indicated that there were twenty-seven mills with a capacity in excess of 60 sacks per hour. These twenty-seven mills (19% of the total number) represented at least 35% of the U.K. milling capacity. The evidence from the enquiry is set out in Table 2.1.

2.2. Two groups of factors are responsible for this pattern of concentrated ownership and production: the importance of imports, and the security which large-scale capital resources afford in a period of economic pressure. Table 1.1 makes it clear that the major input item of flour millers is imported wheat. Firms established in port areas close to major consuming areas are able to minimise the costs of transport within the U.K. Furthermore, the ability to buy large amounts of imported grain strengthens the bargaining position of the port miller and spreads the overhead cost of maintaining information facilities on a world-wide scale.

2.3. Each of the large milling organisations has important overseas connections. Overseas purchases are bought directly or through associated companies from such international suppliers as the Canadian Wheat Board. Shipping is sometimes arranged by the firm itself but more often is provided by specialist shippers.[2] An increasing amount of grain is transshipped from European ports in vessels which could be unloaded directly

---

[1] Of the sixty-nine firms, one is in Eire and one does not produce flour.

[2] See Part IV, Chapter 1.

TABLE 2.1

SIZE DISTRIBUTION OF NATIONAL AND INDEPENDENT FLOUR MILLERS BY MILLING
CAPACITY (sacks per hour)

	Under 20	20–59	60 or more	Not known	Total
Independent companies	47	19	5	3	74
Large national groups	20	22	25	—	67
Total	67	41	30	3	141

Source: Information from millers and from the National Association of British and Irish Millers.

into port mills. One large group depended upon a series of small country mills for its milling capacity but has now started to build a very large flour mill at Tilbury to take advantage of the extensive developments being made there by the P.L.A. The same firm also has a part share in the operation of a new mill at Leith. The advantage of such port locations remains of great importance despite the greater use of home-grown wheat by the milling industry. They provide one important reason for the high degree of concentration in the industry.

2.4. Flour milling has been under continuous economic pressure since the 1920's. Only during the war-time period did the decline in the number of mills slow down. Table 2.2 indicates the number of firms whose names disappeared from the list of members of the National Association of British and Irish Millers between 1926 and 1967.[1] To some extent it may understate the true decline in separately owned mills because businesses which become parts of larger groups in this period sometimes retained their identity and individual membership of the Association.

2.5. Firms which wished to withdraw from flour milling were sometimes assisted by the Millers' Mutual Association. This organisation drew its funds from member millers. Firms were allocated a quota of production based on the previous three years' production. If a firm was unable to sell its full quota in a particular year compensation was paid from the fund. If firms were consistently unable to meet their quota they might elect to be bought out by the Association.[2] These payments, which were naturally confined to members of the "Millers' Mutual", were made on condition that no more flour was produced. They enabled the assets of the business to be transferred to some other activity—for example, the manufacture of

[1] We acknowledge with thanks the courtesy of the Association in helping us with this information. Map 2.1 shows the decline in the number of mills between 1928 and 1968.

[2] Firms who exceeded their quota paid a "premium" on their excess production to the Millers' Mutual Association.

animal feedingstuffs. Undoubtedly, this system facilitated the reduction in the number of mills. It also helped to lift the pressure on millers to cut prices if their capacity was under-utilised. This scheme has now been discontinued.

2.6. The fall in the number of mills was accompanied by the formation of industrial groupings. Thus the four large milling companies possess in

TABLE 2.2

REDUCTION IN NUMBER OF INDEPENDENT MEMBERS OF THE
NATIONAL ASSOCIATION OF BRITISH AND IRISH MILLERS SINCE 1926

Region[a]	Number of independent firms in 1926	Number of independent firms in 1967	Net reduction in numbers
Eastern	78	22	56
South-eastern	67	11	56
East Midland	23	7	16
West Midland	27	5	22
South-western	35	2	33
Northern	10	2	8
Yorkshire and Lancashire	47	8	39
Wales	7	0	5
Scotland	12	4	8
Ireland	5	2	3
Total	309	63	246

[a] The regions in England correspond to those of the National Agricultural Advisory Service.

Source: Membership Lists of Association.

all some sixty-seven mills, many of which were formerly independent businesses. These large groups have been in a relatively strong position partly because of their bargaining power in relation to agricultural merchants, grain importers and bakers and partly because of their strong capital situation. The introduction of new capital has resulted in mills being modernised, obsolete plant being scrapped and the competitive position of the group being enhanced. Independent millers have had to face increasingly severe competition from the more powerful national companies.

2.7. This very dominant position has, in itself, been a source of some concern to the larger millers. Conscious of the very important political and emotional overtones of the price of bread they have for many years advised the Government of any intention to raise flour prices. More recently the activities of the industry have twice been scrutinised by the Prices and Incomes Board. The effect has been to place these firms in a "cost-plus" pricing position so far as flour is concerned. The effective extent of any monopoly power they may have is thus much more apparent than real. One paradoxical consequence may result. Several of the representatives of large firms in the Survey commented on the importance they attach to the survival of the existing independent millers. If these firms were to disappear it might be more difficult to resist claims for some more comprehensive national control of the milling industry. This attitude contrasted sharply with the expectation of many of the smaller milling firms. In the Survey 75 % of the independent millers said that they anticipated further integration within the industry.

2.8. It should be noted that the scope for further concentration is limited by the large proportion of the trade already in the hands of the national groups. Such further concentration as does occur seems more likely to stem from the disappearance of independent companies than from an aggressive policy of acquisition by the national groups. In a position of surplus capacity few national groups would be interested in adding plant, much of which is obsolete, to their existing equipment. A diminution in the number of small flour mills would further reduce direct sales by farmers to flour millers. Large national groups purchase most of their home-grown grain through merchants. The relationship between small millers and farmers is based on propinquity and a mutual understanding of each other's problems. Its disappearance might diminish the benefit accruing to the farmer from producing wheat of especial suitability for flour production.

2.9. Any damage which arises from this development seems likely to be more than outweighed by the tendency for flour milling to become part of an integrated cereal-using industry. The large national groups are now deeply committed to the production of animal feedingstuffs and bread and to the organisation and management of merchant businesses. Within the groups, great flexibility in using home-grown grain already exists. As a result of involvement in merchanting better knowledge is available concerning the home-grown cereal crop and incentives can be provided to farmers to produce grain of suitable characteristics for the market. At this stage there already exists, among some of the national groups, a co-ordinated policy towards the utilisation of home-grown grain. As experience is gained and greater understanding of husbandry and technical processes achieved it seems likely that the benefits in terms of a larger market for home-produced grain will be considerable. The powerful position of the large national groups must always call for vigilance to safeguard the interests of the community; it also provides a real opportunity to promote an efficient and modern home cereal-producing industry.

Geographical Location of Flour Mills

2.10. The geographical distribution of mills is shown on Map 2.1. A more detailed regional breakdown in terms of the regions used by the National Agricultural Advisory Service is shown in Table 2.3. It is clear that the independent mills tend to be more heavily concentrated in the grain-producing areas while those which form part of national groups are more evenly dispersed through the country.

2.11. Even where mills are situated close to grain-producing areas they are seldom far from port facilities. Of the seventy-four mills owned by independent companies, twenty-eight are at ports and a further eleven are within 20 miles of a port. The sixty-seven mills owned by large national groups are even more firmly tied to port locations; thirty-eight mills are actually situated at ports and thirteen are within 20 miles.

2.12. One of the large national flour millers suggested that the cost per ton of moving bulk flour was twice that of moving bulk grain. The siting of bakeries close to flour mills reduces the distance for deliveries of flour, while the location of mills at ports means that only about one-third of the grain comes in by road, the rest coming mainly by sea. To locate mills in areas of home wheat production would not only involve moving imported grain from the ports but would also add to the distance over which flour has to be moved. If more home-grown wheat is used in future, mills which are well placed to receive domestic as well as imported raw materials will enjoy some small advantage. However, the dominant position of imported wheat in the grists used for bread flours means that this advantage is unlikely to undermine the strong position of the port millers. A further factor which reinforces this view is that most of the inland mills are relatively small. They are unable to exploit economies of scale which the larger port millers already enjoy. It is significant that the one large national group which has hitherto relied mainly on country milling capacity is building two new mills at port locations which will in due course replace some of its older inland mills.

2.13. The need to transport flour depends upon the siting of the mills. In the Birmingham conurbation there are no large mills, and flour or bread has to be brought into the area. Other areas of dense population are nearer to port mills but flour still has to be moved from these mills to bakeries distributed through the area. A great deal of flour is now moved in bulk, perhaps as much as half. Large lorries which can be loaded and unloaded by pneumatic methods enable flour to be transported over fairly long distances. Such equipment represents a substantial capital investment, and it is essential that it should be fully employed. The regular pattern of flour consumption assists this process, bakeries having to be supplied on a day-to-day basis. A number of millers expressed concern

Q

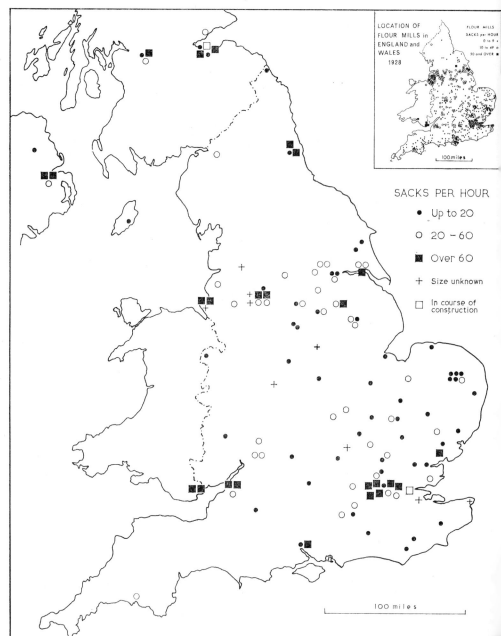

MAP 2.1. Location and size of flour mills in the U.K., 1968. (Inset: Location of mills in 1928).
Source: National Association of British and Irish Millers (and 1928 report).

TABLE 2.3

THE GEOGRAPHICAL DISTRIBUTION OF FLOUR MILLS IN THE UNITED KINGDOM

Regions[a]	Independent companies' mills						Large national groups' mills						
	Sacks per hour				All ind. mills	Of which ind. mills at ports	Sacks per hour			All national mills	Of which national mills at ports	All mills	Of which all mills at ports
	Less than 20	20–60	Over 60	Capacity unknown			Less than 20	20–60	Over 60				
East	20	7	1	1	29	12	4	5	4	13	6	42	18
South-east	10	—	1	—	11	2	7	1	3	11	5	22	7
East Midland	4	2	—	1	8	—	4	1	—	5	—	13	—
West Midland	3	2	—	1	6	1	—	—	3	3	3	9	3
South-west	1	1	—	—	2	—	2	3	3	8	6	10	7
Northern	1	1	—	—	2	2	1	—	2	3	3	5	5
Yorks. and Lancs.	5	3	—	—	8	4	2	10	5	17	8	25	12
Wales	—	—	—	—	—	—	—	—	2	2	2	2	2
Scotland	2	2	2	—	6	6	—	1	2	3	3	9	9
Northern Ireland	1	1	—	—	2	1	—	1	1	2	2	4	3
Total	47	19	5	3	74	28	20	22	25	67	38	141	66

[a] The regions in England correspond to those of the National Agricultural Advisory Service. See Appendix C.

at the possible effects of the recent Transport Act.[1] They feared that pressure to use rail services would add to the costs of delivery and make it difficult for them to guarantee regular supplies of flour. Bakers currently keep very small stocks of flour but if deliveries of flour cannot be assured, capital will be needed for storage to sustain a larger margin of safety for baking. A related problem stems from the specialised nature of bulk-flour lorries. For reasons of design and hygiene they cannot be used to carry other materials. If the effect of the Transport Act is to move part of the load from road to rail it may be more difficult to keep such vehicles fully employed. Effectively this will increase the total cost of moving flour, even if selected long journeys could be made more cheaply by rail.

2.14. Manpower statistics relating to the industry are scarce and somewhat confusing, but from discussions we have been able to have with reliable authorities we understand that the labour force in flour milling has declined from 15,000 to under 10,000 in the last 10 years. This reflects the declining number of mills and the increasing mechanisation of the industry. In the actual milling process relatively few men are engaged. To the casual visitor one of the more striking features of a tour of a flour mill is the small number of workers engaged in production. Such economy in manpower is more difficult in transport or office procedures, but even here new methods must tend towards an overall reduction in the number employed.

2.15. An important aspect of employment in flour milling is the 24-hour nature of milling operations. Mills commonly operate for between 90 and 120 hours a week. This requires continuing supervision by responsible staff. Workers are not always prepared to accept the inconvenience of shift working and recruitment presents some problems, especially in rural areas.

Diversification of Millers' Interests

2.16. An important characteristic of the four large national companies which dominate flour milling, as indeed of many independent millers also, is the extent of their involvement in other areas of economic activity. Two of these areas, the manufacture of compound feed and the production of bread or other flour products, are closely related to the milling industry. Compound feed manufacture makes use of the major by-product of flour production, wheatfeed.[2] It also enables millers to spread the cost of administration, transport and grain handling and storage facilities over a

[1] See Part III Chapter 3 where problems associated with the movement of grain are discussed.

[2] The production of flour uses about 70% of the volume of material present in wheat; the remaining 30% is sold as wheatfeed. The 70% is the *extraction rate*.

wider range of activities. Three out of the four large national flour producers are also large-scale manufacturers of compounds. Even so they are not always able to use all the wheatfeed they produce and must sell the surplus to specialist compounders and others. The market for wheatfeed is particularly volatile. During the past year prices have ranged from £21 10s. per ton to £26 0s. per ton. This price variability is a reflection of the need for millers to dispose of their wheatfeed. An accumulation can cause milling operations to come to a halt. If stored for too long it will go rancid. Rather than face a damaging interruption in production millers prefer to cut prices until the market absorbs the necessary volume of wheatfeed. The existence of compound mills adjacent to flour mills owes much to this need to dispose of the by-products of flour production.

2.17. It has been estimated that the four large national flour milling groups control more than half of the bakery production in the U.K. In particular they own most of the newer plant bakeries in which bread can be produced at lower cost than in small-scale bakeries. This competitive advantage means that many traditional small-scale family bakers have been forced out of business. The large national flour millers are often better placed to provide the capital needed for innovation in bread production than older-type bakers. As a result the process of vertical integration from flour milling into bread production has assumed great importance. It gives the miller an assured outlet for his flour despite rapidly changing market conditions.

2.18. As well as being involved in the bakery and animal feedingstuffs industries, the large national flour groups have also become important owners of agricultural merchant businesses. The pressures behind this development arise more from their interest in compound feeds than in flour and have been more fully discussed in Part III, Chapter 2 and Part V(a), Chapter 5. Here it should be noted that if greater use is made of home-grown wheat these firms are well placed to use their merchant businesses to ensure satisfactory supplies. One notable example of this already exists. The wheat varieties Maris Widgeon and Troll have been found by some millers to be superior in some respects to other home-grown wheat for the production of flour. One large national group has attempted to buy a greater proportion of these varieties through offering contracts to farmers via its merchants for samples which attain certain minimum protein levels.

2.19. Flour milling and bread production are relatively contracting sectors of economic activity. The market for animal feed is somewhat restricted, being closely related to the number of farm livestock and the technology of animal feeding. The major national flour millers are all involved in other economic activities where prospects for growth seem better. Some of these activities have direct relation to flour milling. The manufacture of pet foods or frozen foods and the operation of super-market chains all have some connection with the manufacture and distribution of bread, flour and other cereal products.

CHAPTER 3

Some Technical Constraints
on the Use of Home-grown Wheat

3.1. Wheat is very far from being a simple homogeneous product. Apart from variation in such physical characteristics as moisture, cleanliness and colour, it varies too in its chemical constituents. These variations have placed home-grown wheat in some respects in a disadvantageous position compared with imported grain.

3.2. Wheat is bought by millers at a maximum of 18 % moisture content. It is stored by them at 12–14 %. The danger of damage to wheat through careless or too hasty drying on farms was referred to by many of the millers who were interviewed. Bad drying can destroy the gluten content of wheat, making it useless for flour milling. On the other hand, a crop stored damp may begin to sprout, to develop mildew or to suffer insect infestation. For any of these reasons the crop may prove unacceptable for milling.

3.3. Cleanliness is very much in the hands of the farmer. A crop which is too little screened, as are many which are sold direct from the combine, may contain considerable amounts of weed, straw, animal residues or other forms of trash. Apart from the fact that such useless material has been bought at £20 to £25 per ton, the miller must undertake elaborate cleaning operations if his product is to meet required standards of hygiene. Farmers vary greatly in the care with which they clean their wheat. In a difficult harvest season the problems of attaining a satisfactory standard of cleanliness are often aggravated. If farmers are to be encouraged to produce clean, dry samples of grain suitable for milling, the difference in price for acceptable wheat should be sufficient to offset the extra costs involved. There are grounds for believing this is not the case.[1] The wheat deficiency payment scheme affords support for wheat which is of "millable"[2]

---

[1] It should be noted that the availability of small quantities of specially cleaned wheat will not enable the miller to use it separately in his grist. Where it has to be mixed with a quantity of less clean wheats he cannot justify payment of a premium.

[2] See definition in Chapter 4.

quality. Merchants who refuse this designation deprive the farmer of the benefit of subsidy. They may well be reluctant to impose so severe a penalty on their customers, whose business is of continuing value. A more stringent definition of millable wheat might make it possible to reward farmers who do produce good samples. At the same time it would be necessary to ensure that wheat of feeding quality was not entirely excluded from subsidy.

3.4. Greater care in production and market presentation might remove many of the physical disabilities of home-grown wheat; its chemical limitations are more intractable. Flour for bread-making must attain certain characteristics of "strength" if it is to produce an acceptable load. Weaker flours result in bread of poor conformation and keeping quality. Further, in the baking process they are unable to absorb as much water, and this diminishes the output of bread per sack.[1] Stronger flours rely on imported hard wheats. These may be supplemented by "filler" wheats or soft wheats, the extent to which substitution can take place being limited by the basic need for a stronger flour. Strength is related to the protein level and the protein quality of the wheat.[2] This must reach $11-11\frac{1}{2}\%$ for satisfactory bread flours. English wheats often have protein levels as low as 8%. They sometimes reach more than 11% but this varies with weather, husbandry practices and variety. The uncertainty of quality between different years and different producers is in itself a major deterrent to the use of home-grown grain for bread flours.

3.5. Two approaches are currently being pursued to produce in this country wheat which is more satisfactory to the millers. First, new varieties with higher protein levels are being developed by plant breeders. Second, studies have been made which suggest that the application of nitrogenous fertiliser late in the growing season may improve protein content Each approach has its problems. The variety which has received most publicity for the purpose of bread-making is Maris Widgeon. It illustrates many of the problems in improving wheat quality. An analysis of sixty-three samples of commercially produced wheat and twenty samples produced in N.I.A.B. trials showed protein levels varying from 7% to more than 11%. Less than half the flours had more than 10% protein, and a satisfactory bread-making flour could be made from only twenty-three of the eighty-three samples.

3.6. Any attempt to use 100% home-grown wheat in bread flours is out of the question. Without considerable modification of consumer taste

---

[1] See Part I, Chapter 1, p. 8.

[2] Millers use the terms "extensibility, resiliency, viscosity and stability" to characterise protein quality.

patterns the loaves so produced would prove unacceptable. A more realistic possibility is to increase the proportion of home-produced wheat used in bread from its present 20%. Here relatively small increases in protein level are of great significance. Greer[1] states that at 8% protein the maximum use of home-grown wheat is 20%; this rises 33% for 10% protein and 50% for 11% protein. The importance of such shifts may be demonstrated when it is realised that an improvement in the use of home-grown wheats in bread flour from 20% to 30% would yield a market for an additional 360,000 tons of domestic wheat.

3.7. The possibility of achieving higher protein levels through the late application of nitrogenous fertiliser is influenced by climatic and technical factors. The relevant period during which a worthwhile effect can be attained is short. If the weather is wet it may be impossible to spread the fertiliser. Even in favourable conditions the process of spreading may well damage a growing crop resulting in lower yields which are incompletely offset by better-quality standards. Both fertiliser manufacturers and producers of agricultural machinery are familiar with this problem and some progress may be expected in the longer term.

3.8. A more immediate contribution to the use of home-grown wheat in bread-flour production stems from changing techniques in the bread making industry. The Chorleywood bread process developed by the British Baking Industries Research Association has substituted mechanical working for conventional slow fermentation as a means of making dough.

3.9. Rather more than half of the bread baked in this country is produced using the Chorleywood process. Bakers who employ this method use grists containing about 30% soft wheats compared with the 25% soft wheat content of grists for flour which is to be used for the conventional fermentation process. Some of the millers interviewed stated that the Chorleywood process could in fact use successfully grists containing as much as 35% soft wheat. Thus the use of soft wheat might be increased in future both by more bakers turning to the newer method and by an increase in the proportion of soft wheats used to make bread flour for this purpose. Some bakers may continue to use conventional fermentation processes because they are short of capital or because they believe the resulting loaves more nearly match the tastes of their customers. However, it seems reasonable to anticipate that at least 75% of the bread flours made will in the relatively near future be designed for the Chorleywood method.

3.10. The current utilisation of soft wheats is about 1 million tons.[2]

[1] E. M. Greer, The wheat variety Maris Widgeon, *Ceres,* Vol. 1, 1968,H.-G.C.A., p. 14.

[2] Based on the assumption that 50% of the bread is baked by the Chorleywood process and the grists for this contain 30% soft wheat compared with 25% for the remainder of the bread baked after conventional fermentation of the dough.

Q*

If we assume that the proportion of bread baked by the Chorleywood process might rise to 75% and that the grist used contained 35% soft wheat an extra demand for 200,000 tons of soft wheat would arise.

3.11. The report of the Prices and Incomes Board[1] showed that about 20% of grain used in flour milling was of domestic origin. This accounts for some 700,000 tons of the one million tons of soft wheat used by the industry. If full advantage were to be taken of the Chorleywood process, on the assumptions employed in the previous paragraph total soft wheat usage would be about 1,200,000 tons. Thus there exists a potential extra market of 500,000 tons for the type of grain which can be grown in this country.

3.12. The exploitation of this potential depends on the availability of domestic wheats at competitive prices. It is clear from the fact that substantial amounts of soft wheat have been imported in recent years that these conditions have not been met. If price can be kept competitive and deliveries ensured, the potential expansion of the market for home-grown wheat in flour milling should be readily achieved.

3.13. Bread flours represent about 70% of the output of the milling industry. In addition flour is made for biscuits, for household purposes and for a variety of special needs such as the manufacture of pie casings.

3.14. The technical requirements for biscuit flours differ from those of bread. A weaker, more extensible flour is needed. This permits 90% of the grain used for the biscuit flour to be of domestic origin.[2] Demand for cakes and biscuits, especially for the more expensive types, seems likely to grow slightly in the U.K. Further expansion may flow from increased exports of biscuits, which in 1967 were worth £8·0 million. Further liberalisation of trade with Europe or membership of the European Economic Community could well extend this market, but the significance of such possibilities should not be exaggerated. An expansion of 10% in biscuit production would add only 40,000 tons to the demand for home-grown wheat.

3.15. About half the wheat used in the production of household flour is grown in Britain, though the proportion varies considerably from year to year according to the quality of the home-grown wheat available. This is a declining market and there seems little prospect that the present utilisation of about 350,000 tons of home-grown wheat will grow in the future.

3.16. The consumer in the U.K. has become accustomed to a loaf which will keep for several days, which may be bought sliced and wrapped and which is white in colour. A market exists for other types of bread but the large white loaf is still the most common form of bread bought.[3] In

---

[1] National Board for Prices and Incomes, Report 53, *Flour Prices*, Cmnd. 3522.
[2] Greer, *op. cit.*
[3] See Table 1.2.

continental European countries softer wheats are more widely used in bread production and the baking industry is made up of smaller-scale units. Although bread of the continental type is attractive while it is fresh, it stales relatively quickly into a hard, unpalatable lump. These characteristics make it unsuitable for large-scale manufacture and distribution, especially for sale through supermarkets, where shelf life is of great significance. If housewives were prepared to shop every day for such bread, it might be possible to use a softer type of wheat bread flour. The loaves produced would be of a different conformation and colour than those currently preferred. Such a development would facilitate the wider use of domestic wheat in flour but it would do so at a cost. Extra cost would arise in distribution—not least in the need for more frequent expeditions to buy bread. It would also arise in the form of lost satisfaction among consumers who would find the bread they bought less suitable for sandwich making, less palatable after it had been in the larder for a few hours and possibly less attractive in its shape.

3.17. In some European countries millers are required by law to use a certain proportion of domestic grain in their grist. It has sometimes been suggested that similar requirements might be imposed on British flour millers. Such a measure seems out of place in the circumstances of the U.K. Its major function, to protect domestic wheat producers, is already adequately discharged by the system of cereal deficiency payments. It threatens to impose a rigidity upon the flour and baking industries which may frustrate developments both in the technology of those industries and in the distribution of bread. Finally it would have the effect of imposing an inferior product on consumers. One important side-effect might well be to reduce the total amount of bread sold and thus limit the aggregate market for wheat.

3.18. This examination of some technical aspects of the use of wheat for flour milling suggests that there could well be a worth-while increase in the uptake of home-grown grain by the flour-milling industry. Provided the price relationship between home and imported grain is satisfactory some 500,000 tons might be absorbed into bread flour and a much smaller increase, say 20,000 tons, into biscuit flours. The realisation of this potential is to an important degree outside the control of the millers; it hinges on the production and marketing decisions of farmers and of bakers. In the longer term, approved types of wheat may enable domestic grain to fill an even larger fraction of the grist for bread flours. Again flour millers can do relatively little directly to bring about such changes. The introduction and establishment of better varieties depend very much on the work of the plant breeder and the farmer. The most that the flour miller can be expected to do is to offer a price incentive for grain which more completely meets his needs.

3.19. In the manufacture of biscuit and speciality flours the independent country millers are at least disadvantage, being able to use local supplies for their product. For other flours the technology of production seems to favour the larger firms. The basic physical process of milling has changed little since the introduction of roller mills about 100 years ago. Changes have taken place in the speed at which the job is done, the length of roller contact through which wheat must pass and the techniques of handling grain and flour. Increased speed of production reduces the number of plants required to attain a given level of production. Pneumatic handling of flour and grain reduces the labour force needed for purely manual operations. Bulk handling further cuts the labour cost while the introduction of multi-wall paper bags eliminates the problem of dealing with returned bags. Each of these innovations demands capital. Most of them operate at lower cost where they can be applied to larger volumes of materials. Many small businesses cannot afford new equipment, cannot utilise it on a large enough scale to attain low costs and cannot make any real economies through reducing their already small labour force.

3.20. This discussion of some technical aspects of flour production emphasises the problems of making use of home-grown wheat. It would be wrong to conclude the section without reference to the efforts which many manufacturers are making to decrease their dependence upon imports. Their success in this direction varies. Comparison is not simple because of the variety of flour products made by different manufacturers and the greater ease with which some can shift unwanted wheat into animal feed production. Each year the millers give an undertaking to the Government to attempt to use a certain amount of home-grown grain. This amount has risen from $1\frac{1}{4}$ million tons in 1960 to $1\frac{1}{2}$ million tons in 1968.

3.21. To increase the proportion of home-grown grain does not necessarily mean a corresponding saving in foreign exchange. The economic complexities of import substitution fall outside the scope of this chapter but one aspect is of immediate relevance. If for a given quality of bread flour a higher proportion of soft home-grown grain is used in the grist, there must also be some increased use of harder (imported) varieties, and less of the intermediate (filler) wheats. These hard wheats come mainly from North America, while filler wheats are supplied principally by Australia and Argentina. The relative merits of saving imports of hard wheats or filler wheats must include an assessment of our trading and currency relationships with the countries concerned.

CHAPTER 4

The Marketing of Millable Wheat [1]

The Buying of Wheat

4.1. Home-grown wheat destined for flour production is normally marketed through agricultural merchants. Most flour millers buy some part of their requirement direct from farmers but this represents only a small part of the wheat sold for flour production. The smaller independent millers in the Survey bought a rather higher proportion direct from farms, and in some cases this was accentuated because the miller was also a merchant-compounder. Even so none of the millers interviewed bought all their home-grown grain directly from farmers. Local variations in the amount of wheat available and the time at which it moved from farms meant that at least part of the grain was bought through the merchant sector.

4.2. Most larger national groups leave considerable discretion concerning purchases of home-grown grain in the hands of their mill managements. Using information they have collected about import prices, grain stocks and expected future demand, head offices lay down general guide-lines for mill managements. However, within this broad framework local managements carry out their own negotiations with farmers and merchants. The dominant position of the national flour millers means that their commercial practices have a major influence on the market for wheat. If one or other of the national groups concludes that home-grown prices are too high, either in relation to import prices or because of the level of production of home-grown wheat, it may cease to buy for a short period. This may

---

[1] Millable wheat has been defined as follows: "For the purposes of the Cereals (Guarantee Payments) Order 1964, as amended by the Cereals (Guarantee Payments) (Amendment) Order 1966 'Millable wheat' means wheat which is sweet and in fair merchantable condition, reasonably free from sprouted or smutty grains, commercially clean as regards admixture and tailing and commercially free from heated or mouldy grains or objectionable taint, and capable of being manufactured into a sound sweet flour fit for human consumption having regard to the customary methods employed in the milling industry for cleaning and conditioning wheat; and which has a moisture content not exceeding 18%."

exert a perceptibly depressing effect on wheat prices at home. Equally any attempt to buy an unusually large amount of home-grown grain may force its price up. One advantage of owning merchant businesses is that it makes it possible to buy home-grown grain more circumspectly without "upsetting" the market.

4.3. The extent of this bargaining power is limited first by the alternative use for wheat in compound feed production and second by the international market. If the wheat price falls appreciably, animal-feed manufacturers may find it an increasingly attractive substitute for barley or maize. The effect could be to leave flour millers short of home-grown grain. Import prices cannot be ignored. A miller who continued to pay higher prices for home-grown wheat in a situation in which imports had become cheaper would give his rivals a competitive advantage. Over short periods, especially allowing for the fact that only one-fifth to one-third of the grain used is home-grown, such price disadvantages may be tolerated. If they were to persist on a longer-term basis they would seriously damage the profitability of the user of home-grown grain

4.4. The technical requirements for wheat used for flour milling are more exacting than for grain used in compound manufacture. Millers commonly rely on merchants to select suitable wheat. On arrival at the mill it is subjected to a series of tests. These vary from mill to mill but usually include a test for moisture, protein content, and a trial milling on a laboratory scale. The larger mills, including the large independent mills, usually carry out a rather more elaborate pattern of testing than smaller producers.

4.5. Millers were asked if they would favour a grading system for home-grown wheat. In spite of many complaints about bad grain presentation by farmers only one firm was in favour of a grading system; 8% thought that while grading was an attractive idea it would prove unworkable or prohibitively expensive. The majority of those interviewed took the view that merchants already did a satisfactory job selecting grain. No practicable grading scheme would eliminate the necessity to test samples for milling. The marginal benefits from grading would in this situation be insufficient to offset the costs involved.[1]

4.6. The success with which flour millers buy grain is closely related to their knowledge of both the grain market and the market for flour. The larger national companies have elaborate communication systems. These provide up-to-the-minute data on international grain prices, shipping rates and supplies and prices of home-grown cereals. Traditional methods

---

[1] The same reasons must evidently have been operative for a long period. It is noteworthy that the detailed recommendations of the 1928 Report for a grading system were not acted upon by the trade. (See Part I, Chapter 1 of the present Report.)

still have a role. The large corn exchanges provide for the personal interchange of information and assist buyers and sellers to form a view of market trends. Published data about production and stocks are supplemented by information from wholly-owned merchant subsidiaries. Longer-term developments are explored by market research staff or hired consultants.

4.7. This elaborate array of information services is not possessed by the smaller millers. For them published data, informal contacts with merchants, brokers and local knowledge are more important, and will often give them an equally effective basis for trading. Frequently, they follow the lead established by national companies, although some of the larger independent firms take their own view of the market, especially in relation to home-grown grain.

4.8. There can be little doubt that the ability of the major national groups to absorb and interpret a vast amount of commercial data gives them an important competitive advantage. So far as they are concerned information made available by public bodies such as the H.-G.C.A. arrives late and they sometimes know that it does not give a completely accurate picture. Less well-equipped members of the industry, as well as merchants and farmers, may derive greater benefit from such impartial sources of public information. But no matter how extensive such services become, they cannot overcome the advantage of the large firm. This arises not only because such firms have sound data about the market but also because they alone know about their own intentions. The strong position of the large national groups in the home market means that the actions of any one of them may have an effect on the ruling level of price. Real advantage comes not so much from knowing what the price is but rather from having good reasons to judge what it will be in the next few weeks. No administrative intervention which leaves a free market in existence can place the small producer on a par with large national groups in this respect.

4.9. As mentioned earlier, the price at which flour is sold has become a political as well as a commercial issue. In 1964 and 1967 the flour millers' intention to raise the price of flour was referred to the Prices and Incomes Board. After due consideration of changes in costs, approval was given for some increase in price. From the individual miller's point of view a collective action of this kind diminishes the likelihood that an increase in his flour price will result in lost sales to competitors. Its implications for the economy as a whole are more complex. On the one hand, public scrutiny of pricing policies seems well justified where an industry is in a potentially monopolistic condition. Flour milling still retains some element of competition but open or tacit agreements if they were to exist among the four largest firms could severely attenuate this. On the other hand, cost-plus pricing arrangements are likely to make the structure of the industry more rigid. There is also a danger that the industry may become less and less in touch with changes in consumer demand. The effect of price competition is to make the businessman careful to match his product to consumer needs. This, in an increasingly affluent society, may mean selling more expensive and sophisticated goods rather than simply cutting prices on existing lines. If a ceiling price were to be fixed this type of market development might be frustrated, involving loss not only to the miller but also to the community.

Seasonality and Quality of Wheat Supplies

4.10. The production of flour shows very little seasonal variation either in total quantity or in its use of home-grown wheat (see Table 4.1). Millers' stocks are also fairly stable through the year at 2–3 months' requirements. This means that the home-grown crop, which becomes available around September, is delivered fairly evenly over a 12-month period. The crop is stored mainly on farms;[1] the marketing may take place

TABLE 4.1

MILLERS' PURCHASES AND STOCKS OF WHEAT, AND PRODUCTION
OF FLOUR, U.K., 1967 ('000 tons)

	Wheat bought for milling	% home-grown	Stocks	Flour produced
January	475	32·0	1150	346
February	365	32·6	1177	267
March	358	31·6	1190	262
April	515	32·2	1165	377
May	384	31·5	1159	281
June	396	31·1	1146	291
July	449	28·1	1055	330
August	357	27·2	1090	262
September	408	32·1	1110	300
October	489	34·8	1060	361
November	401	34·4	984	294
December	371	32·6	957	273
Total	4968	31·9	——	3644

Stocks = wheat, and flour in terms of wheat equivalent, at end of month.
Note: Flour produced = 73·3% of wheat bought.
Source: Monthly Digest of Statistics.

[1] See Part II, Chapter 5, p. 116.

on a spot or forward basis. The regular demand for wheat by millers makes forward purchases attractive. By this means the need to commit capital to storage is reduced, yet supplies remain secure. In the Survey six out of twelve independent millers said that they had been able to reduce storage as a result of the Home-Grown Cereals Authority forward contract bonus scheme. Similar points were made by two of the large national milling groups. The extent to which forward transactions can be carried out depends on the reliability of farmers in making and keeping contracts. The H.-G.C.A. scheme has provided an incentive both to make and to keep contracts but the forward contracting method does not depend upon this scheme. Before the scheme operated some of the larger millers already bought on contract, despite the fact that contracts were not always honoured by farmers. The scheme has improved this aspect of marketing, but of the millers interviewed many believed it to be a contributory cause of a new problem. Wheat sales from farms were relatively small in the late autumn, while too much was contracted forward for April–June.

4.11. It was argued that the official statistics (see Table 4.1) did not fully reflect the autumn shortage of wheat for flour milling. Millers have a fairly good idea of the amount of wheat produced at home and if this is not offered at prices which seem reasonable in the autumn they turn to cheaper imported supplies. One large firm reported that it had done this in the autumn of 1967. In addition to the forward contract scheme, which provides a mechanism whereby farmers can assure themselves of an outlet later in the harvest year, the Cereal Deficiency Payments Scheme also rewards them for storing. For reasons discussed more fully later (Part VIII, Chapter 4), once the grain is in store farmers may be reluctant to sell until the end of the harvest year when maximum advantage accrues from storage incentives and bonuses on forward contracts. Moreover, for farmers who grow sugar-beet the late autumn period may be a bad time to move grain; labour is then fully stretched in harvesting the sugar-beet crop. It was suggested by four of the large millers that a monthly bonus period would improve the flow of grain. The removal of the system of storage incentives under the C.D.P. scheme would enable the rate of delivery from farms to be more responsive to market requirements.

4.12. The varying characteristics of home-grown wheat and the relatively stringent needs of millers provide a basis upon which price differentials for quality might be established. Reference has already been made to the incentive scheme operated by one large company for the production of Maris Widgeon and Troll. This provides a bonus over market price of 30s. per ton for wheat with a protein level of 10% or more. Other millers, too, will pay more for quality but the extent of these differentials is usually too small and their incidence too uncertain to encourage farmers to concentrate on wheat quality—especially if this means a reduction in quantity. An intractable aspect of this problem arises in relation to the development of improved wheat varieties. Small quantities of a new wheat, which is itself of good quality, do not enable the miller to adjust his milling process to exploit their superiority, He is thus not justified in offering a higher price. On the other hand, farmers are unlikely to grow the new wheat, even if it is superior for milling purposes, unless its yields compare well with traditional trusted varieties. To make a shift he requires a noticeable price incentive. Even where new varieties are preferable, their extra value to the miller seldom justifies more than a marginal increase in price. Thus improvement of the home-grown wheat crop as a basis for flour production may tend to lag. It might be sensible, within the framework of the subsidy scheme, to offer some extra reward for promising new varieties in order to encourage their exploitation.

CHAPTER 5

The Profitability of Flour Milling

The Price of Wheat

5.1. The enquiry by the Prices and Incomes Board into the flour milling and baking industries in 1964 included an analysis of the cost structure of each activity. The information published in their report[1] forms the basis of the following discussion. Since 1964 the levels of wheat prices, labour costs and other charges have risen but there seems no reason to believe that the underlying pattern of costs has changed.

5.2. Two prices series dominate the economics of flour milling—those for wheat and for wheatfeed. Purchases of wheat and other raw materials account for almost 90% of the selling price of flour. On the other hand, rather more than 20% is recovered by sales of wheatfeed.

5.3. Although some 80% of the wheat used in bread flour is imported, the overall dependence of the flour industry on imports is much smaller— about 66%. Thus although the major element determining the cost of flour is the price of imported wheat, home-grown grain does play a significant role. This is especially true for those millers whose major concern is biscuit as distinct from baking or bread flour.

5.4. The price level of imported wheat is largely outside the influence of even the larger national flour millers. Apart from changes in world demand and supply conditions such as arose when harvests were poor in the U.S.S.R. and mainland China, the activities of governments, or of quasi-governmental authorities such as the Canadian Wheat Board, are of major importance. Since November 1967 the devaluation of the pound sterling has added to the cost of imported wheat.[2] However, although the overall price level must be accepted, importing firms can make small but significant reductions in cost through wise buying policies. To do so they

---

[1] National Board for Prices and Incomes, Report No. 3, *Prices of Bread and Flour,* Aug. 1965, Cmnd. 2760.

[2] See National Board for Prices and Incomes, Report No. 53, *Flour Prices,* Jan. 1968, Cmnd. 3522.

467

must assemble a wide range of data concerning day-to-day prices in various international markets, changing shipping rates and the technical character-istics of wheat from a variety of sources. Table 3.1, Part IV sets out the main countries of origin of U.K. wheat imports in 1967. There seems no reason to believe that any important cut in the costs of imported wheat could be achieved by modification of the pattern of imports or of the structure of the industry. Indeed the concentration of the industry into a few, relatively large, firms enables each of them to set the heavy overhead costs of comprehensive information services against an adequate volume of trade. A less concentrated industry might find it difficult to sustain the range of information services necessary to successful buying in world markets.

5.5. Reference has already been made to the limited influence of even large flour millers on the prices paid for domestic wheat. A decision to withhold purchases does exercise some downward pressure on market prices but this is likely to be of a short-run nature. The most important contri-bution which home-grown grain can make to reduced milling costs is through an improvement in its quality. Wheat of higher protein content could permit cheaper home-grown supplies to be substituted for imports. Since devaluation this possibility is of added significance. The better presentation of wheat in terms of the consistency of samples, the moisture content of the crop and the cleanliness of grain would increase the fraction of the domestic crop which is suitable for milling. To some extent quality

TABLE 5.1

AVERAGE MONTHLY WHEATFEED
(MIDDLINGS) PRICES PER TON
AUGUST 1966–JULY 1968

	1966/7	1967/8
	£ s.	£ s.
August	20 0	23 5
September	19 6	23 16
October	19 10	24 9
November	21 11	25 9
December	23 10	25 11
January	25 4	25 10
February	25 10	25 14
March	25 10	25 14
April	25 16	25 12
May	23 6	25 12
June	24 4	23 5
July	23 2	22 0

Source: Farmer and Stockbreeder. London Mer-chants' prices ex-store. Subject to additions for transport, handling, store and credit charges.

differentials in prices paid by millers may foster progress in this direction but a great deal depends on appropriate action by farmers.

5.6. There are a number of grades of wheatfeed of differing nutritive value and an even larger number of names used to describe these by-product types. All of them are used for animal feed. The price at which wheatfeed is sold varies considerably over the year, rising to a peak in February–March (Table 5.1). Instability arises from changing prices of other animal feedingstuffs, changes in availability of other types of feed and changes in the number of farm livestock. The close relationship between wheatfeed and compound feed production has encouraged many millers to become involved in the production of feedingstuffs. To some extent this provides a secure outlet for wheatfeed. Even so most millers produce more wheatfeed than they can use in compound production. The prices they receive for the surplus can have a noticeable effect on the costs of making flour.

Return on Capital

5.7. Mills vary greatly in their level of capital costs. This is demonstrated by the fact that although, on average, independent millers have a lower margin on turnover than national companies, their return on capital is rather higher. The broad impression of relatively modern and elaborately equipped mills owned by large national groups and rather older, more traditional mills belonging to independent companies is an oversimplification. Some independent millers have very modern and efficient equipment. However, the shortage of capital resources and depressing market prospects of many independent firms explain a certain reluctance to embark on new investment.

5.8. The existing capital commitment of the port millers may in itself be a source of inflexibility. Even where premises are old the machinery is often relatively modern and represents an asset which is difficult to move. New port developments such as those at Tilbury or the projected new terminal at Liverpool may attract less milling capacity than their operational advantages suggest. In the short run at least, firms which are already heavily committed in established sites will find it difficult to write off existing assets despite the lower landed cost of grain at the new terminals. (See also Part IV, Chapter 3, p. 356.)

5.9. The average profit margin in flour milling reported by the Prices and Incomes Board in 1965 was 10·4% on sales. The independent companies enjoyed rather smaller profits on sales, but their profit per unit of capital employed was said to be substantially higher than the $17\frac{1}{2}\%$ per annum of the large national groups. Two aspects merit attention. First, in the light of their falling share of the market it seems paradoxical that the independent companies should earn higher profits than the national groups. Second, the average level of profits seems high in an industry afflicted by surplus capacity.

5.10. The apparent prosperity of the independent companies is deceptive. A high return on capital arises as much from the writing down of the value of old equipment as from the income received from flour sales. If this equipment had to be replaced at current costs the level of the profits would be much less reassuring. In addition the relevant figure for assessing the long-term prospects of many independent companies, as of other small family businesses, may be the total income accruing to the family rather than return on assets employed. On this basis the relatively low level of sales and the expectation that the existing level may be even more difficult to sustain in future may explain the decision of many millers to leave the industry during the past 20 years.

5.11. The average return on capital among the national groups is higher than that ruling in many other sectors of economic activity. Satisfactory processes of economic adjustment appear incompatible with this stage of affairs. Orthodox economic analysis suggests that, confronted by surplus capacity, profits should fall until enough resources are driven from the industry to bring about a balance between its capacity and the demand for its services. The resources realised should, in suitable economic conditions, find alternative uses in which their value to the community is greater. However, monopolistic practices can frustrate this development. Agreed pricing policies or market-sharing arrangements can prevent competition forcing out the least efficient firms and can sustain the average level of profits for the industry as a whole.

5.12. The concentrated nature of flour milling might facilitate agreements among the firms concerned to avoid a mutually uncomfortable price war. Such agreements are illegal under the Restrictive Trade Practices Act of 1956, but it is clear that even in the absence of any conscious attempt to restrain competition each of the firms engaged in flour milling is likely to be strongly influenced by the behaviour of the large national groups. The pattern of costs, in which wheat prices are of such great importance, means that when one firm faces rising costs and wishes to preserve its margins by raising prices, others are in a similar situation and are likely to take similar action. Thus an effective system of price leadership is established. This affords considerable protection to weaker firms so long as their relative cost structures, compared with the rest of the industry, remain unchanged.

5.13. The extent to which arrangements of this type could increase flour prices is, of course, limited by the fact that flour can be imported duty-free from Commonwealth countries.

5.14. To some extent the separation of flour milling from baking presents an unrealistic picture of the profitability of the industry. The profit on capital of the major groups engaged in bread baking was only $6\frac{1}{2}\%$ in 1964. For national groups this profit level must be read in conjunction with that from flour milling. Without a share in the baking industry it would be difficult, or even impossible, to maintain flour sales. If the volume of flour sold were to fall then the profitability of the group would be substantially reduced. Viewed as a whole the level of profits from flour milling and bread baking does not seem excessive. The fact that historical accident results in statistics which show high profits in one part and low profits in another part of what, in economic reality, is a single industrial complex is not a good reason for legislative intervention.

5.15. Undoubtedly, the structure of the flour-milling industry has given grounds for public comment from time to time, but there is little evidence to indicate that the national groups have abused their position. Indeed, the practice, which existed before the Prices and Incomes Board, of giving early warnings to the Government of anticipated price increases, demonstrates the responsibility with which the national flour millers have regarded their position in the market. The existing powers of government, via the prices and incomes legislation, provide ample machinery to ensure that the public interest is safeguarded. If these powers are used wisely there seems no reason for any major intervention in the flour-milling industry.

PART V

Utilisation of Grain in the U.K.

(c) Malting

CHAPTER 1

Malting by
Brewer-maltsters and Sale Maltsters

1.1. Malting is the process in which grain (usually barley, though wheat is very occasionally used) is germinated under carefully controlled conditions for a specific length of time, then dried. The resultant product is malt. It resembles the original grain in appearance but is very different in chemical properties. The principal uses of malt are in brewing and distilling; much smaller amounts go for use in human food and in vinegar production, and some is exported.

1.2. The malting industry in the United Kingdom is divided into three distinct sectors: sale maltsters, i.e. firms making malt for sale on the open market; and brewer- and distiller-maltsters, forming part of breweries and distilleries respectively, who make malt for their own use. Of these three groups, the distillers have many special characteristics, both economically and geographically, and they are dealt with separately in the following chapter. Whenever the term "maltster" is used in the present chapter, it refers to a sale maltster or brewer-maltster.

1.3. Around 1·1 million tons of barley were sold for malting (not including direct uptake by distillers) from the 1966/7 harvest. Malting thus came second only to stockfeed as a sales outlet for that year's crop, and accounted for about one-eighth of total barley production. In addition, about 3000 tons of wheat were taken up by maltsters in the same year.

The Survey of Sale Maltsters and Brewer-maltsters

1.4. As part of the programme of the Cereals Survey, a study of sale maltsters and brewer-maltsters was carried out in the winter of 1967/8. A sample of firms were visited and a set questionnaire was completed at each interview; the findings given in this chapter are based mainly on the summarised results. In addition, a brief postal questionnaire containing questions on throughput and location was sent to all the firms who had not been interviewed.

475

1.5. The interview sample was chosen from trade lists supplied by The Maltsters' Association of Great Britain and The Brewers' Society. It was known in advance that there were a few very large concerns which together accounted for a large proportion of the malt production in the U.K.; certain of these were excluded from the lists and visited separately, to make sure that their attitudes and practices were taken account of. Next, to avoid double-counting, firms known to be subsidiaries of these large concerns were struck out of the lists (and the relevant details were gathered at the head office of each group). The remainder of the interview sample was chosen by taking one in two of the remaining addresses in the list. The trade bodies had allocated the firms in their lists to approximate size-groups, according to the amount of malt made; where a firm in the original sample did not complete a questionnaire (e.g. because it was no longer malting) it was replaced by randomly selecting a firm of the same type (sale or brewer-maltster) and size-group.

1.6. A total of forty-one interviews were carried out with firms selected in this way. None of the sale maltsters interviewed was a subsidiary of another sale maltster; a few maltsters of both types were subsidiaries of brewers, but on no occasion was there double-counting between one malting enterprise and its subsidiary. Thus the interview sample covered forty-one separate malting enterprises; the firms interviewed accounted for about two-thirds of the total production of malt for brewing and sale in the U.K. in 1966/7. The postal questionnaire was sent to all other brewers and sale maltsters who were shown on the trade associations' lists as currently making malt. In total, twenty-five firms sent back positive returns to the postal enquiry. As in the interview survey, the postal returns were vetted to exclude the possibility of double-counting between enterprises. In addition there were five firms (one in the interview and four in the postal sample) who declined to give specific information but confirmed that they were making malt. All other firms on the trade lists were found to have ceased malting, or to have amalgamated with another firm since the trade lists were issued (in which case information for both firms was gathered on one form for the merged enterprise). Thus, at the time of survey, there were seventy-one separate enterprises making malt in the U.K. and sixty-six of these were covered by the interview and postal enquiries. Though the malt production figures for the five non-responding businesses are, of course, not known exactly, it is possible to estimate them roughly by taking the average output shown by Survey results for the type and size-group into which these businesses were placed in the trade lists; on this estimate, the businesses which did respond to the postal and interview enquiries accounted for some 95% of all malt production by brewers and sale maltsters in the U.K. in the year under review.

Structure of the Industry

1.7. The word "structure" implies something static and unchanging, and to this extent it is a misleading term to use in the context of the malting industry. The structure of the industry is, in fact, continually changing; the overall movement is towards greater concentration in larger and larger enterprises, with small maltsters either amalgamating with other concerns or going out of the industry. There are, unfortunately, no figures which would show the precise extent of these changes over a run of years. Trade lists must inevitably be out of date even by the time they are published, and, more seriously, they do not always show whether one firm is a subsidiary of another. However, a very rough idea of the speed of change, in the brewer-maltsters' sector at least, can be gained from their trade list: the list used in selecting the sample related to the year 1965/6, and showed eighty-one separate brewing businesses as making their own malt. By the time the survey was carried out—some 2 years later—there were only forty-three such businesses. The fall in numbers occurred both because some breweries were taken over by others, and because certain breweries ceased making their own malt. Very often, both these changes go together: when two breweries amalgamate, it is common for one of them to cease malting and draw its requirements from a single malting serving both breweries. No comparable figures are available for the sale malting side of the industry, but it is probable that the changes in it—though still in the direction of more concentration and larger units—have been less impressive in scale than those in the brewing sector; whereas sale malting firms can gain any advantages of amalgamation in the field of malt production only, breweries amalgamate for many additional reasons of productive and marketing advantage. Probably the most important stimulus towards concentration in sale malting in recent years has been the emergence of over-capacity following the cut-back in demand for malt by distillers (see next chapter). Within individual firms, both in brewing and in sale malting, the production of malt is being further rationalised by the closing down of some maltings—especially the older ones—and the concentration of production in larger and more modern plant. Again, no precise figures exist to show the extent of this change over the past few years. Because such rapid changes are taking place, any description of the structure of the malting industry can only be a snapshot of the situation at one moment of time—in the case of the description which follows, the winter of 1967/8.

1.8. Malting is commonly regarded as an industry in which the bulk of production is concentrated into a few hands, with a large number of smaller firms sharing the remainder of the market. This picture was confirmed by the Survey, though the degree of concentration was not as great as had been suggested by various sources; it had been asserted, for example, that one single firm accounted for over half of the country's total production. In fact, this concern made just on a quarter of the total malt produced by sale maltsters and brewer-maltsters in the production year covered by the Survey.[1] The top four concerns together, all of them large sale maltsters, accounted for one-half of this production total. Table 1.1 shows the size distribution of the sixty-six malting firms responding to the Survey (in compiling the table, firms were ranked in ascending order of size, then

TABLE 1.1

SIZE DISTRIBUTION (BY TONNAGE OF MALT PRODUCED) OF SIXTY-SIX MALTING FIRMS SURVEYED, WINTER 1967/8

	No. of firms			Average production per firm (tons)	Percentage of total production [a]
	Brewer-maltsters	Sale maltsters	All maltsters		
Groups of eleven firms in ascending order of size:					
First group	8	3	11	325	—[b]
Second group	9	2	11	900	1
Third group	7	4	11	1719	3
Fourth group	6	5	11	3119	5
Fifth group	4	7	11	8549	12
Sixth group	3	8	11	53,776	79
Total sample	37	29	66	11,398	100

[a] Total refers to production by U.K. brewer-maltsters and sale maltsters; it excludes malt production by distillers.   [b] Less than 0·5%.

grouped for convenience in size-groups of eleven firms each). The table gives a striking illustration of how production was concentrated among the large concerns, with a long string of small firms accounting for relatively little of the malt made. Among these smaller firms, brewer-maltsters were more numerous than sale maltsters, while the opposite was true of the larger firms.

1.9. So far in this section, the size structure of the industry has been described in terms of malt production. Since production of malt is in a more or less constant ratio to barley bought, the same structural picture applies to grain buying in the industry. Usually, even among large malting

[1] If the production of malt by distillers, as well as that by sale maltsters and brewer-maltsters, is taken into account, the proportion of the total accounted for by this concern is about one-fifth.

firms, buying is the direct responsibility of one man, with or without the help of assistants. A few of the very largest groups split buying up geographically, but nevertheless pursue a closely co-ordinated policy on buying and price fixing. Therefore, each separate enterprise covered in the Survey corresponds to one independent buying unit, and the sizes of these units vary in the way described above. The possible effects of this purchasing structure on the market for malting grain are discussed later in this chapter. Meantime it is only necessary to make a point with regard to description; because of the widely varying sizes of malting businesses, it would often be misleading to quote "numbers of firms" against sample results without regard to their size, and this has usually been avoided in the description which follows.

Location of Maltings

1.10. Map 1.1 shows the location of maltings in the U.K. at the time of survey (winter 1967/8). It also indicates the approximate size of these maltings and whether they belonged to sale maltsters or brewer-maltsters. A very few firms either did not respond to the survey or declined to allow the locations of their maltings to be published. Of the seventy-one separate enterprises making malt in the U.K. at the time of survey, the map shows maltings belonging to only sixty-four. None of the excluded firms were among the largest in size, and the general picture given in the map remains accurate.

1.11. Maltsters taking part in the interview sample were asked what had led to their maltings being sited where they were. Of the maltings belonging to brewer-maltsters, most are in, or close to, the brewery; it is also common for a malting to adjoin an old brewery, which may now be used as a store or depot. The siting of the breweries depended in turn on several factors. Perhaps the most important was the availability of a suitable water supply (this accounts partly for the concentrations of breweries around Burton and Edinburgh). Obviously, it is also important that a brewery should be near population centres; and another necessity is an outlet for effluent from the malting and brewing processes. Most sale maltsters' premises were sited in or near grain-growing areas.

1.12. Very few of the firms in the survey reported any problems of location. In a fair number of cases, maltings were near railway lines which were now closed; some were near canals which had previously brought in imported grain, but which had become useless to the maltings now that only home-grown grain was used. But no significant difficulties with transport were reported, even in these cases. A few firms were having trouble over effluent disposal, where new regulations against river pollution were preventing them from discharging effluent into a river without expensive purification plant.

1.13. It was not practicable to ask exactly what quantities of grain came to each malting from different areas of the country. However, maltsters

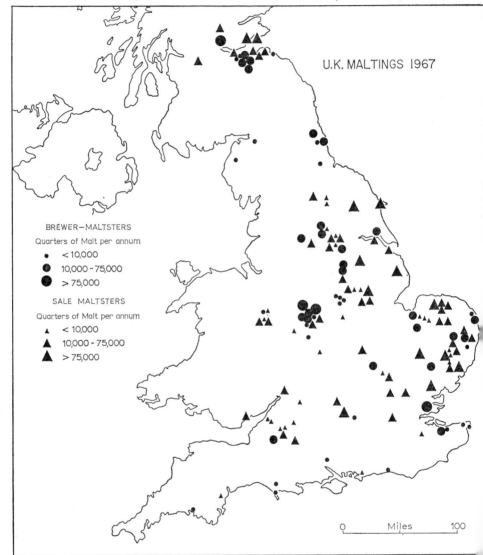

MAP 1.1. Location and size of maltings belonging to brewer-maltsters and sale maltsters in the U.K., 1967.
Source: Information supplied by the trade.
Note: Maltings in Scotland linked with the distillery industry are shown in Part V(c) Map 2.2.

taking part in the interview sample were asked to state from what counties they usually bought grain, and their answers give a broad picture of the geographical pattern of buying. The main centre for the growing of malting barley in England has traditionally been East Anglia, particularly Norfolk. As might be expected, the maltings surveyed in East Anglia took virtually all their grain from their own localities. Also important for malting barley production is an area comprising the other East Coast counties, from the East Riding of Yorkshire southwards, together with a few counties immediately adjacent to these. Maltings in this area again tended to buy almost wholly from within its boundaries. It is interesting to note that, of the maltings covered in the interview survey, those outside East Anglia but within this "second-rank" group of counties very seldom bought any East Anglian barley; the indication is that the maltsters concerned found the quality of their local barley perfectly adequate or at least high enough not to justify the extra cost of bringing grain from East Anglia. Moving outside the "second-rank" counties, further towards the north and west of England and Wales, the survey showed a very different pattern of buying: overall, only a minority of businesses bought grain locally, and when local grain was bought it usually made up a relatively small part of total purchases (though there were a few relatively small firms who made a policy of local buying). For most firms surveyed in the north and west of England, all or most of their barley came from East Anglia or the other eastern and south-eastern counties. It might have been expected that a maltster in a district where not much malting barley was grown would buy from his nearest growing area in the east or south. Frequently, of course, this was the case. But almost as often, distance seemed to be no object; for example, purchases of East Anglian barley were reported by firms whose maltings were in Yorkshire, Nottingham, Northumberland, Staffordshire and Devon (more than one firm in several of these counties). Of the Scottish maltsters interviewed, all had their premises in the east of Scotland—itself no mean area for malting barley—and all bought their full requirement within that area, except for one firm which bought from East Anglia.

1.14. Of the firms interviewed in the Survey, the great majority had all their barley brought to their premises by road; almost always the maltster would buy the grain at a delivered price and the transport would be arranged by the merchant or farmer. A few firms used rail transport for longer hauls.

Grain Buying

1.15. As Table 1.2 shows, sale maltsters took about four-fifths of the barley bought for malting in the Survey year—not surprisingly, in view of the

R

fact that the large firms which account for the lion's share of production are all sale maltsters.

1.16. One striking feature of these results is the very small proportion of barley imported; almost 99% of the barley used in the year was home-grown. No brewer-maltster, and only a few sale maltsters, used any impor-ted barley. Some firms had never done so. More commonly, it was stated that imported barley had made up most or all of total purchases until the beginning of the Second World War, but that imports had then virtually

TABLE 1.2

ESTIMATED TONNAGE OF BARLEY PURCHASED BY SALE MALTSTERS
AND BREWER-MALTSTERS IN THE U.K. IN THE SURVEY YEAR[a] [b]
('000 tons)

	Sale maltsters	Brewer-maltsters	All maltsters [d]
Bought direct from farmers	140	6	146
Bought from merchants	670	233	902
Imported (direct or via importers)	11	—	11
Bought from other sources	[c]	—	[c]
Total purchases[d]	821	239	1059

[a] Figures are totals for all firms responding to the Survey, plus an estimate for firms making malt but not responding. The adjustment has been carried out separately for sale maltsters and brewer-maltsters, and involves additions of about 4% and 10% respectively. [b] The Survey year corresponded to July–June 1966/7 for most firms; a minority of firms quoted other year-ends. [c] Less than 500 tons. [d] Totals are calculated before rounding and may differ slightly from the sum of the rounded figures.

ceased and had never regained their importance. Before the war, it had been a tradition among many brewers that foreign barley—especially Californian—was essential in the production of brewing malt.[1] Wartime restrictions cut this trade off abruptly, and—according to those who took part in the Survey—brewers then found that they could very well do with-out foreign barley. Since then, relatively more favourable prices for the British crop have helped to prevent a return to imported grain, and there is no evidence that this will change in the foreseeable future.

1.17. It is also noticeable from Table 1.2 that maltsters buy relatively little grain direct from farmers; most is bought through merchants. Half of the firms in the Survey bought no barley from farmers. Of the firms which

[1] The 1928 Report (pp. 149–50) estimated that 25% of the barley used by maltsters and brewers at that time was imported. The Report expressed concern about this level of usage and about an apparent tendency for the use of imported grain in malting to increase. This, at least, is one problem which has since found its own solution.

did, most bought only a small proportion of their total needs in this way—most often because of long-standing ties of goodwill with local farmer customers. The only firms buying substantial amounts of grain direct from farmers were some of the larger sale maltsters, these being large enough concerns to maintain a specialised department for the purpose. In addition, a few of the smaller maltsters made a deliberate policy of buying from farmers direct, so as to save paying the merchants' commission. But the great majority of maltsters—whether or not they bought some of their barley from farmers—felt that buying from merchants had many advantages and that merchants' charges were very reasonable in relation to these. It was often pointed out that buying much grain direct from farmers would mean dealing with many relatively small orders, making necessary a large buying force which would be under-employed for most of the year. Many maltsters felt that merchants could be relied on more than farmers to meet quality requirements, and made the point that if a merchant delivers a parcel of low-quality grain the maltster can reject it without argument, whereas this is not always found to be so with farmers. In many other ways, the merchant is regarded as providing a marketing service which well justifies his commission. He arranges transport, usually sending in grain at a delivered price; he takes care of the collection of grain from farms; he meets the maltster's delivery schedule, usually buying forward to do so; he often has facilities for cleaning, drying and storage which can be used to the maltster's advantage. Not surprisingly, the Survey showed that purchases direct from farmers were most common in southern and eastern areas, where farms tend to be larger and better equipped to perform services like these. But in the U.K. as a whole there is no doubt that—to quote the grain buyer of one large firm—"merchants form the basis of the marketing system" for the malting industry.

1.18. Of the firms interviewed, only two—both among the very largest concerns—had grain grown for them on contract, and in both of these cases contract growing had been only very recently introduced. The objectives were to get an assured supply of grain (particularly later in the season) at a reasonable price and of specified variety. The great majority of the firms interviewed had never given much consideration to contract growing and had no plans to start it, since the more usual sources of supply were found perfectly adequate. Among the firms which did comment on the subject, the general feeling was that contract growing involved the maltster in too much of a gamble—quality could fall very seriously in a bad harvest year. It is probably no accident that those firms making contracts with growers were very large sale maltsters; where large quantities of malt are made, differing in type, it is relatively easy to divert lower-quality grain to the making of lower-grade malt, or to dilute it by bulking it up with better grain. Clearly, these alternatives are not open to the smaller sale maltster or the brewer-maltster making malt to a strict brewing specification, and it is unlikely that contract growing will ever become popular except among the very largest sale malting firms. At the same time, because of the very large grain requirements of these firms, contract growing could attain relatively great importance in the market if it becomes an established part of their buying practice.

1.19. In buying grain from merchants, the practice among the biggest firms surveyed was to divide total purchases between a large number of merchants, shopping around for the best price for each quality of grain required. The very largest maltsters might take grain from over one hundred different merchants in a season; more commonly among the larger firms, the practice was to buy from a few dozen merchants each year. Small maltsters, on the other hand, very often dealt with fewer than half a dozen merchants, or even with only one, and these were often the same merchants year after year. In cases like these, the merchant or merchants were virtually performing the functions of a buying department for the maltsters concerned—taking care of their requirements on quality, collection and delivery, drying and storage, and for all practical purposes on pricing as well, since under these arrangements the merchant will usually quote a price which the maltster will accept without much haggling. This method of buying has obvious advantages for the small maltster, who does not have staff to spare for shopping around and does not need a wide range of differing types of grain. Goodwill is often built up over many years' dealing with the same merchants; at the same time, the maltster retains the ultimate sanction of withdrawing his business if the merchant should give him a bad deal.

1.20. Most usually, merchants call at maltsters' offices, bringing samples of the grain they have to offer. Another common practice, especially during the harvest season, is to send samples to the maltster through the post; the maltster evaluates the samples and the deal is completed by post or over the telephone. Alternatively, merchants may be contacted at grain markets or exchanges.

1.21. The use of markets and exchanges is most common among larger maltsters. Small firms often reported that staff time was not available for visits to markets; but this was not considered a handicap, since it was felt that grain could quite well be bought from visiting merchants. As often as not among firms who did send buyers to exchanges and markets, the visits were as much to pick up information—to "get the feel of the market" on a season's price and quality—as to carry on trade. A few firms, however, had trading stands on grain markets, where dealings with merchants were conducted. The use of markets and exchanges by maltsters has declined steadily for many years, and this trend is still continuing. Reasons suggested during Survey interviews were that chemical testing of grain has gained in importance, this being much easier to do if samples are brought into the maltster's office; and that maltsters are now fewer and larger on average than they were 10 or 20 years ago, making it easier for merchants or their representatives to travel round.

Market Information and Pricing

1.22. As in other sectors of the cereals market, the most important source of market information used by maltsters in day-to-day pricing is direct contact with others in the market—with merchants principally (on their visits to maltsters' offices, at markets, or over the telephone) and also with other maltsters, brewers and farmers. In the larger malting and brewing groups, there is day-to-day contact between buyers in different parts of the group. Apart from informal contacts, merchants provide a very important source of market information through the prices they ask for the samples they offer. Most maltsters use published market intelligence to some extent. The most popular published source is the *London Corn Circular,* and the *H.-G.C.A. Weekly Bulletin* is taken by many firms. These sources, however, are usually used as a general indication of price levels rather than as a basis for day-to-day pricing.

1.23. The beginning of the harvest season is a particularly difficult time for fixing prices. Quite often, the grain buyer will tour his buying areas to get an idea of the likely crop; estimates of supply are gained also from the various information sources already mentioned, as are estimates of likely demand and stocks. A rough starting price for the coming season may then be arrived at by adjusting the previous season's price in the light of these factors. Often, a buyer will go into the market early in the season, offer rather less than his estimate of price and see what sellers' reactions are. Some firms watch the price of the first barley bought for feed and offer a premium over this for malting grades. Many smaller firms leave pricing to their merchants, at the beginning of the season as at other times. The likely price of malt is not the most important consideration in fixing the buying prices of grain, because the malt price is itself determined largely by the barley price.

Forward Dealing

1.24. About half the firms taking part in the Survey of Maltsters bought barley forward. Forward contracts were almost always made with merchants, rather than direct with farmers. Very few of the firms who bought forward had a programme of even deliveries throughout the season; the almost universal practice was to fill all available storage with grain bought spot at harvest, then to buy forward to cover the rest of the year's requirement. Of the firms not buying forward, there were a few who actually had enough storage to cover the whole year's production of malt and who took in all the barley they needed at harvest time.[1] More usually, where storage was less than adequate for the year's needs, the store was filled at harvest and the remainder of the total requirement was bought spot at the same time for later delivery, being held at a storage charge by the merchant or, less often, by the farmer. Very few maltsters gave a reason for the practice of filling the store at harvest; those who did comment said that they preferred to dry and store the barley themselves rather than to rely on farmers' drying and storage. Because so few firms contracted forward with farmers directly, very little comment was offered on the H.-G.C.A. Forward Contract Scheme—too little to attribute any opinion on the Scheme to maltsters as a whole.

1.25. Table 1.3 shows how deliveries vary from month to month as a result of the buying policy described. It is almost universal practice among maltsters to "carry over" a certain amount of barley from one cereal year to the next. Enough is carried over to keep the maltings in operation during the few weeks between the resumption of malting and the arrival of the new season's grain on the market, plus a further period of a month or so during which the new grain, after drying, is stored temporarily; this is done for technical reasons connected with germination of the grain (i.e. "dormancy").

1.26. As a rule, sale maltsters contract forward for the sale of all or most of their malt. A common complaint among the medium-sized and smaller firms of sale maltsters interviewed was that brewers and distillers did not make their forward contracts until after the harvest season was over; typically, a contract would be made in the period from October to December to cover forward purchases of malt for the whole year running from the following January. Thus, the maltster had to rely on guesswork when he was buying barley at harvest. Sometimes, it was stated, the larger brewers would give an indication of their probable next year's purchases before the harvest began; but this would not include any details of how purchases were to be spread through the year. This latter point would be specified in the purchase contract itself—but, according to some of the sale maltsters interviewed, the brewer did not always keep to the dates arranged. A fairly

---

[1] No precise figure exists of the total storage capacity for barley in brewer-maltsters' and sale maltsters' premises, but the survey suggested a total for the U.K. of around 750,000 tons. This represents *maximum* available capacity; the amount in store at any one time will be less.

TABLE 1.3

ESTIMATED TONNAGES OF HOME-GROWN BARLEY
PURCHASED IN EACH MONTH FROM JULY 1966 TO
JUNE 1967 BY MALTSTERS IN THE U.K.[a] ('000 tons)

	Sale maltsters	Brewer-maltsters	All maltsters [d]
1966: July	1	1	2
August	106	25	131
September	217	47	263
October	106	41	147
November	87	26	113
December	59	25	84
1967: January	75	25	100
February	58	23	81
March	49	9	58
April	30	8	38
May	19	4	23
June	4	2	6
Total [d] [e]	810	235	1045

[a] [d] As for Table 1.2.   [e] Slight differences from
the purchases of home-grown barley shown in Table
1.2 are due to differences in year-end.

general feeling among the small and medium-sized firms of sale maltsters
was that, in the field of sale contracts, some brewers tended to take advantage of their strong buying position.

Grain Quality

1.27. In malting, probably more than in any other grain-using industry,
it is essential that the grain should meet strict standards of quality. Nowadays, laboratory testing of barley is almost universal; of the firms surveyed,
only a very few of the smallest maltsters carried out no chemical or physical
tests on the grain. Practice in grain testing is very similar throughout the
industry, with only small differences between firms—for example, in
whether sample or bulk, or both, are subjected to a given test; in whether
off-loading of the grain is allowed before a certain test has been made; or
in the technical details of the tests employed. But a general and very brief
summary of quality standards can be made as follows:

Nitrogen content must not exceed a certain level, which depends on
the intended use of the malt: the best pale-ale malt might contain a
maximum of 1·40% of nitrogen, while malts for the food trade would
be acceptable at 1·70%. Malt for other uses would usually fall between
these limits (see table, Appendix G, p. 802).

Moisture content is less important now than formerly, because efficient driers are available. However, when grain has to be dried it loses weight, and the drying process costs money; therefore a sliding scale of price adjustments has been agreed between maltsters' and merchants' associations, applying to grain delivered at more than a given moisture content.

Germination is of obvious importance in malting; the firms surveyed reported, as limits below which they would reject grain, germination percentages varying between 98% and 94%.

Other features such as size, appearance, smell, freedom from damage, admixture, mould or taint, are judged by hand evaluation.

1.28. Nearly all the large and medium-sized firms of maltsters taking part in the survey reported rejection rates of 4% or less for barley from the 1967 harvest (only two firms were above this, at 5%, and 7%). These figures, of course, apply to rejection of loads of grain which have actually arrived at the malting, and not to rejection at the sampling stage. The proportion of all *samples* rejected—either on first inspection or on testing—will of course be far higher than the figures for rejection of grain at the malting, though how much higher is not known. It was generally agreed that rejection rates in 1967 were lower than usual, because of the good harvest conditions. Rejection was for a variety of reasons; sometimes the grain did not correspond to the purchase sample; sometimes nitrogen was too high, germination was too low (because of bad drying) or the grain was mouldy. A striking result of the Survey was that small maltsters, whether malting for brewing or for sale, all reported that they had had to reject no barley, or only insignificant amounts, from the 1967 harvest. Many said that the same had been true in earlier years. It may be that testing standards are less stringent among some of the smaller firms; but a large part of the explanation must lie in the small firms' common practice of buying from a small number of trusted merchants, who can select good parcels of grain for these relatively small buyers. This, indeed, was stated by a number of the smaller firms interviewed.

1.29. Most maltsters said they preferred grain not to have been dried on the farm, since they had grave doubts about the germination of farm-dried grain. Some firms actually insisted that no grain coming in should have been dried by farmers. The majority did accept it, subject to germination tests; sometimes a firm specified that grain dried on the farm should be declared as such. In malting it is almost universal practice to dry all incoming grain down to a low moisture content (usually 12%) whether or not it has been dried previously. This is done to ensure safe storage—particularly important for carry-over barley—and for other technical reasons.

1.30. Asked for their comments on the introduction of an official quality-grading scheme for barley, a large majority of maltsters replied that it would be undesirable or impossible, or would have no effect on their buying policy. A very wide variety of reasons were given for these views, but in general the doubts were the same as those expressed by merchants: there was too much variation from season to season and sample to sample; no precise standards could be laid down; a scheme would mean centralised collection and official testing, which would add greatly to costs; maltsters and their merchants already did all the grading that was necessary.

Production Aspects

1.31. Table 1.4 shows how much malt was produced for various uses in the year under review. Of sale maltsters' total production, over one-half went to brewers and a further one-third to distillers; and just over 10% was exported.

TABLE 1.4

ESTIMATED TONNAGES OF MALT PRODUCED BY SALE MALTSTERS AND BREWER-MALTSTERS IN U.K. IN THE SURVEY YEAR[a] [b] ('000 tons)

	Sale maltsters	Brewer-maltsters	All maltsters [d]
Malt for brewing	319	175	494
Malt for distilling	191	2	193
Malt for vinegar manufacture[d]	10	[c]	10
Malt for food manufacture[d]	22	—	22
Malt for malt extract	9	—	9
Malt for export	63	—	63
Total all uses[d]	614	177	791

[a] [b] [c] As for Table 1.2.   [d] On a few returns, tonnages for vinegar and for food manufacture were not separated; an estimated split has been applied in these cases.

1.32. It is outside the brief of this Report to examine technical features and developments in malt production, except in so far as these affect the marketing of barley. In summary, there are three main systems of malting:

(a) *Floor maltings:* the traditional system, in which the germinating barley is spread on a large floor and is turned by hand. More recent modifications include mechanical turning and air-conditioning of the floors (the latter allows greater control of the process and lengthens the productive season). Many existing floor maltings are in very old buildings.

(b) *Box maltings:* the grain is spread in long, walled compartments; turning machines run over the grain, on rails along the walls. (A variant of this is the Wanderhaufen system, in which the grain moves along the box on a conveyor-belt.) The building is air-conditioned.

(c) *Drum maltings:* the grain is germinated inside a large drum. Turning is accomplished by rotating the whole drum. Only the interior of the drum is air-conditioned.

1.33. At the time of survey, floor maltings accounted for about 36% of total malt production by maltsters and brewers (it was not possible to estimate how much of this came from the older floors and how much from the newer mechanical or air-conditioned floor maltings). Box maltings produced another 36%, excluding the output of Wanderhaufen installations; if the latter is included, box maltings count as the most important single system, with a share of 47% of total output. Drum maltings accounted for the remaining 17%.

1.34. Where a malting firm owned several malting installations using the same malting system, it was not always possible to tell from Survey returns exactly how that firm's production was divided between these, so that figures of average size of installation must be approximate. "Installation" here is used to mean "the total amount of malting capacity on a given system at one particular malting establishment"; it does not imply that the whole "installation" is necessarily contained in one building. On this basis, there were just over 100 floor malting installations in the U.K. at the time of survey, producing on average about 16,500 quarters (about 2500 tons) per year; there was a very wide range of size in floor maltings, from under 1000 quarters to over 100,000 quarters per year. There were twenty or so drum malting installations, averaging around 50,000 quarters (7500 tons) annual production; most of the drum maltings in fact were rather smaller than this, but there was one very large drum malting (over 300,000 quarters) which raised the average figure. Box maltings, excluding Wanderhaufen maltings, also numbered around twenty; their average production was about 105,000 quarters (16,000 tons) annually. The largest units of all, with an average annual production of some 135,000 quarters (20,000 tons) were the Wanderhaufen installations, of which there were only about half a dozen in the U.K.

1.35. As might be expected, the floor malting was the most common system reported among small maltsters; to some extent, this obviously reflected small throughput and lack of capital, though many brewers in the small and medium-sized groups felt that the traditional system was still the only one that could produce the kind of malt they wanted for their beer. The larger companies usually had several maltings, using a variety of systems.

1.36. The malting system used has an effect on the seasonality of production. Air-conditioned maltings can operate all the year round (some of them do, though most close for a few weeks annually for holidays and repair) whereas maltings without air-conditioning must stop for 2 or 3 months in summer, for technical reasons connected with air temperature. In addition, the system of malting—in theory at least—sets the quality of grain which can be used; the traditional system is limited to grain of high and even quality, whereas the mechanised systems can handle rougher barley.

1.37. In practice, however, malting systems in themselves do not have any noticeable effect on the seasonality of deliveries or the quality of barley bought. Maltsters pointed out that the timing of deliveries is determined by the policy of filling storage at harvest, rather than by the seasonality of production; and that the quality of grain bought for malting depends first and foremost on the sort of malt one is aiming to produce, rather than on the production system.

Future Developments

1.38. Of the maltsters who commented on the future structure of the industry, all thought that concentration was bound to continue. A few respondents, in both brewing and sale malting, suggested that the biggest groups were deliberately selling at uneconomically low prices at present, in order to force smaller maltsters out of business, and felt that if this happened the big concerns might then increase malt prices. One brewer said that, in this event, his company would not hesitate to build a new malting for its own use.

1.39. Maltsters were also asked what changes, if any, they intended making in their own malting enterprise in the immediate future. The most striking result was that among small brewer-maltsters, over half of the firms interviewed had plans to close down their maltings within the next 5 years. These were all traditional floor maltings; typically, firms using such maltings were allowing them to run down, and because of low overhead costs were often still making malt cheaper than they could buy it. When the maltings finally became unserviceable, they would be closed and malt would be bought in. As time goes on, it seems certain that this situation must apply to more and more of the traditional maltings; probably, as one maltster said, "the floors must go—in 20 years at the outside". This development will probably have little effect on the seasonality of grain purchases by maltsters, since, as has already been stated, this is not determined by production systems as such. However, the swing away from brewers' malting in small floor maltings and towards the buying-in of brewing malt from sale maltsters is likely to affect the grain market in various ways. The geographical flow of grain will change, with collection points becoming fewer and larger. Further, the sale maltster relates quality and price in a very different way from the brewer-maltster. To the brewer, malting represents only a relatively small proportion of total production cost (about 10%, according to one brewer interviewed) and variations in the price of barley thus make relatively little difference to the cost of production of his end-product; therefore many brewers have traditionally gone for barleys of very high quality, even at relatively high prices. The sale maltster, on the other hand, must always buy as keenly as possible, because a slight difference in barley price means a big difference to his profits. Therefore it is to be expected that the more brewing malt is made by sale maltsters, the less easy it will be to get a premium for malting barley of very high quality.

Economic Assessment

1.40. It is not the purpose of this Report to discuss the economics of malt production and sale, except in so far as these affect the marketing of grain. As far as the grain-marketing side of the malting industry is concerned, we would say first of all that, in our view, the survey of maltsters revealed very few economic faults. The systems of buying and pricing reported by maltsters seemed to us to be rational and well suited to the circumstances of the different types of firm.

1.41. There are a few topics which, we think, deserve a second look from the economic point of view:

Structure of the Industry (paragraphs 1.7–1.9)

1.42. If, as seems likely, concentration in the malting industry is going to continue, one must ask whether there is any possibility of one buyer, or a few buyers, getting so big that they can dictate the price of malting barley— or, indeed, whether this situation has arisen already. We have not found any evidence that price dictation is going on at the moment, and we think that it is, in fact, not a danger for the future. Despite the predominance of a few very large firms in the market the malting industry is only in a position analogous to the compound feed industry, in which there is no price dictation. If large firms tried to depress grain prices unduly, the first people to be hit would be their merchants—and, as we have seen, the loss of merchants' services would be a severe handicap to these firms. Most important of all is the fact that barley grown for malting can quite easily be diverted to use for feed or for export, so that the feed or export price forms a "floor" for malting barley prices (supplies of malting barley would, of course, begin to dry up even if the malting price were only pushed partway down towards the feed price). For this reason, we suggest that any significant degree of price dictation for malting barley would be virtually impossible even if the market were dominated by a single firm.

1.43. On the opposite side of the coin, it is obvious that concentration should decrease costs in marketing by reducing the number of sale and delivery points. While accepting this, we see no reason why any steps are necessary to hasten the process of concentration and rationalisation in malting; these are already going ahead rapidly in response to commercial pressures within the industry.

Location and Transport (paragraphs 1.10–1.14)

1.44. At first sight, it might seem irrational to haul large quantities of barley from the east and south of England to maltings a long way distant, when large acreages of barley are grown just beside the maltings concerned; yet this is what many maltsters and brewers continue to do. In our view, it is not so easy to reach a conclusion on this subject as might be thought at first. The usual reason for these long hauls is that the maltster wants certain quality features which, in his opinion, are found only in grain from the east and south. As long as he and his customers are willing to pay the cost of transport from these areas for the sake of getting the quality they want, there is no economic justification for changing to closer sources —assuming that grain of the same quality could not, in fact, be found nearer the malting. To a large extent, of course, quality depends on weather and soil conditions, in which the east and south have advantages. But from comments made during the Survey, we think that more local grain could be taken by maltsters in northern and western areas if storage and drying on farms could be made more reliable. For this reason and others (see below) we would recommend that more publicity and advisory work be undertaken by government and trade bodies with the aim of improving farmers' practice in drying and storing malting barley (see also the Statement by the Maltsters' Association, Appendix G, p. 800).

Forward Dealing and Seasonality of Purchases (paragraphs 1.24–1.26)

1.45. Another seemingly irrational practice, on the surface at least, is that of filling barley stores at harvest and buying forward (or ordering forward) only as much grain as cannot be stored then. As compared with a programme of even deliveries through the season, filling the store at harvest must tie up the maltster's storage space and capital (particularly the capital represented by the grain itself), and must mean a rush of buying activity over one very short period of the year. However, we accept that given present circumstances in the market, maltsters have good reasons for stocking up at harvest despite these disadvantages. One important reason for the practice is distrust of farm drying and storage; we have already recommended that this should be improved. Further, a maltster is very often storing in old buildings which have been written down virtually to zero, so that capital costs on storage buildings are very low; and drying costs are no deterrent, since grain will almost always be dried down in any case as part of the malting process. Stocking up at harvest allows the maltster to buy cheap and be sure of the quality of his grain. We therefore do not consider that any change in maltsters' seasonal buying practice is needed at present. In future, when old storage buildings have gone out of use, the economic picture will be different; but we believe that maltsters' buying policy will alter as necessary to fit the situation without any outside guidance.

1.46. We find it easier to criticise another aspect of forward dealing, namely the timing of the contracts which some brewers and distillers make for malt purchase (paragraph 1.26). There could obviously be some saving in marketing costs if sale maltsters could buy to known requirements rather than on guesswork. We admit that savings would probably be small (buying on guesswork is, after all, the rule in most sectors of the cereals market). But we can see little or no difficulty for brewers and distillers in making their malt contracts a few months earlier in the year, to come just before harvest rather than after—though we realise that while amounts and delivery schedules could be fixed at this time, the price might have to be left open until after the harvest was in. We recommend that this change should be encouraged by the trade bodies.

Quality and Grading (paragraphs 1.27–1.30)

1.44. With regard to the possible introduction of an official grading scheme, we are convinced by the objections which maltsters offered to us, and would regard such a scheme as unworkable and unnecessary. As has already been mentioned, the Survey gave a strong impression that unreliable farm drying and storage constitute a pressing problem in connection with grain quality in malting, at least in some parts of the country, and we have suggested that the situation can best be dealt with by publicity and advisory work. It has been suggested to us that if maltsters want better-quality grain from farmers, they should offer a higher premium for it. We do not agree, however, that a higher price for malting grain would necessarily improve farm drying and storage. Our impression is that the difference between good and bad drying and storage lies mainly in technique, rather than in resources used. In other words, where difficulty exists, it lies not in considerations of cost and return but in lack of knowledge and skill.

CHAPTER 2

Malting by Distillers

Introduction

2.1. The key raw material in whisky distillation is barley malt. In the malting process enzymes are synthesised within the grain. When the malt is used for distilling, these enzymes convert the barley's starches into fermentable sugars, which in turn are converted to alcohol by the addition of blended yeasts.

2.2. In malt whisky production, the barley itself is relied on for total provision of convertible starches. Because it is well within the capacity of the synthesised enzymes of a good barley malt to convert all its convertible starch, the important criterion of a good malting barley is a high content of convertible starch or extract. Extract potential is, in general terms, inversely related to the nitrogen content of barley. Therefore a good malting barley for malt whisky distilling should, like barley for brewing malt, have a low nitrogen content, with a capacity for rapid vigorous germination, but primarily it should have a high convertible starch content.

2.3. In grain whisky production, as opposed to that of malt whisky, the surplus enzymes of the barley malt are utilised to convert starches from another cheaper source, namely maize. Because barley malt with particularly vigorous enzymic action can convert up to $5\frac{1}{2}$ times its own weight of convertible starch, maize (which has a much higher starch content than barley) is used in the ratio of as much as 85 parts by weight of maize to 15 parts by weight of barley.

2.4. Barley malt for grain distilling must therefore be characterised by a high enzymic activity or "diastatic power" (D.P.) and its content of convertible starch is only of secondary importance. Home-grown barleys are of characteristically lower D.P. than Canadian six-row barleys, and for that reason the latter have been used almost exclusively until very recently by grain distillers for their malt requirements.

2.5. Malt whisky distilling is characterised by long-established brands with individualistic reputations, these whiskies being produced by 107

495

small self-contained distilleries dispersed particularly in the Highlands and Islands of Scotland (see Map 2.1). The dispersed and frequently isolated nature of their locations arises partly from their historical origin in days when distilling was illegal and partly because of the reliance of whisky manufacture on the occurrence of water, in sufficient quantity, with certain desirable characteristics of softness and flavour associated with peaty soils.

2.6. Grain distilleries, fourteen in number, and relatively newer introductions (many being post-war) are larger than malt distilleries and produce some 70% of total whisky output. Relying largely on imported grain, these distilleries are mostly found in the Central Lowlands of Scotland, or alternatively, at small coast ports such as Girvan, Fort William, Invergordon and Montrose (see Map 2.1).

2.7. All 121 distilleries in Scotland are owned by thirty-seven firms. Twenty-six of these firms, representing ninety malt distilleries and eleven grain distilleries, co-operated with the Survey team and the findings of the report as stated in the following sections are based largely on the consolidated information thus collected. The Survey of Distillers was carried out during the winter of 1967/8. At each visit, a set questionnaire was completed very similar to that used in the Survey of sale maltsters and brewer-maltsters (the results of which have been summarised in the previous chapter). An important point of difference in the two sets of results is that in the Survey of Distillers—unlike that of sale maltsters and brewers—it was not possible to gain virtually complete coverage of all producing firms. Neither was the Survey sample of firms large enough to allow reliable estimates of results for the industry as a whole to be worked out from sample results. Therefore, the figures quoted below as having been derived from the Survey of Distillers must not be assumed to reflect the exact position for U.K. distillers as a whole. Nevertheless, in most cases the Survey results are the only information currently available on the topics concerned.

Trends in Cereal Utilisation

2.8. Table 2.1. gives a breakdown of utilisation of the three main categories of cereal used in distilling in the U.K. since the war, and shows also the trends in production of malt whisky and grain whisky. It is important to realise the basis of the figures shown for "malted cereal utilisation". These include not only the tonnages of barley actually bought by distillers for malting, but also the tonnages of whole cereal which went into the malt bought by distillers from sale maltsters; and both these tonnage figures include imported as well as home-grown barley. It is helpful to describe

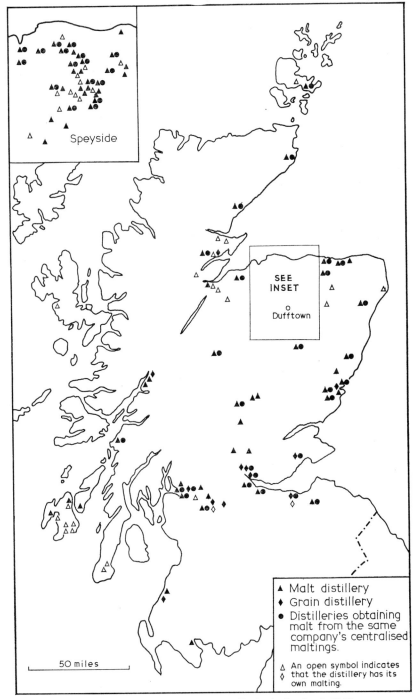

Speyside

SEE INSET

o
Dufftown

▲ Malt distillery
♦ Grain distillery
● Distilleries obtaining
malt from the same
company's centralised
maltings.

△ An open symbol indicates
◇ that the distillery has its
own malting.

50 miles

MAP 2.1. Location and type of distilleries in Scotland, 1968.
Source: Information supplied by the trade.

Table 2.1

GRAIN USED IN DISTILLING IN U.K. AND PRODUCTION OF CEREAL SPIRIT

Year ending 30th Sept.	Cereal utilisation				Total cereal spirit production (million proof gallons)		
	Malted cereal ('000 tons whole cereal equivalent)			Unmalted cereal ('000 tons)			
	Malt distil- ling	Grain distil- ling	Total malted		Malt spirit	Grain spirit	Total spirit
1946–50 Average	80	33	113	73	8·4	12·0	20·3
1951–5 Average	126	45	171	118	13·3	19·1	32·4
1956	147	95	206	149	15·1	27·1	43·0
1957	171	63	234	177	18·2	31·4	49·6
1958	194	66	260	182	20·6	33·7	54·2
1959	211	63	275	209	22·4	36·6	59·0
1960	233	78	311	256	25·1	44·6	69·7
1961	244	82	326	287	26·5	49·4	75·7
1962	248	93	341	333	27·1	57·0	84·1
1963	264	105	369	375	28·9	65·2	94·1
1964	299	135	434	460	33·0	82·0	115·0
1965	352	156	508	553	39·4	95·8	135·2
1966	401	173	574	565	44·9	97·8	142·7
1967	460[a]	140[a]	600[a]	462[a]	51·4[b]	80·6[b]	132·0[b]

[a] Estimated from figures of spirit production in 1967. [b] Extracted from *Statistics Relating to British Made Potable Spirit*, William Birnie, C.A., Inverness, January 1968. *Source:* Derived from statistics contained in Annual Reports of the Commissioners of Her Majesty's Customs and Excise: H.M.S.O.

more fully the nature of the three main categories of cereal distinguished in the table:

(i) Malting grain used in *malt* distilling is predominantly home-grown barley, with very infrequent and declining instances of the use of Australian barley.

(ii) Malting grain used in *grain* distilling is the "high D.P." malting barley described earlier and is mainly imported Canadian barley, except for a very recently increasing proportion of selected home-grown malting barleys. As discussed later, devaluation may result in the increased use of home-grown barleys by this sector of the industry if encouraging results are obtained from experiments currently being carried out by some distillers and maltsters.

(iii) Non-malting grain is, nowadays, almost entirely imported maize. In former years, when the price difference between local rye or feed barley and imported maize was greater, there was a somewhat greater use of these home-grown cereals, but it can be assumed that reference to non-malting grain in the Report signifies imported maize.

2.9. Whisky production has increased rapidly in the post-war period, as shown in Table 2.1, leading to a similar increase in the use of cereals. In fact, cereal utilisation by the distilling industry has shown a rate of expansion unequalled by any other major cereal-using industry.

2.10. However, for the last 20 years, whisky production has exceeded consumption to such an extent that there are now on hand stocks for $10\frac{1}{2}$ years' requirements at the present level of demand, compared with the minimal stocks of some 3 to 4 years' production needed for maturing whisky. The expansion of production has been markedly greater for grain whisky than for malt whisky. The result of this has become visible in a cutting-back of production of grain spirit, and hence a drop in the utilisation of maize and high D.P. barley, in 1967. At the time of survey, in 1967/8, many grain-distilling firms were found to be working at two-thirds capacity or less (though it was not safe to generalise, since at least one grain distillery was still working to full capacity). Malt whisky distillers, on the other hand, showed little sign of cutting-back production; most reported capacity orders, and some were expanding production.

The Purchase of Barley for Malting

2.11. As explained above, it is not possible to use the results of the survey of distillers to arrive at a figure of total barley purchases by the distilling industry. It is possible to calculate a very rough estimate for 1966/7 as follows:

(a) From Table 2.1, the estimated total utilisation of barley (i.e. barley bought and the whole-grain equivalent of malt bought) was 600,000 tons. This is equivalent to 450,000 tons of malt, since 1 ton of barley makes 15 cwt of malt.

(b) In the previous chapter (Table 1.4, p. 489) it was estimated that sale maltsters and brewer-maltsters together sold some 193,000 tons of malt to distillers.

(c) By deduction, therefore, it can be estimated that distillers made 257,000 tons of malt, out of the 450,000 they used in 1966/7. This in turn would indicate that distillers bought around 340,000 tons of barley—home-grown plus imported—in that year for the production of malt.

2.12. The most consistent trend revealed in the Survey has been the proportionately greater use of Scottish-grown barleys, as opposed to English Proctor and imported Australian barley, by malt distillers over the last 10 years, because of a considerable improvement in the quality of the Scottish crop. However, should bad weather cause a poor season, there is a marked reversion to the more expensive English supply. It was generally stated that English barley was £2 to £3 10s. per ton dearer, delivered, than Scottish barley for comparable specifications; much of this differential, of course, would be due to transport costs.

2.13. Of the malting barley purchased by firms interviewed in the Survey, some 30% was from England and was mostly Proctor; 45% was of Scottish origin, being mainly Ymer, Freja, Baldric and other varieties; and about 22% was imported Canadian barley for use in grain malt production. Australian barley made up a mere 2% of the total.[1] Only three companies still insisted on English Proctor. It was invariably said that Australian barley was excellent for malting because of its low moisture content, uniformity, cleanness and rapid vigorous germination. However, price difference between it and home-grown barley had increased in recent years, and this tendency, together with the rise in volume and quality of the home-grown crop, had almost stopped the use of Australian barley.

2.14. Criteria for selection of good malting samples were fairly standard throughout the industry, but degrees of tolerance and the extent to which distillery managers relied on merchants to supply to specification varied considerably. Generally speaking, the smaller distilleries—like smaller breweries—relied more on the good faith of their merchants. Most large firms have been at pains to provide themselves with up-to-date laboratories for sample testing, but much visual examination is still relied on by the smaller operators. The quality testing carried out covers very much the same features as the testing done by brewers and sale maltsters (see Chapter 1, p. 487). Similar standards of germination and moisture are called for; nitrogen content can be rather higher in barley for distilling—more particularly grain distilling—than in barley for brewing. A general comment was that Scottish-grown barleys had improved greatly in the last 10 years, though English barleys still tended to be more reliable and uniform. In Scotland there were local problems of lack of uniformity and a tendency

---

[1] This sample result, of 75% of purchases home-grown and the remainder imported, can be checked against other estimates. The Ministry of Agriculture's statement of the utilisation of cereals, 1966/7 (provisional) gives 222,000 tons as the amount of home-grown barley taken up by distillers. Comparing this with the estimate of 340,000 tons for *total* barley purchases by distillers, arrived at in paragraph 2.11 above, the indication is that about 65% of the total was home-grown barley and the remaining 35% imported. Thus the picture is broadly the same on both estimates; the differences between them reflect the difficulties of sampling and estimation already mentioned.

to too high nitrogen content. Several companies disliked or would not buy farm-dried grain; the majority of companies would buy it subject to laboratory germination tests, or, in the case of two firms, if the grain had been dried under their own supervision on the farms.

2.15. Attitudes to a possible statutory grading scheme were very mixed. About half the companies were against the suggestion, mainly for reasons of the varied individual preferences of distillers and the inability to reach agreement on acceptance criteria for grading. The other half gave qualified support to a grading scheme, one firm expressing the need to grade for corn size, at present a very variable characteristic giving rise to difficulties in even germination at malting. Another firm supported a standardised bushel-weight grading system. One large firm would welcome a standardised grading scheme to enable avoidance of present embarrassing situations over pricing that arise between merchants and distillers, due to the different criteria against which distillers buy at present.

2.16. The commonest method of grain purchase was from merchants visiting distillery offices with samples. Fewer than half the companies bought grain direct from farmers, and very few indeed relied primarily on this system of buying. The main reasons for not buying from farmers were the general inability of the distilleries to dry and store farm grain bought at harvest-time, coupled with the general dislike for farm-dried grain. The larger firms stated also that they were not administratively equipped to deal with large numbers of small producers. Two firms, however, expressed a preference for purchases ex-farm, since buying in this way afforded them greater selectivity, and both firms were expanding this method of purchase. One of these firms was expanding a system of contracting with farmers to grow certain acreages; selected seed, advice and supervision were provided by the firm.

2.17. Purchasing of grain at corn exchanges and markets had declined rapidly because of the time-consuming nature of such visits and the increasing competitiveness of merchants' visiting representatives. Only six firms now attended corn exchanges or markets, and Edinburgh Corn Exchange was the only one specified.

2.18. It was apparent that malting barley merchants' representatives were very numerous and diligent in visiting distilleries canvassing for orders between September and January. These representatives came largely from Scottish east coast firms but also from Yorkshire, Cambridge, Norfolk and Berkshire. In the north of Scotland there was growing activity by Inverness and Aberdeenshire merchants, representing the barley-growing areas of Ross-shire, Moray, Black Isle and further south (Fife, Angus, Lothians).

2.19. The most prevalent pattern of barley purchase was to contract forward for regular delivery as from December up to the end of that crop year, i.e. to June or July; if prices appeared attractive and storage was available, barley would be over-ordered for the latter months to cater for a carry-over to December, when the succeeding year's grain would in turn become available for use in malting (having been stored during the "dormancy period" of a month or two after harvesting). However, it was equally common practice to buy forward in June for the carry-over period to December; in this case the purchases would usually be of English barley, which is available 1 month to 6 weeks earlier due to the earlier harvest in the south and the shorter dormancy period of the English crop.

2.20. Very little spot or speculative buying of barley was reported, since firms are anxious to determine their raw-material costs well before their calendar-year production season in order to arrive at an appropriate contract price for their whisky, which is, in turn, all sold forward.

2.21. Because most distilleries need to carry 3 to 4 months' supply of grain for the period between 2 years' crops, there was generally sufficient storage capacity at distilleries or maltings for 25–33% of annual usage. Only two companies with their own drying facilities were able to, and in fact did, buy and take delivery of a full year's supply at harvest time.

2.22. Because only seven firms interviewed bought grain direct from farmers, there was little general comment on the H.-G.C.A. forward contract scheme. One firm, which purchased close on 20,000 tons of barley undried, direct from farmers, believed that the scheme was a good one but that at present very few farmers had the skill or equipment to dry grain to the exacting requirements of the distilling industry. Another firm buying direct from farmers disliked the scheme because, they alleged, it created artificial shortages of grain at certain times of the year, and because it was preferable to buy most farmers' grain undried. Two other firms buying direct from farmers favoured the scheme because it partly solved their storage problems. In general, because of the predominant tendency for distillers to buy their barley from merchants, there was little knowledge of the finer points of the scheme.

2.23. Distillers' main sources of market intelligence were the quotes made by the many merchants visiting them. It was stated that merchants quoted a similar price for barleys of similar type during the major buying season of September to December. As in the brewing and sale malting industries, distillers often formed close trading relationships with a few chosen merchants who came to know their requirements very accurately, and they tended to deal with these same few merchants each year, although they were visited by many more. Some large merchant firms provided periodical market reports, but these were generally of more interest to those firms buying imported maize or barley.

The Purchase of Imported Grain

2.24. Thirteen of the firms interviewed bought imported cereals in 1967. Five were malt distillers whose purchases of imported grain consisted only of Australian barley. Four of these bought from merchants in small tonnages only, for use during the gap between the two seasons' crops. The fifth firm bought from Leith and Hull shippers, buying each year 2000–3000 tons to meet all of its requirements between the 2 years' crops.

2.25. As already stated, maize is the cheapest known source of convertible grain starch. It has been suggested to us that if good home-grown barley is to be competitive with maize as a non-malting grain, the barley must be at least £7 per ton cheaper delivered. The most attractive maize at present is South African White Dent. During recent years of shortage of South African maize, North American Yellow No. 2 and No. 3 grades were used, but these are considered inferior to South African White as they have a higher moisture content and lower extract.

2.26. Because devaluation has increased the price of South African and North American maize by up to £3 per ton, firms have had a closer look at other possible sources, e.g. Spain, which has also devalued, but because of problems of unassured regular supplies and variable qualities, it is unlikely that a better alternative to present sources can be found.

2.27. South Africa is probably a larger supplier of maize than North America at present, but substantial tonnages are still purchased from the U.S.A. Because of a certain tendency for poor grading, variability and impurities in the North American grades, there has been a recent development in buying selected maize direct from U.S. merchants, which, according to one firm, fully justifies the slightly higher price.

2.28. Apart from these direct negotiations most maize appears to be purchased through Glasgow and Edinburgh brokers and shippers, and some English shippers. Prices are normally c.i.f. Glasgow or Leith and it is the responsibility of the distilleries to arrange collection from these points. Distilleries operating from the port of Glasgow make use of the Clyde Port Authority granaries, and may draw daily from their own stocks held there. There are similar arrangements at Leith.

2.29. There appeared to be little speculative buying of maize. Most distilleries dealt with at least two or three brokers or shippers. It was stated that the larger brokers and shippers provided useful trade newsheets and the *Financial Times* was frequently stated to be a useful guide.

2.30. In general, little attempt was made to hold large stocks of maize, due to the undesirability of tying up capital and the relative ease of forward buying of part shipments for delivery to the ports three, four or more times a year. Further, it was stated that granary rentals could amount to considerable sums over a period.

2.31. The five grain distillers who also imported Canadian barley for malting normally bought "Canadian No. 2 Feed", and it was said that this grade did have frequent shortcomings due to its imprecise definition and hence undue variability. Whilst such barley was normally bought from a large Scottish maltster/broker, there were instances of a dearer selected Canadian barley being bought direct from Canadian granaries, f.o.b. a Canadian port. The normal method of purchase was to buy forward for regular delivery at anything between 3 months and 12 months hence, and stocks carried were generally kept low.

Production and Purchase of Malt

2.32. Traditionally, malt whisky distilleries incorporated their own floor maltings. With the desire to increase production rapidly after the Second World War, some distilleries installed the more efficient and labour-saving Saladin box or other form of mechanical malting unit. A more customary form of post-war development was to expand only the distilling capacity of the site and to purchase additional malt requirements from large sale maltsters. Today many small distilleries no longer use their floor maltings, and their mechanical malting capacity can often supply only a part of their requirements. Many grain distilleries, too, do not have adequate malting capacity for their needs; in fact, some of the newer ones have no maltings at all.

2.33. Our survey suggests that there is a tendency for smaller "one-distillery" companies to rely on sale maltsters for malt supplies, due largely to the importance of economies of scale in malt production and the considerable capital outlay on buildings and installations required in maltings. (One small distillery had estimated that the capital outlay for a modern malting with a capacity of 20,000 quarters per annum would be £27,000.) In terms of overall costs, also, it would seem that floor maltings have no advantage over purchasing from sale maltsters. The larger concerns state frequently that they can save £1 per quarter on the price of purchased malt, being able to produce for around 120s. per quarter at 1967 barley prices, and there is a general move among the larger operators to expand and centralise their malting capacity to measure up to their increased requirements. Even so it is only now that this expansion is beginning to catch up with the increase in the distilleries' utilisation of malt. Only two large distillery companies stated that it was not their policy to develop maltings, and one of these had a strong financial commitment in a large firm of sale maltsters.

2.34. The map of distilleries (Map 2.1) distinguishes those with their own maltings from those without, and also shows which of the latter depended on malt from centralised maltings. Map 2.2. shows the locations of all distilling maltings, including both the distillery maltings and the centralised maltings, and indicates the malting systems used (for a brief description of these systems, see the previous chapter).

2.35. Among the *grain* distillers interviewed during the Survey, all the firms using home-grown barley reported that its use for grain malt production was a recent innovation and would be increased in the future if economically justified. Two firms had not yet used home-grown barley malt, but both indicated that trial lots would be used in 1968.

2.36. The main reason for switching over to home-grown barley malt was one of price, and a general increase in the ability to adjust the production process to make better use of less efficient malt. Malt from Canadian barley cost, at the time of the Survey, around 160s. per quarter delivered, if purchased from a maltster. From 1966 up to the time of writing (1968), local barley malt had dropped in price from around 148s. to as low as 135s. per quarter delivered, because of expanded production by sale maltsters and increased malt production by the large distillery companies. It has therefore become much more attractive to the grain distiller to substitute local barley malt for that made from Canadian barley. Devaluation reinforces this situation, although up to March 1968 the delivered price of malt from imported Canadian barley had not been increased by maltsters.

2.37. The effect of using home-grown barley malt, with its lower D.P., is to increase the cost of each unit of fermentable sugars, and thus of each gallon of spirit produced. Several distillery companies and associated maltsters are experimenting with various local barleys in an endeavour to identify the characteristics of higher D.P. types. Nitrogen content is related to high D.P. but it has yet to be established that nitrogen content alone can indicate a high D.P. value. Present indications are that the varieties Maris Baldric and Cambrinus are promising, but that seasonal and geographical factors also affect D.P. potential. There is also at present a general preference for English-grown barleys.

2.38. The Survey suggests that approximately 40% of *grain* malt was purchased from sale maltsters and that the remainder was produced by the distilling companies themselves. Some companies aimed to produce all their requirements, whereas others preferred to rely entirely for supplies on sale maltsters.

2.39. Of the twenty-six malt and grain distillery companies giving information only two reported that they purchased any malt "spot", and this only in small irregular amounts. The remainder revealed a very consistent practice of placing orders for malt in August to December for the whole, or the greater part of, the succeeding calendar year. There are two main reasons for this policy. First, as most companies work to a very full

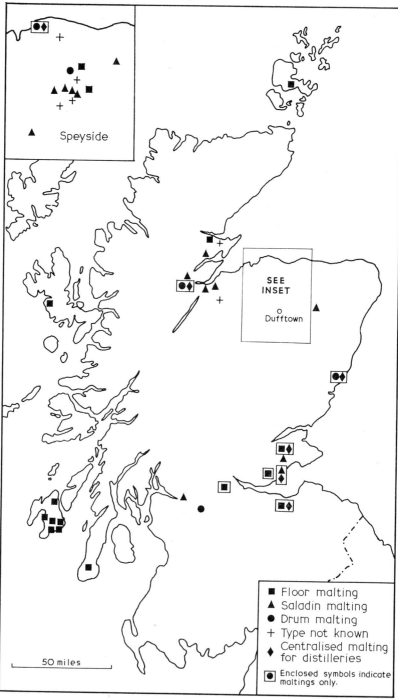

Speyside

■ Floor malting
▲ Saladin malting
● Drum malting
+ Type not known
♦ Centralised malting for distilleries
▣ Enclosed symbols indicate maltings only.

SEE INSET

○ Dufftown

50 miles

MAP 2.2. Location and type of distillery maltings in Scotland, 1968.
Source: Information supplied by the trade.

programme to meet forward orders for spirit and have to ensure an even flow of raw material into the distillery, regular delivery rather than stockpiling at any time of the season is the general practice of purchased malt intake. Second, by contracting for supplies for a 12-month or shorter period, distillers get better terms from the sale maltsters.

Handling and Transport

2.40. Firms participating in the Survey were very conscious of the labour-saving advantages of bulk, as opposed to bagged, handling of malt and barley, and there has been an obvious move in that direction. However, the cost of installing bulk handling facilities has deterred some of the smaller operators from making a switch. Although several firms felt that suppliers might in future discriminate against bagged deliveries by charging a higher rate, there was no evidence that this had yet happened. Distilleries in the islands of Islay and Orkney still invariably received their raw materials in bags mainly because of a lack of bulk handling facilities at the small ports used by ships to these islands. Furthermore, because of the hygroscopic nature of malt and the risk of both barley and malt absorbing excess moisture, handling in bags was safer. Some firms insisted that malt be delivered in polythene bags.

2.41. Transport systems used varied greatly depending on the geographical location of distilleries, their size and the existence of railway sidings. Malt and barley were invariably bought on a delivered basis, and transport methods were left to the discretion of the supplier in many cases. In the case of grain distilleries, the companies were themselves all responsible for moving their maize from the docks to the distilleries or maltings. Three companies relied almost entirely on hired rail transport, one firm hired rail and road, and three companies used road transport exclusively— on a contract-hire basis in two instances and with its own transport in the other. Some grain distilleries were concerned at the cost of transporting maize, and were investigating the possibility of using their own transport (at present used for hauling spirit only) or the possible use of coasters from Glasgow or Leith to a nearer harbour. The main problem arising in the latter possibility, however, is that of devising a speedy method of unloading coasters, as demurrage charges can be high.

2.42. The dispersed nature of malt distilleries and their remoteness from rail facilities forced many to rely on road transport. However, it was frequently stated that rail facilities, even where these existed, were not used because of the extra handling costs involved in transferring the barley to road transport at the end of its rail journey. Nevertheless, most malt and barley from England is moved north by rail (in enclosed hopper trucks if in bulk) and, in most cases, transhipped on to road transport. A large distillery company, a road haulage firm and British Rail have co-operated to develop such a scheme to service distilleries and maltings in the north-east of Scotland, and it is likely that increasing use will be made of it, since it is geared to meet the special needs of distillery companies; attention is paid to prompt delivery, separation of bulks, and use of suitable hoppers and bulk transporters.

2.43. Scottish barleys, on the other hand, are mostly moved by road transport with no transhipment. For the shorter distances involved, and the need for timely, regular deliveries, there is no doubt that road transport provides the more satisfactory service.

2.44. Transport costs were obviously an important factor affecting distillers' decisions on where to buy their barley. Scottish barley held a considerable advantage in this respect, because its transport costs could be as low as 10s. to 12s. per ton for many users, as compared with an average transport cost of £2 10s. 0d. per ton for barley from the south. The island distilleries were faced with considerably higher transport charges; it was established, for example, that the average cost of transporting barley from the Ayrshire coast to Islay by chartered coaster was 52s. 6d. per ton.

2.45. The commonest method of transporting malt and barley to Islay was by 100- to 200-ton coasters from Ardrossan or other Ayrshire coast ports to Islay, but firms made use of direct Dutch coaster services from King's Lynn to Islay and instances were reported of malt being sent by coaster from Northern Ireland to Islay. Distilleries in the north-east of Scotland had from time to time received English barley by coaster from King's Lynn delivered to places such as Inverness, Lossiemouth and Burghead, but these routes were not developed, probably because of the lack of speedy unloading facilities and the high demurrage charges associated with the use of coasters.

2.46. Several companies expressed concern about likely implications of the Road Transport Act (see Part III, Chapter 3). If road transport costs increased, it was said, many firms would be in difficulty, since transport charges featured largely in their costs; in the purchase of malt and grain requirements, in the disposal of spirit, and in the movement of empty wooden casks for spirit to the distilleries.

The Future

2.47. As has been mentioned, production of grain whisky has been running ahead of effective demand for some years past, and large stocks have built up. There is already evidence of a cut-back in production of grain whisky, and it is to be expected that this will prevail for several years to come; how long it is impossible to say. As a consequence of this trend, the utilisation of maize and high D.P. barley, and the malt made from the latter, can be expected to lag for the next few years.

2.48. In their efforts to protect their profit margin, grain distilleries will strive to increase efficiency and reduce raw-material costs. Devaluation has aggravated their problem, and one major change is likely to be a fairly rapid replacement of Canadian barley by home-grown barley in distilling malt production.

2.49. In malt whisky distilling also, although there is no strong evidence of a fall-off in demand for spirit, it is likely that increasing attention will be given to reducing production costs. There is already definite indication that the utilisation of Scottish barley, with its lower transport costs, is on the increase, and the 1967 season's excellent crop gave a boost to this trend. It was frequently stated that much remains to be done to educate farmers in the particular needs of the distilling industry and that barley quality can still be improved. Attention was drawn to the undesirable tendency for some farmers to grow a dual-purpose barley, and it was emphasised that the need to pay a premium for malting barley would continue. This latter argument will, however, be partly countered by the increasing skills in malting technology and the ability to exert more control over the malting and subsequent processes.

2.50. The smallest independent distilleries will continue to move out of malting. Nevertheless, because of the present development of new maltings, a larger proportion of the industry's malt requirements will be produced in Scotland by the distilling companies themselves or their subsidiaries, and this factor will further emphasise the swing to Scottish barleys.

2.51. Bulk handling will continue to develop and to replace handling in bags. The new Transport Act will have some bearing on the use of road and rail transport in the future, and present researches by some firms might give rise to greater use of coastal shipping facilities.

2.52. At the time of writing, the main worries of the industry are the overall recession in grain distilling, and the effects of present and future government taxation policies on demand for whisky. The consequences of the Transport Act could affect small isolated distilleries particularly adversely. General fears of higher cereal prices as a consequence of possible entry into the European Common Market add to a prevalent air of anxiety within the industry.

PART V

Utilisation of Grain in the U.K.

(d) Grain for Industrial Uses and for Cereal Breakfast Foods

CHAPTER 1

Grain for Industrial Uses and for Cereal Breakfast Foods

Grain for Industrial Uses

1.1. Apart from a negligible quantity of wheat (5000 tons),[1] the main cereal used for industrial purposes is maize. Some 500,000 tons of maize are used per annum for the manufacture of glucose, dextrose and starch, with corn oil and animal feeds being obtained as valuable by-products (Table 1.1).

1.2. In the manufacture of starch and glucose the separation of the

TABLE 1.1

DISPOSAL OF GRAIN FOR INDUSTRIAL USE AND CEREAL BREAKFAST FOODS IN THE U.K., 1946–67 ('000 tons per annum)

July/June years	Industrial use		Cereal breakfast foods		
	Wheat	Maize for starch and glucose	Wheat	Oats (oatmeal etc.)	Maize (corn flakes)
1946/7 to 1950/1 (average)	3	130	[a]	294	47
1951/2 to 1955/6 (average)	6	184	[a]	194	87
1956/7 to 1960/1 (average)	5	294	[a]	154	98
1961/2	5	357	[a]	136	106
1962/3	5	383	[a]	141	113
1963/4	5	445	[a]	126	151
1964/5	5	(568)[b]	[a]	123	[b]
1965/6	5	(637)[b]	[a]	129	[b]
1966/7 (prov.)	5	(652)[b]	[a]	118	[b]

[a] No statistics are available but the total is estimated to be about 75,000 tons for 1966/7.
[b] Maize for corn flakes has not been separately recorded since 1963/4.
Note: In addition to the uses shown in the table, about 25,000 tons per annum of barley is used for pearl and pot barley, and about 15,000 tons per annum of rye for flour and meal and rye-based products.
Source: Commonwealth Secretariat: *Grain Crops;* and M.A.F.F.

[1] Probably sales of wheat flour unfit for food—see E. N. Greer, New uses for milling products, *Milling,* Vol. 110, No. 4.

S

constituent parts of the grain takes place wet and the process is called "wet-milling". The output of starch is about half that of glucose and dextrose. The starch is used for such purposes as packaged foods, paper, textiles and adhesives; glucose is used in confectionery, jams and preserves, beer, soft drinks and baking; and dextrose in pharmaceuticals, cider and soft drinks, confectionery and baking. The use of glucose in the form of wort syrups in brewing is a likely growth factor.

1.3. The maize used for starch and glucose comes mostly from America, with some from Mexico, South Africa and France. There is a tendency to switch to South African White Maize, but this is unlikely to go very far (unless there is a big price advantage) because the yellow corn yields (as a by-product) the coloured corn oil most favoured by the housewife. One large Manchester-based company dominates the industrial maize market. It is situated on the Manchester Ship Canal and can receive imported maize direct in vessels of up to about 10,000 tons. The other wet-millers, which are very much smaller, are located in London and a substantial proportion of their imports of maize are transhipped via Rotterdam.

1.4. From the viewpoint of the marketing of home-grown grain, the main interest in this topic lies in the possibility of home-grown grain being substituted for imported materials such as maize, corn starch and tapioca as a source of starch. Only a few thousand tons of wheat are used in starch manufacture including some wheat starch which is produced as a by-product of starch-reduced bread manufacture. However, Canada has some large plants for the separation of wheat starch, whilst Australia uses only wheat for this purpose and discourages the import of maize with a high tariff. Starch can also be made from barley but little seems to be known as to whether this would be economic or whether the product would be commercially successful. The possibility of considerably more home-produced grain being used as a base for starch manufacture in this country certainly cannot be discounted, and the H.-G.C.A. might well stimulate further research in this field. Another possibility worth looking into is that cereal flour might be used as a filler in paper manufacture; some 50,000 tons of starch are at present used in paper manufacture. Neither of these possibilities may be economic at present but if cereal production increases faster than demand, and grain prices fall as a result, such outlets for home-grown grain may assume greater importance in the future.

Cereal Breakfast Foods

1.5. Cereal breakfast foods are made either from oats (oatmeal and porridge oats), wheat, or maize. The popularity of porridge has declined in the last 20 years, not only in Britain but in most parts of the world, whilst other breakfast foods have expanded their sales. This decline is likely to continue, although the recent introduction of new types of cereal of the "instant porridge" variety may go some way to arrest the decline in terms of value of sales, if not in quantity. No statistics are available for the production of cereal breakfast foods made from wheat but after conversations with manufacturers we estimate the annual total to be about 75,000 tons. The principal grain used in cereal breakfast foods is maize for corn flakes, and Table 1.1 suggests that about 200,000 tons of maize per annum are used for this purpose.

1.6. The oats used for porridge are mostly home-produced, but the wheat used for cereal breakfast foods is almost wholly imported. One manufacturer of cereal breakfast foods gave as the reason for this that hard wheats are needed of a type not produced in the U.K.; another, that imported white winter wheat from Australia, unlike English wheat, is available at moisture contents of 8–10% and at uniform quality and variety. In the latter case there would seem to be no basic reason why English wheat could not fulfil the requirements, especially if a merchant, co-operative, or producers' marketing group could assure regularity of supplies of the required variety, grade and moisture content. Imported wheat has to be stored up to 8 weeks (the period between shipments) and a lot of capital is tied up this way which could be released if English grain were bought on forward contract—there would also, of course, be a substantial saving in freight costs. It may be that imported wheat continues to be used simply because when English wheat was last used during the 1939–45 War (when a proportion of English wheat had to be used compulsorily) it could not then satisfy the manufacturer's requirements. If a merchant, co-operative, producer marketing group, or indeed any such marketing agency, could organise the supply of grain for this purpose the use of English wheat might become a practical possibility and one that would be well worth investigating.

1.7. Maize used for corn flakes is, of course, all imported and comes either from the United States or Argentina. Plate maize is not used at all and, in fact, it would result in a different product from that which consumers now expect, so that it is unlikely to be used even if the price is relatively attractive. Wheat cannot be substituted for maize to obtain an identical product. The principal manufacturer of corn flakes in Britain (there are only two and one is small) has introduced "Wheat Flakes", but as a completely different product complementary to corn flakes rather than as a substitute. There is no possibility, therefore, of substitution of wheat for maize in the production of "corn flakes".

PART VI

The World Grain Situation

Introduction

This report is primarily concerned with cereals marketing in the U.K. However, the domestic grain market is very much influenced by what is happening in the international grain trade, and developments in cereals marketing in countries supplying grain to the U.K. (or competing with the U.K. for grain export markets) often have a direct impact on the U.K. grain situation. Therefore the world grain situation cannot be ignored, yet equally one cannot hope to cover such a vast subject adequately in the limited space available. What has been attempted in the following chapters is a selection of those aspects of cereals marketing in other countries, and those features of the international grain trade, which seem most relevant to the U.K. situation. Thus the European chapter describes the cereals marketing arrangements in three countries only (the Netherlands, France and West Germany) since their general farming conditions most nearly resemble those in the U.K.; this chapter also includes an account of the E.E.C. regulations for grain. The other country chapters, covering the U.S.A., Canada and Australia, deal with aspects of particular interest to the U.K. grain trade, e.g. systems of grading, the operation of the Wheat Boards, and the importance of the U.K. as a destination for these countries' exports. The chapter on the international cereal situation looks at the present problems of world trade and assesses the prospects for the future.

Grain Marketing in some
Western European Countries

The Netherlands

1.2. In recent years the total grain harvest has been averaging about 1·8 million tons, of which around 1 million tons are feed grains. The Netherlands also import $4\frac{1}{2}$–5 million tons of grains, mostly for animal feed. Exports have been rising partly as a result of the growth of the transshipment business and partly through increased exports of Dutch grain, particularly feed wheat. Total exports are now in the region of 2 million tons a year. The quantity of grain retained on farms has fallen recently (from 505,000 tons in 1962/3 to 243,000 tons in 1965/6). The change is mainly attributable to increasing specialisation of farm production. Few livestock farms in the Netherlands now produce sufficient quantities of grain so as to make on-farm mixing economic. Due principally to E.E.C. regulations (discussed later) there has been a rather similar decline in the amount of grain used for compound feeds. By contrast the industrial use of grain has expanded over the past few years and has now reached nearly 500,000 tons, about half of which goes to the starch and spirits industries.

1.2. The main users of grain in the Netherlands are rather highly concentrated. In flour milling there are two large private groups, and the Consumers' co-operative also has an interest in milling. These three take up some 85% of production. The Dutch brewers operate a central buying organisation responsible for purchasing all their raw materials, including grain. The compound feed industry is also concentrated. Production is divided roughly equally between the private sector and the co-operatives, which are organised in two centrally controlled groups. The private sector chiefly consists of about a dozen large compounders each producing over 20,000 tons a year, of which Ut Delfia (a Unilever company) is the largest, and over 1000 small-scale compounders. At present the output of the private sector is divided roughly equally between the large and small compounders,

521

but the former are tending to increase their share of the trade. Competition is fierce, both on price and services, including credit, and in this respect the small local compounder is at a disadvantage.

CO–OPERATIVES AND MERCHANTS

1.3. The grain trade in the Netherlands is divided roughly equally between the co-operatives[1] and the private merchants. Due largely to the fact that co-operatives have a bigger share of compound feed manufacture (about 45%) than of flour milling (about 10%) they handle a relatively greater quantity of feed grains than of wheat, in the marketing of which the merchants predominate.

1.4. Co-operation is a well-established feature of life in the Netherlands, but its particular strength is in agriculture, where it has flourished for nearly a century. The movement has developed "naturally", that is to say without special grants or credits as in France, and without even any special taxation advantages. Its development has been much assisted by the parallel growth of the two Dutch agricultural credit banks which, despite a growing diversity of business, still regard the support of agricultural co-operation as their first responsibility.

1.5. There are some 900 agricultural co-operatives in the Netherlands, but very many of these do not operate commercially. They have been merged into larger co-operatives, but still retain their identity and often function as social clubs. In this way the "grass roots" flavour and the sense of farmer participation, important ingredients in co-operation, have been retained.

1.6. There are two central co-operatives, C.E.B.E.C.O. and C.I.V., and practically all Dutch agricultural co-operatives are affiliated to one of them. There are important differences between the two organisations which affect grain marketing. The C.I.V. co-operatives, being mostly in livestock farming areas, are basically requirement societies. Much of the grain bought from members is retained by the co-operatives for manufacture into compound feeds. One of the C.I.V. regional co-operatives owns one of the largest plants in Europe, located at Veghel, which is capable of producing about 600,000 tons of compounds a year. In the C.I.V. organization there are only two co-operatives (at Sas van Gent and Rosendaal) which deal in

---

[1] Agricultural co-operatives in Western Europe have been established in a number of different ways. In general they correspond to the agricultural marketing and requirement societies set up in the U.K. under the Industrial and Provident Societies Acts, except that members are usually liable, in whole or in part, for their debts, whereas in the U.K. members are only liable to the extent of their shareholding. For a fuller account of agricultural co-operation in the E.E.C. see *La Co-opération Agricole dans la Communauté Economique Européenne*.

grain on any scale. So far as imports are concerned C.I.V. acts solely as the buying agent of the four regional co-operatives.

1.7. C.E.B.E.C.O. covers more of the arable part of the Netherlands. Consequently it handles a much larger proportion of grain, about 35–40% of all Dutch production. There are some 350 members of C.E.B.E.C.O. but, after allowing for the amalgamations among co-operatives described earlier, there are only about fifty dealing in wheat and about twenty-five in barley. As concentration in the co-operative movement continues, these numbers are tending to decline. Most Dutch farmers producing grain are members of a co-operative, but they are not obliged to sell their grain exclusively to it. Shortly before harvest time a representative from the co-operative calls on each farmer to talk about grain, and to try to persuade the farmer to sign a contract to sell his output to the co-operative. Normally the co-operative is only interested in buying the whole of a certain crop, expressed in hectares rather than tons to allow for the uncertainty of yields, but the farmer can sell his wheat to the co-operative and his barley elsewhere.

1.8. At present there are four bases on which the co-operative will purchase grain. The farmer can accept a fixed price either at harvest time or, if he has storage facilities, whenever he delivers his grain. This, of course, is the normal method adopted in the U.K. In the Netherlands it is referred to as the "today's price" system. Alternatively, the farmer can deliver his grain to the co-operative and then give instruction for sale at the prevailing price later on, and he would then be paid this price less costs of storage, handling, cleaning, drying, etc. Finally, there are two alternative "price-pooling" systems. With the first the individual farmers retain some control over the timing of sales. With full pooling the co-operative is given complete discretion over when and how much to sell. Farmers who agree to pooling are paid about two-thirds of the expected price soon after they deliver their grain to the co-operative. A further payment is made in January or February, and a final payment, based on average realisations, after the end of the season.

1.9. Practically all grain bought by the co-operatives, with the exception of quantities retained for seed or for manufacture into compound feeds, is then sold to the C.E.B.E.C.O. head office in Rotterdam. At present there is usually a "gentlemen's agreement" between C.E.B.E.C.O. and the co-operatives affiliated to it that the latter will not sell grain elsewhere. The current procedure is for each transaction to require the approval of the co-operative that owns the grain. C.E.B.E.C.O. wishes to introduce a system whereby discretion on sales is handed over to the centre, and the co-operatives will simply advise what quantity of various grains they have for sale. The system, which would be a further pooling of pooled grain, would be accompanied by a drive to get all farmers to agree to sell on a pooled basis. Clearly such a system would greatly strengthen the marketing strength of C.E.B.E.C.O., which claims that about 75% of co-operatives are already prepared to agree in principle to it. On the other hand, it would further weaken farmer control.

1.10. By buying imported grains exclusively for its members, by the "gentlemen's agreements" referred to earlier, and by the influence that it has on the credit policy of the agricultural banks, C.E.B.E.C.O.'s grain marketing has already taken on a strongly centralised character. It is now hoped that further benefits can be obtained for members by giving the central organisation full control over grain marketing, a step which the management justifies by pointing to the existing concentration in grain buying in the Netherlands for milling and brewing and, to a less extent, for compound feeds.

1.11. These developments in co-operative grain marketing have presented the private merchants with a highly competitive situation. Their share of the total market has been retained at the cost of large investments in grain silos, which are commented on in the following section. Most merchants complain about shrinking profit margins. A number have gone out of business. At least one has made links with an international shipper, a move which provides access to both capital and better sources of market information. Some of the surviving merchants now collaborate more closely together. But there seems to be some despondency about how long the merchants will be able to sustain the low return on capital employed which competition with the co-operatives and the E.E.C. grain system itself induce.

1.12. Most merchants deal on the "today's price" basis, and the drive towards complete pooling by the co-operatives may give them some temporary advantage, since not all farmers favour it. To estimate the effect of this is difficult, particularly as some merchants, mostly in Groningen, have already been forced to introduce a pooling system. The traditional business of the Dutch grain merchant has been further upset by the entry of the international shippers into the domestic grain market. Having established offices in Rotterdam (or in one case Amsterdam) it was a rather logical step in the conditions of the Dutch market for shippers to begin selling imported grains direct to local mills, thus by-passing the merchants, and even operating on country grain exchanges.[1] These developments justify the rather pessimistic attitude to the future of many Dutch grain merchants.

1.13. Concentration has also occurred at the centre of the grain trade in Rotterdam. Shippers' agents have become a rarity. One firm of Rotterdam

---

[1] Groningen, the most important country exchange, is regularly attended by representatives of international shippers. This market is unusual in that most farmers sell their grain through agents who work on a commission of about 1% paid by the farmers.

brokers does a very substantial business, but the number of brokers has declined and there are now no more than three or four of any importance. Similarly there are now far fewer large importer-merchants, who need to have good inland trade connections, and also storage facilities, in order to be able to survive in competition with the shippers. In 1965/6 the total number of grain importers in the Netherlands was 111. Many of these imported only very small quantities. Only thirteen imported more than 75,000 tons (about 2% of total imports). This figure includes both the international shippers and the two central co-operatives (who together accounted for about 20% of all imported grain). The number of genuine importer-merchants importing more than 75,000 tons is probably no more than three or four.[1] The number of firms engaged in exporting grain is also falling. In 1965/6, 125 firms were concerned with Dutch grain exports (about 1·1 million tons) compared with 149 in the previous year. But of these 125 only twenty-three exported more than 5000 tons.

1.14. *Grain storage.* Some evidence on grain storage in the Netherlands is contained in a report by the Commodity Board published in 1964. This report showed that storage capacity totalled about 2 million tons, of which 73% was controlled by private firms, including millers/compounders, and 27% by co-operatives. Unfortunately the evidence from this report is now largely out of date. In recent years the co-operatives have very greatly added to their storage capacity, including some silos erected largely to assist foreign business, such as one at Zwijndrecht with a capacity of 38,000 tons and another of 23,000 tons at Delfzijl in Groningen. Merchants have also added to their capacity. After taking account of port storage facilities there appears to be some surplus grain storage capacity in the Netherlands.

1.15. The cause of this excess capacity can be traced to conditions at harvest time. The small-scale structure of Dutch farms and good rural communications have favoured the construction of central grain drying and storage installations. Outside the province of Groningen where there is a maximum of 100,000 tons of on-farm storage capacity, farmers are dependent on facilities away from their farms. In the highly competitive situation between co-operatives and merchants, contracts depend on the ability of either side to store the grain offered by farmers. There is a rush at harvest time to secure grain without, in general, bothering too much about the price, which is often based on subsequent realisations. But in order to secure the grain both sides must have sufficient storage. Hence the tendency towards surplus capacity.

[1] In international trade grain is very frequently bought and sold many times before it reaches a consumer. The total of the turnover of the shippers and importers would be very many times greater than the total shown in the import statistics.

1.16. Normally farmers bring their grain to the central store in their own transport, using sugar-beet and potato trailers adapted for grain. In some areas there is some road congestion, but this does not seem to be a big problem. Transport from the central store is by barge or coaster, most silos being located on canals, or by trucks, which are usually hired by the co-operatives and merchants on contract. Rates for both types of transportation are freely negotiated and vary through the season, being highest usually at the time of the sugar-beet harvest.

1.17. *Government and the grain trade.* The main link between the agricultural industry and the trade is provided by the Commodity Boards. The Commodity Board for Grains, Seeds and Pulses consists of thirty-six members who are appointed on the nomination of the organisation they represent (farmers, millers, merchants, etc.) for a period of two years. The Board has a general responsibility for ensuring that circumstances exist which promote the orderly marketing of grains, seeds and pulses. In this task it acts partly on its own initiative and partly in collaboration with the Government. It has a right to express its views to the Minister of Agriculture or to the principal government advisory body, the Social and Economic Council. It has powers to impose levies to cover its own administrative costs and to raise funds for research purposes. Its regulations, which must be approved by the Minister of Agriculture, are legally enforceable. The Board has a permanent staff of about seventy and offices in The Hague. In practice the Board's main activities can be summarised as follows:

(i) Acting as an agency for the Commission of the E.E.C. the Board works closely with Brussels, providing daily information about prices, issuing import and export licences and collecting levies and paying restitutions. It would be consulted if intervention buying of grains became necessarily under the Community's common grain regulations.

(ii) Collaborating with the Government. The Board works in an advisory capacity and represents grain producers, processors and traders at international meetings, along with representatives of the Dutch Ministry of Agriculture.

(iii) Stimulating research. The Board makes research grants to the Dutch Grain Centre Foundation at Wageningen.

(iv) Providing information to the trade. The principal publications of the Commodity Board are its very comprehensive Annual Report and periodical reports on special subjects within its field, such as the report on facilities for grain storage mentioned earlier. However, the Board does not have sole responsibility for providing trade information. A good deal of statistical data is published regularly by the Central Bureau of Statistics and the Agricultural Economics Institute.

SUMMARY

1.18. The main features worthy of note in the Netherlands system of grain marketing, apart from those which relate to membership of the Common Market and operation of its arrangements, are:

(i) a strong tendency towards concentration among end-users, co-operatives and merchants;

(ii) a flourishing co-operative system of organisation, with two central or "apex" co-operatives, having a substantial share of the total compound feed manufacture;

(iii) grain storage taking place mainly off the farms, with a tendency to excess capacity and vigorous competition between co-operatives and merchants to obtain supplies at harvest time;

(iv) the development of "price-pooling" systems of marketing, with varying degrees of farmer and co-operative control over the timing of sales.

France

1.19. France is much the largest grain producer in Western Europe, the total harvest amounting to nearly 30 million tons. Wheat is the most important crop: recently production has been running at about 13 million tons. Barley production has been increasing rapidly and reached about 9 million tons in 1967, roughly double the average of the 1950's. By contrast, the quantity of oats grown in France has declined, as elsewhere in Europe, and now amounts to only about $2\frac{1}{2}$ million tons. The biggest increase has been in maize, production of which, stimulated by new varieties and improvements in irrigation, now exceeds 4 million tons.

1.20. The number of farms growing cereals and contributing to the "collecte", the total quantity sold off farms, is very large, but over the past 30 years there has been some concentration of production. In 1936 O.N.I.C. (Office National Interprofessionel des Céréales, whose activities are described later) has estimated that about one and a half million families were dependent in whole or in part on the cultivation of wheat. By 1949/50 the number of farmers contributing to the "collecte" had fallen to below one million, and by 1964/5 to 730,000. The importance of large farms as suppliers of wheat is illustrated by analysis of deliveries of wheat in that year by quantity. The 12,500 farms (only 1.7% of the total) which delivered more than 100 tons contributed just over 25% of the whole "collecte". In the same year 2.8% of the 212,000 farmers who sold all or part of their barley contributed 27.4% of the total "collecte".

1.21. The relative importance of grains in total arable production has slightly increased over the past 10 years and now constitutes about one-third of its value. The Fifth Plan (1966–70), which is based on the assumption that farm incomes will rise at a more rapid rate than non-farm incomes (4·8% annually over the period of the Plan compared with 3·3% for non-farm), forecasts that soft-wheat production would rise to $15\frac{1}{2}$ to 16 million tons, representing a considerable slow-down in the rate of expansion of production. Total cereal production, according to the Plan, should remain at about the same level in 1970 as at the beginning of the Plan, i.e. about 30 to $31\frac{1}{2}$ million tons.

1.22. Given average weather conditions it seems almost certain that the Plan's forecasts for grain production in 1970 will be considerably exceeded. Average yields are still low—wheat about 24 cwt. per acre in the mid-1960's—and the E.E.C.'s common grain prices should continue to stimulate production. Finally, it is worth stressing that not only has production been increasing but that the "collecte" has been increasing still faster. This has been particularly striking in the case of feed grains. In 1950 only 15% of production was sold off farms compared with nearly half in 1966. The average total "collecte" immediately after the Second World War (1946–50) was about 5 million tons. By 1966 the total "collecte" of grains had increased threefold to 15 million tons. This is the quantity, plus about $1\frac{1}{2}$ million tons of imported grain (mostly hard wheat and maize), which the French grain trade has to market either domestically or for export.

1.23. *Co-operatives and merchants.* The present structure of the French grain trade was profoundly influenced by the establishment in 1936 of the Office National Interprofessionel du Blé which developed 4 years later into O.N.I.C. Faced by a disastrous decline in grain prices since 1929, Léon Blum's Popular Front government decided to intervene more directly in French grain marketing. The Office du Blé had three main features. First, its structure was "interprofessionnel", that is to say it was composed of producers, merchants, processors, consumers and the government departments concerned, charged with arriving at decisions (for instance, on the price of wheat) for which common interests are lacking. Delegation by the State of powers of this kind was an important innovation in French agricultural policy. Second, it was given a monopoly on imports and exports of wheat and made responsible for forecasting and dealing with surpluses. Third, the law of 18th June 1936 also established the principle of the system of *organisme stockeur*, which has greatly affected the structure of the trade.

1.24. Part of the blame for the breakdown of grain markets in the years 1929 to 1936 was attributed to the activities of the French flour millers who, under the pressure of excess capacity, forced down producer wheat prices. Some blame was also attributed to the private grain merchants, not only for failing to provide sufficient grain-storage capacity, which might have mitigated the extreme swings in grain prices, but also for taking, or so it was suspected, an undue share of the prices paid by the grain users. The *organisme stockeur* was set up in order to insulate the flour millers from grain producers and in order to ensure that rules about minimum prices were observed. Both co-operatives and private merchants were eligible for approval as *organismes stockeurs*, and such approval was normally forthcoming. In order to help them with their financing of grain, *organismes stockeurs* were allowed to obtain special rediscount facilities from the Banque de France via the Crédit Agricole provided that the contracts had been approved and stamped by the Office du Blé. But from 1936 until 1952 approval was limited to contracts made by co-operatives, who thereby gained a great advantage to add to their existing advantage of access to credit facilities from the Crédit Agricole on favourable terms for the construction of grain silos, grain processing plants, etc. Very largely as a result of this advantage, co-operatives now control 80% of the wheat "collecte" and about 65% of the coarse grains "collecte".

1.25. Practically all French agricultural co-operatives trading in grains are affiliated to one of the two central co-operative grain organisations. These are the Union Nationale des Coopératives Agricoles de Céréales (U.N.C.A.C.)[1] which is often called the Groupe MacMahon from its address in the Avenue MacMahon, and the Fédération Nationale des Coopératives de Céréales (F.N.C.C.) also known as the Groupe Lafayette. These two compete actively against each other but make use of a Federation, to which each belongs, when they want to present a joint viewpoint for the benefit of the co-operative movement. For a number of years they have worked together on agricultural research projects, and it is not impossible that they might ultimately merge. The two groups do about the same amount of grain trading.

1.26. Just before the Second World War there were about 1300 agricultural co-operatives in France which dealt in grain. The number has fallen to about 800. The tendency to concentrate has been especially strong in the South-west where the co-operatives face most competition from the private sector.

1.27. Until very recently most grain bought by local co-operatives from their members has remained within the co-operative network until sold by the central offices of the co-operatives to the millers, the compounders, etc., or shipped for export. Local co-operatives have sold to the regional co-operatives who, in turn, have normally sold grain on to Paris. The

[1] See also the statement by the Agricultural Co-operative Association in Appendix G.

regional offices have also acted as brokers on behalf of the local co-operatives. This centralised system is now tending to break down. While the larger export business is, as one would expect, still handled from Paris, the regional and local co-operatives also handle exports, this typically consisting of barge and rail traffic into the Rhine Valley, or short-haul business to Spain or Switzerland. They are also increasingly selling to the private sector, either direct (through brokers) to private processors or to independent grain merchants. The regional and local co-operatives still depend greatly on market information from their central offices.

1.28. Despite the disadvantages which they have suffered there are still about 2000 private merchants in France who are trading in grain. The number in 1936 has been estimated by O.N.I.C. at about 2650. At the summit of the trade are the international shippers, some of whom are French by origin and all of whom have offices in Paris. The shippers are increasingly making links with the larger country merchants, many of whom do some exporting, particularly intra-E.E.C., on their own account. In 1964–5 there were only 151 merchants (about 7% of the total) doing more than 20,000 tons a year, and probably less than fifty of these were private merchants. At the bottom of the scale are the many small private and co-operative merchants. It is expected that their number will continue to fall. Now that the private trade suffers from fewer disadvantages it may regain some ground previously lost to the co-operatives. It obtains re-discount facilities, making use of O.N.I.C.'s *aval* with the Sociétés de Caution Mutuelle. The terms are now much the same as those which co-operatives get from the Crédit Agricole.

1.29. Unlike the Netherlands, very little grain business, either domestic or international, is done without the use of the brokers. They provide one of the main sources of market information. The brokers' position is strengthened by the diversity of the French grain market, the large number of merchants and processors and the many areas in which grain is produced. Nevertheless, expressions of discontent are frequently heard about the unusual hold on the French grain market that brokers retain.

1.30. *Grain storage.* O.N.I.C. makes periodical censuses of grain storage capacity in France. The most recent, in January 1966, included the following:

Farm storage	3,357,000 tons
Storage through *organismes stockeurs*	8,663,000 tons
Transit storage	1,771,000 tons
Storage in processing industries	659,000 tons
	14,450,000 tons

1.31. It will be seen that the storage provided by the *organismes stockeurs* is much the most important, representing about 60% of total capacity. It is doubly important because it is at this stage that most of the drying and cleaning of grain is done. Both the co-operative and the private *organismes stockeurs* have greatly increased their grain storage capacity over the past thirty years, the co-operatives from 1,800,000 tons to 6,313,000 tons and the merchants from 800,000 tons to 2,350,000 tons. Of the total of 8·7 million tons of grain storage capacity only a little over 1 million tons consists of inland silos. Three million tons of capacity is in the ports and a further 4¼ million tons located on rivers, both of these being basically intended for international trade. Of the 650,000 tons of storage in the hands of processors about 450,000 tons belongs to millers and compounders and the remainder to other processors, mostly starch manufacturers.

1.32. The forecasts of the French Fifth Plan have already been mentioned. These suggest that the total cereal harvest in 1970 will be a little over 31 million tons and that the "collecte" will amount to 20 million tons. It was earlier suggested that these forecasts may well be exceeded. Present grain storage capacity is scarcely sufficient for the current "collecte", even after taking account of the length of the harvesting period and the fact that large-scale export business is usually done immediately after the harvest. The shortfall may become severe by 1970 and the Plan therefore calls for silo construction during the Plan period of 800,000 tons. This figure is for the whole of France. Certain areas, especially in the Centre and South-west, are particularly short of grain-storage capacity and need relatively more new silos. With the exception of the Paris Basin area most grain drying is still undertaken by the *organismes stockeurs* (now *collecteurs agréés*), most of the larger ones having drying facilities. The number with such facilities is now about 1100, and the total drying capacity is about 6600 tons an hour for a moisture reduction of 4 points. In addition it is estimated that some 2·6 million tons can be dried by ventilation or refrigeration. The Fifth Plan also calls for a major extension of drying facilities amounting to 1500 tons an hour, to be carried out during the Plan period. The object is that drying facilities should be sufficient to deal with not less than a quarter of the total grain harvest, or about 8 million tons.

1.33. *Government and the grain trade.* In France there has been a long history of government intervention in grain marketing which goes back to the twenties. A wide variety of techniques have been employed including tariffs and quotas on imports, controls on grain exports, regulations on the use of domestic grain by the millers, taxation raised through the *passage obligatoire*, intervention buying by O.N.I.C., subsidies on denaturation of wheat, quality controls on flour and other products based on grains, and the provision of cheap rediscount facilities through O.N.I.C.'s guarantee, a privilege which, as has already been shown, was reserved to the co-operatives until 1952.

1.34. In addition, O.N.I.C. has tried to improve marketing practices by issuing model forms for grain contracts, by improving arbitration practices and by making occasional studies of matters of special interest to the trade, such as grain-storage capacity. O.N.I.C. also makes available reports on crop conditions obtained from its representatives in each department and from its thirteen regional offices, production forecasts, and statistics of actual grain production and yields. It could hardly be expected that O.N.I.C. would escape criticism for the way that it has carried out these responsibilities. But the criticism which has come from the Chambres d'Agriculture, the F.N.S.E.A. (Fédération Nationale des Syndicats d'Exploitations Agricoles), the principal farmers' union, the co-operatives and others has, by farming standards, been mild. Perhaps part of the explanation for this lies in the fact that these organisations and the important and politically influential A.G.P.B. (Association Générale des Producteurs de Blé et autres Céréales) are closely involved in the Central Council of O.N.I.C., and therefore have to take some responsibility for its activities.

Germany (Federal Republic)

1.35. During this century the German grain market has been just as much insulated from the outside world as the French. The role of the Government in protection of its domestic market is commented on later in this section. Prices for all grains were kept at very high levels, even during the latter stages of the Community's transition period, the German Government declining to adapt grain prices gradually to those prevailing elsewhere in the Community. As a result German prices had to be reduced abruptly for the first season of harmonised Community grain prices. The anticipated reduction in income for German farmers will anyway be cushioned by direct subsidies, paid partly from the Agricultural Fund, up to 1970, originally the last year of the transition period. But, fortunately, the 1967 harvest was an exceptionally good one amounting to about 18 million tons. On the basis of previous years' experience not more than half this quantity, or about 9 million tons, will enter the market, the remainder being held back on farms. Despite the record harvest in 1967 imports remained high. The Statistisches Bundesamt has reported that imports for the calendar year 1967 totalled about 6·6 million tons (a little up on the previous year), of which about 2 million tons were bread grains.

1.36. While the total number of mills has fallen from 12,000 in 1955 to under 4000, the German milling industry still remains remarkably unconcentrated, though three or four groups associated with grain importing and feed compounding interests are becoming of increasing importance. Of the 250 larger mills represented by the Arbeitsgemeinschaft Deutscher Handelsmühlen, about eighty have a capacity of over 10,000 tons a year.

Since 1962 there has been no compulsion on the mills to take up a certain proportion of the home crop. Hard wheat imports have recently constituted 25–30% of total supplies. German compound feed production is similarly fragmented, though concentration is now taking place fairly rapidly. There are about half a dozen firms with annual output exceeding 100,000 tons a year which distribute compounds over most of Germany. Some of these have close links with the major grain importing companies. There are also about twenty firms which are manufacturing on a regional rather than a national scale. Then there are the many hundreds of small local compounders operating over a very small area and often buying grains direct from farmers. The future for these firms, which are probably responsible for about a quarter of the compound feeds produced by the private sector, seems extremely uncertain. In particular it is difficult for them to compete with the technical advice which the larger firms provide. Finally, the co-operatives have very greatly increased their capacity for compound feed production since they started in 1956. Nearly all the fourteen Hauptgenossenschaften (see below) have compound feed plants. Production now totals about 2 million tons a year, roughly a third of total output. It is largely as a result of the expansion in the co-operative sector that national manufacturing capacity currently greatly exceeds output of compounds.

1.37. *Co-operatives and merchants.* Of the total quantity of grain marketed some is sold to co-operatives, some to private merchants and some direct to processors, mostly the small local flour millers and compounders mentioned earlier. The proportion purchased directly by the latter, 16·4% of the total in 1966/7, has remained about the same as ten years earlier. But the co-operatives have greatly expanded their share of the market. Comparing 1956/7 with 1966/7 the co-operatives' share has increased from 40% to 47·1%, while that of the private merchants has fallen from 44·3% to 36·5%.

1.38. Co-operatives (Genossenschaften) cover the whole of the Federal Republic. There are several thousand of them, most combining marketing with requirements business, though a few specialise in wine or fruit and vegetables. On the commercial side they are grouped under fourteen regional co-operatives (Hauptgenossenschaften) which buy the bulk of the grain gathered by the local co-operatives. The regional co-operatives are linked through a central organisation at Frankfurt, but this does virtually no trading and does not provide the strong leadership of, for instance, C.E.B.E.C.O.'s head office in Rotterdam.

1.39. The political head of the German agricultural co-operatives is the Deutscher Raiffeisen Verband based on Bonn, which has regional co-operative associations corresponding with the Länder. These are responsible for the supervision of the activities of co-operatives under German law. As a centre for political negotiation the Raiffeisen Federation is much more important in commercial terms than is the corresponding organisation in Frankfurt. The Federation represents the whole agricultural co-operative movement, the membership of which totals over 4 million, in all negotiations with government agencies. It is a very powerful political organisation. In addition to carrying out its general supervisory responsibilities towards the regional co-operatives the Federation also provides them with regular and comprehensive information on commodities, including grains. In this respect the commodity section of the Federation has been very active since the common agricultural policy came into force. The Raiffeisen Annual Reports provide comprehensive information on all aspects of agriculture in Germany. The Federation is responsible for training co-operative managers and is closely concerned with research on co-operation which is carried out at five university institutes including one, at Giessen, exclusively concerned with agricultural co-operation.

1.40. Most grain bought by the local co-operatives is sold on to the regional co-operatives. Local managers are urged to deal exclusively within the co-operative organisation, and are actively discouraged from attending markets and in general from developing contacts with the private sector. The regional co-operatives, however, necessarily deal with the whole grain trade. Some of the grain they buy is used in their own compound feed plants. The remainder is sold either direct to mills (there is one mill, at Mannheim, owned by a consumers' co-operative), compound feed manufacturers and other processors, or to merchants in the private sector at the wholesale level. Co-ordination in grain marketing between the regional co-operatives has yet to be achieved.

1.41. At grassroots level the private grain trade consists of the Landhändler, a small-scale agricultural merchant, supplying fertilisers, compound feeds and other farm requirements and purchasing grain from farmers. Traditionally the Landhändler has depended on close links with his customers who often rely on him for market, and even husbandry, advice. In most parts of Germany there exist associations, called Landhandelsvereinigungen, to which the Landhändler sell most of their grain, the remainder being sold to local mills or compounders, or to the wholesalers. The latter, numbering some 250, provide important trade channels for fertilisers and compounds, manufacturers of which find them a useful means of marketing with the minimum of credit risk. So far as grain is concerned the traditional wholesalers have already lost business to the Landhandelsvereinigungen, who now duplicate their role, but they remain important customers for the importers/shippers for maize, sorghum, etc. Their future in a world of specialisation and concentration appears to be very uncertain. In the struggle between the co-operatives and the Landhandel it seems inevitable that further gains will be made by the co-operatives, who are expanding into egg-packing stations, broiler production and slaughterhouses. Many firms of private merchants have been bought by co-operatives, the ex-proprietors becoming (if they are the right age) co-operative managers. Against increasing managerial professionalism, and the well-known advantages of agricultural co-operatives in E.E.C. (good access to credit, community links and a modest tax advantage) the private trade is likely to lose more business in the future. Some of the international shippers and the few large grain-importing companies that remain have penetrated the German domestic market. Two of the major shippers have established a number of offices in the interior, which are active in both selling imported grains direct to compounders and millers and trading between them and co-operatives, wholesalers and the Landhandelsvereinigungen. The importers trade similarly; no less than three of them have representation in Bonn to act as listening-posts and to facilitate negotiation at government level. Links between importers and the grain-processing industry have already been mentioned. The four largest German grain importers, one of whom also acts as a shipper, all have milling and/or compounding interests.

1.42. *Grain storage.* Wherever possible imported grain, whether from third countries or from within the Community (such as French wheat and Dutch barley), is moved direct to the consumer without intermediate storage. Domestic grain is stored at harvest time in silos owned by the co-operatives and the private merchants or in the facilities owned by the independent silo companies, the Lagerhausgesellschaften. Most of the latter's silos are old and badly placed and their business has consequently suffered. Currently their silo accommodation is being used by the intervention agency, the EVSt, in Frankfurt (see below). Clearly the 1967 harvest put strains on storage capacity. Co-operatives and merchants were able to manage the greatly increased quantity of grain by temporary expedients, such as using disused mills, and by employing facilities of the Lagerhausgesellschaften which might otherwise have been empty. In view of this, it is perhaps not surprising that the grain trade in Germany more than elsewhere in the Community has recently been complaining that monthly price increments to take account of storage costs are not sufficiently high.

1.43. *Government and the grain trade.* Soon after the handover by the Occupation Powers the Federal Government set up a number of Import and Storage Agencies (Einfuhr- und Vorratstellen, often referred to as EVSt), on which producers and merchants were represented, including one for grains. This agency remained largely responsible for grain marketing in Germany up to the commencement of the common agricultural policy and still exists, having become the intervention agency acting on behalf of the Commission in Brussels. In the pre-E.E.C. period its method of operation basically consisted of buying from importers at a "take-over price" grain from certain stated countries of origin (often as a result of bilateral agreements) and selling to the trade at a "disposal price" which corresponded with the price (usually higher) at which the Government wished to maintain the domestic market, the process being known as skimming. In addition, EVSt intervened in the market in some years on a substantial scale. In 1960/1 purchases amounted to over 2 million tons, but in the last 2 years of the transition period they were on a much smaller scale. Increased intervention became necessary in the first year of harmonised prices (1967/8;, when purchases by EVSt as a community intervention agency exceeded 1 million tons.

E.E.C. Grain Regulations and their Effects on the Market

1.44. The common market for grains came into force on 1st July 1967, at which date all impediments to intra-Community trade in grains were removed. The regulations establishing this common market which were agreed in principle as early as 1964 are, as one would expect, rather complicated. For the present purposes an outline only of these regulations will be given.

1.45. The basic aim of the E.E.C. grains policy is to provide farmers with a stable market for their products at prices which are, at least at present, far above those that would prevail without protection. These prices, which are established each year by the Council of Ministers acting on proposals from the Commission, are known as indicative (or target) prices, and represent the levels to which wholesale prices should tend in the area of greatest deficit in the Community, for which Duisburg was chosen. These prices are maintained in two ways. First by the establishment of threshold prices equivalent to the indicative prices at Duisburg less the cost of transport to Duisburg from Rotterdam, the closest port for third countries' grain. The threshold prices are implemented by variable levies which make up the difference between the threshold prices and the lowest c.i.f. prices as determined by the Commission on the basis of market reports from Hamburg, Frankfurt (EVSt), Antwerp, Rotterdam and Paris (O.N.I.C.). As the Community is still in overall deficit for grains this has so far been the more effective means of market support. The second method is support buying by intervention agencies acting on behalf of Brussels. For this purpose a basic intervention price is established at Duisburg which, for wheat, is about £3 2s. 0d. per ton below the indicative price, together with derived intervention prices which apply at other grain centres and which basically relate to the Duisburg intervention price according to the cost of transporting grain from the point in question to Duisburg. For the 1967/8 season 90% of all levies on grains (and other agricultural products) was handed over to the Guarantee Fund. This money is used for structural reform (up to a limit of about £118 million), to finance losses on sales of commodities, including grains bought through market intervention, the cost of export restitutions, etc. The latter are made necessary by the difference between Community and world prices. Both levies and restitutions are common to the whole Community. The original idea was that the two should normally be the same and that restitutions should not exceed current levies. This still applies to third countries' grain imported and then exported. But the Commission now uses its powers over restitutions to regulate the market, restitutions being varied according to conditions of supply and demand and to the country of destination of the grains.

1.46. The procedure for fixing levies on imported grains consists in the importer applying to the Community grain agency in his country (O.N.I.C., EVSt, etc.) for a licence to import a certain quantity of a stated type and quality of grain, the application being made usually by telex. The importer has the option either to fix the levy at the time, in which case the current day's levy is stamped on the licence, or to wait until the grain arrives and pay the levy ruling at the time of arrival. A levy fixed in advance is good for the current month plus the three following months. This levy is always the same for the month in question plus 1 month, but for the next 2 months it may differ in the future market and the threshold prices do not keep in step after allowing for the monthly changes in the threshold prices. If such a difference exists the importer has to pay a fixation premium representing the difference between current and future levels. The levy and fixation premium for each type of grain is announced daily. They can be changed every day, but small variations in the spread between world market prices, c.i.f., the Community and the threshold prices are permitted without a change in the levy. In the case of maize, for instance, c.i.f. offers can rise or fall by U.S. $0·60 (about 5*s*.) per metric ton compared with the last levy fixation without the levy being changed. No fixation premium is charged if the difference between current and future month's prices is less than 12½ cents (about 1*s*.) per ton. There are some special provisions concerned with the fixation premium, including powers for the Commission to increase it above the normal level if the volume of fixations is tending to disturb the Community's grain market. An importer has to lodge immediately a bank guarantee to cover his fixation, the cost of which is normally about ½% a year. There are complicated regulations dealing with failure to use a licence. The penalty for this consists of 50 cents (about 4*s*. 3*d*.) per ton if the day levy on the last day of validity of the licence is as high or higher than the fixed levy which is about to expire, or this sum plus the difference between the day levy and the fixed levy if this day levy is lower than the fixed levy. The procedure over restitutions, which can also be fixed in advance, is much the same and there are similar penalties. Strictly speaking, licences cannot be bought or sold, but in fact they are so traded and this provides much of the flexibility of the Community's grain trade. Regulations also exist on a number of other points, including denaturation of wheat, quality differentials, procedures for intervention buying, and arrangements for subsidising grain stocks at the end of the season, designed to deal with the problem of the changeover from the high prices at the end of the season to the base price of the new season which would normally be lower.

1.47. Before turning to the main effects these regulations have had on the grain trade, three general comments on their application to the market must be made. First, the regulations have produced a generally stable market but at the cost of insulating the Community from trends in world prices. This is a major disadvantage. The competition between substitutable feed grains is fossilised. If the price of one grain increases and that of another (substitutable) grain falls, the Community should be able to benefit by buying less of the former and more of the latter. But in fact the action of the levies, one falling and the other rising, neutralises this, the difference in levy-paid prices remaining the same. In this respect a fixed tariff is preferable to a variable levy.

1.48. Second, the exclusion from protection by levies of a range of secondary products, suitable for mixing in compound feeds, has given a powerful incentive to their use at the cost of grains, which, as a result of the threshold prices and levies, are relatively more expensive. Duties on oil-cakes and on soya-bean and other meals under the common external tariff are mainly nil. Other ingredients for compound feeds such as beans, maize gluten and rape seed extracts have been favoured by the protection given to grains, and their use has consequently increased.

1.49. Third, there is the difficult problem of intervention in the grain market. The Community's policy, which is designed to favour first and foremost the producer, can be implemented in two ways: by maintaining or even increasing the amount of protection against imports, or, alternatively, by propping up the domestic market through buying grain at the intervention level and then moving it off the market. Both of these two methods will be used in the future, but it is still uncertain which will become more important. One of the peculiarities of the present system has been shown by the large-scale purchases of grain, mostly wheat, which have been made during the first season of harmonised prices in Germany, although the country is still in overall deficit for grains. Most of the grain bought through intervention has come from the South. The high transport costs from this area to the main deficit areas in the Ruhr make grain from this source uncompetitive with French production. Theoretically, Bavaria should be exporting to the East, but there is the difficulty that its customers in that direction are state-trading countries. Differences in intervention prices will anyway have to be revised as the transport costs which form the basis of their derived calculation are now out of date. It seems likely that the whole strategy of intervention will have to be reviewed for grains as for other agricultural commodities.

1.50. Some of the effects on the trade of the common agricultural policy in grains have already become apparent. The system prevents large and unpredictable fluctuations in price. Losses due to wide swings in the market are now ruled out, but so also are large profits. High prices leading to surplus production depress prices down to the intervention level. If the market stays at, or very close to, this level there is less scope for profitable trade activity. Average profit margins in the trade are somewhat difficult to ascertain, but it is clear that the trend has been downwards. The trade must view with some concern the possibility of increasing intervention in the market. At present merchants are usually involved in intervention buying because the agency will only accept lots of 100 tons. But if intervention became common there seems no good reason why farmers should not organise themselves into small groups to meet these minimum requirements. At present merchants are generally involved in sales made by the intervention agency which are by tender. Here again it would be possible for the trade to be by-passed through mills dealing direct with the agency, although by so doing they would lose the benefit of credit terms.

1.51. Due to the grain-storage practices in Common Market countries described earlier, the monthly increments over the basic target, intervention and threshold prices are of great interest to the trade. There is no general agreement about their appropriate level, which is crucial to the success of policies directed towards orderly grain marketing. The trade in Germany, including the co-operatives, considers that the monthly increments are too small, no doubt being influenced by the relative inefficiency of their storage compared with France and the Netherlands. But even in Germany unanimity is lacking, as the mills and merchants with milling interests have a different viewpoint. Monthly price increments is the type of subject which comes up before the Consultative Committee in Brussels which consists of representatives of all grain interests, producers, processors, merchants, etc. The diversity of these interests severely limits the effectiveness of the Committee.

1.52. In general the E.E.C. regulations seem to have accentuated the changes in the structure of the grain trade, already taking place in member countries. By increasing the complexity of trading the regulations have given a further fillip to concentration. The unified market has put a premium on good communications, an advantage enjoyed by the international shippers, the few large firms of merchants, and the co-operatives. If the latter succeed in making effective links across frontiers they should benefit by the common grains policy. As their commercial importance increases, the co-operatives should also improve their political position. In the future the main trends in the structure of the trade in the E.E.C. are likely to be concentration among merchants, brokers, etc., increased penetration of domestic markets by the international shippers and continuing severe competition between the co-operatives and private merchants, from which the former seem likely to emerge with a larger proportion of the total market.

CHAPTER 2

Wheat Marketing in Australia

(*Contributed by* F. O. GROGAN, *Agricultural Adjustment Unit, University of Newcastle upon Tyne*)

2.1. Since the Second World War, Australian wheat growers have received certain price guarantees from the Government. These guarantees are implemented by the Australian Wheat Board, a statutory body, which, under the Wheat Industry Stabilisation Plan, is the sole constituted authority for the marketing of wheat within Australia and for the marketing of wheat and flour for export from Australia for the period of the plan.

2.2. In this chapter the Wheat Industry Stabilisation Plan, which incorporates the price guarantees, will be described with a brief account of earlier marketing experience. Following this the Wheat Board's marketing procedures and policies will be outlined and the Australian marketing system will be compared with that in the United Kingdom.

The Wheat Industry Stabilisation Plan

2.3. Wheat is a major crop in all the Australian mainland States. Australia first became a consistent exporter of wheat in the 1870's and today 75% or more of a "normal" crop is exported. Prior to the First World War Australian wheat farmers sold their wheat to local millers—there are still about 100 flour mills scattered through the Australian States—or to wheat merchants, often big firms with overseas associations. The price paid for the wheat was, nominally, closely related to the world price at the time of sale and particularly to the price on the Liverpool Corn Exchange because of the predominance of the British market in Australian overseas sales.

2.4. During the First World War, compulsory pooling was introduced to allocate available markets and available shipping equitably between the States and between individual growers. After the 1921/2 harvest compulsory pooling lapsed and responsibility for marketing reverted to growers and merchants, with some pooling continued on a co-operative basis. The degree of support received for these pools from growers, and the degree of success achieved by the pools, varied in the different States.

2.5. A continuing feature of the free-market system had been income

541

instability for wheat growers as a result of fluctuations in the world price for wheat. This instability was increased by the considerable variations in the size of the crop from season to season due to climatic conditions. This situation led to continued pressure by growers for government intervention to change the marketing system. Such pressure was a recurrent feature of Commonwealth and State politics.

2.6. The problem of income instability reached a climax during the 1930's when world wheat prices were extremely depressed, with large surpluses in exporting countries.

2.7. The financial position of Australian wheat growers became parlous and there was much parliamentary consideration of measures which might be taken, by intervention in the marketing system or by other means, to alleviate their situation. Much of the thinking on the subject, as on the similar problems of other Australian rural industries during this period, was along the lines of "orderly" marketing, i.e. means by which the prices received by primary producers could be, even if only partially, insulated against low prices and fluctuations in world markets. Two-price schemes, under which the grower would receive for that portion of his produce consumed on the home market a price divorced from the fluctuations of the world market, figured prominently in this thinking. These schemes were regarded as analogous to the protection which Australian secondary industries received by way of tariffs.

2.8. A Royal Commission into the Wheat, Flour and Bread Industries in 1934/5 examined in great detail the costs under the competitive system of the time and discussed the advantages and disadvantages of controlled marketing. It recommended that, subject to a referendum of growers, a system of controlled marketing, not unlike the present one, be set up.

2.9. No such major steps were taken at the time although a number of lesser measures of assistance were implemented including, in 1938, a flour tax to enable a stable price to be paid to growers for wheat sold for human consumption in Australia.

2.10. Following the outbreak of the Second World War in 1939, however, the Australian Wheat Board was established under wartime legislation to acquire and sell all wheat grown in Australia, other than that retained for use at the farm where it was grown (i.e. for seed or feed). During this period of wartime control various measures of price support were introduced.

2.11. After the war it was clear that wheat growers had no wish to go back to a free-market system and in 1948 the first of a series of 5-year wheat stabilisation plans was introduced under legislation following negotiations between the Commonwealth and State Governments and the Wheat Industry. While the plans have differed in details the underlying principles have not changed.

2.12. The term "stabilisation", as commonly used in Australia, usually includes elements aimed at price support as well as at dampening price or income fluctuations, and the stabilisation plans have all included

(a) a specified price at which wheat is sold for consumption in Australia, and

(b) a guaranteed minimum price for a stated quantity of exports in any season.

2.13. The guaranteed price is directly related to the cost of producing wheat in Australia as ascertained by surveys carried out by the Bureau of Agricultural Economics. The home consumption price is equal to the guaranteed price plus a small loading to cover the cost of shipping wheat from the mainland to Tasmania. During the currency of a stabilisation plan the guaranteed price is adjusted from year to year in accordance with movements in an index of wheat production costs as estimated by the Bureau and examined by a Wheat Cost Index Committee. This Committee is composed of a representative of the Australian Wheatgrowers' Federation and a representative of the Australian Agricultural Council with the Director of the Bureau as chairman.

2.14. To finance the price guarantee which currently applies to 150 million bushels[1] of export wheat, the legislation provides for a Wheat Prices Stabilisation Fund. Growers contribute to the Fund by means of a levy, not exceeding 15 cents per bushel, on any price excess above the guaranteed price derived from all exports. There is an upper ceiling to the growers' contributions, at present $60 millions, any excess being returned to growers on the first-in-first-out principle. The Government underwrites the Fund in order to meet the guarantee when growers' contributions are exhausted.

2.15. The 1948 5-year plan, and subsequent plans authorised under legislation in 1953, 1958 and 1963, provided for the continuation of the Australian Wheat Board to implement them. The current plan nominally expires with the handling of the 1967/8 harvest. (See also paragraphs 2.58–2.63.)

2.16. Since 1948 there have been a number of changes in the composition of the Wheat Board. Currently it comprises two growers' representatives from each of the five mainland States, a commercial member, a financial member, a member representing flour-mill owners, an employees' representative, and a chairman. The members, including elected growers' representatives, are appointed by the Minister for Primary Industry.

2.17. The Board has its head office in Melbourne with an office in each State and it appoints licensed receiving agents in each State who receive wheat on its behalf at a large number of delivery points throughout the wheat-growing areas. In each State there is a single bulk-handling authority

---

[1] 4 million tons, using a conversion rate of 37·3 bushels per ton.

which is a licensed receiving agent for the Board but is not necessarily the only one. The bulk-handling authority, at the delivery point, issues the grower with a weigh-bridge receipt setting out the weight and grade of wheat delivered and has responsibility for storing and/or transporting the wheat to sub-terminal depots or to terminal silos at ports for loading on to ships. Alternatively, the wheat may in some cases be delivered by growers to local mills or merchants who, as licensed receivers, receive it on behalf of the Board which pays the growers. The bulk-handling authority has responsibility for carrying out such cleaning, grading or phyto-sanitary processes as the Board might direct or approve.

2.18. Shortly after the grower delivers his wheat he receives a first advance payment which the Board is able to make because of arrangements for finance with the Reserve Bank under guarantee by the Commonwealth Government. The level of this first advance, while it bears no fixed relation to the guaranteed price, is, of course, related to the Board's expectation of overall realisation for the particular harvest. It is determined by the Government and for eleven consecutive seasons to 1966/7 has been $1·10 per bushel on a free-on-rail-at-ports basis, with the appropriate rail freight from each grower's siding to the port being deducted from the first advance made to him. During this period the guaranteed price has varied between $1·41 and $1·64 per bushel (approximately £28 10s. 0d. sterling per ton). As the sale of the crop proceeds the Board progressively makes further payments to growers. But the final level of payments, and the liability of the Government to the Stabilisation Fund, can only be known accurately after payment for the whole crop has been received because it is only then that the average price received for exports can be known.

2.19. The timing of the second and subsequent payments depends on the timing of receipt of monies due to the Board. With export credit sales some payments are not received until 12 months after shipment, which may be 18 months or more after sale.

2.20. In a complex marketing and production situation any assessment of the effects of one variable such as the stabilisation scheme is not easy and would necessitate some assumptions as to the situation that would have existed in the absence of the scheme. Obviously, the scheme has implications regarding allocation of resources and income distribution that are important both for the national economy and the wheat industry. However, the scheme is only one part of a wider policy of government intervention in agriculture and it is not proposed to discuss these wider issues here.

2.21. Table 2.1 sets out statistics regarding wheat acreage, production, and export and local prices since 1948/9. The following observations may be made.

TABLE 2.1

AUSTRALIAN WHEAT STATISTICS

Year	Million acres	Yield per acre (bushels)	Production (million bushels)	Home consumption price (cents per bushel) (a)	Average export return (cents per bushel) (a)	Average return to grower (cents per bushel) (a)	Value of production ($ million)
1947–8	13·9	15·9	220	60·6	175·3	143·1	315·0
1948–9	12·6	15·2	191	66·7	142·0	112·8	215·1
1949–50	12·2	17·8	218	66·7	162·3	130·1	283·9
1950–1	11·7	15·8	184	78·3	167·8	126·2	232·5
1951–2	10·4	15·4	160	100·0	172·9	142·4	227·4
1952–3	10·2	19·1	195	119·2	170·4	149·8	292·4
1953–4	10·8	18·4	195	141·3	140·3	120·7	238·9
1954–5	10·7	15·8	169	141·3	125·4	119·8	202·0
1955–6	10·2	19·2	195	134·6	128·8	120·0	234·5
1956–7	7·9	17·1	134	137·9	142·3	125·1	168·2
1957–8	8·8	11·0	98	143·3	138·7	129·9	126·7
1958–9	10·4	20·7	215	146·7	131·9	131·7	283·3
1959–60	12·2	16·3	199	150·0	132·3	134·5	270·0
1960–1	13·4	20·4	274	153·3	133·9	136·3	373·1
1961–2	14·7	16·8	247	158·3	142·9	144·4	356·9
1962–3	16·5	18·6	307	159·6	135·7	139·4	427·8
1963–4	16·5	19·9	328	145·8	142·9	137·3	450·2
1964–5	17·9	20·6	369	146·7	133·8	134·9	497·5
1965–6	17·5	14·8	260	153·3	140·9	141·1	366·4
1966–7	20·8	22·4	467	156·5	144·7	141·0	657·9
1967–8	22·7	12·2	277	165·5	N.A.	N.A.	N.A.

Value of production has been calculated by multiplying the quantity produced by the average return to grower.
(a) f.o.r. ports.   *Note*: £1=2·14 Australian dollars.
Source: Department of Primary Industry.

T

2.22. Over the period since the Wheat Stabilisation Plan was introduced the export prices received for Australian wheat have shown some decline.

2.23. The guaranteed price is based on periodic surveys made by the Bureau of Agricultural Economics of the cost of producing wheat in Australia. In the years between the surveys, in order to protect wheat-growers' incomes against rising costs, the guaranteed price is adjusted annually in accordance with an index of movements in the individual cost elements. Except in 1955/6 and in 1963/4, when the cost-of-production formula was altered, the home consumption price has continued to rise.

2.24. In the early years of the scheme, when the guaranteed price was below the export price, growers contributed to the Stabilisation Fund but in more recent years when the guaranteed price has been above the export price and the growers' contributions have been exhausted, the Government has had to contribute to the Fund. If one assumes that in the absence of the scheme Australian consumers would have had to pay the export price for wheat, then in the early years the consumers gained whereas in the later years the wheat growers gained, through the sales on the home market which were made at the home consumption price.

2.25. There seems little doubt that over recent years the price guarantees have acted as an incentive to farmers in their production policies. These guarantees, in conjunction with the better average yields and, in some cases, probably with economies of scale, have protected profits in wheat growing from the full impact of rising costs. The greater predictability regarding returns has probably encouraged farmers not only to increase acreage but to make the investment necessary to utilise the best technology.

2.26. Perhaps the main comment to make on the scheme is that it has been a contributory factor in a major expansion of Australian wheat exports and foreign exchange earnings. During most of the period concerned the level of export returns in relation to cost of production (as officially assessed) was favourable and while some expansion in wheat production might have taken place in the absence of the stabilisation arrangement it can be questioned whether it would have been on the same scale.

2.27. As a corollary it may be remarked that if the relation of export prices to cost of production becomes less favourable (with implications for the government's financial liability under the guarantee), and especially if marketing problems should arise, the plan may come under more critical scrutiny than in the past.

Grading

2.28. The bulk of Australian wheat is still sold on an f.a.q. (fair average quality) basis and this has been the case for some 70 years.

2.29. The description "fair average quality" refers to the milling quality of the grain and not to the strength or other characteristics of the flour made from it. Australian wheat has been traditionally bought mainly for blending or for use as a filler wheat of medium strength, and has been valued for its ease of milling, dryness, and high yield of flour of good colour and flavour. The f.a.q. basis of sale has apparently been satisfactory to many or most buyers and, despite criticism that the system did not encourage improvement in the crop, it has persisted for many years.

2.30. Each season, for each Australian State, or in some cases for regions within a State, the declared f.a.q. is, in effect, a carefully bulked, thoroughly mixed, sample of the season's wheat drawn from delivery points throughout the State or region. The weight of a bushel of the f.a.q. wheat is determined by the committee which in each State declares the f.a.q. and a sample is forwarded as soon as possible to the various potential buying countries.

2.31. Recently markets have become more discriminating and, partly on account of new baking techniques, have been more concerned with the exact specification of the wheat in terms of protein, flour quality and baking performance. To meet the requirements the Wheat Board has moved towards the identification and segregation of a number of special grades of wheat in addition to the f.a.q. which comprises the bulk of the crop. Each of the special grades is specified according to the hardness of the grain, the protein content, the dough qualities and the milling quality as indicated by the bushel weight. The specifications for the 1967/8 crop and the methods used in testing them are set out in Newsletter No. 213a issued by the Bread Research Institute of Australia, an independent research organisation, and published in the Wheat Board's *Wheat Australia* of April 1968. In 1964/5, 1965/6 and 1966/7 the premium and semi-hard grades comprised about 8 or 9% of total deliveries. The minimum specifications for f.a.q. or other grades with which each load of wheat delivered by growers must comply are determined by the Wheat Board and supervised by the State bulk handling authorities.

2.32. The characteristics of the special grades are related to the wheat varieties and to the areas where they are grown. Australian wheat breeders have been successful in their endeavour to combine high yield per acre with desirable flour strength. Changes in husbandry practices, especially the introduction of leguminous pastures into the crop rotation, have contributed to general improvement especially with respect to yield and protein content. Certain areas produce soft wheats of low protein content and these find special markets for biscuit-making and cake flours. As a safeguard for the quality of the crop, the relatively small number of organisations engaged in breeding new wheat varieties have voluntarily accepted a system of testing and registration of new varieties before these are disseminated.

Storage and Handling

2.33. The generally hot and dry Australian summer weather and the low moisture content of the grain enables the harvest to be carried out over a period of several weeks from late October onwards, the actual starting date depending on the district concerned. The possibility of summer rain or hailstorms is an incentive to reduce to a minimum the period of harvesting, which is carried out with combine harvesters. The threshed grain is bagged from the harvester or, more commonly, discharged in bulk into motor trucks for transport to delivery points at railway sidings.

2.34. The receival, storage, transport and shipment of the harvest were all revolutionised by the introduction of bulk handling over the period since the 1920's. Today, all except a small proportion of the wheat deliveries are in bulk, with major economies in labour as well as savings through the elimination of sacks. In each State, as mentioned above, there is one bulk-handling authority, in some cases owned and operated by the Government concerned and in others by growers' co-operatives. Since considerable investment is necessary to provide the bulk-storage facilities and handling equipment, the Governments have taken the view that duplication of these facilities would be uneconomic and the authority in each State has been given sole rights on a statutory basis. Although systems vary somewhat from State to State with consequent variations in the capital and operating costs, the Wheat Board averages these charges before deducting them from the market realisations for each season's harvest.

2.35. The dryness of Australian wheat simplifies storage in that artificial drying is rarely necessary. But the warm climatic conditions encourage insect infestation which is a major problem in storing wheat under Australian conditions. The bulk of the harvest is shipped from some sixteen ports, to which it has to be transported over considerable distances by rail or road.

2.36. By and large, farmers prefer to deliver their wheat direct from the harvester to the railway siding thereby avoiding double handling and the cost of on-farm storage and at the same time establishing an early entitlement to the first advance payment. The Board, for its part, tries to accept all wheat as farmers choose to deliver it. Occasionally the pressure on storage facilities and on available rolling stock has been such that in some localities some curtailment in deliveries has had to be imposed.

2.37. The wide fluctuations from year to year in the size of the harvest have increased the problems of storage for the Wheat Board. For example, during the period 1947/8 to 1967/8 the size of the crop varied from 98 million bushels in 1957/8 to a peak of 467 million in 1966/7. Over the same period, end-of-season carryover varied from 14 million to 93 million with an average of approximately 40 million bushels. The Board and the bulk-handling authorities are assisted in planning the handling and selling of the crop each season by a system of crop forecasting, based on acreage and yield estimates, made by the State Departments of Agriculture and by the Board itself. These forecasts, reviewed throughout the growing season, reach an accurate level at the commencement of harvesting.

2.38. The recent spectacular increase in the crop lends added force to the question whether storage facilities on a scale adequate to cope with peak deliveries at all points in peak production years are justified. Provision of storage on this scale could mean, of course, that in some years the facilities at some points would not be fully utilised. This problem may lead to some re-examination of the costs of centralised compared with on-farm storage and of the economics of staggered deliveries. A related problem and subject of controversy is that of finding the considerable capital sums necessary for providing additional storage. The service charges for such capital and the costs of storing and handling the crop are debited against the grower's return.

2.39. Bulk storage for a total of 428 million bushels was available at ports, in the country and at mills in 1965/6. This was not much below the record Australian crop of 467 million bushels in 1966/7. The stores, other than at mills, are controlled and operated by the respective bulk-handling authorities, although some storage facilities are owned by the Wheat Board, and some bulk storage at mills, particularly in Victoria, is leased by the bulk-handling authority. The Board, however, has no power to provide permanent storage, which, like transport facilities, is a matter for the States concerned. Loading facilities at the ports are highly mechanised with loading rates of up to 3000 tons per hour.

2.40. The average cost of handling for the 28th pool (season 1964/5) was, for bulk wheat, 5·136 cents per bushel, with a further 16·694 cents for rail freight and 1·087 cents for bank interest and charges and administration.

Selling Procedures

2.41. In general the Board aims to supply orders from the nearest suitable source in order to save unnecessary freight costs. However, millers naturally have preferences as to the wheat they want and sometimes for a particular growing area. Within limits the Board tries to satisfy such preferences. Also, growers may deliver their wheat to mills or merchants and may receive a premium payment, but the mills or merchants, in such cases, act as licensed receivers. On delivery to such licensed receivers the wheat becomes the property of the Board. The grower receives from the Board a first and subsequent advance at the same times and on the same conditions as any grower delivering f.a.q. wheat to a bulk-handling authority.

2.42. The Board, as sole exporter of wheat from Australia, sells on f.o.b., c.i.f., or c. & f. terms. Where the Board sells on c.i.f. or c. & f. terms it arranges its own chartering of ships. While it is still sometimes possible to get vessels of about 10,000 tons, the general tendency now is for shipment in large bulk carriers. Since early 1967 the Board has shipped wheat to Europe only in large vessels with trans-shipment through Amsterdam or Rotterdam. A 48,000-ton vessel was recently loaded for Europe. All wheat exported is subject to government inspection and certification that it is equal to the particular Standard or Standards of the State and season concerned. With the sole exception of sales made to United Kingdom millers all export sales by the Wheat Board are on the basis of weight, quality and condition of the wheat being final at the time of shipment.

Marketing Patterns and Policies

2.43. The pattern of Australian wheat marketing has varied over time as a result of changes in the demand for wheat of various qualities in total as well as in particular markets, and of fluctuations in the domestic crop in importing countries. A noteworthy factor has been the development, under support policies, of wheat production in countries which traditionally had been large importers of wheat but which have become self-sufficient or even exporters. Perhaps the most striking of these changes has been the decline in the importance of the United Kingdom and Europe generally as outlets for Australian wheat. This is illustrated in Table 2.2.

2.44. Australia makes shipments of wheat to a large number of countries but the only destinations in 1967 which individually accounted for 5% or more of shipments were China, India, Pakistan, Japan, Singapore and Malaysia, the Middle Eastern countries (Iraq, Iran, Lebanon and Saudi Arabia) and the Netherlands. Direct shipments to the U.K. dropped to only 4%, while the U.S.S.R. had dropped out altogether.

2.45. The Chairman of the Australian Wheat Board sees world demand for wheat as falling into three fairly distinct categories:
  (a) the affluent communities where *per capita* consumption of bread is steady or declining. Among Australia's markets, Western Europe is the most important example;
  (b) those economies where bread and bread products still have a high, and in some cases, a growing dietary status; the most important of these are the Japanese, African and Middle East markets;
  (c) those countries short of food, where increasing use is being made of wheat as a substitute for other energy foods, including rice; impor-

TABLE 2.2

DESTINATION OF AUSTRALIAN WHEAT AND WHEAT FLOUR EXPORTS

Importing countries	1938/9(a) Wheat 000 tons	%	1938/9(a) Wheat flour 000 tons	%	1952–3 Average Wheat 000 tons	%	1952–3 Average Wheat flour 000 tons	%	1965 Wheat 000 tons	%	1965 Wheat flour 000 tons	%	1966(e) Wheat 000 tons	%	1966(e) Wheat flour 000 tons	%	1967(e) Wheat 000 tons	%	1967(e) Wheat flour 000 tons	%
U.K.	863	51	106	16	499	39	93	13	630	10	33	8	497	11	23	8	300	4	14	4
Netherlands	—		5	1	33	3	2	—	(b)		(b)		(b)		(b)		376	5	(b)	
U.S.S.R.	(b)		(b)		(b)		(b)		1105	17	(b)		156	4	(b)		—		(b)	
Japan	12	1	—		37	3	3		419	6	(b)		362	8	(b)		508	6	(b)	
India	185(c)	11	1(c)	—	238	18	40	6	319	5	(b)		276	6	(b)		850	11	(b)	
Pakistan					65	5	(b)		54	1	(b)		123	3	(b)		605	8	(b)	
China	284	17	194	30	(b)		(b)		2733	41	(b)		1459	35	(b)		2822	35	(b)	
Singapore and Malaya	(b)		60	9	(b)		119	16	105	2	60	14	334	8	14	5	459	6	1	—
Iraq, Iran, Lebanon, and Saudi Arabia	(b)		(b)		(b)				283	4	(b)		362	8	(b)		508	6	(b)	
Other countries	347	20	281(d)	44	408	32	471(d)	65	896	14	337(d)	78	739	17	257(d)	87	1557	19	371(d)	96
Total	1691	100	647	100	1280	100	728	100	6544	100	430	100	4308	100	294	100	7985	100	386	100

(a) Twelve months commencing July 1938.
(b) Included in "Other countries".
(c) Including Pakistan.
(d) Mostly to countries round the Indian Ocean, particularly Ceylon.
(e) Provisional.
Source: Commonwealth Secretariat, *Grain Crops*, 1954, 1955, and No. 12.

tant Australian markets in this category are Malaya, Singapore, Hong Kong, China, India and Pakistan.

2.46. The Board's selling policies aim at meeting the special needs and opportunities of these different markets, for example by the economies of large-scale shipments to Europe and through the freight advantage which Australia enjoys in the Asian market, where China has become a major buyer.

2.47. The Australian Government attaches considerable importance to maintaining outlets for wheat exports. Trade agreements containing provisions relating to wheat and flour are in force with the United Kingdom, Japan, Malaya and Ceylon. When Australia agreed in 1964 to the U.K. imposing levies on cereals offered for sale in U.K. below an agreed schedule of minimum import prices, it insisted that the new system was not to affect the provisions of the existing Trade Agreement under which both Governments affirm the intention that the annual sales of Australian wheat and flour to the U.K. on commercial terms amount to not less than 28 million bushels of f.a.q. wheat (including wheat equivalent of flour). In the case of the other three countries mentioned the arrangements have succeeded in maintaining or increasing sales of Australian wheat and/or flour.

2.48. As regards the various International Wheat Agreements, Australia, as an exporter, has consistently supported the principle of greater stability and predictability in world wheat marketing and especially limitation of the fluctuations in the world wheat price. The negotiation within the Kennedy Round of an International Grains Arrangement which, in addition to a Wheat Trade Convention, incorporated a Food Aid Convention providing for multilateral disposal of $4\frac{1}{2}$ million tons of wheat on special terms to needy countries was strongly supported by Australia. The problems of concessional wheat sales on a bilateral basis, because of their possible impact on normal commercial markets, have been of continuing concern to Australia.

Wheat Marketing in the United Kingdom and in Australia

2.49. A major difference between the two systems is that in Australia, by arrangement between the Commonwealth and State Governments, wheat for home consumption is sold at the guaranteed price (plus a small loading) whereas in the United Kingdom there is no necessary relationship between the guaranteed price and the price paid for home-grown wheat by local users.

2.50. Although there is incentive to individual growers in the U.K. to obtain the highest price possible for their wheat because the deficiency payment is calculated on the national average realisation and not on the price actually received by the individual grower, relatively low prices for home-grown wheat do not unduly penalise U.K. growers because of the deficiency payment. In this situation it might be suggested that a low U.K. domestic price by tending to depress the price of imported wheat could, without hardship to U.K. growers, benefit consumers and the U.K. trade balance. However, the system of minimum import prices introduced by the British Government in 1964 would seem to indicate that any such price-depressive effects are not an aim of official policy.

2.51. The price of wheat grown in the U.K., compared on a quality basis with that for imported wheat, is governed by a complex marketing and utilisation pattern including such factors as the relative bargaining strength of the growers as sellers and the millers and compounders as buyers, and of course competition from overseas suppliers. In so far as internal factors affect the price of home-grown wheat there are obvious implications for competing wheats of comparable quality.

2.52. As regards quality, British wheat is predominantly soft wheat used by millers and compounders for blending purposes and for animal feeding-stuffs. It is sold on the basis of a submitted sample with acceptable moisture content and cleanliness as the main desiderata. In essence, this procedure is not very different from the Australian practice of selling on the basis of fair average quality. Recently some varieties stronger than the predominantly soft wheats have been successfully grown in the U.K. However, any development towards the segregation of special grades for separate sale is likely to depend, as in Australia, on the combined economics of the yield of the special varieties and of any price premium which they may command.

2.53. In Australia the bulk of the crop is exported. This fact, in conjunction with the system of centralised marketing and the virtually universal practice of handling the grain in bulk, has facilitated a system of co-ordinated and centralised receival, transport, storage and shipment of the wheat in bulk aimed at maximum economy, speed and convenience.

2.54. In the U.K. all the wheat produced is consumed within the country either as flour or in feedingstuffs. There are a very large number of millers and compounders of widely varying size scattered throughout the country, some near ports and some inland. In this situation the transport costs associated with the location of a particular mill or compounder can be extremely important. On the grower's side, the fact that he often buys feedingstuffs or requisites from the merchant to whom he sells his wheat can have a bearing on his choice of outlet and on the price at which he sells.

2.55. The problems of distribution and storage are further complicated by the need to dry practically all of the U.K. crop, a need which does not arise in Australia.

T*

2.56. Information regarding the economics of centralised drying and storage, as well as the marketing advantages through handling larger quantities than some individual growers produce, is likely to become available through the activities of the cereal drying and/or marketing groups. On the evidence currently available it would appear open to question whether there would be very significant advantages under U.K. conditions in the highly centralised storage facilities typical in Australia.

2.57. Any generalisations regarding the applicability of the system in either country to the circumstances of the other would be difficult, if not impossible, without a good deal of further study. The different nature of the grain trade in each country, especially with regard to location and destination of the domestic production in the two countries, reduces the relevance of Australian experience as far as the U.K. is concerned. On the other hand, circumstances have changed so greatly with regard to domestic cereal production in the U.K., over recent years, that some, if not all, of Australian experience could have relevance with respect to the long-term development of cereal marketing arrangements in the U.K.

Postscript

2.58. Since this Chapter was written, the Australian Government has announced details of a new wheat stabilisation plan to apply for five seasons commencing with the 1968/9 crop.

2.59. A major change from the previous arrangements will be that cost of production, as traditionally estimated from survey data collected by the Bureau of Agricultural Economics, will no longer be used for setting the guaranteed price for exported wheat. However, the home consumption price which will apply to approximately 60 million bushels sold for all purposes on the domestic market, will be based on a cost of production derived from the 1967 survey with a yield divisor of 20·25 bushels. This home consumption price for 1968/9 will be 1·70 dollars per bushel for f.a.q. bulk wheat free on rail at ports plus an amount to cover cost of shipment to Tasmania. The present loading is 1·5 cents per bushel subject to variation as necessary.

2.60. The guaranteed price will apply to exports up to a maximum of 200 million bushels in any one season and for 1968/9 will be $1·45 for f.a.q. bulk wheat free on board vessel. This price is considered by the Government to be consonant with world wheat trading conditions and prospects and it takes into account the floor put under prices by the International Grains Agreement. Both the home consumption price and the guaranteed price will in subsequent years be annually adjusted on the basis of index movements in cash costs (including interest actually paid) and in rail freight and handling charges.

2.61. There will continue to be a wheat prices stabilisation fund with a ceiling of $80 million. The fund will be financed from an export charge which will apply when export returns exceed an amount equal to the guaranteed return plus 5 cents per bushel. The charge will not at any time exceed 15 cents per bushel. If export returns for a season are less than the guaranteed price then the deficit, on up to 200 million bushels, will be made up from the fund. If the balance in the fund is insufficient or if, as at present, there are no growers' moneys in the fund, the deficit will be made up by the Commonwealth.

2.62. The guaranteed price for the 1967/8 season was $1·64 per bushel f.o.r. ports. After allowing for fobbing costs of about 3·5 cents per bushel, the 1968/9 guaranteed price of $1·45 per bushel f.o.r. vessels represents a reduction of about 22·5 cents per bushel.

2.63. The Minister for Primary Industry in announcing the new plan referred to prosperity in the Australian wheat industry and to a significant rise in land values. He said that continuation of the old principles of price guarantee could lead to unreasonable and unbearable levels in the Commonwealth's subvention to the industry and also in the home consumption price. The effects of this would be to endanger the stability of the industry—the principal objective of the Stabilisation Plan.

CHAPTER 3

Grain Marketing in Canada

3.1. The Canadian grain production and marketing system is large and complex and has an important position in the Canadian economy. In recent years about 30% of farm cash receipts from the sale of farm products has been derived from the sale of grains. (About 5·5% of the gross domestic product has been attributable to agricultural production and 8% of the labour force is employed in agriculture.) Exports of grains and grain-based products have accounted for about 65% of Canadian agricultural exports, and agricultural exports account for about 20% of total exports.

3.2. The average annual production of each major grain crop for the most recent 5-year period is shown in Table 3.1. The distribution of the grain production according to the major regions—British Columbia, Prairies, Eastern Canada and the Maritimes—is also shown in Table 3.1 together with the breakdown by provinces in most cases. Total Canadian

TABLE 3.1

CANADIAN GRAIN PRODUCTION (5-YEAR AVERAGE 1962/3 TO 1966/7)

('000 long tons)

Province	Wheat	Rye	Oats	Barley	Maize	Mixed grains	Total four feed grains	Total all grains
British Columbia	77	2	58	87	—	4	149	226
Prairies:	17,497	328	4135	4341	4	436	8916	26,742
Alberta	4018	90	1612	2524	—	258	4395	8503
Saskatchewan	11,422	172	1424	1376	—	83	2883	14.477
Manitoba	2057	66	1099	441	4	95	1639	3762
Eastern Canada:	463	37	2061	179	1327	898	4375	4876
Ontario	443	35	1227	151	1232	769	3379	3857
Quebec	14	2	684	13	5	70	772	788
Maritimes	6	—	150	15	—	59	225	231
CANADA	18,037	367	6254	4607	1241	1338	13,440	31,844

Note: Because of rounding, totals may differ slightly from the sum of individual figures.

grain production averaged 31·8 million long tons, with 84% of this production located in the prairie provinces of Alberta, Saskatchewan and Manitoba.

3.3. Total wheat production averaged about 18 million long tons, with 97% located in the prairie region. The province of Saskatchewan alone accounted for 63% of the total Canadian wheat crop. The bulk of the 17·5 million tons of wheat produced in the prairie region is of the Hard Red Spring type, with the exception of about 1·0 million tons which is of the Amber Durum type and a very small amount in Alberta which is of the Soft Red Winter type. Some 3% or about ½ million tons of Canadian wheat is grown in the Eastern Province of Ontario and is a soft white winter type not unlike English wheats.

3.4. In the most recent 5-year period, an average of 13½ million tons of feed grains per year—oats, barley, maize and mixed grains in that order of volume—have been grown in Canada, with the overall trend being upward as feed grain production is expanded to meet a growing livestock demand. Some two-thirds of these feed grains are grown in the prairies, with the bulk of the remainder grown in Ontario—although some feed grains are produced in every province. The main feed grains produced in the prairie region are oats and barley. In Eastern Canada, oats has been the largest crop, but maize production has increased rapidly in Ontario in the last 5 years. Mixed grains, which are usually a mixture of half oats and half barley, are grown solely as a livestock feed, with Eastern Canada producing twice as much as Western Canada. Some 95% of Canadian barley is grown on the prairies, with 66% of the oats and 32% of the mixed grains, but virtually no maize is grown there.

Supply and Utilisation

3.5. Canada is virtually self-sufficient in grains, with the exception of small imports of about ½ million tons of maize from the United States. The utilisation of each grain varies and there is a distinct difference between wheat, which is grown primarily for export, and all the other grains. The composition of supplies and their utilisation are shown in Table 3.2 as annual averages over the 5 years 1962/3 to 1966/7.

3.6. In the case of wheat, nearly three-quarters of the average crop is exported. Of the rest, about 9% is consumed in Canada for wheat products or industrial use, 6% goes for seed (excluding seed exports) and 7·5% is the residual for feed, waste and loss. The actual quantity fed varies with wheat prices, export opportunities and the quality of the crop.

3.7. It is interesting to note that although a relatively small proportion of the Canadian wheat crop is processed for human food in Canada, it is

TABLE 3.2

SUPPLY AND UTILISATION OF GRAINS: AVERAGE 1962/3 TO 1966/7

('000 long tons)

	Wheat	Rye	Oats	Barley	Maize	Mixed grains	Total four feed grains	Total all grains
Supply								
Carryover	12,167	170	2022	1936	142	—	4100	16,437
Production	18,037	367	6254	4607	1241	1338	13,440	31,844
Imports	—	1	—	1	596	—	597	598
Supply	30,204	538	8276	6544	1979	1338	18,137	48,879
Utilisation								
Human food	1552	11	85	4	471[a]	—	903[a]	2552[a]
Industrial use	46	40	—	343	[a]	—	[a]	[a]
Seed	1091	20	337	235	8	—	580	1691
Feed, waste, loss	1357	96	5507	2870	1362	1338	11,077	12,530
Total domestic use	4046	167	5929	3452	1841	1338	12,560	16,773
Exports	12,996	178	233	840	9	—	1082	14,256
Change in stocks	+1005	+23	+73	+317	−12	—	+378	+1406
Utilisation as %								
of production								
Human food	8·6	3·0	1·3	0·1	38·0[a]	—	6·7[a]	8·0[a]
Industrial use	0·2	10·9	—	7·4	[a]	—	[a]	[a]
Seed	6·0	5·4	5·4	5·1	0·6	—	4·3	5·3
Feed, waste, loss	7·5	26·2	88·0	62·3	109·8	100·0	82·4	39·3
Total domestic use	22·4	46·6	94·7	74·9	148·3	100·0	93·4	52·6
Exports	72·0	48·5	3·7	18·2	0·7	—	8·1	44·8
Change in stocks	+5·6	+6·3	+1·2	+6·9	−1·0	—	+2·8	+4·4

[a] Industrial use included with human food.

the largest quantity of any grain used this way. When industrial use is added (this covers mostly distilling, brewing, gluten and starch manufacture; breakfast cereals are covered in human food), the total quantity of grains processed as human food amounts to 2½ million tons or 8% of production.

3.8. Barley is the most important coarse grain export and this has averaged just under a million tons over the last 5 years. When the four coarse or feed grains are added together, about 80% of production (11 million tons) is fed to livestock.

3.9. An important aspect of the Canadian grain situation is the relatively large carryover stocks which are maintained from one year to the next. The 5-year average carryover stocks amount to 12 million tons of wheat, 2 million tons of oats and 2 million tons of barley. Such a level is not excessive in relation to an annual total utilisation which averages 17 million tons of wheat, 6 million tons of oats and 4 million tons of barley and taking into account the variable nature of crop production in the Canadian Prairies. In the same 5-year period, wheat production varied from a high of 22 million tons to a low of 15 million tons after a very small crop in 1961. Oats ranged from 7·5 million tons down to 5·3 million, and barley from 6·5 million tons to 3·6. Over the 5-year period there has been a slow build-up of carryover stocks in all grains except maize, largely to take account of the steady increase in total utilisation.

3.10. The location of these stocks helps to explain their relatively large size. More than half the total of oats and barley are stored on farms, since this is where they are fed. One-fifth of the wheat is also carried over from one year to the next on farms—some will be seed, some will be fed, but most will be marketed by farmers. The rest of the stocks are either in the marketing channels or stored on the way to export positions or Eastern Canadian feeding areas. Very large quantities of wheat are held in the country elevators and the terminal and port elevators at the Lakehead, Montreal, Vancouver and other Canadian export ports.

3.11. The size and location of the various sectors of the livestock industry have important implications for the Canadian grain industry. In terms of cereal consumption, slightly more than half the livestock industry is located in Eastern Canada while 97% of the wheat and 66% of the feed grains are produced in Western Canada. Grain is therefore moved from Western Canada to feed the livestock in Eastern Canada, as the East is not self-sufficient in feed grains.

3.12. The difference, as between Western and Eastern Canada, in the relative importance of cereal growing has meant that marketing systems for grain have developed along different lines in these two main areas. They are discussed separately in paragraphs 3.13–3.20 below.

Marketing Western Grain

3.13. In Western Canada the marketing system starts with the country elevator. In 1967 there were 5032 licensed public country grain elevators with a combined storage capacity of about 10½ million tons, located at some 2000 railway stations. Over 98% of this storage is controlled by twelve firms, either farmer co-operatives or private companies. Some of the stations have only one elevator each; others, where grain handlings are heavy, may have several competing with one another for the farmer's patronage.

3.14. Most of the older elevators were constructed to hold about 500 to 800 tons of grain, but large numbers with greater capacities have since been built, several having space for well over 2500 tons. Frequently, the original capacities have been increased by building additional storage bins, called "annexes", additional to the elevators. The total capacity of these country elevators and their annexes is about 9·4 million tons. Additional grain-storage facilities amounting to about another million tons are available in the larger urban centres in the prairie region including the elevators belonging to flour millers, maltsters and other grain processors, and the six interior "terminal elevators" (see below).

3.15. Grain moves to Eastern Canada and to the Pacific by rail, and most of that used by processors or stored in the interior terminals is received from country elevators by rail, although some is received directly from growers by truck. Whether grains are moved to overseas markets through Fort William or Port Arthur, through the Pacific Coast ports of Vancouver, Victoria or Prince Rupert, or by way of Churchill on the western shore of Hudson Bay, they must first be hauled for considerable distances by rail.

3.16. In several parts of Canada where grain is transferred from rail to water transport, or from lake vessels to rail transport or ocean-going ships, large terminal elevators have been built. In recent years the average annual shipment of grain to these elevators has been almost 13 million tons. These elevators are primarily designed to unload rail wagons or vessels efficiently, to store grain safely and to ship it out promptly as required. They also provide facilities for cleaning, drying and weighing grain.

3.17. Most of the grain is cleaned before it is put away in the storage bins, and when grain with a high moisture content is received it is dried. All grain-drying operations are under the strict supervision of the Board of Grain Commissioners (see below) whose regulations require the use of specific drying temperatures which preclude the possibility of damage to the quality of grain. Normally only small quantities of grain are dried, and little if any is dried on the farm.

3.18. Remuneration for the market service provided by the elevator companies is determined largely by the handling and storage charges permitted on the three major grains and consequently by the competitive success of the individual company in securing the patronage of the producer. Annually, owners of country elevators sign a handling agreement with the Wheat Board under which they agree to accept delivery and make payment to the producers for the grain on the basis of the authorised initial payment. For this service the elevator companies are authorised to make a deduction from the purchase price of the grain of a handling charge which has been agreed upon with the Board. Storage rates are also provided for in the same agreement. Charges are also made by the Board of Grain Commissioners for the services of weighing, grading and inspecting grain entering and leaving terminal elevators.

Marketing Eastern Grain

3.19. There are no large, highly organised chains of country elevators in Eastern Canada, but many small groups or single enterprises operated by farmer co-operatives, milling companies and elevator companies. These elevator operators are closer to the role of the corn merchant in England than are the elevators in the Western system. In the East the elevator caters for producers selling winter wheat for delivery to flour mills or for export, maize for livestock compounds or industrial use, soya beans for crushing, white beans for export, or oats or barley for industrial use or for feed. The elevator operator may dry the grain if required, store it at the farmer's request or buy it outright from the farmer. In the case of feed grains the elevator operator may also grind or mix the farmer's own grain as a service, but more grain goes direct to the feed plants (compounders) for mixing as prepared feeds. The farmer may sell to the elevator for movement to the feed plants, or the farmer may sell and deliver direct. Most of the industrial plants and mills are in Toronto and Montreal so that the elevators will sell and move the grain by rail (sometimes by road) to these cities. Exports of the grains are made from the ports on Lake Erie and Lake Huron as well as from Toronto and Montreal.

3.20. During World War II, Canadian agriculture was forced to make radical changes in the production pattern by curtailing shipments and increasing production and exports of livestock to fulfil its contracts with the U.K. In order to increase livestock production at the price in the U.K. contract, the Federal Government introduced a subsidy on Western grain freighted to the East for feeding to livestock. This feed freight assistance has been changed in detail many times but has continued over the years. The current objectives are to offset the transfer costs, and maintain price stability, for prairie feed grains used for livestock production in Eastern Canada and British Columbia. In the 25 years since freight-assisted shipments began, between 2 to 3 million tons have been moved each year in this way.

Elevator Operators and the Exchange

3.21. Farmers' co-operatives have played an important role in the development of the Canadian grain-marketing system. Each of the three prairie provinces has its own wheat co-operative—Saskatchewan Wheat Pool, Alberta Wheat Pool and Manitoba Pool Elevators. Farmer-controlled through annually elected delegates and district directors meeting monthly, the Pools own and operate a country elevator system and terminal elevators at the Lakehead and on the West Coast. Private trade elevators are also important.

3.22. The farmer-owned elevator companies function in a manner very similar to the other private and public grain companies. Extremely keen competition exists among the elevator companies and all are members of the Winnipeg Grain Exchange. The Exchange lays down conditions under which trading in grain shall be conducted on the Exchange which provides two types of market, a physical or spot market for immediate delivery and a futures contract market for delivery at a forward date. The wheat futures market in Canada was suspended in September 1943, but there are futures markets for rye, barley and oats as well as other commodities. Facilities available include quotations from other important markets, information on crop conditions, etc., and a place where buyers and sellers can meet daily.

The Canadian Wheat Board

3.23. The Canadian Wheat Board, established under the Canadian Wheat Board Act of 1935, has responsibility for orderly marketing of western grains entering inter-provincial and export trade. The control of the Board extends from the issuing of producers' delivery permit books to the loading of the grain on ship for export.

3.24. A key part of the market system for prairie grain is the delivery quota. There is no acreage control or limitation and the delivery quota was adopted by the Wheat Board to provide producers in its area with an equitable opportunity to utilise the nation's storage facilities when grain marketings are in excess of commercial storage and handling facilities. At the beginning of each crop year, each producer has an initial quota which might be 100 units of grain consisting of either 300 bushels of wheat, 500 bushels of barley, 500 bushels of rye, 1000 bushels of oats, or any combination of these grains amounting to 100 units. As the space becomes available at specific delivery points (country elevators), general quotas are established ranging from 1 to 10 bushels per "specified acre". Specified acreage consists of each permit holder's acreage seeded to wheat, oats, barley, rye, summer fallow and eligible acreage seeded to cultivated grasses and forage crops. In years of very large crops, the delivery quota will prevent producers marketing their harvest and they will hold the grain on the farm in storage bins and, on occasion, in piles in the field. In view of the yield variation from year to year, many producers like to have full storage bins.

3.25. The entire commercial supplies of wheat, oats and barley from the Wheat Board's area are placed in annual marketing pools by grades. The Board then tries to effect an orderly movement of grain from western producers to the domestic user and foreign buyer. The pool permits a uniform per-bushel return, subject to transport costs, to all producers for each grade regardless of when the grain is delivered. The Wheat Board closes the pool at the end of the crop year, but it may be 6 months or more before the financial position of the pool can be determined depending on the progress of disposal of the grain. Occasionally, remaining stocks may be transferred to the subsequent pool in such a way as to give neither a profit or loss to either pool. Any additional funds from the Board's operations of the pool are distributed to producers as a final payment, often more than 18 months after the grain was harvested, in accordance with the quantity and grade of grain delivered.

3.26. Because producers are often restricted by the quota system, the Wheat Board is empowered to make advance payments to producers on farm-stored wheat, oats and barley up to a maximum of $3000 per producer. The Board recovers the money, which is interest-free, when the producer actually delivers the grain.

3.27. The Wheat Board makes handling agreements with elevator companies to act as its agents in the handling and sale of wheat, oats and barley in Western Canada, specifying maximum handling and storage charges. The Board, through its control over movement, is able to regulate not only the quantity of grain moving to the terminals but also the grade and type of grain, by allotting rail wagons to elevators where the desired grain is located. The Board sells wheat either directly or, in most cases, through the grain exporting companies, acting as agents for the Board, to any customer in the world at the Board's quoted prices. (The activities of the Wheat Board in the export trade are described in more detail in paragraphs 3.41–3.42 below.)

3.28. In Eastern Canada the only body with a similar role to the Canadian Wheat Board is the Ontario Wheat Producers' Marketing Board, which regulates the marketing of winter wheat grown in that province.

The Board of Grain Commissioners

3.29. With the passage of the Canada Grain Act of 1912, the Board of Grain Commissioners was established to maintain strict grade standards and control the movement of grain with the dual purpose of maintaining the competitive position of Canadian grain in world markets and upholding the interests of the producer and the merchandiser. The three-man Board has no power or duties with respect to price. Its concern is rather with the maintenance of quality, and it operates a staff of 800 persons stationed across Canada. It has jurisdiction over the grading and weighing of grain, the deduction made for dockage and shrinkage, elevator shortages or overages, the deterioration of any grain during storage or treatment, and the refusal or neglect of any person to comply with the provisions of the Canada Grain Act. On a fee basis the Board provides official inspection, grading and weighing of grain, and registration of warehouse receipts. The fees received are not far short of the cost of inspection, grading and weighing, but do not cover research and statistics. It licenses annually all operators of elevators in Western Canada, and of elevators in Eastern Canada that handle western grain for export, as well as all parties operating as grain commission merchants or as grain dealers. The Board also manages and operates the Canadian Government elevators. Its Statistics Branch compiles and publishes basic statistics relating to storage and handling of grain, within the elevator system, and also has responsibility for administering, collecting and recording the 1% levy under the Prairie Farm Assistance Act.

3.30. The Board of Grain Commissioners maintains a Grain Research Laboratory. As a centre for basic and applied research on the traditional uses of cereal grains, this laboratory is renowned throughout the world. In co-operation with the Research Branch of the Canadian Department of Agriculture, extensive tests are conducted on new hybrids developed by plant breeders. Only a few of the total number of varieties examined show sufficient improvement over accepted varieties to survive the rigid tests to qualify for licensing.

Handling, Grading and Quality Standards

3.31. All grain in Canada is handled in bulk and has been so handled ever since the Prairies first produced grain in excess of local demand. Thus from the time the combines enter the fields and empty their tanks into the trucks for delivery to farm storage facilities until the huge port elevators at Montreal or Halifax or Churchill or Vancouver discharge it into ocean grain vessels, it flows as bulk grain and never in sacks except for pedigreed seed. This not only facilitates handling by modern methods but means that grain can be mixed to give relatively uniform grades. Inspection and grading is made easier at all times and sampling can be systematically and mechanically undertaken. Quality control is a prime objective in servicing both the domestic and export markets, and bulk handling facilitates the attainment of the high grain quality for which Canada is recognised.

3.32. As already stated, the grading of grain in Canada is centrally controlled by the Board of Grain Commissioners. The main factors on which grading is based are:
(a) moisture,
(b) bushel weight,
(c) total foreign material, including other cereal grains,
(d) grain of classes other than the type being graded,
(e) soundness of grain.

Other than moisture and bushel weight, which require testing and measuring, the grading is based on visual inspection. This visual inspection has been developed very thoroughly over many years, being undertaken with the aid of special equipment and guides, and the margins by which each grade is now specified are quite narrow. For example, Western Canadian Red Spring Wheat is graded on a number series, with the highest standards for No. 1 and a larger tolerance of foreign material and other wheats as the grade gets lower. This is illustrated by Table 3.3.

TABLE 3.3

GRADING STANDARDS FOR WESTERN CANADIAN RED SPRING WHEAT, 1966/7

Name	Total foreign material			Wheats of other classes and varieties	
	Test weight (lb/bushel)	Other cereal grains	Other seeds	Not equal to Marquis	Contrasting classes
		(%)	(%)	(%)	(%)
No. 1 Manitoba Northern	64·1	0·15	0·05	0·2	0·1
No. 2 Manitoba Northern	62·7	0·3	0·15	1·5	0·2
No. 3 Manitoba Northern	61·4	0·45	0·15	6·0(4·0)[a]	0·5
No. 4 Manitoba Northern	60·4	0·65	0·15	(5·0)[a]	1·0
No. 5 Manitoba Northern	59·4	0·8	0·15	(7·5)[a]	2·5[b]

[a] Wheat of other classes.   [b] Durum.

3.33. In addition, in 1966/7 there were three grades of Canada Western Garnet, six for Canada Western Amber Durum, four of Alberta Winter. The care with which wheat is graded is brought out by the definition of the lowest grade, No. 6 Wheat: any variety of spring or winter wheat excluded from the higher grades on account of frosted or otherwise damaged kernels; minimum bushel weight, 51 lb; maximum foreign material other than wheat including cereal grains, about 3%; maximum limit of durum wheat, 6%.

3.34. For other cereals, grades are similarly established—there are six grades for feed oats and four higher grades called Canada Western; there are three grades for feed barley and six higher grades divided between six-row and two-row types; there are six grades of rye.

3.35. Since there are climatic differences between the Prairies and Eastern Canada, different grade standards are established for wheat, oats, barley, rye and maize grown in the East with the prefix "Canada Eastern", but the same basic principles are followed. The detailed practices differ as the volumes are smaller and the marketing system is not the same.

3.36. Normally, all but a small percentage of the wheat samples can be fitted into these grades. But excessive moisture at harvest may cause exclusions. If a sample contains between 14·5% and 17·0% moisture, the word "Tough" is added to the grade designation, and for more than 17·0% "Damp" is used. Other sub-grades are established for excessive other grains or weed seeds or other foreign material, the word "Rejected" and the cause added to the grade designation.

3.37. Before the beginning of each crop year, the Board of Grain Commissioners must establish a representative and highly qualified Committee on Grain Standards to select and settle standard samples. As soon as possible after 1st August, the Board collects samples of the new crop and from these prepares tentative standard samples of the statutory grades and other commercial grades as required. After reports on the milling and baking qualities have been obtained, the Grain Standards Committee is convened. The standard samples established by the Committee represent the minimum of each grade for that year. The Committee further prepares, under even more rigid conditions, standard export samples of the first nine statutory grades and of all commercial grades considered advisable.

3.38. The basic quality of Canadian grain production is secured and maintained through the licensing of varieties for seed. Most basic plant breeding for new varieties is carried out by the Federal Department of Agriculture in many research stations across Canada. No hard red spring wheat varieties are licensed unless they are judged equal to or better than Marquis, an excellent variety of wheat produced at the turn of the century, using ten basic criteria associated with milling and baking requirements. Similarly for the other grains, basic standards of quality are used for the approval of new varieties. These criteria are aside from improvements looked for in better varieties, such as freedom from rust, disease resistance, high yields, shorter growing season and so forth.

Exports

3.39. In recent years Canadian Pacific seaboard ports have exported over 5 million tons of wheat a year, St. Lawrence ports over 8 million in the peak years of 1963/4 and 1965/6, the Atlantic seaboard about 1 million tons, the port of Churchill just over $\frac{1}{2}$ million tons, and $\frac{1}{4}$ million tons annually have been exported direct from the Lakehead. While the trend has been for higher exports from most areas, the volume and proportion shipped from the Pacific have risen most. Atlantic seaboard shipments have fluctuated and do not seem likely to exceed 1 million tons. Churchill also seems limited in volume, and the amount shipped via the St. Lawrence Seaway appears to have stabilised. This is probably because of the size limitations on ships moving through the locks at a time when the use of larger bulk carriers has been increasing in the ocean movement of grain.

3.40. In the case of oats and barley and rye, where the volumes have been much smaller—barely $\frac{1}{4}$ million tons of oats, 1 million tons of barley, and 200,000 tons of rye—a larger proportion has been shipped from the West Coast to Pacific destinations or moved direct from the Lakehead.

3.41. Through its pricing policies, the Canadian Wheat Board endeavours to maintain a strong competitive position in all markets. In order to accomplish this, the Board quotes separate asking prices daily for wheat in store at the terminals at the Lakehead, Vancouver and Churchill. At current freight costs—open or closed navigation on the St. Lawrence—the Lakehead prices are immediately convertible to Montreal or other St. Lawrence ports or Atlantic seaboard prices. The price at any one of these three terminal positions may fluctuate freely and independently of the price quoted for the other two. The Board takes into account not only internal costs of moving wheat to these positions but also the various ocean freight rates involved and any variations in these rates. Prices are quoted daily for each grade and type of wheat, oats and barley.

3.42. The Board offers wheat to exporters on a deferred pricing basis. The buyer has the right to declare the final price up to 8 market days after the date of call on shipment from St. Lawrence or Atlantic ports and from 15–22 market days from date of loading from Pacific Coast ports depending on destination of shipment.

Production of Compound Feeds

3.43. In recent years there have been slightly less than 900 manufacturing establishments in Canada whose primary activity is the production and sale of livestock feeds (including pet foods). In addition, the flour-milling industry, the breakfast-food manufacturers, the brewers, distillers and other processors such as starch, malt, meat, fish and vegetable oil industries, also produce a small proportion of livestock feeds as a means of utilising the by-products of their principal activity.

3.44. In 1964 feed manufacturing plants employed about 8000 persons, of whom 4500 were involved in manufacturing. The value of the product is nearly $400 million (about £154 million) and the value added about $100 million (£38 million). In terms of size, eight plants each sell over $5 million worth of feed products of their own manufacture, sixty-five sell between $1 and $5 million, and 797 less than $1 million. At present, the trend is towards a slight decline in the number of establishments.

3.45. Feed plants are located in nearly every major trading centre across Canada. A few larger companies operate almost on a national basis with several centralised plants which specialise in the production of pre-mixes and a large number of smaller plants located in the livestock areas, either company-owned or operated on a franchise basis, which process the pre-mixes, grains and other ingredients into complete feeds and supplements. There are also smaller regional companies and co-operatives operating in a similar fashion. Some of the larger companies are subsidiaries of United States feed companies or subsidiaries or divisions of Canadian companies with a feed operation to utilise by-products, such as the meat-packing industry and flour-milling industry.

3.46. In the past, most farms in Western Canada ground and mixed their own livestock ration by combining their own grain (oats or barley and occasionally wheat) with a commercial protein supplement. It is only in recent years that livestock producers in the West have begun purchasing complete rations from feed manufacturers in large quantities. This contrasts with the situation in Eastern Canada, where the volume of complete feeds sold has been much larger. The complete feeds in Eastern Canada are based on western grains shipped to the East under feed freight assistance, or on grain produced direct from Eastern farms or through the local elevator or using imported U.S. maize. In the West complete feeds will be based on grain purchased direct from the farms and outside the Canadian Wheat Board delivery quota system—"non-quota" grain as it is called.

3.47. Feed plants usually offer a variety of services to farmers in conjunction with the sale of feed. Some plants have mobile feed-mixing units which move from farm to farm and custom-mix feeds to the farmer's specification, using the farmer's own grain. Many plants also custom-grind and mix feeds on the mill premises, using the farmer's own grain or the mill grain and the mill concentrate or supplement. Other services offered by feed mills are feed testing, farm audits and specialised animal nutrition advice. A recent development has been the establishment of "one-stop" farm supply centres where producers can obtain all manner of farm inputs, such as fertilisers, pesticides and animal health products.

3.48. The Canadian livestock producer does not purchase as large a volume of compound feeds as does his British counterpart, but makes wider use of his own grain both in direct feeding and in custom-mixing. In a number of cases, however, the production of livestock (especially broilers and eggs in Ontario) has been to a large degree integrated with a feed plant operation. This takes the form of actual ownership and management by the feed plant or more commonly contract and credit operations by feed companies.

3.49. Thus the Canadian producer of feed grains sells only about 3 or 4 million tons of the 13½ million grown. While there may be a trend to use more complete feeds purchased from feed plants, it seems likely that livestock feeders will continue to grow and feed a high proportion of their requirements. While the Canadian farmer does have several choices in his method of obtaining a balanced ration for his livestock, the producer of feed grain for sale does not have a very strong bargaining position and in general does not find a high degree of organisation in the marketing channels for his feed grain.

Future Developments

3.50. In the area of production, Canada will remain an efficient, low-cost producer of a high-quality product—hard spring wheat—and production is likely to continue its steady increase. This will be partly through further technological advances in new varieties, more fertiliser and better cultivation practices, and partly through larger acreages. At the same time, further changes in the production pattern may lead to the concentration of wheat in the centre and east of the Prairies. This could result in greater consistency in quality, and possibly higher average protein levels.

3.51. Proportionally to wheat, the emphasis will move to feed grains, with maize as the front-runner and some increase in oats and barley. The potential for increased production, both of the traditional feed grains on the Prairies and of maize and barley in the East, is very considerable. Fertiliser, drainage and modern crop practices are constantly providing higher yields per acre. Until maize varieties with a shorter growing season were introduced, production of maize was limited, but in the last 6 years it has increased rapidly. There may be some increase in barley relative to oats, but it is unlikely that there will be a wholesale switch to barley of the sort which has taken place in the U.K.

3.52. Over the next decade or so, consumption in Canada for wheat products will decline a little, but consumption of other grains, especially maize, is likely to increase for breakfast foods, industrial products and distilling and brewing. Consumption of livestock products, except milk and butter, will rise very considerably. In consequence, a large rise can be expected in the demand for feed grains. In terms of exports, this seems to indicate larger volumes of wheat, but with a continued emphasis on more wheat to the Far East and to other areas of developing countries and less to Europe, except perhaps for durum wheat. Wheat shipped under Food Aid programmes will also increase.

3.53. As the use of bulk carriers increases, the efficiency of the ports becomes even more important to the vessel charterers. This may lead to fewer and larger ports, possibly operating a longer season with use of deeper-water harbours and longer use of ice-breakers during the year.

3.54. The chain of events from the production to the consumption or export of Canadian grains is a long one, involving the research stations of the Department of Agriculture, the farmers and co-operatives, the Wheat

Board, the railways, the traders, shippers, terminals, the Departments of Trade and Commerce, the Board of Grain Commissioners and others. A closer integration of the development of all these participants in the Canadian grain economy could be brought about by the establishment in the near future of a National Grains Council. Such a Council would provide a forum for all these parties and organisations to put forward their views and to study and survey the future needs of the grain market.

CHAPTER 4

Grain Marketing in the U.S.A.

The Production and Disposal of Grain

4.1. The supply of wheat in the United States market in 1967 was 53 million tons, of which nearly 12 million tons had been carried over from the previous year, and 41 million tons were current production. Of this total, 40% (21 million tons) was exported, 26% (14 million tons) was milled into flour, and about 8% (4 million tons) was used for seed or feed. The remaining 26% (14 million tons) was carried over.

4.2. The supply of feed grains in 1967 totalled 212 million tons, of which 37 million tons had been carried over from the previous year and 175 million tons was current production. Of this total 59% (125 million tons) was used for feed (i.e. on the farm of origin or manufactured), 11% (23 million tons) was exported, 7% (15 million tons) was used for the food industry or for seed, and the remaining 23% (49 million tons) was carried over.

4.3. Information about the marketing channels through which the wheat and feed grains passed has been published for the years up to 1963/4 in a Report published in 1966 by the United States Department of Agriculture[1] and much of what follows in this chapter has been based on this report. For wheat all but a very small proportion of the crop which is kept back for seed or feed is sold off the farm and, in 1963/4, 80% of it passed through country elevators and most of this also passed through the sub-terminal or terminal elevators (see below). A much smaller proportion of the total production of feed grains is sold off the farm but this has increased from about 25% in 1939 to nearly 50% in 1963/4. As with wheat, the country elevators were the major outlet for off-farm sales of feed grains and in 1963/4 they handled 81% of the total volume. A substantial part of this grain was manufactured into animal feeds at or near the country elevators but the rest moved on to sub-terminal and terminal elevators, which handled 51% of the total volume of off-farm sales.

[1] *Agricultural Markets in Change,* U.S. Dept. of Agriculture, Economic Research Service, Agricultural Economic Report No. 95.

The Country Elevators

4.4. The most important difference between the grain marketing system in the U.S.A. and that in the U.K. is that whereas in this country virtually all the grain sold off the farms moves from the farm direct to the processor (although a merchant has usually acquired the title to it and arranges the transport), in the U.S.A. nearly all the grain entering the market moves from the farm at harvest time to intermediate storage sites. Generally there are two such intermediate stages, the country elevator and the sub-terminal or terminal elevator. It is far less common for grain to be dried or stored (for ultimate sale) on farms in the U.S.A. than in the U.K. and the farmer's involvement in the marketing process generally ends when he transports the grain by lorry to the nearest country elevator at or soon after harvest and sells it to the elevator operator. At the country elevator the grain is graded, blended and stored and, increasingly in recent years with the introduction of faster methods of grain harvesting, dried. The elevator operator then decides whether to sell the grain immediately to the sub-terminal or terminal elevator operator or to keep it in his silos until it is required later in the season. In 1963 there were 7650 country elevators in the U.S.A. and they varied in size from a few hundred tons to over 25,000 tons, the typical size being several thousand tons.

4.5. Although in most instances the country elevator operator buys the grain outright from the farmer, he may sometimes accept the grain for storage on a rent basis, the sale to be negotiated later, or he may simply hire out storage facilities and not come into ownership of the grain at all. The country elevator operator's primary objective is not to store grain for long periods but to move it onwards as soon as disposal can be arranged at the sub-terminal or terminal elevator or a processor can be found to buy it. At harvest time there is frequently more grain to be stored than he can accommodate. Queues of lorries wait in the elevator yard until space can be found, which may take time because there is also heavy pressure on rail services at this time and railway wagons may not be available to move the grain already in the store. Thus it sometimes happens that the grain has to be stored temporarily on the ground, either at the farm itself or at the elevator. The country elevator therefore fulfils a dual role of a transit silo with a fairly rapid turnover of grain and also a silo where grain can be stored for longer periods. The "transit silo" function is particularly characteristic of those areas where there is a succession of crops coming to maturity at different times. Then the elevator operator may be anxious to clear his store of one crop so as to allow space for the next. In addition to his function as a link in the grain-marketing chain the country elevator operator also carries out some of

the trading activities of a typical merchant in the U.K. Thus he will sometimes stock feeds, fuel, coal, baler wire and other agricultural requisites
which farmers can load on to their lorries after delivering their grain.

4.6. Country elevator operators have shared in the general increase in
grain business resulting from the expansion in cereals production in the
last decade or two. However, there is a tendency for some of the bigger
cereal farmers who have large vehicles and elevators, and who are not too
far from market outlets, to by-pass the country elevators and to take their
grain direct to the sub-terminal or terminal elevator or even straight to the
processor. By circumventing one link in the marketing process in this
way these farmers can sometimes obtain a better price for their grain. This
has led to a slight reduction in the number of country elevator operators
since before the war, but nevertheless they remain the principal point of
first sale for most cereal producers.

4.7. Surveys carried out in 1950 and 1951 in two wheat-surplus areas,
Oklahoma and North Dakota, showed that even where the farmer already
had bins available for farm storage it was still more economic for him to
sell all his grain to the country elevator.[1]

4.8. In 1954 the United States Department of Agriculture made a survey
of off-farm commercial types of grain storage in which the total capacity
for the country as a whole, both private and public storage, was estimated
to be 78 million tons. In addition the Commodity Credit Corporation
owned or leased emergency storage space amounting to about 25 million
tons, mostly in small metal bins located mainly in the Corn Belt. Corn
(i.e. maize) is too damp to store in a metal bin immediately after it is
harvested and the traditional method of storage was to put it into a crib
where further curing could take place; recently, however, the tendency in
recent years, particularly since the spread of the picker-sheller, has been
for the corn to be dried and then stored in metal bins since this facilitates
harvesting and reduces losses in store.

The Sub-terminal and Terminal Elevators

4.9. The sub-terminal elevators occupy an intermediate position between the country elevators located near the producer and the terminal
markets located near the processors. They are all located away from the
metropolitan areas and are the only large grain-handling facilities in their
immediate vicinity. They tend to be larger than the country elevators, but

[1] *Where and How Much Cash Grain Storage for Oklahoma Farmers,* U.S.D.A. Farm
Credit Administration Bulletin No. 58, May 1950; and *Where and How Much Cash Grain
Storage for North Dakota Farmers,* U.S.D.A. Farm Credit Administration Bulletin No.
61, May 1951.

it is function rather than size that distinguishes them from other elevators. Grain moves from the country elevators to the sub-terminal elevators (or arrives direct from farms as mentioned above) and is then sold to terminal elevator operators or direct to processors and exporters. The storage requirements of the Commodity Credit Corporation have given a stimulus to the construction of a number of large new sub-terminal elevators in recent years, and the operators of these have in turn developed their grain-selling activities.

4.10. Terminal elevators are located in the principal grain markets, and some of the more important ones are Chicago, Duluth, Kansas City, Milwaukee, Omaha, Houston, Portland, New York, Baltimore and Charleston. The capacity of terminal elevators may vary from a few thousand tons to over 1 million tons. The operators of terminal elevators buy grain from a variety of sources; some comes from the sub-terminal elevators or country elevators and some from the grain merchants. They in turn sell grain to processors, exporters and sometimes to terminal elevator operators elsewhere.

4.11. Terminal elevators, like country elevators, may be owned by independent firms, integrated terminal companies controlling a chain of such terminal elevators, or by farmers' co-operatives. There were 650 terminal elevators in the U.S.A. in 1963, an increase of 250 over 1948, most of the increase having taken place in the North Central and South Central regions.

4.12. The elevator system as it has developed in the United States owes its origins to the great distances from the main grain-producing regions to the principal consuming areas. This is reflected in the predominance of rail and water transport of grain from country elevators onwards. Unlike the U.K., where virtually all the internal transport of grain is by road, motor transport still accounted for only 41 % of total movements of grain in the U.S.A. in 1963, although as indicated above, the importance of motor transport has been tending to increase (much of this being first-stage movements from the farm to the country elevator or river terminal); in 1958, for example, the proportion was only 30 %. Because distances in the U.K. are so much smaller it is unlikely that an elevator system like that in the U.S.A., linked mainly to rail and water systems of transportation, would ever be appropriate.

The Grain Merchants

4.13. The grain merchants in the U.S.A. (generally known as "cash merchants"), unlike their counterparts in the U.K., seldom buy grain direct from the farmers. They more usually act as intermediaries between the elevator operators and grain processors and other grain users, or they may have grain elevators of their own. They also perform a variety of functions for the elevator operators such as putting them in touch with potential buyers (i.e. acting as grain brokers), advising them on market trends, handling futures transactions, assisting with the grading, weighing and transportation of grain, and obtaining finance. Apart from the relatively few who trade mainly as brokers the cash merchants are of two kinds; the "terminal receivers" who deal largely in cash grain from country elevator operators with whom they are geographically linked and who seldom take up a position on grain; and the "merchandisers" (including the big shippers), who are the largest firms as a rule and who frequently engage in the export of grain as well as terminal elevator operation on their own account. The merchandisers take up a position on the market and frequently operate "cross-country", i.e. they cover several interior grain markets. The big farm co-operatives also perform many of the functions of the cash merchants.

The Grading of Grain

4.14. A Federal system of grain inspection and grading was established in the U.S.A. as far back as 1916 (Grain Standards Act) and has been modified by subsequent legislation. Under this system all grain shipped in inter-state or foreign commerce to or from a point at which an inspector licensed under the Act is located must be officially inspected and graded if the grain is marketed by grade. Under the Act classes and sub-classes, as well as grades within each class, were fixed for each type of grain. The grain is inspected and graded by inspectors licensed by the Federal Government and located at approximately 140 inspection points in the U.S.A.; there is provision for appeal. The Report referred to earlier comments:[1]

Accurate, definite, and reliable grade determination is essential to many phases of marketing. With accurate grading, the farmer obtains a more exact price than when price was based on the "average for the crop" in his locality. The producer of better than average quality receives benefits in higher returns. Market reports are likely to be vague and misleading unless they are based on precise grades. Accurate grading makes possible modern bulk grain warehousing and commingling of grain with assurance of fairness to the owners of any part of it. Futures contract trading would be impossible if grain were not graded for delivery on contracts. Transactions for the purchase of grain from distant grain markets would be much more hazardous without the use

[1] *Agricultural Markets in Change,* Economic Research Service, U.S.D.A., 1966, p. 226.

U

of grades. Accurate grading is also of particular importance in the export market. The exporting countries capable of supplying a uniform product have gained buyer preference for their grain.

4.15. This passage has been quoted in full because it summarises succinctly the reasons why grading has become so firmly established in the American (and indeed Canadian) grain trade. The lack of any similar system in the U.K. is another major difference between the grain trades in the two countries. Again, the basic reason is to be found in geography. Sellers and buyers in the U.K. have never been far removed from each other and the grain has traditionally been kept on the farm until it has been required by the trade, i.e. it never required to be assembled or bulked at intermediate points of the marketing chain as in the U.S.A. Thus the selling of grain has always been a matter of an individual deal between a smaller seller (selling perhaps 50 to 100 tons) and a buyer who is often dealing with thousands of similar sellers; there was no point in the process at which grading would have been economically feasible. This may have been the fundamental reason why the proposals of the 1928 Report evoked little or no response from the grain trade. However, circumstances today are changing. Such developments as the rise of a substantial barley export trade, the growth of producer marketing groups, the increasing rate of concentration of cereals production into the hands of a relatively small number of large farmers, and the decline of the corn exchange and the increase in selling by telephone, would all lead one to expect a more lively interest in the possibility of cereals being traded on the basis of recognised quality standards. (See, however, Part IX, para. 87.)

4.16. In the United States system of grading there are seven classes of wheat: Hard Red Spring; Durum; Red Durum; Hard Red Winter; Soft Red Winter; White; and Mixed Wheat. The sub-classes of Hard Red Spring Wheat are (a) Dark Northern Spring, (b) Northern Spring, and (c) Red Spring. Sub-classes of Durum Wheat are (a) Hard Amber Durum, (b) Amber Durum, and (c) Durum. Red Durum Wheat has no sub-classes. Hard Red Winter Wheat sub-classes are (a) Dark Hard Winter, (b) Hard Winter, and (c) Yellow Hard Winter. Sub-classes of White Wheat are (a) Hard White, (b) Soft White, (c) White Club, and (d) Western White. In addition to these classes there are five numerical grades of wheat and "sample grade" as shown in Table 4.1. There are also classes, sub-classes and grades for the other main cereal crops.

4.17. Not all sales of grain in the U.S.A. are made on a graded basis; some grain is sold on the basis of a type sample. In other cases the grade system is used as a basis, but the processor specifies certain other requirements such as a certain protein content and this is not part of the grade specification.

TABLE 4.1

NUMERICAL GRADES AND SAMPLE GRADE AND GRADE REQUIREMENTS FOR ALL
CLASSES OF WHEAT EXCEPT MIXED WHEAT

Grade	Minimum test weight per bushel		Maximum limits of						
			Defects					Wheat of other classes[a]	
	Hard red spring wheat (lb)	All other classes (lb)	Heat-damaged kernels (%)	Damaged kernels (total) (%)	Foreign material (%)	Shrunken and broken kernels (%)	Defects (total) (%)	Contrasting classes (%)	Wheat of other classes (total) (%)
1	58·0	60·0	0·1	2·0	0·5	3·0	3·0	0·5	3·0
2	57·0	58·0	0·2	4·0	1·0	5·0	5·0	1·0	5·0
3	55·0	56·0	0·5	7·0	2·0	8·0	8·0	2·0	10·0
4	53·0	54·0	1·0	10·0	3·0	12·0	12·0	10·0	10·0
5	50·0	51·0	3·0	15·0	5·0	20·0	20·0	10·0	10·0

Sample grade: Sample grade shall be wheat which does not meet the requirements for any of the grades from No. 1 to No. 5 inclusive; or which contains stones; or which is musty, or sour, or heating; or which has any commercially objectionable foreign odour except of smut or garlic; or which contains a quantity of smut so great that any one or more of the grade requirements cannot be applied accurately; or which is otherwise of distinctly low quality.

[a] Red Durum Wheat of any grade may contain not more than 10·0% of wheat of other classes.

Source: Official Grain Standards of the United States, U.S. Dept. of Agriculture, Agricultural Marketing Service, 1964.

The Grain Exchanges and the Futures Markets

4.18. Chicago is the leading grain market in the United States, a pre-eminence which it owes to its strategic position at the southern end of Lake Michigan, a natural focus of routes. Other important grain markets are Minneapolis (Spring Wheat) and Kansas City (Hard Winter Wheat). The grain exchanges have developed their own rules, but they are also regulated by State and Federal laws. Thus the Commodity Exchange Act authorised limitations on both the amount of speculative trading in one business day and the amount of the speculative net position that may be held at any time by any one person or by any group of persons acting as a unit (hedging transactions being excluded from such limitations). Dealings in grain at exchanges may be either in actual (i.e. "cash") grain, in which the trading is done in specific lots of grain, or in futures contracts, which do not relate to specific lots of grain but to grain of a specified class and grade. The Chicago Board of Trade is the leading grain futures market in the U.S.A. and accounts for the major part (about 85 %) of the trading in grain futures, the Kansas City and Minneapolis markets sharing the remainder about equally.

4.19. The elevator operator in the U.S.A. has to decide which are the best market outlets, and in doing so he has to study a range of factors such as the movement of prices, the availability of markets, transport costs to different markets and the special needs of particular buyers. When shipping grain to a terminal market the country elevator operator is usually represented by a commission agent or cash grain receiver, but he retains the ownership of the grain until it is eventually sold and this usually means until it arrives at the terminal market. This is a rather risky process so far as the country elevator is concerned, since prices may change during shipment, and an alternative method of selling is the "deferred shipment", i.e. the grain is sold for a price fixed at the time of the contract but shipment is deferred until the buyer wants the grain (the buyer paying the storage costs). The risk is thus transferred to the buyer as soon as the contract is signed, with the grain still lying in the country elevator. Under this system the transport is sometimes arranged by the buyer and sometimes by the seller. To help them meet the heavy risks sometimes associated with their storage function the elevator operators make considerable use of the futures markets for hedging. The concentration of large quantities of grain in intermediate stores, with the consequent risks involved, may help to explain why the futures markets are used so much more in the U.S.A. than in the U.K.[1] In this country the risks are borne by the farmer, but he has the assistance of seasonal guarantees for wheat, and seasonal incentives for barley, and in any case the grain in store will often represent only a part of the farmer's total production so that he is less concerned than if it represented his total worth. Therefore the need for British farmers to hedge their risks is far less than it is for the elevator operator. The trading in grain futures in the U.S.A. has been carried to the point where the risks of major price changes are generally carried by people other than those who own the grain. James S. Schonberg[2] has described the process as follows:

> Purchases of cash grain are offset by simultaneous sales of futures contracts; similarly, sales of cash grain are offset by purchases of futures. This results in a balanced market position at all times. The grain contracts may represent grain in various physical locations and differ materially from the offsetting futures contracts, but one contract is protection against the other from major fluctuations in price, no matter which way the price of that particular kind of grain moves.

4.20. The operators of the sub-terminal and terminal elevators carry fewer long-term risks and they are typically short-term hedgers, i.e. their purchases of grain are offset by sales of futures which usually take place within a short time after the cash purchase.

4.21. The futures markets in the U.S.A. depend heavily upon the large number of "speculators", people often with little or no knowledge of the grain trade as such who speculate on the commodity markets more or less as an alternative to speculating on the stock exchange. This type of speculator is virtually non-existent in the U.K. grain futures market. Another important feature of the United States futures markets is their

---

[1] See also Part IV, Chapter 2, para. 2.18 *et seq.*
[2] James S. Schonberg, *The Grain Trade—How it Works,* 1956, p. 41.

dependence upon an adequate flow of information about the market. Regular crop reports are issued by the Department of Agriculture, with elaborate precautions to prevent prior leakages, and these have an important impact on the commodity markets. Other statistics relate to grain stocks at terminal and sub-terminal markets, agricultural prices, foreign crops and markets, grain exchanges and weather. Many comparable statistics are published by the Ministry of Agriculture, Fisheries and Food in the U.K., but these seem to be prepared and published with less regard to their impact on the market than is the case in the U.S.A.

4.22. To the extent that centralised grain stores in the U.K. would be performing similar functions to those of the country elevators in the U.S.A., it follows that the organisers would need to consider the use of the futures markets in Britain to spread their risks in a way that perhaps the individual farmer would seldom consider necessary. Moreover, they might have available the expertise in marketing that is needed to operate the futures markets and which few farmers would claim to have. Any extension of group-selling or group-storage in Britain is therefore likely to be associated with an increase in trading on the futures markets.

Grain Exporting

4.23. The grain-export business is conducted by large export firms and shippers who have specialised knowledge of ocean transportation, foreign market methods and customs, and the operation of foreign exchange markets. They buy their grain generally from the grain dealers at the larger markets, and sometimes from the country and terminal elevator operators; they also use the services of brokers. The bigger firms often have vertically integrated organisations reaching back to the country elevators in the grain producing districts.

4.24. Grain intended for export is usually bought from interior sellers either on a "delivered to port" basis or "f.o.b." vessel at port. Some exporters have their own terminal elevator facilities and can store the grain until it is sold or whilst awaiting shipment, but others buy only to fulfil orders already received and for immediate shipment. All grain coming into these elevators is inspected by licensed inspectors as it is unloaded from the rail wagons. When the grain is loaded into a vessel the Federal licensed grain inspector again makes an official inspection; the grade assigned at this stage is the final one and forms the basis of the export sale.

4.25. In 1966/7, 49% of wheat exports were made under government programmes and 51% were commercial sales. In contrast most exports of feed grains move through commercial channels without government assistance. In 1964/5 private firms accounted for 90% of all feed grain exports.

The Grain Processors

4.26. Unlike their counterparts in the U.K., the processors in the U.S.A. try, whenever possible, to hedge their purchases of grain and their sales of grain products by counterbalancing sales and purchases in the futures market. Although some grain is bought by processors at country elevators, and some direct from farmers, the bulk is purchased in the terminal markets. The great advantage of this system is that the processor can obtain a large quantity of grain of the specific type and quality he needs at any time convenient to him. As in the U.K., the maltsters buy a high proportion of their grain requirements for the whole year at harvest time, but most other processors are in the market buying grains throughout the year. The processor's main task is to buy grain of the right type and quality to meet his requirements. He is helped by the Federal grading system, and many processors are satisfied to rely wholly on this system. But others, particularly the millers and maltsters, are more exacting in their requirements, and they will frequently rely more upon the appraisal of a skilled grain buyer who, through years of experience in the trade, can determine certain qualities in the grain that are not satisfactorily described by the United States grade designations. Some processors submit grain to tests under conditions similar to those the grain will encounter in the process itself, e.g. milling. The Federal grading system is thus for many flour millers merely a first "filter", i.e. it allows them to reject grain that is obviously unsuitable, but they still apply their own criteria of quality so far as their own particular process is concerned.

4.27. As in the U.K., a process of concentration has been taking place in the cereal-using industries, particularly the flour industry. In 1963 there were only about one-third as many establishments (with twenty or more employees) making flour as there had been in 1939. The cereal breakfast-foods industry has always been highly concentrated in the U.S.A., as in the U.K., and it continues to grow roughly in parallel with the increase in population. The biggest changes have taken place in the animal-feeds industry (called in the U.S.A. the "Mixed-feeds Industry"). These changes include: the increase in popularity of formula foods, integration of the feed industry with livestock production, increase in direct sales to producers, more bulk delivery of feeds, and a steady growth in on-the-farm mixing and custom-mixing for the farmer.

4.28. A process of decentralisation of production, combined with concentration of ownership, has been taking place, with the total number of feed-mixing installations increasing and the average size of mills decreasing. In 1963/4 the number of commercial feed manufacturers was estimated to be over 9000. Between 1939 and 1963 the number of establishments increased by about 90%, the greatest increases being in the South Atlantic and South Central states, the number in the Middle Atlantic states actually declining.

4.29. The percentage of feed delivered in bulk varied from about 36% in the West North Central and East North Central Regions to between 68% and 76% in the South Atlantic, East South Central and Pacific Regions. Methods of feed retailing have been changing. Retail dealers are still important, but increasingly feed firms are making direct sales to farmers, a system which has hitherto been used to only a limited degree. Many firms now feel that direct sales enable them to meet competition and changing market conditions better, and this trend is likely to gather momentum in the future.

4.30. Another trend has been the shift by many livestock producers to on-the-farm mixing and the buying of feeds through organised buying groups. Many large livestock and poultry producers have gone over to on-the-farm mixing because the local merchants were not able to meet their needs and because they were able to cut costs that way.

4.31. Vertical integration has moved a long way in the poultry industry, the impetus having come mainly from the feed manufacturers, the hatcheries, poultry processors and others in the chain between the producer and the consumer. Some indication of the extent to which vertical integration has proceeded is provided by Table 4.2, which is based on a survey of firms in the North Central Region (eleven states centred on Iowa). The degree of vertical integration is measured by the percentage of grain procured from other plants owned by the same company and the

TABLE 4.2

ACQUISITION AND DISPOSITION OF GRAIN AND PRODUCTS FROM COMPANY-OWNED PLANTS AND OUTSIDE FIRMS, NORTH CENTRAL REGION, 1960

Type of plant	Acquisition of grain		Disposition of grain		Disposition of products	
	Company-owned plants (%)	Outside firms (%)	Company-owned plants (%)	Outside firms (%)	Company-owned plants (%)	Outside firms (%)
Merchandising	12·0	88·0	19·1	80·9	—	—
Feed manufacturing	7·2	92·8	3·4	96·6	27·4	72·6
Flour milling	12·0	88·0	9·7	90·3	22·2	77·8
Oilseed processing	17·0	83·0	28·2	71·8	11·1	88·9
Dry milling and cereal manufacturing	0·4	99·6	0·3	99·7	21·6	78·4
Distilling and alcohol	23·9	76·1	—	—	42·0	58·0
Malting and brewing	5·8	94·2	—	100·0	8·3	91·7
Total merchandising and processing	10·7	89·3	18·5	81·5	17·6	82·4

Source: Market Organisation of Grain Industries in the North Central Region, L. B. Fletcher, University of Missouri Research Bulletin, No. 847, 1964, p. 20.

percentage of grain and processed products shipped to other plants owned by the same company.

Future Prospects

4.32. The carryover of wheat in June 1967 was the smallest for 15 years, but the carryover at 30th June 1968 was 27% up on the previous year, whilst it is forecast that the carryover at 30th June 1969 will be about 30% higher than the 1968 level. This would bring the mid-1969 carryover to about 19 million tons compared with an average annual carryover of nearly 36 million tons during the period 1959–63.[1]

4.33. The feed grain picture is rather similar. The carryover in 1968 at 49 million tons was 12 million tons greater than in 1967, and it is estimated that the carryover at the end of the 1968/9 season will be greater still. The progressive decline in carryover stocks from 69·3 million tons in 1964 to 37·1 million tons in 1967 has therefore been reversed.[2]

4.34. The significance of these forecasts, so far as the U.K. grain market is concerned, is that the United States is likely to be looking for every opportunity of expanding its exports of grain in the years immediately ahead.

[1] These figures are taken from *Wheat Situation,* August 1968, U.S.D.A. Economic Research Service.

[2] From *Feed Situation,* August 1968, U.S.D.A. Economic Research Service.

CHAPTER 5

The International Cereal Situation

5.1. The prices received by British cereal producers and the level of British cereal production have in the past been considerably affected by developments on the world market. This was particularly marked in the 1880's and 1890's, when the opening up of new grain-producing areas in the U.S.A. led to a fall in world cereal prices, and also in the depression of 1920/1 and the more severe depression of the early 1930's, which led to a change in Britain's traditional policy of free trade and the introduction of a policy of subsidising home production of cereals.

5.2. Since the Second World War, the annual average prices received by British cereal growers have been guaranteed by the Government and isolated to a large extent from prices on the world market. This situation is not likely to be radically changed in the foreseeable future. Nevertheless, whenever British guaranteed prices have stood well above world prices, the Government has been under pressure to lower them. Thus developments in world production, trade and consumption—which underlie trends in world prices—have more than an academic relevance to the British situation.

World Grain Production

5.3. Cereals account for 71 % of the world's harvested area; they provide 53 % of man's supply of food energy when consumed directly and a large part of the remainder when consumed indirectly in the form of livestock products. Measured in terms of calories or tonnage, cereals dominate world trade in foodstuffs.[1]

5.4. The proportion of the total food supply (in terms of calories) obtained directly from cereals varies considerably from region to region, reflecting mainly the level of income and thus the consumption of livestock products; in 1958 it ranged from 20·5 % in North America to 67·7 % in Asia.

[1] Lester R. Brown, *Man, Land and Food,* U.S. Department of Agriculture, Foreign Agricultural Economic Report No. 11, 1963.

U*

TABLE 5.1

PERCENTAGE OF TOTAL CALORIE SUPPLY DERIVED FROM GRAINS: 1958

	Rice	Wheat	Maize and other grains	Total from grains	From other foods	Total
North America	0·8	17·7	2·0	20·5	79·5	100
Oceania	0·6	25·5	0·5	26·6	73·4	100
Western Europe	1·1	30·8	4·9	36·8	63·2	100
Latin America	7·4	14·0	18·3	39·7	60·3	100
E. Europe and U.S.S.R.	0·7	41·7	10·6	53·0	47·0	100
Africa	2·7	10·6	31·5	44·8	55·2	100
Asia	41·1	13·0	13·6	67·7	32·3	100
World	21·2	19·6	12·6	53·4	46·6	100
Developed regions	0·9	31·7	6·3	38·9	61·1	100
Less-developed regions	32·4	12·8	16·5	61·7	38·3	100

Source: Lester R. Brown, *Op. cit.,* table 12.

5.5. Whereas the *direct* consumption of cereals *per capita* declines with rising incomes, the *total* consumption—including the cereals transformed into livestock products—rises, as it takes from 3 to 7 lb. of cereals to produce a pound of animal products. Thus the total supply of cereals per head of population in the U.S.A. is roughly five times as high as in India (Appendix H, Table 19).

5.6. The world's two most important cereals—considered in terms of direct human consumption—are wheat and rice. In terms of production and trade, however, wheat is the leading cereal. The world production of wheat exceeds that of any other cereal, and it is by far the most important cereal entering into international trade (Table 5.2).

5.7. In recent years both the production and exports of wheat have been running at record levels; exports have risen above 50 million tons a year, nearly a quarter of total production. A much lower proportion of the other major cereals enters into international trade. The production of maize is now a close second to that of wheat, but exports are less than half those of wheat. World production of milled rice is roughly half that of wheat, but only a very small proportion of rice production is traded internationally. Wheat and rice constitute, however, the world's two major food grains, for rice is used almost exclusively for human consumption,whereas most maize is used for animal feed. Barley is used both for brewing and for feeding to livestock; it is the only other cereal of importance in international trade.

5.8. Wheat and barley can be grown under a wide range of climatic conditions, but are for the most part grown in regions having a temperate

TABLE 5.2

WORLD CEREAL PRODUCTION AND EXPORTS[a]

	Million long tons			
	Average 1937–9		Average 1965–7	
	Production[d]	Exports	Production[d]	Exports
Wheat	143·8	18·0[b]	260·3	57·0[b][e]
Maize	109·9	9·0	207·9	23·9[e]
Barley	43·4	2·2	95·3	7·1[e]
Oats	63·2	0·8	45·7	1·4
Rye	43·0	1·0	32·6	0·5
Rice[e]	67·2	7·7	111·5	6·1[e]
Total	470·5	38·7	753·3	96·0

[a] Excluding China.   [b] Including wheat equivalent of flour exports.   [c] In milled rice equivalent.   [d] In this table and in subsequent tables based on Commonwealth Secretariat publications, production figures relate in the Northern Hemisphere to the harvest in the calendar year (or the first year where two are shown), and in the Southern Hemisphere to crops harvested in the latter part of that year and the first part of the following year. Export figures relate to calendar years (or the first year where two are shown).   [e] Average 1965/6.

Source: Commonwealth Secretariat, *Grain Crops.*

climate. Maize requires a mild to warm climate, while rice can be grown only in a tropical or sub-tropical climate. These climatic requirements affect the distribution of world production (Appendix H, Table 15). Wheat production is divided more or less equally between four regions: the Soviet Union; Eastern and Western Europe; North America (i.e. the U.S.A. and Canada); and the rest of the world. Barley production is somewhat more concentrated in Europe and the Soviet Union. In the production of maize the U.S.A. is dominant, producing more than half the world's output, whereas the production of rice is concentrated in South and South-east Asia.

5.9. The yields of cereals vary considerably between countries as a result of both climatic differences and the level of agricultural technique (Table 5.3).

5.10. British yields of wheat and barley are among the highest in the world, being virtually double those in the U.S.A., Canada and Australia, which are limited by low rainfall. But agricultural techniques also play a role; the fact that British yields are substantially higher than French, or that the Japanese yield of rice is three times the Indonesian, is the result more of differences in agricultural technique than climate. It is clear that in some of the countries with relatively low yields there is considerable scope for increasing them, although this will require considerable changes in

TABLE 5.3

YIELD PER ACRE IN CERTAIN COUNTRIES
AVERAGE 1961/2 TO 1965/6[a]

	Cwt per acre			
	Wheat	Barley	Maize	Rice
United Kingdom	32·2	28·6	—	—
France	23·3	22·3	24·0	20·6
U.S.A.	13·5	15·3	33·2[b]	24·9
Argentina	12·2	9·9	14·7	19·0
Canada	11·0	13·2	37·4	—
Australia[c]	9·7	9·0	15·1	33·4
U.S.S.R.	7·7	9·0	17·7[b]	12·9
India	6·7	6·9	7·9	7·8
Japan	—	—	20·3	28·4
Indonesia	—	—	7·3[d]	9·1
Thailand	—	—	15·4	8·0

[a] See footnote [d] to Table 5.2.   [b] Maize harvested for grain.   [c] Per sown acre.
[d] Average for less than full period shown.
Source: Commonwealth Secretariat, *Grain Crops.*

techniques and substantial investment in agriculture and ancillary industries such as fertiliser production.

5.11. Considerable changes have occurred in the regional pattern of
trade in cereals since before the War (Fig. 5.1). The first big change has
been that North America has emerged as the dominant exporter of cereals
(although the comparison with the 1934-8 period is a little misleading,
since this period contained 2 years of extreme drought, which caused
exports to fall below the normal level). The second change has been that
Eastern Europe (including the U.S.S.R.), Africa and—most strikingly of
all—Asia have changed from net exporters to net importers. Total grain
imports into Asia in 1964/5 to 1965/6 approached those into Western
Europe, which did not differ greatly from their pre-war level.

5.12. Of the most important cereals in terms of world production—
wheat, barley, maize and rice—only wheat and barley (and an insignificant acreage of maize) are grown in Britain. However, Britain is a
substantial importer of maize, and since wheat, barley and maize are
all to some extent substitutable as animal feeds, their prices are related.
World trade in rice, on the other hand, is largely divorced from that in the
other three cereals and has little direct importance for the United Kingdom.
We will therefore concentrate attention on world cereal trade under the
two main headings of (a) wheat and (b) coarse grains, meaning mainly
maize and barley. The coarse grains traded internationally are used
predominantly as feed grains, but in the developed countries wheat is
also a feed grain of considerable importance.

NET REGIONAL TRADE IN ALL GRAINS

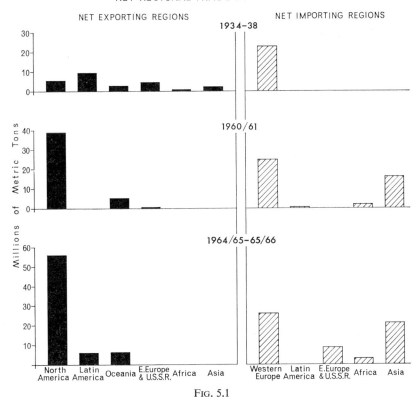

FIG. 5.1

Source: Lester R. Brown, *Man, Land and Food*, Fig. 21 (updated).

The World Wheat Trade

5.13. World production of wheat has approximately doubled since the immediate pre-war period, while exports have risen from 18 million tons to about 50 million tons (Table 5.4).

5.14. This substantial increase in the world wheat trade does not indicate a corresponding increase in sales on commercial terms; the greater part of the increase has consisted of disposals on concessional terms to developing countries, in other words, food aid.

5.15. The world's leading producers of wheat are the U.S.S.R., the U.S.A., China, Canada, France and India in that order (Table 5.5).

5.16. In terms of exports, however, the situation is very different, with the U.S.A. a clear leader, followed by Canada, Australia, Argentina and France (Table 5.6).

TABLE 5.4

WORLD WHEAT ACREAGE, PRODUCTION AND EXPORTS[a]

	Area	Average yield per acre	Production[d]	Exports[b][d]
	million acres	long tons	million long tons	million long tons
1937/8–1939/40	368·3	0·39	143·8	18·0
1957/8–1959/60	443·7	0·48	211·5	33·5
1961/2	440·8	0·47	207·7	45·2
1962/3	452·0	0·52	232·8	41·3
1963/4	451·7	0·47	213·4	48·9
1964/5	472·2	0·52	247·1	55·9
1965/6	470·4	0·50	236·4	55·0
1966/7[c]	468·9	0·59	278·9	58·9
1967/8[c]	485·8	0·55	265·7	46·3

[a] Excluding China.   [b] Including wheat equivalent of flour exports.
[c] Provisional.   [d] See footnote [d] to Table 5.2.

Source: Commonwealth Secretariat, *Grain Crops* and *Grain Bulletin.* International Wheat Council, *Review of the World Wheat Situation 1966/7*, table 1.

TABLE 5.5

ACREAGE AND PRODUCTION IN MAIN WHEAT-PRODUCING
COUNTRIES 1965–7

	Area[a] (million acres) 1967	Production[d] Average 1965–7 (million long tons)
U.S.S.R.[b]	174·9	77·9
United States	59·0	37·0
China	67·5[c]	27·5[c]
Canada	30·1	18·5
France	9·8	13·3
India	32·5	11·2
Italy	9·9	9·4
Australia	22·7	9·0
Argentina	14·6	6·5
Other countries	132·2	77·5
World total	553·3	287·8

[a] Harvested area where figures for both sown and harvested are reported.   [b] Sown area.   [c] The Commonwealth Secretariat's rough estimate for China is about 65 to 70 million acres and 25 to 30 million tons.   [d] See footnote [d] to Table 5.2.

Source: Commonwealth Secretariat, *Grain Crops,* No. 12, tables 12, 13.

5.17. Since the 1920's these five countries between them have supplied the bulk of the world's wheat exports, although their shares of the world market have varied from period to period (Appendix H, Table 17); it is only in recent years that France has emerged as a major exporter. In recent years the U.S.A. and Canada together have supplied around 60% of world wheat exports, and have dominated the world market (Table 5.6).

5.18. There have been dramatic changes in the pattern of imports since 1939 (Tables 5.7 and 5.8 and Fig. 5.2). In the pre-war period, virtually all world wheat imports were taken by Western Europe, of which the U.K. took roughly half. In that period Asia and the U.S.S.R. had not emerged as significant importers of wheat and, indeed, the U.S.S.R. was a net exporter. Since the early 1960's, imports into Western Europe have declined to roughly half the pre-war level. Within Western Europe as a whole, imports by the U.K. have declined slightly, while the European Economic Community has become a net wheat exporter.

TABLE 5.6

EXPORTS OF WHEAT[a]

	Thousand long tons			
	Average 1961 to 1963	1965	1966	1967
United States	17,918	18,541	23,438	18,286
Canada	10,211	12,534	15,401	10,145
Australia	5790	7141	4723	8516
Argentina	1896	6570	4998	2046
France	2216	4695	4186	2812
U.S.S.R.	4832	2001	3172	n.a.
Others	2275	3501	3065[c]	4520[c]
Total	45,138	54,983	58,983	46,325[b]

[a] Including wheat equivalent of flour exports.   [b] Excluding U.S.S.R.   [c] Incomplete.

Source: Commonwealth Secretariat, *Grain Crops,* No 12, table 15, and *Grain Bulletin.*

5.19. Whereas imports into Western Europe have been declining, imports into other countries have increased dramatically. These new importing countries fall into three main groups: Japan, the one major Asian country that can be classified as "developed"; the developing countries, especially those of Asia and South America; and the centrally planned countries, the U.S.S.R. and China. Japan's imports, which are paid for commercially, have risen to around 3·5 million tons in recent years or about three-quarters of her total consumption. Imports into the developing countries of Asia, i.e. excluding Japan and China, have risen even more rapidly,

NET REGIONAL TRADE IN WHEAT AND WHEAT FLOUR

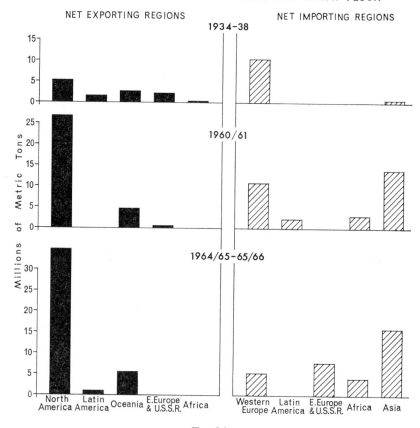

NET EXPORTING REGIONS   NET IMPORTING REGIONS

Fig. 5.2

Source: Lester R. Brown, *Man, Land and Food*, Fig. 20 (updated).

although this had only been made possible by food aid programmes. Imports of wheat into these Asian countries began to exceed imports into Western Europe in the early 1960's, and in 1965/6 they reached a level of 13·6 million tons. Imports into Central and South America (notably Brazil) had also risen to 5 million tons in 1965/6. Imports into China first became important in 1961, and have been running at 5–6 million tons in the last few years—in other words, somewhat more than imports into Western Europe. Substantial imports into the U.S.S.R. also commenced in the early 1960's; they have fluctuated considerably since then but have been over 8 million tons. The post-war period has thus seen a complete change

TABLE 5.7
NET GRAIN IMPORTS OF THE MAIN IMPORTING COUNTRIES

	Average 1934-8	Average 1954/5	Average 1960/1	1963	1964	1965
				Million metric tons		
WHEAT AND WHEAT FLOUR						
Developed Countries						
Western Europe[a]	11·2	9·6	10·7	6·2	3·6	4·7
of which E.E.C.	3·6	2·3	3·7	0·1	−2·0	−1·5
United Kingdom	5·6	5·2	4·7	4·6	4·2	4·7
Japan	—	2·2	2·7	3·8	3·5	3·4
Developing Countries						
Asia[b]	0·2	2·5	9·1	10·2	11·8	13·6
of which India and Pakistan	−0·4	0·4	4·4	5·7	7·4	8·7
Central and South America	−2·1	3·0	4·4	4·3	4·9	5·0
Centrally Planned Countries						
U.S.S.R.	−0·6	−0·4	−1·1	8·5	1·4	8·1
China (People's Republic)	0·8	—	3·3	5·2	5·0	6·3
COARSE GRAINS						
Developed Countries						
Western Europe[a]	9·9	10·7	15·5	18·2	17·8	23·5
of which E.E.C.	5·4	5·7	8·2	9·7	9·8	13·6
United Kingdom	4·3	2·6	4·6	4·1	3·8	3·6
Japan	0·2	1·0	2·2	4·6	5·1	5·2
Developing Countries						
Asia	−0·8	−0·6	−0·7	−1·2	−0·5	0·7
of which India and Pakistan	—	—	0·2	0·1	0·1	1·1
Central and South America	−6·8	0·0	0·5	0·0	0·2	−1·3
Centrally Planned Countries						
U.S.S.R.	−0·7	−0·3	−0·9	−0·2	−0·6	−0·2
China (People's Republic)	—	—	0·9	0·7	0·2	−0·2

Negative figures indicate net exports.
[a] Post-war figures excluding Yugoslavia.   [b] Excluding Japan.   [c] 1 metric ton = 0·98421 long ton.
Source: F.A.O. *World Grain Trade Statistics.*

in the pattern of world wheat imports, with the traditional market in Western Europe declining, while imports into Asia and Africa have risen from virtually nothing to a point at which they take the bulk of the greatly increased volume of world trade (Table 5.8).

5.20. The rise in imports to the developing countries has, however, been made possible only by exports under special concessional programmes. These programmes have expanded enormously since the war. As can be seen in Table 5.9, 19 million tons per year have been exported in recent years under these programmes, or over one-third of total world exports. Thus world sales in the commercial market have remained fairly constant since the pre-war period at around 20–25 million tons; the increase in the world wheat trade has consisted mostly of the expansion of concessional sales.

TABLE 5.8

THE CHANGING PATTERN OF WHEAT IMPORTS[a]

Importing country	Thousand long tons			
	Average 1937–9	1965	1966[c]	1967[c]
United Kingdom	5076	4665	4278	3902
German Fed. Rep.	1224[f]	1653	1583	1824
Eastern Germany	—	1206	1329	n.a.
Brazil	963	1847	2382	2440
Italy	852	912	1150	842
Japan	94	3595	3855	4064
India	87[b]	6466	7659	6265
Pakistan		1760	986	2160
China	47[e]	5661	5740[d]	3968[d]
Egypt	2	2045	2239	n.a.
U.S.S.R.	n.a.	6673	7904	n.a.

[a] Including wheat equivalent of flour imports.   [b] Undivided India.   [c] Provisional.   [d] Shipments reported by exporting countries.   [e] One year only. [f] Undivided Germany.

Source: Commonwealth Secretariat, *Grain Crops* and *Grain Bulletin.*

5.21. The bulk of these concessional sales have been made by the U.S.A. under the "Food for Peace" (more recently "Food for Freedom") programme authorised by Public Law 480, passed in 1954; in the last few years Canada has also made substantial sales on special terms, including outright grants and long-term credits. Wheat sold on concessional terms has in recent years come to make up the greater part of U.S. exports; in 1964/5 and 1965/6 sales under special programmes were running at a record level of over 15 million tons out of total exports of just over 20 million tons. In 1966/7, however, there was a sharp drop in

TABLE 5.9

WHEAT AND FLOUR EXPORTS UNDER SPECIAL PROGRAMMES

Million metric tons—wheat equivalent

	United States		Canada		Australia		Total	
	Total exports	Exports under special programmes	Total exports	Exports under special programmes	Total exports	Exports under special programmes	Total exports	Exports under special programmes
Average 1930–9	2·0	—	5·5	—	3·1	—	10·6	—
Average 1945/46–49/50	11·3	7·0	6·9	—	2·3	—	20·5	7·0
Average 1955/56–59/60	12·2	8·4	8·0	0·6	2·6	—	22·8	8·9
1960/1	18·0	12·6	9·4	0·9	5·1	—	32·5	13·6
1964/5	19·6	15·3	11·8	4·0	6·5	0·2	37·9	19·4
1965/6	23·4	15·9	14·8	3·3	5·7	0·2	43·9	19·4
1966/7 (P)	20·2	11·3	14·8	4·8	7·0	0·2	42·0	16·5

(P) Provisional.
Source: International Wheat Council, *Review of the World Wheat Situation,* 1960/1 and annually to 1966/7.

concessional exports to 11·3 million tons. This curtailment of the food aid programme by the U.S.A. reflected a sharp decline in the level of wheat stocks (Table 5.10) combined perhaps with a certain disenchantment regarding the results of economic aid that had become prevalent in the U.S.A. as well as other developed countries.

TABLE 5.10

WHEAT SUPPLIES IN THE U.S.A.

	Million long tons			
	Initial stocks	Production	Domestic consumption	Exports
Average 1946/7 to 1950/1	5·9	31·7	22·0	7·6
1963/4	32·0	30·6	15·5	23·1
1966/7	14·3	35·1	18·5	19·7
1967/8	11·4	40·8	13·9[a]	15·9[a]

[a] July 1967 to March 1968.
Source: United States Department of Agriculture, *Wheat Situation.*

5.22. At the same time, there have been some moves by other developed countries to share some of the burden of food aid with the U.S.A. The most important development of this type has been the signing of a Food Aid Convention embodying the G.A.T.T. agreement resulting from the Kennedy Round negotiations in 1967. Under this agreement, the developed countries have agreed to supply 4·5 million tons of cereals annually to developing countries as food aid; the cost is allocated in fixed shares between the developed countries, and can be paid either in the form of grain or in cash.

5.23. This agreement represents a significant development in acceptance of food aid as a part of the international trading pattern. Before 1939 any system in the world grain trade other than commercial sales would have been hard to envisage. After the War there were the problems of rehabilitating devastated Europe, and later of dealing with food shortages in the developing countries as their population increase outstripped their food production. The U.S.A. reacted to these problems by launching programmes of food aid, which were at first the subject of controversy among economists; those critical of the programmes argued that they were an inefficient form of aid and that it would be more satisfactory to grant aid in the form of money, allowing the recipients to buy what they needed in the cheapest market. More recently, however, it has come increasingly to be felt that, given the political limits set to monetary aid, programmes of food aid are a valuable supplement to other forms of aid. However, food aid has for the most part been carried out on a national basis by the U.S.A. and, to a lesser extent, by Canada. The agreement to set up an international food-aid programme which emerged from the Kennedy Round negotiations in 1967 may represent the beginning of a permanent international action by the developed countries to provide food aid for the less fortunate countries of the world.

Demand and Supply in the Commercial Market

5.24. The commercial market for wheat—as distinct from food aid—has gone through three phases since the War. Up till about 1950 there was the period of post-war scarcity brought about by the setback to production during the War. Then, in the 1950's, production began to outstrip demand. A collapse of prices on the world market was prevented by an increase in stocks in the U.S.A. and Canada, but these stocks reached an exceptionally high level at the end of the decade and there were fears of a continuing chronic wheat surplus. In the event, however, the situation once again changed rapidly. The increased exports to Asian countries in the 1960's caused a sharp decline in North American stocks. Thus stocks in the U.S.A. fell from 35 million tons at the end of 1961/2 to 14 million tons at the end of 1965/6, a level which the U.S. authorities considered necessary to meet normal eventualities, such as a poor crop (Table 5.10 and Appendix H, Table 18). At the end of 1966/7 they fell even below this level to 11·4 million tons and the authorities became concerned at the level of stocks.

5.25. The accumulation and subsequent disposal of stocks in the U.S.A. and Canada—the only countries to maintain large stocks in relation to output—had a stabilising effect on world prices. As a result, wheat prices on the world market have not been characterised by the enormous fluctuations that have characterised the prices of some other primary products. A contributing factor to this stability has been the International Wheat Agreement, under which the main importing and exporting countries have agreed to maintain a "floor price" and a "ceiling price" on the world market. However, the control exercised by the U.S.A. and Canada on supplies coming on to the world market—by means of stock changes and acreage control—have since the early 1950's held the world price within the price range permitted by the Agreement (Table 5.11).

5.26. At the end of 1967 world stocks were starting to rise slightly, but for a time at least the problem no longer appeared to be that of the bulging granaries of the early 1960's.[1] The commercial wheat market of

---

[1] However, the slower export trade after the 1967 harvest resulted in larger stocks.

the world was in reasonable balance, and seemed likely to remain so for some years. As the International Wheat Council commented in its *Review of the World Wheat Situation for 1966/67* (p. 12), "It seems that in round terms world trade will lie between 50 and 55 million tons for several years to come. Given such a level of world trade and the normal evolution of world production, it seems unlikely that additions to world stocks will reach sufficient proportions to give rise to undue anxiety."

TABLE 5.11

OPEN MARKET EXPORT PRICES FOR
WHEAT COMPARED WITH I.W.A. MAXIMUM AND MINIMUM

(No. 1 Manitoba Northern Wheat in store Fort William/Port Arthur)

	U.S. $ per bushel		
	I.W.A.		Average open market price
	Maximum	Minimum	
1962/3	2.02½	1.62½	1.82
1963/4	2·02½	1.62½	1.88
1964/5	2.02½	1.62½	1.84
1965/6	2.02½	1.62½	1.86
1966/7	2.02½	1.62½	1.96
1968/9[a]	2.35½	1.95½	

[a] New International Grains Arrangement prices relate to grain at Gulf ports. The maximum and mimimum prices of the new reference grain, U.S. Hard Red Winter No. 2 (ordinary) are U.S. $2.13 and $1.73 per bushel.

Source: Trends and problems of the World Grain Economy, International Wheat Council, 1966, and *Review of World Wheat Situation, 1966/7.*

The World Coarse-grain Situation

5.27. World imports of coarse grains display a quite different pattern from world imports of wheat. The grains which have been classified as coarse grains are maize (together with milo and sorghums), millet, barley, rye and oats. Of these, maize and millet are used extensively for human food in some tropical regions and rye is an important bread grain in Germany and some other European countries; maize is also used in Western Europe for corn flakes and distilling. However, the coarse grains entering into international trade—of which maize is by far the most important—are for the most part used as feed grains in the production of meat and dairy products, and as these are inevitably relatively expensive forms of food, they are mainly consumed in the developed countries. Thus the world imports of coarse grains are almost wholly taken by Western Europe and Japan (Table 5.7). This trade is on purely commercial terms; it is only in recent years that small quantities of coarse grains have been supplied to developing countries as part of programmes of food aid.

5.28. World trade in feed grains has increased even more rapidly than world trade in wheat in the last 10 years, having risen from roughly 16 to roughly 32 million tons; this reflects the rising demand for animal products that has accompanied rising real incomes in Western Europe and Japan. As with wheat, a tendency for production to rise faster than consumption caused stocks of maize to rise to high levels in the U.S.A. in the early 1960's. Subsequently, however, rising demand at home and abroad, and some slowing down of the rate of increase in production (to 1967/8) partly due to acreage restrictions, has caused stocks to fall to a more manageable level (Table 5.12). Thus the world market for coarse grains, like that for wheat, is at present roughly in a state of balance, taking one year with another. Looking to the future, the question is whether this balance in the commercial market is likely to remain or whether there will be shortages in the developing countries, which are not at present reflected in the commercial market.

TABLE 5.12

SUPPLIES OF MAIZE IN THE U.S.A.

	Million long tons			
	Initial stocks	Production	Domestic consumption	Exports
Average 1946/7 to 1950/1	11·2	70·2	65·1	2·3
Average 1956/7 to 1960/1	36·8	86·0	76·6	5·2
1961/2	50·2	90·6	89·5	10·4
1965/6	28·7	102·1	93·1	16·7
1966/7	21·0	102·9	91·7	11·7
1967/8	20·6	118·1	51·1[a]	8·3[a]

[a] First two quarters of season only.

Source: Commonwealth Secretariat, *Grain Crops* and *Grain Bulletin.*

Prospects for the World Cereal Situation

5.29. Projections of the world food situation, and, more narrowly, the world cereal situation, are extremely problematical and any precise figures need to be treated with considerable reserve. It is, however, possible to point out certain trends which have been in operation in the recent past and are likely to continue to a greater or lesser degree for the remaining decades of this century.

5.30. The dominating factor in the world food situation is that world population is increasing at a rate unprecedented in history. Between the beginning of the Christian era and 1600 world population rose from 250 million to 500 million. By 1900 it had risen to 1500 million and by 1960 to 3000 million. It is estimated that by the year 2000 another 3000 million people will have been added to the world population (Table 5.13). Thus world food production in 2000 will have to be twice what it was in 1960 if even present consumption levels are to be maintained; moreover, it is estimated that at present two-thirds of the world's population, although not starving, has a diet which is inadequate in some respect.[1] Thus to maintain and, if possible, improve nutritional standards, an unprecedented increase in world food production is needed, at a time when there are no more fertile virgin lands like those opened up in North America and Australasia at the end of the nineteenth century.

TABLE 5.13

U.N. POPULATION DATA AND PROJECTIONS[a]
(medium variant[a])

	Population ('000 million people)			
	1950	1960	1980	2000
Less-developed regions (inc. China)	1·6	2·0	3·1	4·7
More-developed regions	0·9	1·0	1·2	1·4
World	2·5	3·0	4·3	6·1

	Rate of population increase (% per annum)		
	1950–60	1960–80	1980–2000
Less-developed regions	2·0	2·1	2·1
More-developed regions	1·3	1·0	1·0
World	1·8	1·9	1·8

[a] High, medium and low projections were made; the figures here relate to the medium projection.
Source: World Population Prospects, U.N., New York, 1966.

5.31. This situation is assessed with differing degrees of optimism and pessimism by various economists. The most optimistic assessment is that represented by Mr. Colin Clark, Director of the Agricultural Economics Research Institute, Oxford, who foresees no problems in raising food production to feed the world's increasing population, until the time when excess population can emigrate to other planets.[2] At the other extreme

[1] *World Food Budget 1970*, U.S.D.A., 1964.
[2] The earth can feed its people, *World Justice*, Vol. 1, Louvain, 1959.

are the prognostications of the Food and Agriculture Organisation of the United Nations, whose publications in recent years have warned of the danger of widespread famine in many parts of the world within the next few decades. Most economists would probably take the view that even though modern technology may make possible the increased food production needed to feed a world population double the present number, the speed of the needed increase raises serious problems.

5.32. The global problem of world food supply is aggravated by the imbalance between the developed and developing regions of the world. The less developed regions—which comprise most of Asia, Africa and South America—have a population growth rate double that of the more developed regions. But grain production—which can be taken as representative of total food production—has in recent decades been higher in the developed than in the less-developed regions. As can be seen in Table 5.14, world grain production *per capita* rose by 7% between 1934–8 and 1960/1, but this concealed a rise of 26% in the developed regions and a fall of 3% in the less-developed regions. The F.A.O. figures of *per capita* consumption in most countries of the less-developed regions show a slight rise in consumption since 1960/1, but in some of the most populous countries, such as India, there appears to have been little, if any, improvement in diet since before the War.

TABLE 5.14

INDICES OF WORLD GRAIN PRODUCTION, 1960/1
1934–8 = 100

Geographic regions	Total production	*Per capita* production
North America	200	144
Latin America	142	84
Western Europe	131	119
E. Europe and U.S.S.R.	124	105
Africa	154	108
Asia	141	98
Oceania	220	151
World	147	107
Economic Regions		
Developed regions	151	126
Less-developed regions	142	97
Political Regions		
Non-Communist countries	159	115
Communist bloc	129	96

Source: Lester R. Brown, *Man, Land and Food*, tables 17, 18.

5.33. The projections that have been made of future world food needs have been based on two main approaches. One approach is to set a certain nutritional standard and then, given the expected population increase, work out the increase in production or imports needed to maintain this standard. A detailed investigation of this type by the United States Department of Agriculture[1] estimates the food supplies needed to (a) maintain current *per capita* consumption levels in the less developed world, and (b) achieve modest rises in consumption levels. On the former assumption, total grain supplies in the less-developed regions would have to rise from 448 million tons in the 1957/8–1960/1 base period to 697 million tons in 1980, and 1102 million tons in the year 2000. It is assumed that total grain imports into these regions will rise from 15 million tons in the base period 1957/8–1960/1 to 35 million tons in 1980 and 68 million tons in the year 2000, being limited by, among other factors, poor transport and distribution facilities in the less-developed countries. This then gives the production needed in the less-developed countries to maintain current *per capita* consumption; it will have to rise from 433 million tons in 1957/8–60/1 to 662 million tons in 1980 and 1034 million tons in 2000 (Table 5.15).

TABLE 5.15

SUPPLIES IN LESS-DEVELOPED REGIONS
NEEDED TO MAINTAIN *per capita* CONSUMPTION

	Million metric tons		
	1957/8–60/1	1980	2000
Production	433	662	1034
Net imports	15	35	68
Total supplies	448	697	1102

5.34. When, on the other hand, it is assumed that *per capita* consumption rises by 10% by 1980, and by a further 10% by 2000, the figures are as given in Table 5.16, with a required production of 1253 million tons in 2000.

5.35. Although economically unsophisticated, this estimate serves to bring out the very large increases in grain output that will be necessary if the less-developed regions are to maintain their existing meagre standards, let alone improve them. It also brings out the marginal nature

[1] Lester R. Brown, *Man, Land and Food*.

TABLE 5.16

SUPPLIES IN LESS-DEVELOPED REGIONS
ASSUMING RISE IN CONSUMPTION LEVELS

	Million metric tons		
	1957/8– 60/1	1980	2000
Production	433	732	1253
Net imports	15	35	68
Total supplies	448	767	1321

of the contribution made by imports from the developed world. Even if the assumed level of imports is doubled, the general size of the production required is not greatly altered. This is not to decry the importance of imports (which would probably have to be on concessional terms). In certain coastal regions, and in periods of famine, imports could be of considerable importance, but the bulk of the developing world's increased grain needs will have to come from domestic production.

5.36. A different approach, which can be applied to both the developed and the less-developed regions, is to estimate changes in real income, and calculate from the projected real income the quantity of cereals that will be demanded by consumers. The future production of cereals is then calculated on the basis of historical trends or likely technical developments, and the difference between the quantity demanded and the amount produced will indicate the quantity of imports or exports. This approach has been used in a number of recent projections.[1]

5.37. This approach, however, makes a number of arbitrary assumptions, and the surpluses or deficits that it indicates do not have as unambiguous a meaning as might appear at first sight. For example, the projections of both demand and supply are based on the assumption that prices (in real terms) remain unchanged, usually at the level in the base period chosen. But if a calculation on this basis showed, for example, that a less-developed country would have a grain deficit at a certain date, it would not necessarily follow that the country would be able to pay for this level of imports; this would depend on its general balance of payments situation. If the country were not able to pay for this level of imports, and were unable to obtain them on concessional terms, internal prices would be forced up, limiting demand and perhaps stimulating

[1] *Agricultural Commodities—Projections for 1975 and 1985*, F.A.O., Rome, 1967. Martin E. Abel and Anthony S. Rojko, *World Food Situation*, Foreign Agricultural Economic Report No. 35, U.S.D.A., Washington, 1967.

supply. For this reason, the figures of cereal surpluses and deficits given by these projections refer neither to the trade that is likely on commercial terms nor to the food aid that would be needed to maintain certain consumption levels. Any estimate of future commercial trade would have to bring in more variables, such as the outlook for the less-developed countries' exports, and this would involve an even greater element of crystal ball gazing.

5.38. These qualifications must be borne in mind when examining the results of the projections that have been mentioned which, not surprisingly, come to somewhat differing conclusions. The report by the United States Department of Agriculture makes three projections for production in 1980 in the less-developed countries, on the basis of (a) historical trends in production, (b) a moderate improvement over this rate, and (c) a rapid improvement. On the other hand, a single projection is made for the developed and Communist countries. Communist Asia and Eastern Europe, including the U.S.S.R., are assumed to be slight net exporters (which they are not at the moment). The combination of the three projections for the less-developed countries with the single projection for the rest of the world gives three different outcomes (Table 5.17) with the net imports of the less-developed countries ranging from 34·3 million tons to 5·8 million tons. The imports of the "developed free world" are estimated to be 73·2 million tons. As the export surpluses of the main grain exporters—notably the U.S.A.—exceed total world imports on all three assumptions, it follows that there would be excess capacity in world grain production ranging from 34 million tons to 68 million tons. Thus the problem would be of surplus rather than scarcity.

5.39. The projections made by the F.A.O. for 1975 also imply a surplus of grain. Two projections are made for 1975 on the assumptions of a high or low rate of increase in real *per capita* income, which would affect the level of demand, and in some countries also the level of production. On these two assumptions, the total world production of grain will rise to 944 or 991 million tons in 1975, compared with 696 million tons in 1961–3 (Table 5.18). Demand will in both cases rise somewhat less, so that there would be a world "surplus" of 6·4 million tons or 27·8 million tons respectively. These projections assume that imports into the developed countries would rise on either income assumption—from 32·8 million tons to 37·7 or 40·7 million tons—but that imports into the developing countries would either rise from 17·2 to 36·7 million tons or fall to 14·1 million tons.

5.40. In its predictions for 1985 the F.A.O. is rather cautious. It estimates that grain consumption as human food will rise from 322 million tons in 1961–3 to 458–487 million tons in 1985, depending on the growth of

TABLE 5.17

ALL GRAINS: U.S.D.A. PROJECTIONS FOR 1980

Million metric tons

	Historical trend[a]		Moderate improvement in production[b]		Rapid improvement in production[b]	
	Production	Net imports	Production	Net imports	Production	Net imports
Less-developed countries	381·5	34·3	399·3	29·7	430·8	5·8
Developed countries	767·0	−77·6	767·0	−77·6	767·0	−77·6
Grain exporters	(430·0)	(−152·0)	(430·0)	(−152·0)	(430·0)	(−152·0)
Other developed free world	(106·8)	(73·2)	(106·8)	(73·2)	(106·8)	(73·2)
Eastern Europe and U.S.S.R.	(230·2)	(1·2)	(230·2)	(1·2)	(230·2)	(1·2)
Communist Asia	183·5	9·0	183·5	9·0	183·5	9·0
World total	1332·0	−34·3	1349·8	−38·9	1381·3	−62·8

[a] The rate of growth of grain production during the 1970's is assumed to be the same as in the period 1954–66 except that some allowance is made for the accelerated agricultural development expected in India and Pakistan as a result of the emphasis on "self help" in the late 1960's.   [b] Less developed countries only; constant projections for other countries.

Source: World Food Situation, table 9, p. 19.

TABLE 5.18

ALL GRAINS: PRODUCTION, CONSUMPTION (ALL USES) AND TRADE
1961-3 and F.A.O. Projections for 1975

Million metric tons

	1961-3 Average			1975					
				Low income assumption			High income assumption		
	Production	Consumption	Trade	Production	Consumption	Trade	Production	Consumption	Trade
Developing countries									
(a) Importers	120·1	136·2	17·2	162·8	199·5	36·7	193·3	207·4	14·1
(b) Exporters (Argentina)	15·3	8·4	−6·4	26·3	11·6	−14·7	26·3	12·0	−14·3
Developed countries									
(a) Importers	120·5	153·9	32·8	163·1	200·8	37·7	165·0	205·7	40·7
(b) Exporters	209·3	162·9	−55·1	288·2	217·0	−71·2	288·2	220·6	−67·6
Centrally planned countries	230·8	239·1	9·1	303·6	308·7	5·1	318·1	317·5	− 0·6
World total	696·0	700·5	−2·4	944·0	937·6	−6·4	990·9	963·2	−27·7

Note. A negative figure for "trade" indicates exports.
Source: F.A.O. *Agricultural Commodities: Projections for 1975 and 1985,* Vol. 1, table 3, p. 86.

population and income (Table 5.19). However, the amount of grain used for human food in 1961–3 was less than the amount fed to live-stock—378 million tons was fed to livestock as against 322 million tons consumed direct. Thus the demand for livestock products will have a major influence on the total demand for cereals.

TABLE 5.19

ALL GRAINS: CONSUMPTION IN 1961–3 AND PROJECTION FOR 1985

	Million metric tons	
	1961-3	1985
Use for human food		
Developing countries	110·8	204·2–215·2
Centrally planned countries	130·3	162·0–178·8
Developed countries	81·1	88·9–97·3
World	322·2	458·1–486·9
Use as animal feed	378·3	525·5[a]
World (all uses)	700·5	983·6–1012·4

[a] Based on estimated increase in consumption of animal products (F.A.O.), assuming constant average food conversion.

Source: F.A.O. *Agricultural Commodities: Projections for 1975 and 1985*, Vol. 1, table 7, p. 94.

5.41. The F.A.O. estimates the total world increase in the consumption of animal products between 1961–3 and 1985 as 38·9%. If we assume that the amount of grain fed to livestock would have to rise by the same amount—assuming in effect a constant average feed conversion rate—and add this amount to the estimate for human consumption, total world consumption would rise from 700 million tons to 984–1012 million tons, depending on the growth of population and income.

5.42. The conclusions of the U.S.D.A. and F.A.O. studies that there will be a world surplus, rather than a shortage of grain, differs markedly from that of other studies, and must be considered alongside the more alarmist estimates. But two things must be pointed out. Firstly, these assumptions do not indicate that diets would be adequate. On the contrary, the F.A.O. projection implies that, even on the optimistic assumption, there would be extensive malnutrition in many parts of the world. Secondly, the projections rest on certain assumptions which, in the nature of things, can be little more than informed guesswork; changed assumptions would give very different results. For example, it is assumed that Communist Asia and the U.S.S.R. and Eastern Europe will be slight net grain exporters. However, these areas have in the 1960's become net grain importers, and if it were assumed that their imports would continue, or even increase, this would alter the world trade projections considerably.

5.43. What is likely to happen to grain imports into the Communist countries is, in fact, one of the great imponderables in the situation. Grain production in these countries faces certain physical limitations—the area of fertile land, especially in the U.S.S.R., is not as great as the size of the countries would suggest. More important, however, is the general experience that a Communist system is particularly unsuited to agricultural production. It seemed for a time as though Communist China was an exception to this rule, but it now appears that the Cultural Revolution has done nothing to raise agricultural productivity (although any statements about China are complicated by the fact that no statistics have been published for 10 years!). Perhaps the Communists will eventually solve their agricultural difficulties, but this is obviously a matter on which predictions are difficult. All that can be said is that the agricultural policies pursued in the Communist countries—and also in some other countries, notably India—will very considerably affect the world grain situation.

Conclusions

5.44. It seems only too likely that over the next few decades the less-developed countries will have difficulty in raising food production sufficiently to match the increases in their population. However, they are also likely to continue to face the balance-of-payments problems that they face already, and in many cases their staple food is rice, rather than the cereals, such as wheat, which can readily be supplied by the developed countries. For these reasons it is unlikely that the shortages in the less-developed countries will lead to shortages on the world market, at least up till 1980. The developed countries will no doubt continue, and possibly expand considerably, programmes of food aid of the type pioneered by the U.S.A., but the size of these programmes is limited both by the willingness of the developed countries to assist the less-developed countries and also by transport problems. Thus the demand and supply situation on the commercial world market for cereals is unlikely to change dramatically. The most likely outcome during the next 5 or 10 years is a continuation of the slightly precarious balance which now characterises the commercial market.

PART VII

Price Behaviour in the U.K. Cereals Market

CHAPTER 1

Price Behaviour in the U.K. Cereals Market

1.1. The behaviour of market prices for cereals is, in the last analysis, the most significant feature of the marketing system to all those who participate in it, farmers, merchants, shippers, millers, compounders and merchants, and of course the H.-G.C.A. and the Government. The way cereals prices move over the season, and between seasons, is a measure of the final outcome of all the many factors influencing the cereals market. Those factors are so numerous that it is seldom possible to look at a price series and say with absolute confidence that this or that feature was the result of this or that cause. Nevertheless the price series exhibit certain characteristics that occur so regularly that the likely cause can be suggested, especially if these price movements can be correlated with other relevant factors such as the supply of cereals on the market. It is with relationships of this sort in the pattern of price behaviour that this Part is mainly concerned.

1.2. The monthly prices for home-grown wheat, barley and oats in England and Wales between 1955 and 1968 as reported to the Ministry of Agriculture, Fisheries and Food under the Corn Returns Act are shown in Fig. 1.1. Several features are worth noting:

(a) Market prices were moving downwards during the period 1955 to 1961 and reached a low point in the summer of 1961 when the prices of all three cereals fell to about 17*s*. per cwt. From that low point there was a strong recovery in 1962, and since that year prices have fluctuated less than previously—mostly within the band 20*s*. to 23*s*., with a slight trend upwards.

(b) The fluctuations within each cereal year were greater in the period 1955 to 1961 than in the period 1961 to 1968. For instance taking the difference between the highest and lowest prices quoted during each cereal year for wheat, the average for the first period was about 5*s*. 6*d*. per cwt (ranging from about 3*s*. to about 7*s*. 6*d*.) and in the second period only 3*s*. (ranging from about 2*s*. to about 4*s*. 6*d*.). Barley and oats prices have generally shown a similar reduction in fluctuation within the season.

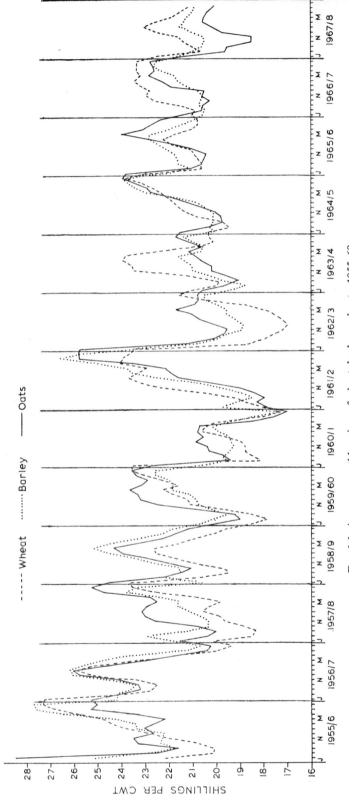

FIG. 1.1. Average monthly prices of wheat, barley and oats, 1955–68.
Source: Monthly corn returns.

(c) In most years the lowest price occurs at harvest time, and normally there is then a period of rising prices up to about March. By this time the prospects of either a surplus or a deficiency in the remaining three months of the cereal year are becoming clearer, and from this point prices tend either to continue to rise (as in 1965) or to turn downwards (as in 1966). In one or two years the price at the end of the cereal year has actually been lower than at harvest.

(d) The prices of the three cereals generally move in the same direction, and reach their high and low points at about the same time. However, the rate of change varies with different cereals at different times with the result that they change their relative positions; thus the wheat price will tend to be higher than the barley price in one season, and lower in another, whilst in some seasons, or parts of seasons, the oats price rises above those of the other two cereals.

1.3. The most outstanding feature of price behaviour in the period reviewed is the contrast between the period 1955 to 1961, and the period since 1961. It was in fact the progressive decline in prices during the late 1950's, culminating in the low point of 1961, that led directly to the introduction of the minimum import price system and the setting up of the Home-Grown Cereals Authority. The relative stability of prices since 1962, despite the rapid increase in the barley acreage, certainly suggests that the measures taken have had some degree of success, although how much of this can be attributed to the minimum import price system introduced in 1964, how much to the H.-G.C.A. and its forward contract system, and how much to the improved international grain situation, it is impossible to say.

1.4. The relative stability of prices during the 1960's, compared with the second half of the 1950's, may have been due in part to the considerable investment in storage facilities stimulated by the Farm Improvement Scheme (see Part II, Chapter 5). Some indication of the effect of this increased storage capacity on the pattern of marketing is given by a comparison of the statistics of stocks of grain on farms in the periods 1956/7 to 1958/9, and 1965/6 to 1967/8. Table 1.1 shows that in the earlier period only 45% of the wheat and 32% of the barley still remained on the farms at end-December, whilst in the period 1965/6 to 1967/8 the percentages were 57 and 54 respectively. At end-March the average stocks of grain on farms were reported as being 21% and 11% of wheat and barley production in the first period, but as much as 28% and 19% in the second.

1.5. Another interesting feature is the uncertainty surrounding the movement of prices during the cereal year. Bearing in mind that every farmer with any storage capacity has to take a decision how much of his grain to store, and for how long, this uncertainty can make his decision very much a gamble. It is probably this feature of the market system that disturbs the farming community more than any other. Thanks to the way the deficiency payments system operates, farmers' receipts as a whole are not normally affected by the level of market prices, but the system still allows the market price to have a considerable influence upon the individual farmer's return. In the case of barley, marketing the crop at a bad time can cost the farmer as much as the subsidy itself is worth.

1.6. The decision how much and how long to store may not have to be

TABLE 1.1

WHEAT AND BARLEY STOCKS ON FARM OF ORIGIN (ENGLAND AND WALES) AS PERCENTAGE OF TOTAL PRODUCTION

	Wheat		Barley	
End of	Average 1956/7 to 1958/9	Average 1965/6 to 1967/8	Average 1956/7 to 1958/9	Average 1965/6 to 1967/8
	% of production			
October	60·5[a]	70	47·5[a]	74
November	52	64	40	64
December	45	57	32	54
January	36	47	24	41
February	29	38	17	30
March	21	28	11	19
April	11[a]	19	4[a]	11

[a] Two-year average 1957/8 to 1958/9.

Source: M.A.F.F.

altogether a gamble since it is possible to detect a definite pattern in the movement of prices during the season. There appears to be an alternation from year to year. Thus at the end of the 1955/6 season prices were high, but at the end of the following season they were low. Then in 1957/8 they were high again, and at the end of the following season they were low. This alternating pattern repeats itself throughout the period under review, and we have called it the "Scissors Effect". A likely explanation is that if prices are high at the end of a season those farmers who did not store are encouraged to do so in the next season, but when this increased quantity comes to be sold at the end of the following season the price is depressed and as a result some farmers decide not to store in the next season. Thus the cycle is repeated. Figure 1.2 demonstrates how this cycle operates in the case of wheat. The left-hand side shows the monthly fluctuations in total deliveries of grain, imported and home-grown, from 1962/3 to 1967/8. The right-hand side compares the monthly deliveries with the monthly price of home-grown wheat, and shows that when the deliveries in the latter part of the cereal year are relatively low the price tends to be high, and when they are high the price tends to be low, and that these opposite movements occur in alternate years. This supports the suppositions (i) that the volume of deliveries in the latter months of each cereal year has an effect upon the price level at that time, and (ii) that this price level in turn influences the volume of end-year deliveries in the following season. It seems therefore that the market prices still have an important influence upon producers' decisions as to whether and for how long to store grain even though the guarantee system includes a seasonal guarantee as in the case of wheat. More research is needed to substantiate the hypothesis suggested here, and indeed better knowledge of the price relationships in the cereals market in general might go far to enable a marketing authority like the H.-G.C.A. to improve the stability of the market.

1.7. Another way of looking at the "Scissors Effect" is to consider the extent to which the market recompenses the farmer for storing his grain to the end of the season. Table 1.2 compares the average August/September price with the average price in the following April–June period, the difference representing the "premium" awarded by the market to cover storage costs.

1.8. Although over the thirteen years as a whole the market yielded a "premium" of about 1*s*. 6½*d*. per cwt, in 5 years out of 13 there was no return to speak of, or even an actual loss. Thus the farmer cannot always rely on the market rewarding him for storage, and this may help to explain

TABLE 1.2

ESTIMATED "PREMIUM" FOR STORAGE IN THE
MARKET PRICE FOR BARLEY, 1955/6 TO 1967/8

	Average price Aug./Sept. (per cwt)		Average price April/ May/June (per cwt)		"Premium" for storage (per cwt)	
	s.	*d.*	*s.*	*d.*	*s.*	*d.*
1955/6	22	3	27	1	4	10
1956/7	23	10½	21	10	−2	0½
1957/8	22	1	23	2	1	1
1958/9	21	9½	22	5		7½
1959/60	19	10	22	4	2	6
1960/1	19	7	18	4	−1	3
1961/2	18	10½	25	9	6	10½
1962/3	20	4	20	3	−	1
1963/4	19	2	20	11	1	9
1964/5	20	2	23	7	3	5
1965/6	21	3½	21	5		1½
1966/7	20	6½	22	5½	1	11
1967/8	20	8½	20	10		1½
Average:					1	6½

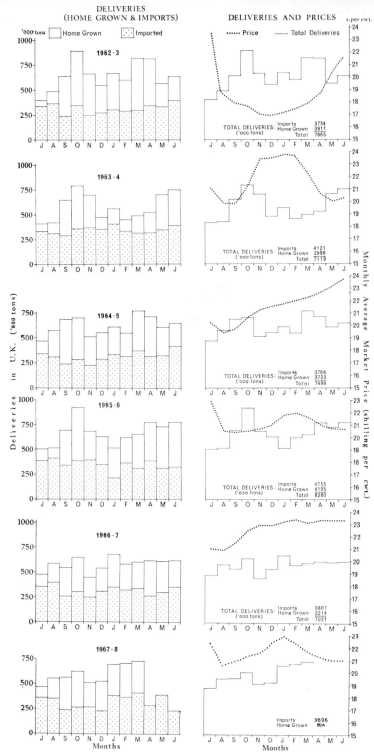

FIG. 1.2. Comparison between tonnages of wheat delivered (imports and home-grown) and prices, 1962/3 to 1967/8.

Source: Prices: Corn Returns. Tonnages (E. & W. only): deliveries under the C.D.P. Scheme (M.A.F.F. statistics).

why he tends to react rather strongly to experience in the adverse years. The seasonal incentive payments for barley, introduced in 1963, may sometimes just tip the scales and make storage worthwhile even in years when the market itself gives little premium, but there is still a large element of doubt about how profitable the whole storage operation will be, especially when grain is stored to the end of the season.

Movements in International Cereals Prices

1.9. The inclusion of imports in Fig. 1.2 is a reminder that prices on the domestic market do not reflect only the fluctuations in home-grown supplies but also the supply and price of imported grains. The way international prices move is therefore of great significance to an understanding of prices in the U.K. market. A study of international cereals prices over the period 1955–65 reveals a numbers of interesting features.[1] The broad price trends have been similar to those in the U.K.; Figs. 1.3 and 1.4 show that the trends were downwards from 1955 to 1958, roughly level until 1961, and thereafter generally upwards. The different grains however have not shown the same trend. Hard wheat, maize and sorghum prices have remained relatively stable and have shown a low degree of inter-year variability. Soft French wheat has shown a less stable pattern, its price movements being more closely allied to changes in the prices of coarse grains than to those of hard wheat. It is significant, in view of the depressive effect of low-priced imports of French wheat in the latter half of 1967/8, that the price of soft French wheat has moved closely with that for wheat in the U.K. market (Fig. 1.1). In contrast to hard wheat, maize and sorghum, a high degree of variability has been exhibited by the prices of barley, oats and rye.

1.10. These differences in the stability of the international prices of cereals may be attributed to a number of factors. In the case of wheat the role of the International Wheat Agreement in stabilising wheat prices should be acknowledged. The Agreement, by effectively assuring the major wheat exporters of a large stable market with an announced price range, has created an environment in which the two largest exporters, Canada and the U.S.A., have been prepared to hold large stocks of wheat, and to control the size of these stocks in such a way that the realised price of Manitoba

---

[1] This brief summary is based on an article by Mrs. A. K. Binder, Relationships in World Prices of Coarse Grains and Some Implications for an International Grains Agreement, Food and Agriculture Organisation of the U.N., *Monthly Bulletin of Agricultural Economics and Statistics,* Vol. 15, No. 9, Sept. 1966, pp. 1–15.

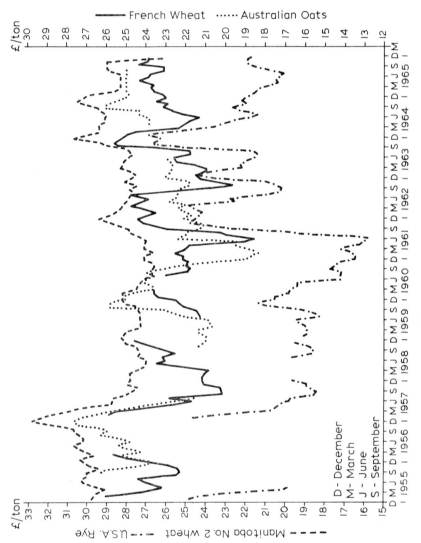

Fig. 1.3. Movements in international cereals prices, 1955 to 1965.

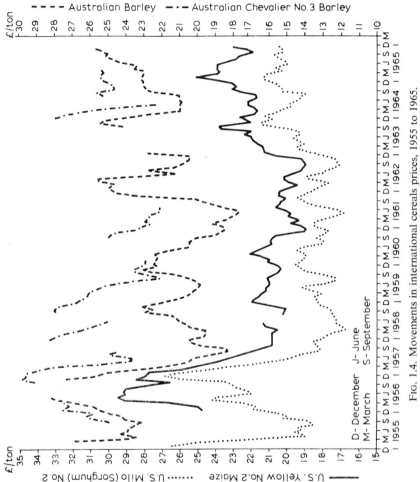

Fig. 1.4. Movements in international cereals prices, 1955 to 1965.

No. 1 wheat has remained within the established price range since 1953/4.[1]

1.11. Maize and sorghum price stability is largely due to the dominant role played by the U.S.A. in the markets for these two grains. In 1966 United States exports accounted for 64% of world maize exports and 81% of world sorghum exports. During the same year the U.S.A., as the only major holder of stocks, carried stocks of maize and sorghum equivalent to 88 and 109%, respectively, of the volume of world trade in these grains, a reduction of more than half from the stock levels of 1961, when the U.S.A. carryover of these grains was at its peak. As a result of this dominance the price of maize from other exporting nations has remained close to U.S.A. prices at all times, and the price of sorghum has maintained a steady relationship with the price of maize (on average 10% below).

1.12. The volatility of the international prices of barley, oats and rye is due mainly to the fact that no country dominates the export market for these grains to the extent that the U.S.A. (and Canada in the case of wheat) does for the grains previously mentioned, or keeps large buffer stocks.

1.13. It is interesting to note the implications of the observed price movements of cereals for the substitutability of cereals for one another in different uses. While milling wheat and high-quality malting barley prices might be expected to have some independence of coarse grains prices, it would seem reasonable to expect coarse grains prices to move closely together. That the prices of coarse grains relative to one another may change from year to year suggests some limits to the substitutability between them. This is not wholly surprising in view of the inflexible nature of demand for some industrial uses, and of the large demand for compounding with its built-in resistance to changes in the composition of proprietary feeds.

The Relationship between Imported and Home-produced Cereals Prices

1.14. If price series of imported cereals and home-produced cereals are compared certain well-defined seasonal patterns in the price differentials can be detected. The fact that such strong patterns exist, in spite of somewhat erratic year-to-year patterns in the prices of imported cereals, suggests that domestic prices are strongly affected by the prices of grain on the international market. The existence of a seasonal pattern of price differentials appears to be attributable to seasonal differences in the degree of self-reliance of the U.K. market for those types of grain which are domestically produced; the degree of self-reliance being high in the immediate post-harvest period, but diminishing as the cereal year progresses

---

[1] See: International Wheat Council, *International Wheat Prices*, Secretariat Paper No. 1, Dec. 1961, p. 35.

and as the importation of qualities of grain competitive with domestic supplies steadily increases.

1.15. In order to study the relationship between imported and home-produced prices the monthly U.K. price was subtracted from the monthly import price for each grain, for each month in the period January 1955 to December 1965, and the resulting price differentials were plotted.[1] The results are summarised in Fig. 1.5 which shows the average differential in selected periods of years for five types of cereals produced in the U.K.

1.16. A clear basic similarity in the pattern of the price differential is evident for all five classes of cereals. After the U.K. harvest, as Fig. 1.1 shows, the prices of domestic cereals fall rapidly and there is a large positive differential of imported over domestic grains; in the case of malting barley U.K. prices fall rapidly after harvest so that the negative differential is eliminated (Fig. 1.5D). A peak differential is rapidly reached, in August for both wheats, September for feeding barley, November for feeding oats, and December for malting barley, after which domestic prices improve steadily relative to the prices of imported cereals so that the price differential declines (or increases in the case of malting barley).

1.17. In the cases of milling wheat and malting barley there have been major differences in the seasonal pattern of the price differential in some years. Figure 1.5c indicates that, while six of the years exhibited the "normal" seasonal pattern, in four years the prices of U.K. milling wheat fell relative to the prices of imported wheat in the later months of the cereal year. The reason for this difference in these four years appears to be attributable to the high level of unsold stocks of wheat on farms towards the end of the season in these years. In the three years 1958/9, 1960/1 and 1963/4, for instance, unsold stocks of wheat on farms at the end of April were equivalent to 15, 14 and 17% of the year's crop respectively, as opposed to an average of 8% in the other years. In two of these years, 1956/7 and 1960/1, and to a lesser extent in 1958/9, the high levels of these unsold stocks may themselves be attributed to the relatively large wheat

---

[1] The data used in this analysis were as follows: the domestic cereals prices were "Cereals Prices Paid to Growers" (excluding cereals deficiency payments) as recorded at the ports of Bristol, Hull, Liverpool, and London (M.A.F.F. Statistics). For the prices of imported wheat, barley and oats of feeding quality, the average market prices of ex-store lots of imported grain at the ports of Bristol, Hull, Liverpool, and London have been used, and these are recorded on the same basis as the series used for domestic grains prices. The price of Manitoba Northern No. 2 wheat c.i.f. U.K. ports, was used for the price of imported milling wheat, Argentine Plate c.i.f. for barley, and North Sea ports for the price of imported malting barley (see A. K. Binder, op. cit.). This latter series of prices is rather low for barley of malting quality, but as it can be shown that the prices of barley from all overseas sources move closely together it seems reasonable to accept it as an appropriate indicator.

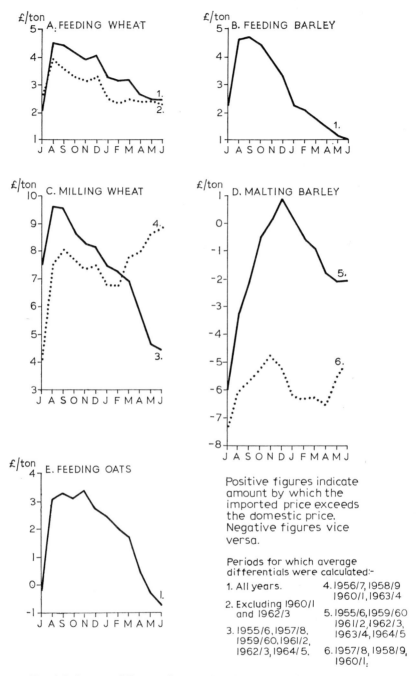

£/ton A. FEEDING WHEAT

£/ton B. FEEDING BARLEY

£/ton C. MILLING WHEAT

£/ton D. MALTING BARLEY

£/ton E. FEEDING OATS

Positive figures indicate amount by which the imported price exceeds the domestic price. Negative figures vice versa.

Periods for which average differentials were calculated:-

1. All years.
2. Excluding 1960/1 and 1962/3
3. 1955/6, 1957/8, 1959/60, 1961/2, 1962/3, 1964/5.
4. 1956/7, 1958/9 1960/1, 1963/4
5. 1955/6, 1959/60 1961/2, 1962/3, 1963/4, 1964/5
6. 1957/8, 1958/9, 1960/1.

FIG. 1.5. Average differences in monthly prices between imported and U.K. cereals—averages of various periods between 1955 and 1965.
Source: Unpublished Econometric Study of Aspects of the U.K. Cereals Market by D. R. Colman, Department of Agricultural Economics, University of Manchester.

harvests, and they reflect also the "Scissors Effect" described earlier. In 1963/4, difficulty was experienced in selling the crop of domestic wheat, unsold stocks towards the end of the season were large, and prices fell in the latter part of the season in comparison with prices of imported wheat as farmers tried to reduce their holdings of wheat. With the malting barley price differential the case is different. It is not that the basic pattern of the price differential varied, but that in three years, 1957/8, 1958/9 and 1960/1, the amount by which the price of U.K. malting barley exceeded that of imported supplies was markedly greater than in other years (see Fig. 1.5D). This major difference resulted from the extremely low prices which were obtained for barley in international markets during the period from August 1957 to September 1961, and from the fact that prices for domestic malting barley held up well, except in 1958/9 when they dropped following a particularly large increase in the size of the barley crop over that of previous years.

1.18. Needless to say the use of average relationships, as in Fig. 1.5, obscures the degree of year-to-year variability in the seasonal price differential, and if this variability were extremely high the conclusions to be drawn could be misleading. However, for milling wheat, malting barley and feeding barley the variation about the various seasonal price patterns is low. For both feeding wheat and feeding oats the variability is large, and in the case of the former the removal of the price differentials for 1960/1 and 1962/3 causes a marked fall in the average price difference (see Fig. 1.5A). The similarity in the shape of the average seasonal price differential of these two grains to those of the grains whose price pattern exhibits relative stability, would however appear to justify both their presentation and the derivation of general conclusions for all grains.

1.19. It is of importance to note that a general price differential pattern occurs between home-produced and imported grains, in spite of large year-to-year variations in the pattern of international prices. It can be seen from Figs. 1.3 and 1.4 that in some years international grain prices rise, and in others fall, steadily throughout the year. But, nevertheless, U.K. grains prices differ from these in a systematic way as indicated in Fig. 1.5. It is possible to conclude, therefore, that there is a relationship between world and U.K. cereals prices, which causes them to differ from one another in a predictable and relatively stable manner. A possible explanation of this relationship is as follows. In the immediate post-U.K. harvest period, there are ample supplies of domestic grains, and these can be substituted for imported grains of similar quality. The limits of substitution are however low; for example, soft domestic wheats cannot replace hard wheats in milling, and maize is only replaceable to a limited extent by barley or wheat. There will be room, however, for the replacement of some imported soft and filler wheats by domestic wheat, and in the years before the U.K. became a net exporter of barley there should have been a displacement of some imported barley. Both of these substitutions do appear to have occurred, so that imports of barley and French wheat in the months after harvest have in general been less than the monthly average—this for a period from 1955 to 1965.

1.20. As a general working hypothesis it appears reasonable to suggest that in the months following harvest the supply of domestic grains is high relative to the demand for them, and thus domestic cereals prices are generally depressed during this period. The "floor price" will be represented by producers' reluctance to sell grain below certain prices, given that they expect prices to rise later in the season and that they are eligible for storage incentive payments. The lower cereal prices fall at harvest time the higher the inducement to improvise means of grain storage, a move which would reduce the excess of grain supply.[1]

1.21. During the post-harvest period the degree of interaction between the domestic and international cereals market will be low since imported cereals will tend to be of types which are not directly competitive with U.K. grains. It is probable however that market expectations about cereals prices later in the season will depend upon international cereals prices, and therefore that the "floor price" of domestic cereals will be a function of the international prices of grains.

1.22. As the cereal year proceeds there will be an increase in the interaction between the two markets. Later in the year the supply of domestic cereals declines relative to the demand for them, with the result that the prices of domestic grains harden progressively, so that at the end of the season they approach the international prices for comparable grains. During this period millers, compounders and other users will be purchasing grain of similar quality from both domestic and foreign sources, so that the prices paid for them will be much the same.

1.23. It is of interest to note that, despite much larger domestic crops of both wheat and barley in recent years, the prices of domestic cereals in the period of post-harvest flush have not been lower relative to imported grain prices than they were in earlier years. A probable explanation for this is that the rapid development of on-farm storage facilities, and the operation of the seasonal incentive schemes for wheat and barley, have done much to reduce the relative excess of domestic cereals supply in the post-harvest period and have mitigated the effect of larger crops.

---

[1] An additional factor helping to depress prices at this period will be the movement of non-storable grain into the market.

1.24. The discussion above has treated the U.K. cereals market as if it were uniform, and has analysed the market situation employing prices measured at the ports. In fact the market is not uniform and it is useful to distinguish between country and port markets. Country markets are insulated from port markets by the costs incurred in transporting grain to and from the ports. The effect of this insulation will be to make country markets less susceptible to the influence of international grain markets. In consequence it should be expected that prices of domestic cereals in country areas will in the post-harvest period fall below prices at the ports and that they may well rise above port prices at the end of the cereal year.

The Operation of the Guide Price

1.25. The low level of prices of home-grown grain at harvest time has been a characteristic feature of the cereals market for many years. Indeed, it is only to be expected that prices at harvest will normally be lower than later in the season because of the cost of storage and because there will always be farmers who must sell promptly to obtain cash. However the harvest price is often lower than these reasons alone would justify and attempts have been made by various means to strengthen the price at this time. One of these was the setting up of the Working Party on Feeding Barley in 1956 under the chairmanship of Mr. Charles Norman (now Sir Charles Norman, deputy chairman of the Home-Grown Cereals Authority) with the objective of issuing "indicated values", or as they later came to be called, "Guide Prices". These were announced weekly and covered three months forward; when there did not appear to be any need for them, e.g. if prices were buoyant towards the latter part of the cereals year, they were temporarily discontinued. The Guide Prices were based at first on the values of imported barley, but as the tonnage of imported barley declined this became increasingly unrealistic and as from 1962 they were based on the prices of competitive imported grains such as maize. From 1963/4 onwards Guide Prices were issued for millable wheat as well as for barley. In 1965/6 the Home-Grown Cereals Authority took over responsibility for issuing the Guide Prices. The success or otherwise of the Guide Prices is considered later (Part VIII), but it is of interest in the present context to study the relationship between the market prices and the Guide Prices over the twelve years since they were first introduced.

1.26. As one would expect, the Guide Prices have tended to be fixed above the market prices for the greater part of the season, although in some years, when Guide Prices were issued for the latter part of the season, they were actually below the market price. The most important period of the year, so far as the Guide Prices are concerned, is August to November. The market prices and Guide Prices for barley and the differential between them, for the period August–November in the years 1956 to 1968 have averaged as follows:

	Average market price (£ per ton)	Average Guide Price (£ per ton)	Average differential	
			(£ per ton)	as % of av. market price
August	18·7	19·9	1·2	6·4
September	19·0	19·95	0·95	5·4
October	19·3	20·1	0·8	4·1
November	19·4	19·9	0·5	2·6

Conclusions

1.27. Prices between 1962/3 and 1967/8 were relatively stable and it would seem that this was mainly due to the improvement in the international grain situation since it was the imports of low-priced grain in the years before 1962 that were mainly responsible for the decline in prices during that period. Fluctuations within the cereal year have also been reduced and part at least of the credit for this must go to the seasonal incentives for barley, introduced in 1963, and the operation of the H.-G.C.A.'s Forward Contract Scheme. The system of minimum import prices has also prevented prices from falling too low in certain periods. Another important stabilising influence on the U.K. cereal market has been the relative stability of the prices of imported wheat and maize, and this is due primarily to the control exercised by the Canadian and United Stated governments. On the other hand the imports of European soft wheats have had an unsettling effect on the U.K. market, particularly in the latter part of the cereal year—the total quantities of such grain involved, however, are not large. The main problem of orderly marketing would seem to be finding some means of achieving a more even trend in prices during the year so that the farmer's decision as to when to sell his crop is less of a gamble than it is today.

PART VIII

U.K. Policy for Cereals

CHAPTER 1

Changes in Government
Cereals Policy since 1953

1.1. This Report has been mainly concerned hitherto with a descriptive analysis of the production, marketing and use of cereals, and questions involving agricultural policy have been only lightly touched upon. In this Part, however, we take a detailed look at agricultural policy in relation to cereals and suggest various ways in which we think it might be improved.

1.2. The basis of the present agricultural price support system for cereals was laid down in 1953/4 when the Government decontrolled grain prices and reverted to the pre-war system of deficiency payments (Cmd. 8947). For wheat and rye (both grown primarily for sale), these were related to tonnages sold, and for barley and oats (mainly feed crops) to acreage grown. The wheat guarantee was related to a seasonal scale "to encourage growers to hold their wheat off the market during the harvest peak"; the total spread over the season was 5s. per cwt. The guaranteed prices laid down for the 1954 harvest were 30s. 9d. for wheat, 25s. 6d. for barley, 24s. 0d. for oats, and 25s. 0d. for rye.

1.3. For some years after the War there was a heavy emphasis in government policy on the need for more home production of food to save foreign exchange. However, as the years went by and wheat surpluses began to build up throughout the world, leading to higher cereal deficiency payments, government policy was modified. The differential in favour of wheat was gradually whittled away until by 1957/8 it had disappeared and the Government was doing all it could to encourage the growing of more feed crops and "taking the plough round the farm". Increasingly the emphasis in government policy statements came to be placed on cost reduction through improved management, with markedly less emphasis on increased production.

1.4. The cut in wheat guarantees, combined with increases in the barley guarantees, new barley varieties and new techniques of production, set in motion the rapid increase in barley acreage, at the expense mainly of oats, that has characterised the 1960's (Appendix H, Table 1). Between the years 1955/6 and 1960/1 the barley subsidy leapt from £10·4 million to £33·6 million, whilst the wheat subsidy remained at about £18 to £20 million. The increasing domestic output of barley plus substantial imports of very low-priced European barley caused prices to fall to abnormally low levels, and when in 1963/4 the total subsidy bill for cereals rose to almost £80 million the Government decided it was time to protect the Treasury from this open-ended commitment. Minimum import prices and standard quantities were introduced to achieve this objective, and at the same time the Government decided to set up the Home-Grown Cereals Authority, charged with the task of improving the marketing of home-grown cereals.

1.5. The minimum import prices, which were fixed at £20 for barley and £22 10s. for European milling wheats (£20 10s. for denatured wheat) provided an effective means—should the need arise—of preventing unduly low-priced imports from undermining the market. At the same time standard quantities were introduced for wheat and barley respectively which caused the guaranteed prices to be reduced when production exceeded these quantities. The standard quantities were to be "determined at each Annual Review taking account of the growth of the demand for cereals for home consumption and animal feed and any changes necessary as a result of reviewing the balance between home production and imports".[1] Their introduction was seen as a means of attaining "such part of the nation's food and other agricultural produce as in the national interest it is desirable to produce in the United Kingdom", which was one of the objectives set out in the 1947 Agriculture Act. Although it is uncertain how effective this scheme has been in influencing farmers' decisions, it has certainly had a direct impact on the Exchequer cost of the cereal subsidies (Appendix H, Table 2) and must inevitably have had some impact on cereal growers. The Home-Grown Cereals Authority introduced a scheme of forward contracts financed by a compulsory levy.

1.6. The stabilisation of the grain market, after the alarms of the early 1960's, and the changed international grain situation have led to a marked change in government policy towards cereal production in the last few years. The large stock-piles of grain in Canada and the U.S.A. were rapidly reduced by substantial buying on the part of Russia and China, and as a concomitant of this our overseas suppliers adopted a liberal attitude to the continued expansion of cereal growing in the U.K.—despite the fact that imports into the U.K. had fallen below the level agreed with our overseas suppliers in 1964. This general relaxation in attitude towards the continued rapid increases in cereal production found expression in the

---

[1] *Annual Review and Determination of Guarantees, 1964,* Cmnd. 2315.

National Plan in 1965, which set modest targets for a continued expansion in the cereal acreage to sustain the increased numbers of livestock called for in the Plan. Although the National Plan as a whole is now virtually a dead letter, the agricultural component of it remains more or less government policy. The removal of standard quantities from wheat altogether at the 1968 Annual Review, together with the increase in the standard quantity for barley and the increase in the guaranteed prices for wheat, barley and oats, indicated a desire to give some further encouragement to a selective expansion of cereal production.

1.7. This brief review of the main trends in government policy since 1953 has highlighted only the main features; it has not covered the many changes in the "mechanics" of administering the subsidy schemes. Some of these, such as the introduction of a seasonal incentive scheme for barley in 1961, the target indicator price, the crop adjustment formula and the definition of millable wheat, are considered later.

CHAPTER 2

The Likely Supply and Demand
Situation for Cereals in the Next Five Years

2.1. Before moving on to consider current and future problems of policy in relation to cereals, it is first necessary to look at the likely demand and supply situation in the next few years, to decide, for instance, whether policy has to be formulated in relation to an expected surplus of home-produced grain over estimated demand, or a deficiency, or perhaps a position of balance. This chapter attempts to take such a forward look, covering the five years from 1967/8 to 1972/3, and suggests a range of possibilities representing respectively the lowest, the "most likely" and the highest increases in demand or supply.

2.2. Since the U.K. still imports about 40% of its cereal requirements it is not meaningful to talk of a "surplus" or "deficiency" of home-produced grain without making some assumption about the proportion of imported to home-produced grain. What follows is based on the assumption that the present proportion will remain unchanged. This is not so artificial as it may seem at first sight, since imported grain is often of a different kind from home-produced and performs a different function in flour milling, animal feed manufacture or distilling. Undoubtedly some substitution of home-produced for imported grain will continue to take place and this is discussed in the following chapters. To the extent that substitution takes place, a surplus is less likely to occur.

A. The Demand for Grain in the Next Five Years

2.3. The total U.K. usage of grain in 1966/7, the latest full year available, was as follows:

633

	Usage (million tons)	Proportion home-produced
		(%)
Animal feed	11·8	73
Human food	7·4	41
Industrial	0·5	—
Seed	0·7	100
Exports	1·1	100
	21·5	62

Taking each of these items in turn:

(a) ANIMAL FEED

2.4. Using the statistics of utilisation of crops published by the Ministry of Agriculture, and import statistics (for maize, and estimated imports of feed wheat) it can be estimated that the total consumption of cereals for animal feeds in the U.K. increased by about 335,000 tons per annum over the last 6 years. This period was one of rapid expansion of broilers and of beef fed partly on barley, but these expansions are now levelling off. Pig production rose during the period, but declined after 1965/6. Production of eggs has remained at about the same level for the last 4 years and is unlikely to increase to any significant extent. The dairy and beef herds are likely to increase, and pig output is likely to expand at a rate as great as, or greater than, the rate experienced during the period under review, given appropriate government support.

2.5. The recently published report of the E.D.C. for Agriculture puts forward a possible programme of livestock expansion over the 1966/7 level which would require an additional $2\frac{1}{2}$ million tons of feed grains per annum by 1972/3. Since we are concerned with the five-year period 1967/8–1972/3 this should be adjusted to 2,085,000 tons. The E.D.C. was not concerned with the immediate prospects of achieving the increases postulated, but merely with their feasibility; we can therefore take their figure as the "highest" for our purpose.

2.6. The "most likely" figure would seem to be a continuation of the trend of the past 6 years. Table 1.6 in Part V(a), showed that poultry rations accounted for over 40% of total manufactured concentrates in 1966/7 and apart from an increase expected in broilers and turkeys, this sector as a whole is unlikely to show any substantial increase in the next 5 years. The down-turn in pig production hardly affected the increase per annum based on the last 6 years, and the latter should therefore be adequate to reflect the likely upturn in pig production. As to the E.D.C.'s suggestion of an increase in the dairy and beef herds, we regard this as dependent on adequate incentives being offered at Annual Reviews and experience over past years suggest that incentives of the order necessary to achieve this scale of expansion are unlikely to be offered. On balance, therefore, we pitch our "most likely" estimate at the straight projection forward of the past trend.

2.7. In making this assumption we are also assuming that the U.K. population will continue to grow at about the rate of the last few years.[1] The income elasticity of demand at retail for livestock products ranges from 0·1 for beef and bacon to 0·8 for poultry meat other than broilers—say 0·5 for all livestock products. This means that an increase in average income per head of, say, 10% will lead to an increase in demand for animal products of 5%. We have no reason to postulate any increase or decrease in the impact of rising incomes on demand.[2]

2.8. As to our "lowest" estimate, we pitch this at a continuation of the present level of consumption of animal feeds plus ¾% per annum to allow for an increased human population and an equal amount to allow for a continued increase in real incomes per head, i.e. we assume no increase in livestock numbers other than that required to meet these two needs.

2.9. Our estimates, on these various assumptions, are as follows:

Lowest increase	Most likely increase	Highest increase
	('000 tons)	
930	1675	2085

On the assumption that the proportion home-produced remains at the 1966/7 level of 73% the increased requirements of home-produced cereals by 1972/3 would be:

Lowest increase	Most likely increase	Highest increase
	('000 tons)	
660	1230	1520

(b) HUMAN FOOD

2.10. The main cereals used for human food are wheat for bread and other flour products and breakfast foods, barley for brewing and distilling, maize for distilling, glucose and breakfast foods, and oats for milling. Since

---

[1] The Registrar-General's projection to 1980, published in the Annual Abstract of Statistics 1967, indicates an increase of 5 million in the population of the U.K. from the present level of about 55 million.

[2] Elasticities of demand are discussed at greater length in Appendix D.

maize is wholly imported it can be ignored for the purpose of this exercise.

2.11. Bread consumption per head is slowly declining, but this is to some extent offset by an increase in the consumption of cakes and biscuits.[1] The use of home-grown wheat for flour milling has increased by an average of about 40,000 tons per annum over the last five years (1961/3 average to 1965/7 average) and we estimate that this is likely to continue.[2]

2.12. The demand for barley for brewing and distilling has been increasing over the last few years by about 65,000 tons per annum and we expect this rate of increase to continue.

2.13. The effect of these various trends (assuming $+20\%$ and -20% for highest and lowest estimates) is as follows:

ESTIMATED INCREASE IN ANNUAL CEREAL REQUIREMENTS FOR
HUMAN FOOD IN 1972/3 COMPARED WITH 1967/8 (U.K.)

	Lowest increase	Most likely increase	Highest increase
	('000 tons)		
Wheat	160	200	240
Barley	260	325	390
Total	420	525	630

ESTIMATED INCREASE IN HOME-PRODUCED REQUIREMENTS ASSUMING THE
PERCENTAGE OF HOME-PRODUCED REMAINS AT 1966/7 LEVEL

	% home produced	Lowest increase	Most likely increase	Highest increase
		('000 tons)		
Wheat	31	50	62	74
Barley	96	250	313	376
Total		300	375	450

(c) INDUSTRIAL USES AND SEED

2.14. The prospects for increased industrial uses of grain are discussed in more detail in Part V(d). There is scope for research into the use of grain for

[1] This has been a long-term trend. In the report of the British Association 1881, the *per capita* consumption of flour in the U.K. was given as 280 lb per annum. By 1965 this had fallen to 155 lb. See also Appendix D, Table 1, and Part V(b), Table 1.1.

[2] The percentage of home-produced wheat used in flour milling has been between 31% and 33% for the last five years. Since the production of U.K. wheat has been increasing the tonnage used by U.K. millers has also been increasing in parallel.

industrial purposes, but we do not expect any significant change in this item in the next five years.

2.15. The demand for seed is a reflection of the increases in cereals acreages postulated for the next five years. These are set out below as part of the discussion on the supply of cereals. Assuming that seed represents about 5% of production on average, the extra seed requirements will be as follows (i.e. assuming the U.K. does not enter the Common Market—an assumption that is common to all the estimates in this chapter: the U.K.'s entry into the Market would bring about so many consequential changes that the value of an exercise of this sort would be very questionable):

ESTIMATED INCREASE IN ANNUAL SEED REQUIREMENTS IN
1972/3 COMPARED WITH 1967/8

	Lowest increase	Most likely increase	Highest increase
	('000 tons)		
(All home-grown)	—	90	190

(d) EXPORTS

2.16. As indicated in Part IV, Chapter 3, the export trade in grain grew rapidly from its previous level of about 100,000 tons per annum to over 700,000 tons in 1965/6, and to over 1 million tons in 1966/7. In 1967/8 total exports have slipped back to about the 800,000 tons mark, and prospects for the future are not bright. For the reasons discussed more fully in the above section of the Report we forecast the following reductions in exports by 1972/3:

ESTIMATED REDUCTION IN ANNUAL EXPORTS OF CEREALS FROM THE
U.K. IN 1972/3 COMPARED WITH 1967/8

	Biggest reduction	Most likely reduction	Least reduction
	('000 tons)		
(All home-grown)	−500	−250	No change

Summary: A. The Demand for Home-grown Cereals

2.17. Bringing together the above estimates we have the following total increase in demand for home-grown cereals in 1972/3:[1]

---

[1] The F.A.O. have recently made detailed projections of the increase in demand for cereals in the U.K. between 1965 and 1975, and their results are discussed in Appendix D. It is of interest to note, in connection with the estimates made in this chapter, that the F.A.O. projection to 1975 represents roughly an increase in demand for cereals for all purposes of about 1½ million tons between 1967 and 1972, which agrees very closely with our far less sophisticated estimates.

ESTIMATED INCREASE IN ANNUAL DEMAND FOR HOME-GROWN
CEREALS IN THE U.K. IN 1972/3 COMPARED WITH 1967/8

	Lowest increase	Most likely increase	Highest increase
	('000 tons)		
Animal feed	660	1230	1520
Human food	300	375	450
Industrial and seed	—	90	190
Exports	−500	−250	Nil
Total	460	1445	2160

B. The Supply of Grain in the Next Five Years

2.18. In Part II, Chapter 1, we have set out in detail the results of a survey of N.A.A.S. officers and their counterparts in Scotland and Northern Ireland in which they were asked to estimate the likely changes in yields and acreages of cereals in the five years from 1967 to 1972. Although the survey also asked for an estimate to be made assuming the U.K. entered the Common Market this has not been used for the reasons stated above.

2.19. These estimates show the following expected increases in cereals production:

ESTIMATED INCREASE IN ANNUAL PRODUCTION OF CEREALS
IN THE U.K. IN 1972/3 COMPARED WITH 1967/8[a]

	Lowest increase	Most likely increase	Highest increase
	('000 tons)		
Wheat	83	523	1031
Barley	58	1170	2441
Oats	−106	132	378
Total	35	1825	3850

[a] The F.A.O. report (see Appendix D) projected an increase of 4·3 million tons between 1967 and 1975 or about 2¾ million tons between 1967 and 1972. This is rather higher than our "Most likely" figure. It is, perhaps, to be expected that the N.A.A.S. would adopt a cautious attitude to forecasting increases in acreages and yields. Past experience has shown that the Ministry's yield forecasts tend to be too low when the yield rises above average and too high when it falls below it (See A. H. J. Baines, Estimating Cereal Yields, *Agriculture*, August 1966).

C. Relationship between Demand and Supply over Next Five Years (Home-grown Cereals)

2.20. The above estimates of demand and supply may now be brought together as follows:

	Lowest increase	Most likely increase	Highest increase
	('000 tons)		
Estimated demand	460	1445	2160
Estimated supply	35	1825	3850
Estimated surplus of deficit	−425	+380	+1690

The figures for estimated surplus or deficit shown under "lowest increase" and "highest increase" respectively do not represent the most extreme possibilities. It is conceivable (but unlikely) that the lowest increase in demand could coincide with the highest increase in supply and vice versa. In such circumstances a surplus of over 3 million tons or a deficit of over 2 million tons could arise.

D. Conclusion

2.21. The U.K. grain market appears to be in a position of "uneasy balance", and this is likely to remain roughly the position in the next few years but with a tendency for supply to increase rather faster than demand.[1] Bearing in mind the fluctuations in production due to weather and other factors (see Part II, Chapter 4) there are bound to be years when a substantial surplus of grain has to be disposed of. Equally, there will be years when domestic supplies are short, but as imports are usually readily available this seldom causes any difficulty. These "seasonal surpluses" are likely to become more serious in the years ahead unless there is a significant substitution of home-produced for imported grain. Thus the U.K. can be said to be entering a period of "incipient surplus", i.e. troublesome surpluses in some years but not in others, with a tendency for the surplus situation to get steadily worse as production rises faster than demand. This is the background to the discussion that follows on agricultural policy in relation to cereals.

[1] T. E. Josling, in *The United Kingdom Grains Agreement (1964): An Economic Analysis,* Institute of International Agriculture, Michigan State University 1967, arrives at a similar conclusion: "From the empirical relationship of Chapter VI, it would appear that cereal production is expanding, at constant prices, by some 520 thousand tons each year, and that use is increasing at about 450 thousand tons a year. The disparity has increased in the last four years" (p 150).

CHAPTER 3

Summary and Critique of Current Cereals Policy

3.1. The conclusion reached in the previous chapter, that the U.K. cereal market is in a situation of uneasy balance, with surpluses to be expected at least in some years, does not represent an altogether new situation. Strange as it may seem now in the light of the rapid expansion in cereal production in the last few years, there was a great deal of anxiety 5 or 6 years ago about the disposal of barley surpluses (although production was then only half its present level) and indeed this led, as was indicated earlier, to the introduction of the Minimum Import Price system and the setting up of the Home-Grown Cereals Authority (Chapter 1, page 630). In the event, the increasing production of grain was absorbed in various ways, e.g. by increases in livestock populations, by the use of an increasing proportion of home-grown grain by animal feed manufacturers and flour millers, and by a much larger export market.

3.2. The present and likely future situation, therefore, is not markedly different from that which has existed for some years now. However, there are signs that surpluses of grain are likely to recur more frequently, and on a larger scale. Moreover, the possibility of an increasing production finding satisfactory outlets seems less promising in the next 5 years than in the past 5 for the following reasons:

(a) The international grain market, which moved from a position of surplus at the turn of the decade to a position of impending deficit accompanied by a sharp run-down of stocks in the mid-1960's, has now swung back again to a surplus position. The great unknown factor is the extent to which the U.S.S.R., China and India will continue to be buyers of grain. There are signs that they may not be such big buyers as in the past. Meanwhile the Common Market countries are continuing to step up their grain production and find it difficult to adopt price policies aimed at checking this increase.

(b) It seems unlikely, as was indicated in Part IV, Chapter 3, that the U.K. grain export trade will recover the high level reached in 1966/7.

641

(c) Bulk grain carriers are being used for an increasing proportion of grain moving in world trade (Part IV, Chapter 3) and the consequent reduction in freight rates improves the competitive position of the exporting countries.

(d) Substitution of home-produced grain for imported has been going on for some years already and there is a limit to the extent to which this process can continue without a major change being required in consumer habits, e.g. from the present type of loaf to a Continental type.

3.3. These issues will be discussed at greater length shortly but they are outlined here to support the assertion that the main policy problems that face the U.K. in the cereals field are associated with the threat of over-production. It is true that there are some people who argue that the growth in world population is such that a situation of world food shortage is more likely than one of food surplus. This issue is discussed in Part VI, Chapter 6, where it is concluded that world grain production is likely to exceed the "effective demand" for some years yet. In this context, of course, the growth of government-financed food aid programmes in the last 20 years is significant but it does not alter the basic situation.

3.4. We turn now to the implications of this situation of contingent recurrent surpluses for cereals policy, with barley particularly in mind. There are two main approaches, firstly to take measures to stop surpluses from developing (what one might call the *ex ante* approach) and secondly to dispose of surpluses once they have arisen (the *ex post* approach). We will first consider policies of the "surplus-preventive" type.

The Supply Control Approach

(i) ENCOURAGING BREAK CROPS

3.5. If more profitable break crops were available there is no doubt that this would encourage many cereal farmers to move from a continuous cropping system for barley to a more traditional rotation system. This is the purpose of the action already taken by the Government in this direction (e.g. the subsidy on beans) and the future is likely to see further stimulus given to other break crops also. However, the problem is often not so much persuading farmers to grow a break crop as finding suitable market outlets for the product, and we would like to see more attention paid to the problems of marketing such break crops as oil-seed rape and dried grass, as well as beans.

3.6. There is always a risk that the policy of encouraging profitable break crops will eventually be self-defeating, in that it might encourage farmers who had moved out of cereals, because of falling yields and disease problems, to come back again. In our view this risk is one that has to be taken.

(ii) ENCOURAGING THE SUBSTITUTION OF WHEAT AND MAIZE
GROWING FOR BARLEY

3.7. In Part V(a), Chapter 2, the problem of substitution of home-produced grains, like wheat and barley, for imported grains like maize and sorghum has been discussed, and the conclusion was that there is a possibility of a considerable increase in substitution taking place, even in the short run. Wheat, in particular, can readily be substituted for maize in pig and poultry rations with potential benefits to the U.K. balance of payments. More English wheat could be used in bread making but only to any significant extent if the public would accept a different type of loaf—this possibility has already been discussed in Part V(b), Chapter 3. It may not be an easy matter, however, to increase wheat production at the expense of barley. There is evidence that most farmers already sow as much autumn wheat as they can, since this is the most profitable grain crop. They are usually inhibited from putting in wheat more than say, 2 or 3 years in succession on the same land (except on heavy land as in North Essex) for fear of disease and low yields. The extra wheat, therefore, may have to be mainly spring-sown wheat, but this is considerably less profitable than autumn-sown wheat, and a wider differential in the guaranteed price than now exists may be necessary if there is to be much substitution for barley.[1] Nevertheless, the current policy is in the right direction, and the suggestion we make below, regarding the payment of the wheat subsidy on an acreage basis, may give a further incentive to some farmers to grow wheat rather than barley.[2]

3.8. As to the growing of maize, even after making allowance for our fickle climate we are rather surprised that so little progress appears to have

[1] See, for example, How can we best grow more wheat? by Edward Bullen, Director of Boxworth Experimental Husbandry Farm, *British Farmer,* 10th February 1968: "A further possible way to grow more wheat is an increase in the spring wheat acreage, but there is a substantial gap between the yield potential of winter and spring wheat. Indeed heavier yields can usually be achieved from Capelle or Champlein sown as late as mid-February than from orthodox spring wheats. N.I.A.B. data suggested that Opal was about 13 per cent lower yielding than Capelle: this was of course before disease devastated Opal. New varieties of high potential will soon be available for both winter and spring, but it still seems that Cambier or Ranger will be at least 10 to 15 per cent higher yielding than Kolibri."

[2] For further discussion of relative acreages of the three cereals, see Economic Development Committee for Agriculture, *Agriculture's Import Saving Role,* Part II, paras. 26–9.

been made with this crop in the U.K., whilst in France it has become one of the main cereals in the space of only a few years (Part VI, Chapter 1, page 527). The acreage of maize in the U.K. at present is of the order of 15,000 (*Farmers Weekly,* 28th June 1968), while production in France is over 4 million tons. There are some problems still to be overcome in the production, harvesting and marketing of maize, but it seems that these are being tackled at the moment mainly by a few enthusiasts at various experimental farms or agricultural colleges, or by private innovators who have faith in the future of this crop, and who have established the Maize Development Association to stimulate its production. We would like to see a far greater injection of effort and capital in this field, since the potential reward is great even allowing for a proportion of bad years. The likelihood of a more serious barley surplus situation developing than has been experienced hitherto should lend a new sense of urgency to research and experiment in maize growing in the U.K. The Home-Grown Cereals Authority should be empowered to take a leading part in this.

(iii) REDUCING THE RELATIVE PROFITABILITY OF CEREALS *vis-à-vis* OTHER FARM ENTERPRISES

3.9. If the above approaches do not succeed, the main alternative in the direction of preventing over-production of feed grains, notably barley, is to change the relative profitability of the crop by reducing guaranteed prices, or by increasing them less than for other commodities.

3.10. This is the approach that was used for wheat in the late 1950's when wheat surpluses were building up in the main producing countries, and it was reasonably successful. The cuts in the guaranteed price of barley in the 1960's, however, failed to stop the increase in output of that crop. One of the problems with this approach is that the Government is not free to arrive at price decisions purely on the grounds of stimulating the most desirable pattern of production. The Government has recognised for a long time, and certainly since the 1947 Agriculture Act, an obligation to give a measure of support to farm incomes. This does not extend to individual commodities (except as covered by the long-term assurances set out in the 1957 Agriculture Act) but if the guaranteed prices for cereals are reduced it may be difficult to find commodities whose guaranteed prices can be increased, since commodities like eggs and milk are already tending to be over-produced. Poultry meat and horticulture are not guaranteed; price policy for pigs is complicated by the need to take account of the current phase of the pig cycle; so that beef and sheep are really the only remaining major commodities for which higher guarantees might be appropriate. It seems likely that if the threatening surplus situation materialises and the other lines of policy discussed previously and in the rest of this chapter fail to cope with the problem, the only ultimately effective solution will be to cut the guaranteed prices for cereals. If this creates problems of income support for agriculture, alternative methods other than through commodity price guarantees may have to be found. The system of standard quantities is admittedly designed in part to discourage excess production, but its effects on cereal growers are so indirect and delayed that it is far less effective as a device for supply management than it is as a device to restrict the Exchequer's commitment.

3.11. Ironically, it may happen that if guaranteed prices for cereals are reduced, some farmers who have not yet moved to the limit of continuous corn-growing may be impelled to do so, seeing opportunities for economies of scale and simplification of their farming system. However, there comes a time when this is self-defeating and declining yields force farmers to look for more profitable alternatives. There are some who would turn the argument on its head and say that rather than being reduced the cereal guarantees should be increased, since this would remove the pressure on cereal growers to turn to continuous cropping. We believe, however, that if this policy were adopted it would simply attract more farmers into cereal growing and this would more than offset any saving on the continuous cropping side.

(iv) DIRECTLY CONTROLLING PRODUCTION, E.G. BY FARM QUOTAS

3.12. We mention this alternative more for the sake of logical completeness than because we regard it as a choice in any way comparable with the others discussed in this chapter. In fact the use of this alternative would only be justified if all else had failed. Possible methods would be a quota system, as for sugar beet and potatoes, control of entry into cereal growing, or a differential pricing system, i.e. so that those who produced above a pre-determined level received a lower price. None of these would be easy to operate and they would mean the virtual disappearance of the free market at least in anything like its present form. We regard this approach as a last resort because it would impose a rigid pattern of production on an industry that is in a state of rapid technological change, and because such a system would inevitably put a brake both on technological innovation and structural adjustment. This is a high price to pay for supply management, and other solutions to the problem should be tried first.

The Market-widening Approach

(i) ENCOURAGING EXPORTS

3.13. The remarkable growth in the export trade in grain in the last few years has already been commented upon in Part IV, Chapter 3. However, it was suggested there that the future of the U.K. grain export trade is an uncertain one. The countries in relation to which the U.K. is best placed geographically, mainly those on the European seaboard, are precisely those where competition from the Common Market countries is likely to be greatest. The 1967/8 season showed how France, for instance, was prepared to use restitutions to secure a competitive advantage over U.K. grain. There is little doubt that the Common Market countries will continue to pay higher prices for grain than the situation really demands, and that they will try to dispose of the surpluses abroad to the detriment of our own export trade.

3.14. Nevertheless, the merchants, shippers and brokers continue to search for export outlets for U.K. grain with considerable success. Not only are exports a useful safety valve in the event of over-production in a good season (and the U.K., with its lack of storage facilities off the farm, badly needs a safety valve of this kind), but also they put the U.K. in a better position, in the event of our joining the Common Market, to profit from the rapid increase in trade which can be expected to develop across the North Sea and the English Channel.

3.15. There are some who favour direct subsidies to exports, but these could cause difficulties for the U.K. with our G.A.T.T. partners and might in any case merely add impetus to a vicious circle of export subsidisation and import restriction which in the end is no help to anybody. If it is felt that assistance to export should be given, our preference would be for aid in the improvement of wharfing facilities at export ports and provision or modernisation of grain-handling equipment. Only if it were proven that our main competitors were gaining a perpetual unfair advantage through devices like restitutions should direct assistance to exports be considered, and then not on a selective basis if possible, but in such a manner as to affect all exporters alike. We have a highly competitive grain export trade, and as far as possible it should be left to operate freely.

(ii) INCREASING DOMESTIC CONSUMPTION OF GRAIN

(a) *The Livestock/Feed Balance*

3.16. The most important variable in the demand/supply equation for cereals is the demand for livestock feed. Apart from the question of whether to use home-grown or imported grain, a compounder's production plans are virtually dictated by the demand for his feeds, and this in turn is very much influenced by weather (an open spring can bring a marked reduction in the demand for cattle feed) and by the number of livestock to be fed. Thus the Government's policy for livestock has a direct relevance for cereal growers, just as, in reverse, the Government's policy for cereals has a direct relevance for livestock producers, since they are the main users. An attempt was made in the National Plan, and more recently in the Report of the E.D.C. for Agriculture, to relate these two elements of the situation more closely than hitherto, but still the relationship, in terms of agricultural policy, tends to be a loose one. Inevitably perhaps, government policy for agriculture is a compromise between many conflicting pressures, and the desirability of keeping policy for cereals and livestock in close harmony with each other cannot always be the over-riding consideration.

3.17. The evidence of discordant tendencies is apparent in the acreage and livestock statistics. The substantial increases in livestock populations needed to consume the additional cereals likely to be produced do not seem to be getting under way. As grain production moves increasingly into surplus and marketing problems begin to dominate the scene, so the Government will be forced to pay more attention to this question of the livestock/feed balance. The N.F.U. of Scotland have said in their statement (Appendix G, page 758) that the problem is more one of under-production of livestock than of over-production of grain. Whether or not one accepts this view, the Union is surely right in emphasising the need to keep the two policies dovetailed as closely as possible. Whilst there would be little point in expanding domestic livestock production simply to absorb surpluses of grain without regard to relative costs, if the economies of scale still to be obtained in grain production are substantial (e.g. larger fields, bigger combines, etc.—see Part II, Chapter 4) it may be that a parallel expansion of livestock to enable these economies to be achieved would be sensible.[1] Again, it may be more economic to convert grain into barley-beef in the U.K. than to export the grain and import the beef. Another example is the impact of the feed formula (designed originally to help the livestock producer) on the marketing of grain (see Part V(a), Chapter 1). Issues of this kind need to be worked out not in isolation but as part of a comprehensive livestock/feed policy.

---

[1] The report of the E.D.C. for Agriculture suggests that in the predominantly cereal-growing areas, "70 per cent of any expansion would be absorbed by the present number of combines" and "it may be possible to dispense with labour at a faster rate than before".

(b) *Substituting Home-grown for Imported Grain in U.K. Cereals Industry*

3.18. It may seem odd at first sight that a country like Britain, which imports about 8 million tons of grain, could be in a "surplus" grain situation. But of course these imports include maize (about 4 million tons) and hard or filler wheats (about $3\frac{1}{2}$ million tons) for which U.K.-grown wheat is not directly substitutable. The problems of substituting the English product for imported grain in flour milling and feed compounding have been discussed in Part V, a and b. We conclude that there is a reasonable chance that 1 million tons[1] of English grain might be substituted for imported grain in animal feeds, and possibly another half million tons of English wheat might be used in flour milling without basically altering the character of the English loaf (Part V(b), Chapter 3). Any further substitution would entail major changes in farmers' techniques of livestock production (e.g. a return to breeds of poultry more capable of making effective use of low-energy feeds), or in millers' flour milling practices. Such changes are not impossible but they are unlikely to come soon enough to affect the situation in the near future. We are therefore thrown back on substitution of the "technically feasible" kind referred to earlier.

3.19. How is this substitution to be brought about? Basically this imported grain is used rather than domestic grain because it is cheaper. The recently announced higher minimum import prices and the higher wheat prices resulting from the new International Grains Arrangement may stimulate some substitution although the cost of all imported grains would be increased as a result and this would work to the disadvantage of our balance of payments. We doubt whether the present or likely near-future grain surplus position will be serious enough to justify government intervention to force millers and compounders to use a certain proportion of home-grown grain (as is done in some European countries), but such a step might ultimately become necessary.

(c) *Discouraging Imports*

3.20. From the point of view of disposing of domestic surpluses of grain, the level of imports is a vital factor. The Government has discouraged the inflow of very cheap grain by its minimum import prices, and there is some reason to think that some of the largest buyers of grain in the U.K.

---

[1] The E.D.C. Report estimated that about 1 million tons of home-produced grain might be used to replace imported maize, sorghum or feed wheat and another $\frac{1}{2}$ million tons of home-grown quality wheat could replace imported filler wheat (*Agriculture's Import Saving Role*, Part II, para. 12).

exercise a degree of self-imposed restraint in buying foreign grain if this is likely to damage the domestic grain situation. However, there is a limit to the extent to which any one firm can act individually, especially if its competitors have no such scruples. Indeed there are some who say that rather than trying to protect the domestic price, certain of the big buyers sometimes import grain for the purpose of breaking the domestic price if they think it is too high. Farmers as a whole certainly do not feel that the present policy for imports gives adequate protection to the interests of the domestic producer, and they quote the impact of imports of French wheat early in 1968 on the U.K. grain price as an example of how even small offers of cheap imports can completely upset the home market.[1] We would not advocate an outright protectionist policy, but it seems to us that in an "ex post" situation of grain surplus and of ineffective supply/demand management within the U.K., imports might have to be subject to a greater degree of control than exists at present. We consider this issue more fully in the next chapter in relation to a discussion of the role of the Home-Grown Cereals Authority.

3.21. There is one aspect of the grain imports situation that calls for special comment, in addition to the mention it has already received in Part V(a), Chapter 7; this is the substantial quantity of grain imported directly into N. Ireland from North America. At present most of the grain entering N. Ireland—one of the principal grain-deficit areas in the U.K. (Map 3.2, page 216)—comes directly to Belfast across the Atlantic in 12–15,000 ton ships. It would obviously save imports if British grain could be used in N. Ireland instead, but this does not happen at present because of the heavy cost of transporting the grain across the Irish Sea (estimated to add up to £3 per ton to the cost of grain landed in Belfast). It may be that when the new grain terminal at Liverpool is complete and grain comes across in 50,000–60,000 ton ships, which cannot be accommodated at Belfast, the situation will change. However, this is a possible development in the future and many observers argue that steps should be taken immediately to assist the transport of British grain across to N. Ireland so that it can replace imports.[2] We recommend that a study should be made

---

[1] A comparison between the trend of the U.K. soft milling wheat price in the period January–June 1968 and the imported West European milling wheat price shows that they kept closely in parallel.

[2] See, for example, *Financial Times,* 22nd May 1968. "Northern Ireland will almost certainly buy at least an extra 300,000 tons of British barley a year if Whitehall can find a way of making up some of the £3-a-ton transport costs that apply at present." See also Norman Hicks in *Farmers Weekly,* 31st May 1968: "The demand for barley in Britain could be usefully boosted by subsidising the cost of shipping it there—few people realise that it costs more to ship to Belfast from the grain growing areas of England than from

of the existing transport facilities, particularly to establish why the freight rates are so high; but whether a subsidy on the freight costs would be justified would depend upon whether the benefits to the balance of payments and the reduction in deficiency payments outweighed the desirability of manufacturing the grain into feed as near as possible to the place of origin of the grain. There would certainly be no point in stimulating an *expansion* of the N. Ireland feed manufacturing industry based on heavily subsidised freights, but there would seem to be good sense in substituting British grain for the imported barley and some of the imported maize and sorghum if this could be achieved by reducing the freight costs, and a study of this possibility seems to be called for.

(d) *Measures to Increase the Uptake of Home-produced Grain*

3.22. These generally involve attempts to match the U.K. product more nearly to market requirements, e.g. in terms of the quality of grain, moisture content, seasonality of deliveries, type of grain and assured supplies of the quantities required. Merchants, co-operatives and producer marketing groups of various kinds are now much more alert than they were to the need to stimulate production of the right quality, type and condition of grain to supply specific outlets. We would like to see this process accelerated. The self-discipline in production techniques exercised in varying measure by producer marketing groups in particular (but not confined to them—merchants can achieve similar results through contract arrangements) is a relatively new feature in grain marketing and opens out opportunities for a new approach such as contracting to sell an assured quantity of grain of a specified quality over a whole year. These groups can offer some of the advantages of imported grain. They may also in turn develop central driers and stores, and the necessary process of "bulking" the grain would facilitate more sophisticated grading techniques and more efficient systems of transport from store to mill than are possible at present. Facilities offered by groups might well encourage users hitherto accustomed to use mainly imported cereals to take up a higher proportion of U.K. grain. However, research is needed to establish exactly what users' requirements are, what they will be in the future and how best they can be met by domestic producers.

Canada. I suspect that the cost of such a subsidy would be more than recovered in the reduction of deficiency payments on barley resulting from raising demand by some 300,000 tons a year." See also *Annual Report of the Home-Grown Cereals Authority, 1967–8* (para. 61).

(e) *Finding New Uses for Domestic Grain*

3.23. In Part V(d) we have already commented upon the possibility of more domestic grain being used in the production of starch and starch derivatives, and for breakfast foods manufacture. Whilst the possibilities of substituting U.K. grain for imported in this sector seem to be small, they are certainly worth investigating more methodically than has been the case hitherto. Uses for straw, wheatfeed and other by-products are also worth further consideration.

(f) *Carrying-over Grain Stocks*

3.24. In any year a certain amount of grain has to be carried over by the cereal-using industries between the final disposal of one crop by farmers and the harvesting of the succeeding crop, to ensure continuity of manufacture. However, bearing in mind the tendency observed in recent years for production to increase faster than demand, the situation may well arise following an abnormally bountiful harvest that the market will not be able to absorb the whole crop before the next harvest ensues, or at least not without a serious collapse in market prices. Some carryover in excess of normal requirements might therefore be desirable. This is a regular feature of grain marketing in most of the major grain-producing countries of the world, but has hitherto been effectively prevented in the U.K., so far as carryover on farms is concerned, by the regulation that all grain must have left the farm by 21st July if it is to qualify for cereal subsidy (in the case of wheat) or incentive payments (barley). This was introduced because of the difficulty of ensuring that the new season's crop does not qualify for the higher guaranteed prices or incentive payments relating to the tail-end of the old season's crop.

3.25. A small carryover may indeed be helpful in the sense that it can act as a strategic supplementary reserve in the event of a bad harvest, thereby contributing to orderly marketing. But almost inevitably, it seems, a modest carryover of this kind develops in due course into a "stock-pile", perhaps because it is politically simpler to take this way out than to reduce the guaranteed price. It is true that the huge American stock-piles have served a valuable humanitarian purpose in that they have enabled supplies of grain to be sent to India and other countries which needed them but were not in a position to buy; but famine relief measures, even if they persist, do not remove the necessity to prevent unduly expensive stock-piling which results from a collective failure of the nations trading commercially in grain to adjust supply to demand.

3.26. A realistic assessment of the U.K. grain situation suggests that ome capacity for carryover is going to be needed, at least in years of good harvests. The problem is, who would be responsible for the policy and where would the grain be stored? The Home-Grown Cereals Authority would seem to be the obvious body, but, as indicated in the following chapter, the Authority is at present specifically barred from carrying over grain it has acquired for market-support purposes to the following season. This policy is unworkable should a surplus situation develop and we suggest ways later in which it may have to be altered.

Conclusions

3.27. In this chapter our aim has been to study the elements of the situation from the approach first of preventing surpluses, and second of disposing of surpluses once they have arisen. The question naturally arises as to which of these approaches is the right one. In our view it is not necessary to make a choice; they are complementary, not mutually exclusive, approaches. We expect to see both approaches adopted contemporaneously in the future, as they have been in a rather unco-ordinated manner in the past. At first the emphasis may be more on the "market-widening" approach, which involves the grain trade more than the Government; but if the surplus situation worsens, the emphasis will switch more to the "supply control" approach and this involves primarily the Government. As the problems arising out of surpluses become more acute, so the pressures on both the trade and the Government will increase, and there will be a cry for more effective marketing. The role of the Government, through its subsidy policy, and the way the Home-Grown Cereals Authority reacts to the situation, are key issues, and it is to these that we now turn.

CHAPTER 4

Systems of Price Support and the Role of the H.-G.C.A.

Systems of Price Support

4.1. There are a few who would argue that there should be no guaranteed prices for cereals at all, and that the free market system should be allowed to operate without government interference. This attitude may seem superficially attractive at a time when the existence of large surpluses of grain in many of the producing countries would enable the U.K. to import her grain very cheaply. Such a policy would of course cause hardship among farmers but, so the supporters of this viewpoint would argue, the end result would be a switch of resources from agriculture into some other sector, say manufacturing industry, and this would be to the advantage of the nation as a whole in the long run. There is some logic behind this point of view but we do not think that such a drastic policy would be acceptable on social grounds or consistent with the spirit of the 1947 Agriculture Act, nor is it immune from criticism on economic grounds, i.e. it would cause such a loss of confidence in home agriculture that general efficiency would certainly suffer. We therefore accept that the necessity for some measure of government support for agriculture will continue, but the degree of support should be such as to ease the strains of desirable structural adjustments without insulating farmers from market pressures to such an extent that those adjustments never take place. We believe that successive governments since 1947 can claim to have achieved some success in this. Our main concern in the first part of this chapter is therefore not whether the Government should be assisting cereal producers, but whether the system of price support which has evolved is the best one.

1. WHEAT SUBSIDY PAID ON A TONNAGE-SOLD BASIS

4.2. Early in the Survey we gave particular attention to the question of whether the wheat subsidy should be paid on a tonnage-sold basis, or, like the barley subsidy, on the acreage grown. This issue had already been a subject of debate for some years, and we decided to make a rather detailed investigation into the likely impact on farm incomes such a change would make, and to assess the potential savings in transport which might result.

4.3. The results of this investigation are given in more detail in Appendix E. Briefly the arguments for and against the change may be summarised as follows:

Arguments for changing to the acreage basis for wheat:

(i) It would enable those farmers who wished to feed their wheat on their own farms to do so without losing the subsidy; this would allow more flexibility in planning home-mixed rations and would save the transport costs now incurred in taking straight wheat off the farm and bringing an identical product back. We estimate that the total aggregate saving in transport might be anything up to £4 million (England and Wales) depending on the extent to which farmers in fact decided to feed wheat rather than sell it.

(ii) The change might encourage farmers to switch from barley to wheat, particularly in the west where yields tend to be lower.

(iii) The change would marginally assist the hard wheats whose yield tends to be less than the soft wheats.

(iv) There would be administrative savings.

Arguments against changing to the acreage basis for wheat:

(i) It would reward both inefficient and efficient producers equally.

(ii) Farmers not accustomed to growing wheat might "farm for the subsidy", i.e. secure the acreage payment but take little trouble over the crop.

(iii) It would mean the ending of the present seasonal guaranteed prices, which would be replaced by an average guaranteed price for the year as with barley.

Our views on the three above arguments against a change are:

(i) For most producers the gross receipts would be altered by something less than £1 per acre (see Appendix E, Table 6). It is very unlikely, therefore, that the change would cause any producer to relax his efforts to obtain maximum yield. Of the 65,000 wheat producers in England and Wales, about 45,000 would be virtually unaffected by the change as their yields are at or close to the national average.

(ii) The subsidy now represents only about 13% of the guaranteed price compared with 32% in 1960/1. This argument has therefore lost a lot of its force.

(iii) We accept that the withdrawal of the seasonal guaranteed price would be a loss to wheat producers but we doubt whether this feature of the wheat subsidy system can anyway long survive the conditions likely in the years ahead, since it insulates producers too much from the requirements of the market.

4.4. For the above reasons, and because we believe it is important wherever possible to cut away artificial limitations of this kind on the free operation of the market system, we propose that the wheat subsidy should be changed to an acreage basis, thus allowing wheat to "find its own level" as a feed grain on the farms where it is grown.

VIEWS OF THE FARMERS' UNIONS

4.5. The three farmers' unions were invited to answer the question: "If the wheat subsidy were paid on an acreage basis, as with barley, to what extent would cereals farmers who at present sell off their wheat crop be likely to retain it on the farm to feed to livestock?" The National Farmers' Union answered that the quantity of wheat likely to be retained for feeding on the farm would probably be too small to justify making the change, but the National Farmers' Union of Scotland and the Ulster Farmers' Union both said that the change would benefit the livestock producers in their countries. See Appendix G, pages 759 and 762.

2. SEASONAL INCENTIVES

4.6. When cereal prices were decontrolled in 1953/4 it was decided to encourage the spread of marketing of wheat throughout the year by offering a higher guaranteed price as the season advanced. It seems to have been taken for granted that even though the price would tend to rise through the season in a free market, the premium for spring delivery would not be sufficient to justify a farmer's cost of storage, drying and interest. The seasonal incentives may also have been designed to encourage cereal farmers to erect driers and stores to cope with the growing quantity of grain that was being unloaded on to the market at harvest time as a result of the increasing use of combines. With the old binder and rick system, grain was threshed at intervals during the post-harvest period and there were fewer gluts (see Part I, Chapter 1 and Part II, Chapter 1). If this indeed was the purpose, it would seem that what must have been intended as a short-term incentive has remained a permanent feature of the scheme, and this in spite of the further encouragement to farm storage given under the Farm Improvement Scheme since 1957. Indeed when the barley expansion got under way in the early 1960's, the Government introduced a seasonal incentive scheme for that crop also. The incentives have varied over the years as follows:

WHEAT

	Harvests			
	1954–64[a]	1965 and 1966	1967	1968
	(cumulative, per ton)			
	s. d.	s. d.	s. d.	s. d.
				Oct. 21 8
July–Sept.	—	—	—	Nov. 30 0
Oct./Nov.	+30 0	+30 0	+25 0	Dec. 38 4
Dec.–Feb.	+60 0	+60 0	+45 0[b]	Jan. 46 8
March/April	+85 0	+80 0	+65 0[c]	Feb. 55 0
May/June	+100 0	+90 0	+80 0[d]	Mar. 63 4
				Apr. 71 8
				May/June 80 0

[a] With slight changes up or down, between seasons.   [b] Dec.–Jan.   [c] Feb.–March.   [d] April–June.

BARLEY

	Harvest		
	1963 and 1964	1965 and 1966	1967 and 1968
	(cumulative, per ton)		
Disincentives	s. d.	s. d.	s. d.
Sept./Oct.	−15 0	−15 0	−10 0
November	−15 0	—	+ 6 8
Incentives			
December	—	+10 0	+ 6 8
January	+20 0	+20 0	+10 0
February	+20 0	+20 0	+13 4
March	+30 0	+25 0	+16 8
April	+30 0	+25 0	+20 0
May/June	+30 0	+25 0	+23 4
Total spread	+45 0	+40 0	+33 4

4.7. Whereas in the case of wheat the seasonal incentives are part of the guaranteed price (i.e. the deficiency payment is worked out for each period separately) with barley they are not. Thus the barley incentive (or disincentive) is not directly related to the average year guaranteed price for barley, and it was originally intended to be self-financing; the aim was that the money collected in disincentives for sale at harvest time would cover the cost of the premium to be paid on later sales. In fact, however, only about two-thirds of the funds needed have come from the disincentive payments and the rest has been found by a deduction from the acreage payment.

4.8. There has been a great deal of discussion as to the appropriate level of the seasonal incentives; whether it is right that the incentives for barley should be so much less than those for wheat; and whether the barley scheme should be self-financing. But in our view the time has come to reconsider whether these incentives now serve a useful purpose at all. They were begun at a time when storage facilities were inadequate to cope with the flow of grain and when a special incentive was needed to encourage farmers to put in grain stores. It is a tribute to the success of the scheme, together with the Farm Improvement Scheme, that today there appears to be adequate storage on most farms, and a special incentive for this invest-ment is no longer necessary. The great drawback of the present scheme is that it introduces an undesirable element of rigidity into cereals marketing. For example, the wheat incentive certainly exceeds the variable costs of storage, and indeed virtually insulates the wheat farmer from having to consider the market price when he sells (except insofar as he tries to "beat the average" during a particular guarantee period). The result is that wheat farmers sell wheat when it happens to suit them, without having to consider when the market really needs it. There is also a dearth of U.K. grain a few weeks after harvest and often a surplus at the end of the cereal year, and as grain production rises this will become an even more urgent problem. It is surely desirable for the maximum quantity of U.K. grain to be taken up by millers and compounders fairly early in the season, allowing imports to make up the deficiency at the end of the cereal year. It is a criticism of the present storage incentives that they cause this desirable pattern of marketing to be reversed. In recent years grain has been impor-ted in November and December when the bulk of the U.K. crop was lying in store and farmers saw no reason to release it. Another criticism of seasonal incentives in general is that they cause some farmers to sell their grain (to get the benefits of the incentives) rather than store for feeding on the farm, since if they do not sell they forsake any incentives. But the main point is simply that the seasonal incentives are now unnecessary. They are a complication in the subsidy system than can be done away with. Farmers would have to study the market rather more than at present, but this would be desirable in the interests of better marketing. Once the grain was in the store, farmers would respond to the market signals far more readily if they had no incentive "steps" to think about. Moreover, there would then be no reason for deliveries of grain to be bunched at the beginning of each incentive period as they are now, with all the incon-venience this causes to the trade in general. Apart from simplifying the subsidy schemes the abandonment of seasonal incentives would ease the administrative burdens of the Agricultural Departments (the new wheat incentive scheme with its eight periods must call for still more clerical work than before), and would save both farmers and merchants a great deal of trouble. If the objection is raised that this would leave the cereal farmer exposed to the fluctuations of the market during the cereal year, our reply is that, in moderation, adjustments in response to such fluctuations are desirable for efficient marketing, and that our proposals below for a system of import levies will remove the risk of excessive price fluctuations brought about by low-priced imports. We would therefore like to see the seasonal incentives abandoned completely, not only for wheat but for all cereals.

4.9. This may be thought to be too radical a change. If so, at least the following modifications to the existing scheme would bring about some improvements in cereals marketing:

   (i) Remove the barley disincentive altogether since it acts as too heavy a deterrent on the sale of grain at and soon after harvest. This would mean abandoning the self-financing element of the scheme as it is at present.

  (ii) Start paying the barley incentive in October, thus bringing barley more into line with wheat.

 (iii) Reduce the spread of the wheat incentive to the barley level—the wheat incentive at present more than covers the variable costs of storage.

 (iv) Stop paying incentives after May—this should help clear the home crop before the new harvest is ready and thus prevent too great a carry-over by processors, which will depress prices in the following harvest.

3. STANDARD QUANTITIES AND TARGET INDICATOR PRICES

4.10. The 1964 Annual Review heralded the ending of the open-ended guarantee to which the Government had hitherto been committed. By fixing a standard quantity the Government both limited its own liability and signalled to producers as a whole its ideas of what constituted a desirable level of output. The standard quantity for wheat was abolished at the 1968 Annual Review. The standard quantity adjustment for barley caused the deficiency payment in 1966/7 to be reduced by about 17%. The Crop Adjustment Formula was introduced in 1964 to deal with the fluctuations in the acreage of winter wheat due to weather conditions, but this was also abolished at the 1968 Annual Review. The standard quantity for barley as a percentage of production has been as shown overleaf.

4.11. Another feature of the package of new arrangements for cereals introduced at the 1964 Annual Review was the Target Indicator Price. This was a device similar in its objectives and methods to the standard quantity

	Harvest			
	1964	1965	1966	1967
	('000 tons)			
Production of barley	7242	7840	9004	8773
Standard quantity[a]	6500	6750	7467	7948
S.Q. as % of production	89·8	86·1	82·9	90·6

[a] After adjustment under the crop adjustment formula.
Source: Hansard, 17th June 1968, and M.A.F.F.

but related to prices rather than production. Thus if the market is weak and the average market price is below the T.I.P., the deficiency payment is calculated on the difference between the T.I.P. and the guaranteed price, not on the difference between the average market price and the guaranteed price. The Target Indicator Price was 20s. 6d. for wheat and 19s. for barley for the 1966 and 1967 harvests, and in July 1968 these were raised to 21s. 6d. and 20s. 8d. respectively.

4.12. We would not criticise the concept of standard quantities, which, apart from their role in safeguarding the Exchequer, also have the objective of signalling overall policy intentions to producers. It is probable, however, that the latter function would be better served by changes in the guaranteed price, since these are likely to be a far more effective means of supply management, partly because they operate before a crop is grown whereas the effects of standard quantities are not felt until long after the event, and partly because they are more easily understood. Once again, we cannot fail to be impressed by the widespread complaints about the complexity of existing arrangements.

4.13. Hitherto the Target Indicator Prices have had little impact or meaning in the cereals market, as they were set well below normal realised price levels and served as a safety net to keep the Exchequer commitment in check in case of a serious collapse of the market. However, we are concerned at the much more significant role the T.I.P.s are likely to have now that they have been pitched at a level not much below the expected realised price. (This applies more to the barley T.I.P. than to that for wheat.) If farmers suspect that the average realised price at the end of the year will fall below the T.I.P. (with a consequent reduction in their deficiency payments) they will exert pressure on the H.-G.C.A. to intervene to support the market. Hitherto, the H.-G.C.A. has not been subjected to heavy pressure of this kind but with the higher T.I.P.s it will take only a relatively small fall in the market price to cause the farmers to urge intervention. This will make the Authority's task very difficult and may force it to change its market role from that of intervention only in support of a very depressed market (it has powers to do this but has not yet had to use them) to market intervention to counteract even a small fall in prices—a very different matter.

4.14. We do not regard this new role for the Target Indicator Prices as an appropriate one, and indeed it strikes at the root of the deficiency payments system. The great advantage of this system is that it allows the market to operate freely and efficiently, whilst assuring farmers of a guaranteed return. By fixing higher Target Indicator Prices the Government is diminishing the producers' confidence in the system and is likely to generate pressure for more intervention which could ultimately destroy the deficiency payments system.

4.15. If the Government wishes to alter the guarantees on cereals, which is virtually what the effect of a high T.I.P. would be (assuming the realised price is below the T.I.P.) the proper place for this is the Annual Review. The Target Indicator Price should either revert to its original function of providing a safety net for the Exchequer, or it should be abandoned altogether. If the proposals we outline later give an adequate assurance against a collapse of prices in the market, then even a safety-net T.I.P. is unnecessary and we would prefer to see it abandoned altogether.

4. MINIMUM IMPORT PRICES

4.16. One of the most important features of the 1964 package was the new system of minimum import prices covering straight and manufactured cereals. These minimum import prices remained unchanged until 1st August 1968 when they were increased to take account of devaluation. The minimum price for denatured wheat was increased from £20 10s. per ton to £22 19s., for European milling wheats from £22 10s. to £25 5s., for barley and oats from £20 to £21 19s. and for maize from £21 to £22 19s.

4.17. The minimum import prices have provided a means of preventing the more extreme fluctuations in price that occurred before 1964 (see Part VII), but they have been subjected to a great deal of criticism because the exporting country gets the benefit of the higher price whereas with a levy system (as operated by the E.E.C. for instance) the levy accrues to the Exchequer of the importing country. We regard the present minimum-import price system as basically a temporary and emergency device which, if applied promptly, could be useful over short periods. However, it is not a system acceptable in the long term and since some system of controlling the price of imported grain is likely to be necessary as a permanent feature of the U.K. grain market (because of the tendency to over-supply in world markets) we favour a change to a system of levies such as would provide for payments to be made to the Exchequer by importers to bring the price of grain landed in the U.K. up to a "floor" price. We discuss such a system in more detail later.

5. COMMITMENTS TO OVERSEAS SUPPLIERS

4.18. The 1964 package was designed in part to reassure our overseas suppliers that the U.K. was not aiming at self-sufficiency for cereals. This policy was directly acknowledged when the Government gave an undertaking at the 1964 Annual Review that cereal imports would not fall below the average of the three years to June 1964, and at the same time overseas suppliers were offered "the opportunity of securing a fair share in the growth of the market". In the event, imports tended to fall below the agreed level, but this has not caused much difficulty with our suppliers, who have been preoccupied with their changing trading relationships with other countries. These access agreements have now been dropped and we welcome this since it seems to us that in the rapidly shifting pattern of trade in the international cereals market, the idea of a fixed "quota" of imports has become unrealistic, and U.K. cereals policy cannot be dominated by considerations of this sort. The U.K. no longer accounts for the major share of the exports of Commonwealth countries. Table 2.2, Part VI shows, for instance, that the U.K.'s share of Australia's wheat exports dropped from 39% in 1952/3 to 4% in 1967. This changing pattern of international trade in cereals is being reflected in a less contractual relationship with our main suppliers. It is time to recognise that if the U.K. imports 200,000–300,000 tons less wheat from Canada or Australia this is not going to be of critical importance for those countries, whereas it might make all the difference to the stability of the U.K. cereals market.

6. MILLABLE WHEAT

4.19. In our view the fiction that all wheat entering the cereal trade in the U.K. must be "capable of being manufactured into a sound sweet flour fit for human consumption" should be dropped, since half at least of the wheat grown in the U.K. is now used for animal feed. The criterion does not, in fact, act as a barrier, since it is virtually ignored, but that itself is a good reason for abandoning the concept. If the wheat subsidy is put on to an acreage basis this issue will no longer arise, but if the tonnage system is retained then the way "millable wheat" is defined should be revised to correspond more nearly to current practice in the trade.

7. OPERATION OF THE FEED FORMULA

4.20. The guaranteed prices for pigs and eggs are linked to the prices of feeds. Thus if feed prices rise the guaranteed prices rise, and vice versa if the feed prices fall. The effect of this system is virtually to insulate these livestock producers from fluctuations in the prices of feed grains. This is a disadvantage when it comes to disposing of a large U.K. crop of barley, since if the price is lowered, and there is a consequent fall in the price of manufactured feeds, the guaranteed prices for pigs and eggs fall as well and there is no effective encouragement to pig and poultry producers to take up more feed. This system is, therefore, badly designed for the kind of situation likely to prevail in the U.K. cereal market in the years ahead, and we would like to see it amended so that only part of the change in the feed price is reflected in the guaranteed price, thus retaining some security for the producer whilst allowing scope for grain price reductions to encourage a higher take-up of grain.

CONCLUSION ON SYSTEMS OF PRICE SUPPORT

4.21. The deficiency payment system of price support has proved broadly satisfactory, and only for one period during the last 14 years (1961 to 1963) has it seemed to be going seriously wrong—and then it was the Exchequer, not the farmers, who bore the brunt. It has enabled the free market for grain in the U.K. to be maintained largely untrammelled by direct intervention by any government or quasi-government body, and in our view it has contributed very largely to the efficient grain market that the U.K. now has. These are not benefits to be lightly discarded, and indeed, we do not suggest that the deficiency payments scheme should be abandoned; rather, as indicated earlier, we would prefer to remove impediments to its proper functioning. We believe it should be possible to operate a flexible levy system on imported grain without the essentials of the deficiency payments scheme being changed at all. If, however, the U.K. joins the Common Market, then the deficiency payments scheme may have to be abandoned, and this would have to be accepted for the sake of the wider benefits to be obtained from our joining.

The Role of the Home-Grown Cereals Authority

4.22. The H.-G.C.A. was established in 1965 with the basic objective of "improving the marketing of home-grown cereals". It comprises 23 members, i.e. 5 independents, 9 farmers, and 9 merchants or users (compounders, millers and maltsters). The administrative costs of the Authority are borne equally by the Exchequer and by cereal-growing farmers.

4.23. The Authority grew out of the troubles of the early 1960's, and was one of the package of measures taken by the Government in 1964/5 to prevent a repetition of these events (the others included minimum import prices, standard quantities and target indicator prices). The main specific task it was given was the introduction of the Forward Contract Bonus

Scheme. The primary object of this Scheme was to enable home-grown grain to compete more effectively with imported grain by giving users a greater assurance of continuity of supply. The first scheme was introduced in 1965/6 and by 1966/7 up to 80% of marketings in some months were being made on forward contract. The money out of which the contract bonuses are paid is acquired by compulsory levy on cereal farmers.

4.24. The Authority's other non-trading powers and responsibilities include: operating a "bonus on delivery" scheme; making or guaranteeing loans on forward contracts; improving market intelligence; and research and development. This present Survey was commissioned by the Authority as part of its research and development responsibility.

4.25. The Authority also has reserve trading powers but it has not yet had occasion to use them. The circumstances in which the Authority may start trading, and those in which it would have to stop trading, are tightly controlled by Ministers, and if the Authority trades and is holding stocks of grain at the end of a cereal year (30th June) it must sell those stocks by 31st July immediately following.

The Authority's Non-trading Powers

THE FORWARD CONTRACT BONUS SCHEME

4.26. At first sight the willingness of farmers and merchants to participate in this scheme may be taken as a sign that they approve of it, but this would not be a correct inference. Many farmers participate because, if they do not, they lose the levy deducted from their subsidy payment; whilst merchants participate because they virtually have no choice if the farmers insist. A substantial proportion of farmers who participate dislike the scheme (page 142) whilst two-thirds of the merchants disapprove of it (page 261). As to the end-users, they are less directly involved, but it would seem that whereas some of them make use of the assurance of a forward delivery to reduce their stocks of grain, others find it more difficult than before to buy grain when they want it because the grain has been "locked up" on forward contract. This, together with the seasonal incentive arrangements, has contributed to the frequent dearth of grain in the two or three months following the harvest, and some millers and compounders have had to resort to imported grain during this period.

4.27. The assertion sometimes made that the relative stability of the grain market in the last three years is due almost entirely to the forward contract scheme is probably an exaggeration, changes in the international cereals market being of greater influence; but the scheme has certainly helped to maintain stability. On the other hand, there has been a cost. Not only has the scheme involved a lot of paper work, but it has introduced an undesirable rigidity into the cereal market and has made it difficult for supply always to respond rapidly to demand. In 1967/8 the Authority went some way to remedy this by reducing its rates of bonus by 20% to damp down the high level of forward contracting. For some reason, very little use has been made of the Type B contract which allows a little more flexibility, although the bonus is currently only 8s. per ton instead of 10s. for the three-month contract. The Authority recognise the need to keep a significant proportion of grain after harvest on the spot market (say 40%) and they aim to pitch their bonuses at a level to achieve this.

4.28. The publication of quantities forward contracted, and whether they are at open or fixed prices, strengthens the already strong position of the big cereal buyers. Whilst it may be difficult to prove one way or the other, some influential cereal growers certainly claim that the big buyers arrange their imports of European soft wheats to depress the U.K. market and so reduce the price at which grain on "open price" is eventually sold. The Forward Contract Scheme as at present arranged seems to favour the user more than the seller, although there is very little evidence that in itself it has encouraged cereal users to take home-grown grain rather than imported grain.

4.29. On balance we feel that the Scheme is worth continuing, but on a diminishing scale as a statutory scheme, the object being that it would be replaced in due course by voluntary forward contracts negotiated independently between buyers and sellers. The Authority might still have a role to play as an "Umpire" in these contracts—e.g. the parties might agree to refer questions of breach of contract to it for arbitration or advice—but the present levy and bonus system would eventually be abandoned. We certainly would not wish this recommendation to be taken as meaning that we do not value forward contracting as a system for selling grain; very much the reverse. The present scheme has served a very useful role in introducing farmers generally to the system of forward contracting (although some farmers were selling grain on forward contract before the scheme started). However, it is by nature inflexible, and having become familiar with the idea of forward contracting, farmers, merchants and end-users should be left free to decide for themselves what kind of contracts to make.

THE GUIDE PRICE

4.30. At the time when it was introduced the Guide Price filled an important need. There was a serious lack of storage capacity and the increasing use of combines was bringing more grain on to the market at harvest time than the market required, with the inevitable result that prices at that time tended to be abnormally low. The Guide Price showed what the prices of competitive imported grains were, and usually these were considerably higher at harvest time than the domestic levels. Thus it was brought home to farmers that if they persisted in flooding the market at harvest time they would receive a lower price than if they withheld supplies until demand was stronger. The Guide Price takes its place alongside the Seasonal Incentives, the Farm Improvement Scheme and, more recently, the Forward Contract Scheme as one of a group of measures which, taken together, have succeeded in persuading farmers to phase the deliveries of their grain more evenly over the year.

4.31. However, circumstances have changed gradually since the Guide Price was first introduced and it is now questionable whether it is needed—at least in its present form. These changes are as follows:

(a) Storage capacity on farms has been vastly increased since the mid-fifties and is now adequate to cover the major part of the crop. Sometimes lack of storage is still a reason why farmers sell in the autumn but usually the farmers concerned have made a conscious decision not to take advantage of the storage grants available, i.e. it suits their particular farming systems to sell at harvest time despite the lower prices then.

(b) There is better information available to cereal growers than there once was. For instance, the H.-G.C.A.'s own weekly bulletin, which is reproduced in part in the weekly farming press, provides an excellent market intelligence service that was not available in the mid-fifties. Moreover, cereal farmers tend to be larger and more price-conscious sellers than they were, and are better able to negotiate a reasonable price with their merchants.

(c) The problem of estimating what the Guide Prices should be has been exacerbated by the changes taking place within the industry. Thus the U.K. no longer imports any substantial quantities of barley for animal feeding, whilst an increasing proportion of wheat production goes for animal feeding rather than flour milling. It is not altogether realistic to relate the U.K. barley Guide Price to the prices of imported maize, since maize and barley are not always directly substitutable. Nor would it be adequate to take the export price of barley, as some would advocate, since so many special factors apply in the export business, which in any case tends to be confined to the areas nearest the coast.

4.32. It seems to us that in the circumstances of today the friction generated among those engaged in the cereals market by the Guide Prices in their present form is likely to exceed whatever benefits the system can still achieve. The truth is that no accurate forecast can be made of what will be the "proper" price for grain in the near future because no-one knows in advance what will be the factors influencing the market, e.g. how much will be sold on the export market, to what extent compounders will use barley rather than maize, or whether substantial quantities of Canadian wheat will find an outlet in the U.S.S.R. or other countries. When a Guide Price is issued in July for the forthcoming harvest, this is necessarily very much a guess because the amount of the harvest and its quality are not closely predictable. No doubt the Guide Price is a highly informed guess but it is still no more than that, and its authors have never made extravagant claims for it. The trouble is that farmers seldom interpret the Guide Price in this spirit; they tend to quote the Guide Price as an authoritative statement of what they should receive for their grain and they will often hold out for this price. This may be one of the factors contributing towards the relative dearth of grain on the market during the late autumn, because the Guide Price in the period following the harvest is invariably higher than the market price (Part VII, p. 626).

4.33. In the last analysis, prices are determined by supply and demand and a Guide Price can have only a marginal effect, e.g. by possibly causing some farmers to defer their sales from one period to another. Most farmers today are adequately equipped to decide when to sell without the assistance of the Guide Prices in their present form. We suggest therefore that the Guide Prices be given a rather different character. There would be a good case for including them (in the form of realistic price ranges rather than as exact single figures) in a comprehensive discussion about likely price levels and market trends which might form part of a regular "Market Situation and Outlook" statement by the Authority. These statements might also be supplemented by one or two Market Outlook Conferences held at and soon after the harvest. There will always be a need for market guidance in the weeks before the harvest, but what is required is more an open and expert discussion of market prospects than a schedule of prices; there would be no need to imply that farmers should "hold out" for any of the prices discussed in the statement. In this way the "Guide Price" could be absorbed into the Authority's on-going function of supplying market intelligence and informed comment on market prospects, and one would hope that it might lose the quasi-political overtones it has unfortunately acquired over the years.

RESEARCH AND DEVELOPMENT

4.34. The Authority has an important role in the general field of research and development and in our view this could well be extended. We have not made a detailed study of the research field but in the course of our enquiries we became aware of important gaps in knowledge in the broad field of cereals marketing and these gaps the Authority is best equipped to fill. Moreover, it can look at problems of production and utilisation from a marketing point of view. In this context we are astonished that at present the Authority is barred from conducting research in matters that appear to relate to production more than marketing. Marketing problems and decisions begin with production, and for the Authority to be inhibited from investigating such problems as the breeding of wheat and barley for particular uses or with particular qualities (or indeed of maize varieties suitable for the U.K. climate) seems to be an absurd frustration. A break-through on this front could do more than any specifically "marketing" research to bring about great changes in the U.K. cereals market. There is a whole field of applied research which the Authority should be covering, and it should have the funds to stimulate the kind of pioneering studies that few cereal growers, merchants, or small users could, or would, finance from their own resources. The Authority should act as a "power-house" collecting and disseminating information and ideas on all aspects of the cereals industry. We trust that the present Report, for example, will not be conceived as a once-for-all study, like that of a Royal Commission, but rather as a springboard for a continuing process of investigation and informed debate.

4.35. There is one development in particular in which the Authority might well take the lead. We have already indicated in Part III, Chapter 4, that we support the views expressed by the great majority of merchants that a statutory grading scheme for cereals is impracticable in the U.K. at present. But this is not to say that progress cannot be made on a voluntary system of classification of grain according to its principal characteristics. There is a great deal of research needing to be done to establish what are the characteristics that are required by the various end-users, and how they can best be defined and measured. In due course, once an agreed list of "specifications" has been published, it is possible that the larger grain sellers, including the producer marketing groups and the co-operatives, may adopt them as a basis of trading by specification rather than by sample. This is a development the Authority should foster by every means in its power. The more cereal producers are acquainted with the specific requirements of end-users, and begin to adapt their production and marketing processes with this end in view, the better placed home-produced grain will be to compete with imported.

4.36. We would recommend that the considerable funds needed to support this enlarged research and development role (we envisage expenditure of up to £$\frac{1}{2}$ million per annum) should be made available by the

Government direct and not by levy, since the programme should not be orientated towards any one section only (i.e. the cereal growers if they provided the levy income) but towards all the sectors in the grain trade. Moreover, the money might need to be focused towards particular individuals (e.g. merchants experimenting with new bulk containers or farmers trying out new varieties), whereas if the funds were raised from cereal growers only it might be difficult to secure agreement on their disposal in this way.[1] A large part of the money might be used to finance innovations or pioneering techniques, i.e. we envisage an "AMDEC" or "Central Council" function but specifically for the cereals industry and with the information so obtained being disseminated within the industry by the Authority. Typical fields in which we feel more research is needed are:

> productivity studies in merchanting,
> cereals classification and grading projects,
> centralised storage and drying facilities,
> marketing of break crops,
> vertical integration and contract farming,
> role of market intelligence,
> palletisation and transport rationalisation,
> economics of form-mixing,
> development of export facilities,
> business management techniques.

4.37. There may also be a case for an extension of the Authority's statistical and market intelligence functions, but as indicated in Part III Chapter 4, this is a field in which more research is needed. Already the Authority has provided a valuable core of statistical knowledge, and its market reporting system is widely used. It might improve its market reporting service if it were allowed to take over the collection of Corn Returns from the Ministry of Agriculture, whilst it would greatly improve the statistical background to cereals marketing if it could persuade the Ministry of Agriculture to improve both the present crude system of estimating yields at harvest time, and its estimate of stocks of grain on farms during the year.

4.38. The Authority has an important role in influencing national policy towards cereals production and marketing. Necessarily this function must go on largely unacknowledged, and even unrealised by many, but it is one of the most important of the Authority's functions. On many occasions the

---

[1] We feel it is a weakness of the way the Authority is at present financed that only the farmers and the Government contribute to the administrative costs; the merchants and end-users have no financial stake.

Authority makes representations to the Ministry when it feels that the cereals market is likely to be adversely affected by policy decisions—or the lack of them. In the nature of things, however, much of this work has to go on behind the scenes.

4.39. The Authority may have a future role to play in facilitating the provision of capital (through the banks and other established agencies) for medium- to long-term investment, e.g. by agricultural merchants to improve marketing facilities. Examples are: new hangar-type buildings for pallet storage and handling, blower-equipped bulk lorries, wharfage facilities for imports and exports, and containers for transporting bulk grain. Such assistance could take the form, for instance, of feasibility studies.

4.40. Finally, in this review of the Authority's non-trading functions, it has been suggested by many people that the Authority should be given the responsibility for administering all the cereal subsidy schemes, subject to the over-riding control of the Ministry through the Annual Review. The main object of making this change would be that all payments for cereals, e.g. seasonal incentives as well as forward contract bonuses, would emanate from the one source. This would simplify matters from the farmer's point of view and ultimately might enable the seasonal incentives to be dove-tailed into the forward contract bonus scheme, since both are basically aiming at the same thing. If our earliet recommendations as to the abolition of the seasonal incentives and eventually of the forward contract bonuses are accepted, we see no point in making this change, since it would be administratively simpler for the Ministry to continue making the deficiency payments as part of its wider responsibility for a range of payments to farmers. However, if the seasonal incentives are retained, there is something to be said for the Authority becoming responsible for their adminis-tration, but only if this carries with it not merely the responsibility for making the payments but also the responsibility for deciding at what level the seasonal incentives should be pitched. Indeed, this latter function is the key one, and if it is administratively cheaper for the Ministry to make the payments it is more sensible for them to continue to do so.

4.41. We have discussed so far only those functions that the Authority is at present exercising, or could take over from the Ministry, but there are many other functions that the Authority could assume as pait of its overall responsibility for efficient marketing. For instance, it could use the powers it already has to introduce a "Bonus on Delivery" Scheme of its own, or it might initiate a scheme for assisted exports to dispose of surpluses and to avoid having to fall back on its trading powers. It might undertake nego-tiations with overseas buyers on behalf of grain exporters where this was desirable (e.g. with centrally controlled economies). The point to remem-ber is that the Authority has been in operation for only three years and it must not be assumed that the present pattern of its activities is necessarily the pattern for the future. It has been preoccupied with establishing the operation of its basic functions, including the contract bonus scheme, but in future it should be able to direct its efforts towards a wider and ultimately more promising range of possibilities.

The Authority's Trading Powers

4.42. Important as they are, we realise that the non-trading powers of the Authority are of less immediate interest to farmers and merchants and others in the grain trade, than are its trading powers—actual or potential. Indeed, it is by its effectiveness in maintaining a stable cereals market that the Authority must ultimately stand or fall. We do not dissent from the commonly accepted attitude, and we suggest ways in which the Authority would be empowered to sustain a stable cereals market.

4.43. It is clear, from the very circumscribed trading powers given to the Authority in 1965, that the Government did not envisage the Authority intervening in the cereals market, except in an acute emergency. And by insisting that all grain held by the Authority should be sold by 31st July, the Government ensured that it would not become a "carry-over" body comparable in any way with the Commodity Credit Corporation in the U.S.A. If it intervened at all, it could be only to give a very short-term boost to the market.

4.44. These trading powers are virtually useless. In the circumstances pertaining in the U.K. today, with total storage capacity probably adequate to accommodate most of the crop needing to be stored at harvest, the most likely glut period is no longer at harvest but in the spring. But the Authority is powerless to intervene effectively in the spring if everyone knows that it must dispose of its stocks by 31st July. Moreover, under the strict regulations governing the Authority's decision as to when to intervene, it is almost certain that rather than intervening to *prevent* a price collapse, the Authority would be intervening *after* the collapse had occurred, so that its losses on the transaction would be very great indeed—and that means the farmers' losses, since they would be recovered by a levy on cereal producers.

4.45. If the Authority is to intervene effectively it must have more flexibility as to when to intervene; and it must, if need be, have the power to carry grain over from one season to another. We are not suggesting that the Authority should be encouraged to intervene more often (it has not had to intervene at all as yet), but simply that if everyone knows its intervention powers are virtually useless, then it ceases to be effective at all. If its intervention powers were realistic, the mere threat of intervention by the Authority would probably often be sufficient to save the price from collapsing altogether.

THE CASE FOR A CEREALS MARKETING BOARD

4.46. In our discussions with cereal growers we have been left in no doubt that they do not have much faith in the Authority as it is at present constituted. They say that it has only the shadow of authority over the cereal market, not the substance—it lacks "teeth".

4.47. Many of the cereal producers we have met feel keenly that the farmer is at a complete disadvantage as a very small seller compared with the big buyers, the national compounders, millers and maltsters. And now that some merchants are owned by these same firms (Ranks Hovis McDougall alone handled $1\frac{1}{4}$ million tons of home-grown grain through their own merchant businesses in 1967)[1] some of them feel even more at a disadvantage. It tends to be the larger growers who feel this way, since the small farmer is happy to sell through his merchant in the time-honoured fashion; he knows that his quantity of grain is altogether insignificant, and it is often, in any case, only a minor part of the output of his business. Thus it is probably the top 10,000 or so cereal growers who particularly want some strengthening of their position *vis-à-vis* the buyers. But apart from this vague feeling of unease, there is no clear-cut view as to what is wanted. Usually what most cereal farmers seem to want is an assured price for their crop and an escape from the hazards of having to guess how the market will behave. As indicated in Part VII, they can lose up to 50% of the profit on the crop if they make a serious mistake in selling at a period when heavy imports are depressing prices, and it is this kind of risk they want to avoid. Some farmers are joining the producer marketing groups that sell the grain of all the members and "pool" the revenues so that the risks are shared. Presumably a Cereals Marketing Board would operate like one of these groups but on a national scale, and all growers (over a minimum cereals acreage) would have to sell their grain to, or at least through, the Board. This would not necessarily mean the end of the present farmer/merchant relationship if merchants acted as agents of the Board (as in Ontario, Canada) but it would completely change the nature of the relationship, since the merchant would no longer be acting as a principal but as an agent on a commission basis.

4.48. We do not believe that the majority of cereal growers want a compulsory Cereals Marketing Board of this kind. Many growers, in fact, confess that they like the element of chance involved in selling grain and would not like to see it disappear altogether. Nor is it likely that such a Board would be as efficient as the present system, which is extremely flexible and which has proved its great adaptability in recent years. Nor do we believe that such a Board would be able to secure higher prices for

[1] Chairman's Review of 1967—*Annual General Report 1967*, published January 1968.

growers unless it could also control imports. But no Government of this country is likely to allow a producer-controlled marketing board to possess such powers so long as a substantial proportion of the nation's food is imported. In our view what most cereal producers want is not a highly controlled and rigid selling organisation but an assurance that unrestricted imports are not going to knock the bottom out of the cereals market and drag prices down to levels that entail a high Exchequer subsidy and with it the likelihood of reduced guarantees in the future.

THE CASE FOR GREATER CONTROL OF IMPORTS

4.49. We have given much thought to the question of controlling imports, and we feel that the cereal growers have a good case for a greater but not an absolute influence to be exercised over imports by a marketing authority. The experience of the early part of 1968, with some French grain being offered at a price less than half the price prevailing in France and dragging U.K. grain prices down to a low level (see Part VII), is surely the negation of efficient marketing. If the Authority had been in a position of real power it would have prevented these imports (it made representations to the Government and did succeed in getting French restitutions slightly reduced), so that they did not adversely affect the disposal of the U.K. crop. We have already proposed a variable import levy system for cereal imports, and now we suggest that the Authority should have a considerable share in the decision as to when and how these should be imposed. The Government is unlikely to surrender to the H.-G.C.A. its overall responsibility in negotiations with other countries (despite the fact that the H.-G.C.A. is representative of all sections of the industry) but it can certainly give a far more important role to the Authority in this field than it has done hitherto. It is no more possible to keep "home-grown cereals" in a separate pigeon-hole than it is to keep marketing separate from production. To retain the confidence of the producers the Authority needs "teeth" such as a degree of influence over the import levies would provide. It would be the responsibility of the Government and the non-farming representatives on the Authority to ensure that the powers were used only to promote good cereals marketing. The farming community as a whole would have no direct incentive under the deficiency payments scheme to urge the use of these powers to raise domestic prices once they were above the target indicator prices (if they continue to exist), since the Exchequer makes up the deficiency in any case.

4.50. There are some who would go further than merely replacing the minimum import prices with variable levies; they would favour changing from the deficiency payments system to the complete apparatus of the

Common Market, with threshold prices, indicative prices, intervention prices, and restitutions. We do not favour this approach for several reasons. If levies are to be set high enough to provide the funds needed to sustain farmers' incomes, this would deny to the livestock farmers in the U.K. the benefits of relatively low priced imports, and it would greatly upset our trading relationships. Then again, there would have to be a great deal of intervention buying to support prices at these high levels. Lastly, there is the simple argument that if a method is working reasonably well, why switch to another and untried one unless it promises a worthwhile reward or is part of a more comprehensive re-arrangement (e.g. entry into the Common Market)? We therefore favour retaining the present deficiency payments scheme whilst using import levies merely to put a floor in the market.

4.51. Some might object that a system of variable levies is contrary to the provisions of G.A.T.T., but our answer would be that the proposed levies are in effect little different from the present minimum import prices— only they can be applied more sensitively. In any case other European countries have introduced schemes involving much greater control of imports, and G.A.T.T. has accepted these.

Conclusions

4.52. The foregoing suggestions may be summarised as follows:

(a) Retain the present Deficiency Payments Scheme as the basic means of price support.

(b) Replace minimum import prices with a levy system. The prices to which the levies would relate would generally be pitched at such a level that in a normal year most imports would enter the U.K. at prices above this level, i.e. levies would be payable on only a small proportion of total imports during that part of the year that the import price fell below the "floor" level. In a year of low U.K. grain production the levy system would probably be inoperative for most of the year, but in a year of high U.K. output it would "bite" for some part of the year depending upon the level of import prices.

(c) The "floor" level prices would be announced at the beginning of the cereal season, i.e. at harvest time and would normally remain unchanged. However, if it became apparent that they had been pitched too high or too low in relation to the requirements of the U.K. market they could be adjusted.

(d) Prices of imported grains seldom have a strong seasonal element in them—as the U.K. prices have—because the marketing agencies in the exporting countries average out the storage costs over their

Y

operations as a whole. Since it is important that the market price in the U.K. should yield a return to cover the costs of storage it would be necessary to quote a "stepped" floor price for the season, the price rising in (say) four steps of about 10*s*. per cwt each giving a spread of £2 over the season as a whole.

(e) If these proposals proved successful the more extreme fluctuations in price would be eliminated, and in virtually all years the market price would yield some return for storage. Thus growers would no longer require the elaborate and complex systems which have been devised over the years to give them (and the Exchequer) a measure of security in the marketing of their crops. Our proposals make possible a dismantling of these arrangements and a simplification of the whole cereals subsidy scheme, which has become altogether too complicated. We envisage the early abandonment of seasonal incentives and seasonal guaranteed prices, target indicator prices, and forward contract bonuses. The merchanting sector is highly competitive and the restraints imposed by schemes of this kind can be an impediment to the efficient operation of the market. With these restraints removed, producers could still be left with the basic protection of the variable levy and the "stepped" floor price and these should give them adequate confidence.

(f) The Home-Grown Cereals Authority would have a key part to play, with the Government, in helping to fix the level of the "floor" price and keeping the situation constantly under review. It would also have a role in improving the commercial efficiency of the grain market (through its market outlook and information activities) and in improving technical efficiency through its powers to grant-aid promising innovations and to initiate research.

PART IX

Summary and Conclusions

CHAPTER 1

Summary and Conclusions

1. The findings of the Survey are preceded in **Part I** by a discussion of the principles of efficient marketing. Although the title of this Report refers to production, marketing and utilisation these are not thought of as wholly separate activities but are closely linked together in the whole operation of matching supply to demand. Marketing is no less "productive" than farming or manufacturing if it facilitates the movement of grain of the right kind to the right place at the right time. A marketing system is efficient if it achieves this with the least possible waste of resources. To the extent that there is inefficiency, this may be due to low standards of performance within the existing market structure or to defects in the structure itself.

2. Price fluctuations are sometimes regarded as symptoms of an inefficient marketing system; but some movement in prices must be a normal feature in a market in which supply and demand cannot be accurately predicted. The desire for "orderly marketing" must therefore be accommodated to the realities of an ever-changing situation. Stabilisation is a worthy objective but it should not be pursued to the point where rigidity prevents necessary adjustment.

3. **Part II** describes the present situation and trends in cereal production in the United Kingdom, and gives particular attention to those features which have special relevance for marketing.

4. The outstanding feature of cereals production in the past ten years has been the rapid expansion of the acreage of barley, from 2 million to 6 million acres. To some extent this was offset by the continuing decline in oats, but total cereals expanded by about 2 million acres. This expansion was accompanied by a significant increase in yields per acre, so that total cereals production reached $14\frac{1}{2}$ million tons in 1967 compared with 8 million tons in 1957. The technical and economic reasons for this growth in production are examined in detail, and the question arises: will this rate of growth continue in the years to come?

5. The opinion of the Advisory Services was sought on this question in a nation-wide survey. The results suggested that both acreage and yield

will continue to increase but at a slower rate than in the past decade. For the 10-year period 1967 to 1977 an increase of some 3·2 million tons of cereals (wheat, barley and oats) is forecast—assuming that the United Kingdom remains outside the Common Market—compared with the increase of 6·4 million tons in the previous decade. Wheat is expected to increase rather more than barley in percentage terms.

6. Even this more modest rate of growth in production could lead to important marketing problems, as there are already some pressures on the market and difficulty has been experienced in finding suitable outlets for barley in some years.

7. In 1967 there were about 175,000 growers of cereals in the United Kingdom. Many of these grow only a small acreage. For instance, in England and Wales over half the growers have less than 30 acres of cereals each. On the other hand, 10,000 farmers grow more than 200 acres each and these account for nearly half of total production. Marketing problems have different aspects according to whether one is considering the many small producers—for whom cereals are seldom an important enterprise—or the relatively few large producers to whom grain marketing is of vital concern.

8. A study of trends shows that there is a marked tendency for production to be concentrated in fewer hands. The number of growers has been falling while the acreage per grower has risen steadily. In 1954 the "top" 30,000 growers accounted for 60% of the acreage in England and Wales; by 1967 they covered more than 75%.

9. Two randomly selected samples of 600 cereal farmers each were interviewed in July/August 1967 and October/November 1967 respectively, and a further sample of 202 cereal farmers in selected areas were contacted on five different occasions in order to obtain information about their production and marketing. Tables are presented to show some of the characteristics of cereal farmers, their choice of varieties of each kind of cereal, factors influencing the yields they obtained, their methods of estimating yields, their decisions to grow more or less cereals than previously and their attitudes to cereal growing in the future. On the whole the larger farmers were more inclined to expect an increase than a decrease in their cereals acreage, while the reverse was the case for the smaller farmers. This confirms the trend towards concentration already noted.

10. The importance of the influence of weather on cereals production in the United Kingdom is noted. The inverse relation between autumn rainfall and autumn sowing is evident, and has implications for marketing because it affects the relative proportions of wheat and barley and the average yield of wheat. It is also noted that variations in yield from year to year have usually been in the opposite direction to variations in rainfall during the summer ripening period.

11. Because of the increase in yields, average gross returns per acre of wheat increased from £38 to £41 in the past ten years, although prices were generally falling. For barley the corresponding increase was from £32 to £36 per acre. Mechanisation has been a major factor in cost reduction, and this in return has accelerated the process of concentration into larger units of production. Farmers have been able to maintain a fairly constant relationship between costs and returns, though the profitability of both wheat and barley appears to have reached a peak in 1962/3.

12. Yield is still undoubtedly the major determinant of profitability in any individual year, and yield fluctuations are of crucial importance for marketing. In the last decade, year-to-year fluctuations in total United Kingdom cereals production due to weather alone have averaged about half a million tons either way. This indicates the need for considerable flexibility in the marketing system, and research into the causes and means of combating yield variation is desirable.

13. With the spread of combine harvesting, much larger quantities of threshed grain are available at harvest time, and the farmer is in a weak selling position unless he can store his grain. Normally it comes off the combine too moist for safe storage, and some drying or other method of preservation is necessary. Furthermore, larger combines and fewer men on farms means that provision must be made for handling grain in bulk.

14. An investigation was made into the characteristics of the various systems of drying and storage, and their capital and running costs. Capital costs vary considerably between the different systems, but the average is just over £14 per ton, for 450 tons capacity. Estimated total costs under the different systems range between 16*s.* and 49*s.* per ton stored per annum. The chilling system generally appears to offer low costs but it cannot be widely adopted unless marketing arrangements are changed.

15. The great majority of farmers in all parts of the country acknowledge that the grain merchant serves a useful function, and many consider that any great reduction in the number of merchants would be to their disadvantage. They appreciate the reciprocal trading services offered by merchants.

16. Many of the farmers interviewed in 1967 said that they had not heard of the Home-Grown Cereals Authority, but these were mainly farmers growing less than 100 acres of cereals. Of those who were aware of the Authority, 9% regarded its formation as highly necessary and desirable and 46% as quite necessary and desirable; 32% felt it was not really needed and 13% said that it was totally undesirable. Favourable comments were more widespread among farmers who had sold grain on forward contract. Eighty-three per cent of those whose cereal acreages exceeded 300 acres claimed to have registered a forward contract with the Home-Grown Cereals Authority at some time.

17. Just over half of the cereal farmers interviewed seemed to be satisfied with the Cereal Deficiency Payments system. An appreciable number wanted to see higher rates of subsidy, but there were others who favoured abolishing subsidies altogether. However, there appeared to be a widespread lack of knowledge or understanding of the details of the system.

18. Considerable dissatisfaction was expressed by many of the larger farmers regarding the extent to which the price they receive for their grain reflects its quality; but most of these did not think that a grading scheme would help.

19. **Part III** is concerned with the merchant sector of the marketing system. After the problem of defining a merchant business had been tackled, a survey was undertaken in the summer of 1967. Some 260 interviews took place with a random sample of businesses from an estimated "population" of 2050 in the United Kingdom.

20. The businesses in the sample were classified into subsidiaries of national companies (25), co-operatives (20) and independent firms (215). Of these independent firms, 42 had links with another firm or firms (by ownership, management or financial arrangements). About half the merchants in the survey were "general merchants", but among the rest it was possible to distinguish those whose business was mainly in grain or mainly in feed, and those who were mainly concerned with selling compound feeds which they had manufactured themselves. Much of the subsequent analysis takes account of these classifications.

21. The size of merchant businesses was examined in terms of turnover, labour force and purchases of grain. There was a wide range of turnover, with the majority of firms having a turnover of less than £500,000 a year, but with firms of £1 million turnover or more accounting for nearly 60% of the total turnover of the sample while they represented only 14% in number. In national terms this suggests that under 300 firms account for around 60% of the trade of the merchant sector.

22. In terms of turnover, the relative importance of co-operatives and of subsidiaries of national companies far exceeds their numerical ratio to the total number of merchant businesses.

23. As regards purchases of grain (as distinct from total turnover), concentration has proceeded to the point where one-third of the firms account for nine-tenths of total purchases. This would represent about 700 firms in the United Kingdom as a whole.

24. The diversification of businesses is analysed, details being given of other agricultural activities besides dealing in grain, and of non-agricultural activities (garden sundries, coal, builders' materials, etc.).

25. The dispersed geographical distribution of merchant businesses means that most farmers have a choice between several merchants to deal with. More than half the merchants in the sample traded within a radius of

20 miles from base. This keeps down transport costs in relation to volume of trade. However, independent firms work in smaller trading areas than do subsidiaries of national companies.

26. The study of the structure of the merchant sector reveals the strong element of family control and the relatively low numbers of hired managerial and executive staff.

27. The smallness of many businesses precludes storage in quantities likely to achieve economies of scale. If more merchants were to undertake extensive storage operations, the structure and capital base of the sector would have to be changed.

28. The main changes taking place in the structure of the merchant sector are the continuing decline in the number of merchants (there was a net disappearance of one-third of the membership of N.A.C.A.M. between 1947 and 1967); the tendency for farmers to establish trading groups; and the growing concentration of the trade, associated with the takeover of merchant businesses by national manufacturers of animal feedingstuffs.

29. Changes in technology have tended to increase economies of scale in the merchant sector. Improvements in transport include bulk movement of grain in specially-designed lorries and improved handling of grain. They require considerable investment and this increases the pressures of competition. Another factor is the greater complexity prevailing in the manufacture and use of compound feeds.

30. Merchants expressed varying attitudes to farmers' trading groups. Most were prepared to deal with groups on the same terms as with any other customer; some were resolutely opposed to group trading; while a few had promoted groups. The larger the merchant's business, the more common it was to buy from farmers' groups. The prospects for extension of group trading are discussed.

31. It is suggested that the reduction in the number of independent merchants (as national firms take them over or come to special terms with them) will reinforce the already strong tendency of the smaller independent firms to follow prices set by the national companies.

32. At first sight the size structure of the merchant sector may appear to be incompatible with the needed development of new techniques, services and skills. However, many small businesses have considerable capacity to survive because of their specialisation of functions, their intimate knowledge of their customers, their flexibility in handling grain (assisted by a considerable volume of transactions between merchants) and the possibility of spreading overhead costs over several activities, enhanced by the reciprocal nature of the agricultural merchant's trade.

33. The study of the problems of the internal transport of grain includes an analysis of grain-surplus and grain-deficit areas of the country and the

Y *

resultant flows of grain. The relative advantages for merchants to own or to hire lorries are discussed. Most merchants make use of some hired transport; frequently they own enough lorries to deal with demand during the slackest period and use hired transport for the rest. Availability of return loads is an important factor in long-distance operations.

34. Many merchants do not know their transport costs at all accurately and it is suggested that more attention should be given to this.

35. Some of the inefficiencies in the transport system are caused by inadequate facilities for collection and delivery on farms, particularly on the smaller farms. Some merchants are considering imposing penalties for unreasonable delay at farms. It is suggested that if this is done it should not be absorbed in the price but should be a specific charge, so that the farmer may be encouraged to improve his facilities.

36. Among questions studied in the Survey of Merchants was the extent to which merchants, when buying grain from farmers, vary the purchase price between one farmer and another. About one-third of the firms interviewed made no such variation, but among the rest prices were varied for large buyers of feeds, for customer loyalty, where the customer owes money, to attract new customers, etc. However, these price variations were generally not more than 10*s.* per ton.

37. Merchants generally are not in favour of introducing a statutory grading scheme for grain, though a minority of one in three thought such a scheme would be desirable. It is pointed out that although such schemes work successfully in the U.S.A., Canada and other countries, growing conditions are vastly different in the United Kingdom.

38. When merchants were asked to comment on the Home-Grown Cereals Authority's Forward Contract Scheme, adverse comments outnumbered favourable comments by almost two to one. The most common complaint was that the Scheme created too much paperwork and other administrative tasks for the merchants.

39. The frequency and volume of grain sales between merchants are examined. Generally, grain moves between merchants only when temporary stresses of demand or supply cause it to do so.

40. On price determination, the survey showed that most merchants when selling grain to end-users had usually to take or leave the price quoted by the buying firm. However, the large merchant firms reported quite frequently that they had scope for some price negotiation. The level of the import price was considered to be the main determinant of the feed grain price near the main port centres. Sales to other merchants or to farmers leave more room for bargaining; but knowledge of the "going price" is spread through the market by a sensitive system. More importance is attached to personal contact by telephone and at markets than to published market

intelligence. To increase the *amount* of official market intelligence might therefore bring no perceptible improvement in the efficiency of marketing; but further research could probably indicate worthwhile changes in the *kind* of market information supplied.

41. An analysis of the management problems of merchant businesses and of their profitability concludes that the pressure on poor-quality management is very real and that the economic environment has become more rigorous in the merchanting sector.

42. A special study of the economics of centralised group storage suggests that on-farm and centralised storage are fairly evenly balanced in economic terms, but that centralised storage appears to have a slight advantage. It is therefore suggested that government subsidies to storage should not be confined to the farm. Each proposal for central storage arrangements should be considered on its merits, with due regard to local requirements and conditions. If the number of centralised stores operated by farmers became substantial the farmers might for the first time have some collective strength as sellers of grain. Such stores might also facilitate the operation of a system of season-to-season carryover of grain which might in some circumstances be necessary; and they could facilitate exports.

43. **Part IV** is concerned with the firms and institutions which handle the importing and exporting of grain. "Shippers", despite their name, are primarily international traders in grain and only secondarily movers of it. Traditionally shippers in the United Kingdom have been mainly engaged in selling imported grain to the large port-based millers and compounders and the port merchants. More recently they have taken an interest in the United Kingdom export trade and this has brought them into closer touch with corn merchants. Another recent development has been the buying up of some agricultural merchant businesses by the big shippers, and the reasons for this move are discussed.

44. A number of factors are analysed which have brought about the concentration of importing into the hands of a very few large firms; but it appears to remain a highly competitive market.

45. In a section on the cereals futures market in the United Kingdom the conclusion is reached that there is considerable potential for the use of this market—particularly by merchants—as insurance against price risk, and that a programme of information and education is needed to demonstrate the system and its benefits. It is also suggested that much more comprehensive and detailed statistics about the operation of the futures markets should be produced.

46. A section on the United Kingdom grain export trade considers the problem of deciding whether heavy capital investment should be made to

"tool up" for a permanent trade of this kind (wharf-side silos, bulk-handling equipment, larger loading berths, etc.). The likelihood is that some such investment will take place on a limited scale at certain ports. Demands for intervention to discourage shippers from competing with each other to sell United Kingdom grain abroad are not viewed with favour.

47. **Part V** deals with the utilisation of grain in the animal feedingstuffs industry, flour milling, malting and other industrial uses.

48. Cereals represent a large part of the concentrated feedingstuffs used in the United Kingdom, and demand has increased with livestock numbers. The expansion of poultry numbers has had a particularly marked effect on supplies of manufactured compound feeds.

49. Changes in the distribution of livestock between the different regions in the United Kingdom have affected the marketing pattern. The eastern port areas, around London and Hull, have shown larger increases in demand than the other areas of Great Britain, because of their rapid expansion of pigs and poultry numbers. This eastward shift of feed requirements may continue.

50. Total compound production shows a marked seasonal pattern, being highest in the winter months of October and March (though for poultry feeds there is a trough in the January–March period). Relating this pattern to the date of harvesting the United Kingdom crop it is clear that the amount of storage required is less than would have to be provided if demand were constant throughout the year. This seasonal demand creates serious problems of under-utilisation of capacity for the compound manufacturing industry.

51. The feedingstuffs industry provides by far the largest outlet for home-grown cereals. Imported wheat and barley have been increasingly replaced by home-grown supplies, and now imported maize also is being partly replaced by domestic grain. The patterns of utilisation change rapidly from month to month, with variations in the size of the home crop, the availability and price of imports and the changes in the export trade. Despite technical constraints on substitution within compounds, the feedingstuffs industry has maintained considerable flexibility in its pattern of utilisation.

52. The number of factories producing concentrated feedingstuffs has fallen by about 30% in the past 15 years, while the total quantity manufactured has substantially increased. The traditional distinction between "port" and "country" manufacturers is becoming less clearly defined. In the course of time seven "national" compounders have emerged who, although still vastly stronger in the port areas, have spread their manufacturing and distribution activities into the country areas. A more useful classification for many purposes is therefore that between "nationals" and "non-nationals". In 1966/7 the nationals supplied 60% of all the

compounds manufactured in Great Britain. Those firms manufacturing relatively small tonnages of compounds tend to diversify their business activities to a much greater extent than the larger compounders. For many small manufacturers, grain merchanting is a major activity.

53. Since the mid-1950's economic considerations have tended to favour the location of new plant away from the port areas.

54. The national compounders rely mainly on merchants for their purchases of home-grown supplies; for imports they generally do not operate through merchants but directly. Small compounders rely more on direct purchases of grain from farmers. Most of the smaller manufacturers sell other compounds as well as their own. In this way they can expand their turnover without increasing their plant size. However, a diminishing role is visualised for the small merchant-compounders.

55. Various processes of integration of manufacturing firms are described and discussed. A factor influencing the national companies towards diversifying their manufacturing points away from the ports and into the heart of the cereal-growing areas is that a certain radius (perhaps 50 miles) is the maximum that can be economically supplied by a factory relying mainly on home-grown grain.

56. The Surveys of Farmers and of Merchants included questions about mixing of feed on farms. Although this practice has been on the increase, at least until fairly recently, and some 2 million tons a year may be involved, there does not appear to have been any real adverse effect on merchants' trade. Farm-mixing may tend to increase as the size of farms increases, unless there are technical changes in compound production or in animal husbandry which might make purchased compounds more attractive to the farmer in future.

57. Flour milling is a relatively static industry in terms of total production. Although more use is made of home-grown wheat than formerly, two-thirds of the grain used in flour milling in the United Kingdom is imported. Bread flour manufacture is the largest part of the industry and the part which uses the smallest proportion of home-grown wheat, for technical reasons which are discussed in a separate section. The biscuit trade represents a growth market for home-grown wheat but it uses only 10% of the total consumption of flour and flour products in the United Kingdom.

58. Annual fluctuations in the home crop of wheat are taken up by feed use rather than by changes in the quantity used for milling. An interesting positive relationship, however, appears to exist between the quantity of home-grown wheat used for milling and the yield per acre. This needs further investigation but it seems likely that in years of high yield per acre, quality has also been good from the miller's point of view.

59. Flour milling is a highly concentrated industry. Four firms produce over two-thirds of all flour used in the United Kingdom. Because the major raw material is imported wheat, the large firms are established in port areas and close to centres of population.

60. The number of independent member-firms of the National Association of British and Irish Millers fell from 309 in 1926 to 63 in 1967. Most of those who remain anticipate a further process of integration. The reasons and consequences of this, and of the geographical location of flour mills, are discussed.

61. Examination of the technical considerations affecting the use of British grain in flour milling leads to the conclusion that there exists a potential market for an additional 500,000 tons, depending upon the delivery of suitable varieties at competitive prices. The realisation of this potential is largely outside the control of the millers; it hinges on the production and marketing decisions of farmers and bakers.

62. The millers interviewed were not in favour of introducing a grading system. They thought that merchants already did a satisfactory job of selecting grain, and in any case no practicable scheme would eliminate the necessity of testing samples for milling according to procedures now generally in use.

63. Millers will often pay more for quality, but the extent of these differentials is usually too small and their incidence too uncertain to encourage farmers to pay great attention to quality—especially if this means a reduction in quantity. To shift to a new variety a farmer normally requires a noticeable price incentive, but this the miller cannot offer unless he can feel assured of sufficient quantities of the variety to enable him to adjust his milling process. It is suggested that it might be suitable, within the framework of the subsidy scheme, to offer some extra reward for promising new varieties.

64. The Report includes results of a special study of brewer-maltsters and sole maltsters. At the time of the survey there were 71 such enterprises making malt in the United Kingdom. Around 1·1 million tons of barley were sold to them for malting from the 1966/7 harvest. This use was second only to stockfeed as a sales outlet.

65. The malting industry, like the feed and flour industries, is undergoing a process of concentration. Over half of the small brewer-maltsters interviewed had plans to close down their maltings within 5 years. When brewers amalgamate it is common for one of them to cease malting. Individual firms have closed down some of their maltings to concentrate production in larger and more modern plant.

66. Maltsters buy most of their barley from merchants. A negligible proportion is imported. Purchases direct from farmers are not uncommon in

southern and eastern areas of England, where farms tend to be larger and better equipped for cleaning, drying and storage. Most maltsters divide their total purchases between a large number of merchants, shopping around for the best price for each quality of grain required. The use of markets and exchanges by maltsters has declined; more usually samples are brought or sent to the maltster's office. Laboratory testing of barley is almost universal.

67. Most maltsters said they preferred grain not to have been dried on the farm, since they had grave doubts about the germination of farm-dried grain.

68. A large majority of maltsters were not in favour of a grading scheme for barley. There was too much variation from season to season and from sample to sample; no precise standards could be laid down; the scheme would be costly to administer; and maltsters and their merchants already did all the grading that was necessary.

69. The Report suggests that more local grain could be taken by maltsters in northern and western areas if storage and drying on farms could be more reliably carried out. It is recommended that more publicity and advisory work be carried out to encourage this.

70. It is also recommended that brewers and distillers should make their malt contracts a few months earlier in the year—as far as amounts and delivery schedules are concerned—so that maltsters could buy to known requirements.

71. In reporting on the distilling industry it is noted that the production of grain whisky has been running ahead of demand for some years. In malt whisky distilling the utilisation of Scottish barley is on the increase: its quality has improved and its transport costs can be as low as 10*s*. to 12*s*. per ton for many users, compared with an average of £2 10*s*. for barley from the south. Bulk handling will continue to develop and some firms may make greater use of coastal shipping.

72. **Part VI** includes studies of grain marketing in certain other countries so that comparisons may be made and the relevance of their experience to the United Kingdom situation may be considered. With regard to the Common Market countries the conclusion is reached that the main trends are likely to be further concentration among merchants, brokers, etc.; increased penetration of domestic markets by international shippers; and continuing severe competition between the co-operatives and the private merchants from which the former seem likely to emerge with a larger proportion of the total market.

73. On the world market there are many imponderables both on the supply side and on the demand side, but it is tentatively concluded that there are unlikely to be commercial shortages of grain, at least up till 1980. The most likely outcome during the next 5 or 10 years is a continuation of the rather precarious balance which now exists.

74. The way cereals prices move, both within a season and between seasons, is of vital significance to all engaged in the marketing of grain. In **Part VII** a study of price behaviour is made covering the period 1955 to 1968 and several important features are noted. Prices were generally falling in the late 1950's and reached a low point about 1961, since when they have been relatively stable, whilst within-season price fluctuations have also diminished. This change is ascribed partly to the buoyancy in the international grain market during the period following 1961 when Russia and China were large buyers of cereals, and partly to developments in the domestic grain situation such as the effect of increased storage capacity on farms, seasonal incentive payments for barley, and the Home-Grown Cereal Authority's forward contract system.

75. Certain consistent patterns of price behaviour on the domestic grain market are identified. Prices generally rise between harvest and about March, by which time the prospects of either a surplus or a deficiency in the remaining three months are becoming clearer, and from this point they tend either to continue rising or to fall. There appears to be an alternation from year to year, a year when end-season prices are high being followed by one in which end-season prices are low. The reason for this appears to be that farmers' storage decisions at harvest time are very much influenced by the price-levels at the end of the immediately preceding season, but the need for more research to throw light on these price movements is noted. This pattern may also be a reflection of the farmer's uncertainty as to when to sell his barley, since a study of prices over the last 13 years shows that in 5 of those years there was no premium for storage in the market price or even a penalty, i.e. the price at the end of the season was lower than at harvest.

76. A study was made of the patterns of movement of international grain prices and their relationship to prices on the domestic market. Maize, sorghum and hard wheat prices have fluctuated less, over the last 13 years, than have those of other cereals such as soft French wheat, and this is ascribed partly to the role of the International Wheat Agreement and partly to the dominant position of government or quasi-government selling agencies in the marketing of hard wheat, maize and sorghum. A consistent relationship is found between international grain prices and those on the domestic market. At or soon after harvest time the difference is always greatest, with international prices above those on the domestic market, but the differential declines as the season advances so that by the end of the season the domestic prices usually approach those of imported grains. Because of the effect of the alternating pattern on the domestic market, noted above, this relationship is less marked in some years than in others, but it is present in some degree in all years. It is suggested that the existence

of this differential is due to the absence of complete substitutability between imported and home-produced grain, but again this is a field in which more detailed research is needed to establish the full causes.

77. The discussion on policy in **Part VIII** is preceded by a brief review of changes in government policy for cereals since 1953. The decline in prices of the late 1950's and early 1960's led to the establishment of the Home-Grown Cereals Authority and to a "package" of measures designed to protect the domestic market from a collapse of grain prices. These included the system of minimum import prices on the one hand and the standard quantity system on the other. In the last year or two the Government has given some encouragement to a selective expansion of cereals, in line with an increased livestock population, but with the emphasis on more wheat rather than barley. As a prelude to the discussion of current and future policy for cereals, a study is made of the likely supply and demand situation for cereals in the next 5 years. The National Agricultural Advisory Service estimates of increased cereals production, referred to earlier, form the basis of the projection of future supply (an increase of 1·8 million tons by 1972) and various projections of demand are made covering requirements for livestock, human feed and industrial uses. On the assumption that the proportion of home-grown to imported grain will remain as at present it is estimated that demand will rise by about 1·4 million tons by 1972. Thus supply is likely to outrun demand by about 400,000 tons per annum unless increased substitution of home-grown for imported grain takes place. The conclusion is reached that the United Kingdom is in a situation of "uneasy balance", i.e. uneasy because there appears to be an underlying tendency for supply to increase faster than demand and because, as was shown earlier, seasonal fluctuations of about half a million tons on average are to be expected and in years of good harvests there could be difficult problems of disposal. This is the background for the discussion of present and future policy for cereals.

78. Two main types of policy measures are noted—those designed to prevent surpluses of grain from increasing, and those designed to widen the market so that the grain can be disposed of once it has been produced. Included among the "supply-control" group are such measures as the encouragement of break crops (the need for more research into market outlets for these crops is noted); the encouragement of wheat or maize in substitution for barley (again more research on maize growing in the United Kingdom is needed); and the reduction of the relative profitability of the cereal or cereals in question by adjusting the guaranteed prices or the standard quantities at Annual Reviews. The latter might be necessary in the future, as it has been in the recent past, if other measures were unsuccessful; and if surpluses become such as to threaten to swamp the market this would be the only sensible policy. Consequent problems of income-support might

have to be solved by other means. Included in the group of "market-widening" measures are: the encouragement of exports; stimulating the increased domestic consumption of grain by livestock (policies for livestock and for cereals need to be integrated more fully); encouragement to the substitution of home-grown grain for imported (not by such measures as government control of the grist but by increasing the competitive advantage of the United Kingdom grain. In particular, a study should be made of the possibility of reducing the freight cost of taking grain from Britain to Northern Ireland so that it can compete with imported feed grains); the discouraging of imports (by tighter import controls as described below); and improving the presentation and quality of home-produced grain and matching these more closely to users' requirements. Although not exactly comparable with the other "market-widening" measures, the possibility of carrying-over grain may be included in this group because it might enable a seasonal surplus in one year to be carried over to offset a possible seasonal deficiency in the next. The conclusion is reached that both supply-control and market-widening policies are needed concurrently, but if surpluses develop on an increasing scale the supply-control approach must inevitably become predominant and the role of the Government and the Home-Grown Cereals Authority in this connection is likely to be an increasingly important issue in the years ahead.

79. The various systems of price support and their relevance to supply-control are discussed in Chapter 4 of Part VIII. The present system of cereals deficiency payments, combined with a floor level of prices below which imports are not allowed to enter, is accepted as the one best fitted to United Kingdom conditions. However, a number of changes in the method of implementation are proposed. The most significant of these is the suggestion that the minimum import prices be replaced by a levy system designed to operate in a broadly similar manner so as to ensure certain floor level prices—though not, as in the E.E.C., as the main vehicle for maintaining farmers' incomes. The floor levels would be pitched a little higher than the minimum import prices so that instead of being a safety device to prevent a complete collapse of domestic prices induced by cheap imports they would be a more sensitive regulator capable of being used to sustain a measure of stability on the domestic market. The new-style floor price would be varied during the season if it became clear that the level had been pitched too high or too low, and it would be "stepped" so that the market in all except the most exceptional years could be expected to yield some premium for storage. With this modest degree of protection to producers it is suggested that many of the elaborate schemes introduced since 1953 to protect farmers (and the Exchequer) could be dismantled and the whole cereals support system greatly simplified. Thus it is suggested that

seasonal incentives, target indicator prices and eventually forward contract bonuses could be abandoned once the new scheme was in operation; but standard quantities would remain as a useful protection to the Exchequer in the event of over-production by United Kingdom farmers.

80. Another simplification of the cereal subsidy schemes which is proposed is the abandonment of the tonnage basis of payment for wheat and the substitution of an acreage payment as for barley. This change is proposed partly in the interests of simplification, partly because the present degree of isolation from market pressures afforded wheat producers by the seasonal guaranteed prices is not compatible with efficient marketing, but mainly because it is a waste of national resources for farmers to have to move their wheat off farms and then buy wheat back again simply because the subsidy system requires it. The economic implications of the proposed change for individual cereal farmers are stated in some detail in Appendix E, where it is shown that the total receipts of most cereal growers would hardly be affected.

81. The role of the Home-Grown Cereals Authority is discussed in the context of these proposed changes in import control and subsidy schemes. It is suggested that the Authority should share in the responsibility of determining when and how the variable import levies should be imposed, keeping the situation constantly under review. Its concern with home-grown cereals should not preclude it from preparing and publishing at appropriate times a well-informed analysis of the total grain situation and outlook as it affects the United Kingdom, including imports of grains of all kinds.

82. With total storage capacity now probably adequate to accommodate most of the crop needing to be stored at harvest, the most likely glut period is no longer at harvest but in the spring. If the Authority's trading powers are to enable it to intervene effectively it must not be so inflexibly restricted by having to dispose of all stocks by 31st July, as it is at present.

83. The Authority should play a more important role than at present in stimulating innovation and research over the whole field of cereals marketing. It should be able to make financial grants to assist firms or even individuals who will undertake pioneer developments or innovations, and it should conduct and commission a wide range of research investigations. The results should be made generally available to all in the industry. Up to £500,000 per annum could be allocated to this function, and it is suggested that the necessary funds be found by a direct government grant since the present levy system would clearly be inappropriate. The Authority could also expand its market intelligence role, although the need for more research into consumer-requirements in this field is emphasised, and it is suggested that the Guide Price could be absorbed into this wider role.

84. Of first importance among the national statistics which the Authority should be concerned to improve are the reports of ex-farm prices and the estimates of total production of wheat, barley and oats. The returns of prices which are collected under the provisions of the Corn Returns Act of 1882 leave much to be desired. They are limited to transactions in the 172 prescribed towns in England and Wales (though there are special provisions covering offers and acceptances by telephone or teleprinter); they do not distinguish between grain of different qualities or uses; they may or may not include transport or other charges; they are not confined to "spot" transactions, but include some forward purchases; and although they purport to relate to sales made during the week in question, irrespective of date of delivery, it is known that some merchants fail to observe this. Price statistics are needed which, while being nation-wide in scope, relate to a more precisely identifiable concept than the present "Corn Returns average". Due account should be taken of the quality, location and stage of marketing to which each quoted price relates. Different averages can be calculated subsequently for different purposes.

85. Accurate and timely estimates of production are of the greatest importance. The government departments having responsibility for agricultural censuses in the various parts of the United Kingdom must continue to produce the acreage statistics, but the whole question of yield estimation seems to call for joint review by all those concerned. It is a source of disturbance and uncertainty to markets when the national estimates have to undergo substantial revision months after the harvest. More accurate measurement at the farm at harvest time is needed, including more knowledge of moisture content.

Conclusions

86. Although this Report has been primarily descriptive in nature, certain suggestions and recommendations have been made where it is thought that the present situation can be improved. These are listed as follows, no attempt being made to rank them in order of importance, but taking the order in which they appear in the Report:

87. Most farmers (Part II 6.34), merchants (Part III 4.25) and end-users (Part V(a) 4.7, Part V(b) 4.5, Part V(c) 1.30, Part V(c) 2.15) think that a statutory grading scheme for cereals would be unnecessary or unworkable. However, the use of standard specifications on a voluntary basis might be feasible and worthwhile, particularly if grain were bulked in centralised stores (Part III 7.20 and Part VIII, 4.35).

88. Merchants who penalise farmers for imposing unreasonable delays in delivering grain to lorries should do so directly and not as an adjustment to the price (Part III 3.32).

89. Government assistance for grain storage should not be confined to storage on the farm—other proposals for centralised storage, e.g. from merchants or processors, should be considered on their merits (Part III 7.21).

90. The futures markets are less used than they profitably could be, and a programme of information and education is needed (Part IV, 2.28).

91. The present system of standardising coastal freight rates is unlikely to lead to maximum efficiency (Part IV 3.19).

92. Some extra reward might be offered, within the framework of the subsidy scheme, to encourage farmers to grow promising new varieties of wheat suitable for bread making (Part V(b) 4.12).

93. If brewers and distillers would make their malt contracts a few months earlier each year it would help the maltsters (Part V(c) 1.43).

94. Better drying and storage of malting barley might enable maltsters, especially in the northern and western areas, to use more local grain, and publicity and advisory work should be directed to this end (Part V(c) 1.44).

95. If assistance to grain exports were thought necessary this should preferably take the form of assisting improvements in the "infrastructure" (i.e. wharfage facilities, etc.) rather than direct aid to exporters (Part VIII 3.15).

96. A study should be made of existing transport facilities between Britain and Northern Ireland with a view to making British grain more competitive with grain imported into Northern Ireland (Part VIII 3.21).

97. The wheat-subsidy system should be changed from a tonnage basis to an acreage basis (Part VIII 4.2 and Appendix E).

98. The system of seasonal incentives for both wheat and barley should be abandoned and instead a stepped levy system introduced designed to ensure that in a normal year the market price will yield a seasonal premium to cover storage (Part VIII 4.8 and 4.52).

99. If this proposal is thought to be too radical the existing incentive schemes should be amended so that: the barley disincentive period disappears, barley incentives begin in October, the spread of wheat incentives is brought closer to those for barley, and all seasonal incentives stop at the end of May (Part VIII 4.9).

100. Target Indicator Prices should be abandoned (Part VIII 4.15).

101. The present Minimum Import Price system should be abandoned and replaced by a stepped levy based on a "floor price" slightly above the present M.I.P. level. The Cereal Deficiency Payments Scheme should be retained (i.e. assuming Britain does not join the Common Market) (Part VIII 4.17 and 4.21).

102. The definition of "Millable Wheat" should be revised (Part VIII 4.19).

103. The feed formula system should be revised to allow the grain price to have some impact on livestock producers' decisions (Part VIII 4.20).

104. The H.-G.C.A.'s Forward Contract Bonus Scheme should be gradually phased out, to be replaced by voluntary forward contracting arrangements as far as possible (Part VIII 4.29).

105. The Guide Prices should be absorbed into the Authority's expanding Market Situation and Outlook responsibility, and should become more a range of prices than exact figures (Part VIII 4.33).

106. The Home-Grown Cereals Authority should expand its research and development role and should have funds available to grant-aid innovations in cereals marketing. These funds (up to £$\frac{1}{2}$ million per annum) should be made available direct by the Government since the levy on farmers is clearly inappropriate (Part VIII 4.36).

107. The Home-Grown Cereals Authority should aim to secure improvements in two important series of statistics; the price statistics derived from the Corn Returns and the estimates of production, particularly the average yield at each harvest (Part VIII 4.37 and paras 84 and 85 of this Part).

108. The Report identifies the following specific fields in which more research is needed: causes of yield variations and means of combating them (Part II 4.19), user requirements for grades and qualities of grain (Part VIII, para. 3.22), kinds of market intelligence needed (Part III 4.54), behaviour of market prices (Part VII 1.6), marketing of break crops (Part VIII 3.5); growing of maize (Part VIII 3.8); use of cereals for industrial purposes (Part V(d) 1.4).

109. The H.-G.C.A. should have powers to carry over grain from one season to another (Part VIII 4.45).

Appendices

APPENDIX A

Factors Affecting the
Geographical Distribution of Cereal
Acreages and Cereal Yields in the U.K.

I. Distribution of Acreages

A.1. The factors affecting the geographical distribution of cereals are so numerous and interrelated that no simple explanation can be expected, even if satisfactory data were available. In reality, there is generally a poverty of suitable data; for example, there is no reliable information about regional differences in the incidence of cereal diseases, in yields of spring- and autumn-sown crops or in dates of ripening and harvesting, although each of these is likely to be of some importance in any full explanation of cereal distributions. Again, although there is a wealth of climatic data, essential parameters are either not available or are presented only as mean values for calendar months; the great variability of the British climate is also a complicating factor and average values have generally been used to minimise the effects of annual variations. Similarly, soil data for the whole country are almost completely lacking and recourse has had to be made to the land classification map.

TOTAL CEREALS

A.2. Cereals can be and have been grown in most parts of the U.K.; only in the uplands (broadly the areas under rough grazing) does shortness of growing season and frequency of precipitation actually prohibit cereals and even some of these areas could grow a crop of a kind. No firm altitudinal limit can be given, partly because of the intricate interplay of physical and human forces which have fixed the present levels of the moorland edge and more generally because it varies throughout the country, being much lower in the more oceanic climates of western Britain than in the more continental east and on smaller uplands than on larger. Cereals are successfully grown and harvested above the thousand-foot contour in areas as different as Exmoor, the Peak District and the Lammermuirs, although the proportion of the cereal acreage to be found at such elevations is very small.

A.3. Thus, except in respect of the moorlands, these distributions are not determined by whether particular cereals can be grown at all, but rather whether they can produce grain of acceptable yield and quality and whether they enjoy a comparative advantage over other crops and enterprises. There is abundant historical evidence to show that cereals (admittedly of different varieties, grown in very different ways and with much lower yields than at present) have been far more widely grown in the past. The distribution can be expected to be related not only to physical factors—climatic parameters, such as onset of growing season, available moisture, temperatures above the minimum for cereal growth and weather during harvest; to relief, especially slope and roughness, which affect suitability for mechanisation; to soil conditions, especially drainage, ease of working and acidity; to farm and field size and availability of harvesting, drying and other equipment—but also to a whole range of economic factors which determine the availability of markets and the comparative level of returns to be obtained from other enterprises. In detail, of course, much will depend on the standards of management, abilities and preferences of individual farmers, and because of the small-scale relief, complex geology and oceanic climate of the U.K. there is great local variety within quite small distances.

A.4. Yet there are certain broad regional tendencies which can be observed and can be measured statistically. Climatic controls are clearly important and more than two-thirds of the cereal acreage is grown in counties with an average summer rainfall of 15 inches or less and three-quarters in those with a summer water balance[1] of 0 inches or less. The proportion of the crops and grass acreage under cereals is strongly correlated (negatively) with both annual rainfall and summer rainfall and with water balance; multiple regression of cereal acreages on county climatic data identifies annual rainfall as the only significant variable at the 5% significance level. Again, few cereals are grown where the summer accumulated temperature[2] is less than 1000°(C) day-degrees and two-thirds of the cereal acreages are to be found in counties with an accumulated temperature of 1500 day-degrees or over, although the relationship with cereal acreage is not statistically significant.

A.5. It also seems likely that cereal distributions are related to the existence of favourable weather at times of seeding and harvest and to the date at which harvests can take place. The later growth begins, the later is likely to be the date of harvest (although differences in day-length will be a complicating factor). Growth begins approximately a month later in the lowlands of the Moray Firth than in south-west England, and since drying power decreases with lateness of harvest (with potential drying power in September approximately half that in August), it becomes increasingly difficult for cereals to reach an acceptable moisture content. It is true that grain-driers can reduce the importance of this factor, but they are found mainly in drier eastern parts where farms are large and large acreages of cereals are grown. In wetter districts, where farms are generally small, the capital cost of installing drying equipment may be a disincentive and lack of sufficient drying equipment a bottleneck at harvest time. Unfortunately, there is very little reliable information on date of harvesting. Records for seed trial grounds of the National Institute of Agricultural Botany, which give dates of both ripening and harvesting, generally show gradients of progressively later dates from south to north, although there are considerable variations from year to year in the ranking of individual stations and in the dates of harvest. Thus for all stations the average date of harvest for winter wheat (Cappelle-Desprez) ranged from 7th September in 1962 to 23rd August in 1963; at the extremes, an Edinburgh average for the 5 years 1962–7 of 6th October compared with a Cambridge average of 2nd August. For spring barley (Proctor) averages for all stations range from 7th September in 1963 to 24th August in 1967, and 5-year averages from 20th August at Wye to 5th September at Edinburgh and at High Mowthorpe, Yorkshire. For spring oats (Condor) there was a similar range of years and dates, with 5-year averages ranging from 24th August at Sparsholt (Hants) to 26th September at Edinburgh. Late harvests not only lead to lower yields and quality, but also interfere with harvesting of root crops and with the preparation of ground for crops for the following year.

A.6. The distribution of relief also controls the distribution of cereals, and the importance of altitude is confirmed by the negative correlation between cereal acreages and height. The higher proportion of the crops and grass acreage to be found on steep or uneven land in western and northern areas is also a factor, especially where cultivable

---

[1] *Water balance* is defined as the difference, measured in inches, between rainfall and estimated potential evaporation.

[2] *Accumulated temperature* is defined as the cumulative total of the number of degrees Centigrade by which the mean daily temperature exceeds 6°C (42°F); it is expressed in terms of *day-degrees*.

land is confined to steep-sided valleys and where the marks of glaciation are still fairly fresh. Land quality, which subsumes elevation, slope and soil, is also a significant variable and two-thirds of the cereal acreage is to be found in counties, mainly in eastern England, with 30% or more of their crops and grass acreage in class 1A, 2A, and 2AB land; if a modern land-classification map were available this proportion would be appreciably higher, since much of what was classified in 1939 as medium-quality land on the chalk would undoubtedly be upgraded. Although there are regional tendencies in land quality, notably the high proportion of gently sloping land with medium-textured soils in eastern England, the intimate pattern of soil parent material and relief makes it likely that the influence of soils on cereal distribution will be felt primarily at a local level. If suitable data were available, it is probable that poor drainage and heavy texture could be shown to be major factors, depressing yields, hindering cultivations and making for a low proportion of land under tillage, as in such areas as the Weald of Kent. Such conditions are especially important for cereal distribution where they coincide with unfavourable climatic conditions such as high rainfall or late onset of temperature favourable for growth, as on the boulder-clay areas of Northumberland.

A.7. Economies of scale in cereal growing suggest that regional differences in farm size play an important role in the distribution of cereals and are at once a cause and an effect; for the trend towards enlargement is most marked where farms (as measured by crops and grass acreage) are already large, viz. in the main arable areas. There is a discernible gradient of farm size across the U.K., with farms of 300 acres and over accounting for more than 50% of crops and grass acreage in south and east England (especially on the chalk areas), compared with only 9% in Wales and an even smaller proportion in Northern Ireland, where small farms predominate. The cereal acreage of a region is strongly correlated with the proportion of crops and grass acreage occupied by holdings of 300 acres and over and, with annual rainfall (negatively correlated) and tillage, it accounts for 91% of variation in the cereal acreage at the 5% significance level. In view of higher operating costs on small fields and their lesser suitability for mechanised farming, especially for large combines, differences in field size are also relevant to cereal distribution. Although satisfactory data are lacking, there are clearly marked regional variations in field size, with small fields predominating in the west and north and large fields in the main arable areas; for example, average field size is four times larger in Essex than in Devon. The distribution of field sizes is broadly similar to that of farm sizes, and the prevalence of small farms and fields in northern and western counties reinforces the physical obstacles to cereal-growing there. Conversely, where favourable climate, relief, soil and farm and field size occur together, as in East Anglia, the proportion of the crops and grass acreage devoted to cereals is likely to be very high.

A.8. Since cereals occupy the greater part of the tillage acreage, their distribution is broadly related to conditions favouring tillage, notably ease of ploughing. Sometimes, however, physical controls operate negatively through their effect on other crops; the proportion of the tillage acreage under cereals is high (exceeding 85%) in areas like the chalk downs where few other cash crops are possible, but low in areas like the uplands where cereal growing is more difficult and in counties like Lincolnshire (Holland) where a large acreage of sugar-beet, potatoes, vegetables and other cash crops is grown. Stoniness and depth of soil similarly exert their influence negatively; thus, few crops other than cereals and grass are grown on the very stony soils derived from clay-with-flints.

VARIETIES

A.9. In one sense, the distributions of cereals shown in Part II, Maps 1.1 and 1.2, are highly generalised, for what is shown as a single crop comprises a large number of different varieties. Unfortunately, there are no official data to show either the distribution of these varieties or their relative importance, but it is possible to examine the latter on the basis of information collected by Messrs. Peter Darlington and Partners Ltd. Data for 1966 show the dominance of winter wheats in Scotland (80%+) and in southern, eastern and midland England (60%+) and the relatively greater importance of spring wheat in south-west England, Yorkshire and Lancashire and Wales, although the total acreage of wheat there is small, as it is in Scotland. Cappelle-Desprez was the principal variety nearly everywhere, with Champlein prominent as a second variety in Wales and north Scotland; Kloka was the leading spring wheat, except in Wales and south-west England. The proportion of spring barley never fell below 75% and increased north-wards, exceeding 95% everywhere north of the line from Carmarthen Bay to the Wash. Maris Otter was the dominant winter barley. Although Proctor was still the leading spring variety, spring barleys display a complex pattern of regional preferences, with Proctor in eastern and much of south-west England, Impala elsewhere in southern England, Zephyr in the midlands and north-east England, Vada in Yorkshire and Lancashire and in north-east Scotland, Ymer in other parts of Scotland and Rika in west and south-west Wales. The importance of spring varieties of oats shows a similar northward gradient, with winter oats dominant in the south and only spring varieties being grown in northern England and Scotland; Peniarth was the leading winter variety of oats and Condor the leading spring variety, although its place was taken by Astor and Forward in north-west England and Scotland. Each of these varieties has its own distributions and requirements and this regionalisation ought properly to be taken into account in any attempt to explain the distribution of the individual cereals.

WHEAT

A.10. Wheat is the least widely distributed of all the cereals and is usually considered the most difficult to grow. It is mainly concentrated in the cereal-growing counties of eastern England, from Kesteven to Essex. Some 81% of the acreage is found in counties with a summer rainfall of 15 inches or less, 90% in counties with a water balance of 0 inches or less and 84% in those with an accumulated temperature of 1500 day-degrees and over; nearly all the wheat acreage is grown in areas with 1200 day-degrees or over. Wheat also shows a stronger association with good-quality land than either barley or oats; 77% of the wheat acreage occurs in counties with 30% or more of their crops and grass acreage in land of classes 1A, 2A and 2AG. There is a positive correlation between wheat yield and wheat acreage, indicating the concentration of this crop in the areas best suited to it. Wheat acreage is also correlated strongly with tillage and with the proportion of land in farms of 300 acres and over, although the latter is present in multiple regression equations only in respect of data for England and Wales. There is also empirical evidence that its strong rooting system makes wheat a preferred crop on heavy soils and it certainly occupies the largest proportion of the cereal acreage in such areas as the clays of the English Midlands and of south Essex, although in absolute terms the acreage under wheat is often quite low. Quality is likely to suffer from late harvests and this may well be a factor in the small acreage of wheat grown north of York; a late harvest is also likely to conflict with the needs of the potato harvest in Scotland. Wheat also competes for land with other crops for, apart from the clay lowlands, the physical conditions favouring wheat are also suitable for the growth of cash crops other than cereals; the acreage under wheat is consequently less in such areas than might be expected if the physical requirements of wheat alone were being considered.

BARLEY

A.11. The production of barley is less localised than that of wheat and larger acreages are grown in western and northern parts of the U.K. Even so, the greater part of the acreage is also enclosed by the 1200 day-degree isoline and the most important areas lie in counties east of the line from the Tees to Lyme Regis Bay. The barley acreage also shows similar negative correlations with rainfall and water balance and positive associations with good land, large farms and arable cropping in general. Barley yields, too, like those of wheat, are positively correlated with the barley acreage, suggesting that most of the barley is grown in those areas where conditions are most favourable. High soil acidity is likely to be a factor restricting the extent of land under barley in the north-west, although farmers' preferences and the exclusion of barley from the list of approved crops under the winter keep scheme are also important. On the other hand, the ability of spring barley to be sown late and still produce a satisfactory crop is a factor which favours the wide extension of barley-growing. Barley is particularly prominent in those arable areas where other cash crops cannot easily grow because of shallowness or other unfavourable soil conditions, and where its lower susceptibility than wheat to certain cereal diseases has often enabled several successive crops to be grown with impunity; barley occupies the highest proportions of the acreages of crops and grass, of tillage and of cereals on the chalklands, although the reason which made it the leading crop there in the 1930's—the need for soil of low fertility and the demands of the maltsters—no longer apply. Conversely, in south-west and eastern Scotland the acreage of barley may be restricted by farmers' preference for a low pH on their potato fields.

OATS

A.12. Oats are the most widely grown of the cereals and Part II, Map 1.1 shows a surprisingly uniform distribution; but when account is taken of the proportion of the tillage and cereal acreages occupied by oats, a very different situation emerges, with oats as either the leading or a prominent cereal in much of the north and west and of very minor importance elsewhere. The conditions governing the distribution of oats are quite different from those relevant to wheat and barley, since oats tend to be a residual legatee, occupying those areas in which, for a variety of reasons, wheat and barley are not grown to any extent. This view is confirmed by the absence of any significant correlation between oat yields and oat acreages, except in England and Wales, where the correlation is negative. In a sense, the distribution of oats (or more strictly, the areas where they predominate) indicates indirectly the physical controls on wheat and barley. Most of the oats acreage is found in areas with 1000 day-degrees or over and with a water balance of less than $+5$ inches but the acreage under oats is positively correlated with height and negatively with temperature and sunshine variables, suggesting that it is principally the deterioration of temperature and sunshine conditions which leads to the replacement of barley and wheat by oats. Conversely, there are no significant correlations with factors such as proportion of the crops and grass acreage under tillage, cash crops or large farms which are such a notable feature of the wheat and barley distributions. In so far as those major cereal diseases favoured by damp conditions are more likely to occur in the western districts, the lesser susceptibility of oats may be a factor, as their greater tolerance of high soil acidity certainly is. Apart from attracting winter keep subsidy and often being preferred by more conservative farmers in upland districts, the positive correlation with altitude confirms the view that—at least until quite recently—the growing of oats was being increasingly confined to the margins of cultivation. The fact that oats can be cut green (50% of the crop is said to be so harvested in Merionethshire) and their suitability for harvesting by binders (which in the mid-1960's still outnumbered combine harvesters in most counties of the north and west) are also relevant; barley is reputed to be a difficult and unpleasant crop to harvest by binder, although modern short-strawed varieties are said to dry more quickly than other cereals. The prevalence of small fields and steep slopes in these areas is also likely to be a factor hindering substitution of combines for binders. The survival of oats in eastern counties is probably related to their lower susceptibility to the soil-borne and air-borne fungal diseases which affect wheat and barley; on the other hand, they are susceptible to attack by cereal root eelworm.

II. Distribution of Yields

A.13. If the distribution of cereal acreages is controlled by a variety of factors, this is even more true of regional differences in cereal yields, which are much more likely to be affected by management, a conclusion supported by the low proportion of county variation in yield which is explained by physical variables. Such variables may work in opposite directions, so that yields in south-eastern counties may be depressed by inadequate soil moisture and those in north-western counties by excessive rainfall. Unfortunately, data on cereal yields are based only on estimates by crop reporters and are probably more variable and less reliable than crop acreages, even at county level; indeed, the Department of Agriculture and Fisheries for Scotland discontinued publication of county estimates in 1965. Furthermore, yield estimates are available only for each cereal as a whole and do not distinguish yields on autumn-sown and spring-sown crops, although the former are likely to be appreciably higher. In view of the varying proportion of the individual cereals which are spring- and winter-sown, this gap of knowledge is of some importance. An additional difficulty, which would arise even if accurately measured yields were available, is that cereal yields vary greatly from year to year, depending on the sequence of weather conditions, the occurrence of major outbreaks of disease and many other factors. It is therefore preferable to consider average yields rather than those for a single year, when weather conditions may be exceptional.

A.14. The most striking feature of the maps of cereal yields (see Part II, Maps 1.6, 1.7 and 1.8) is their relative uniformity, as compared to the marked differences which characterise the distribution of cereal acreages. It is true that the range of yields is large, but there is a marked clustering around the mean. The second major feature is the similarity of pattern and of causation in the distribution of yields of the three cereals; in multiple regressions of cereal yields on relevant factors, seven out of ten variables are related to all three cereal yields at the 0·1% level of significance and these variables occupy broadly similar positions for each cereal when ranked according to the strength of the correlation. A major factor in all cereal yields is the importance of farm size, for the proportion of land in holdings of 300 acres and over is significant in each correlation and appears in each multiple regression at the 5% level of significance. Yield is also positively correlated with proportion of good quality land and negatively with that of medium quality, annual rainfall and water balance, and summer water balance.

WHEAT

A.15. The pattern of wheat yields shows a concentration of high values in northern England and south-east Scotland, and above-average values in most of lowland England and northern Scotland; the lowest values occur in Wales, south-west Scotland and Northern Ireland. Of course, the yields in western and northern counties are in respect of small acreages only and it is probable that wheat there is carefully managed and planted only where it is likely to do well. For this reason, it should not be supposed that if larger acreages of wheat were grown in these counties their high average yields would be maintained. According to unpublished results from the National Wheat Survey, the only major soil factors affecting yields were drainage and to a lesser extent soil parent materials. Yields from poorly and excessively drained soils were respectively 30·7 and 31·4 cwt/acre, compared with 38·5 on moderately drained soils, while those soils derived from clayey parent materials and those from sandy and coarse materials averaged 34·4 and 32·6 cwt/acre respectively, compared with 37·4 from silty and from fine loamy materials. Thus, even if wheat is the preferred crop on heavy land, it does not do as well there as on soils of medium texture. The National Wheat Survey also showed correlations with farm size and land quality. A contributory factor, which was not included in the statistical analyses because of lack of data, is fertiliser treatment. Fertiliser applications on land in crops and grass generally are much lower in the north and west than in the south and east, as are applications on cereals, although the Ministry of Agriculture's Surveys of Fertiliser Practice[1] show that, apart from nitrogen (applications of which

---

[1] *Survey of Fertiliser Practice 1966 (Preliminary Report)*, M.A.F.F., 1967.

are almost twice as heavy in arable districts as in upland and grass districts of the west and north) actual dressings per acre on cereal crops are not dissimilar, the chief difference being that a lower proportion of cereal fields are treated in the upland and grassland areas.

BARLEY

A.16. Barley shows a smaller range of county average yields than does wheat; many of the counties with high yields are again in northern England and Scotland. Lower yields are found in many of the counties where large acreages of barley are grown, notably those containing a substantial acreage of chalk outcrop. The lowest yields occur in south-west England, parts of Wales, south-west and north Scotland and Northern Ireland. Barley yields are strongly correlated with proportion of land in holdings of 300 acres and over, and this factor accounts for more than 30% of total variation in multiple regression equations at the 5% level of significance. Barley yields are similarly positively correlated with proportions of good-quality land and negatively with medium-quality land, rainfall and water balance.

OATS

A.17. Apart from the fact that oat yields are almost everywhere lower and show a smaller range of values than those of wheat and barley, the principal difference between the distribution of oat yields and those of the other cereals is that whereas the latter are positively correlated with wheat and barley acreages respectively, oat yields are not correlated with acreage under oats. Instead, and in contrast with the distribution of the acreage under oats, oat yields are positively correlated with farm size, total cereal acreage and proportion of good-quality land. The implication is that, as in the case with wheat and barley, yields are associated with farm management factors, as least in so far as these can be inferred from the data available, and this view is again suggested by the lower proportion of oats which received fertiliser treatment in western counties.

APPENDIX B

Survey of Farmers:
Description of Methodology

B.1. The purpose of the Survey undertaken by Market Investigations Ltd. was to collect data on the growing and first-stage marketing of cereals in the United Kingdom. Two approaches were used: the National Study and the Area Study.

The National Study

SAMPLING

B.2. The study was designed to be based on a total of 1200 interviews in Great Britain; the total being divided into two matched sub-samples of 600 each, described as Phase I and Phase II. Because of the highly skewed distribution of farms in terms of their acreage, it was decided that the basic design should incorporate over-sampling of farms growing more than 300 acres of cereals with corresponding weighting at the tabulation stage.

B.3. Practical considerations indicated that a measure of clustering would be desirable and it was, therefore, agreed that 12 interviews be taken in each of 50 sampling areas in each Phase. The 50 sampling points required for each Phase were distributed between the countries forming the United Kingdom on the basis of 1966 total acreage, which resulted in the following distribution by country: England and Wales 41; Scotland 7; Northern Ireland 2. In the case of England and Wales, sampling points were then selected at random with a probability proportionate to cereal acreage by taking a first stage sample of N.A.A.S. districts and within each N.A.A.S. district thus selected choosing two parishes, using tables of random numbers. Where more than one sampling point appeared in one N.A.A.S. district, the number of parishes selected was twice the number of sampling points. Five additional contiguous parishes were then aggregated around each of the parishes selected. In each case, the first selected parish was the focus of the before harvest interviewing, and the second that for the after-harvest Phase. Cereals Deficiency Payments cards were then extracted from Home-Grown Cereals Authority records for all farms in these two clusters of six parishes which had made a Cereals Deficiency Payment return in 1966. These farms were then listed by computer in ascending order of holding number within parish within county, using the M.A.F.F. Enumeration System. Farms were selected in three size-groups:

(i) Up to 99 acres of cereals in 1966

(ii) From 100 to 299 acres of cereals in 1966

(iii) 300 acres of cereals and over in 1966.

703

Two names were issued for every one interview required. Interviewers were required to complete two separate quotas, one for farms with 100 acres and over and the other for farms with up to 99 acres of cereals in 1966. Every large farm name which appeared in the enumeration (that is over 300 acres of cereals in 1966) was issued to interviewers, and the balance of names required was made up with a sample of the farm names available in the 100–299 size group together with an entirely separate sample of the farm names available in the up to 99 acre size group. A similar procedure was applied in Scotland but based on counties (not N.A.A.S. districts) as first stage sampling units. In Northern Ireland the fact that virtually all cereal growers grew less than 100 acres of cereals obviated any need for disproportionate sampling. The sample was based on counties as first stage units, and on farms growing cereals within clusters of parishes within the two counties selected.

SAMPLING POINTS

B.4. The sampling points used were as follows.

One Primary Parish for each Phase (England):

Counties: Berkshire, Buckinghamshire, Cambridgeshire, Devon, Durham, Gloucestershire, Hertfordshire, Huntingdonshire, Lancashire, Lincolnshire (Holland), (Kesteven), Northamptonshire, Northumberland, Oxfordshire, Rutland, Staffordshire, Surrey, Warwickshire, Wiltshire, Yorkshire (North Riding), (West Riding).
(Wales): Pembrokeshire.

Two Primary Parishes for each Phase (England):
Counties: Hampshire, Kent, Suffolk.

Three Primary Parishes for each Phase (England):
Counties: Essex, Norfolk, Yorkshire (East Riding).

Four Primary Parishes for each Phase (England):
County: Lincolnshire (Lindsey).

Total number of Primary Parishes for each Phase: 41.

One Primary Parish for each Phase (Northern Ireland):
Counties: Down, Tyrone.

Total: 2.

(Scotland): Counties: Aberdeen, Angus, Berwick, Midlothian, Perth, Roxburgh, West Lothian.

Total: 7.

INTERVIEWING

B.5. The topics covered in the two phases of interviewing included the following:

	Phase I (before 1967 harvest)	Phase II (after 1967 harvest)
Cereals grown and growing practice	*	*
Cereal yields, 1966	*	—
Yield expectation, 1967	*	—
Cereal yields, 1967	—	*
Storage facilities	*	*
Storage procedure, 1966	*	—
Storage procedure, 1967	—	*
Drying facilities	*	*
Drying procedure, 1966	*	—
Drying procedure, 1967	—	*
Milling and mixing	*	*
Usage and disposal of 1966 crop	*	—
Usage and disposal of 1967 crop	—	*
1967 contracts	*	*
Attitudes to H.-G.C.A.	*	*
Attitudes to price policy	*	*
Attitudes to cereal growing	*	*
Membership of farmers' clubs etc.	*	*

* Indicates that the topic was covered.

B.6. Fieldwork was conducted by trained investigators with experience of agricultural market research. Interviewing was carried out at the following times:

Phase I—July/August, 1967

Phase II—October/November, 1967

In general, the later month noted above refers to interviews in the North of England and in Scotland.

FIELD PERFORMANCE

B.7. As noted above, each interviewer was issued with twice as many names and addresses of farms as interviews were required, and instructed to work through the lists (in two size-groups) systematically until the required number of interviews had been achieved in the appropriate size groups. The allowance for additional names was well in excess of requirements and contact was attempted overall with only three out of four of the names issued. Of the 940 interviews attempted in Phase I, and 836 in Phase II, 590 and 582 respectively were successfully completed. The rest either refused, were away, could not be contacted or could not be retained in the sample for some other reason.

z

INTERVIEWS ACHIEVED

B.8. The actual numbers of interviews achieved with different sizes of farm were as follows:

Cereal acreage of farm	Phase I			Phase II		
	England and Wales	Scotland	N. Ireland	England and Wales	Scotland	N. Ireland
1–99	174	37	23	176	39	22
100–299	178	33	—	180	28	—
300 and over	132	13	—	122	15	—
Total	484	83	23	478	82	22

WEIGHTING PROCEDURES

B.9. As noted earlier, the sample was deliberately chosen so as to over-represent larger cereal farms. Effectively the aim was to under-sample farms having less than 100 acres of cereals by a factor of 6, and to over-sample farms with 300 acres of cereals and over by a factor of 3. Perfect performance would thus have implied that weights of 18, 3 and 1 would be applied respectively to farms having less than 100 acres of cereals, 100–299 acres of cereals and 300 acres and over, in order to bring the sample into line with the known distribution.

B.10. Minor variations in sample fulfilment between different districts, however, resulted in a distribution which was not perfectly in accordance with the sample design. Moreover, limitations on the number of sampling points in Scotland and in Northern Ireland also resulted in a measure of imbalance in the actual number of interviews achieved in these two areas. To correct for these factors and to ensure that the data presented were in accordance with the known distribution of cereal farms, weights were calculated for England, Wales, Scotland and Northern Ireland separately and applied to all results at the tabulation stage.

The Area Study

PURPOSE

B.11. The purpose of this Survey was to examine the cereal growing and marketing practice of a sample of farms in six selected areas of the United Kingdom at intervals over a period of 12 months, in order to provide more detailed information on certain aspects of cereal marketing and to check certain aspects of data provided on a longer memory recall in the National Study.

B.12. The six areas specially investigated were as follows: Devon (South Hams), Hampshire (downland around Whitchurch), Norfolk (around Hickling), Northampton-shire (around Thrapston), Yorkshire, East Riding (north of Driffield) and East Lothian (south of North Berwick).

METHOD

B.13. After initial selection of areas of interest by the University of Nottingham Department of Agricultural Economics, rough areas were defined to County N.A.A.S. officers who then provided Market Investigations Limited with lists of units of at least 6 parishes which were both reasonably homogeneous and typical of the area.

B.14. For each parish thus nominated, Cereal Deficiency Payment cards were selected for all farms which grew cereals in 1965. From these cards, lists were prepared for each parish giving the cereal acreage of all cereal producers in that parish. Sufficient parishes were then selected from those enumerated by the N.A.A.S. County Advisory Officers to yield the number of interviews required in each area.

INTERVIEWING

B.15. Informants were interviewed, as far as possible, five times over the period of nearly a year. The calls were distributed as follows:

Phase	Date	Type
First	May/June 1967	Personal
Second	July 1967	Telephone
Third	October 1967	Telephone
Fourth	February 1968	Telephone
Fifth	April 1968	Personal

B.16. The Second, Third and Fourth Phase interviews were carried out primarily by telephone. Certain recalls were made by personal interview. A certain number of farmers in each area were "lost" as members of the panel at either the second or third stage. In this case, they were reapproached personally at the fourth stage and as much data as possible was collected then pertaining to the third (and if necessary to the second) stage.

FIELD PERFORMANCE

B.17. In the First Phase, interviews were attempted with 389 farms and 234 interviews were successfully achieved, the remainder having refused to co-operate (52), were away (33), could not be contacted (37), or had moved or could not be traced (33).

B.18. Not all farms recruited at the First Phase were retained throughout the year. The number of interviews achieved at each phase of the study were as follows:

Area	Phase				
	First	Second	Third	Fourth	Fifth
Devon	40	39	39	37	37
Hampshire	40	38	38	37	37
Norfolk	40	38	34	34	34
Northamptonshire	38	34	30	29	29
Yorkshire (E. Riding)	40	40	40	40	40
E. Lothian	36	32	31	25	25

B.19. Tabulations were based only on those farms for which data were available for all five phases. The structure of the sample achieved by size of farm was as follows:

Size of farm (total acres crops and grass)	Devon	Hants.	Norfolk	Northants.	Yorks., E.R.	East Lothian	All farms
Up to 99 acres	23	6	26	10	13	3	81
100–299 acres	14	13	7	16	10	9	69
300 acres and over	—	18	1	3	17	13	52
	37	37	34	29	40	25	202

B.20. The topics covered in each phase of interviewing included the following:

Phase	Topics
First	1965 Cereal growing practice 1966 Cereal growing practice 1966 Yield Drying, general Drying, 1966 crop Storage, general Storage, 1966 crop 1966 Usage and disposal 1967 General growing practice and intentions 1967 Fertilisers and sprays purchases 1967 Disposal General
Second	1967 Crop, satisfaction with growing 1966 Disposal (contracts) 1966 Disposal (others) 1967 Crop disposal (contracts) General
Third	1967 Crop (harvesting) 1967 Crop disposal (contracts) 1968 Crop (intention) 1968 Beans (intention)
Fourth	1967 Crop disposal (intentions) 1967 Crop disposal (spot sales) 1967 Crop disposal (contracts) 1967 Crop contracts, delivery 1968 Crop disposal (intentions)
Fifth	1967 Crop disposal (contracts) 1967 Crop contracts, delivery 1967 Crop disposal (spot sales) 1968 Cereal sowing 1967 Crop drying Comparison: 1967 and 1968 cereal growing Milling and mixing Attitudes to H.-G.C.A. Attitudes to price policy Grain marketing Attitudes to merchanting

APPENDIX C

Regions used in the Report

(A) Advisory Regions

(i) N.A.A.S. Regions, England and Wales

South Eastern

Berkshire
Buckinghamshire
Greater London—part
Hampshire
Isle of Wight
Kent
Oxfordshire
Surrey
Sussex, East
Sussex, West

Yorks. and Lancs.

Lancashire
Yorkshire, E. Riding
Yorkshire, W. Riding

Eastern

Bedfordshire
Cambridge and Isle of Ely
Essex
Greater London—part
Hertfordshire
Huntingdon and Peter-
    borough
Lincolnshire (Holland)
Norfolk
Suffolk

West Midland

Cheshire
Herefordshire
Shropshire
Staffordshire
Warwickshire
Worcestershire

South Western

Cornwall
Devon
Dorset
Gloucestershire
Wiltshire
Somerset

Northern

Cumberland
Durham
Northumberland
Westmorland
Yorkshire, N. Riding

East Midland

Derbyshire
Leicestershire
Lincolnshire (Kesteven)
Lincolnshire (Lindsey)
Northamptonshire
Nottinghamshire
Rutland

Wales

Anglesey
Breconshire
Caernarvonshire
Cardiganshire
Carmarthenshire
Denbighshire
Flintshire
Glamorgan
Merioneth
Monmouthshire
Montgomeryshire
Pembrokeshire
Radnorshire

(ii) College Advisory Regions, Scotland[1]

West of Scotland Agricultural College

North Argyll
South Argyll
Islay, Jura, Gigha and
  Colonsay
Bute, Dunbarton and
  Renfrew
Lanark
Stirling and Clackmannan
West Perth
Ayr
Dumfries
Kirkcudbright
Wigtown

North of Scotland College of Agriculture

Aberdeen Region

Aberdeenshire
Kincardineshire
Banffshire
Orkney
Shetland

Highland Region

Inverness
Moray and Nairn
Ross and Cromarty
Sutherland
Caithness
Skye
Lewis and Harris
Uist and Barra

Edinburgh and East of Scotland College of Agriculture

Perth, East
Kinross
Angus
Fife
Mid and East Lothian
West Lothian
Peebles
Selkirk
Roxburgh
Berwick

(B) Farm Management Survey Regions, England and Wales

Northern

Cumberland
Derbyshire
Durham
Lancashire
Northumberland
Westmorland
Yorkshire, N. Riding
Yorkshire, W. Riding

Eastern and South Eastern

Cambridge and Isle of Ely
Essex
Greater London
Hertfordshire
Huntingdon and Peter-
  borough
Kent
Lincolnshire, Holland
Lincolnshire, Kesteven
Lincolnshire, Lindsey
Norfolk
Nottinghamshire
Suffolk, East
Suffolk, West
Surrey
Sussex, East and West
Yorkshire, E. Riding

[1] The College Advisory Regions were the only regions used for Scotland in the Report.

Central and Southern

Bedfordshire
Berkshire
Buckinghamshire
Cheshire
Gloucestershire
Hampshire
Herefordshire
Isle of Wight
Leicestershire
Northamptonshire
Oxfordshire
Rutland
Shropshire
Staffordshire
Warwickshire
Wiltshire
Worcestershire

South Western

Cornwall
Devon
Dorset
Somerset

APPENDIX D

The Future Demand and
Supply Position for Cereals

A. Demand for Cereals

TRENDS IN *per capita* CONSUMPTION

D.1. The direct human consumption of cereals has remained virtually constant since before the War. In a period when population increased, this implies a reduction in *per capita* consumption. A decline of this type is, in fact, a long-term trend in the U.K. and other developed countries. It has been estimated for example that *per capita* flour consumption in the U.K. fell from 280 lb per annum in 1880[1] to 221 lb in 1946–50 and 155 lb in 1965 (Table 1).

TABLE 1

TOTAL CONSUMPTION OF FLOUR PER HEAD
(IN BREAD, CAKES, ETC)

lb per annum

	Average 1946–50	Average 1956–60	1965
Canada	159	140	136
Australia	201	181	168
United Kingdom	221	172	155
West Germany	213	170	153
U.S.A.	141	120	115

Source: Commonwealth Secretariat *Grain Crops.*

D.2. This reduction in the direct human consumption of cereals reflects both a reduction in calorie intake as work becomes more sedentary, and the switch from cheap foods such as bread and potatoes which accompanies rises in real income. The change in *per capita* consumption since 1956, as revealed in the family budgets analysed by the National Food Survey, is shown in Table 1.2, Part V(b). It can be seen that total bread consumption fell from 51·08 oz per person per week in 1956 to 40·02 oz in 1967. There was a slight rise in the consumption of biscuits and cakes, but the total consumption of cereals fell from 70 oz in 1956 to 58 oz in 1967.

D.3. It seems likely that the reduction in human consumption of cereals still has a long way to go in the U.K. The British *per capita* calorie intake is—after the Irish—the highest in the world, and the level of cereal consumption is still high by American standards. If

[1] Report of the British Association, 1881.

713

Z*

it is assumed that flour consumption is directly correlated with real income, then projections of future consumption can be made, using the methods discussed below. However, it should not necessarily be assumed that cereal consumption is merely a function of income; it may also be a function of tastes, fashions, and of the quality of the product. If this is the case, consumption should be susceptible to influence, either through advertising, or through the quality of (in particular) bread. With the introduction of large-scale mechanised baking, the tendency in the U.S.A., and to some extent in Britain, has been for bread to become very standardised and to lose the distinctive quality it once had. This trend is not, however, inevitable; West Germany and the Scandinavian countries have been able to retain the traditional quality of their bread—mainly rye or wholemeal wheat bread—while adopting modern techniques.

D.4. It is bread consumption that largely determines flour consumption; in 1966, 66% of flour produced was used for making bread. Moreover, it has been the fall in bread consumption that has accounted for most of the fall in flour consumption. Thus the advertising campaign for bread, undertaken in recent years by the milling and baking industries, is understandable, and is probably justified.

ELASTICITIES OF DEMAND *per capita*

D.5. The *per capita* consumption of any foodstuff can be expressed as a function of income and price—which are measurable—as well as of the more intangible factors such as taste and habit. The relation of consumption to income and price is indicated by the "income elasticity" and "price elasticity" respectively. There is a considerable amount of empirical evidence regarding these elasticities, although it has to be treated with considerable caution when used to project future consumption.

D.6. The income elasticity for cereal products as a whole is negative in the U.K., and in countries with comparable levels of real income: this indicates that consumption falls as income rises. However, the income elasticity can be expressed in two forms—the income elasticity of quantity purchased and the income elasticity of expenditure. The income elasticity of quantity purchased represents the percentage change in the quantity of a commodity that is purchased when the consumer's real income rises by 1%. For example, if consumption rises by 1% when income rises by 2%, the elasticity is 0.5. The income elasticity of expenditure is similar, but measures the change in *expenditure* on a foodstuff as income rises. This would differ from the other type of income elasticity if the rise in income led the consumer to change to dearer qualities of the foodstuff, or to more expensive forms of packaging and processing, such as sliced and wrapped bread.

D.7. Estimates of income elasticities for foodstuffs are made regularly by the National Food Survey, and the 1965 figures for cereal products are given in Table 2. The figures for bread are negative, indicating that a rise in real income leads to a fall in consumption and expenditure. The figures for "cakes and biscuits" and "other cereals" on the other hand, are positive, indicating that consumption rises as income increases. However, the total consumption of these items is small compared with the consumption of bread. If the 1965 income elasticities of quantity purchased are weighted by the average quantities consumed in 1966, an aggregate elasticity of -0.15 is obtained. This would indicate, for example, that a 10% rise in real income *per capita* would lead to a 1.5% decline in the *per capita* consumption of cereals.

D.8. Price elasticities of demand (as distinct from income elasticities) represent the proportionate change in quantity purchased when the price changes. For example, if the price elasticity is 0.5, this indicates that a price rise of 1%—in relation to the general price level—will cause the amount purchased to fall by 0.5%. Since a price rise would not lead to a *rise* in the amount purchased, price elasticities are either zero or negative.

D.9. It is difficult to calculate price elasticities in the case of cereal products, because commodities like bread and cakes are not subject to large swings in price. There is therefore very little evidence to go on. However, some figures have been calculated by the National Food Survey, and are given in Table 3.

TABLE 2

INCOME ELASTICITIES OF DEMAND, U.K. 1965

	Income elasticities of expenditure	Income elasticities of quantity purchased
CEREALS:		
Brown bread[a]		
Unwrapped	0·40	0·36
Wrapped	−0·22	−0·20
White bread		
Large loaves, unwrapped	−0·10	−0·11
Large loaves, wrapped	−0·51	−0·51
Small loaves, unwrapped	0·12	0·10
Small loaves, wrapped	−0·01	−0·01
Wholewheat and wholemeal bread[a]	1·01	1·01
Malt bread	0·38	0·43
Other bread	0·16	0·12
Total bread	−0·20	−0·25
Self-raising flour	−0·12	−0·10
Other flour	−0·34	−0·38
Total flour	−0·18	−0·18
Buns, scones and teacakes	−0·31	−0·30
Cakes and pastries	0·16	0·09
Chocolate biscuits	0·56	0·53
Other biscuits	0·08	−0·02
Total cakes and biscuits	0·13	0·03
Puddings	−0·05	−0·16
Oatmeal and oat products	−0·38	−0·53
Breakfast cereals	0·26	0·23
Rice	−0·05	−0·10
Cereals, flour base	0·24	0·26
Other cereals	0·21	0·02
Total other cereals	0·18	0·14

[a] Certain proprietary brown breads were classified as "wholewheat and wholemeal bread" in 1958, but as "brown bread" subsequently.

Source: Household Food Consumption and Expenditure: 1965, Ministry of Agriculture, Fisheries and Food (Appendix E).

TABLE 3

PRICE ELASTICITIES OF DEMAND 1958–1963

	Elasticity	Standard error
Bread	−0·20	0·07
Flour	−1·36	0·76
Cakes and pastries	−0·92	0·26
Chocolate biscuits	−0·93	0·44
Other biscuits	−0·03	0·18
Oatmeal and oat products	−0·86	0·26
Breakfast cereals	−0·27	0·40

Source: Household Food Consumption and Expenditure: 1963, Ministry of Agriculture, Fisheries and Food.

The standard error column gives an indication of the extent to which the figure calculated for the elasticity, which is based on a small number of observations, might be incorrect. Broadly speaking, it is highly probable that the true figure will be within two standard errors, and almost certain that it will be within three.

D.10. In the case of bread, the elasticity given is −0·20, which means that if the price of bread went up 10%, the amount purchased would drop 2%. However, this estimate might be incorrect: all we can say is that the true figure is likely to be within the range 0·06 to −0·34 (twice 0·07 on either side of −0·20). This indicates that the range of error is rather large, and the same applies to most of the other cereal products. It is therefore difficult to say much about the reaction of the amount purchased to changes in price. If the prices of bread, cakes etc. should rise substantially more than the general price level this would cause some fall in the quantity purchased, but the effect is unlikely to be quantitatively important.

D.11. It has been stated above that the direct *per capita* consumption of cereals in the U.K. is likely to decline still further. But this does not mean that the total demand for cereals will decline—even apart from rises in population. On the contrary, the switch from cereals to animal products necessitates a higher total *per capita* supply of cereals, because of the losses entailed in transforming cereals into animal products. The extent of this rise in the demand for cereals will depend on various factors. It will be affected by the relative increase in the consumption of beef, dairy products and mutton as compared with eggs, pigmeat and poultrymeat, since cattle and sheep can obtain a large part of their food requirements from grass and fodder crops. It will also be affected by the amount of cereals (as compared with grass and fodder crops) fed to cattle and sheep. Finally, it will be affected by the improvement in the feed conversion ratio; in the case of broilers and pigs this ratio (the liveweight gain per lb of feed) has been improved considerably in recent years, and further improvements may be expected.

D.12. As far as meat is concerned, some indication of the likely change in *per capita* consumption can be obtained from recent figures of income elasticity of demand for these products (Table 4). The figures for most types of meat are positive; this indicates that a rise in real *per capita* income which in the longer run should be possible in spite of the present gloomy economic situation—would lead to a rise in the consumption of these types of meat. The highest elasticities—apart from those for rabbit and game—are for pork and chicken, and both pigs and poultry derive the bulk of their food from cereals. It therefore seems likely that the consumption of these two types of meat will continue to increase. Insofar as this meat is home-produced—and it seems likely that most of it will be—this will increase the demand in the U.K. for cereals as animal feed.

TABLE 4

INCOME ELASTICITIES OF QUANTITY PURCHASED, 1965

Meat and meat products:	
Carcase meat:	
Beef and veal	0·10
Mutton and lamb	0·21
Pork	0·31
Total carcase meat	0·18
Other meat:	
Corned meat	−0·31
Bacon and ham (uncooked)	0·11
Bacon and ham (cooked incl. canned)	0·14
Cooked chicken	0·87
Liver	0·08
Offals (other than liver)	0·37
Broiler chicken (uncooked)	0·42
Other poultry (uncooked)	0·82
Rabbit, game and other meat	0·62
Sausages, uncooked, pork	0·24
Sausages, uncooked, beef	−0·58
Other meat products	−0·07
Total other meat and meat products	0·08

Source: Household Food Consumption and Expenditure: 1965, Ministry of Agriculture, Fisheries and Food.

INCREASE IN POPULATION

D.13. In addition to the likely increase in *per capita* cereal requirements, there is the likely increase in total population. The Registrar-General's latest estimates for 1970, 1975 and 1980 (Table 5) indicate an increase of 5·6 million in the population of the U.K. by 1980 compared with 1965.

TABLE 5

PROJECTED FUTURE POPULATION OF THE U.K.
(million)

1965	1970	1975	1980
54·4	56·0	57·4	60·0

Source: Annual Abstract of Statistics 1967.

PROJECTED CEREAL REQUIREMENTS IN 1975

D.14. The most authoritative published estimates of future British cereal requirements are contained in a recent report by the Food and Agriculture Organisation of the United Nations.[1] This report makes detailed calculations of the demand for, and the supply of, cereals in the U.K. in 1975 together with more tentative estimates for 1985.

D.15. The F.A.O. first makes estimates of the future growth of the U.K. population and of real income (Gross Domestic Product at constant prices). For the 1975 population, its estimate is virtually the same as the Registrar-General's figure given above. For

[1] *F.A.O. Agricultural Commodities—Projections for 1975 and 1985.*

real income, it makes both a low and a high low estimate for 1975 and 1985, as shown in Table 6. The high estimates are based on the assumption (also shared by the National Plan) that the growth in *per capita* real income will be very much higher in the future than in the past. For the "high" figures, the compared annual growth rates from 1965 to 1975, and from 1975 to 1985, are 3·4% and 4·0% respectively, as against an actual growth rate of 2·0% over the period 1950 to 1963. The "low" figures—at 2·1% and 2·2% respectively —assume only a very small improvement over the past growth rate.

TABLE 6

1965 AND PROJECTED *per capita* REAL
INCOME, U.K.

1965	1975		1985	
	Low	High	Low	High
	U.S. $ at 1961–63 prices			
1373	1688	1916	2097	2820

*Source: F.A.O. Agricultural Commodities—
Projections for 1975 and 1985*, Vol 2, Table 1, p.10

D.16. The F.A.O. study then decides on income elasticities of demand for the main food groups; details are given in Table 7. The F.A.O. elasticities for meat are on the whole higher than the 1965 figure of the National Food Survey, and this would give a higher figure for the cereal stockfeed requirements. One complication is that the elasticities tend to fall as income rises, meaning that consumption levels off once a sufficiently high level of income is reached. This can, however, be taken care of by a mathematical function. Given, therefore, estimates of population, income and income elasticity of demand, and assuming that prices remain constant, it is possible to estimate future consumption of the various foodstuffs. The *per capita* figures are given in Table 8, from which it can be seen that a continual fall in cereal consumption and a rise in the consumption of meat is forecast.

TABLE 7

INCOME ELASTICITY OF DEMAND, U.K.

Cereals	−0·2
Beef and veal	0·3
Mutton and lamb	0·3
Pork	0·4
Poultry	0·8
Total meat	0·4
Eggs	0·3
Milk and milk products excluding butter	0·0
Butter	0·1
Calories	0·0
Total proteins	0·06

*Source: F.A.O. Agricultural Commodities—
Projections for 1975 and 1985*, Vol. 2, Table
1.8, p. 28, 29.

TABLE 8

1965 AND PROJECTED 1975 AND 1985 *per capita* CONSUMPTION

Kg. per year

Food Group	1965 (Actual)	1975		1985	
		Income Assumption			
		Low	High	Low	High
Cereals	80·4	77·6	76·1	75·2	72·9
Starchy roots	96·0	89·5	86·3	84·3	79·1
Sugar	48·8	48·0	47·5	47·3	46·5
Vegetables	59·3	62·9	65·2	66·7	71·9
Fruit	56·9	62·8	66·4	68·9	77·4
Meat, total	71·7	76·2	78·9	80·7	86·3
Beef and veal	23·6	24·9	25·6	26·0	27·8
Mutton and lamb	11·3	12·0	12·4	12·7	13·7
Pork	9·5	10·3	10·8	11·1	12·2
Poultry	6·9	8·0	8·7	9·1	10·7
Other	20·3	21·1	21·4	21·7	22·4
Eggs	15·5	16·3	16·8	17·0	17·9
Fish	19·0	19·8	20·3	20·6	21·7
Milk	199·1	199·1	199·1	199·1	199·1
Fats and oils	26·6	26·7	26·8	26·8	27·0
Calories (no.)	3275	3277	3282	3288	3312
Proteins	89·3	90·2	90·9	91·4	93·2
Animal protein	54·0	56·0	57·1	57·9	60·3

Source: F.A.O. Agricultural Commodities—Projections for 1975 and 1985, Vol. 2, Table B, p. 126.

D.17. The total estimated consumption of beef, pigmeat, poultrymeat and eggs is shown in Table 9 together with estimates of home production in 1975. It can be seen that home production is estimated to increase roughly in line with the increase in consumption, so that the deficit to be filled by imports remains roughly the same.

D.18. The projected increase in the production of meat and eggs would lead to an increase in cereal requirements for animal feed. As shown in Table 10, col. 6, total cereal requirements for non-food purposes are estimated at between 18·3 and 19·0 million tons in 1975, according to whether the low or high income assumption is realised. This represents a rise of 3–4 million tons over the 1967 level. The human consumption of cereals in 1975 is estimated to be 6·3 million tons or 5·7 million tons according to the low or high income assumption—in this case a higher income means a lower consumption. This level of "food use" represents either a continuation of the present level, or a fall of about 0·6 million tons. The total consumption for both food and non-food uses is estimated to be 24·6 on the low or 24·7 million tons on the high income assumption. Thus the total consumption is not greatly influenced by income, and represents an increase of some 2 million tons over the 1967 level.

B. Supply of Cereals

D.19. The estimation of cereal production in 1975 is even more problematical—although less complex—than the estimation of consumption. It is bound, in effect, to be a guess based on present production trends. The F.A.O. estimates a figure of 18·9 million tons. This represents a rise of 4·3 million tons compared with the record crop of 1967/8, a rise to 10·5 million acres as compared with 9·4 million acres in 1967, and a rise in

TABLE 9

ESTIMATED PRODUCTION AND CONSUMPTION OF LIVESTOCK PRODUCTS, U.K. 1975

Food group	Base period 1961–3 average			1967/8 production (forecast)(a)	1975 projected low income assumption (high income assumption in brackets)		
	Production	Net imports	Total consumption		Production	Total demand	Deficit(b)
		'000 tons		'000 tons		'000 tons	
Beef	837·0	548·8	1385·8	906	1015	1574 (1619)	559 (554)
Pigmeat	753·0	630·9	1383·9	769	985	1621 (1695)	636 (675)
Poultrymeat	351·0	8·7	359·7	455	455	460 (496)	6 (6)
Eggs	770·0	54	824	n.a.	910 (936)	936 (962)	26 (26)

(a) *Annual Review and Determination of Guarantees*, Cmnd. 3558 Appendix I.   (b) A deficit would correspond to imports unless either production or demand were altered by official action.

Source: *F.A.O. Agricultural Commodities—Projections for 1975 and 1985*, Vol. 1, pp. 151, 153, 154, 160.

TABLE 10

U.K. PRODUCTION AND CONSUMPTION OF GRAIN

	(1) Area '000 acres	(2) Yield cwt/acre	(3) Production	(4) Net imports	(5) Total consumption	(6) Non-food uses	(7) Food uses (gross)
			'000 tons				
All grains							
1961–3	7843	27·1	10,800	8901	19,788	13,506	6282
1975 (low income)	10,549	35·4	18,963	5654	24,617	18,282	6335
1975 (high income)	10,549	35·4	18,963	5734	24,697	19,003	5694
Wheat							
1961–3	2004	30·0	3061	4449	7626	2653	5573
1975 (low income)	2718	39·0	5390	2558	7948	2656	5292
1975 (high income)	2718	39·0	5390	2434	7824	2773	5051
Coarse grains							
1961–3	5839	26·1	7739	4452	12,162	11,453	709
1975 (low income)	7831	34·1	13,573	2726	16,299	15,626	673
1975 (high income)	7831	34·1	13,573	3300	16,873	16,230	643

Source: F.A.O. Agricultural Commodities—Projections for 1975 and 1985, Vol. 1, Tables 9–17, pp. 96–104.

average yield from 31 to 35 cwt per acre. All of this is quite feasible, if perhaps at the upper limit of the likely increase, assuming that there is no substantial rise in the prices of cereals in relation to other crops. Assuming that this increase in production takes place, net imports would fall to 5·6 or 5·7 million tons compared with the present level of around 8 million tons. If home production should rise less than the F.A.O. calculates, imports would be higher than 5·6 million tons, but it is extremely unlikely that they would be higher than the present level. Thus the level of cereal imports is likely to fall by 2 million tons between now and 1975.

THE NATIONAL PLAN

D.20. A somewhat different approach to the question of future cereal requirements has been provided by the Department of Economic Affairs in the National Plan. The "National Plan for 1965 to 1970"[1] calculated that the demand for cereals for human and industrial use would grow by £12 million (at 1964 import prices) between 1964 and 1970, and that the demand for cereals for animal food would "increase substantially" (p. 138). The Plan's proposals for cereals were as follows:

"In the case of crops the most important additional requirements would be for cereals. This would be due almost wholly to the cereals needed for increased livestock production. Additional output of cereals would be possible through increased acreages of both wheat and coarse grains and a continued increase in yields. The additional acreage would be at the expense of fodder crops and grains; and there are limits beyond which it would not be economic to expand. Nevertheless, it would be technically feasible to increase home production of cereals to meet the additional requirements expected" (p. 139).

The Plan went on to say that:

"The expansion of meat and milk production will increase considerably the demand for cereal feed and, consistently (*sic*) with our international commitments, a substantial part of these additional requirements should come from home sources. The Government would expect home agriculture to be able to meet a major part of the additional demand expected by 1970, totalling some £200 million on food for human consumption. It would also supply much of the cereals required for the increase in livestock production" (p. 141).

D.21. It need hardly be added that many of the targets (or forecasts) of the National Plan are unlikely to be achieved. Indeed, there have been many criticisms from economists of the Plan's confusion between targets and forecasts, and of the detailed way in which the Plan was prepared; one might add that the solecisms of its prose style hardly suggest great clarity in the underlying thought. However, in the case of agriculture in general, and cereals in particular, the idea that increased demand should be met from home production was both a reasonable target and a forecast that is likely to be realised.

C. Conclusions

D.22. In recent years there has been a fall in direct *per capita* consumption of cereals in the U.K., together with a rise in indirect consumption *via* animal products. This trend is likely to continue for the next decade. Between now and 1975, the demand for cereals for human consumption is likely to remain fairly constant, with increasing population balancing a falling *per capita* consumption. The demand for cereals for stockfeed—and hence the total demand—is, however, likely to rise by 3 to 4 million tons. Home production is likely to rise by at least this amount, and probably more; thus imports are unlikely to rise and are more likely to fall by up to 2 million tons. The fall in imports will affect both wheat and coarse grains, but substantial quantities of hard wheat (around 2·5 million tons annually) will be required for the foreseeable future.

[1] H.M.S.O. Cmnd. 2764, 1965.

The Economic Implications
of a Change in the Basis of the
Wheat Subsidy from Tonnage to Acreage

E.1. As the wheat subsidy scheme is arranged at present, all wheat has to move off the farm (except seed wheat under special arrangements) before the subsidy can be claimed. The subsidy is based on the tonnages sold, and if wheat is not sold it is not eligible for the subsidy. Those farmers who grow wheat, keep livestock, and have the necessary feed preparation equipment on the farm, would probably keep some or all of their wheat for feeding on the farm if it were not for these arrangements. The subsidy system is thus preventing the natural development of on-the-farm feeding (i.e. in so far as this is economic for the farmer—it is always so), and is using resources of transport, labour, storage, etc., that could be put to other uses. This Appendix sets out the results of an exercise aimed at estimating what the national saving in resources might be if the subsidy were changed to an acreage basis, and what might be the effect on producers' net incomes.

E.2. From information supplied by farmers and merchants it can be estimated that the cost to the farmer of having to buy in wheat rather than using his own, is about £2 to £3 per ton. The potential overall saving can be illustrated by an actual example. In the course of our enquiries we met a farmer who grows 1000 acres of wheat and has a poultry unit of 80,000 laying birds. He could use all, or nearly all, the wheat he produces for his own poultry flock if it were not for the way the subsidy system is at present operated. As it is, he must sell all his wheat and buy exactly similar wheat (but not his own) back again at £3 (this was the farmer's estimate) per ton extra cost, i.e. about £4500 per annum. Admittedly he has been saved storage costs, but a substantial part of the extra £4500 could be avoided. This case is exceptional as there are not many units of this size in the country, but even if the potential savings on other farms are smaller, they would still be worth having.

E.3. An attempt was made to estimate what the total potential saving might be. This involved estimating—(a) the tonnage of wheat grown on farms with livestock, and (b) of this total how much might be consumed by the livestock on these farms. With the co-operation of the Ministry of Agriculture, which is gratefully acknowledged, a random 1 : 4 sample of the 2000 non-horticultural farms in the Farm Management Survey (E.W.) for 1965, was drawn and each of the 256 farms in this sub-sample that grew wheat was classified into one of the four categories:

(1) Wheat, but no pigs, poultry or cattle.
(2) Wheat and cattle but no pigs or poultry.
(3) Wheat and cattle and pig and/or poultry.
(4) Wheat and pigs and/or poultry but no cattle.

The likely wheat requirements of the three above classes of livestock were then estimated for each farm and expressed in terms of wheat acreage. It was assumed that 0·6 acre of wheat would be needed per 100 birds, and 2 acres per sow plus pigs. On the basis of these figures the equivalent wheat acreage for each farm with pigs and poultry was calculated. This is the acreage of wheat that would be potentially available for feeding on the farm if the subsidy system were changed. In addition there would be a sizable additional acreage that might be fed to cattle but no attempt was made to estimate for each farm the acreage-equivalent of cattle numbers, as was done with the intensive livestock.

E.4. The results of this analysis (raised to the England and Wales level) are given in Table 1. They may be summarised as follows:

Of the total wheat acreage in England and Wales (2,432,500 acres on 65,373 farms in 1965, the year to which the data relate):[1]

109,700 acres were on 2300 farms that had no cattle, pigs or poultry (generally the farms in the middle size ranges).

742,800 acres were on 15,400 farms that had cattle but no pigs and/or poultry (well distributed throughout the farm size groups).

1,326,600 acres were on 38,700 farms that had cattle and pigs and/or poultry (well distributed by farm size). Of this total, the pigs and/or poultry represented a feed requirement of 316,300 acres of wheat.

253,400 acres were on 8950 farms that had pigs and/or poultry but no cattle (these tended to be either the smallest or the largest farms). Of this total, the pigs and/or poultry represented a feed requirement of 62,400 acres of wheat.

E.5. These results suggest that, contrary to some opinion, nearly all the wheat-growing farms have some livestock on them. However, the livestock would not be able to consume all the wheat grown on these farms. Thus the production from 378,700 acres of wheat (316,300 + 62,400) out of over $1\frac{1}{2}$ million acres, could potentially be mixed on the farm for feeding to pigs and poultry, provided, of course, that the farmer was prepared and equipped to mix his own feed. At the national average yield of wheat of 32·4 cwt per acre this would represent 613,490 tons. To this total must be added the proportion of the output of wheat on farms with cattle that might potentially be fed to those cattle. This can only be a rough estimate but it was assumed that the wheat requirements per cow might be 4·4 cwt per annum, and per follower 2·3 cwt per annum. On average each wheat-growing farm with cattle in the F.M.S. sub-sample had 30 dairy cows, 32 cattle one year and over and 28 cattle under one year.[2] On this basis the wheat requirements per farm would be about $13\frac{1}{2}$ tons per annum. For the 54,100 wheat-growing farms with cattle this would represent a total of 730,300 tons. This figure is less exact than that for pigs and poultry because no attempt was made to estimate how much wheat might be consumed by the cattle on each farm. Thus the potential tonnge of wheat which might be fed on farms could be anything up to about $1\frac{1}{4}$ million tons. At an estimated average saving in transport and handling of £2 to £3 per ton there is a potential total saving of up to £2$\frac{1}{2}$ to £4 million. These figures make no allowance for any increase in the acreage of wheat that might result from a change in the basis of the subsidy, but on the other hand they assume that all farmers with livestock capable of using wheat would in fact feed it on the farm—obviously many would prefer to buy compounds as at present. Thus they represent the maximum potential saving rather than the saving likely to be realised in practice.

E.6. The potential saving per farm on average would be as shown in Table 2: These figures show that if farmers took advantage of a change in the subsidy arrangements to feed wheat to livestock on the farm there could be useful savings in costs (before allowing for storage costs).

E.7. Apart from the saving in transport costs and the benefits accruing from allowing the farmer to decide for himself whether to feed the wheat on the farm or to sell it, the other advantages that might follow from the change are:

---

[1] Comparable figures for 1966 are 2,171,356 acres on 55,801 farms, and for 1967, 2,222,535 acres on 53,574 farms.

[2] The F.M.S. farms tend to be rather larger than average and these figures may therefore be on the high side.

TABLE 1 ANALYSIS OF WHEAT-GROWING FARMS IN ENGLAND AND WALES 1965/6
(Raised results from a sample of 256 wheat-growing farms in the Farm Management Survey)

Farm size group (acres)	Wheat acreage size group (acres)	Acreage of wheat on farms that:						Total wheat acreage
		Have no cattle, pigs or poultry	Have cattle but no pigs or poultry	Have cattle and pigs and/or poultry		Have pigs and/or poultry but no cattle		
				Total acreage of wheat	Feed requirements of pigs and poultry expressed as acres of wheat	Total acreage of wheat	Feed requirements of pigs and poultry expressed as acres of wheat	
0–49¾	0–49¾	—	8064	38,708	33,547	33,870	19,354	80,642
50–99¾	0–19¾	—	15,448	60,934	31,450	9440	5961	85,822
	20–99¾	39,743	22,320	14,558	14,558	—	—	76,621
100–149¾	0–19¾	2947	11,786	44,199	13,702	—	—	58,932
	20–149¾	—	46,274	69,411	35,414	28,922	4049	144,607
150–299¾	0–19¾	1957	23,479	36,524	26,020	3261	181	65,221
	20–29¾	—	21,052	56,923	12,026	—	—	77,975
	30–49¾	—	53,120	92,960	18,364	19,919	11,121	165,999
	50–69¾	—	73,845	54,645	12,587	19,200	8682	147,690
	70–299¾	10,247	53,287	53,287	2181	88,128	6020	204,949
300–499¾	0–49¾	—	54,222	27,111	8721	—	—	81,333
	50–69¾	—	37,511	42,300	14,110	—	—	79,811
	70–99¾	—	33,261	105,326	24,939	—	—	138,587
	100 and over	—	70,648	184,774	35,526	16,303	4840	271,725
500 and over	0–99¾	—	38,253	48,686	3171	—	—	86,939
	100–199¾	54,775	114,530	79,674	1155	—	—	248,979
	200 and over	—	65,670	316,684	29,873	34,335	2231	416,689
Total		109,669	742,770	1,326,704	316,344	253,378	62,439	2,432,521

TABLE 2

AVERAGE POTENTIAL SAVING ON FEED COSTS PER FARM IF WHEAT WERE FED
TO LIVESTOCK ON THE FARM OF ORIGIN

	Saving on feed for pigs and poultry	Saving on feed for cattle	Total saving
	(£ per farm per annum)		
15,400 farms with wheat and cattle but no pigs or poultry	—	27	27
38,700 farms with wheat and cattle and pigs and/or poultry	16	27	43
8950 farms with wheat and pigs and/or poultry but no cattle	139	—	139

(a) It might encourage farmers to switch some of their production from barley to wheat—the proposed change would help towards this objective insofar as it encouraged livestock farmers growing cereals to choose wheat rather than barley. This would apply particularly to the western parts of the country where yields are below average and the tonnage basis works to the disadvantage of wheat.

(b) It is widely accepted that there is a need for more hard wheats of the Maris Widgeon type to be grown. These tend to be somewhat lower-yielding than the soft wheats, and hence attract a lower subsidy per acre as well as giving a lower value of production. An acreage basis of subsidy payment would help slightly to offset this disadvantage.

(c) There would be some administrative economies if wheat were put on the same basis as barley, and this would go some way towards streamlining the existing over-elaborate cereal subsidy schemes.

E.8. One of the main arguments likely to be raised against making the change is that it might discourage productive efficiency, since all would benefit alike regardless of yield. Although all wheat producers would certainly benefit alike, it does not follow that there would be any decline in productive efficiency. Whilst the subsidy remains a relatively small proportion of total returns it will always pay a cereal farmer to aim for maximum yields. There would, of course, be a reduction in the value of the wheat crop to the farmer who at present gets high yields, and an increase to the farmer with low yields, but the impact on any one farmer's returns is likely to be small. An attempt was made to calculate what this might be as follows:

(a) First a study was made of 789 wheat-growing farms in the total Farm Management Survey Sample to obtain the yield distribution of wheat by size of wheat acreage grown per farm (Table 3A).

(b) From this was calculated the percentage distribution by yield of (i) wheat acreage, and (ii) total holdings with wheat, for the sample farms.

(c) These percentages were then applied to the total wheat acreage, and the total number of holdings with wheat in England and Wales (M.A.F.F. statistics), to obtain the national yield distribution within each wheat acreage group (Table 4).

(d) It was now possible to estimate the net increase or decrease in net income resulting from a change from the tonnage basis of subsidy to the acreage basis (Table 5). The average subsidy per ton (£3 12s. 6d.) was taken from Table B of the 1966 White Paper *Annual Review and Determination of Guarantees 1966* (Cmnd. 2933). To estimate what the subsidy per acre would have been if the acreage basis had been in operation the total subsidy was divided by the acreage grown (£5 17s. 5d.). By comparing the acreage figure with the tonnage figure appropriate to each yield it was possible to estimate the net benefit or loss per acre, and by multiplying by

TABLE 3A

ANALYSIS OF YIELDS

(Sample of 789 wheat farms in the Farm Management Survey in E. and W. 1965/6)

WHEAT

Yield (cwt per acre)	\multicolumn{9}{c}{Acreage grown}	Total	Cumulative total
	0–49¾	50–99¾	100–149¾	150–199¾	200–249¾	250–299¾	300–349¾	350–399¾	400+		
19 and under	14									14	14
20	14	2								16	30
21	11	1								12	42
22	10	4								14	56
23	11	4		1						15	71
24	17	3	1	1						22	93
25	13	4	2	1	1					22	115
26	21	9		1	2					33	148
27	13	7	2	1	2					24	172
28	32	13	3	5	2	1		1		56	228
29	16	11	4	2						35	263
30	34	12	5	1	1					53	316
31	24	10	4	2	2		1			41	357
32	31	18	6	1	4	2		2		54	410
33	21	12	3	7		2				40	464
34	24	7	6	1	1		1		1	50	504
35	33	13	6	1				1	1	60	554
36	11	19	5	3						33	614
37	24	13	4	2	1	2				41	647
38	14	9	3	2	2	2	1	1		25	688
39	19	6	1		2					28	713
40	2	4								5	741
41	8	2	3							12	746
42	7	1								9	758
43	5	2			1					7	767
44		2									774
45 and over	13	1			1					15	789
Total	466	189	66	31	18	9	2	6	2	789	

Median $= \dfrac{790}{2} = 395 = 32$ cwt

Lower quartile $= \dfrac{790}{4} = 197\tfrac{1}{2} = 28$ cwt

Upper quartile $= \dfrac{790 \times 3}{4} = 592\tfrac{1}{4} = 36$ cwt

Quartile deviation $= \tfrac{1}{2}(36 - 28) = 4$ cwt

Source: Farm Management Survey.

TABLE 3B

ANALYSIS OF YIELDS

(Sample of 966 barley farms in the Farm Management Survey in E. and W. 1965/6)

BARLEY

Yield (cwt per acre)	Acreage grown									Total	Cumulative total
	0–49¾	50–99¾	100–149¾	150–199¾	200–249¾	250–299¾	300–349¾	350–399¾	400+		
19 and under	7	3	—	1	—	—	—	1	—	12	12
20	23	—	1	—	—	—	—	—	—	24	36
21	3	—	1	—	—	—	—	—	—	4	40
22	14	2	3	2	1	—	1	—	—	18	58
23	7	5	1	1	1	—	—	—	—	16	74
24	11	5	6	2	3	1	—	—	—	19	93
25	30	10	5	2	1	1	—	—	1	47	140
26	17	9	6	4	1	1	—	—	—	37	177
27	11	13	6	2	3	3	—	—	1	32	209
28	33	11	8	5	3	4	3	—	2	59	268
29	17	29	10	9	3	1	1	1	2	48	316
30	125	10	12	7	1	2	2	—	—	185	501
31	19	14	9	4	2	2	1	1	1	52	553
32	35	16	7	4	3	1	—	—	—	70	623
33	30	16	8	6	1	—	1	2	1	61	684
34	23	13	9	7	2	1	1	—	—	58	742
35	30	9	16	3	3	—	—	—	—	66	808
36	19	8	2	—	—	—	—	—	—	57	865
37	8	3	2	2	—	—	—	—	—	21	886
38	11	4	3	1	1	—	—	—	—	23	909
39	6	1	1	—	—	—	—	—	—	12	921
40	28	—	—	—	—	—	—	—	—	36	957
41	4	—	—	—	—	—	—	—	—	6	963
42	1	—	—	—	—	—	—	—	—	2	965
43	—	—	—	—	—	—	—	—	—	—	—
44	—	—	—	—	—	—	—	—	—	—	—
45 and over	1	—	—	—	—	—	—	—	—	1	966
Total	513	199	118	62	27	16	12	8	11	966	966

Median $= \dfrac{967}{5} = 483\tfrac{2}{5} = 30$ cwt

Lower quartile $= \dfrac{967}{4} = 241\tfrac{3}{4} = 28$ cwt

Upper quartile $= \dfrac{967 \times 3}{4} = 725\tfrac{1}{4} = 34$ cwt

Quartile deviation $= \tfrac{1}{2}(34 - 28) = 3$ cwt

Source: Farm Management Survey.

TABLE 4

DISTRIBUTION OF WHEAT ACREAGE AND HOLDINGS WITH WHEAT IN ENGLAND AND WALES BY SIZE OF WHEAT ACREAGE AND YIELD (E. AND W.) 1965/6

Yield range (cwt per acre)	¼–49¾ %	¼–49¾ Acreage ('000)	¼–49¾ Holdings ('000)	50–99¾ %	50–99¾ Acreage ('000)	50–99¾ Holdings ('000)	100–199¾ %	100–199¾ Acreage ('000)	100–199¾ Holdings ('000)	200–299¾ %	200–299¾ Acreage ('000)	200–299¾ Holdings ('000)	300 and over %	300 and over Acreage ('000)	300 and over Holdings ('000)	All sizes %	All sizes Acreage ('000)	All sizes Holdings ('000)
14 and under 16	0·8	6·6	407	—	—	—	—	—	—	—	—	—	—	—	—	—	6·6	407
16–18	0·8	6·6	407	—	—	—	—	—	—	—	—	—	—	—	—	—	6·6	407
18–20	1·3	10·6	661	—	—	—	—	—	—	—	—	—	—	—	—	—	10·6	661
20–22	5·4	44·2	2746	1·6	9·6	141	—	—	—	—	—	—	—	—	—	—	53·8	2887
22–24	4·5	36·8	2288	4·2	25·1	369	—	—	—	—	—	—	—	—	—	—	61·9	2657
24–26	6·4	52·4	3254	3·7	22·1	325	4·1	23·4	177	11·1	23·4	101	—	—	—	—	121·3	3857
26–28	7·3	59·7	3712	8·5	50·8	746	4·1	23·4	177	3·7	7·8	34	20·0	46·8	102	—	188·5	4771
28–30	10·3	84·3	5237	12·7	75·8	1117	13·4	76·5	578	18·5	38·9	168	10·0	23·5	52	—	299·0	7152
30–32	12·5	102·3	6353	11·6	69·4	1021	12·4	70·9	535	7·4	15·6	67	—	—	—	—	258·2	7976
32–34	12·0	98·2	6100	15·9	95·1	1399	16·5	94·3	712	14·8	31·1	134	20·0	46·8	102	—	365·5	8447
34–36	9·7	79·4	4932	10·5	62·8	924	17·5	99·9	754	22·3	46·9	201	20·0	46·8	102	—	335·8	6913
36–38	9·4	76·9	4779	17·0	101·9	1496	15·5	88·6	669	3·7	7·8	34	10·0	23·5	52	—	298·7	7030
38–40	8·1	66·3	4118	7·9	47·2	695	9·3	53·1	401	7·4	15·6	67	20·0	46·8	102	—	229·0	5383
40–42	4·5	36·8	2290	3·2	19·1	281	4·1	23·4	177	7·4	15·6	67	—	—	—	—	94·9	2815
42–44	3·2	26·2	1627	1·6	9·6	141	3·1	17·7	134	—	—	—	—	—	—	—	53·5	1902
44–46	1·5	12·3	763	1·6	9·6	141	—	—	—	—	—	—	—	—	—	—	21·9	904
46–48	0·8	6·6	407	—	—	—	—	—	—	3·7	7·8	34	—	—	—	—	14·4	441
48–50	0·8	6·6	407	—	—	—	—	—	—	—	—	—	—	—	—	—	6·6	407
50–52	0·7	5·7	356	—	—	—	—	—	—	—	—	—	—	—	—	—	5·7	356
Total	100·0	818·5	50,844	100·0	598·1	8796	100·0	571·2	4314	100·0	210·5	907	100·0	234·2	512	—	2432·5	65,373

Source: Based on Farm Management Survey and June Census Statistics.

TABLE 5

EFFECT ON INDIVIDUAL FARM INCOMES OF A CHANGE IN THE WHEAT SUBSIDY FROM TONNAGE PAYMENTS TO ACREAGE PAYMENTS

Average yield (cwt per acre)	Subsidy per acre on a: Tonnage basis (£ s. d.)	Acreage basis (£ s. d.)	Net benefit or loss (£ per acre)	¼–49¾ Av. acreage=16 No. of farms	¼–49¾ Benefit or loss (£ per farm)	50–99¾ Av. acreage=68 No. of farms	50–99¾ Benefit or loss (£ per farm)	100–199¾ Av. acreage=132 No. of farms	100–199¾ Benefit or loss (£ per farm)	200–299¾ Av. acreage=233 No. of farms	200–299¾ Benefit or loss (£ per farm)	300 and over Av. acreage=451 No. of farms	300 and over Benefit or loss (£ per farm)
15	2 14 5	5 17 5	+3 3 0	407	+50	—	—	—	—	—	—	—	—
17	3 1 7	5 17 5	+2 15 10	407	+45	—	—	—	—	—	—	—	—
19	3 8 11	5 17 5	+2 8 6	661	+39	141	+140	—	—	—	—	—	—
21	3 16 1	5 17 5	+2 1 4	2746	+33	369	+116	—	—	—	—	—	—
23	4 3 5	5 17 5	+1 14 0	2288	+27	325	+91	—	—	—	—	—	—
25	4 10 8	5 17 5	+1 6 9	3254	+21	746	+66	177	+177	101	+312	—	—
27	4 17 11	5 17 5	+ 19 6	3712	+15	1117	+42	177	+129	34	+227	102	+440
29	5 5 1	5 17 5	+ 12 4	5237	+10	1021	+17	578	+81	168	+144	52	+271
31	5 12 4	5 17 5	+ 5 1	6353	+4	1399	-7	535	+33	67	+59	—	—
33	5 19 7	5 17 5	- 2 2	6100	-2	924	-32	712	-15	134	-25	102	-49
35	6 6 11	5 17 5	- 9 6	4932	-8	1496	-56	754	-63	201	-111	102	-214
37	6 14 1	5 17 5	- 16 8	4779	-13	695	-81	669	-111	34	-194	52	-376
39	7 1 5	5 17 5	-1 4 0	4118	-19	281	-106	401	-159	67	-278	102	-541
41	7 8 7	5 17 5	-1 11 6	2290	-25	141	-130	177	-207	67	-363	—	—
43	7 15 11	5 17 5	-1 18 8	1627	-30	141	-155	134	-255	—	—	—	—
45	8 3 1	5 17 5	-2 5 8	763	-36	—	—	—	—	—	—	—	—
47	8 10 5	5 17 5	-2 13 0	407	-42	—	—	—	—	34	-617	—	—
49	8 17 7	5 17 5	-3 0 2	407	-48	—	—	—	—	—	—	—	—
51	9 4 11	5 17 5	-3 7 6	356	-54	—	—	—	—	—	—	—	—

Source: Derived from Farm Management Survey and June Census Statistics.

the average size of farm in each wheat acreage group to estimate the net benefit or loss per farm. The results on a per acre basis are given in Table 6 and on a per farm basis in Table 7.

E.9. Table 6 shows that for the majority of wheat produced the change in the basis of the subsidy would not alter the gross receipts more than about £1 per acre. This is small relative to the effect of yield itself on gross receipts. Thus, an increase in yield from $27\frac{1}{2}$ cwt per acre to $32\frac{1}{2}$ cwt per acre can increase gross receipts by about £5.[1] This can be illustrated by Fig. 1. The effect of the change in the basis of subsidy is clearly small compared with the effect of yield over the range 25–40 cwt per acre (which accounts for 84% of the acreage). In the light of these figures it is difficult to believe that a change in the basis of the subsidy would lead any of the high-yield producers to produce less efficiently, or cause any of the low-yield producers to relax their efforts to increase yields.

E.10. As to the effect of a change from the tonnage basis to the acreage basis on incomes per farm, Table 7 shows that of the 65,373 farms in England and Wales with wheat in 1965 two-thirds (44,041) would be virtually unaffected by the change (i.e. the net income would not be changed by more than £25 per annum), whilst another 27% would be affected to the extent of plus or minus £25–£99. The remainder, about 3000 farms, would be affected by more than £100 per farm (1321 better off and 2603 worse off). Of course, all those farmers who had livestock enterprises and chose to feed their wheat would stand to gain up to £2/£3 per ton fed. This could more than offset the loss in income on many farms, and add to the gain on others. We conclude from this analysis that although a small number of farmers could be significantly worse off with an acreage basis there would be many more who would be better off. But the important point is that these are compensating changes within the farming industry, whilst the saving resulting from wheat not having to be sold off and then bought back again is a real saving in national resources. If the acreage basis can lead to a more rational use of agriculture's, and merchants', resources we do not think that it should be rejected because some farmers will be a little worse off and some a little better off.

E.11. It is sometimes said that cereal producers on marginal farms are "farming for the subsidy", i.e. so long as their yield of barley is just sufficient to satisfy the Ministry's requirement that the crop must be grown for a profit they are content to put the minimum of effort and expenditure into the crop and take the subsidy as the main reward. It is suggested that if the acreage basis were introduced for wheat, farmers might begin to "farm for the subsidy" with this crop as well. However, this idea gained currency at a time when the subsidy was as much as £8 or £9 per ton. Now the subsidy is a relatively small proportion of the value of the crop and it would hardly pay the rent let alone other costs (Table 8). The statistics show a steady decline in the proportion of total cereals grown on a small scale, and the advice we have received from cereal farmers is that with costs of production as they are today, very few farmers are likely to be "farming for the subsidy". Perhaps a few small cereal producers, using family labour with the minimum of field operations and virtually no machinery, may still be farming in this way but these can be regarded as a small and fast-disappearing group and on the whole this argument can be discounted.

E.12. Some millers, in particular, have expressed the view that anything which is likely to lower the quality of English wheat coming on to the market will diminish the proportion of English wheat to imported that millers are prepared to take. It is important to keep a high quality standard for English wheat because the imported wheats come in with reliable grading and homogeneous quality, and if the English product has a high admixture, is badly dried, or comes in "penny packets" of variable quality, it will not be wanted for milling.

E.13. These are sound arguments but it seems likely that a change to acreage payments is likely to improve the position, rather than worsen it, by discouraging poor quality wheat from coming on to the market. At present virtually every ton of wheat

---

[1] The increase in net profits would be proportionately much greater, because of the large element of fixed costs.

TABLE 6

EFFECT ON GROSS RECEIPTS PER ACRE OF WHEAT OF THE ACREAGE BASIS COMPARED WITH THE TONNAGE BASIS

| Yield per acre | Gross receipts per acre | | | | | | Estimated acreage grown (England and Wales) | Estimated tonnage grown |
| | Tonnage basis | | | Acreage basis | | | | |
	Market return	Subsidy	Total	Market return	Subsidy	Total		
(cwt)	£ s. d.	£ s. d.	£ s. d.	£ s. d.	£ s. d.	£ s. d.	('000)	('000)
Under 20	18 7 11	3 3 7	21 11 6	18 7 11	5 17 5	24 5 4	24	21
20 and under 25	23 13 0	4 1 8	27 14 8	23 13 0	5 17 5	29 10 5	176	198
25 and under 30	28 18 1	4 19 9	33 17 10	28 18 1	5 17 5	34 15 6	548	754
30 and under 35	34 3 2	5 17 10	40 1 0	34 3 2	5 17 5	40 0 7	792	1287
35 and under 40	39 8 3	6 15 11	46 4 2	39 8 3	5 17 5	45 5 8	696	1305
40 and over	44 13 4	7 14 0	52 7 4	44 13 4	5 17 5	50 10 9	197	375
Total	34 1 6	5 17 5	39 18 11	34 1 6	5 17 5	39 18 11	2433	3940

TABLE 7

EFFECT ON INDIVIDUAL FARM INCOMES OF A CHANGE IN THE BASIS OF THE WHEAT SUBSIDY PAYMENTS FROM TONNAGE TO ACREAGE

Number of farmers who would be:

Acreage of wheat grown per farm	Significantly better off by: (£ per annum)					Significantly worse off by: (£ per annum)					Unaffected to any significant extent (i.e. incomes up or down by less than £25)	Total all wheat farms (England and Wales)
	25–99	100–199	200–299	300 or more	Total	25–99	100–199	200–299	300 or more	Total		
¼–49¾	6509	—	—	—	6509	3560	—	—	—	3560	40,775	50,844
50–99¾	2188	510	—	—	2698	3115	563	—	—	3678	2420	8796
100–199¾	1113	354	—	—	1467	754	1070	311	—	2135	712	4314
200–299¾	67	168	34	101	370	—	235	67	101	403	134	967
300 and over	—	—	52	102	154	102	—	102	154	358	—	512
Total	9877	1032	86	203	11,198	7531	1868	480	255	10,134	44,041	65,373

Source: Derived from Farm Management Survey and June Census Statistics.

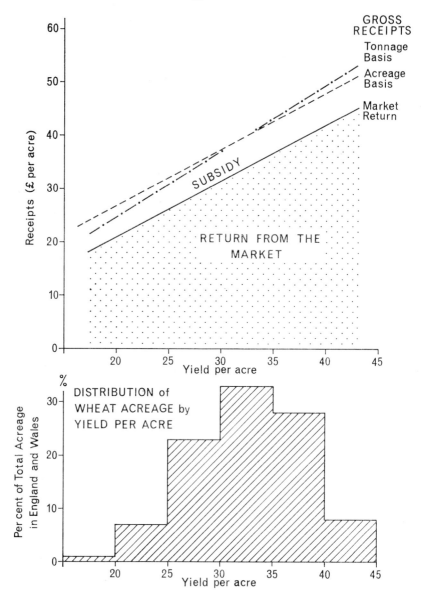

FIG. 1.1. Comparison between the tonnage basis of subsidy for wheat and the acreage basis.

TABLE 8

SUBSIDY AS A PROPORTION OF TOTAL RETURN
1959/60 TO 1967/8

	1959/60	1960/1	1961/2	1962/3	1963/4	1964/5	1965/6	1966/7	1967/8
	s. d.	*s. d.*	*s. d.*	*s. d.*	*s. d.*	*s. d.*	*s. d.*	*s. d.*	*s. d.*
Wheat (per cwt)									
Market value	20 4¾	18 10½	21 6¼	17 11	21 4	21 3¾	21 0	22 6¼	22 4
Unit subsidy	6 10½	8 0½	5 2¾	9 5¼	5 5¾	4 8¾	3 7¾	2 10	3 7
Total return	27 3¼	26 11	26 9	27 4¼	26 9¾	26 0½	24 7¾	25 4¼	25 11
Subsidy as % of total return	25	30	20	35	20	18	15	11	14
Barley (per cwt)									
Market value	20 5½	19 5¾	19 11¾	19 8	20 5¼	20 8¾	21 9¼	20 11½	21 2
Unit subsidy	8 6½	9 3¼	7 7¼	7 11	6 2¾	5 4	3 0¾	3 7½	3 2¼
Total return	29 0	28 9	27 7	27 7	26 8	26 0¾	24 10	24 7	24 4½
Subsidy as % of total return	29	32	28	29	23	20	12	15	13

Source: Table B, White Paper, *Annual Review and Determination of Guarantees 1968*, Cmnd. 3558.

produced has to be sold and some of it is of very indifferent quality; much of this wheat would never enter the market if the acreage system were introduced.

E.14. A change to the acreage basis for wheat would mean replacing the present short-period guaranteed prices with one guaranteed price for the year as a whole. This would be an important consequence of the change, and it is the one that we have found most difficult to evaluate. At present the wheat producer knows that whatever the price of wheat will be during the season he will receive a deficiency payment, based on the difference between the average price in each one-month or two-month period and the guaranteed price for the period. The seasonal steps are cumulatively so substantial that in effect many producers virtually ignore the price factor and store their wheat for as long as they conveniently can. This does not mean to say that they all store to the end of the season, although an increasing number do, but that their decision as to when to sell is dictated mainly by such factors as the need for capital, or storage space, etc. This system is undoubtedly a great asset to the wheat producer as it largely removes any anxiety about price. Even if the price were to fall by a pound or two per ton at the end of the season when the wheat is released from store, the farmer still benefits from the deficiency payment related to the high seasonal guaranteed price for that period. Thus, in a sense, he is immune from the consequences of his own actions. If there were an average-year guaranteed price (as with barley) the deficiency payment would only make up the difference over the year as a whole—and over such a long period the price could move up or down quite sharply. The farmer could not be certain what the price levels would be several months hence and this would introduce a note of uncertainty that is not there now, so far as the farmer is concerned. The farmer who "played the market" best might stand to gain by selling his grain when the price was high and drawing a deficiency payment reflecting the lower prices received by those farmers who were not so skilful and sold when prices were down. Confidence in the future would be diminished and the steady movement of wheat off the farm might be prejudiced.

E.15. The present system is undoubtedly a great asset to wheat producers and not unnaturally they would be reluctant to lose it. But the question has to be asked whether it is best in the national interest. Is it reasonable that the wheat producer should be almost entirely insulated from the market price in this way? Merchants are already complaining that with the system as it now stands they are almost powerless to attract wheat off farms—they cannot offer a price sufficiently attractive to offset the seasonal guarantees. At present the market price tends to be effective only in the very short term. It may influence the farmer when he decides on which day or in which week to sell; it has little influence over the longer term. So far the system seems to have worked reasonably well, but there are some warning signs on the horizon that it may eventually break down. Already in some years a high proportion of wheat is kept in store until the end of the season, and prices then are depressed as a result (see Part VII).

Appendix E

E.16. With the present system the price of wheat early in the season would have to rise considerably before it would give the farmer any incentive to sell rather than store. This can be illustrated with reference to the 1966/7 season. Table 9 shows the market prices together with the guaranteed prices for each month (these are only estimates but they are near enough for this purpose) and the variable costs of storage. Only a variable charge is included because once the farmer has built storage facilities he will disregard his fixed costs in deciding at what point in the season to sell his wheat. Only the cost of the drying required to reduce the moisture to a level adequate for storage has been included. The miscellaneous costs have been purposely pitched on the generous side to cover every possible eventuality. The table is not intended in any way as a detailed study of the

TABLE 9

RELATIONSHIP BETWEEN THE MARKET PRICE AND THE EXPECTED RETURN AFTER STORAGE
(1966/7 SEASON)

	Market price(a) (per cwt)	Seasonal guaranteed price(b) (per cwt)	Variable costs of storage			Guaranteed price less cost of storage (per cwt)	Plus H.-G.C.A. contract bonus	Total expected return after storage
			(1) Interest (per cwt)	(2) Drying	(3) Misc.			
	s. d.	*s. d.*	*s. d.*	*d.*	*d.*	*s. d.*	*d.*	*s. d.*
Aug.	20 10	23 0	2	3	1	22 6	—	22 6
Sept.	21 5	23 0	4	3	2	22 3	—	22 3
Oct.	22 6	24 6	6	3	3	23 6	—	23 6
Nov.	22 10	24 6	8	3	4	23 3	3	23 6
Dec.	22 9	25 10	10	3	5	24 4	6	24 10
Jan.	23 2	25 10	1 0	3	6	24 1	7½	24 8½
Feb.	23 4	25 10	1 2	3	7	23 10	7½	24 5½
Mar.	23 1	27 0	1 4	3	8	24 9	7½	25 4½
Apr.	23 3	27 0	1 6	3	9	24 6	7½	25 1½
May	23 3	27 6	1 8	3	10	24 9	7½	25 4½
June	23 3	27 6	1 10	3	11	24 6	7½	25 1½

(a) Corn Returns.   (b) Broomhall's Yearbook.   (1) Taking 7½% per annum.   (2) Assuming 5s. per ton to dry from 18% to 16% for storage.   (3) Approximate estimate to cover wastage of the crop, insurance, handling expenses, loss of value through drying.

storage problem but merely to illustrate the fact that so long as the expected return later in the season, after allowing for the variable costs of storage, (about £1 5s. 3d. for March–June) and including the H.-G.C.A. contract bonus, exceeds the guaranteed price in the immediate post-harvest period (23s. 9d. average Sept./Oct.), the post-harvest market price becomes completely ineffective. The market price would have to rise by about 3s. or 4s. per cwt in Sept./Oct. to persuade a farmer, who had intended to store, to dispose of his crop earlier. Until the market price rises above the guaranteed price, of course, it does not affect the income of the average producer since the increase will be counterbalanced by a decrease in the deficiency payment.

E.17. If millers are not prepared to pay the high prices necessary to attract wheat out of store soon after harvest they have no alternative but to turn to imported wheats, to the disadvantage of our balance of payments. There is no doubt that the shortage of domestic wheat for sale in November and December leads to more wheat being imported than would otherwise have been the case. Thus, in a year when wheat is in short supply the acreage basis would generally lead to more flexible marketing than the tonnage basis.

E.18. To sum up, the present system certainly gives valuable assurances to the farmer, but it seems to us that it insulates him from the market in a way that is likely to lead to inefficient marketing and that the implications of this are becoming increasingly important as production continues to rise. On balance, therefore, we favour a change in the basis of the wheat system to acreage. The basic reason why we take this view is that it seems desirable, in a situation like the present when cereal acreage is still expanding and seems likely to go on expanding for some time in the future, to let wheat "find its own level" as a feed grain on farms. We have not included a detailed analysis of the potential role of wheat as a component of animal feed rations for on-the-farm mixing, but the advice we have received is that it could form a substantial part of rations for pigs and poultry and to a lesser extent for cattle. The present subsidy system distorts the natural pattern of wheat utilisation in that wheat has to be sold to qualify for the subsidy, and it involves unnecessary transport and handling which could amount at the maximum to £4 million. In our view, if it is economical for farmers to keep their wheat on the farm for feeding to livestock rather than to sell it, then it is undesirable that they should be forced to sell it simply to claim the subsidy. We therefore suggest that consideration be given to the possibility of changing the basis of the wheat subsidy to an acreage basis.

APPENDIX F

Details of Storage Costs Estimates

Building costs for five central stores of varying size have been synthesised from actual building quotations using the following method.

(a) The dimensions are arrived at by calculating the volume available assuming 10 ft retaining walls with grain (at 46 ft³ per ton) heaped to its natural angle of repose (35°). The length of the building has then been increased by about 15% to allow space for the equipment and to make some allowance for not fully utilising the theoretical capacity.

(b) Following from this, the framework has been charged at £28 per foot run and the cladding (double skin with vapour barrier and insulation) at £0·134 per ft². All five buildings have the same span (100 ft) and are merely lengthened to accommodate the larger tonnages.

(c) As the 100-ft span is not suitable for a 2000-ton building, the costs for the co-operative store have been synthesised from a 70-ft open building on a £0·525 per ft² basis.

(d) Foundations have been estimated at £2 per yd², an above-average cost intended to cover excavation for the drier, intake pits and provide an approach road.

(e) The cost of grain walling has been calculated using a price of £6 per foot run.

(f) The cost of the drier(s) includes grain handling equipment, erection and electrical work and is based on the information collected during the Survey of Machinery Agents.

(g) Pre-drying storage is provided for 1 day's drying, assuming a 50-day intake period.

(h) Other equipment includes ventilating ducts to condition the grain whilst in store and a cleaner.

739

2000 TON CO-OPERATIVE STORE

Dimensions 70 × 85 × 16 ft

BUILDING	£	£/ton
Structure	3124	1·56
Foundations	1400	0·70
Grain walling	2016	1·01
	6540	3·27

EQUIPMENT	£	£/ton
Drier and grain handling	7392	3·70
Pre-drying storage (50 tons at £10)	500	0·25
Weigher (not weighbridge)	600	0·30
Other equipment	1148	0·57
	9640	4·82
TOTAL COST	£16,180	£8·09

OPERATING COSTS	£	£/ton
Haulage	672	0·34
Fuel	440	0·22
Electricity	200	0·10
Repairs (1 % total cost)	162	0·08
	1474	0·74

Cost of on-farm drying and storage 2000 tons at £10 per ton=£20,000

OPERATING COSTS	£	£/ton
Fuel	300	0·15
Electricity	200	0·10
Repairs (1 % total cost)	200	0·10
	700	0·35

DISCOUNTED COSTS

	Co-operative	On-farm stores
Capital cost	£16,180	£20,000
	£	£
10-year total of discounted operating costs	10,352	4916
Plus capital cost	16,180	20,000
	26,532	24,916
Less salvage value at 20% (£3236 discounted)	1644	(£4000 discounted) 2032
	24,888	22,884

5000 TON CENTRAL STORE

Dimensions 100 × 120 × 20 ft

BUILDING	£	£/ton
Structure	6713	1·34
Foundations	2667	0·53
Grain walling	2460	0·49
	11,840	2·37

EQUIPMENT	£	£/ton
Drier and grain handling	7605	1·52
Pre-drying storage (100 tons at £10)	1000	0·20
Weighbridge	2500	0·50
Other equipment	1900	0·38
	13,005	2·60
TOTAL COST	£24,845	£4·97

OPERATING COSTS	£	£/ton
Labour	2400	0·48
Haulage	2060	0·41
Fuel	1100	0·22
Electricity	500	0·10
Repairs (1% total cost)	248	0·05
	6308	1·26

Cost of on-farm drying and storage 5000 tons at £10 per ton=£50,000

OPERATING COSTS	£	£/ton
Fuel	750	0·15
Electricity	500	0·10
Repairs (1% total cost)	500	0·10
	1750	0·35

DISCOUNTED COSTS

	Central store	On-farm stores
Capital cost	£24,845	£50,000
	£	£
10-year total of discounted operating costs	44,301	12,290
Plus capital cost	24,845	50,000
	69,146	62,290
Less salvage value at 20% (£4969 discounted)	2524	(£10,000 discounted) 5080
	66,622	57,210

10,000 TON CENTRAL STORE
Dimensions 100 × 220 × 20 ft

BUILDING	£	£/ton
Structure	11,469	1·15
Foundations	4889	0·49
Grain walling	3480	0·35
	19,838	1·98

EQUIPMENT	£	£/ton
Drier and grain handling	10,962	1·10
Pre-drying storage (200 tons at £10)	2000	0·20
Weighbridge	3000	0·30
Other equipment	2800	0·28
	18,762	1·88
TOTAL COST	£38,600	£3·86

OPERATING COSTS	£	£/ton
Labour	3300	0·33
Haulage	4370	0·44
Fuel	2200	0·22
Electricity	1000	0·10
Repairs (1% total cost)	386	0·04
	11,256	1·13

Cost of on-farm drying and storage 10,000 tons at £10 per ton = £100,000

OPERATING COSTS	£	£/ton
Fuel	1500	0·15
Electricity	1000	0·10
Repairs (1% total cost)	1000	0·10
	3500	0·35

(For Discounted Costs see Table 7.5 of Part III.)

15,000 TON CENTRAL STORE
Dimensions 100 × 315 × 20 ft

BUILDING	£	£/ton
Structure	15,988	1·06
Foundations	7000	0·47
Grain walling	4500	0·30
	27,488	1·83

EQUIPMENT	£	£/ton
Drier and grain handling	18,460	1·23
Pre-drying storage (300 tons at £10)	3000	0·20
Weighbridge	3000	0·20
Other equipment	3700	0·25
	28,160	1·88
TOTAL COST	£55,648	£3·71

OPERATING COSTS	£	£/ton
Labour	4090	0·27
Haulage	6930	0·46
Fuel	3300	0·22
Electricity	1500	0·10
Repairs (1% total cost)	556	0·04
	16,376	1·09

Cost of on-farm drying and storage 15,000 tons at £10 per ton = £150,000

OPERATING COSTS	£	£/ton
Fuel	2250	0·15
Electricity	1500	0·10
Repairs (1% total cost)	1500	0·10
	5250	0·35

DISCOUNTED COSTS

	Central store	On-farm stores
Capital cost	£55,648	£150,000
	£	£
10-year total of discounted operating costs	115,009	36,871
Plus capital costs	55,648	150,000
	170,657	186,871
Less salvage value at 20% (£3275 discounted)	5654	(£30,000 discounted) 15,240
	165,003	171,631

20,000 TON CENTRAL STORE
Dimensions 100 × 415 × 20 ft

BUILDING	£	£/ton
Structure	20,744	1·04
Foundations	9222	0·46
Grain walling	4272	0·21
	34,238	1·71

EQUIPMENT	£	£/ton
Drier and grain handling	21,924	1·10
Pre-drying storage (400 tons at £10)	4000	0·20
Weighbridge	3500	0·17
Other equipment	4600	0·23
	34,024	1·70
TOTAL COST	£68,262	£3·41

OPERATING COSTS	£	£/ton
Labour	5100	0·25
Haulage	9740	0·49
Fuel	4400	0·22
Electricity	2000	0·10
Repairs (1% of total cost)	683	0·03
	21,923	1·09

Cost of on-farm drying and storage 20,000 tons at £10 per ton = £200,000

OPERATING COSTS	£	£/ton
Fuel	3000	0·15
Electricity	2000	0·10
Repairs (1% total cost)	2000	0·10
	7000	0·35

DISCOUNTED COSTS

	Central store	On-farm stores
Capital cost	£68,262	£200,000
	£	£
10-year total of discounted operating cost	153,965	49,161
Plus capital cost	68,262	200,000
	222,227	249,161
Less salvage value at 20% (£13,652 discounted)	6935	(£40,000 discounted) 20,320
	215,292	228,841

25,000 TON CENTRAL STORE
Dimensions 100 × 515 × 20 ft

BUILDING	£	£/ton
Structure	25,500	1·02
Foundations	11,444	0·46
Grain walling	5280	0·21
	42,224	1·69

EQUIPMENT	£	£/ton
Drier and grain handling	24,738	0·99
Pre-drying storage (500 tons at £10)	5000	0·20
Weighbridge	3500	0·14
Other equipment	5500	0·22
	38,738	1·55

TOTAL COST	£80,962	£3·24

OPERATING COSTS	£	£/ton
Labour	5370	0·21
Haulage	12,500	0·50
Fuel	5500	0·22
Electricity	2500	0·10
Repairs (1% total cost)	810	0·03
	26,680	1·07

Cost of on-farm drying and storage 25,000 tons at £10 per ton = £250,000

OPERATING COSTS	£	£/ton
Fuel	3750	0·15
Electricity	2500	0·10
Repairs (1 % total cost)	2500	0·10
	8750	0·35

DISCOUNTED COSTS

	Central store	On-farm stores
Capital cost	£80,962	£250,000
	£	£
10-year total of discounted operating costs	187,374	61,451
Plus capital cost	80,962	250,000
	268,336	311,451
Less salvage value at 20% (£16,192 discounted)	8226	(£50,000 discounted) 25,400
	260,110	286,051

AA*

APPENDIX G

Statements by Associations

G.1. In the course of the Cereals Survey, several trade Associations offered to prepare statements to assist in the work, and asked for guidance on those aspects of the broad field of study on which their comments would be most valuable. It was felt that not only these bodies, but all the main organisations within the cereals industry, should be given the opportunity of contributing to the Survey. With this end in view, the organisations listed below were invited to prepare statements for publication in the Cereals Survey Report. The various organisations were left free to frame their statements in any way they chose, but each was given a list of general questions as a guide to the aspects which seemed particularly important from the Survey's point of view. This Appendix contains these questionnaires and the statements which were submitted in reply.

G.2. The replies received have been arranged in the following order, corresponding with that adopted in the report itself.

G.3. Except where otherwise stated, submissions have been reproduced verbatim. The lists of questions sent to the various organisations are also printed in full; each questionnaire is immediately followed by the replies received from the bodies to which it was sent.

A. The Farmers' Unions

Questions addressed to the National Farmers' Union, the National Farmers' Union for Scotland, and the Ulster Farmers' Union.

A. PRODUCTION TRENDS

1. What are the main factors likely to affect the supply of the main cereals from U.K. farms in the next 5 years, and what do you expect the effect of these factors to be?

2. What are the likely trends in relation to mixing of feed on farms?

3. If the wheat subsidy were paid on an acreage basis, as with barley, to what extent would cereals farmers who at present sell all their wheat crop be likely to retain it on the farm to feed to livestock?

B. MARKETING OF GRAIN

4. What are the main difficulties facing the farmer in the marketing of his grain, and how could the marketing system be improved? In particular, to what extent is the farmer a weak seller? What difference has the formation of farmers' groups made in this respect?

5. What factors influence the farmer as to whether to store or sell his grain and what is their relative significance? Do the present Cereals Deficiency Payments system and the H.-G.C.A. Forward Contracts Scheme work satisfactorily so far as the farmer is concerned? To what extent do such factors as shortage of storage facilities, lack of capital to finance spring cultivation, indebtedness to merchants, cause the farmer to act in a way that may be contrary to his best interests in the marketing sense, e.g. by causing him to sell during the harvest period rather than later?

6. Does the price the farmer receives for his grain adequately reflect quality differentials? If not, how might this be achieved?

7. To what extent does grain fetch a lower market price than it otherwise would because of such factors as inefficient drying, poor storage or lack of cleaning?

8. To what extent has the process of concentration in the agricultural merchanting industry affected the farmer's position as a seller of grain and a buyer of requisites?

9. In choosing how, where and when to sell his grain does the farmer have adequate means available of informing himself as to the market situation? Does the N.A.A.S. play any part in this? Should it do more, and if so in what way?

10. Are there any other features of the grain market which have not been mentioned in your replies to the above questions but which you consider to be of special importance?

Statement by the National Farmers' Union

QUESTION 1. *What are the main factors likely to affect the supply of the main cereals from U.K. farms in the next 5 years, and what do you expect the effect of these factors to be?*

749

The main factors are, (a) profitability depending mainly on government policy, (b) the volume and nature of demand, (c) technical factors affecting yield per acre, (d) availability of capital and labour, and (e) availability of land.

(a) In recent years the profitability both absolute and relative of cereal-growing has declined sharply. This has been reflected in the slowing down of the expansion in acreage which occurred in the first half of this decade. Unless this trend of profitability were to be reversed, further expansion would be unlikely. The fundamental influence upon profitability is the Government's determination as to prices and standard quantities at successive annual price reviews: the overall picture since the mid-1950s has been one of reduced guaranteed prices despite continually increasing costs, and, since 1964, with the additional restrictive effect of standard quantities upon the overall return per acre. For 1968, the important step of removing the standard quantity for wheat has been taken, together with a significant increase in guaranteed price, but a standard quantity for barley has been retained, and the increases in guaranteed prices for both barley and oats are small. While these most recent indications of government policy suggest some recognition of the potentially important contribution which can be made to the national balance of payments through import-saving by increased production of cereals in the U.K., they do not clear the way for realisation of that potential, even to the extent of achieving the target set in the national economic plan of 1965.

Devaluation has increased the prospective cost of cereal imports. World market prices, expressed in sterling, must be likely to rise (quite apart from the effect of the international agreement on grain prices that is projected as a result of the G.A.T.T. "Kennedy round" discussions). Under the present system of deficiency payments, supported by minimum import prices at a reasonable level, higher market prices should lead to the possibility of greater encouragement of U.K. production without increase in Exchequer cost. (U.K. entry into the European Economic Community, with adoption of the Community's current agricultural policy, would, in the short term at least, significantly stimulate cereals production in the U.K.—since our productive efficiency compares most favourably with the standard upon which E.E.C. price levels are based).

Providing that profitability per acre recovers from recent depressed levels, continued expansion in cereal acreage can be foreseen. On present government policy, some expansion in wheat acreage appears likely, but whether this proves to be a net increase in cereal acreage or to be at the expense of other cereals depends upon future government policy.

(b) Demand for cereals must be distinguished as between human and animal consumption.

For human consumption the main outlets are for flour-milling wheat, malting and distilling barley, and milling oats.

Milling wheat imports are substantial, more than 2 million tons: while present milling and baking techniques continue, home-grown wheat cannot be expected to replace them fully: these techniques cost the country dearly in foreign exchange. Nevertheless, with some modification in baking processes and the developing use of new home-grown milling varieties such as Maris Widgeon, it should be possible to replace fully half a million tons of these imports from home-grown supplies—provided that market premia give the required incentive—overall demand for milling wheat is not likely to change significantly in the next 5 years.

Demand for barley for malting is now almost fully met from home-grown sources: but for distilling purposes some substitution of imports would be possible with a modification of processing techniques. Export trade in malting barley, and in malt has been developed: its maintenance depends on a number of factors, including the trade policy of the importing countries.

The main utilisation of home-grown grain is for livestock feeding, either through manufactured compounds, or straight on the farm. To feed the existing livestock population, substantial imports of feed wheat and sorghums are still needed: these can be entirely replaced from home sources, as could also a high proportion of current imports of maize. The replacement may more readily take place through increased wheat production, since wheat is in greater demand as a high energy feed than barley: but there is room for substitution of barley without loss of quality, as experience, when deliveries of maize have been interrupted, has clearly shown.

Expansion of U.K. livestock population should take place in the next 5 years, and the demand for cereals should increase correspondingly, particularly in the production of pig and poultry meat.

Thus, both in terms of an increase in total requirements of cereals, and of potential replacement of imports within the existing demand, there can be no doubt that there should be ample outlet for expanding home production.

(c) Yields per acre of all home-grown cereals have increased substantially in the past 20 years, although, with the exception of oats, the upward trend has been less marked in the last 3 or 4 years. Many technical factors (apart from yearly climatic variations) affect yield trends: it is to be expected that the trend will continue upwards, albeit at the reduced rate of the most recent past.

Much of the striking rise in yields since the last war has been due to the introduction and rapid adoption of new varieties: further progress in this direction is foreseeable in the next 5 years, particularly with winter wheat and oats, from new varieties already in the pipeline. There is evident need for higher yielding disease-resistant varieties of spring wheat, the acreage of which must be expected to increase: it is hoped that the work of plant-breeders will be directed to this end.

Usage and methods of application of fertilisers have had considerable impact on yields: continuing development can be looked for, provided that increasing costs are adequately recouped. Straw shortening, either through new varieties or through chemical treatment, could increase the response to additional fertiliser application.

As with fertilisers, the increased use of sprays for weed control in recent years has had considerable impact on yields; the impact in the next 5 years will be less, although further progress in the control of grass weeds should be expected.

Much attention has been focused on the effect of diseases and pests upon cereal yields, particularly in certain areas and in certain cases of so-called monoculture. Their increased incidence, especially of foliar diseases, is not to be ignored, but it must be anticipated that techniques of husbandry will be so adapted as to keep them under control, e.g. through the use of more tolerant varieties, the development of chemical treatments, and the adjustment of cropping rotations.

"Break" crops must become of increasing importance: under present conditions increases in acreage of potatoes and sugar-beet are unlikely: that of peas will be determined by the processing demand: that of field beans should increase substantially in the "wheat" lands: the downward trend in oat production has already been reversed, and the oat acreage may (if prospects for utilisation improve) continue to increase somewhat in the arable areas in view of the crop's suitability as a break between other cereal crops: rapeseed has potentiality, particularly in conditions where beans are not a practicable proposition, and is a useful import-saver: experiments in the development of maize production are in progress.

It is vitally important for expansion of cereal production that the overall price arrangements determined by the Government, under the Agriculture Acts, should take into account the need for a suitable rotation, and should produce a proper balance between cereals production and livestock population.

(d) Increased cereals production, especially on farms where they are already grown, demands less capital than expansion of production of other farm commodities. It is, however, far from negligible—the need for drying and storage installations, in addition to cultivating and harvesting equipment exemplifies this. Until recently, the supply of capital for investment has not been a limiting factor: but the fall in profitability has now begun to bring into question the justification for new investment. Profitability has indeed declined to the point where even replacement of existing assets has become a strain on resources.

Unless the downward trend in the profitability of cereal production is reversed, the supply of capital required for expansion will be jeopardised.

Providing that the necessary investment in mechanised production and handling of grain on the farms is forthcoming, it is to be expected that an adequate labour force will be available for expanding production.

(e) The potential expansion in cereal production which should make an invaluable contribution to the national balance of payments should arise partly from a continuing increase in yields per acre, but will also require a substantially greater acreage. Total cereal acreage has increased by over one million acres between 1962 and 1965. A further increase of similar proportions can, given the right conditions, (including encouragement of improved field drainage), be envisaged in the next 5 years. It would occur through a gradual extension of the mainly arable areas towards the west and north. Structural changes occurring through the amalgamation of holdings will create larger units which are more readily adaptable to cereal cropping than before.

Inevitably the grassland acreage will be reduced, and more intensive use of grass will be called for. This is feasible, provided that the necessary incentive for the production of livestock and livestock products is available.

Summary of reply to Question 1

During the next 5 years a considerable expansion of cereal production will be required, (a) to reduce national expenditure on imports of produce which can be grown in the U.K., and (b) to meet expanding demand. Technically this expansion is entirely possible. If profitability is restored, the necessary capital investment will be forthcoming. The expansion of acreage should occur mainly in wheat, although some lesser increase in supplies of barley and oats could be expected particularly from a continuing increase in yield per acre.

QUESTION 2. *What are the likely trends in relation to the mixing of feeds on the farm?*

In addition to the availability of the labour and capital required for the development of farm-mixing and to a careful assessment of its relative profitability, there are other considerations which will influence individual decisions, e.g. the relative complexity or simplicity of the required ration (taking into account the use of pre-mix additives), and the proportion of the total ingredients which can be supplied from the farm itself as compared with the proportion that must be brought in. These considerations must in turn be affected by such factors as the extent to which different types of livestock production expand in future, and the volume and variety of grain or pulse production on farms which also carry a significant livestock population. The development of grouping for production and marketing purposes could also be a relevant factor.

It is therefore extremely difficult to forecast a future trend. There is some evidence of increased farm-mixing in recent years, but insufficient evidence to justify a firm prediction either that this will continue or that it will die away.

QUESTION 3. *If the wheat subsidy were paid on an acreage basis, as with barley, to what extent would cereals farmers who at present sell all their wheat crop be likely to retain it on the farm to feed to livestock?*

Wheat production is much more concentrated in the predominantly arable areas than is the production of other cereals, i.e. in areas where livestock population is lower than average. Within those areas it tends to be grown on farms where stock numbers are at a minimum. The scope for retention of wheat on the farm of origin is thus much less than it is for barley or oats.

In this factor lies one of the main historical reasons for the tonnage basis of the wheat deficiency payments scheme, as opposed to the acreage basis for barley and oats.

A tonnage basis for barley or oats could draw off large quantitites from the farm of origin for which a ready use can be found there: this is not the case with wheat. For example, wheat is particularly in demand as a feed for poultry—but large poultry units are not usually to be found in association with large-scale arable production.

The normal principle of the guarantee arrangements for farm commodities is that the farmer is rewarded for the quantity that he produces. This principle is fully observed by the tonnage scheme for wheat; the tonnage basis is indeed the most satisfactory from the angle both of the progressive farmer and of the national interest. The farmer has every incentive to go all out for maximum production.

Clearly there would be some retention of wheat on the farm of origin, if the basis of the scheme were to be changed from tonnage to acreage: but, having regard to the areas and farms on which wheat is mainly grown, it is considered that the percentage of the total crop that would be so retained would be very small. It is believed that this could continue to be the case even if the wheat acreage is increased. The increase in farm retention would not be large enough to justify discarding the tonnage basis.

Notes

1. The percentages of total U.K. acreages of the main cereals in 1967 that were grown in the M.A.F.F. Eastern, South-eastern and East Midland Regions were:

Wheat	67%
Barley	48%
Oats	20%

2. Wheat comprised 24% of the total U.K. acreage in 1967, but 33% of the total cereals acreage in the Eastern, South-eastern and East Midland regions.

QUESTION 4. *What are the main difficulties facing the farmer in the marketing of his grain, and how could the marketing system be improved? In particular, to what extent is the farmer a weak seller? What difference has the formation of farmers' groups made in this respect?*

In 1963 a special Cereals Marketing Study Group, set up by the N.F.U., considered that farmers faced a number of difficulties in marketing their grain, e.g. (a) a continuing tendency for the number of processors, and of independent merchants, to diminish, resulting in an increased disparity in commercial strength between individual farmers as sellers, and those to whom their grain must be sold; (b) a need for many to sell at or soon after harvest in order to meet outstanding financial commitments; (c) insufficiency of market intelligence, leading to marketing decisions based less on an informed assessment of market requirements than on a mixture of instinct and perhaps inaccurate trade rumours; (d) the non-availability of forward contracts which could enable the farmer to keep the market more regularly supplied; (e) the ever-present possibility that the home market could be depressed by unreasonably low-priced imports.

Since 1963 the introduction of minimum import prices, and the establishment of the Home-Grown Cereals Authority, have led to changes in the marketing situation. Some of the difficulties that were then diagnosed have been eased (their present impact is commented on in greater detail in later answers).

The individual farmer's main aim must be to sell at the right price, whether "spot", or "forward", i.e. one which will bear satisfactory relationship to the average market price, as calculated for deficiency payments purposes. Collectively, farmers' aims should be to ensure that average market prices are properly in line with the prices of comparable imported grain: in the setting of a market that is freely open to imports, subject to the "floor" provided by minimum import prices, this is a fair objective, from the angle not only of sellers, but of buyers, and of the Exchequer. The indications in recent years have been that, despite increasing home production, this collective aim of selling at prices in line with those of competing imports has more nearly been achieved than it was a few years ago. The very fact of this collective achievement means that it is less easy for the individual to beat "bogey" (i.e. the average market price) by a significant amount. The individual consequently needs more access than ever before to the best possible market intelligence, he needs as close and confident a working relationship with his buyer as possible, and he needs to think in terms of a pattern of forward contracts for selling much of his grain.

The larger the selling unit the less difficult it is to establish these prerequisite marketing conditions, particularly when farmers are faced with the continuing decline in the number of independent competitive buyers: for this reason, as well as for possible economies in drying and storage, farmers' cereal marketing groups have been formed in many parts of the country. Some sell direct to processors, some through nominated merchants—their trading pattern is varied. Group-selling implies marketing in larger lots than would otherwise be offered: this should reduce distribution costs, and also be more attractive to end-users, who have in the past adversely compared the small size and varying quality of home-grown grain with imported grain. Reports of group operations are not, however, available for long enough to enable a full evaluation of their contribution to be made: but the keen support of their members, despite the sacrifice of some individual freedom of action, demonstrates that they are providing a needed service: a genuine local sense of need for such service is vital as a stimulus for a successful group.

In general terms, an outstanding feature of cereal marketing in the U.K., as compared with many other countries, is the absence of any significant carry-over of stocks at the end of the cereal year. Statutory arrangements (whether the deficiency payment schemes, or the powers and activities of the H.-G.C.A.) are geared to the movement off farms, and to the absorption by users, of the whole of one year's crop before the next harvest is gathered. In the context of an import requirement still running at 8 million tons annually (and allowing for the fact that some quantity of these imports, e.g. hard wheat, cannot, in present circumstances, be replaced from home production), such arrangements are sensible, and, indeed, desirable. With the increasing home production that should progressively reduce the need for imports, more flexibility will be required: as home supplies approach nearer to total demand, the farmer's selling position could be gravely weakened if the marketing of home-grown grain is statutorily constricted in this way. While the primary aim should be to ensure that first place in the market goes to home grain so that no carry-over arises, an increasingly important secondary aim must be, if a carry-over becomes necessary, to ensure that it can emerge without the collapse of market prices that might well occur under present arrangements. The method should be, as in other major producing countries, through suitable intervention facilities for a statutory authority to carry buffer stocks.

QUESTION 5. *What factors influence the farmer as to whether to store or sell his grain and what is their relative significance? Do the present Cereals Deficiency Payments system and the H.-G.C.A. Forward Contracts Scheme work satisfactorily so far as the farmer is concerned? To what extent do such factors as shortage of storage facilities, lack of capital to finance spring cultivation, indebtedness to merchants, cause the farmer to act in a way that may be contrary to his best interests in the marketing sense, e.g. by causing him to sell during the harvest period rather than later?*

Broadly, grain can be stored in three places; on the farm of origin, at the point of processing, or at some intermediate point.

Intermediate storage is clearly the most expensive in so far as it means two transport costs and an extra in-and-out handling cost. Additional storage for British grain at the processing point was not readily provided as the use of combine-harvesters developed; it would have committed the processor to purchase British grain to an extent that he was not then prepared to be committed.

Grain storage policy in the U.K. has rightly been to encourage storage on the farm of origin. In pre-combine-harvester days grain was stored on the farm in stacks and put on to the market as required. The new situation has naturally developed from the old. The pattern of movement of grain from British farms is necessarily very different from that in many predominantly exporting countries, where it flows along defined channels of communication to one or two ports, and where stores have been erected along those channels.

Stimulus for the provision of farm storage has been provided by government financial incentives, including the deficiency payment arrangements. Substantial capital investment by farmers was necessary: this could not have taken place without reasonably firm assurances of an increased return from stored grain. In this way the deficiency payments incentive schemes have, beyond doubt, contributed to better marketing. In their absence, marketing would have been chaotic, as use of the combine-harvester spread.

The individual farmer has considered (and, as grain production continues to increase, must continue to consider), (a) whether the financial incentives justify the investment required for farm storage; (b) for what proportion of his production storage is required, bearing in mind his likely need for cash realisation of some of his crop as soon as possible; (c) how disadvantageously failure to provide storage will place him at the mercy of the market at harvest time; and (d) to what extent the inevitable streamlining of his harvesting operations—with his available labour force and the required degree of bulk handling—requires a ready and immediate place to store grain as it comes off the field.

These are the factors which determine whether a farmer provides storage capacity— the paramount consideration being the degree of assurance of financial return upon his capital investment. Farm storage has, in fact, increased in line with increased production, and seems, in national terms, now to be adequate to avoid an undue amount of forced selling at harvest. The incentives have thus been effective, and have worked satisfactorily. The incentives must not be abandoned: if they were to be eliminated or substantially reduced, those who have invested capital upon their assurance would be seriously let down.

The H.-G.C.A. contract bonuses do not affect significantly the farmer's decision to store rather than to sell at harvest. They help the decision as to when to sell after the basic decision to store is taken.

Not all farmers have been induced to provide storage by the incentive arrangements: some have deliberately decided, rightly or wrongly, that they will continue to sell at harvest.

The need for ready cash, and indebtedness to merchants, are a significant factor in these decisions, particularly when, as at present, Bank Rate is high. For individuals these are important considerations, although in general they are less important than they were a few years ago—merely because more farm storage exists, and, since it exists, will be used rather than left empty. The crucial test is whether the true value of home-grown grain is realised at harvest in comparison with imported grain. Until this happens— and it has not yet happened—the marketing situation cannot be accepted to be satisfactory.

QUESTION 6. *Does the price the farmer receives for his grain adequately reflect quality differentials? If not, how might this be achieved?*

In a freely operating market, the measurement of a quality differential can scarcely be measured other than by variations in the price at which the producer of a quality article is prepared to sell and the buyer is prepared to buy. If the producer cannot obtain a price which will recompense him for the extra cost incurred in producing it, he will cease to produce it: if the buyer has an abundant supply of the desired quality available to him, he will not need to offer a large premium in order to obtain his requirements.

A typical example of this contest in operation has been malting barley. On one side has been the continually increasing production of barley, giving, theoretically at least, a greater volume of supply from which to choose: on the other has been the natural reaction of the producer if confronted by lower malting premia—to go for yield at the expense of malting quality. After allowing for inevitable year-by-year variations in the overall quality of barley, it seems that buyers and sellers have reached some degree of agreement as to the adequacy of reward required to ensure the required supply of malting barley—since the market has been adequately, but not over-adequately, supplied. The farmer-seller is not, however, always well enough informed as to the requirements of the maltster with whom he is rarely in direct contact in his marketing.

With seed grain also, since there is an adequacy of supply, if would seem that the market price adequately reflects the quality differential.

It is with milling wheat that the greatest problem arises. In recent years the premium paid to the farmer for wheat of special milling quality has been insufficient to induce his special attention to specifically milling varieties or quality. New varieties are, however, now available which are more qualitatively competitive with certain imported wheats, which are usually available only at significantly higher prices than ordinary British wheats. There is thus scope for a market premium for these newer varieties: and there is also necessity for such a premium if they are to be produced in sufficient quantity, and to the desired quality, instead of newer types of higher-yielding wheats which are also becoming available. If anything like the maximum potential amount of import-saving is to be achieved through increased production of wheat for flour-milling, a more assured and adequate premium must be available to the farmer.

This premium should come from the market itself. It should not be artificially contrived through some modification of the deficiency payments scheme: any such arrangement would (a) run counter to the general principle of such schemes, and (b) be complicated and difficult to administer.

Two aids to the provision of an adequate premium are suggested: firstly, the maintenance of adequate price differentials in the structure of minimum import prices as between different imported qualities; and secondly, the progressive development of contracts entered into before the grain is grown and harvested. Such contracts would provide, as do many existing contracts for seed grain, for a premium over the market price, say, for feed grain at the time of delivery. Neither buyer nor seller would thus be committed to an absolute level of price, which could not be reliably forecast months ahead; but the necessary premium to encourage the production of the required quality would be agreed in advance. Contracts of this kind, even if arranged through an intermediary merchant, would have the effect of bringing the producer and the processor into greater awareness of each other's requirements. They are particularly desirable where the utilisation of new varieties needs to be developed quickly, because they will lessen the necessity for the producer to rely solely on a speculative estimate of the eventual return from the "spot" market.

An important instance of the failure of the existing market to reflect quality differentials is its response to deliveries of grain that are drier than the customary standard for the particular grade. Processors traditionally make unfavourable comparisons between the moisture content of British grain, and the lower moisture content of many imported grains—and they normally seek price allowances if British grain exceeds a particular level of moisture content, e.g. 18% for milling wheat, and 16% for feeding grains. Market comparisons of the value of British grain in relation to imports are usually related to these nominal "standards" of moisture content, i.e. an assumption is made that British grain is no drier than these "standards". In practice British grain is increasingly delivered at lower moisture contents, but little or no recognition of this is given in the market price. The producer of feed barley may be confronted with a claim for an allowance if he delivers at 18% m.c., but is most unlikely to receive any increment in price if he delivers at 14% m.c. If the producer is to have the incentive to deliver grain with the highest possible dry-matter-content, he must be assured of either, (a) a proportionately higher market price, or (b) a scale of automatic price increments in parallel with the price allowances claimed by processors on damp grain.

QUESTION 7. *To what extent does grain fetch a lower market price than it otherwise would because of such factors as inefficient drying, poor storage or lack of cleaning?*

Some years ago when farm storage and drying was increasing rapidly to keep pace with the development of combine-harvesting and with the expansion of production, the quality of grain may sometimes have been affected by the novelty of design of equipment and by farmers' inexperience of the drying and storage operations, but there is no statistical evidence, and it is probable that reports of its extent were exaggerated. Nevertheless some buyers, e.g. maltsters, were then cautious in their purchasing policy for farm-dried grain—and it is possible that average market prices may have been less for this reason.

The years of experience now gained by manufacturers of equipment and by farmers in its use have substantially reduced the risk of mishaps occurring during drying and storage. It is not possible to quantify the extent of any problem that still exists, but it seems that buyers have modified their attitude and in general no longer maintain their earlier prejudice against farm-dried grain. No significant effect on average market price levels probably now arises from this factor.

The farmer needs to be assured that the costs incurred in cleaning, drying and storage will be recouped either through his market price or through the pattern of the guaranteed price incentives. He has had some assurance in this respect in recent years, and this has undoubtedly been the main factor in the provision of efficient facilities on farms. Removal of this assurance could have serious marketing effects.

QUESTION 8. *To what extent has the process of concentration in the agricultural merchandising industry weakened the farmer's position as a seller of grain and a buyer of requisites?*

The reduction in recent years in the number of independent merchants, substantial though it has been, has not had an adverse effect on the farming industry either as sellers of grain or as buyers of requisites. Particularly on the requisite supply side there has, in fact, been a sharpening of competition, prompted to a considerable extent by increased activity by farmer-owned businesses.

There could be cause for apprehension if the concentration of independent merchanting interests continued too far, and if a situation of widespread vertical integration developed in the control of too few firms. For this reason the existence and progressive development of farmer-owned businesses is an important safeguard for the farming industry.

QUESTION 9. *In choosing how, where and when to sell his grain does the farmer have adequate means available of informing himself as to the market situation? Does the N.A.A.S. play any part in this? Should it do more, and if so in what way?*

The N.A.A.S. does not play any part in disseminating market information, nor should it, as a government agency, do so: its officers are not trained or qualified to perform this kind of function.

The H.-G.C.A. has instituted and is continuously developing a comprehensive and reliable service of market information which is made available to and through the press, as well as to individual farmers on a subscription basis. It is a matter for some regret that more farmers do not subscribe and thus assure themselves of a regular and up-to-date service.

QUESTION 10. *Are there any other features of the grain market which have not been mentioned in your replies to the above questions but which you consider to be of special importance?*

(a) Comment is sometimes made on the requirement for the farmer who feeds all or some of his barley on his own farm to make a contribution in respect of this barley towards market stability through the operation of the deficiency payments incentive scheme and the H.-G.C.A. levy. The general objective of these schemes is to include a pattern of marketing which will return the best possible market price, thereby keeping down the cost of deficiency payments: in this way a safeguard is provided against the use of the argument of high cost to the Government as a reason for depressing the guaranteed price. This objective is in the interests of the self-feeder and of the seller alike. On this ground it is reasonable that the self-feeder should make some contribution (and in the case of the H.-G.C.A. he can also benefit from the Authority's market information services). Present levels of contribution can be accepted as fair.

(b) Concern is felt at the impact on distribution costs of grain of recent and prospective increases in transport costs. These will have the effect of making home-grown grain less competitive with imported grain at the ports, where a high proportion of processing capacity is located.

Statement by the National Farmers' Union of Scotland

QUESTION 1. *What are the main factors likely to affect the supply of the main cereals from U.K. farms in the next 5 years, and what do you expect the effect of these factors to be?*

ANSWER. Acreage and yields have been increasing steadily and we expect this general trend to continue, though the rate of yield increase may diminish somewhat. The fact that labour costs are a lesser proportion of total production costs in cereals than in almost any other farm commodity contributes largely to the trend towards increased acreage.

Government policy and action is likely to influence this trend more than any other external factor. It is assumed that this will continue to follow the requirements of the National Plan. Overall policy, however, has many aspects and we believe that those most relevant to cereals production can be outlined thus:

Direct action. Price Review determinations in regard to guaranteed prices and the standard quantity for barley are obviously of direct influence on production. This Union has opposed the standard quantity system since its inception and maintains that this should be completely discontinued in the interests of import saving. Official policy since 1966 has been to favour wheat. While the encouragement given to wheat in the 1967 and 1968 Price Reviews is welcome, prices for barley and oats must also be maintained in line with costs. Natural factors impose a definite limitation on the extent to which wheat can be increased in Scotland. On the cost side, fertiliser price levels are likely to be important.

Indirect action. In this Union's view, it is of basic importance that demand for cereals should increase in step with production and we believe that recently this has not always been achieved. The result which has been termed overproduction of barley has been, in fact, underproduction of livestock products. It is the Union's view that action is clearly needed to encourage the livestock sector and, in particular, to ensure expansion and long-term stability in pig production. Further, although certainly such encouragement might induce some farmers to revert from cereals to livestock production, we think that this would in no way harm the cereals sector or have more than a marginal effect on the trend of increase of cereals production.

Import substitution is of immediate and particular importance and the 1966 dock strike showed clearly that barley is a more acceptable substitute for imported maize and sorghum in compound rations than has hitherto been acknowledged. Likewise, the introduction of Maris Widgeon has for the first time brought a prospect of a reduction of wheat imports. Devaluation has strengthened the argument that expanded home production is needed in the national interest. Provided market prices reflect the anticipated increase in the level of minimum import prices, this can be achieved without any significant increase in the cost of deficiency payments.

Developments in research cannot be forecast but will be of great importance. Of particular importance to Scotland is research in the breeding of new varieties of oats and barley, both for malting and feeding.

Although the question of disease incidence and the need for break cropping is of more limited significance in Scotland than in England, it is nevertheless important in those areas where lengthy periods of unbroken cereals cropping are practised—notably the east and south-east—and the need here for assistance to break crops is no less than is England's. It is this Union's view that under Scottish conditions, and in the interests of efficient land utilisation, this should take the form of a subsidy on crops of rape and roots.

The question of Common Market entry is perhaps outside the scope of this paper. Suffice it to say that this Union concurs with the general view that entry would stimulate cereals production at least in the short-term.

Although forecasting is difficult, we expect in Scotland to see a continuation of the trend towards increased grain production. In the short-term, an increase is likely in wheat output in south-east and eastern counties, but this must clearly be limited by natural factors. Barley is likely to continue to expand though more slowly than in the recent past, while the future for oats may depend partly on new varieties, partly on whether the increase in the crop in England which took place in 1967 is maintained.

QUESTION 2. *What are the likely trends in relation to mixing of feed on farms?*

ANSWER. It is believed that the proportion of farmers employing mechanised feed mixing systems on farms in Scotland is at present small, though increasing. This system is likely to continue, particularly as farm unit sizes grow and if home grown proteins, such as beans, become available, but this growth may be a long-term rather than a short-term feature.

QUESTION 3. *If the wheat subsidy were paid on an acreage basis as with barley, to what extent would cereals farmers who at present sell all their wheat crop, be likely to retain it on the farm to feed to livestock?*

ANSWER. This Union has not taken any formal decision on the question of whether or not the wheat subsidy should be based on acreage as is the case with barley. The idea is attractive from many points of view, not least in that it would enable a producer of feed wheat to retain his crop for feeding on the farm without depriving himself of deficiency payment as at present. This would undoubtedly be attractive to many farmers with pig and poultry enterprises.

QUESTION 4. *What are the main difficulties facing the farmer in the marketing of his grain and how could the marketing system be improved? In particular, to what extent is the farmer a weak seller? What difference has the formation of farmers' groups made in this respect?*

ANSWER. Increasingly, merchants are looking for regular supplies of assured quality. Especially in the quality field, e.g. of malting barley and milling oats, climatic vagaries put the Scottish producer at a disadvantage as compared with his English counterpart and buyers who may feel obliged to buy from, for example, East Anglia, on quality grounds in years when the quality of the local crop is not high, are often loth to accord the local crop an appreciably bigger market even in years when it is of good quality.

The early crop is often at an advantage, particularly in the case of milling oats, and it may be appropriate to point out that because under present C.D.P. arrangements, grain must be removed from the farm of origin not later than 21st July, there is a bigger gap between crops in terms of the later Scottish harvest, than is the case further south. The extension of the final date to 15th August in Scotland would be equitable.

The concept of contracting the sale of a crop at the growing stage is a relatively new one which would seem to offer scope for assuring a market at an earlier stage than hitherto, and may be particularly applicable to malting barley and milling oats.

Although group marketing is as yet relatively little developed in Scotland, there is some evidence that it can strengthen the individual producer's position, if by this means a sufficient tonnage can be regularly made available to a selected, efficient and progressive merchant, to give him incentive to obtain and keep the business. Early experience is against the concept that a group of farmers can normally engage, to advantage, in direct sales to end users.

QUESTION 5. *What factors influence the farmer as to whether to store or sell his grain and what is their relative significance? Do the present C.D.P. system and the Home-Grown Cereals Authority forward contracts scheme work satisfactorily so far as the farmer is concerned? To what extent do such factors as shortage of storage facilities, lack of capital to finance spring cultivation, indebtedness to merchants, cause the farmer to act in a way that may be contrary to his best interests in the marketing sense, e.g. by causing him to sell during the harvest period rather than later?*

ANSWER. Lack or shortage of storage facilities is of course a compelling reason to sell, and one which is normally evident to a buyer. This is naturally more often the decisive factor in selling in the case of small holdings than of large.

There is almost always a need for cash income at harvest time so that depending on his scale of operation, the question is how much of his crop a farmer must sell at harvest. Traditionally in the fattening areas of Scotland, cattle courts are filled in the autumn and this, rather than the spring cultivation, is the source of need of cash. Indebtedness, especially when as at present the Bank Rate is high, is also a potent factor.

A serious criticism of the Cereals Deficiency Payments Scheme is its extreme complexity. Evidence of the fact that a significant proportion of farmers fail to understand it adequately is provided by the numbers who annually, for one reason or another, lose benefits to which they would have been entitled. Often these are the very men whose need is greatest. The Scheme is also frequently cited as a cause of unresponsiveness in prices. For all this, it is difficult to see how it could be improved other than in detail, except in that the remaining standard quantity on barley should be abolished. One point of detail affecting the barley incentive scheme is of concern, namely that the transition from early disincentive period to later premium period should be less steep.

Individual experience in regard to the Authority's Forward Contract Scheme has been mixed. The Scheme has certainly induced a welcome acceptance of the practice of contracting forward a proportion of the crop and registered contracts have been found acceptable as security for loans. Although the grower feeder has resented a liability to pay levy, this Union accepts that there are grounds for requiring him to contribute at a reduced rate and believes that the present apportionment as between acreage and tonnage is about right. Devaluation has regrettably introduced an upsetting factor this season in its clearly disadvantageous effect on many farmers forward contracted in the early part of the season.

QUESTION 6. *Does the price the farmer receives for his grain adequately reflect quality differentials? If not, how might this be achieved?*

ANSWER. It is commonly accepted that quality premia become most clearly established in an over-supplied market. Nevertheless, there has been little if any evidence of premia in the feeding barley market in recent years and it is this Union's opinion that if a greater degree of price differentiation could be induced, based on bushel weight, this would be to the good.

Some of the difficulties facing the would-be seller of malting barley have been discussed in answer to question 4. Additional to these, there has been an unfortunate customary reticence on the part of maltsters, brewers and distillers to state what exactly their requirements are. Accepting that quality requirements are becoming ever higher, and that these must be subject to regular review in the light of harvest conditions, it would be helpful if these buyers would make known their requirements in terms of quality at harvest each year. Recent discussions give grounds for optimism that a common Code of Practice for nitrogen testing of malting barley will be agreed for 1968.

QUESTION 7. *To what extent does grain fetch a lower market price than it otherwise would because of such factors as inefficient drying, poor storage or lack of cleaning?*

ANSWER. It is our view that lack of cleaning is probably the most important factor in the three mentioned and that inefficient drying is becoming less frequent as experience is gained.

It is not possible to quantify this problem.

QUESTION 8. *To what extent has the process of concentration in the agricultural merchanting industry weakened the farmer's position as a seller of grain and a buyer of requisites?*

ANSWER. Such difficulties as have arisen have arisen because of concentration in end-using processes rather than in merchanting as such. Maltings provide a case in point.

There is an increasing tendency for farmers to buy their requisites from the merchant to whom their grain is sold. Although savings can sometimes be affected by group buying from a main supplier, limited experience to date suggests that success in terms of a reciprocal trade in grain depends very much on the considerations mentioned above in our reply to question 4.

QUESTION 9. *In choosing how, where and when to sell his grain, does the farmer have adequate means available of informing himself as to the market situation? Does the N.A.A.S. play any part in this? Should it do more and if so, in what way?*

ANSWER. N.A.A.S. does not exist in Scotland and the three Colleges of Agriculture, which are together responsible for advisory work, have not so far ventured into the business aspect of marketing of grain. Nor do we think it appropriate that they should do so.

Customarily, cordial relations between merchants and farmers have generally provided an adequate basis for decision in the past, but the Authority's Weekly Information Bulletin has provided a welcome source of information which puts decision-taking on a much more business-like basis. It is a matter for some regret that the number of farmers subscribing to the Bulletin is small.

QUESTION 10. *Are there any other features of the grain market which have not been mentioned in your replies to the above questions, but which you consider to be of special importance?*

ANSWER. (1) The Union is concerned that proposed transport legislation will interfere with the free movement of grain and will add unduly to the costs of transporting grain in a way which will diminish grain values to the producer. We support objections to this legislation which have been lodged by the Scottish Corn Trade Association.

(2) The union is concerned that in the event that the Authority should ever have to engage in direct trading, its power to support a falling market will be found wanting. In particular, the obligation imposed on it to sell any grain bought by 31st July following is unreasonably inhibiting. The Union fears that the Authority's proposal to take immediate possession of grain bought in direct trading and to pay storage charges where the grain is stored at its instruction on the seller's farm may increase the risk to the farmer of liability to rating. Indeed, the question of rating and the effect on it of farm grain storage requires to be clarified.

(3) The Union regard it as vitally important that research work, especially in the plant-breeding field, should be fostered, and, naturally, that breeding in Scotland of varieties suited to Scottish conditions should receive due attention. We welcome the stimulus which the introduction of plant breeders' rights should give to this work.

(4) The continuance of the high standards of the present Scottish Seed Certification Scheme is important and the Union is pleased to note that these will be maintained in a unified British Scheme.

Statement by the Ulster Farmers' Union

A. PRODUCTION TRENDS

QUESTION 1. *What are the main factors likely to affect the supply of the main cereals from U.K. farms in the next 5 years, and what do you expect the effect of these factors to be?*

ANSWER. The financial return on cereal crops in comparison with practicable alternate land uses is the principal factor. Policy in the past 5–6 years has tended to diminish the profit from cereal growing whereas the profitability of grass products (e.g. beef) has improved.

Though the cereal acreage in England has continued to increase, that of Northern Ireland has declined. For example, the barley acreage is down from 184,000 acres in 1965 to 153,000 acres in 1967, due to the extra incentives to grass crops while arable crops remain on a par with the rest of the United Kingdom.

It would appear that although cereal acreage may continue to expand in Great Britain it will continue to decrease in Northern Ireland.

QUESTION 2. *What are the likely trends in relation to mixing of feed on farms?*

ANSWER. The high expense of grinding and mixing machinery has tended to restrict home production of rations to the larger farms. One would assume that as farm sizes increase the number of farms where home grinding and mixing of feeds is practised will also increase. This, of course, will be the case particularly where grain is home grown and where larger numbers of animals are kept.

QUESTION 3. *If the wheat subsidy were paid on an acreage basis, as with barley, to what extent would cereals farmers who at present sell all their wheat crop be likely to retain it on the farm to feed to livestock?*

ANSWER. At present the acreage of wheat grown in Northern Ireland is very low (approx. 2000 acres per annum). This is to a large extent due to the necessity to sell the wheat to obtain the subsidy. This necessity is particularly injurious to Northern Ireland farmers who would be more inclined to feed the wheat to their own stock than would the farmers in, say, East Anglia who are not basically stock-keepers.

Recently the subsidy has barely paid the transport cost of shipping home-grown grain to the mills and carrying imported grain back to the farm. In consequence wheat offers a lower financial return than barley and tends to be a more troublesome crop in our damper Northern Ireland climate.

There is no doubt that it would be a more attractive crop financially if it could be used on the farms where it was grown and the acreage grown would increase on farms where it could be mixed for livestock feeding. No Northern Ireland wheat is used for any purpose other than animal feeding.

B. MARKETING OF GRAIN

QUESTION 4. *What are the main difficulties facing the farmer in the marketing of his grain, and how could the marketing system be improved? In particular, to what extent is the farmer a weak seller? What difference has the formation of farmers' groups made in this respect?*

ANSWER. There are no particularly outstanding difficulties. The N.I. price is decided by reference to the price of imported equivalents plus transport costs. Farmers have little difficulty in finding either a local miller or farmer to purchase their grain, this being basically a grain-using area.

Towards the end of the season the storage incentive payments make it worth while to sell to millers only as farm to farm sales are excluded from payment, which tends to restrict the price available to the vendor without helping the farm user in any way and also often involving extra transport costs. So far the formation of a few groups in Northern Ireland has not altered the position to any great extent.

QUESTION 5. *What factors influence the farmer as to whether to store or sell his grain and what is their relative significance? Do the present Cereals Deficiency Payments system and the H.-G.C.A. Forward Contracts Scheme work satisfactorily so far as the farmer is concerned? To what extent do such factors as shortage of storage facilities, lack of capital to finance spring cultivation, indebtedness to merchants, cause the farmer to act in a way that may be contrary to his best interests in the marketing sense, e.g. by causing him to sell during the harvest period rather than later?*

ANSWER. Many farmers feeding their own livestock on home-grown grain are uncertain as to the extent of their eventual surpluses. Storage incentive payments just about cover bank interest rates, so it is to the general advantage of the feeder to retain his stocks of grain until the end of the season in May or June when the extent of feed requirements is known.

These factors do not, however, apply to grain grown on conacre land (land taken annually). Farmers in this position often do not have the storage facilities for the grain grown on this extra land and rent for the land must be paid in November. Such a man is often a weak seller at harvest time at prices which are always depressed at this time of year.

QUESTION 6. *Does the price the farmer receives for his grain adequately reflect quality differentials? If not, how might this be achieved?*

ANSWER. In our experience the price obtained does normally reflect quality differentials within the limits that quality matters for feeding purposes. In a tight community such as this the normal standards of a farmer are known and these standards reflect the price he will be paid.

QUESTION 7. *To what extent does grain fetch a lower market price than it otherwise would because of such factors as inefficient drying, poor storage or lack of cleaning?*

ANSWER. Grain in Northern Ireland is habitually dried to 14% moisture content as opposed to Great Britain's 16%—this is, of course, reflected in a 10% differential in price for the 2% difference in moisture and on the whole grain is reasonably well stored. Most modern combines give a pretty good sample of grain when properly set and most farmers today rely on their combine to give a good sample without resort to special cleaning.

QUESTION 8. *To what extent has the process of concentration in the agricultural merchanting industry affected the farmer's position as a seller of grain and a buyer of requisites?*

ANSWER. Northern Ireland being predominantly a livestock-rearing area means that the millers and compounders are basically in a seller's position so must give farm prices for grain bought locally—often from the same farmers to whom they are selling compounds. The same conditions tend to provide a larger number of small merchants than is probably the case in other parts of the United Kingdom.

QUESTION 9. *In choosing how, where and when to sell his grain does the farmer have adequate means available of informing himself as to the market situation? Does the N.A.A.S. play any part in this? Should it do more, and if so in what way?*

ANSWER. Both the Union and the farming press publish details of current prices and the grain situation generally. The local equivalent of N.A.A.S. will not become mixed in commerce in any way, so farmers do not get any help from this quarter.

QUESTION 10. *Are there any other features of the grain market which have not been mentioned in your replies to the above questions but which you consider to be of special importance?*

ANSWER. The only aspect untouched is whether it might not be to the national advantage to give special encouragement to grain growing in an area of large grain deficiency. At present the special funds available are wholly devoted to grass or other livestock products all of which are in surplus in Northern Ireland and must be exported or sold locally at prices well below G.B. prices. Current transport costs being what they are it is often cheaper to import from abroad than from G.B.

Encouragement to grow grain in the better lands in the north and east of Ulster might result in rather less depressed prices for milk and beef and a more plentiful supply of feedingstuffs.

B. The Merchants, Co-operatives and Shippers

Questions addressed to the National Association of Corn and Agricultural Merchants, and the Agricultural Co-operative Association Ltd. (incorporating also the views of the Welsh, Scottish and Ulster Agricultural Organisation Societies)

A. USE OF HOME-GROWN CEREALS

1. Do you consider that merchants could do any more than they are doing already to facilitate the disposal of home-grown grain, e.g. for flour milling, compounding or export?
2. What are the main factors influencing the continued development of the grain export trade, and how can exports be stimulated?

B. MARKETING OF HOME-GROWN CEREALS

3. Do farmers as a whole present their grain in a quality, and in a manner, that best suits market requirements? In particular to what extent does grain fetch a lower market price than it otherwise would because of such factors as inefficient drying, poor storage and lack of cleaning?
4. Is the present system of Cereals Deficiency Payments, Seasonal Incentives, and H.-G.C.A. Forward Contracts conducive to a desirable pattern of grain marketing? If not, what improvements do you suggest?
5. To what extent, if at all, would the grading and bulking of grain improve the marketing of U.K. grain? What is the relevance of developments of the Mendlesham (Suffolk) type, and the activities of farmers' groups, in this connection?

C. AGRICULTURAL MERCHANTING

6. What are the economies of scale obtainable in agricultural merchanting, particularly so far as grain and feedingstuffs are concerned, and how do these affect decisions as to the location and economic size of merchant businesses? In particular, is there a potential for groupings of small merchant businesses to achieve the economies of scale?
7. Do you consider that the transport of grain internally, or coastwise, is being carried on as efficiently as possible? If not, what improvements would you suggest?
8. Do you consider that the capital position of agricultural merchants is generally satisfactory? If not, what improvements would you suggest?
9. Are there any other features of the grain market, which have not been mentioned in your replies to the above questions, but which you consider to be of special importance?
10. Is the present number and distribution of branches of the larger co-operatives satisfactory in relation to grain buying and selling? Is there scope for more specialisation between the branches of a large co-operative? (*A.C.A. only.*)
11. Do you consider that co-operatives who compete with other co-operatives within the same trading area are acting rationally in the farmers' best interests? (*A.C.A. only.*)

Statement by the National Association of Corn and Agricultural Merchants

PREAMBLE

N.A.C.A.M. represents all those actively engaged in the marketing of cereals including public companies, such as the large national organisations with milling and compounding interests, private firms and agricultural co-operative trading societies. In practice marketing is carried on in a similar way by all three types of business; there is no fundamental difference in the marketing problems which they face or in the way in which they overcome them and there is therefore no need for this statement to differentiate between them.

Approximately one-half of the members of the Association also manufacture animal feedingstuffs and about two-thirds carry on a seed trade, a number of the latter being specialist seed processors. About one-third of members are engaged in all three processes of feed manufacture, seed processing and drying and conditioning of grain.

DISPOSAL OF HOME-GROWN GRAIN

We have been asked whether merchants could do more than they are already doing to facilitate the disposal of home-grown grain for flour milling, the manufacture of compound feedingstuffs or for other markets, including exports. We do not consider that merchants could do any more than at present. The very competitive nature of the corn trade ensures that the best possible market is found at all times and compels merchants to satisfy their customers, both buyers and sellers.

In order to meet this challenge the merchant must have first-class market intelligence drawing his information from all available sources and making use of the judgement which he has built up with long experience to select the best outlets in any given market situation. The merchant's achievement is demonstrated by the successful marketing of a home-grown cereal crop which nearly doubled in volume between 1957 and 1967.

The country manufacturers of animal feedingstuffs within N.A.C.A.M., known as country compounders, produce about one-third of the compound animal foods manufactured in the United Kingdom. It is clearly in their interest to use the maximum quantity of home-grown cereals in their compounds. The national compounders (meaning those who largely manufacture at port mills), however, exercise considerable influence in the market. It is to be hoped that the substitution of home grown for imported cereals in their manufactured feedingstuffs will increase as much as can be justified on economic and nutritional grounds. There are naturally many factors outside the merchant's control which influence the usage of home-grown cereals. In flour milling, for example, the percentage of home-grown wheat used in the grist depends largely on the technical judgement of the miller. Its use would, however, undoubtedly be stimulated by the production of varieties especially suited for flour milling and we feel that such varieties should attract a realistic premium on the market to enable merchants to encourage farmers to sow them.

We also feel that greater use could be made of home-grown barley in brewing and distilling in place of imported material. We feel some concern, too, at the increasing use of imported materials in the manufacture of starch.

EXPORTS

The continued development of the export trade in home-grown grain is subject quite simply to harvest prospects and the out-turn of crops in other parts of the world and to changes in the overall pattern of the international grain trade. In other words, it will be affected by the balance of supply and demand. This balance itself can be affected by political action, ranging from outbreaks or rumours of war and interference with trade routes (such as the closing last year of the Suez Canal) down to the adjustment of import levies and export rebates by the European Economic Community. But in the last analysis all exports must depend upon the price at which the grain can be offered. Clearly the price at which British grain can be exported is related directly to our domestic demand for grain of the same quality.

Among factors tending to inhibit the export of British cereals is the lack of adequate transit facilities at ports. Some merchants own installations at or near ports allowing them to screen and grade grain by reference to its quality, but there is a great need for more such facilities to be available to exporters. Dock charges vary sharply from one port to another and are often very high for the service offered. The cost of inland transport is also an important factor referred to in more detail on page 768.

The export trade involves considerable risks. Some international firms are prepared to take a view of the market on a scale which individual merchants in this country could not possibly match. This factor alone has a significant influence on the trade. Individual merchants trading on their own produce a fragmentation of the export effort and we feel that a strong argument can be made for the formation of consortia of export merchants in the various port regions or a national consortium. Larger contracts could then be made, a wider view of the market could be taken by sharing the risk and it should also be possible to improve market intelligence on the world grain situation.

PRESENTATION OF GRAIN

Farmers' presentation of their grain has improved substantially over the past 10 years but considerable difficulties do, however, still arise. If, for example, a farmer is anxious for the safety of his crop because of bad weather at harvest he tends to pass the grain through his driers too fast. This fast drying can easily damage the gluten in wheat, making it unfit for milling and can similarly make barley unfit for malting. The germination of seed grain can be badly affected by rapid drying. With the increasing capacity of combine harvesters even those farmers who have good drying equipment find that they cannot cope with the rate of intake of grain in a wet harvest, so that the grain is sometimes allowed to stand in heaps for too long before drying and this undoubtedly causes considerable damage. The value of malting barley can also be harmed by being heaped or placed in a bin without having been cooled. Furthermore, there are still too many cases of insufficient attention to regular precautions against infestation in farm stores.

It can thus be seen that the difficulties which still exist are of two kinds. One is the lack of adequate facilities on the farm and failure to turn to the merchant for the service which is needed. The other is failure to make proper use of equipment which exists on the farm. As a further example the screening of grain leaves much room for improvement: sometimes this is because the farmer has no screening equipment of his own but sometimes he does not use properly the equipment which he has.

The net result of these difficulties is that there are still occasions when grain has to be sold for feeding although the farmer might have hoped at harvest time that it could be sold for a specialised use at a higher price. It must be remembered, however, that farmers sometimes take the view that it is more economic to sell grain for feeding than to incur the additional expense and trouble of presenting it as a higher grade parcel.

STATUTORY SUPPORT SYSTEMS

The storage incentives payable to growers under the Cereals Deficiency Payments Scheme are commendable in principle but lately severe weaknesses have become apparent. The system has resulted in a spasmodic pattern of marketing for wheat with a rush of offers at the beginning of each accounting period which tails off until the next period begins. The deduction period under the Barley Incentive Scheme has led to many abuses as it is an inducement to farmers to avoid normal trading channels to evade the deduction and the tonnage element of the H.-G.C.A. levy. This evasion of payment by some barley growers is clearly against the interests of their fellow farmers.

The Home-Grown Cereal Authority's forward contracts scheme has made farmers more conscious of marketing ahead but the scheme's provisions on delivery are too rigid and have tended to starve the spot market. We are convinced that there should be facilities for a contract to be varied by mutual consent so that delivery could be postponed if the market were glutted or brought forward if a particular cereal were in short supply.

The Association has prepared a memorandum of its views on the deficiency payments scheme which is appended to this statement. Our recommendations on barley to some extent mask the penalty on early sales instead of abolishing it but even so we consider them an improvement on the present arrangements. These proposals for barley and for wheat were put forward at meetings of the Ministry of Agriculture, Fisheries and Food's advisory committees on 30th January 1968, where they were supported in principle by the trade interests represented. They were rejected by the committees, however, because the N.F.U. and the H.-G.C.A. did not accept them. The arrangements for barley could, we feel, be more effectively and equitably operated if the Government financed the barley incentive fund by a direct grant which would entirely separate storage incentives from the deficiency payment on acreage.

The existing deficiency payments system which has evolved over the years has undoubtedly now become far too complex and is fully understood by few farmers. In our view the time has come for a complete reappraisal and a drastic simplification of the system. It would undoubtedly be desirable to incorporate the deficiency payments scheme with the forward contracts scheme as the present separate administration of the two schemes only adds to the complexity of the arrangements.

GRADING AND BULKING OF GRAIN

In Great Britain the grading and bulking of grain has been one of the activities of merchants for very many years. It may consist simply of a merchant arranging deliveries of parcels of similar grade from different farms direct to a single user against a single large contract, as distinct from sales of individual parcels. This practice is followed quite frequently in the sale of milling wheat, feed wheat, feeding barley and some grades of malting barley. Its financial advantage is that the only transport charge incurred is that from the farm to the user's premises.

The merchant may, however, adopt the more sophisticated practice of bulking, grading and drying grain in his own premises when he has guaranteed the delivery of large quantities of a given specification to a specialist buyer. This practice becomes more difficult and more specialised as new varieties of cereals appear on the market in increasing numbers. The flow of new varieties has been stimulated by the Plant Varieties and Seeds Act, 1964, which for the first time enables plant breeders to gather royalties on all sales of their new varieties. We feel that a comprehensive enforced grading system would be quite impracticable in the United Kingdom as local requirements vary considerably in terms of quality. (An article by Mr. C. A. Brooks on this subject was published in the *Agricultural Merchant* in April 1952.) We would refer to the memorandum on the drying and storage of grain which was submitted to you in 1967 and of which a further copy is attached to this statement.[1]

You refer to one type of marketing scheme introduced by maltsters in Suffolk and we have been asked to comment on the role of farmers' groups in the grading and bulking of grain. The Suffolk scheme involves a contract by a farmer through his merchant to supply barley to a local maltster at fixed prices. The maltster is registered as an approved service operator under the Cereals Deficiency Payments Scheme and does not necessarily take ownership of the grain until a later date. This scheme was specifically designed to suit a specialised malting process in purpose built premises and is simply one aspect of a general effort to develop new methods in cereals marketing, some of which are successful while others are not. Such methods should be studied with interest but the particular

[1] See pp. 771-6.

scheme in Suffolk, although it may appear to be a new concept in marketing, really offers little more than the service already provided by other approved service operators.

So far as the activities of farmers' groups generally are concerned, they cannot draw their supplies of grain from a sufficiently large circle to undertake specialised grading and they are indeed bound to take all parcels offered by their members. There is not necessarily any advantage in offering a large quantity and in certain circumstances to do so may actually deter buyers and even depress the market. There appears to be no evidence that farmers' groups can improve the marketing of home-grown grain and a group member is no better off than an individual efficient farmer who trades through his merchant.

ECONOMIES OF SCALE

Economies of scale in agricultural merchanting can be more apparent than real. The optimum size of a merchant's business must depend on the local pattern of agriculture and on its needs. There does appear, however, to be a need for country merchants to co-ordinate their trade, particularly in specialist fields and this is a trend which is developing spontaneously. In many cases the small country business is a most efficient and economic unit. It is interesting in this connection to note the following extract from the I.C.F.C. *Small Firm Survey:*

> "It is arguable that in a modern mixed economy the better small concerns not only operate efficiently in more established sectors, but also include some outstanding innovators and pioneers, and that, above all, they are a nursery from which many of our future business leaders will emerge. Thus, whilst some of the firms in this particular sample may flounder and fall, many more will grow substantial—and some may become the giants and household names of tomorrow."

TRANSPORT

You have asked us whether we consider that the transport of grain internally and coastwise is as efficient as possible. Most home-grown grain is carried by road transport. Merchants are continually improving their lorry fleets and the service given to agriculture can hardly be equalled in any other industry. It is the highly efficient and flexible operation of transport which has played a major part in the rapid development of the trade and its ability to market successfully the very greatly increased cereals crop.

Transport charges have shown negligible increases over the past 15 years but it is disturbing to note that recent and impending legislation is inexorably driving costs upwards. Merchants are deeply concerned to note the tendency to undermine by fiscal measures and new statutory controls the development of this vital factor in cereals marketing, especially as nationalised transport, both road and rail, is not noticeably more efficient or economic than other transport services. The railways have indicated that they are building no new bulk wagons for grain, although their existing fleet is too old and too small. The closure of private sidings, branch lines, and local stations has sometimes added to costs where grain is consigned by rail. This is a regrettable development since it would seem that certain grain traffic could be transported quickly and economically by rail if the railways would only provide the necessary facilities instead of progressively limiting them.

The use of containers for grain transport may have some limited application, particularly where there are no bulk rail facilities at the consumer's premises. This is a possibility which may prove increasingly useful if government pressure against the use of large road vehicles develops.

The developments which we would like to see are the use of larger bulk lorries where practicable and the installation of modern bulk facilities by more farmers, including those who buy grain as well as those who sell it. Regrettable delays constantly occur in the loading or unloading of bulk vehicles at farms because of inadequate equipment. It is, for example, not uncommon for a 10-ton lorry to take $1\frac{1}{2}$ hours to load on a farm.

The Association's booklet on bulk handling successfully foresaw trends in vehicle construction and is in standard use by consultants and contractors for bulk storage facilities and bulk handling equipment. It was last revised in 1964, but although development in this field continues at a rapid pace with the use of larger vehicles only minor revision of the booklet is likely to be required in the near future.

Difficulties also occur in the delivery of grain to consumers, particularly on a Monday morning, for example, when a number of lorries may arrive at a mill simultaneously.

Some users have instituted a partial appointment system; there are difficulties in operating such systems but more could be done to improve them.

So far as coastal shipping is concerned, this can certainly be efficient, but the costs of loading and unloading, dock dues and inland cartage often make it uneconomic. There is also considerable loss involved in waiting time at docks. Above all, however, it is the high Conference freight rates which hinder the development of this traffic.

CAPITAL

The working capital position of agricultural merchants is broadly satisfactory although the very low profit margins available to the trade make it increasingly difficult to finance development. The position of merchants would be improved if farmers themselves had more adequate capital, as this would ease the demands on many traders for substantial credit to their farmer customers.

We feel that there is a case for the establishment of a government-sponsored agricultural finance corporation to make funds available to the whole agricultural industry, including the ancillary trades.

OTHER MATTERS

Agricultural production

The United Kingdom's natural resources must clearly be exploited to the utmost in the national interest and we feel that there should be no restriction on agricultural production. Full government support for the maximum production of meat and livestock produce would increase the demand for home-grown grain. It must, of course, be borne in mind that our climate and local soil conditions govern the amount of wheat which can satisfactorily be grown; but the present system of agricultural support with its standard quantities and generally inadequate encouragement to livestock farmers is not conducive to the achievement of maximum agricultural production.

Unity of the agricultural industry

A fresh look is needed at the structure of the agricultural industry as a whole and there is no justification for drawing artificial lines between different sectors of it. It would be in the interests of the national economy for the grain trade, including growers, merchants and users, to be treated as a single entity. At present both farmers and industrial processers may obtain government grants which are denied to country merchants whose role lies in distribution and we feel that this situation tends to hinder the orderly economic marketing and usage of home-grown grain.

Technical assistance is available from the Government's advisory services to farmers, and to the industrial processors from their own specialist advisers; it is equally important for the country merchant to have access to the best technical advice from government sources. Merchants also obtain information and knowledge from their practical experience in the trade and from other sources and this experience is made readily available to

BB

farmers. Merchants certainly maintain very close contact with the Government's advisory services so that farmers are not given conflicting advice on technical matters and the continuation of this liaison and co-operation is most important. The advice which merchants give to farmers is generally very much welcomed by them.

Market statistics

It would be of great value to all concerned with cereals marketing if reasonably reliable estimates of production and anticipated consumption requirements could be made available to the trade before each year's crop was actually harvested, covering both the home and international market. Such estimates would need to be revised regularly throughout the season as figures several weeks old are of very little value.

Research

Further research on the production and utilisation of home-grown cereals for animal feeding and for industrial purposes could lead to a valuable increase in the scope for consumption at home and could consequently minimise the need for imports. It would seem that this is a function to which the H.-G.C.A. might well devote more time.

Recruitment

The low margins available in the corn trade (the gross margin in cereals transactions excluding transport costs was recently estimated at no more than 2% on average) have contributed to a decline in the recruitment of able young men. This is one of the more serious factors affecting the trade's future development and efficiency.

Registered barley buyers

Under the present Cereals Deficiency Payments Scheme it is accepted that merchants authorised to sign wheat certificates entitling farmers to payment from public funds should be subject to proper investigation before they are so authorised. We are very strongly of the opinion that a similar scrutiny should be applied to those who wish to be registered as barley buyers for the purposes of the scheme.

New cereal varieties

Plant breeders have achieved remarkable increases in cereal yields by breeding improved varieties. We feel, however, that greater emphasis should now be placed on the fitness of new varieties for specialised use; that new varieties of wheat, for example, should be intrinsically valuable to flour millers and that barley varieties should be bred for their value to brewers and maltsters. Greater attention to quality rather than to yield alone will increase the potential usage of the home crop. As already mentioned earlier in this statement, it should of course follow that the produce of such crops should command a realistic premium from users.

Futures markets

In the outline to your survey you refer to the value of futures markets. It is our view that the wheat and barley futures markets have now established themselves as an important feature of orderly marketing.

Support buying

Your outline of the survey also refers to the role of the "buyer of last resort". We feel that the intervention of such a buyer could have a very harmful effect on the market and could cost growers and the Exchequer substantial sums of money, unless some system similar to that in the E.E.C. were adopted, whereby any surplus is exported through the trade with the assistance of export bonuses.

EUROPEAN COMMON MARKET

The nature of the cereals market and the trade's standpoint on many of the questions referred to above would, of course, be radically changed if the U.K. were to join the E.E.C. or if the Government were to adopt a system of cereals support similar to that of the E.E.C. This system, by means of import levies, export subsidies and intervention buying, ensures that cereal prices are maintained at a level high enough to give farmers a satisfactory return from the market alone without any need for deficiency payments from the Exchequer.

The working capital used by both corn merchants and processors would need to be increased very significantly in order to finance purchases of grain at the new high market prices. A system of intervention buying at a pre-announced intervention price would have to be devised. Within an enlarged Common Market British grain merchants would be in competition on their domestic market as well as on the export market with French exporters, and this could well hasten the development of a closer system of quality control such as French merchants already have in their own silos. The export trade would, of course, benefit from subsidies which would be financed by import levies, and if the intervention prices were not pitched too high in relation to import levies it should be possible to encourage processors to use a greater proportion of a competitive home crop, thus reducing the country's dependence on cereal imports.

These are questions of such importance that they would require very careful consideration in the face of any likely change of policy. The Association would in that event wish to modify many of the views expressed in this paper.

CONCLUSION

The basic problem of cereals marketing is to ensure that home-grown grain is moved at the time it is required, from the point where it is produced to the place where it can be most advantageously used, with the minimum of expense in transport or other costs. Generally speaking these other costs would consist of the merchant's gross margin.

This basic operational function has to be performed whatever system is adopted. In spite of the decrease in the number of home buyers through recent amalgamations of millers, maltsters and other users the corn merchant has proved successful in marketing the crop through these and other outlets both at home and abroad. All studies of cereals marketing so far have confirmed that the present system is in general efficient and economic and we remain convinced that there is a very valuable future for the merchant.

Statement by the National Association of Corn and Agricultural Merchants: Supplement

THE DRYING AND STORAGE OF HOME-GROWN GRAIN

Origins of present policy

1. After the Second World War a number of aspects of agricultural policy in the United Kingdom were reviewed, among them the drying and storage of home-grown grain.

2. Apart from a general movement from the major grain-producing areas in the east to the major livestock areas in the west there was found to be no clearly defined pattern of distribution of the cereal crop. Grain of varied qualities was moved by the corn trade from one part of the country to another to meet demand as it arose. The pattern of supply and demand itself would vary from season to season and could not normally be anticipated. This situation remains unchanged today.

3. Largely for this reason it was decided at the time that the most logical place to store grain was on the farm where it was grown. In consequence the grain also had to be dried there.

4. Thus it has become government policy to encourage the drying and storage of home-grown grain on the farm. The encouragement is given by various means, including grants for the erection of silos and the installation of drying machinery, derating of farmers' premises, storage incentives in the deficiency payments scheme and free technical advice.

The changing background to this policy

5. But the background to this policy has materially changed since its inception. The home-grown crop now amounts to $13\frac{1}{2}$ million tons of which $9\frac{1}{2}$ million tons come onto the market. The following table shows the extent of the increase in production.

HOME PRODUCTION OF CEREALS IN THE U.K. ('000 tons)

	1946	1958	1964	1965	1966 (forecast)
GRAIN					
Wheat	1967	2711	3733	4105	3495
Rye	39	21	25	8	11
Barley	1963	3170	7404	8062	8804
Oats	2903	2138	1325	1213	1116
Mixed corn	350	275	101	91	93
Total	7222	8315	12,588	13,492	13,519

(1964 is the base year of the National Plan—see paragraph 6(a) below.)

The pattern of production has also changed significantly. Larger cereal crops are grown in the south and west and in Scotland.

6. Three major factors must be taken into account in any plans for cereals marketing in this country during the next 10 years:
   (a) Production is likely to continue to increase steadily. The National Plan foresees an increase of $4\frac{3}{4}$ million tons between 1964 and 1970.
   (b) The consumption of coarse grains in this country is limited by the number of stock to be fed. Even if the livestock population increases recent experience has shown that it is most unlikely to do so at the same rate as the expansion in cereals production.
   (c) The present consumption of home-grown wheat for flour milling is limited to about 2 million tons from an ideal harvest. Although more wheat is being used in compounding for animal feedingstuffs there is a limit to this outlet set by the number of stock (mainly poultry and pigs) to be fed.

7. The conclusion arising from these factors is that there will continue to be a surplus of production of barley over the demand for home consumption. Substitution of wheat for barley is limited by the fact that the sowing of winter wheat is very dependent on weather conditions. It is thus vital to consider the export market not only as an important outlet for home-grown barley, but also as a means of earning valuable foreign exchange.

8. The export trade in feeding and malting barley has in fact developed considerably over the past 2 years, principally to Western Europe and North Africa. The major part of the expanded tonnage has been feeding barley. The following table of exports from the U.K. over recent years shows the extent of the expansion:

1960/1	122,000 tons	1964/5	110,000 tons
1961/2	336,000 tons	1965/6	667,000 tons
1962/3	182,000 tons	1966/7	1,050,000 tons
1963/4	75,000 tons		(estimated)

Results of present policy

9. After 20 years of encouragement a large proportion of the cereal crop is now dried and stored by farmers who grow it. It is not thought that more than about 15% or 20% of home-grown grain is dried or stored under arrangements made by merchants.

10. Thus for months after the harvest grain lies scattered in thousands of farmers' stores of varying sizes and with varying facilities. Although grain is occasionally dried for a grower off the farm and then redelivered to his store most of the crop is dried on the farm itself as soon as possible after harvesting and does not then leave the farm until it is required by a user. This may be as late as July the following year.

11. Direct financial grants for the installation of drying plant were only available to farmers for the first time last year. One consequence of this is that grain-drying equipment on farms has never kept up, in terms of capacity, with the increasing volume of the crop, particularly as the rate of harvesting has increased enormously with the use of modern combine harvesters.

12. In these circumstances many farmers have, with the best intentions in the world, been unable to avoid drying their grain too hastily and at too high a temperature and thus damaging its quality. Before it is even dried the fact that the grain may have to wait for an excessively long period before reaching the drier after harvesting is an additional cause of loss of quality. Thus large quantities of grain which are potentially of malting or flour-milling quality can be damaged with significant loss in value. Even feeding grain, which is of far greater importance in terms of volume, can have its value seriously reduced if it is allowed to go out of condition while awaiting drying or afterwards.

13. The limited capacity of farm driers in relation to the volume of the crop and the rate at which it is harvested is therefore a source of financial loss both to the individual farmer and to agriculture and the national economy as a whole. In a wet year particularly the loss can be very heavy as many farm driers are unfortunately not capable of bringing the moisture content of wet grain down to a safe storage level in one operation.

14. It is, of course, readily acknowledged that many farmers with large and small acreages have installed efficient drying machinery and do not damage their grain in the process. Their silos are often most modern and effective in maintaining the quality of grain. Even so the fact remains that grain stored on a farm after the harvest presents certain problems of marketing. Neither its quantity nor its quality can be known with any certainty until it leaves the farm. Thus the merchant who has responsibility for marketing it is dealing in something of an unknown factor and is handicapped in his task of realising the grain's maximum value on the market.

15. If this is a surprising conclusion it only needs to be explained that few farms have any facilities for weighing grain in bulk and that it is extremely difficult to estimate the quantity of grain held in a silo unless it has actually been weighed. So far as quality is concerned the sampling of grain held in a bulk store is notoriously unreliable. Particularly if it has not been adequately turned while in store the grain at the top and bottom of a silo may present quite a different sample from the bulk in the middle.

Loss of exports

16. One particularly regrettable consequence of the comparative unavailability of information about grain stored on farms is that export opportunities for barley are lost. The problem is especially difficult with malting barley of which this country possibly produces the finest qualities in the world. Although there is a continuing overseas demand for English malting barley at a good premium, merchants often find it difficult to make firm offers to foreign buyers with any certainty that grain of the required quality will be available at the time of shipment. Even with feeding barley, which now forms the bulk of our exports, quality is an important factor and the same problem exists.

17. If the U.K. enters the Common Market it will be very important to facilitate exports of grain to other member countries in competition with France. Ideally the exporting merchant should have complete control over the grain which he is selling at the time he is exploring the market and this really means that the grain should be in his own store. Complete reliability of quality and availability of supplies, particularly in relation to malting barley, is a necessity for exports on any large scale.

Storage by merchants

18. It has already been seen that the production of grain in the United Kingdom is expected to increase considerably between now and 1970 and probably beyond that date. An expansion in drying and storage capacity will thus be urgently needed. As the maintenance of a substantial export trade will be vital, very careful attention will have to be paid to the methods of drying and the quality of the new storage.

19. It is very likely that farmers will continue to invest in new storage capacity and drying plant, especially in view of the grants which are now available and the continued amalgamation of farming units. There is a substantial argument, however, for at least some of the new storage which will be needed to be in the hands of merchants, particularly those engaged in the export trade. For reasons already given a merchant who has direct control over the crop from the time of harvest is in a better position to offer his customers guaranteed standards and quantities at the times required, and thus has an inherent advantage in obtaining a good price.

20. Even in the domestic market there is an evident and growing demand on the part of users for assured supplies of grain in substantial quantities and at even standards of quality. Any merchant can more easily meet such a demand from grain over which he has immediate control. He could be expected to obtain a better price for it and thus pay his growers more in turn.

21. The basis of this argument is essentially that a merchant does not know the precise quality or quantity of grain available to him until it leaves the farm. If it leaves the farm early in the season to go to a central store then the necessary knowledge is available before the merchant goes into the market. If he goes into the market without that knowledge he may well fail to realise the full potential value of the grain.

22. Another and quite separate advantage of storage off the farm lies in the field of statistics. Grain coming into merchants' stores early in the season would be weighed, and if this information were made available to a central authority a clearer idea of national production would be available earlier than at present.

Location of stores

23. At present the main store capacity off farms is to be found in the ports, but because of the declining importance of imports, *vis-à-vis* home produced grain, port storage is probably of decreasing value. It has been confirmed by the Ministry of Agriculture, Fisheries and Food that almost half the amount of home-grown grain used for milling or compounding is now processed by mills situated inland. Inland storage is therefore of immediate and growing importance.

24. Perhaps the most economic point at which to dry grain is close to the point of harvesting so as to reduce the weight of water which has to be transported with the grain. This is one argument which is currently used to justify the encouragement of farm storage. If it is accepted that some storage should be off the farm in the hands of merchants that storage should nevertheless be as close as practicable to the farm. This argument points to the desirability of erecting local stores serving fairly small districts in the grain-growing areas, easily accessible to the farm trailer which could cart grain straight from the cornfield. Farmers' groups will tend to erect such stores if merchants fail to do so.

25. In the predominantly livestock areas stores should still be established locally in order to take up, after harvesting, the surplus of grain grown in the area for later distribution to local compounders and stockfeeders. At present much of the grain produced in these areas is already stored off the farm, but in port stores. Much unnecessary transportation is involved in taking it into store when it only has to be replaced later in the year by grain brought in for feeding from other districts.

26. The success of local stores operated by merchants is very well illustrated in France where most grain is removed immediately after harvesting to the premises of the local "organisme stockeur". The total tonnage of grain available for the market and the quality of each parcel is thus immediately known to the trade and detailed statistics can be made available to the official authorities. This system has enabled France to build up a very efficient export trade. Orders from other countries, including the U.K., can be fulfilled accurately at very short notice as opportunities occur. As already explained, the British merchant finds it very difficult indeed to guarantee the fulfilment of export orders at short notice.

27. If the U.K. enters the Common Market all tariff barriers between us and the other member countries, including France, will disappear. There would thus be ample opportunity for the export trade to be developed, but this would only be possible on equal terms with the French if export merchants had access to grain in central stores for which they were themselves responsible. On the domestic market, too, the British merchant would find himself in open competition with French exporters who would find it comparatively easy to ship substantial quantities of grain of specified standards to British millers and compounders. A trade in French grain already exists in this country in a comparatively small way and buyers like its even bulk; there would be no barrier to the expansion of this trade once Britain was in the Common Market. Both British and overseas buyers will pay a small premium for a standard bulk of grain to which certain guarantees are attached.

28. It seems clear from these arguments that of the new grain storage which will be necessary in this country to cope with an increasing crop at least a substantial part should be operated in the form of local merchants' stores to complement storage on the farm of origin.

Financial aid to merchants

29. Although some merchants already own and operate grain silos and drying plant and perform specific functions as approved service operators under the Ministry's Cereals Deficiency Payments Scheme, there are a number of financial factors which militate strongly against the merchant in this role.

30. Farmers have for some time received grants for the construction of silos under the Farm Improvements Scheme. The available grants have recently been extended to fixed plant and machinery, including grain-drying equipment. Total grants now obtainable by a farmer are 30% of the capital cost of new buildings and 30% for fixed machinery. In addition, farmers' grain stores are not subject to local rates and they receive in this way a continuing subsidy on their storage and drying operations.

31. Merchants are now eligible for grants of 25% of capital expenditure on new plant and machinery and for initial taxation allowances of 15% on new industrial buildings.

If a merchant's premises are situated in a development area (broadly Scotland, Wales and the extreme north and south-west parts of England) the available assistance is considerably higher: the grant on plant and machinery amounts to 45% and the initial taxation allowance on new industrial buildings can be supplemented by a building grant of 25% or 35%. Most grain, however, is grown outside the development areas and agricultural merchants are thus generally at a severe financial disadvantage, *vis-à-vis* farmers, when investing in drying and storage facilities. A further serious distinction is that merchants' premises are fully rated by local authorities, a burden which adds considerably to the cost of operations in competition with farmers.

32. It is a very serious step for a merchant to consider investing capital in expensive buildings and equipment in such circumstances. As has been argued above, however, it seems desirable in the national interest that a significant part of the new storage capacity which must be erected to cope with an increasing crop should be in the form of local silos operated by merchants.

33. In order to make this development economically feasible in the greater part of the country the existing financial disadvantages facing merchants should be completely removed. The National Association of Corn and Agricultural Merchants therefore proposes that merchants erecting new silos or installing new grain-drying equipment outside development areas should qualify for capital grants on equal terms with farmers. And in order that the operation of these plants should be possible on an equal basis with those owned by farmers it is also proposed that merchants' and farmers' grain silos should receive equal rating treatment.

34. Investment grants should be available either to individual merchants or to merchants' syndicates.

35. It must be emphasised that the Association's purpose is not to discourage the storage and drying of grain by farmers who have the time and facilities to do the job efficiently and economically. It is not claimed that merchants should receive better treatment in this field than farmers. The only purpose of this memorandum is to show that in the interests of improved cereals marketing, and of the national economy as a whole, merchants should be encouraged and enabled to invest in storage facilities which would supplement those available on farms.

Statement by the Agricultural Co-operative Association (incorporating the views of the Scottish, Welsh and Ulster Agricultural Organisation Societies)

CEREALS MARKETING

Agricultural co-operatives handling grain are in nearly all cases members of the National Association of Corn and Agricultural Merchants which has put forward its views on the nine questions posed by Nottingham University.

There are certain special considerations affecting co-operatives in respect of certain of these questions, and there were two additional questions posed by Nottingham University which are of concern to co-operatives alone.

QUESTIONS 1 AND 2. *Use of home-grown cereals*

The question of export deserves special mention by the agricultural co-operatives in that they have during the past year established an export agency particularly concerned with cereals. The experience gained in this operation is too short to be given over much significance, but it can be said with some degree of assurance that some exports have taken place which would be unlikely to have taken place if the agency had not been in being, and the home market must to that extent have been relieved of grain which it would otherwise have had to absorb.

QUESTIONS 3, 4 AND 5. *Marketing of home-grown cereals*

In answering these questions it is necessary to look not only at the actual situation but at the situation as it might have been if different policies had been pursued. In this connection there is some value in examining the situation in other countries, e.g. France, where the Union Nationale des Co-opératives Agricoles de Céréales (U.N.C.A.C.) which started trading in 1947 and began to undertake stockage in 1950 had, 20 years later, captured 80% of the internal and 12–15% of the external trade in wheat and had its own storage capacity amounting to 400,000 tons. U.N.C.A.C. has had some support from the Government in attaining this remarkable position; more important, however, is the fact that it was not confronted by a government policy of promoting storage of grain on the farm through a system of grants for the construction of storage facilities. It is true that since the new scheme of co-operative grants was introduced in 1967 it will be possible for U.K. producers to obtain off-farm grants for co-operative storage (but no greater than the on-farm grants they can obtain through the Farm Improvement Scheme), but only subject to the conditions that (a) such storage will improve their production businesses, (b) the co-operatives sell their grain on a commission basis and not as a wholesaler, (c) all its members are under contract to the co-operative and the co-operative does not have other members which are not under contract to it, (d) the storage erected is no larger than will be needed for the purposes of the contract, (e) it is appropriately sited in relation to the members' farms, which probably means on or near their holdings, and (f) there is a capital contribution by members at least equivalent to the amount received in grant. It seems unlikely that, under such conditions, co-operative grain marketing units of any size are likely to be formed; indeed the scheme could have the effect of breaking down co-operative marketing organisations that already exist. The contrast between French and U.K. government policy has been drawn because the ability to market grain successfully depends to an important extent on having physical control over it; U.K. agricultural support policy has tended, and is tending, to ensure that this control remains with the farmer, who is a small seller and usually a weak seller, whereas French policy has tended to encourage the emergence of the farmers co-operative, which is a large seller and potentially a strong seller. (It should be pointed out that U.N.C.A.C. is a federation of 650 co-operatives which operate to a large extent independently, the influence of U.N.C.A.C. being particularly strong in inter-regional trading and export.)

Quantity is, of course, only one of a number of marketing factors. Small units also tend to be less efficient than large units so far as drying is concerned and have greater difficulty in achieving a level sample.

QUESTIONS 6–9. *Agricultural merchanting*

As pointed out earlier the tendency in government post-war agricultural policy has been to regard the farmer solely as a producer, not as a person who has to sell what he produces. On the former basis it has been reasonable to operate a system of support for agriculture which is founded on the individual need. But in so far as the emphasis in future may be on the farmer as a seller of produce, it will be necessary to adopt a different approach since "individual marketing" is clearly a contradiction in terms; it then becomes necessary to think rather in terms of farmers' co-operatives. Here, too, one must make the point that an increasing part of the consumers' £ spent on farm produce is incurred in expenses beyond the farm gate, i.e. in marketing. It is for this among other reasons that the Agricultural Co-operative Association, in conjunction with the Scottish, Welsh and Ulster Agricultural Organisation Societies, has presented a case to the Minister for assistance to farmers through co-operatives in the form in which such assistance would be of the most value to them, i.e. through the formation of a co-operative financing institution. (A copy of the Maxwell Stamp report commissioned by the co-operatives and their observations upon this is available on request.)

QUESTIONS 10 AND 11. *Structure and distribution of co-operation*

The questions asked here are important but it would be difficult to attempt an answer on the basis of the very limited information available.

During the last few years the agricultural co-operatives have significantly increased their marketing turnover, and this has led to many suggestions as to how this side of their activities should be developed and further co-ordinated. (A fair degree of co-ordination already exists and this has been further improved since the formation of Farmers Overseas Trading Ltd.) The Central Council for Agricultural and Horticultural Co-operation is to grant-aid a survey, sponsored by A.C.A. Ltd. and to be executed by the Plunkett Foundation, into the structure of co-operatives engaged in marketing, which it is hoped will provide an answer to these questions.

Questions addressed to the National Federation of Corn Trade Associations Ltd.

1. Do you consider that any important changes are taking place, or will take place in the near future, in the availability of supplies of imported grains for the United Kingdom, either in respect of the distribution of the trade between the various supplying countries or in respect of the reliability of such supplies and the general stability of the world grain trade?

2. To what extent are home-grown and imported grains interchangeable, and do you expect to see any important changes in this respect, for technical or other reasons? Do you consider that British grain is becoming more competitive or less on the U.K. market, and in relation to what uses?

3. Do you consider that recent changes in the exports of grain from the U.K. are of a temporary nature or are they likely to persist, and if so for what reasons? To what extent have these changes been affected by forward-contract arrangements for home-grown grains?

4. Are producers in the U.K. sufficiently aware of the requirements and opportunities of the export market for grain, particularly as regards quality, timing, regularity and bulk of their deliveries? If not, how do you think this situation could be improved?

5. Are there any other features of the grain market, as it affects the U.K., which have not been mentioned in your replies to the above questions, but which you consider to be of special importance in any attempt to arrive at a fair and balanced view of the market situation, in the short-term or in the long-term?

Statement by the National Federation of Corn Trade Associations Ltd.

QUESTION 1

The question is most comprehensive in its construction, and the main purpose appears prima facie to ascertain whether the Federation foresees in the near future that the supplies of imported grain to the U.K. from usual suppliers could be withheld or placed in jeopardy. Whilst it can be stated categorically that there are no fears in this connection, it is obviously necessary to give qualifying reasons.

Firstly, it should be emphasised that the major world suppliers to U.K. and to other traditional importing markets have efficient planning departments within their official agricultural organisations. These departments are constantly gathering data from all parts of the globe through their Agricultural attachés concerning world agricultural conditions. These exporters are not selling abroad just because they happen to have an excess of grain supplies over their domestic requirements.

On the contrary, countries such as Argentina, Australia, Canada, France, South Africa and the U.S.A. are growing cereals for export (with preference accorded to "cash" markets) as a matter of deliberate policy and to create specific external income.

During the recent Kennedy Round discussions under G.A.T.T., the major grain-exporting countries expressed the wish to institute a set of conditions whereby importing countries would agree to limit production of indigenous cereals in order to ensure that a given percentage of their needs would accrue to the major exporting countries. This facet of the talks was ultimately shelved and did not finally form part of the various understandings reached. It is safe to say, however, that the countries enumerated above are continually anxious to maximise exports, and thereby endeavour to adjust their own crop programming in order to meet foreseeable world demand in

(a) the field of customers who pay cash, and
(b) to satisfy to the extent possible the demands from underdeveloped nations who have not adequate resources to pay in cash.

In connection with the future equilibrium of world grain supply and demand, the following statement in Washington is imputed to U.S.A. Secretary of Agriculture, Mr Orville L. Freeman, see Reuter, on 23rd August this year:

"There would be no lack of productive capacity to feed the world in the years ahead and certainly not before 1980, U.S. Secretary of Agriculture, Orville L. Freeman said today.

He based his predictions on a study just completed by the Agricultural Department's Economic Research Service. The study implied, he said, that the developed countries of the world may be plagued with grain surpluses at least until 1980, despite the increasing needs of developing countries. The pressures of such surpluses could lead to renewed protectionism and special united efforts would need to be intensified further to open up channels of International Trade so that commercial grains could move freely, he added.

The study itself stated that the combined excess food production capacity of all the developed countries in 1980 would be more than adequate to provide for the increased food import needs of less developed nations, even if those countries did not improve their growth of grain production. Under such circumstances, the surplus in 1980 would be around 30 to 34 million tons and would go even higher if developing countries did not increase their output.

Without an increase in their rate of growth in food production they would need to import at an increased rate of nearly 60 million tons annually by 1980, the study predicted.

Such requirements could be readily met through expanded production in developed countries but ways to finance a large proportion of the needs would have to be found.

Furthermore, the less developed countries would have to develop the capability to import and distribute the increased amounts of needed grain, the study added. The study implied, Mr Freeman commented, that the less developed countries needed to increase their self-help efforts and greatly improve their abilities to produce and buy more food.

To this end, he stated, 'We cannot allow food aid to retard efforts by developing countries themselves to increase food production and to accelerate economic growth.' To pour food aid into the developing countries at the massive rate required to raise their *per capita* nutritional intake to desirable levels would tend to depress prices in recipient countries and deprive them of a major incentive for increasing their own food production, he declared. 'The long-run interests,' he said, 'call for careful weighing of requirements for both food and development, and a balance of our own food production in terms of such requirements as well as in terms of expanding commercial trade'."

The reason for quoting this statement *in extenso* is to demonstrate the type of policy thinking by the world's largest supplier of grain.

To take the other part of the question, perhaps unanswered by the foregoing, there are of course important changes constantly happening in the pattern of world supply of grain. Despite "short-term" and "long-term" planning by governments in acreage programming, as well as the vastly improved technology employed in farming in general, the vagaries of climate do play an important role as to eventual annual production in individual countries. Thus, from year to year, it is possible to have quite substantial variations in production even if the acreage sown shows little change. Official statistics readily available will prove the point. However, carryovers from one crop to another plus the current production have been in the main sufficient, not only to ensure domestic requirements in originating countries but also to permit of regular exports to traditional customers abroad. Although there has been an isolated period in the past decade when world stocks have looked to be coming well into equilibrium with demand, generally speaking, stocks from old crops plus expected production from new crops have been more than ample to meet foreseeable needs. By and large, and in the opinion of the Federation, major exporting countries tend to assure stocks and production in excess of immediately foreseeable demand. The "know-how" of the U.K. grain trade is such that a fear of a temporary shortage of one particular class of grain in one part of the world either stimulates advance buying of that particular type or causes substitution by something similar from another part of the world. Broadly speaking, prices of imported grain fluctuate according to availability of supply versus demand despite temporary artificial government attempts to work contrary to that natural law.

In wheat, and up to July 1967, an International Wheat Agreement (renewed periodically over the years) with a minimum and maximum price range has been in existence. At no time has the Agreement militated against normal commercial movement of wheat. A new Agreement will come into force in July 1968, whereby the minimum and maximum prices agreed under the aegis of the "Kennedy Round" have been fixed at an increase of around 23 cents above the minimum/maximum price levels of the Agreement which has just expired. This incentive can only give encouragement to major exporters to make expeditious plans for any extra needs forecast in the near future in the principal importing countries.

As to coarse grains (maize, barley, sorghums and oats), there has never been any overall shortage apparent in world supply, and so confident has been H.M. Government in the continuance of such circumstances still obtaining, that some 3 years ago it instituted what is known as the Minimum Import Price Regulations, whereby levies are imposable on U.K. buyers in the event of imported grains falling below certain c.i.f. levels. There were also Letters of Exchange between H.M. Government and exporting countries willing to co-operate with these Regulations and it is presumed that Professor Britton is fully aware of these Regulations, Letters of Exchange, also the details of the undertakings of the new Wheat Agreement.

QUESTION 2

There are degrees of interchangeability between certain home-grown and imported grains, depending upon a variety of factors. The British milling industry has in recent years conducted considerable research which has lessened their dependence upon imported strong wheats, so that today up to 35% of the total grist for bread making may be English or soft Western European wheats. It must be appreciated, however, that for the remainder of the grist they still depend upon stronger wheats, Manitobas, U.S.A. high protein Hard Winters and Australian Prime Hard Wheats. In wheats utilised for animal-feeding purposes, home grown is wholly interchangeable with imported.

In coarse grains, the U.K. does not cultivate maize and there is a hard core of around 1 million tons of maize utilised per annum for distilling and in the manufacture of starch, glucose, etc. Industrial usage is approximately one-third of total imports of maize.

The possibilities of interchangeability in animal foods from maize to other grains is relatively limited, and under normal circumstances is governed entirely by price factors. In view of the meteoric increase in the harvest of home-grown grains over the past 5 to 6 years, there has already been a tendency to reduce maize usage for feed in favour partially of barley, and more particularly of wheat. It would probably be difficult to change the ratio of usage between these three grains to any further substantial degree without reducing the nutritional efficiency of the rations. This question should undoubtedly be explored with the compounders, however, who alone can give a true picture as to the relationship of the individual feed grains, i.e. barley, maize, sorghum and feed wheat.

As to the second part of the question, the fact that carryovers from one crop to another in the U.K. are minimal tends to show that British grain is competitive on the U.K. market. It must be borne in mind, of course, that the competitiveness of home-grown grain at any particular moment is dependent upon the existing international level of imported grain of a similar description or of another kind which may be used as a substitute.

QUESTION 3

This is not an easy question. The export figure of 1 million tons of home-grown barley in the season 1966/7 constituted a record and the price level involved was undoubtedly the determining feature.

Bearing in mind that exports have been destined primarily to European countries, and around one-third went to E.E.C. countries, considerations relevant to the price level involved are:

(a) Future production (plus or minus) in those countries which have been importing from U.K.

(b) Competitive level of barley from other origins, as also that of other feeding grains.

(c) Increased utilisation (or otherwise) by U.K. users at prices ruling above those obtained for export.

(d) In the case of E.E.C. countries, the level of the variable levies imposed upon imports from third countries.

As a general opinion, it is thought that, based on the foregoing imponderables, there could be substantial variations of quantities exported from year to year but it cannot be foreseen there is a risk of any sudden cut-off down to nil. In reply to the second part of the question, the incidence of forward contract arrangements with the H.-G.C.A. does not appear to be directly related to the total quantity exportable or actually exported.

QUESTION 4

Publicity is given in trade journals, H.-G.C.A., and Board of Trade statistics as to the progress of exports. It is doubted, however, that producers of barley when actually selling know whether the quantities thus sold and actually delivered to the country merchant will be ultimately utilised domestically or for export. The producer sells ex-farm, and the physical movement either domestically or for export is not in his hands. However, in view of market intelligence exchanged at interior markets, any producer is likely to ask questions regarding both domestic and export demand. The producer's indicator is the "guide price" issued by the H.-G.C.A. and the strength of the demand from local country merchants. The trend of the London and Liverpool "futures" markets is also a useful adjunct. It is suggested by the Federation that the pipeline between this country and the importing destination is one of concern to the exporter rather than to the producer—the latter's interest being to maximise his price ex-farm, and any surge or otherwise in domestic or export demand or both must necessarily be reflected in the market price at which the producer is solicited to sell. Each section of the trade has its particular function in the ultimate marketing.

QUESTION 5

At this stage, consideration should be given by Professor Britton to our answers under 1 to 4, in order to further develop our views. It should be said, however, that speaking from the imported side, this section of the grain trade is highly organised and delicately geared from the farmer abroad to the user in this country. Private enterprise has constructed modern interior and seaboard elevators both abroad and in importing countries to keep pace with modern conditions in the grain trade. Export specifications, timing of shipments, and so on have to be rigidly complied with in accordance with world-wide contractual terms and conditions. Shipowners have built larger and more modern ships to cope with the present-day situation and that of the foreseeable future. The international shippers, merchants and the users of imported grain are constantly striving by investment and technical improvements to bring grain to this country and utilise it in the most economic fashion. Sheer weight of competition dictates the most modern methods of storage, marketing, shipment into and discharge from ocean vessels.

C. The End-users
(i) Animal Feed Manufacturers

Questions addressed to the Compound Animal Feedingstuffs Manufacturers National Association and the National Association of Provender Millers

A. USE OF HOME-GROWN CEREALS

1. What are the prospects of using more home-grown cereals in substitution for imported grains?
2. Could the export trade in feedingstuffs be expanded?

B. MARKETING OF HOME-GROWN CEREALS

3. Would compounders favour the introduction of a more precise grading scheme for cereals? To what extent does grain fetch a lower market price than it otherwise would because of such factors as inefficient drying, poor storage or lack of cleaning?
4. What are the relative merits of trading direct with farmers and trading via merchants?
5. Has the H.-G.C.A. Forward Contract Scheme improved the marketing of grain from the compounder's point of view?
6. Are the present cereal deficiency payment arrangements (including seasonal incentives) conducive to good marketing of grain from the compounder's point of view?

C. THE COMPOUND FEED TRADE

7. What are the likely trends in the provender milling and compound feed industry in the next 5 years in so far as they are likely to affect the demand for cereals?
8. What factors influence the location of compound feed factories, and to what extent are current trading patterns bringing fresh thinking on location policy?
9. Is the trend towards bulk feeds and cubed feeds likely to continue, and what effect will this have on the compound feed trade?
10. Are there any other features of the grain market, which have not been mentioned in your replies to the above questions, but which you consider to be of special importance?

Statement by the Compound Animal Feeding Stuffs Manufacturers National Association

INTRODUCTION

In presenting the following answers C.A.F.M.N.A. wishes to stress that the compound industry has to take a balanced view between the interests of the cereal and livestock producing sectors of British agriculture. While compounders are the biggest single purchasers of home-grown cereals their main obligation is to provide livestock producers, who account for about 75% of total U.K. farm output, with the most efficient feeds in terms of economic conversion into livestock products.

782

Flexibility in raw-material supplies assists compounders in achieving their objective. Too great a substitution of home-produced cereals for imported, regardless of quality and price, could mean not only lower profitability to the individual livestock producer but reduced production overall, leading to increased imports of end-food products to make up the deficiency, with no ultimate net saving to the nation's balance of payments.

A. USE OF HOME-GROWN CEREALS

QUESTION 1. *What are the prospects of using more home-grown cereals in substitution for imported grains?*

ANSWER. For an annual production of about 9·5 million tons the compound industry requires, on average, 600,000 tons (approx.) of cereals per month throughout the year to meet its cereal raw material requirements. To assure continuity of production it must have at least 6–8 weeks—sometimes even 12 weeks—supply in the pipeline. If there is the slightest hitch or doubt at any time throughout the season about availability of home-grown supplies at competitive prices, compounders, in the interests of their livestock producer customers who produce about 75% of total farm output, have no option but to turn to the most economic alternative available source of supply, viz. imports.

Despite these limitations, however, the compound industry has continued to increase its off-take of home-grown cereals as the crop has expanded and is now virtually self-sufficient in home-grown supplies of barley. The bulk of barley imports today (0·18 million tons in 1967/8 compared with 1·2 million tons 10 years ago) go to meet the specialised requirements of the Scottish distilling industry, or for livestock feeding in Northern Ireland where domestic production of cereals is insufficient to meet compounders' total needs. If transport costs were lower it might be possible for Northern Ireland compounders to buy more barley from Great Britain than they do at present.

Maize (yellow) is an essential ingredient in the formulation of those high-energy rations which are a "must" for our laying poultry flocks, particularly for the modern hybrid. Barley by itself is no substitute for maize, either chemically, in terms of metabolisable energy, or physically. It is certainly no substitute if the producer is to maintain the standards of high egg production that he requires to make this type of farming economic, and which is essential if the housewife is to have eggs in abundance, and at a cheap price.

Home-grown wheat, if more were available, and available continuously at an economic price, could provide in part a substitution for maize, if not in the highly sophisticated poultry rations, at least in other rations, and it would also replace imported wheat, and sorghum grains including white maize. The possibilities of greater home-grown wheat usage were clearly demonstrated in 1965/6 when, with a larger than usual home crop, off-take by compounders increased considerably.

The growing of higher gluten wheats at home might encourage substitution for imported wheat, and partial substitution for maize, provided always that they were competitive in price with comparable imports.

Tables showing production of compounds, utilisation of home-grown cereals, and imports of feed grains are supplemented to this statement.

QUESTION 2. *Could the export trade in feedingstuffs be expanded?*

ANSWER. Exports of feedingstuffs valued at about £8·5 million in 1967 (of which only part were compounds) are only a minute fraction of total annual output of compounds valued at about £330 million.

Feed manufacturers throughout the world buy their raw materials at roughly similar prices. To transport these to one country and re-export them after processing into compound feeds to another involves double handling and transport costs on a commodity for which raw material content accounts for 80% or more of total cost. Exports of complete compounds must, therefore, be generally uncompetitive with domestic production in the importing country, irrespective of any tariff considerations involved.

A small export trade has, however, developed to the Mediterranean islands (Malta, Cyprus, etc.) and the Arab States in the Near and Middle East, Iceland, West Indies, etc., where there is no established processing industry and where indigenous raw materials, particularly cereals, are either nonexistent or not available in sufficient quantities to meet the needs of livestock producers. No dramatic expansion can be foreseen, however, in this limited, and highly competitive, export market in the years ahead, rather the reverse.

For manufactured protein/mineral/vitamin concentrates the prospects of expanding export markets are much better but this, of course, does not help the demand for home-grown cereals. But even this export market has a relatively short life, since, as the importing countries become more sophisticated in modern nutritional and milling techniques, they will eventually undertake the whole manufacturing process themselves.

B. Marketing of home-grown cereals

Question 3. *Would compounders favour the introduction of a more precise grading scheme for cereals? To what extent does grain fetch a lower market price than it otherwise would because of such factors as inefficient drying, poor storage or lack of cleaning?*

Answer. Any nutritional or manufacturing advantages to compounders arising from a more precise grading scheme for cereals are most unlikely to offset increased costs arising from the double handling involved in intermediate silos between farm and mill. Those companies who sell direct to the farmer would also miss the prospects for reciprocal trading. But above all such a system would involve the provision of additional storage space at mills, at colossal cost, so that the different grades could be identified.

It is, of course, true that grain fetches a lower market price if it is not dried, stored and cleaned properly, but it is not possible to quantify these between the individual deficiencies. On moisture, however, most compounders operate on the basis of 16% moisture, which specification is largely being met by growers and/or grain merchants, for home-grown barley and wheat, with a scale of deductions well known to farmers if the moisture is higher.

Question 4. *What are the relative merits of trading direct with farmers and trading with merchants?*

Answer. By trading in cereals direct with farmers, the compounder saves the merchant's charge but he, himself, may be involved in higher costs through having to deal with a very large number of small quantities which even then may be insufficient to meet his total needs.

The merchant, on the other hand, by dealing in substantial quantities of grain, contributes to the stability of the market, to the operation of the forward market, and to a subjective grading system.

To some extent, the decision whether to trade direct or via merchants is dictated by the method the compounder uses to market his own finished products. For the compounder selling through the merchant, reciprocal trading in grain is the obvious answer. Similarly, the direct seller will tend to deal with the farmer direct, although, as indicated above, supplies, particularly in certain areas of the country where livestock production predominates, may be insufficient to meet his total needs.

Question 5. *Has the H.-G.C.A. Forward Contract Scheme improved the marketing of grains from the compounder's point of view?*

Answer. It is difficult to say.

On the one hand, the scheme has introduced rigidity into the market for cereals by forcing large quantities on to forward contract. This is particularly apparent in the month of October when, due to the operation of the 2-month forward contract rule, it is impossible to contract forward at harvest time for deliveries in that month, leading to shortage then and over-supply in November, the first month for forward contract delivery.

On the other hand, the scheme has encouraged farmers to think in terms of contracting forward which they were reluctant to do beforehand.

QUESTION 6. *Are the present cereal deficiency payment arrangements (including seasonal incentives) conducive to good marketing of grain from the compounder's point of view?*

ANSWER. The differential of £4 per ton between the guaranteed price for wheat at the beginning and at the end of the season under the Cereals Deficiency Payments Scheme discourages regular marketing throughout the season, and completely overrides the effect of the H.-G.C.A. Forward Contract Bonus. As a result, growers are over-encouraged to retain wheat until the latter part of the season, thus creating an artificial demand for imports, which may eventually prove to have been unnecessary, during the earlier part of the season. The Association has persistently campaigned for a reduction in the differential to £3/£3 10s. per ton.

The position is not nearly so difficult with barley, where the "spread" in seasonal incentives is lower than in the case of wheat. Compounders, however, would like to see more barley coming on to the market in the early months of the season and feel that the disincentive to marketing then should now be removed.

For both grains the trade would prefer one overall scheme covering both deficiency payments and seasonal incentives, with changes in the latter operating at much shorter periods, perhaps monthly as in the case of the Common Market cereal regulations.

C. THE COMPOUND FEED TRADE

QUESTION 7. *What are the likely trends in the provender milling and compound feed industry in the next 5 years insofar as they are likely to affect the demand for cereals?*

ANSWER. The level of output of compound feeds in the U.K. from year to year is more or less determined by total livestock numbers. The objectives for livestock production in the National Plan, reaffirmed by the Government at the 1967 Annual Review, provide official guide-lines on which future compound production can be estimated. Total feed necessary to support the increased stocking envisaged compared with 1964—a larger dairy herd, more sows, more poultry—could well total 2 million tons when consequential increases in the life-cycle, e.g. more calves, fattened pigs, chicks, etc., are taken into account. With manufactured feeds representing 50–60% of total consumption of concentrated feeds it would be reasonable to expect that at least 1 million tons of this additional feed would be in the form of compounds. Support for this estimate is given in another section of the Plan which estimated an increase of £65 million (at 1964 prices) in farming expenditure on purchased feed by 1970/1. With manufactured feeds representing about 70% of total purchases by farmers, this would give an estimated increased expenditure of £45 million on compounds, the equivalent of well over 1 million tons.

In view of the setback to livestock production in 1966, expansion, particularly in pig production, is not much above the 1964 deadline. It seems unlikely, therefore, that the targets set in the National Plan will be achieved on time. Given appropriate government backing, however, the livestock targets should be achieved over the next 5 years, with corresponding increases in compound production on the lines indicated.

Any wholesale switch to farm mixing, both on the traditional farms and on the large integrated units, could lead, however, to a somewhat smaller increase in tonnage from commercial feed mills since a larger percentage of production would be in the form of protein/mineral/vitamin concentrates for subsequent mixing with cereals on the farm. But this is unlikely to have much effect on over-all consumption of home-grown cereals since the grain-buying policy of the large integrated producers, the largest potential source of increase in farm mixing, will not be very different from that of the commercial manufacturer.

QUESTION 8. *What factors influence the location of compound feed factories, and to what extent are current trading patterns bringing fresh thinking on location policy?*

ANSWER. The main factors concerned are raw-material supply, location of customer and transport costs. Originally, with a large percentage of raw-material ingredients, both cereals and protein, coming from overseas, mills were established at the main ports. As home production of cereals, particularly barley, has increased, the economics of scale achieved by the large port mills are being closely challenged by smaller mills at strategic points in relation to home-grown cereals and livestock producers. Current thinking is, therefore, tending towards several smaller mills strategically placed and operating as satellites of a larger mill with ready access to imported proteins.

QUESTION 9. *Is the trend towards bulk feeds and cubed feeds (extruded) likely to continue, and what effect will this have on the compound feed trade?*

ANSWER. Bulk delivery of feeds has expanded rapidly in the poultry industry in recent years and is extending to cattle and pigs as production units become larger and more intensive. This means that the compound industry will have to increase its investment in plant and machinery, with corresponding reductions in man-power.

Extruded feeds (i.e. cubes, cakes, pellets, crumbs, etc.) today form the major part of total production of poultry, cattle and sheep foods. For pig foods where "wet" feeding still predominates meals remain more popular. Producers will continue to look critically at the performance of extruded feeds, which cost more than meals, (e.g. there is some evidence of a swing towards mash diets for laying poultry), but overall the indications are that the proportion of extruded feeds to meals will probably continue to increase. The more sophisticated machinery required for extruding means that the compounder can offer the farmer a product which he cannot readily produce himself on the farm.

TABLE 1. COMPOUND FEED PRODUCTION U.K. (million tons)

	1961	1962	1963	1964	1965[a]	1966	1967
Cattle and calf food	3·05	3·21	3·11	3·09	3·32	3·29	3·48
Calf starters ⎱	0·30	0·34	0·04	0·04	0·04	0·04	0·04
Other calf food ⎰			0·34	0·35	0·38	0·35	0·34
All other cattle food	2·75	2·87	2·74	2·70	2·91	2·91	3·11
Pig food	1·85	2·15	2·04	2·04	2·22	1·93	1·94
Pig starters	—	—	0·05	0·05	0·04	0·03	0·03
All other pig food	—	—	1·99	1·99	2·18	1·90	1·91
Poultry food	3·89	3·89	3·89	4·02	4·03	3·94	4·04
For broiler chickens	—	—	—	0·61	0·75	0·81	0·90
For turkeys	—	—	—	0·21	0·27	0·24	0·25
For all other poultry	—	—	—	3·21	3·00	2·90	2·90
Any other compounds	0·17	0·19	0·17	0·17	0·17	0·18	0·17
Total compounds	8·96	9·44	9·21	9·32	9·74	9·34	9·63

[a] 53-week year.

TABLE 2. INTAKE OF HOME-GROWN BARLEY AND WHEAT IN
GREAT BRITAIN BY COMPOUNDERS, PROVENDER MILLERS,
AND DISTRIBUTING DEALERS ('000 tons)

	Barley	Wheat
July 1961/June 1962	1342	732
July 1962/June 1963	1455	1466
July 1963/June 1964	2045	960
July 1964/June 1965	2090	1370
July 1965/June 1966	1790	1700
July 1966/June 1967 (estimate)	2300	1200

Note: No figures available for U.K., quantities of locally grown
barley and wheat going into compounds in Northern Ireland,
however, are very small.

TABLE 3. IMPORTS OF PRINCIPAL GRAIN FEEDS INTO THE U.K. ('000 tons).

	1961/2	1962/3	1963/4	1964/5	1965/6	1966/7	1967/8 (forecast)
Barley	531	292	419	274	192	188	180
Maize	3938	3831	3431	3140	3490	3334	3608
Sorghum	549	394	260	386	512	496	250
Feed Wheat					500	500	500

Note: About 1,100,000 tons of maize imports each year go for non-feeding purposes
(e.g. distilling, etc.) and up to 100,000 tons of imports of Canadian barley are used
specifically for distilling.

**Letter received from the National Association of Provender Millers of Great
Britain and Northern Ireland**

29th November 1967

Dear Professor Britton,

U.K. Cereals Marketing Survey

With reference to your letter of the 19th October, with which you enclosed a questionnaire addressed to C.A.F.M.N.A. and this Association, this has now been considered
by my Committee, and I have been asked to put forward the following replies to your
numbered questions:

1. Little further replacement by imported grains in the provender industry is envisaged.
2. Possibly, under favourable circumstances.
3. We are satisfied with the position as it exists. With regard to the second part of the
question, the market price is discounted to a reasonable extent on account of
factors such as have been mentioned.
4. We prefer to deal with merchants because of the excellent service they give us in
the selection of suitable samples for our requirements.
5. Broadly speaking, we feel the marketing of grain has slightly improved from our
point of view.

6. The scheme appears to us to function reasonably well, but we do feel that an annual review and necessary adjustment should be made in order to maintain an even flow to the market.

7. Provided the producer of livestock can attain a reasonable profit margin on his products, we anticipate an increased demand in direct relationship to the increase in intensive farming.

8. The location in the country of a provender mill is preferred because of—
    (a) proximity to customers,
    (b) increased usage of home-grown cereals,
    (c) availability of labour.

9. This trend is inevitable.

10. No comment.

I trust that these replies may prove of some assistance to you.

Yours sincerely,
JOHN CROSS
(*Secretary*)

(ii) Flour Millers

Questions addressed to the National Association of British and Irish Millers (incorporating the views of the Scottish Flour Millers' Association), and the British Oatmeal Millers' Association

A. USE OF HOME-GROWN GRAIN

1. What are the prospects of more home-grown grain being used in the grist?
2. Would millers be able to use more home-grown wheat if it were bulked into larger lots and graded more exactly than at present?
3. Are there any technical developments coming along in flour milling which are likely to affect the type, quality and condition of grain required?

B. MARKETING OF HOME-GROWN GRAIN

4. To what extent has the H.-G.C.A. Forward Contract Scheme affected the market for home-grown wheat?
5. To what extent does grain fetch a lower market price than it otherwise would because of such factors as inefficient drying, poor storage or lack of cleaning?

C. THE FLOUR TRADE

6. To what extent is there excess capacity for producing flour in the industry, and is the process of vertical and horizontal integration still proceeding? What effect is this process likely to have on the demand for home-grown grain?
7. Are there any other features of the grain market which have not been mentioned in your replies to the above questions, but which you consider to be of special importance?

Statement by the National Association of British and Irish Millers (incorporating the views of the Scottish Flour Millers' Association) [1]

A. USE OF HOME-GROWN GRAIN

QUESTION 1. *What are the prospects of more home-grown grain being used in the grist?*

QUESTION 3. *Are there any technical developments coming along in flour milling which are likely to affect the type, quality and condition of grain required?*

ANSWER. Since the answers to these two questions are in part inter-related it will be more satisfactory to answer them jointly.

The use of home-grown wheat in flour milling and the prospects of the use of greater amounts need to be considered in part against a statistical background.

[1]With regard to the marketing of home-grown grain, see also the paper submitted by N.A.B.I.M in February 1967, printed as a supplement to this submission (pp. 793–5).

(i) The wheat usage figures of the flour milling industry in recent years are as follows:

Year	Total usage	Imported wheat	Home-grown wheat	(iii) as % of (i)
	(i)	(ii)	(iii)	
1964[a]	5,067,162	3,399,802	1,667,360	32·90
1965	4,960,058	3,385,215	1,574,843	31·75
1966	5,089,441	3,461,243	1,628,198	31·99

[a] A 53-week statistical year.

(ii) The extent to which imported wheat and home-grown wheat will be used for any particular grist will vary according to the cost and availability of the wheats themselves and also to the quality and the type of flour to be produced and the baking characteristics sought. There are thus many permutations of wheat mixtures possible at any given time. It is not feasible to be categoric about the actual proportion of home-grown wheat and imported wheat being used or to be used by the individual miller. No information about the exact composition of a grist would be available except from the individual miller, but such would normally be regarded by him as confidential and known only within the confines of his own business.

A reasonable indication of wheat usage, as between home-grown wheat and imported wheat, emerges from an approximate break-down, into its various use categories, of the total amount of flour eaten in this country. The categories are as follows:

	as % of total used	
For bread	approx. 65·9	⎫
For flour confectionery	approx. 6·5	based on
For biscuits	approx. 8·85	information
Domestic use in home	approx. 16·2	from National
Manufactured puddings, etc.	approx. 1·05	Food Survey
Industrial use including brewing	approx. 1·5	⎭

One can draw certain broad conclusions about the grist composition involved in these flour use categories. Thus home-grown wheat could be expected to be used almost entirely on its own for biscuit flour; in the other "non-bread" categories home-grown wheat could be expected to be used in a reasonable proportion in the grists involved. In the bread-use category the average figure of home-grown wheat would be of the order of 20–25%; it has to be remembered that in some areas, because of special consideration of flour characteristics, e.g. Scotland, and because of transportation/availability/cost factors, less home-grown wheat will be used in a bread grist than in others.

(iii) A factor in overall wheat usage is the level of flour consumption (including bread consumption). Both overall consumption and *per capita* consumption are declining and the rate of *per capita* decline more than offsets the effects of the increasing population; in other words, the more mouths there are to feed does not compensate for the fact that the mouths are eating less flour and bread. *Per capita* and overall flour consumption figures in recent years are as follows:

FLOUR CONSUMPTION

	Per capita	Overall
1953	192·7 lb	4,521,000 tons
1960	166·5 ,,	4,038.000 ,,
1961	164·8 ,,	3,964,000 ,,
1962	161·0 ,,	3,923,000 ,,
1963	161·1 ,,	3,963,000 ,,
1964	155·5 ,,	3,856,000 ,,
1965	154·1 ,,	3,926,000 ,,
1966	153·1 ,,	3,859,000 ,,

(iv) Knowing that the use of home-grown wheat is ideal for biscuit flour it is sometimes thought by some people that increased biscuit consumption must have considerable significance in the usage of home-grown wheat by flour millers. That this is not really so is shown by the following figures which record household consumption (on an ounces per head per week basis) of cakes and biscuits, of flour for household use (by the housewife) and, at the same time, of bread.

HOUSEHOLD CONSUMPTION

	oz per head per week		
	Cakes and biscuits	Flour	Bread
1962	12·36	6·22	43·57
1963	12·16	6·52	43·26
1964	12·20	6·07	41·97
1965	12·54	6·09	40·60
1966	12·06	5·95	38·64
1962–6	−0·3 oz =approx. 2½%	−0·27 oz =approx. 5%	−4·93 oz =approx. 11%

(v) It has to be remembered as an overriding factor in flour usage, and also in respect of the wheat grists used for flour, that Commonwealth milled flour is free to enter the U.K. without duty; imported flour from other sources is subject to 10% duty. All imported flour is at present subject to the Minimum Import Price arrangements but these are not a serious barrier, certainly at present wheat price levels, to the entry of flour, especially if it were directly or indirectly subsidised. Figures of flour imports during the last 3 years have been as follows:

	Flour imports in U.K.	Imports as % of total used
1964	276,000 tons	7·2
1965	234,000 ,,	5·9
1966	196,800 ,,	5·1

It remains that no flour miller could so weaken his breadmaking flour (in the use of large quantities of home-grown wheat) as to be unable to compete effectively with the use of imported flour, and in particular of Canadian flour, by bakers.

(vi) With regard to technical developments in flour milling it should be pointed out that the flour-milling industry shares a common industrial research organisation with the baking industry.

In our research programme we are constantly looking at such developments as high-speed dough-making techniques and protein separation (by which the protein-rich fractions of flour from any given wheat, whether weak or strong, may be separated more precisely) which point in the general direction of a higher usage of home-grown wheat. At the same time the industry is seeking actively (again guided by the results of its own and of other related research) to encourage the development and production of higher-quality home-grown wheat varieties which have in them the promise of a higher and more effective usage in flour-milling grists. It is impossible, however, at the present stage, to equate this work to a specific figure of increased usage of home-grown wheat.

(vii) In considering increased use of home-grown wheat by flour millers, some people are inclined to cite the instance of bread in France, bread made from flour incorporating a very high proportion indeed of French wheats. It has to be remembered that the French loaf is quite different in type, in size, texture and quality from that of this country. French bread is produced in very small lots by a very large number of very small bakeries, each serving its own immediate neighbourhood, working round the clock, with the French housewife collecting the bread for consumption oven-fresh, twice and sometimes three times a day. This is necessary, since the 1-day-old loaf is extremely unpalatable and 2-days-old virtually, by our standards, uneatable.

The present pattern of production and distribution and the type of loaf required by the housewife in this country could not be achieved by the grist used by the French miller. To lower our loaf quality would be undesirable and would undoubtedly adversely affect bread consumption levels.

(viii) All these various factors (in (i)–(vii)) taken together make exceedingly difficult forward projection of the exact extent to which home-grown wheat will be used by flour millers in the next few years. Basically, however, home-grown wheat is likely to be, in general terms, the cheapest component of millers' grists and economic considerations alone could be expected to maximise its use at any given time. With all these points in mind we feel we might reasonably foresee that total usage of home-grown wheat in flour milling will have reached *2 million tons a year by 1973*. We must make it clear, however, that if the U.K. had joined the E.E.C. by that time one could expect the whole basis of our agricultural marketing system to have been changed.

QUESTION 2. *Would millers be able to use more home-grown wheat if it were bulked into larger lots and graded more exactly than at present?*

ANSWER. Whereas at first sight it might be assumed that better grading and the ability to buy in larger graded lots would increase the use of home-grown wheat, in practice the larger flour millers (who jointly account for a high proportion of the home-grown wheat used in flour milling) appear already to be receiving bulked and graded home-grown wheat to meet their own individual requirements. Many merchants and farmers' groups now deem it their responsibility and function to make this facility available. It is difficult, therefore, to postulate a considerably larger usage of home-grown wheat if it were bulked into larger lots and graded more exactly than at present. Cost factors (particularly if home-grown wheat is involved in double handling, and extra transportation is involved) figure, of course, in this particular question.

B. MARKETING OF HOME-GROWN GRAIN

QUESTION 4. *To what extent has the H.-G.C.A. Forward Contract Scheme affected the market for home-grown wheat?*

ANSWER. Flour millers feel in general terms that the Home-Grown Cereal Authority's Forward Contract Scheme has helped the marketing of home-grown wheat and has made available home-grown wheat more regularly and with greater assurance to flour millers. This is not to say, however, the flour millers regard the present arrangements expressly as ideal, particularly in so far as they interact with certain aspects of the existing Deficiency Payments arrangements for home-grown wheat. Flour millers feel that there should indeed by a direct relationship between the subsidy arrangements, whatever such may be, and the Authority's Forward Contracts Scheme; they feel that these should be so linked as to jointly contribute to stability combined with a smooth and regular flow of wheat on to the market.

QUESTION 5. *To what extent does grain fetch a lower market price than it otherwise would because of such factors as inefficient drying, poor storage or lack of cleaning?*

ANSWER. Flour millers buy home-grown wheat basically on a clean and dry basis with deductions, where applicable, appropriate to the condition of each particular consignment. They do not buy wheat which for any reason is unsuitable for milling. While farm drying of wheat has improved considerably over the course of recent years, there is still plenty of room for further improvement. Wheat dried too speedily or at too high a temperature can be encountered, and for the flour-miller this is undesirable as it shortens the protein.

C. THE FLOUR TRADE

QUESTION 6. *To what extent is there excess capacity for producing flour in the industry, and is the process of vertical and horizontal integration still proceeding? What effect is this process likely to have on the demand for home-grown grain?*

ANSWER. No survey has been conducted which would enable a precise evaluation of excess flour milling capacity to be made, but it is estimated that by increasing working hours and by increasing the feed of wheat on to mills it would be a simple matter for the industry to produce approximately 25 % more flour than is at present consumed in the U.K.

The rate of vertical or horizontal integration has now substantially reduced but no industry, particularly one facing a declining market, can be insulated from the pressures of economic factors. Whereas integration in the past has almost certainly helped home-grown grain (from a reduction in part of imported flour, which uses no U.K.-grown wheat at all) it is not felt that further integration, if indeed such were to take place, would directly affect the usage of home-grown wheat.

QUESTION 7. *Are there any other features of the grain market which have not been mentioned in your replies to the above questions, but which you consider to be of special importance?*

ANSWER. It should be remembered that the actual cost—and hence the worth—of home-grown wheat will be different to the port miller than to the inland miller. This is often forgotten or overlooked by farmers and others in comparing, loosely, ex-farm prices of home-grown wheat with c.i.f. prices of imported wheat. Port millers must and indeed do use large quantities of home-grown wheat if the crop is to be absorbed satisfactorily. To the ex-farm price of home-grown wheat must be added the cost of handling and overland carriage to bring consignments to the port mill; in the cases of port mills in wheat-deficiency areas, e.g. Liverpool, Newcastle, etc., the cost of transport over the long distances involved from the wheat surplus regions is especially significant. Invariably a higher moisture content (compared with imported) is also involved. It is these factors which have to be kept well in mind in reconciling an ex-farm price of home-grown wheat with the c.i.f. price of imported wheat delivered to the port mill in bulk alongside. At the present time millers—especially port millers—are very much concerned about the increased transportation costs implicit in the Government's new freight transport proposals outlined in the Transport Bill.

Statement by N.A.B.I.M.: Supplement

NOTES ON HOME WHEAT MARKETING

(Paper originally submitted in February 1967)

Agricultural marketing policy has always been of considerable significance to the efficiency and stability of the flour-milling industry, and at all times this has been a subject engaging closely the interests of flour millers. As an instance, in the early thirties flour millers were actively involved (and indeed played a leading part) in the evolution of the Wheat Act and with it the birth of the deficiency payments system.

Our present philosophy of agricultural marketing has stemmed from a conviction that the deficiency payments system has been the right one for this country in the light of all the diverse factors involved, from an inborn desire for freedom to buy our imported wheats as economically as possible and from the need to be highly efficient and to be able to compete effectively with imported flour. At the same time we have been fully aware of the problems and difficulties of growers, and hence of the importance and desirability of market stability and efficiency. Records will show that flour millers have consistently endorsed (and at various times actively suggested) measures to improve the efficiency of cereal marketing and to enhance stability within the marketing system.

It is appropriate to emphasise that over the years the deficiency payments system has proved its effectiveness as an equitable and efficient system of market support for cereals and for wheat in particular. The system has given growers a secure market and a guaranteed return to the general level desired by the Government; it has done this without eliminating the benefits of the free market, or its flexibility in facilitating the disposal of wheat for its various purposes, it has preserved the need to encourage quality, good husbandry and so on. Flour millers have been permitted to continue to buy efficiently and economically, to the direct benefit of consumers; when market prices have been low (and deficiency payments high) the benefits of the low market prices have been passed on to consumers. In addition to being equitable to all interests, the system has operated in harmony with the machinery of the international grain trade and grain markets; it has been effective and efficient in operation and relatively inexpensive administratively. It has interposed no wasteful trading operations or needless double handling between growers and processors.

These merits have in most part been accepted and acknowledged by all interests and even as between political parties. Criticisms which occurred in recent years arose not from any defect in the basic principles of the policy, but from the open-ended application of the system coupled with low market prices and greatly increasing home production. The concern about the rising cost of the open-ended application of the system was remedied by the introduction of Standard Quantities and Target Indicator Prices; concurrently, and to help towards market stability, came the introduction of Minimum Import Prices and the formation of the Home-Grown Cereals Authority.

These and other modifications have come about slowly and in certain respects they reflect recommendations and views which we ourselves have been expressing. If our papers (which were written, of course, in the light of the situation then prevailing) of April 1963 and March 1964 are studied ("The Future Pattern of Home Wheat Marketing", 25th April 1963; "Memorandum of Home Wheat Marketing", March 1964) it will be seen that some of the measures we then advocated have in part been introduced.[1]

We are pleased that the deficiency payments system for cereals has been retained. We insist that over the years this system has worked remarkably well for all concerned. After a relatively short experience, and one not too significant, it still comes hard to millers, who traditionally have been able to buy to the advantage of the consumer economically in the markets of the world, to accept with equanimity the concept of minimum import prices. The feeling persists that this measure on occasions must deprive them of the advantage of parcels of grain at attractive prices; that it is tantamount to asking our suppliers to keep their prices up to their advantage. Since, however, the minimum import prices system is effectively established and is linked so intimately, administratively, with the application of the standard quantity and with the target indicator prices, millers feel, although with some reluctance, that there can be no real hope of turning back the clock to a levy system on imported grain (with, however, a more genuinely free market without a floor) which was suggested in the Memorandum of 1963.

With regard to the formation of the Home-Grown Cereals Authority we are, and always have been, in sympathy with the concept of an improved marketing system. To this end, therefore, we have supported, and will continue to support, in general terms, the efforts of the Authority directed to encouraging forward contracting and regular deliveries of home-grown wheat, towards developing and disseminating better market and statistical information; towards encouraging research and experimental work in some of the technical problems involved in the marketing of home-grown grain generally. In respect of these aims and improvements we feel that perfection will not be achieved quickly and considerable patience is called for, particularly having in mind the diverse

[1]These papers were communicated by N.A.B.I.M. but are not reproduced here.

points of view which the Authority must encounter in dealing with growers, merchants and processors. We feel we should make clear without hesitation, however, our strong dislike of the situation implicit in any introduction of the Authority's reserve trading powers. This is not the place to develop our detailed views on the possible viability, effectiveness and cost of the Authority's potential trading operations but, in general terms, we are opposed to such operations, both on principle and on practical grounds. We are convinced that the use of trading powers would almost certainly produce a situation sooner or later leading to a full-time trading body armed with powers of direction and control of imports—in effect the complete disappearance of a free market.

It is extremely difficult to look ahead to the cereal marketing situation if and when the U.K. joined the European Economic Community, particularly as so many important questions remain to be answered—and some even to be posed. From every point of view, however, we feel that the deficiency payments system as a basic means of cereal support is preferable to the E.E.C. system of target and intervention prices, maintained by substantial levies on imported grains to bring import prices to the predetermined threshold level. To a country such as the U.K. enjoying a loaf of high quality and at extremely economical cost, a support system similar to the E.E.C. with all its ramifications would promise the chill of considerably higher prices for bread coupled with the prospect of poorer quality.

LETTER RECEIVED FROM THE BRITISH OATMEAL MILLERS' ASSOCIATION

10th November 1967

Dear Professor Britton,

Your letter of the 26th October to the Secretary of my Association has been forwarded to me for my attention.

I fully appreciate the information you are trying to obtain, but the questionnaire you sent to us is hardly applicable to the Oat Milling Industry.

I should explain that the Oat Milling Industry is rather a small one; the total output of oat products in the United Kingdom is not much more than 65,000 tons per annum (as against wheat flour of about 4,000,000 tons per annum and animal feeds of 9,000,000 tons).

I think these figures put the problem into its right perspective.

Of the oat crop of some 1,500,000 tons in the United Kingdom, only between 25% and 30% of these leave the farm anyway, and of the amount of grain that leaves the farm, we as an industry consume approximately 125,000 tons per annum.

Regarding the marketing of oats from the farm, one has to be very careful with this as far as the Oat Milling Industry is concerned because the bulk of the grain is used by the industry in the winter, between the early part of August and the end of March each year, and I would refer you particularly to the statistical review which will illustrate this point very forcibly. The average tonnage from August to the end of March was between 12,000–13,000 tons per month, whereas the average from the end of March to the end of August very seldom is more than about 7000 tons per month. If any form of cereal marketing is put into operation, which took the supplies off the market at the beginning, as is indeed the case of barley and wheat this year, then the industry would not have the grain at the right time and would have to probably import grain with disastrous results on the amount of grain left on the farm later on in the year.

Regarding grain quality, this concerns us greatly. Improved harvesting, drying and storage methods would be of greater assistance to the industry. The tendency at the moment is for the miller to store and condition as much grain as he can himself earlier on in the crop year, because he cannot rely upon the quality of storage of the producer.

Finally, we are a very small association with very small funds, but I feel sure that the individual members of my Association would be prepared to co-operate with yourselves to the best of their ability.

Yours truly,

PHILIP LEA (*President*)

(iii) Maltsters, Brewers and Distillers

Questions addressed to the Maltsters' Association of Great Britain,
the Brewers' Society[1] and the Scotch Whisky Association

A. MARKETING OF HOME-GROWN CEREALS

1. To what extent does the quantity, quality and reliability of malting barley at present being produced meet the requirements of the trade? In particular, would it assist maltsters and distillers if grain could be bought in larger lots, and at more precise quality standards, than at present? If these were improved would maltsters and distillers change their present seasonal pattern of grain buying? What is the significance of the H.-G.C.A. Forward Contract Scheme in this connection?

2. Is the present system of buying grain from merchants and farmers working satisfactorily—particularly with regard to the test for nitrogen content?

B. THE MALTING, BREWING AND DISTILLING TRADES

3. What are the likely trends in the malting, brewing and distilling industries in the next 5 years in so far as they are likely to affect the demand for grain?

4. What are the economies of scale in malting and distilling and how do these affect location policy and size of unit? What is the significance for the future of developments like that at Mendlesham in Suffolk?

5. What are the relative advantages and disadvantages of having the malting, brewing and distilling processes separated or integrated?

6. Can profitable outlets for brewers' grains and the other by-products of malting, brewing and distilling be found?, i.e. which yield a return greater than the cost of disposing of them?

7. What potential is there for increasing exports of malt, beer and whisky?

8. Are there any other features of the grain market which have not been mentioned in your replies to the above questions, but which you consider to be of special importance?

Statement by the Maltsters' Association of Great Britain[2]

QUESTION 1

In general the quantity, quality and reliability of malting barley at present produced meets the requirements of the trade. However, conditions of climate and soil which influence barley quality, vary more in the U.K. from year to year and region to region

---

[1] The Brewers' Society replied by letter. The following is the relevant extract from a letter received from the Secretary of the Brewers' Society dated 18th June 1968:

"So far as the general policy of the Society is concerned I think it would be fair to say that we are satisfied with the views put forward by the Maltsters' Association and I think that if you take the enclosed figures together with the Maltsters' report, you will probably have far more information than I could possibly have hoped to produce on my own."

The figures referred to were estimates made by the Brewers' Society of anticipated purchases of barley by brewer-maltsters in the U.K. and were as follows, all figures being in tons:

1967	1968	1969	1970	1971	1972
292,222	278,995	281,407	293,296	294,984	296,723

[2] See also Supplements to this Statement (pp. 798–803).

than other cereal-producing countries. Maltsters for sale must retain flexibility in the areas from which they obtain their material in order to meet their wide range of requirements. Many maltsters do, however, receive a large percentage of their requirements from the same farms every year because, through having a range in quality of requirements, including malt for brewing and non-brewing, they are able to cope with the varying qualities dictated by weather conditions on such farms from year to year. In general, farmers grow varieties of barley that suit their land, but as long as they continue to take the advice of the N.I.A.B. on the most suitable varieties for malting, the situation should remain satisfactory.

It would not help maltsters to buy in larger lots at more precise standards of specification than at present. We are already buying from individual farms at rigid quality specifications. Our present method of buying relatively small lots enables us to fit these into our many and differing standards.

We are equipped to receive different barley qualities direct from farms and these can be stored separately to meet our varying requirements; any wholesale bulking before delivery to us would be detrimental to quality. Furthermore, any bulking of the crop into larger lots before sale to us would only increase the cost of our raw material. Additional handling between the farm of origin and a maltster's premises would increase the cost of our barley by a minimum of 20s. per ton, taking into account extra transport and handling costs involved.

The significance of the H.-G.C.A. forward contracting scheme is that this greatly aids our present system of buying. We can now receive a regular supply of barley in the various standards required in a phased delivery programme covering most of the year.

QUESTION 2

Yes, and as far as nitrogen content is concerned, since the agreement between the N.F.U., and the M.A.G.B., given at First Supplement, has been brought in this season, there have been no disputes as far as we know.

QUESTION 3

Future trends in the Malting Industry will depend on the replies given by the Brewers' Society and Scotch Whisky Association, which we trust will clarify the position.

QUESTION 4

Although economy of size is a distinct advantage for maltsters' costs, local supply of barley is a most important factor. Therefore, barley supply with regard to quality and quantity is carefully taken into account in siting any malting and deciding its optimum size. It is not considered that the malting at Mendlesham, in the context of the answers given here, has any exceptional relevance.

QUESTION 5

The answer to question 4 is directly related to this question, and location of barley supply will not always coincide with the brewing and distilling processes. Furthermore, as explained before, in the event of weather conditions in one part of the country affecting quality, a maltster is able to divert a proportion of certain requirements to other more suitable areas in the country; a too closely integrated organisation could not do this.

QUESTION 6

There are profitable outlets for the by-products of the Malting Industry which more than cover the cost of their disposal.

QUESTION 7

Without doubt, British malt is the best brewers' and pot distillers' material available in the world. We have made representations prior to devaluation for more help in this matter. Export trade in malt could be increased, but the present policy of levies and restitutions in operation within the European Economic Community is severely restricting this trade. The Second Supplement is a copy of the representation to the Ministry of Agriculture on this matter.

QUESTION 8

Further and continual education of farmers in what our requirements are would be most valuable. The Third Supplement gives a short description of such requirements.

Statement by Maltsters' Association: First Supplement

TRANSACTIONS IN MALTING BARLEY

Code of Practice for the Determination of Nitrogen Content

(Recommended by the National Farmers' Union,
Brewer's Society, N.A.C.A.M.
and the Maltsters' Association of Great Britain.)

(a) As inaccurate sampling on the farm can lead to variations in results from tests for nitrogen content, it is of the utmost importance that great care should be taken by the grower in drawing samples for submission to buyers, to ensure that they are truly representative of the bulk. It is therefore recommended that:

  (i) Where possible the produce of different fields should be kept and offered for sale as separate lots.
  (ii) When barley is stored "on the floor", sampling should be carried out at numerous points covering the whole of a heap, and each heap should be offered separately.
  (iii) When barley is stored in bins, it should be thoroughly mixed before sampling, which should be carried out at numerous levels and positions in the bin.

(b) Buyers should in testing purchase samples of malting barley for nitrogen content use the Kjeldahl method as their standard test and also as a check against any other method of determination. If another method of testing is used, the equipment should be checked and calibrated daily against a Kjeldahl test or with a standard sample of barley, the nitrogen content of which is precisely known.

On delivery of a load in bulk or sack to the delivery point buyers must retain the right to test for nitrogen although the visual examination must remain the most important factor in accepting a load.

If, as a result of this test, the nitrogen content appears to be outside the agreed tolerance of 0.05% a fresh sample should be drawn from ten different places in the load. The original delivery sample and this second sample should both be screened over a 2.25-mm screen and tested for nitrogen. The nitrogen content of the load should be determined as the mean average of these two screened samples.

If the nitrogen content, as thus determined, is outside the agreed tolerance, buyers should make every effort to accept delivery for alternative use at the current market price for barley of similar quality to the delivery, and should reject the load only as a last resort.

(c) Buyers should declare, at time of confirmation of purchase, the nitrogen content on which the transaction is based.

(d) In the event of a dispute the purchase samples should be tested on the Kjeldahl method, if originally tested on another method.

Statement by Maltsters' Association: Second Supplement

EXPORT OF MALT

Our principal competitors in the malt-exporting field are Canada, Australia, Czechoslovakia, Poland, East and West Germany, France and Belgium.

In our opinion maltsters in all these countries enjoy greater government assistance in respect of malt exports than we do. In Canada and Australia the system of barley purchasing is greatly helped by the appropriate grain control boards in favour of the malt exporter, and in Czechoslovakia and other Eastern European countries malt is exported at the highest price at which it is possible to make a sale merely to gain foreign currency and normal commerical considerations are not taken into account. In Common Market countries, in addition to restitutions paid by the Brussels Authorities, further support is given where a particular market requires it, Finland being a case in point. In France, where feeding and malting barleys attract the same subsidy, considerable quantities of the cheaper varieties are used in the production of malt and maltsters are thus further enabled to keep their prices at a very low level.

In regard to our traditional markets, we have a number of specified examples to demonstrate the difficulties members are up against. For instance, in Italy the lowest price at which a British maltster can get his malt to a port in north Italy c.i.f. is of the order of 8,000 lire per 100 kg (£46 0s. 0d. per ton), yet Czechoslovakians are able to offer malt at the north Italian border at 7,600 lire (£43 16s. 8d.) and are prepared to come lower. In Venezuela the very lowest that a British maltster can offer malt is of the order of $160 (£57 11s. 1d. per ton) and yet we know that French malt is offered there at $146 (£52 10s. 4d.). In Switzerland, where the price of British malt is Sw.fr.57.75 per 100 kg (£47 15s. 0d. per ton), we know that French malt is being offered at Sw.fr.46 (£38 8s. 10d.). This difference cannot be attributed entirely to the proximity of France to Switzerland. In Africa and the Indian Ocean, we know that Australian maltsters can bring their prices down to more than £4 per ton lower than any British maltster can hope to achieve and still show a profit. Freight charges from this country and Australia are about the same to Southern Africa and the Indian Ocean and no advantage can, therefore, be claimed on that score.

British maltsters are at a further disadvantage on the Continent in view of the specially low through rail rates quoted for international trade from malting to consuming areas or ports. Being an island we do not benefit from these low through tariffs and in any case the road and rail systems in this country do not offer reduced rates in respect of export cargoes.

In no other country has so much money been spent on modernising maltings and closing down inefficient units in order to keep equipment and production methods up to date. It is well known that a healthy competition exists between maltsters in Great Britain where a few efficient and well-organised malting groups compete for the available home trade. These groups of maltsters are themselves products of integration and rationalisation over the years.

Regarding E.F.T.A., an Eftamalt Committee was recently set up and one of its first tasks was to compare barley and malting conditions in the E.F.T.A. countries. This investigation has shown that, compared with other countries, support for the export of malt in this country is relatively small. For example, in Denmark in the year ended 31st July 1967, maltsters received a subsidy of Kr.9 per 100 kg (£4 13s. 6d. per ton) on malt they had exported and in Sweden exporters also receive subsidies amounting to the difference between the domestic and world price for malt. The Austrians also received a tax refund for the export of malt of 5·78% which we regard as a very useful incentive.

It is well known that barley for malting needs to be of a higher quality than barley for feeding and the excellence of British barley is recognised worldwide. Cases arise where continental maltsters buy British barley at low prices which reflect the British growers' subsidy and export rebate and, aided by their own continental subsidy, are able to sell it as malt at prices which effectively undercut British malt in its own traditional markets.

The production of malt involves manpower and materials and it would thus appear to be to Britain's advantage to malt the maximum amount of barley in this country and export it at an average of £43 10s. 0d. per ton f.o.b. rather than export barley at approximately £22 10s. 0d. per ton. The extra amount of revenue which this would bring in for this country is apparent and we are strongly of the opinion that our members should be given every encouragement to support the government's drive for increased production and exports.

Statement by Maltsters' Association: Third Supplement

MALTING BARLEY

Malting is the first of a series of operations beginning with barley grain as it is delivered to the maltster and ending with beer or whisky. The aim of the maltster is to produce malts that will meet with a wide range of brewing and distilling demands. Malt is produced by steeping barley in water and then allowing it to germinate for a limited period. During germination the starch in barley is broken down by enzymic action to soluble sugars which are the brewers' and pot distillers' raw materials. As soon as the period of germination has been completed, the growing grain is transferred to a kiln where it is first dried and then cured. After drying the mass of rootlets is removed without difficulty when malt is screened. Malt is very similar in appearance to the barley from which it is derived except that it has a somewhat paler look, is friable, and has a slightly sweet taste, rather like a biscuit.

Malting barley requirements

The value of malt to the brewer depends to a considerable extent on the amount of fermentable extract he can obtain from a given quantity of malt; this in turn depends to a great degree on the nitrogen content of the original barley. Barley is made up of starch, protein and fibre; protein and nitrogen have a direct relationship, the more the protein the higher nitrogen content. Barley with low nitrogen facilitates the malting process and gives the maximum fermentable extract.

Barley intended for malting should be dry, sweet and free from evidence of heat or pregermination. It should be rotund rather than long or thin and there should be no splitting. It is important that the grains be even in appearance and in ripeness. Finally, while only limited evidence can be gained from the appearance of the grain, barley intended for malting must be capable of germinating at least 96% quickly and vigorously. *Correct conditions of drying and storage are essential if a satisfactory level of germination is to be maintained.*

Cultivation

Early sowing is now recognised as a factor of considerable importance in the production of malting barley. On light soils where summer droughts are not infrequent, the early establishment of the plant and the attendant deeper and fuller root development are unquestionable factors in maintaining that condition of unchecked growth which is a requisite of successful barley production. Some opinion also holds that early sown barleys are less likely to lodge. *Late applications of nitrogen are not recommended,* since without necessarily increasing the yield of grain, they are directly responsible for impairing malting quality.

Varieties of barley

The National Institute of Agricultural Botany determine the malting values of new varieties of barley and publish a most helpful guide for farmers showing the potential of varieties currently offered commercially. The N.I.A.B. leaflet No. 8, 1967, includes seven spring and three winter barleys in the Recommended List suitable for malting. Proctor is still the outstanding malting variety and needs no further description; Maris Badger is as good a malting barley as Proctor.

The earlier ripening varieties, such as Zephyr and Mosane, have shown themselves to be useful malting barleys. Amongst the winter barleys Maris Otter has proved most successful as a malting variety. In these days of many varieties, it is most important that purity of seed is strictly maintained.

Harvesting and storage

It is most disappointing to see the number of samples ruined at harvest by incorrect setting of the combine harvester. Recent research has shown that the drum of the combine harvester requires frequent adjustment in relation to the moisture content of the crop, in order to maintain quality and reduce losses.

The widest concave setting and slowest cylinder speed that will give a satisfactory standard of threshing should be employed, on all occasions. Damage to the grain will reduce germinative capacity and make it unsuitable for malting.

Separation of varieties at the time of storage is most important as a mixture will reduce the value of a bulk.

Correct methods of drying and storage are vital to preserve sound germination and the following general rules apply:

1. The higher the moisture content the lower the drying temperature must be. The grain temperature must never exceed 110°F in drying.
2. During storage in deep silos regular turning is necessary particularly in the very early stages, to avoid a stifling effect on germination.
3. During storage "on the floor", in ventilated heaps strict adherence to the manufacturer's specification is vital.
4. Storage on the floor or in bins with no means of turning or aeration is fatal, unless the moisture content throughout is down to a safe limit in the first case of 14% and a temperature below 60°F, at the earliest possible moment.

Sampling

Inaccurate sampling on the farm can lead to variations in results from tests for nitrogen content. It is of the utmost importance that great care should be taken by growers in drawing samples for submission to buyers, to ensure that they are truly representative of the bulk. It is therefore recommended:

1. Where possible the produce of different fields should be kept and offered for sale as separate lots.
2. When barley is stored in large bulks "on the floor", sampling should be carried out at numerous points covering the whole heap (each sample representing approximately 20 tons). These samples should not be "averaged", but offered as sampled, representing the range of barley in the bulk.
3. When barley is stored in bins, samples should be taken at numerous levels and positions in the bin. This may only be possible as the bin is being either filled or emptied.

Statement by the Maltsters' Association of Great Britain: Fourth Supplement

USER REQUIREMENTS

The end users of malt, i.e. brewers at home and overseas, distillers, food manufacturers and vinegar brewers, have a wide range of requirements in malt.

The high and rigid standards required by the end users encourage keen competition in the Malting Industry, which in turn demands a high standard of raw material.

The normal range of requirements in terms of extract and nitrogen which the end users require the maltster to meet are shown in the table following this statement.

In order to receive supplies of barley economically and to exercise the degree of selection necessary to meet their requirements, maltsters have built their plants in the major barley-growing areas of the U.K. At the moment nearly all maltsters' requirements are met from U.K. barley, for which they pay a premium, depending on quality over the price of barley for feeding. Purchase on sample representing individual fields has proved the most efficient method of selecting our requirements from the mass of barley grown. This principle of buying on sample is considered basic and crucial to our method of selection.

Maltsters have built up and developed efficient means of selecting, buying and handling their barley requirements and drying and storing these according to their various standards. In our opinion, bulking by an independent authority would not be as efficient or economical. Additional handling between the farm of origin and the maltsters' premises can only increase the cost of our raw materials.

There are in existence centrally operated stores, but these are not used to capacity because the double handling involved, and the rates charged for drying, result in the barley handled in that way not being competitive in price with the barley that is either dried and stored on farms, or moved off farms to maltsters for drying and storage.

Considerable investment over the past few years has been made in drying and storage facilities on farms; a proportion of the cost of these facilities has been met by the Government. Now that farmers take care in handling and drying barley, the storage on farms suits the maltster's method of buying on sample.

There will probably be some standardisation of users' grades but any requirements will still be most efficiently met by maltsters themselves selecting suitable barley. Experience in France and parts of Australia has shown that the standard of barley graded or bulked by grain boards has rapidly sunk to a very low standard. It is most doubtful that any Authority without direct financial responsibility would exercise the degree of expert selection that results in the quality at the moment maintained by maltsters themselves selecting and purchasing their own standards. Any such lowering of quality would put maltsters at a grave disadvantage in meeting competition, especially in the export trade, where U.K. malt enjoys a high reputation.

It is generally noticeable that the standard of malt made on the Continent from barley bulked and exported by U.K. merchants is markedly inferior to that made and exported by U.K. maltsters from barley handled themselves.

Identification of varieties is very difficult in modern high-yielding barleys, and there are few people with this expert knowledge. In these days of rapid malting, a mixture of varieties causes grave difficulties; bulking on analysis alone might disregard the varietal factor.

Wholesale bulking and standardisation, as well as rapidly lowering the overall quality, would cause the premium for malting barley to growers to diminish, and this in turn would begin a vicious circle discouraging farmers from meeting our requirements and end in the rapid degeneration of standards of our raw material and resultant end product.

The question of nitrogen analysis which is vital to maltsters, as shown in the specification in the Supplement, has caused some difficulty in the past, but recent agreement

STANDARDS OF EXTRACT AND NITROGEN WHICH ARE MOSTLY GUARANTEED TO USERS

	Max. nitrogen	Approx. % of total requirement
1. *Malt for U.K. Brewing* Extract in brewers lb dry		
(a) 103·5	1·40	5%
(b) 102·5	1·50	20%
(c) 101·5	1·55	30%
(d) 101·0 (L.N. lager)	1·50	5%
(e) Crystal and coloured malts	1·55	5%
2. *Malt for distilling* 101·0	1·60	15%

and special malt for grain distilling with an exceptionally high enzymic activity.

3. *Malt for export* 101·0	1·68	10%
4. *Malt for the food trade* Maximum extract	1·6 to 1·7	10%

There are several other analytical variations within the above standards with which maltsters must comply.

between the Maltsters' Association of Great Britain and the National Farmers' Union has clarified the situation to the satisfaction of all concerned.

Whilst the Code of Practice regarding nitrogen agreed between the Maltsters' Association of Great Britain and the National Farmers' Union allows a fairly wide tolerance on analytical results, such tolerance is not acceptable to the end users of malt. Consequently the maltster must test deliveries and often downgrade at his own expense, provided in such cases the delivery is up to the standard of the purchase on visual evaluation.

If a Central Independent Testing Station were set up, the price of barley for the end user would probably be increased as someone would have to pay for the service which maltsters are already providing free of charge.

The present system of barley buying used by maltsters results in their supplies moving directly off farms to their plants; the Ministry of Agriculture's Premium Payment Scheme and H.-G.C.A. Forward Contract Scheme encourage this form of marketing. It is doubtful whether any alternative system would operate as efficiently and cheaply to the benefit of both farmer and maltster.

Statement by the Scotch Whisky Association [1]

Before commenting on the specific issues raised in the section dealing with the marketing of home-grown cereals it might be helpful to make a general statement outlining the factors governing the attitude of the Malt Whisky Industry to their requirements.

The buying policy of the Malt Whisky Industry is governed by three factors:
1. Quality.
2. Drying and storage facilities.
3. Malting facilities.

For distilling purposes there is required mature grain of an acceptable variety capable of germinating freely and easily under standard conditions and which will modify with a high potential extract of good fermentability. To maintain continuity of supplies most distillers require to arrange drying and storage facilities outwith the distillery. The type of drying facilities most suited to the distiller is one which is capable of reducing moisture content of the grain to 12% with due regard to germination and other requirements. The quality control exercised at the initial stage of purchase is transferred to the stages of acceptance, drying and storage. Modern mechanised malting units underline the importance of uniformity of quality and control is exercised at all stages in the securing of the raw material requirements for distilling purposes.

The question of whether the quantity, quality and reliability of malt and barley at present produced meets the requirements of the Industry depends very largely on the approach by individual buyers to their requirements. The 1967 barley crop is currently estimated at 9·2 million tons and in the circumstances there seems little doubt that supplies have been adequate. As availability of malting barley, in the long term, must depend on sufficient financial inducement to growers, the remedy to the problem, if one exists, must lie with consumers.

In this connection the breeding of new varieties is of prime importance and the introduction of royalty payments must be regarded as a considerable step forward. We feel, however, more emphasis must be placed on selection by a regional basis with desirable malting characteristics sharing equally with the importance of increased yield.

As buyers at present control costs and specification in the formation of large bulks of barley it seems unlikely that this operation could be carried out more cheaply or efficiently by others prior to the initial purchase. As this inevitably would involve further transportation and handling additional costs would be involved. The Barley Incentive Scheme which has been in operation for some years is considered to have substantially

---

[1] The Association preferred to reply to Part A of the questionnaire only.

assisted towards the more orderly marketing of the U.K. barley crop. The H.-G.C.A. Forward Contract Scheme has further consolidated the position and, in addition, has introduced a degree of stability and confidence into forward sales which may in the past have been lacking. Although no statistics are available the Industry is able to make considerable use of the Forward Contract Scheme either directly or indirectly.

The present method of marketing in the U.K. has existed for a very long time and it is doubtful whether any alternative system of marketing would operate as efficiently and cheaply in a free market. In our experience the grain trade accept the testing of malting barley for nitrogen content as an essential element in the assessment of suitable material. We would therefore regard any difficulties that may have arisen as very similar to those experienced some years ago with germination and anticipate that they will be resolved in due course.

APPENDIX H

List of Tables

805

TABLE H.1

ACREAGES AND PRODUCTION OF CEREALS IN THE U.K.

	Prewar (a)	1946	1952	1953	1954	1955	1956	1957	1958	1959	1960	1961	1962	1963	1964	1965	1966	1967
Acreages ('000 acres) As at June Census:																		
Wheat	1856	2062	2030	2217	2457	1948	2293	2113	2208	1929	2102	1827	2256	1928	2206	2536	2238	2305
Barley	929	2211	2281	2226	2063	2296	2323	2622	2755	3059	3372	3828	3987	4713	5032	5395	6130	6027
Oats	2403	3567	2882	2840	2588	2581	2564	2348	2217	2032	1974	1733	1519	1295	1125	1014	907	1012
Rye	16	55	56	68	44	19	26	26	23	14	19	19	18	21	21	18	10	11
Mixed corn	97	458	838	804	602	463	418	336	281	232	203	147	125	99	80	73	73	88
Total cereals	5301	8353	8087	8155	7754	7307	7624	7445	7484	7266	7670	7554	7905	8056	8464	9036	9358	9443
Total arable	13,088	18,980	18,104	18,107	17,864	17,542	17,610	17,524	17,516	17,760	18,051	17,955	18,099	18,212	18,382	18,524	18,484	18,325
Production ('000 tons) July/June years:																		
Wheat	1651	1967	2307	2664	2783	2599	2845	2683	2711	2785	3064	2573	3911	2998	3733	4105	3420	3836
Barley	765	1963	2334	2521	2244	2936	2800	2957	3170	4016	4241	4974	5773	6599	7404	8062	8586	9242
Oats	1940	2903	2772	2821	2440	2709	2486	2145	2138	2187	2058	1822	1747	1438	1325	1213	1102	1340
Rye	10	39	50	66	39	19	25	24	21	13	18	18	17	22	25	21	11	12
Mixed corn	76	350	830	845	555	510	407	325	275	259	219	169	154	118	101	91	93	117
Total cereals	4442	7222	8293	8917	8061	8773	8563	8134	8315	9260	9600	9556	11,602	11,175	12,588	13,494	13,212	15,547
% of total cereals acreage:																		
Wheat	35	24	25	27	32	27	30	28	30	27	27	24	29	24	26	28	24	24
Barley	18	26	28	27	26	31	31	35	37	42	44	51	50	59	59	61	65	64
% of total cereals production:																		
Wheat	37	27	28	30	35	30	33	33	33	30	32	27	34	27	29	30	26	24
Barley	17	27	28	28	28	33	33	36	38	43	44	52	50	59	59	60	65	58

(a) Average of years 1936/7 to 1938/9.

Source: M.A.F.F. Statistics.

TABLE H.2

ESTIMATED UTILISATION OF WHEAT IN THE UNITED KINGDOM

	Pre-war average	1957/8	1958/9	1959/60	1960/1	1961/2	1962/3	1963/4	1964/5	1965/6	(Prov.) 1966/7
June acreage ('000 acres)	1856	2113	2208	1929	2102	1827	2256	1928	2206	2535	2238
Production acreage	1856	2113	2208	1929	2102	1827	2256	1928	2206	2535	2238
Yield per acre (cwt)	17·8	25·4	24·6	28·9	29·1	28·2	34·7	31·1	33·8	32·4	30·6
Crop production ('000 tons)	1651	2683	2711	2785	3064	2573	3911	2998	3733	4105	3420
Disposal of crop ('000 tons)											
Sales for human and industrial use:											
Total	741	1207	1014	1316	1406	1237	1728	1554	1737	1587	1636
Net receipts by millers[a]	730	1178	984	1289	1361	1213	1564	1521	1709	1559	1606
Maltsters	—	5	3	5	6	2	6	3	3	3	3
Exports	3	7	12	4	20	5	136	10	3	4	5
Industrial use	—	5	5	5	5	5	5	5	5	5	5
Waste in distribution	8	12	10	13	14	12	17	15	17	16	17
Usage on agricultural and non-agricultural holdings:											
Total	910	1476	1697	1469	1658	1336	2183	1444	1996	2518	1784
Seed	137	166	148	159	150	175	150	170	194	173	181
Waste	22	34	34	35	38	32	49	37	47	51	43
Stockfeed:											
Sales	{751	1230	1346	1249	1439	1022	1961	1157	1717	2184	1408
Retentions on farms		46	169	26	31	107	23	80	38	110	152
Utilisation in crop year ('000 tons)											
Used for flour	730	1234	946	1290	1280	1262	1515	1461	1759	1488	1621
Products:											
Flour	500	880	677	935	919	910	1097	1055	1278	1070	1180
Offals	219	350	266	351	357	350	426	413	487	423	448
[a] Total receipts by millers	730	1275	1087	1394	1514	1322	1738	1622	1883	1688	1722
Rejects (used for animal feed)	—	97	103	105	153	109	174	101	174	129	116

Source: Ministry of Agriculture, Fisheries and Food.

Note: See page 825 for notes on this table.

Appendix H

TABLE H.3

ESTIMATED UTILISATION OF BARLEY IN THE UNITED KINGDOM

	Pre-war average	1957/8	1958/9	1959/60	1960/1	1961/2	1962/3	1963/4	1964/5	1965/6	(Prov.) 1966/7
June acreage ('000 acres)	929	2622	2755	3059	3372	3828	3987	4713	5032	5395	6130
Production acreage	929	2621	2755	3057	3372	3828	3987	4713	5032	5395	6130
Yield per acre (cwt)	16·5	22·6	23·0	26·3	25·2	26·0	29·0	28·0	29·4	29·9	28·0
Crop production ('000 tons)	765	2957	3170	4016	4241	4974	5773	6599	7404	8062	8586
Disposal of crop ('000 tons)											
Sales for human and industrial use:											
Total	531	1040	1058	1404	1175	1461	1332	1182	1443	2025	2422
Malting	462	790	746	848	875	916	934	890	1098	1026	1078
Flaking and roasting	—	25	25	25	25	25	25	25	25	25	25
Distilling	57	141	111	130	111	137	146	147	168	182	222
Pearl and pot barley	2	25	25	25	25	25	25	25	25	25	25
Exports	2	44	135	355	122	336	182	78	106	738	1035
Waste in distribution	8	15	16	21	17	22	20	17	21	29	37
Usage on agricultural and non-agricultural holdings:											
Total	234	1917	2112	2612	3066	3513	4441	5417	5961	6037	6164
Seed(a)	62	182	198	217	251	261	307	330	351	400	393
Waste	8	30	32	40	42	50	58	66	74	81	86
Stockfeed:											
Sales	{164	932	828	1122	1344	1724	2232	2776	2959	2996	3162
Retention on farms		773	1054	1233	1429	1478	1844	2245	2577	2560	2523
Utilisation in crop year ('000 tons)											
Used for:											
Malting, flaking and roasting	462	796	787	836	880	928	902	936	1052	1048	1086
Distilling	57	144	110	122	116	137	141	140	169	168	217
Pearl and pot barley	2	25	25	25	25	25	25	25	25	25	25
Exports	2	44	135	355	122	336	182	75	109	668	1091
Products:											
Pearl and pot barley etc.	1	16	16	16	16	16	16	16	16	16	16
Brewers' and distillers' grains	109	17	16	17	18	19	19	19	22	23	24
Malt culms(b)	18	30	29	30	32	34	33	34	39	40	43
Offals(b)	1	6	6	6	6	6	6	6	6	6	6

(a) Includes 50% mixed corn seed requirements. (b) From pearl barley manufacturers.

Source: Ministry of Agriculture, Fisheries and Food. *Note:* See page 825 for notes on this table.

TABLE H.4

ESTIMATED UTILISATION OF OATS IN THE UNITED KINGDOM

	Pre-war average	1957/8	1958/9	1959/60	1960/1	1961/2	1962/3	1963/4	1964/5	1965/6	(Prov.) 1966/7
June acreage ('000 acres)	2403	2348	2217	2032	1974	1733	1519	1295	1125	1014	907
Production acreage	2403	2345	2214	2024	1969	1730	1515	1291	1123	1009	904
Yield per acre (cwt)	16·1	18·3	19·3	21·6	20·9	21·1	23·1	22·3	23·6	24·0	24·4
Crop production ('000 tons)	1940	2145	2138	2189	2058	1822	1747	1438	1325	1213	1102
Disposal of crop ('000 tons)											
Sales for human and industrial use:											
Total	131	126	111	201	117	188	140	102	109	109	100
Milling	115	118	99	126	108	110	118	96	105	106	97
Exports	4	4	9	69	6	5	18	3	1	—	—
Waste in distribution	12	4	3	6	3	3	4	3	3	3	3
Usage on agricultural and non-agricultural holdings:											
Total	1809	2019	2027	1986	1941	1704	1607	1336	1216	1104	1002
Seed(a)	210	159	144	139	151	133	114	99	90	81	89
Waste	34	38	37	38	36	32	31	25	23	21	19
Stockfeed:(b)											
Sales to manufacturers	1565	293	285	294	302	279	223	201	193	158	147
Retentions on farms		1529	1561	1515	1452	1260	1239	1011	910	844	747
Utilisation in crop year ('000 tons)											
Used for milling	115	129	96	121	111	113	111	100	105	105	98
Products:											
Oat products	64	72	50	70	63	65	64	58	60	60	59
Offals	31	40	40	38	38	38	38	32	35	36	33

(a) Includes 50% of mixed corn seed requirements.　(b) Includes quantities sold off as feed for horses used in industry or for pleasure.

Source: Ministry of Agriculture, Fisheries and Food.　*Note*: See page 825 for notes on this table.

TABLE H.5

ESTIMATED UTILISATION OF MIXED CORN IN THE UNITED KINGDOM

	Pre-war average	1957/8	1958/9	1959/60	1960/1	1961/2	1962/3	1963/4	1964/5	1965/6	(Prov.) 1966/7
June acreage ('000 acres)	97	336	281	232	203	147	125	99	80	73	73
Production acreage	96	336	280	231	202	146	124	99	80	73	72
Yield per acre (cwt)	15·8	19·4	19·6	22·4	21·7	23·1	24·8	23·8	25·3	24·8	25·7
Crop production ('000 tons)	76	325	275	259	219	169	154	118	101	91	93
Disposal of crop ('000 tons)											
Usage on agricultural and non-agricultural holdings:(a)											
Total	76	325	275	259	219	169	154	118	101	91	93
Waste	1	3	3	3	2	2	2	1	1	1	1
Stockfeed:											
Sales to manufacturers	8	8	7	6	6	6	6	6	6	5	5
Retentions on farms	67	314	265	250	211	161	146	111	94	85	87

(a) For seed see paragraph 4 of notes.

Source: Ministry of Agriculture, Fisheries and Food.   *Note:* See page 825 for notes on this table.

TABLE H.6

ESTIMATED UTILISATION OF STRAW IN THE UNITED KINGDOM

	Pre-war average	1957/8	1958/9	1959/60	1960/1	1961/2	1962/3	1963/4	1964/5	1965/6	(Prov.) 1966/7
Production ('000 tons)											
Total	5514	6484	5955	6001	5711	6198	6535	6150	6683	7133	7167
Wheat	2070	1967	1859	1845	1701	1717	2044	1425	1712	1902	1696
Barley	723	1920	1675	1986	1949	2641	2802	3324	3699	4036	4425
Oats	2594	2247	2134	1932	1861	1690	1556	1289	1169	1095	951
Mixed corn	86	286	226	192	159	125	111	84	77	74	69
Pea	41	64	61	46	41	25	22	28	26	26	26
Disposal ('000 tons)											
Sales off agricultural holdings											
Total	507	377	381	381	368	321	293	267	227	190	158
Wheat straw	285	188	192	146	166	118	117	75	66	55	40
Barley straw	97	150	145	190	159	165	140	157	127	102	86
Oat straw	125	39	44	45	43	38	36	35	34	33	32
Usage on farms											
Total	5007	6107	5574	5620	5293	5860	6242	5883	6456	6943	7009
Fed to livestock	2211	2503	2315	2208	2126	1998	1700	1478	1378	1270	1183
Used as bedding, ploughed in, composted, etc. or wasted	2796	3604	3259	3412	3167	3862	4542	4405	5078	5673	5826

Source: Ministry of Agriculture, Fisheries and Food.   *Note:* See page 825 for notes on this table.

The flow of wheat in the United Kingdom 1966/7

Flows in the Merchant Sector ('000 tons)

	Independent merchants		Co-operatives	Sub-sidiaries of Nationals	Total
	Without links	With links			
	Not making compound feeds				
Bought by merchants:					
Bought from farmers	624	307	76	150	1157
Imported	—	—	—	—	—
Bought from other merchants	57	49	4	12	122
Sold by merchants:					
Sold to other merchants	91	54	13	47	205
Sold to farmers	107	90	30	50	277
Sold to compounders and millers	482	204	41	64	791
Sold to others and for export	4	—	—	—	4
	Making compound feeds				
Bought by merchants:					
Bought from farmers	1265	771	80	553	2669
Imported	165	2	38	23	228
Bought from other merchants	136	143	65	56	400
Sold by merchants:					
Sold to other merchants	65	254	16	50	385
Sold to farmers	175	69	34	78	356
Sold to compounders and millers	852	377	40	352	1621
Sold to others and for export	8	—	—	—	8
Used by merchants:					
Approximate tonnage of home-grown wheat used by merchants in making compounds	281	212	54	126	673

Source: Raised results of Survey of Merchants.
Note: See page 826 for notes on this table.

Other Flows:
  Imports: Total imports 1966/7: 4,110,000 tons. Of this total 500,000 tons was used for animal feed. (*Source:* Annual Review White Paper Cmnd. 3558.)
  Flour Millers: Direct purchases from farmers—approximately 375,000 tons. (*Source:* Survey of Flour Millers; see Part V(b).)
  Compounders: Direct purchases from farmers—approximately 30,000 tons. (*Source:* Survey of Compounders; see Part V(a).)
  Retained on Farms: Retained on farms 152,000 tons, and Waste 43,000 tons. (*Source:* Table H.2.)

CC**

Appendix H

THE FLOW OF BARLEY IN THE UNITED KINGDOM 1966/7

Flows in the Merchant Sector ('000 tons)

	Independent merchants		Co-operatives	Sub-sidiaries of Nationals	Total
	Without links	With links			
	Not making compound feeds				
Bought by merchants:					
Bought from farmers	983	623	80	426	2112
Imported	14	—	—	—	14
Bought from other merchants	601	102	6	348	1057
Sold by merchants:					
Sold to other merchants[a]	508	185	19	269	981
Sold to farmers	163	138	41	51	393
Sold to maltsters/distillers[a]	317	106	2	130	555
Sold to compounders/ millers[a]	296	177	14	97	584
Exported[a]	329	98	8	264	699
	Making compound feeds				
Bought by merchants:					
Bought from farmers	1883	669	128	883	3563
Imported	70	—	—	—	70
Bought from other merchants	304	156	125	299	884
Sold by merchants:					
Sold to other merchants[a]	429	217	19	263	928
Sold to farmers	258	152	60	159	629
Sold to maltsters/distillers[a]	314	29	3	324	670
Sold to compounders/ millers[a]	492	113	22	244	871
Exported[a]	185	28	—	98	311
Used by merchants:					
Approximate tonnage of home-grown barley used by merchants in making compounds	518	234	142	109	1003

[a] The distribution of total sales between these four categories is approximate only.

Note: See page 826 for notes to this table.

Source: Raised results of Survey of Merchants.

Other Flows:

Imports: 1966/7 188,000 tons. (Cmnd. Paper 3558.)

Compounders: Direct purchases from farmers—approximately 55,000 tons. (*Source:* Survey of Compounders; see Part V(a).)

Maltsters and distillers: Direct purchases from farmers: (maltsters only) approximately 150,000 tons. (*Source:* Survey of Maltsters and Distillers; see Part V(c).)

Retained on farms (2,523,000 tons) and *Waste* (123,000 tons). (*Source:* Table H.3.)

TABLE H.9

THE FLOW OF OATS IN THE UNITED KINGDOM 1966/7

Flows in the Merchant Sector ('000 tons)

	Independent merchants		Co-operatives	Sub-sidiaries of Nationals	Total
	Without links	With links			
	Not making compound feeds				
Bought by merchants:					
Bought from farmers	49	66	8	12	135
Imported	3	—	—	—	3
Bought from other merchants	20	11	4	1	36
Sold by merchants:					
Sold to other merchants	21	29	—	1	51
Sold to farmers	22	40	9	1	72
Sold to compounders/ millers	25	12	3	10	50
	Making compound feeds				
Bought by merchants:					
Bought from farmers	86	132	13	63	294
Imported	2	18	—	—	20
Bought from other merchants	23	31	1	13	68
Sold by merchants:					
Sold to other merchants	15	74	1	14	104
Sold to farmers	30	33	5	21	89
Sold to compounders/ millers	25	12	—	11	48
Used by merchants:					
Approximate tonnage of home-grown oats used by merchants in making compounds	35	66	7	29	137

Source: Raised results of Survey of Merchants.

Note: See page 826 for notes to this table.

Other Flows:

Production: 1966/7 1,102,000 tons. (*Source:* Table H.4.)

Retained on farms: 747,000 tons, and *Waste* 19,000 tons. (*Source:* Table H.4.)

(The tonnage bought direct by compounders from farmers is negligible.)

Appendix H

TABLE H.10

THE FLOW OF MAIZE IN THE UNITED KINGDOM 1966/7

Flows in the Merchant Sector   ('000 tons)

	Independent merchants		Co-operatives	Sub-sidiaries of Nationals	Total
	Without links	With links			
	Not making compound feeds				
Bought by merchants:					
Imported	41	—	4	—	45
Bought from other merchants	10	1	5	3	19
Sold by merchants:					
Sold to other merchants	2	—	—	—	2
Sold to farmers	50	1	9	3	63
Sold to compounders	—	—	—	—	—
	Making compound feeds				
Bought by merchants:					
Imported	502[a]	115	239	90	946
Bought from other merchants	63	27	—	21	111
Sold by merchants:					
Sold to other merchants	3	16	1	6	26
Sold to farmers	52	28	83	20	183
Sold to compounders	—	3	—	—	3
Used by merchants:					
Approximate tonnage used by merchants in making compounds	508	93	155	86	842

[a] Of this total about 300,000 tons was maize imported by merchants in Northern Ireland.

Source: Raised results of Survey of Merchants.

Note: See page 820 for notes to this table.

Other Flows:

Total imports 1966/7: 3,334,000 tons.

Utilisation: Cereal breakfast foods (approximately 160,000 tons). Starch and glucose (approximately 490,000 tons). Distilling and brewing (404,000 tons). Animal feed and waste (2,252,000 tons). (*Source:* M.A.F.F.)

Compounders: Maize used by national compounders 1,100,000 tons. (*Source:* Survey of Compounders; see Part V(a).)

TABLE H.11

FLOWS IN THE MERCHANT SECTOR 1966/7—SUMMARY

('000 tons)

	Wheat	Barley	Oats	Maize
Bought by merchants:				
Bought from farmers	3826	5676	429	—
Imported	228	84	23	991
Bought from other merchants	522	1941	104	130
Sold by merchants:				
Sold to other merchants	590	1909	155	28
Sold to farmers	633	1022	161	246
Sold to compounders/millers	2412	1455	98	3
Sold to others and for export	12	1010	—	—
Sold to maltsters/distillers	—	1225	—	—
Used by merchants:				
Approximate tonnage of home-grown grain used by merchants in making compounds	673	1003	137	842

Note: See page 826 for notes to this table.

TABLE H.12

GUARANTEED PRICES FOR CEREALS AND EXCHEQUER COST OF THEIR IMPLEMENTATION

	1954/5	1955/6	1956/7	1957/8	1958/9	1959/60	1960/1	1961/2	1962/3	1963/4	1964/5	1965/6	1966/7	1967/8	1968/9
	s. d.	s. d.	s. d.	s. d.	s. d.	s. d.	s. d.	s. d.	s. d.	s. d.	s. d.	s. d.	s. d.	s. d.	s. d.
Wheat	30 9	30 0	30 0	28 7	28 1	27 7	26 11	26 11	26 11	26 6	26 6	25 7	25 5	25 11	27 5
Rye	25 0	23 3	23 3	22 1	22 1	21 0	21 1	21 7	21 7	21 8	21 8	21 7	21 7	21 9	21 7
Barley	25 6	24 8	26 2	29 0	29 0	29 0	28 9	27 7	27 5	26 8	26 8	25 4	25 4	24 9	25 2
Oats and mixed corn	24 0	23 3	25 0	27 5	27 5	27 5	27 2	27 5	27 5	27 5	27 5	27 5	27 5	27 5	27 10
Exchequer cost:															
Wheat and rye															
(£ million)	24·5	18·2	19·2	22·2	19·3	20·4	18·1	22·0	16·6	30·3	15·9	14·2	13·5	15·9	n.a.
(£ per acre)	(9·80)	(9·25)	(8·28)	(10·38)	(8·65)	(10·50)	(8·53)	(11·92)	(7·30)	(15·55)	(7·14)	(5·56)	(6·01)	(6·90)	
Barley															
(£ million)	8·7	9·2	10·4	17·1	23·5	25·2	33·6	33·2	36·3	36·8	37·4	21·6	29·2	28·8	n.a.
(£ per acre)	(4·22)	(4·01)	(4·48)	(6·52)	(8·53)	(8·24)	(9·96)	(8·67)	(9·10)	(7·81)	(7·43)	(4·00)	(4·76)	(4·78)	
Oats and mixed corn															
(£ million)	Nil	1·4	4·7	11·9	9·8	12·8	11·7	18·1	11·0	10·0	10·0	7·3	6·7	9·8	n.a.
(£ per acre)	—	(0·46)	(1·58)	(4·43)	(3·92)	(5·65)	(5·37)	(9·63)	(6·69)	(7·17)	(8·30)	(6·72)	(6·84)	(8·91)	
Total cereals															
(£ million)	33·2	28·8	34·3	51·2	52·6	58·4	63·4	73·3	63·9	77·1	63·3	43·1	49·4	54·5	n.a.
(£ per acre)	(4·28)	(3·94)	(4·49)	(6·88)	(7·03)	(8·08)	(8·27)	(9·70)	(8·08)	(9·57)	(7·48)	(4·77)	(5·27)	(5·78)	

The figures of "£ per acre" have been calculated by dividing the total Exchequer cost by the total acreage.

Source: Annual Review White Papers.

TABLE H.14

CORN AND AGRICULTURAL MERCHANTS IN GREAT BRITAIN 1947 AND 1967

(Figures based on a comparison of the N.A.C.A.M. List of Members for 1947 and 1967. The table relates to firms, not branches of firms. Thus the closing down of branch offices, or opening of new branches, is not reflected in this list—unless the branches concerned had a different firm name.)

County	Number of firms 1947	Number of firms disappeared 1947–67	New firms entered 1947–67	Number of firms in 1967
ENGLAND				
Bedfordshire	26	11	7	22
Berkshire	35	15	4	24
Buckinghamshire	20	9	5	16
Cambridgeshire	42	14	7	35
Cheshire	64	35	8	37
Cornwall	68	36	13	45
Cumberland	25	9	3	19
Derbyshire	40	22	9	27
Devon	114	56	13	71
Dorset	28	13	6	21
County Durham	25	14	6	17
Essex	51	23	16	44
Gloucestershire	68	34	10	44
Greater London	103	72	34	65
Hampshire	42	20	9	31
Herefordshire	19	10	3	12
Hertfordshire	36	16	10	30
Huntingdonshire	14	7	4	11
Isle of Man	1	0	0	1
Isle of Wight	7	5	0	2
Kent	65	32	12	45
Lancashire	157	97	25	85
Leicestershire	43	28	3	18
Lincolnshire	92	32	32	92
Norfolk	96	43	21	74
Northamptonshire	29	12	5	22
Northumberland	26	17	9	18
Nottinghamshire	46	21	11	36
Oxfordshire	19	7	3	15
Rutland	2	0	1	3
Shropshire	39	21	14	32
Somerset	58	27	8	39
Staffordshire	67	35	10	42
Suffolk	90	40	25	75
Surrey	38	29	8	17
Sussex	63	29	8	42
Warwickshire	58	33	8	33
Westmorland	11	6	0	5
Wiltshire	33	16	5	22
Worcestershire	38	21	6	23
Yorkshire	216	117	44	143
Total England	2114	1084	425	1455

TABLE H.14—*continued*

County	Number of firms 1947	Number of firms disappeared 1947–67	New firms entered 1947–67	Number of firms in 1967
WALES				
Anglesey	3	1	1	3
Brecknockshire	5	2	0	3
Caernarvonshire	5	2	0	3
Cardiganshire	5	3	0	2
Carmarthenshire	8	3	2	7
Denbighshire	15	11	2	6
Flintshire	9	7	0	2
Glamorgan	35	26	2	11
Merioneth	3	3	2	2
Monmouthshire	20	14	4	10
Montgomeryshire	9	6	1	4
Pembrokeshire	8	4	4	8
Radnorshire	2	1	1	2
Total Wales	127	83	19	63
SCOTLAND				
Aberdeenshire	21	4	2	19
Angus	13	5	1	9
Ayrshire	17	8	5	14
Banffshire	6	4	1	3
Berwickshire	2	0	1	3
Bute	1	0	1	2
Caithness	2	1	0	1
Clackmannanshire	1	0	0	1
Dumfriesshire	5	2	2	5
Dunbartonshire	2	2	0	0
Fifeshire	12	7	2	7
Inverness-shire	1	0	0	1
Kincardineshire	2	2	0	0
Kinross-shire	0	0	1	1
Kirkcudbrightshire	3	0	0	3
Lanarkshire	56	36	6	26
Midlothian	29	17	10	22
Morayshire	3	3	0	0
Nairnshire	1	0	0	1
Perthshire	8	3	4	9
Renfrewshire	8	4	0	4
Ross-shire	1	1	1	1
Roxburghshire	2	1	1	2
Selkirkshire	1	0	0	1
Stirlingshire	7	3	0	4
Wigtownshire	3	2	2	3
Isle of Lewis	1	1	0	0
Orkney	0	0	1	1
Total Scotland	208	106	41	143
Total (Great Britain)	2449	1273	485	1661

TABLE H.15

PRODUCTION OF WHEAT, MAIZE, BARLEY AND RICE BY WORLD REGIONS

	Million long tons				
	Average 1937/8 to 1939/40[a]	Average 1956/7 to 1960/1[a]	1965/6[a]	1966/7[a]	1967/8[a][b]
WHEAT:					
Western Europe	27·5	35·2	44·7	39·3	46·4
Eastern Europe [b][c]	14·1	15·3	21·6	22·0	24·3
Soviet Union	33·0[e]	66·0	58·7	98·9	76·1
North America	31·8	45·3	54·1	58·9	58·8
Others	37·4	48·3	57·2	59·8	60·2
Total[d]	143·8	210·1	236·4	278·9	265·7
China[e]	22·4	n.a.	[h]	[h]	[h]
MAIZE:					
Western Europe	5·1	7·2	8·7	10·1	9·5
Eastern Europe [b][c]	12·1	15·0	17·0	22·4	20·4
United States	58·2	86·0	102·1	102·9	118·1
Other America	16·3	21·9	32·2	35·3	32·7
Others	18·1	28·3	34·5	40·8	36·8
Total[d]	109·9	158·5	194·5	211·6	217·5
BARLEY:					
Western Europe	8·1	18·5	29·8	31·5	36·9
Eastern Europe [b][c]	5·7	5·8	7·5	7·6	7·5
Soviet Union	9·0[e]	11·9	20·0	27·4	24·2
North America	7·5	14·3	13·2	15·1	13·5
Others	13·1	18·1	16·5	16·8	18·5
Total [d]	43·4	68·6	86·9	98·5	100·6
RICE: [g]					
India ⎫	27·5	30·5	30·2	30·0	40·0
Pakistan ⎭		9·1	11·7	10·8	11·3
Japan	9·0	10·8	11·4	12·5	14·2
Other Asia	22·8	34·4	42·5	38·7	39·2
America	2·2	5·9	8·2	7·8	9·7
Others	5·7	3·1	6·1	4·9	5·1
Total[d]	67·2	93·8	110·1	104·7	119·6
China	32·8[e]	53·0[f]	n.a.	n.a.	n.a.

[a] These relate in the Northern Hemisphere to harvests in the first of the calendar years shown and in the Southern Hemisphere to crops harvested in the latter part of that year.
[b] Provisional. [c] Including Yugoslavia. [d] Excluding China. [e] Estimate.
[f] Average for less than full period shown. [g] Milled rice equivalent. [h] The Commonwealth Secretariat estimate production to be between 25 and 30 million tons.
Source: Commonwealth Secretariat, *Grain Crops* and *Grain Bulletins*.

TABLE H.16

EXPORTS OF GRAIN FROM THE MAIN EXPORTING REGIONS

	Thousand long tons					
	Average 1937 to 1939	Average 1956 to 1960	Average 1963 to 1965	1965	1966	1967
WHEAT(d)						
Western Europe	835	3221	5664	6820	6304	5281(e)
Eastern Europe and U.S.S.R.(a)	3313	4675	3033	2120	3644	1163(f)
North America	6228	20,467	32,806	31,075	38,839	28,431
Australia	2790	2684	6734	7141	4723	8516
Others	4864	3226	5016	7827	5473	2934(g)
Total(e)	18,030	34,273	53,254	54,983	58,983	46,325
MAIZE:						
Eastern Europe and U.S.S.R.(a)	1008	1182	1508	1283	875	1465(b)
North America	1549	4577	12,609	14,923	15,302	12,388
Argentina	4897	1728	2832	2758	3657	4255
South Africa	532	776	1393	321	46	1969
Others	1034	988	3145	4355	4469	4939(g)
Total(e)	9020	9251	21,486	23,640	24,349	25,016
BARLEY:						
Western Europe	157	1042	1231	2475	3237	3836
Eastern Europe and U.S.S.R.(a)	695	622	2631	2141	470	613(b)
North America	558	3386	2037	2081	2009	1980
Others	812	1638	1166	1205	561	68
Total(e)	2222	6688	7064	7902	6277	6497

(a) Including Yugoslavia.   (b) Excluding U.S.S.R.   (c) Excluding China.   (d) Including wheat equivalent of flour exports.
(e) Only Jan.–June for Spain.   (f) Excluding U.S.S.R. and Czechoslovakia.   (g) Only Jan.–Sept. for Mexico.
Source: Commonwealth Secretariat, *Grain Crops* and *Grain Bulletin*.

TABLE H.17

PERCENTAGE SHARE OF MAIN EXPORTING COUNTRIES[a] IN WORLD EXPORTS OF WHEAT—1900–65/6

	U.S.A.	Canada	Australia	Argentina	Other countries	World
	%	%	%	%	%	%
Average 1900–9	26·0	6·4	4·4	14·1	49·1	100
Average 1910–9	27·5	19·2	8·3	13·4	31·6	100
Average 1920–9	26·4	31·8	10·5	18·4	12·9	100
Average 1930–9	10·6	28·3	16·1	18·3	26·7	100
Average 1945–9	47·3	28·7	9·4	8·7	5·9	100
Average 1950–4	34·1	31·0	10·1	8·4	16·4	100
Average 1955–9	36·1	23·5	7·7	7·5	25·2[d]	100
1960/1	40·6	19·3	8·8	6·5	24·8	100
1961/2	41·4	23·4	13·9	2·3	19·0	100
1962/3	38·4	20·9	11·4	6·8	22·5	100
1963/4[b]	39·2	23·3	13·0	3·7	20·8	100
1964/5[b]	38·3	23·2	12·6	8·6	17·3	100
1965/6[b] [c]	37·6	23·8	9·1	12·7	16·8	100

[a] Including wheat equivalent of flour exports.　　[b] July–June figures.　　[c] Provisional.　　[d] Includes additional estimates of intra-communist bloc exports not fully accounted for in previous years.

Source: Commonwealth Secretariat, *Grain Crops*, and *Review of the World Wheat Situation 1956/66*, International Wheat Council.

TABLE H.18

PRODUCTION, CONSUMPTION, EXPORTS AND STOCKS OF WHEAT
IN THE FIVE MOST IMPORTANT EXPORTING COUNTRIES

Year	Million metric tons			
	Stocks[a]	Production	Domestic consumption	Exports
U.S.A.				
Average 1946/7 to 1950/1	5·9	31·7	22·0	7·6
1954/5 and 1955/6	26·8	26·1	16·6	8·2
1960/1 and 1961/2	37·1	35·3	16·4	18·8
1963/4	32·5	31·1	16·0	23·1
1964/5	24·5	34·9	17·4	19·7
1965/6	22·3	35·8	19·9	23·6
1966/7[b]	14·6	35·7	18·7	20·2
1967/8[b]	11·4	41·5	n.a.	n.a.
CANADA				
Average 1946/7 to 1950/1	2·4	10·5	3·8	6·0
1954/5 and 1955/6	15·5	11·3	4·4	7·7
1960/1 and 1961/2	16·4	10·9	4·1	9·7
1963/4	13·3	19·7	4·3	16·2
1964/5	12·5	16·3	4·0	10·8
1965/6	14·0	17·7	4·3	15·9
1966/7	11·5	22·5	4·2	15·0
1967/8[b]	15·3	16·1	n.a.	n.a.
AUSTRALIA				
Average 1946/7 to 1950/1	0·7	4·9	2·1	2·9
1954/5 and 1955/6	2·6	4·9	1·9	3·2
1960/1 and 1961/2	1·2	7·1	2·1	5·5
1963/4	0·6	9·0	2·2	6·7
1964/5	0·7	10·0	2·7	7·3
1965/6	0·7	7·1	2·3	4·8
1966/7[b]	0·4	12·7	2·4	8·7
1967/8[b]	2·1	7·5	n.a.	n.a.
ARGENTINA				
Average 1946/7 to 1950/1	1·2	5·5	3·3	1·1
1954/5 and 1955/6	1·9	6·5	3·6	3·1
1960/1 and 1961/2	1·2	4·6	3·5	1·5
1963/4	0·7	8·8	3·6	3·5
1964/5	2·4	11·3	3·8	6·3
1965/6	3·6	6·1	3·8	5·6
1966/7[b]	0·4	6·3	3·7	3·0
1967/8[b]	0·4	7·4	n.a.	n.a.
FRANCE				
1954/5 and 1955/6	0·8	10·5	8·2	1·9
1960/1 and 1961/2	2·1	10·3	9·2	1·2
1963/4	3·3	11·0	9·3	2·7
1964/5	2·3	13·8	10·1	4·0
1965/6	2·0	14·8	10·1	4·1
1966/7[b]	2·6	11·3	9·2	2·2
1967/8[b]	2·5	14·4	n.a.	n.a.
FIVE EXPORTING COUNTRIES TOGETHER				
* Average 1946/7 to 1950/1	10·2	52·6	31·2	17·6
1954/5 and 1955/6	47·6	59·3	34·7	24·1
1960/1 and 1961/2	58·0	68·2	35·3	36·7
1963/4	50·4	79·6	35·4	52·2
1964/5	42·4	86·3	38·0	48·1
1965/6	42·6	81·5	40·4	54·0
1966/7[b]	29·7	88·5	38·2	49·1
1967/8[b]	31·3	86·9	n.a.	n.a.
* Excluding France				

[a] Stocks at the beginning of harvest year; U.S.A. and France 1st July, Canada 1st August, Australia and Argentina 1st December. [b] Provisional.

Sources: U.S. Department of Agriculture, *Grain Market News* and *Wheat Situation,* Washington, D.C.; Canada, Dominion Bureau of Statistics, *The Wheat Review,* Ottawa; E.E.C. Commission, Brussels.

TABLE H.19

AVAILABLE SUPPLIES OF GRAIN IN CERTAIN COUNTRIES
(Domestic production plus net imports)

	Million long tons			
	Average 1951/2 to 1955/6	1964/5	1965/6	1966/7
INDIA–PAKISTAN[b]				
Wheat[a]	12·6	21·1	25·0	24·0
Rice	33·4	50·3	42·5	41·4
Others	6·0	7·3	7·9	8·0
Total	52·0	78·7	75·4	73·4
Population	453·2	572·4	589·7	603·7
Supplies per head	0·11	0·14	0·13	0·12
JAPAN				
Wheat[a]	3·3	4·6	4·6	5·1
Rice	9·9	12·1	12·3	12·1
Others	3·4	5·1	5·3	5·7
Total	16·6	21·8	22·2	22·9
Population	86·7	96·9	98·0	98·9
Supplies per head	0·19	0·22	0·23	0·23
UNITED KINGDOM				
Wheat[a]	7·2	7·8	8·7	7·5
Rice	0·1	0·1	0·1	0·1
Others	7·6	12·1	12·3	12·2
Total	14·9	20·0	21·1	19·8
Population	50·9	54·2	54·6	54·9
Supplies per head	0·29	0·37	0·39	0·36
E.E.C.				
Wheat[a]	24·9	26·8	28·5	25·8
Rice	0·6	0·7	0·8	0·7
Others	24·9	36·2	39·5	40·0
Total	50·4	63·7	68·8	66·5
Population	161·3	179·6	181·6	182·3
Supplies per head	0·31	0·36	0·38	0·37
SOVIET UNION[b]				
Wheat[a]	44·5[c]	79·3	63·4	103·6
Rice	0·6[c]	0·7	0·6	0·7
Others	50·9[c]	59·1	47·4	57·0
Total	96·0	139·1	111·4	161·3
Population	235·9	281·0	284·5	287·6
Supplies per head	0·41	0·49	0·39	0·56
U.S.A.				
Wheat[a]	20·5	16·0	12·2	15·5
Rice	1·0	1·0	1·3	1·3
Others	93·7	93·6	105·9	110·9
Total	115·2	110·6	119·4	127·7
Population	160·3	192·1	194·6	196·9
Supplies per head	0·72	0·58	0·61	0·65

[a] Includes flour on wheat equivalent.   [b] Calendar years–first of two stated.   [c] 1954/5 to 1955/6.
Sources: Commonwealth Secretariat, *Grain Crops; United Nations Yearbook.*

Appendix H

TABLE H.20

THE IMPORTANCE OF GRAIN EXPORTS TO THE MAIN EXPORTING COUNTRIES

	Proportion of harvest exported			Relative value of exports to total exports		
	Average 1937/8 to 1939/40	Average 1956/7 to 1960/1	1965/6	Average 1934 to 1938	Average 1956 to 1960	1965
	%	%	%	%	%	%
AUSTRALIA						
Wheat[a]	57	68	70	14·5	8·6	11·3
Barley	18	57	25	0·4	1·5	0·4
CANADA						
Wheat[a]	42	65	90	19·5	10·1	10·7
Barley	14	30	18	0·2	1·5	0·5
FRANCE						
Wheat[a]	5	15	33	1·1	1·7	3·1
U.S.A.						
Wheat[a]	11	44	65	1·7	4·2	4·2
Maize	3	6	16	0·8	1·3	3·1
Barley	4	22	19	0·2	0·5	0·3
ARGENTINA						
Wheat	57	37	[b]	17·4	13·0	24·6
Maize	27	42	38	22·1	10·0	10·3
Barley	43	31	25	n.a.	1·6	1·0

[a] Including flour on wheat equivalent.   [b] Exports exceeded production during the year.

Source: Commonwealth Secretariat, *Grain Crops*.

NOTES TO TABLES H.2—H.6

Acreages

1. The figures relate to agricultural holdings ("farms") which are defined as holdings exceeding one acre in Great Britain and one acre or more in Northern Ireland. The difference between the June acreage and the production acreage of grain crops represents the small area on which the crop failed or was cut green for fodder.

Yields

2. The following table shows the 5-year and 10-year average U.K. yields:

	Average yield (cwt per acre)	
	5 years 1962–6	10 years 1957–66
Wheat	32·5	29·9
Barley	28·9	26·7
Oats	23·5	21·9
Mixed corn	24·9	23·1
Rye	22·1	20·7

Disposal of Crop

3. Under the Agriculture Act, 1947, the crop year is defined as beginning on the 1st of July and the figures of crop disposal relate to the July–June period. Quantities of sales to processors, obtained under the Statistics of Trade Act, 1947, are the total of August to July monthly figures of intake which broadly correspond to the July–June sales off farms. Export figures are derived from the Overseas Trade Accounts of the United Kingdom and similarly cover the August–July period. Estimates are made (as in paragraph 4 below) of the usage for seed and of waste in distribution and on farms. The residual quantity, representing retentions on farms for stockfeed, includes, for wheat and barley, farmer-to-farmer sales of which no separate estimate can be made. For oats and mixed corn, retentions on farms for stockfeed include, in addition, some unrecorded farmer-merchant-farmer sales. No adjustment is made for stocks held on farms at the beginning or end of the period.

Usage on Farms

4. Requirements for seed are estimated on the succeeding year's acreage at the rates shown below. Seed for mixed corn is assumed to be 50% barley and 50% oats. Allowances for *waste on farms* are calculated at the following percentages of the crop: wheat 1·25, oats 1·75, barley, mixed corn and rye 1·00.

	Seed rate per acre (cwt)		
	England and Wales	Scotland	Northern Ireland
Wheat: Winter	1·45	} 1·97	2·125
Spring	1·70		
Barley	1·25	1·65	1·375
Oats	1·50	1·88	1·875
Rye	1·45	1·45	1·500

Utilisation in Crop Year

5. These sections of the tables show the quantities actually used for human and industrial purposes in the July–June period.

Straw

6. Estimates of sales are based on feeding and bedding requirements of horses used in industry or for pleasure, and on information from trade sources. Usage on farms is the residual quantity after deducting estimated sales from total production. In the absence of definite information, the amount fed to livestock is estimated by taking a constant proportion of each crop of straw; in practice the proportion of each crop fed to livestock will vary. The remaining quantity is shown as used for bedding, ploughed in, composted or wasted. No allowance has been made for carry-over of farm stocks from year to year.

NOTES TO TABLES H.7—H.11

The 260 firms in the Survey of Merchants were asked to supply statistics of the grain they bought and sold in the most recent year for which the data were available. Although not all the firms supplied this information (which could not be collected at the interview itself) a sufficient number did so to justify "raising" the results to the United Kingdom level. These raised results are given in Tables H.7–H.11, and the information has also been used in drawing the three flow charts in Part III, Chapter 2. Additional data, based mainly on M.A.F.F. statistics, were also used for the flow charts, and this information is included in Tables H.7–H.10 for the reader who wishes to have the quantitative basis of the charts.

Because of sampling errors, and because not all firms gave data relating exactly to the same 12-month period, the statistics in Tables H.7–H.11 must be regarded as approximate only. Moreover, when the results were raised it was found that they were not always compatible with known national statistics. An example was the raised total of exports which was almost double the known figure. Obviously many firms had entered grain destined for export under this head although it may have been sold in the first instance to another merchant at a port for eventual export. The raised export total was thus adjusted to the correct figure and the balance allocated to "Sold to other merchants"—this adjustment brought the two figures "Sold to other merchants" and "Bought from other merchants" closely into line.

Despite the approximate nature of the estimates it was thought worthwhile including them in the Report because such information has not hitherto been available elsewhere.

Index